Decimal Equivalents

4ths	8ths	16ths	32nds	64ths	To 4 Places	To 3 Places	To 2 Places
				1/64	.0156	.016	.02
			1/32		.0312	.031	.03
				3/64	.0469	.047	.05
		1/16			.0625	.062	.06
				5/64	.0781	.078	.08
			3/32		.0938	.094	.09
				7/64	.1094	.109	.11
	1/8				.1250	.125	.12
				9/64	.1406	.141	.14
			5/32		.1562	.156	.16
				11/64	.1719	.172	.17
		3/16			.1875	.188	.19
				13/64	.2031	.203	.20
			7/32		.2188	.219	.22
				15/64	.2344	.234	.23
1/4					.2500	.250	.25
				17/64	.2656	.266	.27
			9/32		.2812	.281	.28
				19/64	.2969	.297	.30
		5/16			.3125	.312	.31
				21/64	.3281	.328	.33
			11/32		.3438	.344	.34
				23/64	.3594	.359	.36
	3/8				.3750	.375	.38
				25/64	.3906	.391	.39
			13/32		.4062	.406	.41
				27/64	.4219	.422	.42
		7/16			.4375	.438	.44
				29/64	.4531	.453	.45
			15/32		.4688	.469	.47
				31/64	.4844	.484	.48
					.5000	.500	.50

4ths	8ths	16ths	32nds	64ths	To 4 Places	To 3 Places	To 2 Places
				33/64	.5156	.516	.52
			17/32		.5312	.531	.53
				35/64	.5469	.547	.55
		9/16			.5625	.562	.56
				37/64	.5781	.578	.58
			19/32		.5938	.594	.59
				39/64	.6094	.609	.61
	5/8				.6250	.625	.62
				41/64	.6406	.641	.64
			21/32		.6562	.656	.66
				43/64	.6719	.672	.67
		11/16			.6875	.688	.69
				45/64	.7031	.703	.70
			23/32		.7188	.719	.72
				47/64	.7344	.734	.73
3/4					.7500	.750	.75
				49/64	.7656	.766	.77
			25/32		.7812	.781	.78
				51/64	.7969	.797	.80
		13/16			.8125	.812	.81
				53/64	.8281	.828	.83
			27/32		.8438	.844	.84
				55/64	.8594	.859	.86
	7/8				.8750	.875	.88
				57/64	.8906	.891	.89
			29/32		.9062	.906	.91
				59/64	.9219	.922	.92
		15/16			.9375	.938	.94
				61/64	.9531	.953	.95
			31/32		.9688	.969	.97
				63/64	.9844	.984	.98
					1.0000	1.000	1.00

Military Standards 8C

Drafting Technology and Practice

Industrial drafting involves a study of the techniques of graphic communications used in industry. It is a means of developing ideas and transforming them into drawings others can read and understand. It is a language used by designers, engineers, draftsmen, and craftsmen.

Drafting
Technology and Practice

by William P. Spence
Dean, School of Technology
Kansas State College of Pittsburg

CHAS. A. BENNETT CO., INC.
Peoria, Illinois 61614

KP   10   9   8   7   6   5   4   3
77   76   75

Library of Congress Catalog Card Number: 79–178965
PRINTED IN THE UNITED STATES OF AMERICA

ISBN 87002–129–X

PREFACE

This book has been designed to serve as a class text and reference book in drafting technology and practice. The content covers a wide range. The text is designed so that it can be used in a beginning course yet includes sufficient additional technical subject matter that it can serve for advanced classes. It is ideally planned for a broad survey course in which the student is exposed to an extensive range of experiences in applying basic information to a variety of industrial practices.

An important feature of this text is that a large part of it is devoted to specific uses of the graphic language in industrial applications. For example, the student can learn to make a wide variety of electrical and electronics schematics including printed circuits. He can explore the much neglected area of piping diagrams and drawings. In the aerospace industry are unique drafting applications. Simple problems in this industry are included. Technical illustration is finding ever increasing use in many industries. This area is covered in detail. An introduction to architectural drawing is provided to give students the opportunity to have an experience in this area of industry. Structural drawing is a big part of architecture. The basic types of structural drawings

are explained. Everyone uses maps of all kinds, yet students seldom understand their development and how they are made. This text provides a unique experience for students in the area of map design and symbolism. An almost totally neglected area is drafting in the cabinet-making industry. It has definite procedures and standards that should be understood. A very detailed chapter is included for this interesting industry. Considerable emphasis is placed on drawing for production. Experiences are available in the areas of detail and assembly drawings, gears, cams, and section drawings. The basic introduction to descriptive geometry is designed so the student can grasp the fundamentals and apply these to all types of industrial drawing. The rapidly emerging process of computer-aided design and the computer-controlled automatic drafting machine is presented in simple terms.

A large number of practical drawing problems are included. These are carefully selected to give the student a wide range of experiences. The problems selected use objects most students will recognize. This helps relate the subject matter to realistic drawing problems. Also included are a number of problem solving and

original design situations. Students will have the opportunity to try their hand at designing and detailing. This affords them a very real chance to apply what they are studying. The study problems range from very simple to very difficult. This enables the instructor to offer a challenge to students of varying levels of ability.

In addition to using problem-solving situations, it is hoped that drafting problems be drawn on the materials used in industry. Students should draw on vellum, tracing paper, and polyester drafting film.

They should run prints of all drawings made. This not only gives them the opportunity to do some reproduction work, but will quickly show the defects in line thickness and darkness. This will enable the teacher to show the students why they need to pay attention to line quality. It is not enough to simply tell them.

At the end of each chapter is a section called *Build Your Vocabulary*. Carefully selected technical terms have been listed for review and definitions. The lists will help the student to review the key parts of the chapter and clarify the meaning of items that he may have failed to understand.

Considerable space has been devoted to the use of two-color illustrations to help the student understand the purpose of each illustration and perhaps learn the material more easily.

Many drafting instructors prefer tu start their class by teaching the students to sketch. They then proceed to cover some of the basic units as multiview drawing, pictorial drawing, dimensioning, and sections with the students sketching problem solutions. This permits the students to concentrate on the subject matter and not be held back by trying to learn to manipulate drafting tools. The use of basic drafting tools is then introduced and more difficult problems in the basic units

can be solved. For this reason Chapter 3, Technical Sketching, and Chapter 4, Tools and Techniques of Drafting, were placed together early in the text.

The chapter on fastening devices was placed last. This is primarily a reference chapter. When a student needs to draw a bolt, rivet, or other fastener, he can refer to this material. It also contains reference tables detailing stock sizes for fastening devices.

To all the individuals and companies who have contributed materials and advice during the preparation of this text I give my sincere thanks. Certain material from the

American National Standards Institute, Inc., is copyrighted by and used with permission of the American Society of Mechanical Engineers, 345 E. 47th St., New York, N.Y.

Special acknowledgment is made to the contribution of Dr. George Stegman for his assistance in contributing to a number of chapters. Specifically, his assistance in the development of the following chapters—3. Technical Sketching; 7. Dimensioning; 9. Pictorial Drawing; 13. Developments and Intersections; 15. Vector Diagrams; and 24. Charts—is acknowledged and appreciated.

William P. Spence

CONTENTS

Contents

Section One
Introduction

The graphic language has developed into a detailed, highly complex means of communicating ideas. This space vehicle, weighing 6,208,949 pounds, required thousands of drawings. The various parts were built by companies all over the United States. The design features had to be recorded and communicated to thousands of workers in many different industries.

National Aeronautics and Space Administration

Links in Learning

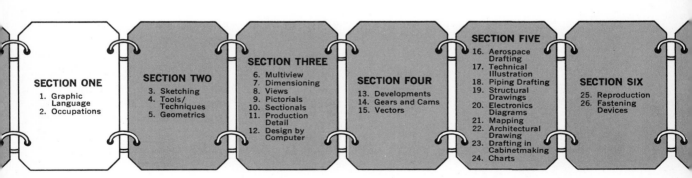

Chapter 1

The Graphic Language

The earliest records of mankind include many picture-type drawings. Man learned to use pictures to communicate with his fellowman. Perhaps the best known form of picture writing is Egyptian hieroglyphics.

The evidence of the use of drawings in early times can be seen in the remains of various civilizations. The pyramids of Egypt and the Roman buildings, aqueducts, and bridges are classic examples. See Fig. 1–1. Projects of this size could not have been built without some type of drawings to guide those building them. Examples of early drawings have been found carved in stone tablets.

As civilization developed, man learned to make materials upon which he could write and draw. For example, in ancient Egypt papyrus paper was developed. Later paper was developed as we know it today. This gave man a lightweight medium upon which to record his ideas. Today paper making is a highly developed industry. Hundreds of different papers are made. Many were developed especially for making drawings for industry.

The story of the development of the graphic language parallels the development of industry. Great strides have been made in the production of material items. The graphic language has changed and developed to keep pace with the increasing demands for drawings. The process of developing new ways and systems for recording ideas with drawings is constantly changing in this age of technological progress.

TECHNICAL DEVELOPMENT THEORIES

Many scientists who study prehistoric times claim that technical developments were slow before 8000 B.C. Man, they claim, first used fire some 300,000 years ago and also made stone tools and pottery in early times. The bow and arrow, they say, was the first composite tool. Some animal drawings have been found which are said to be 35,000 years old. See Fig. 1–2.

From 8000 B.C. to 3000 B.C., some archaeologists say, developments came faster. Some of these probably were sickle harvesting, kiln-fired pottery, brick making, the keystone arch, plows, looms, sailboats, numbering, and sign writing. See Fig. 1–3.

More discoveries probably

1–1. *This aqueduct was built in the 1300's in Italy. Notice the evidence of engineering design through the use of the arch to form the structure.*

Italian Cultural Institute

Over 50,000 YEARS AGO
BOW AND ARROW

About 8,000 YEARS AGO
SICKLE FOR HARVESTING

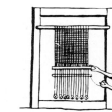

About 8,000 YEARS AGO
LINEN WEAVING
ON WOODEN LOOMS

Over 35,000 YEARS AGO
NATURALISTIC
ANIMAL DRAWINGS

7,000 YEARS AGO
LETTER-LIKE
IDEOGRAMS

6,500 YEARS AGO
COPPER

The DoAll Company

1–2. *According to many archae-ologists, these significant de-velopments took place before 8000 B.C.*

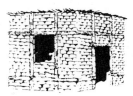

6,000 YEARS AGO
FIRST RECTANGULAR
HOUSES MADE
OF MUD BLOCKS

came between 3000 B.C. and 1000 A.D. Glass making is said to have begun around 1500 B.C. and iron smelting about 1400 B.C. Use of windmills and water wheels to produce power is said to have begun in this period. Paper, block print-ing, and cast type were devel-oped around 100 B.C. See Fig. 1–4.

Rapid developments from 1000 A.D. to 1750 A.D. led to such things as the magnetic compass, clocks, power trans-mission by rope belts, the printing press, clear glass, and the telescope. See Fig. 1–5.

From 1750 A.D. to 1900 A.D.

3,500 B.C.
KEYSTONE
ARCH

3,500 B.C.
IRRIGATION

The DoAll Company

1–3. *Students of prehistoric times estimate that from 8000 B.C. to 3000 B.C. man developed simple tools and machines. He learned to develop symbols and buildings.*

13

**1,500 B.C.
"GLASS"**

**1,400 B.C.
IRON
FROM ORE**

**850 B.C.
STONE
ROADS**

**800 A.D. FOUR-WHEEL
WAGON WITH
TURN-TABLE PIVOT**

**100 A.D.
COMPOUND
PULLEY**

**468 B.C.
CLASSIC
ARCHITECTURE**

The DoAll Company

1-4. *From 3000 B.C. to 1000 A.D. man made great strides in developing machines and using materials.*

**1200 A.D.
POWER TRANSMISSION
BY ROPE BELTS**

**1441 A.D.
GUTENBERG
PRESS**

The DoAll Company

1-5. *From 1000 A.D. to 1750 A.D. technical development continued.*

the complexity of technical developments increased rapidly. See Fig. 1–6. Following are but a few developments:

Spinning machine, 1769
Steam engine useful for manufacturing, 1781
Machine tools, 1776–1800
Cotton gin, 1791
Interchangeable precision parts, 1801
Canning food, 1804–1805
Steam locomotive, 1804
Steamboat, 1807
Telegraph, 1840
Sewing machine, 1845
Assembly line, 1847 (See Fig. 1–7)
Refrigerating machine, 1851
Bessemer steel, 1855
Open hearth steel, 1856
Oil refining, 1859
Telephone, 1876
Electric light, 1879
Steam turbine, 1884
Linotype, 1884
Automobile, 1892
X-ray, 1895

Since 1900 startling technical developments have occurred. These developments required the use of drawings of all types. Without a well-developed graphic language, these breakthroughs probably would not have occurred. Following are just a few examples of the complex type of developments. Compare these with those of earlier periods.

Commercial radio broadcasting, 1920
Major development in plastics, since 1927
Radar, 1936

1769 ARKWRIGHT SPINNING MACHINE

1781 WATT STEAM ENGINE

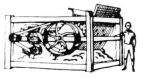

1791 WHITNEY COTTON GIN

1801 WHITNEY DUPLICATE PART PRODUCTION

1804 STEAM LOCOMOTIVE

1807 FULTON STEAMBOAT

1855 BESSEMER STEEL

1859 DRAKE OIL WELL

1876 BELL TELEPHONE

1879 EDISON INCANDESCENT LAMP

1884 MERGENTHALER LINOTYPE

1892 — DURYEA AUTOMOBILE

The DoAll Company

1895 ROENTGEN X-RAY

1-6. *Since 1750 the number and complexity of technical developments increased at a rapid pace.*

1-7. *The Ford assembly line in 1913. The chassis was pushed along a rail by hand. The dash assembly was dropped into position from an overhead rack.*

Courtesy of the Ford Archives, Henry Ford Museum, Dearborn, Michigan

15

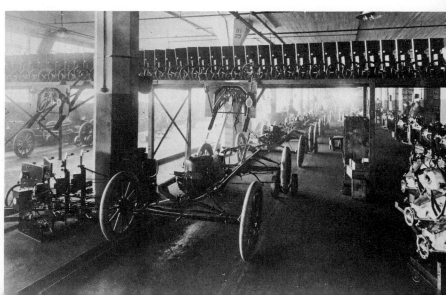

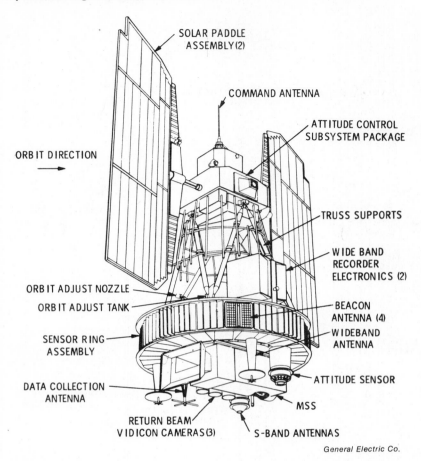

SOLAR PADDLE
ASSEMBLY (2)

COMMAND ANTENNA

ATTITUDE CONTROL
SUBSYSTEM PACKAGE

ORBIT DIRECTION

TRUSS SUPPORTS

WIDE BAND
RECORDER
ELECTRONICS (2)

ORBIT ADJUST NOZZLE
ORBIT ADJUST TANK

BEACON
ANTENNA (4)
WIDEBAND
ANTENNA

SENSOR RING
ASSEMBLY

DATA COLLECTION
ANTENNA

ATTITUDE SENSOR

MSS

RETURN BEAM
VIDICON CAMERAS (3)

S-BAND ANTENNAS

General Electric Co.

1–8. *An earth resources technology satellite used to monitor conditions on the surface of the earth.*

Gas turbine for jet propulsion in aircraft, 1937
Commercial television, 1930
Nuclear energy, 1945
First large-scale automatic digital computer, 1944
Semiconductors, 1949
Laser, 1960

Earth satellite, 1957. Fig. 1–8 shows a more advanced satellite.

Manned moon landing, 1969. See Figs. 1–9 and 1–10.

THE GRAPHIC LANGUAGE

The term *graphic language* refers to the many different ways technical information is communicated with drawings. Through the years many types of drawings have been developed and now find wide use. The chapters in this book show how to use the graphic language. Following is a summary of the more common types of drawings.

Working drawings contain all the information needed to make one part. See Fig. 1–11. There are many types of working drawings. Some of these are casting drawings, forging drawings, and stamping drawings. See Chapters 6 and 11.

Assembly drawings show how various parts of an object fit together. See Fig. 1–12. There are a number of different types of assembly drawings. See Chapter 11.

Pictorial drawings show the object somewhat as it appears if photographed. See Fig. 1–13. Some of the common types are isometric, oblique, and perspective. See Chapter 9.

Schematic drawings show electrical and electronic circuits. See Fig. 1–14. They range from big jobs, such as a high voltage electric substation, to very small projects such as the circuit for a transistor radio. See Chapter 20.

Architectural drawings are used to show how buildings are to be built. See Fig. 1–15. They include a variety of different types. Some of these are building plans, structural drawings, plumbing and electrical plans, and construction details. See Chapter 22.

Maps are made in a variety of ways. See Fig. 1–16. They include such things as a detailed layout of lots in a subdivision. Other types are used in highway construction and geological surveys. Some show the contour of land. See Chapter 21.

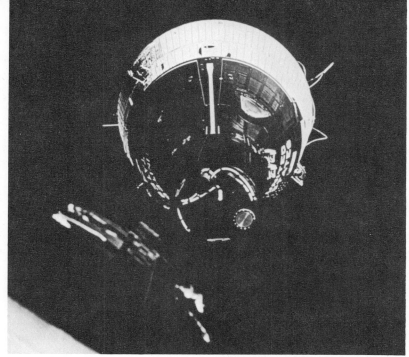

1–9. *A Gemini spacecraft during rendezvous maneuver.*

1–10. *A lunar module and its flight crew on the surface of the moon.*

Production illustrations are a form of pictorial drawing used to present technical information. See Fig. 1–17. They range from very simple drawings used in production to detailed, shaded drawings. See Chapter 17.

Charts are another form of graphic illustration. They are often used to show technical data in a form that a nontechnical person can understand. See Fig. 1–18. There are a large variety of charts in use. See Chapter 24.

A development drawing is a flat pattern. See Fig. 1–19. It is used to give the shape to cut flat material, as sheet metal. The flat piece is then formed to make the finished object or a part. The pipes on a hot-air heating system are designed this way. See Chapter 13.

Vector diagrams are used to solve mathematical problems graphically. See Fig. 1–20. For example, they can be used to find the amount of force that results when several other forces are acting on a single point. See Chapter 15.

DRAFTING STANDARDS

As industry grew and processes and materials became complex, drawings became more and more important. Many companies no longer make all the parts for their products. They employ other companies or subcontractors to make them. Subcontractors use the drawings sent to them by the employing company.

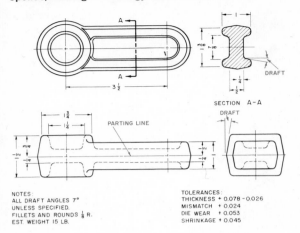

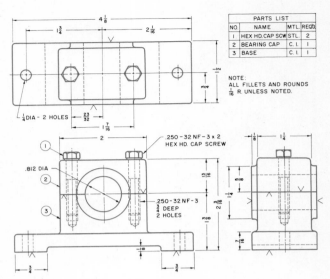

PARTS LIST			
NO.	NAME	MTL.	REQD.
1	HEX HD. CAP SCW	STL	2
2	BEARING CAP	C.I.	1
3	BASE	C.I.	1

NOTE:
ALL FILLETS AND ROUNDS $\frac{1}{16}$ R. UNLESS NOTED.

NOTES:
ALL DRAFT ANGLES 7°
UNLESS SPECIFIED.
FILLETS AND ROUNDS $\frac{1}{8}$ R.
EST. WEIGHT 15 LB.

TOLERANCES:
THICKNESS + 0.078 - 0.026
MISMATCH + 0.024
DIE WEAR + 0.053
SHRINKAGE + 0.045

1-11. *Working drawing.*

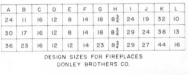

1-12. *Assembly drawing.*

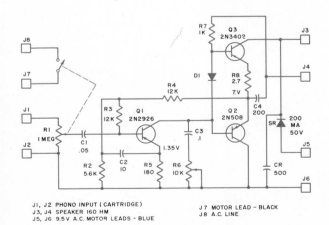

1-13. *Pictorial drawing.*

DESIGN SIZES FOR FIREPLACES
DONLEY BROTHERS CO.

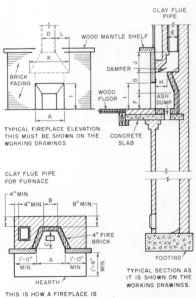

A	B	C	D	E	F	G	H	I	J	K	L
24	11	16	12	8	14	18	$8\frac{3}{4}$	24	19	32	10
30	17	16	12	8	14	18	$8\frac{3}{4}$	29	24	38	13
36	23	16	12	12	14	23	$8\frac{3}{4}$	29	27	44	16

TYPICAL FIREPLACE ELEVATION.
THIS MUST BE SHOWN ON THE
WORKING DRAWINGS.

CLAY FLUE PIPE
FOR FURNACE

THIS IS HOW A FIREPLACE IS
DRAWN ON THE FLOOR PLAN.

TYPICAL SECTION AS
IT IS SHOWN ON THE
WORKING DRAWINGS.

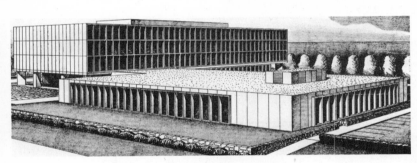

J1, J2 PHONO INPUT (CARTRIDGE)
J3, J4 SPEAKER 16Ω HM
J5, J6 9.5V A.C. MOTOR LEADS - BLUE

J7 MOTOR LEAD – BLACK
J8 A.C. LINE

1-14. *Schematic drawing.*

1-15. *Architectural drawing.*

1–16. *Map.*

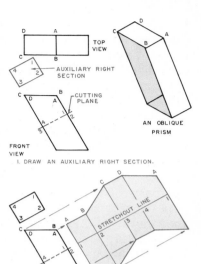

AUXILIARY RIGHT SECTION

TOP VIEW

C D

B

A

AN OBLIQUE PRISM

CUTTING PLANE

FRONT VIEW

1. DRAW AN AUXILIARY RIGHT SECTION.

STRETCHOUT LINE

2. DRAW THE STRETCHOUT LINE.
3. LOCATE THE TRUE LENGTH WIDTHS.
4. LOCATE THE ENDS OF THE CORNERS.

1–19. *Development drawing.*

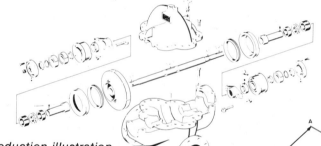

1–17. *Production illustration.*

LOAD CARRYING CAPACITIES
OF SOLID WOOD GIRDERS

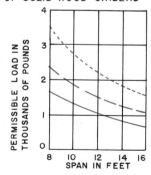

PERMISSIBLE LOAD IN THOUSANDS OF POUNDS

SPAN IN FEET

———— 3 x 6 GIRDER
— — — 4 x 6 GIRDER
- - - - 6 x 6 GIRDER

1–18. *Charts.*

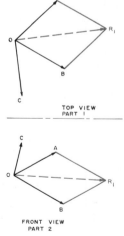

TOP VIEW PART 1

FRONT VIEW PART 2

STEP I. FIRST FIND RESULTANT R_1 OF TWO OF THE FORCES.

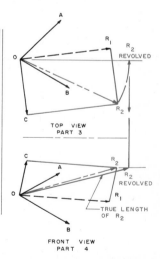

R_2 REVOLVED

TOP VIEW PART 3

R_2 REVOLVED

TRUE LENGTH OF R_2

FRONT VIEW PART 4

STEP 2. NOW FIND THE RESULTANT OF R_1 AND THE THIRD FORCE. THIS IS R_2. REVOLVE R_2 TO FIND ITS TRUE LENGTH.

1–20. *Vector diagram.*

Early it became apparent that drawings and symbols needed to be standardized so they could be understood by the different companies using them.

Some companies have developed drafting standards manuals. These are especially important for large companies having several plants in different parts of the country. It is also necessary for companies that have unique design or manufacturing procedures. In general the basic standards follow those developed on a national basis. National drafting standards are explained in the following paragraphs.

American National Standards Institute. In 1918 five American engineering societies joined together to develop national drafting standards. From this grew the American Standards Association (ASA). In 1966 the Association reorganized as the United States of America Standards Institute (USAS). In 1969 the name was changed to the American National Standards Institute (ANSI). The original organization, ASA, developed a series of drafting standards. They are identified by a letter and number code. Since the reorganization, the designating letters, ASA and USAS, are being changed to ANSI as the standards are revised.

The *ANSI Drafting Manual* is made up of 17 sections. Each is published as a separate publication. See Fig. 1–21.

ANSI Y14.1–1957	Size and Format
ANSI Y14.2–1957	Line Conventions, Sectioning, and Lettering
ANSI Y14.3–1957	Projections
ANSI Y14.4–1957	Pictorial Drawing
ANSI Y14.5–1966	Dimensioning and Tolerancing for Engineering Drawings
ANSI Y14.6–1957	Screw Threads
ANSI Y14.7–1958	Gears, Splines, and Serrations
ANSI Y14.9–1958	Forging
ANSI Y14.10–1959	Metal Stampings
ANSI Y14.11–1958	Plastics
ANSI Y14.14–1961	Mechanical Assemblies
ANSI Y14.15–1966	Electrical and Electronics Diagrams
ANSI Y14.17–1966	Fluid Power Diagrams

American National Standards Institute

1–21. *Standards in the American National Standards Institute Drafting Manual.*

The American National Standards Institute publishes hundreds of standards manuals in addition to the drafting standards. Some that are especially useful in design and drafting are listed in Fig. 1–22.

Society of Automotive Engineers. The Society of Automotive Engineers (SAE) has developed national drafting standards for use in the aeronautical and ground vehicle industries. This is published as *Aerospace — Automotive Drafting Standards.* They conform in general with the American National Standards.

Military Standards. Standards for the preparation of drawings for use in the Departments of Army, Navy, and Air Force were developed by the Department of Defense. The abbreviation used to designate these are MIL-STD.

ANSI Y15.1–1959	Illustrations for Publications and Projection.
ANSI Y15.2–1960	Time Series Charts.
ANSI Y32.2–1962	Electrical and Electronics Diagrams.
ANSI Y32.3–1959	Graphical Symbols for Welding.
ANSI Y32.4–1955	Graphical Symbols for Plumbing.
ANSI Y32.7–1957	Graphical Symbols for Use on Railroad Maps and Profiles.
ANSI Y32.9–1962	Graphic Electrical Wiring Symbols for Architecture and Electrical Layouts.
ANSI Y32.10–1958	Graphical Symbols for Fluid Power Diagrams.
ANSI Y32.11–1961	Graphical Symbols for Process Flow Diagrams in the Petroleum and Chemical Industries.
ANSI Y32.12–1960	Metallizing Symbols.
ANSI Y32.14–1962	Graphic Symbols for Logic Diagrams.
ANSI Y32.16–1965	Electrical and Electronic Reference Designations.
ANSI Y32.17–1962	Nondestructive Testing Symbols.
ANSI Z32.2.3–1949	Graphical Symbols for Pipe Fittings, Valves, and Piping.
ANSI Z32.2.4–1949	Graphical Symbols for Heating, Ventilating, and Air Conditioning.
ANSI Z32.2.6–1950	Graphical Symbols for Heat-power Apparatus.
ANSI Z32.13–1950	Abbreviations for Use on Drawings.

American National Standards Institute

1–22. *Other standards used in design and drafting.*

International Organization for Standardization. Industry has expanded its operations for standardization to other countries. For example, many products designed and manufactured by American companies are also made in foreign countries. The drawings needed to manufacture these products are sent to the foreign companies. Often personnel are sent to work in the overseas plant. This had lead to a need for international drafting standards.

The International Organization for Standards (ISO) has members from 44 countries. The United States is a member of this organization. While no drafting standards manual has been published by the ISO, the representatives reach agreements on standards to be used. These are then shown in the standards publications of each country.

CHANGES IN DRAFTING PRACTICE

Through the years there have been many changes in drafting practices. Emphasis is on accuracy and speed. Drawings must be made as rapidly as possible. Yet they must be perfectly clear and give all the needed information. Speed is no substitute for accurately describing the object.

Drawings are expensive. Attempts are constantly being made to reduce the cost of drawings. The most successful ways are through the use of

mechanical aids. These are described in Chapter 4, Tools and Techniques of Drafting.

Some of these developments include:

1. New drafting papers and polyester drafting film.
2. Electric erasers.
3. Drafting machines.
4. New inking pens.
5. Redesigned drafting tables with easily adjusted top heights and angles.
6. Tables that enable the draftsman to use a comfortable chair.
7. Microfilm.
8. New dimensioning systems.
9. Pressure-sensitive tapes.
10. Templates and mechanical lettering systems.
11. Special typewriters for typing notes on drawings.
12. Use of photography.
13. Automated drafting through the use of computers.

LEARNING THE GRAPHIC LANGUAGE

The graphic language is essential to the successful operation of industry. It is used by many persons not directly involved with the design and production processes. For example, those involved with the service and maintenance of products use drawings. The auto mechanic and air conditioning technician are ex-

amples. Persons in sales work have to refer to drawings of all kinds. For example, a salesman for an elevator company must be able to read building plans so he can understand what the architect wants. He must be able to read working drawings so he can understand how the elevator works. From all of these drawings he figures the cost of the elevator. He then must sell the customer on the values of his product. The customer must also be able to read drawings.

The graphic language is used in some way by almost everyone during their everyday lives. The designing of a house requires an understanding of drawings. Many objects, as a bicycle, are purchased disassembled. A drawing is furnished to show the purchaser how to fit it all together. How to care for a home furnace or outboard motor is shown with drawings. A basic knowledge of this form of communication is needed by all citizens in this technological age. It is a complex language that must be mastered by learning the basic fundamentals. The chapters in this text present in simplified form these fundamentals. The text carries the language beyond fundamentals to some of the more complex means of graphic communications. Careful study and application of principles to the study problems will enable you to develop the skills and knowledge necessary to communicate graphically.

Chapter 2

Drafting and Industrial Occupations

A knowledge of drafting is important to success in many different jobs. Many of these jobs require the ability to read drawings. Some require the employee to make freehand sketches. Drawings are a common way to communicate. Some typical positions requiring this knowledge include machinists and building contractors and those in the various

2-1. *This complex structure, part of a Phillips Petroleum Company refinery, required the services of many kinds of engineers and draftsmen. Knowledge of concrete and steel design was necessary to build the structure. Complex piping and electrical systems are part of the refinery. Patterns had to be developed for piping and tower shells.*

building trades, such as plumbers, masons, carpenters, and electricians. See Fig. 2-1, a refinery where the skills of many trades were required.

Many sales positions require the use of drawings. Salesmen frequently need to make drawings. Persons responsible for production supervision in manufacturing plants use drawings constantly. Those building highways, bridges, and towers work from drawings. Service personnel could not do their job if they could not read drawings. Some of these include radio and television servicemen, appliance installation and repairmen, auto mechanics, aircraft personnel, and power plant technicians. The list of occupations in which a knowledge of drawings is necessary is much longer than those listed above.

Following are some details about occupations in which drafting skill and knowledge is essential. The job titles given are those in general use. They vary some from one company to another.

Women can find successful careers in most of these occupations. The requirements for success, such as manipulative ability, artistic ability, or knowledge of mathematics, are possessed by women as well as men. Great physical strength or exposure to dangerous working conditions are not usually a part of these jobs. Increasing numbers of women are employed in these occupational areas.

THE DRAFTSMAN

The draftsman spends much of his time making drawings. He must be skillful in the use of drafting tools. A knowledge of manufacturing processes is necessary. He must know the materials used in the manufacture of his company's products. See Fig. 2-2.

Generally a beginning draftsman starts as a *tracer*. He copies drawings made by others. As he develops skill and understanding he is given

22

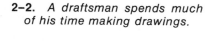

Syndicate Magazines, Inc.

2–2. *A draftsman spends much of his time making drawings.*

2–3. *The draftsman works in a clean, comfortable environment. He must have patience and be accurate in everything he does.*

more difficult jobs. The draftsman's work environment is usually clean. Modern comfortable equipment is generally provided. See Fig. 2–3.

It is not unusual for draftsmen to specialize after they complete a learning period. In large companies draftsmen might work entirely in areas such as electrical, structural, machine, or piping drafting. In these positions they must use some mathematics. They must spell correctly.

A draftsman spends many hours working at a drafting table. He must have the patience to sit long hours and do detailed work. Good eyesight is important.

Draftsmen are trained in many ways. One way is to learn on the job by working with experienced draftsmen. Usually a company expects a new employee to have drafting skills before he is employed. The usual way to get this preparation is in a high school vocational program. Junior colleges offer technical drafting courses. After a beginning draftsman gains experience, he can be promoted to a junior detailer.

A *junior detailer* revises drawings. He corrects detail and assembly drawings. With experience he begins to make simple detail and assembly drawings. Able junior detail draftsmen can be promoted to senior detailers.

A *senior detailer* draws more difficult detail and assembly drawings. He is expected to work with a minimum of supervision, use judgment, and work accurately. Generally he works from layout and engineering design drawings. With experience he is expected to make these drawings. After several years experience he can become a drafting checker.

A *drafting checker* is an experienced draftsman who is responsible for checking the final drawings for errors. After all corrections are made, he signs the drawing in a space provided in the title block. He then is responsible for any errors remaining on the drawing. In a large company the drafting checker is a full-time job. In smaller companies experienced draftsmen check each other's drawings.

A *junior designer* works from engineering notes and specifications. He works with engineering reference material. He must know how to apply tolerances. Sometimes he is given supervisory responsibilities. Many junior designers have college degrees with majors in design and drafting. See Fig. 2–4.

A *senior designer* is usually a part of the engineering staff. He is a thoroughly experienced draftsman. He must be able to work in areas involving mathematics, strength of materials, and kinematics. Kinematics is the study of motion, speed,

Keith Linscheid
Cessna Aircraft Co.

2-4. *College graduates with majors in design and drafting work on engineering design teams. They often start as junior designers.*

and rate of change of speed of the moving parts needed to make a machine operate. The preparation of all types of drawings is his responsibility. He assigns routine drafting jobs to junior draftsmen. He must constantly be aware of the cost of the product being designed.

The *chief draftsman* is in charge of all the drafting in a company. He is usually a part of management. He employs new draftsmen. Anything to do with drafting is under his direction. He sets work schedules, drafting standards, and work loads. Budgeting and purchasing for the needs of the drafting staff are his responsibilities.

THE ENGINEER

The *engineer* must be strong in mathematics and the sciences. Problem solving is his big task. He uses drawings of all kinds. Many of his solutions are developed with freehand sketches. Sometimes he makes instrument drawings to solve problems. Usually he does not make the final finished drawings. This is the work of the draftsman.

Engineers receive their education at the college and university level. The minimum time required is four years. Some engineering programs require five years of college work.

Engineers typically function in one or more activities, such as research, development, design, production, consulting, administration and management, teaching, technical writing, or technical sales and service.

The *aeronautical engineer* performs a variety of engineering work in design, construction, and testing of aircraft and missiles.

The *aerospace engineer* is engaged in research, planning, and development of flight systems and aerovehicles for use in terrestrial atmosphere and outer space.

The *civil engineer* performs a variety of engineering work in planning, designing, and overseeing construction and maintenance of structures and facilities. This includes things such as roads, airports, bridges, dams, and harbors.

The *electrical engineer* is involved in designing, planning, and overseeing manufacture, construction, installation, operation and maintenance of electrical and electronic components, equipment, systems, facilities, and machinery. These items are used in generation, transmission, distribution, and utilization of electrical energy.

The *mechanical engineer* works in planning and design of tools, engines, machines, and other mechanical equipment. He oversees the installation, operation, maintenance, and repair of such equipment. Examples include centralized heat, gas, water, and steam systems.

The *chemical engineer* designs chemical plant equipment and devises processes for manufacturing chemicals and products. Examples include gasoline, synthetic rubber, plastics, and detergents. He conducts research to develop new and improved chemical manufacturing processes. He plans layouts and oversees workers engaged in constructing, controlling, and improving equipment.

INDUSTRIAL ENGINEERING

Industrial engineers determine the most effective methods of using the units of production—manpower, machines, and materials. They may design systems for data processing, and apply operational research techniques to complex organizational, production, and related problems. Industrial engineers develop management control systems to aid in financial planning and cost analysis, design production planning and control systems to coordinate production activities and quality control, and design and improve systems for the distribution of goods and services. Some become involved in plant location studies.

INDUSTRIAL DESIGNERS

Industrial designers are concerned with the appearance and function of a product. The appearance of a product has a lot to do with sales. Industrial designers become involved with color, form, proportion, and texture. They must know about a wide variety of materials and how they are processed. They consult with many persons, including the drafting, engineering, and production staffs. See Fig. 2–5.

Many designers work for large companies. Others set up their own design studios and do work for many different companies.

Artistic ability is important in design. An industrial designer must be able to make and read drawings. He must be very skilled in making layouts of all types. The designer works with marketing and sales staffs. He must understand how business functions.

Industrial designers are educated in colleges and universities. They receive instruction in art, design, materials, and

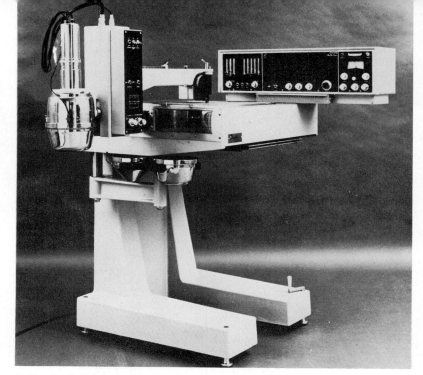

2–5. *Industrial designers work in many different industries. They are concerned with the function and appearance of the product. This is a clinical isotype scanner used in medical diagnosis. It was designed by the staff of the Nuclear-Chicago Co. It was cited for excellence in design by the American Iron and Steel Institute.*

manufacturing processes. The usual degree program requires five years to complete.

TOOL DESIGNERS

Tool designers design fixtures and cutting tools used on machine tools. Fixtures hold the work so it can be machined. The tool designer frequently makes freehand sketches. He uses mathematics and understands manufacturing processes thoroughly. Usually he has a draftsman make any finished drawings that are needed.

Tool designers must have machine shop experience. Some learn by working in in-

dustry, others learn in vocational and college programs. A simple tooling solution is shown in Fig. 2–6A, B, and C. See page 26.

ARCHITECTS

The architect plans buildings of all kinds. He decides how the inside space is to be used. The appearance of the exterior of the building is his concern. See Fig. 2–7. He designs the structure to carry the weight of the building and contents. The building must withstand other forces, as wind. Planning other parts of the building, such as heating,

Heinrich Tools, Inc.

2–6A. *This is an adjustable drill jig. It is used to hold items for drilling on a drill press.*

Home Planners, Inc.

2–7. *The architect is responsible for exterior design as well as the structural and mechanical systems in a building.*

2–6B. *This tooling was developed by a tool designer to fit on the adjustable drill jig. It holds the part in position so that a hole can be drilled in a specific location.*

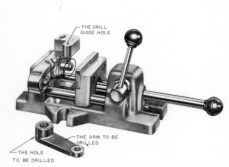

THE DRILL GUIDE HOLE

THE ARM TO BE DRILLED

THE HOLE TO BE DRILLED

2–6C. *The tooling is installed on the adjustable drill jig. The arm at the bottom is placed on the pins. The jig is closed, holding the arm in place. The drill is guided by the hole at the top.*

plumbing, air conditioning and the electrical system are the architect's responsibility. He makes estimates on the materials and cost of the building. The specifications for all

the materials are developed under his guidance. The architect sees that the building is constructed as shown on the plans. He represents the owner in dealings with the contractors.

Some architects open their own offices. Others work for architectural firms. They do work for many companies and individual customers. Others work for governmental agencies or companies large enough to employ their own architectural staffs.

Most architects are prepared in college degree level programs. These are usually five to six years in length. Mathematics and science are important subjects. An architect must pass a state examination to receive a license to practice. Usually several years experience working with a licensed architect is required before licensing.

The field of architecture requires the services of many draftsmen with knowledge of

construction and mathematics. Sometimes they make perspective drawings of the buildings.

BUILDING DESIGN AND CONSTRUCTION TECHNOLOGISTS

Building design and construction technologists work for construction companies and architectural firms. They also work for companies making and selling building materials. They know how to plan residential and commercial buildings. They are familiar with structural design, cost estimating, surveying, and management practices. Often they become supervisors of construction projects. See Fig. 2–8.

Their job differs from the architect. They tend to specialize in areas as specification writing, estimating, expediting, drafting, construction management, inspecting, and structural and mechanical design.

2-8. *Building design and construction technologists often are employed in supervisory positions in the construction industry.*

These technologists receive their education in four-year degree level college programs.

TECHNICAL ILLUSTRATORS

A *technical illustrator* makes three-dimensional drawings. These drawings closely match blueprint specifications. Generally the illustrator uses working drawings. Sometimes he uses photographs and models. He is expected to retouch photographs, use an airbrush, and do some schematic drawing. A knowledge of isometric, dimetric, trimetric, perspective, and schematic drawing techniques is vital. The illustrator must work with speed and accuracy. He must know how to lay out drawings to give the best appearance.

Since the main purpose of a technical illustration is to show a three-dimensional picture of an object, the drawings

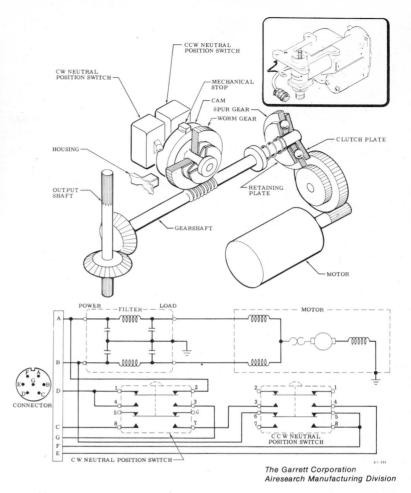

2-9. *This operational schematic drawing is the work of a technical illustrator.*

are not as exact as working drawings. See Fig. 2–9.

Study Chapter 17, Technical Illustration, for more details on the work of the illustrator. Illustrators are taught in post-high school technical programs. Some learn on the job by working for an experienced illustrator.

COMMERCIAL ARTISTS

A *commercial artist* draws and paints illustrations for advertisements, books, magazines, posters, and other such copy. He is involved with design layout. The knowledge of many techniques is important to the artist. He works with pen and ink, watercolor, pastels, scratchboard, tempera, and oils. See Fig. 2–10. He decides on the best visual effect of the media used. He prepares designs and renders details. Artistic talent and creative ability are essential.

Hallmark Cards, Inc.

2–10. *A commercial artist works with many art mediums.*

Fig. 2–11 shows an art department.

Commercial artists are usually prepared in post-high school vocational and technical programs.

CARTOGRAPHERS

A *cartographer* is one who designs and produces maps.

He works from all types of data. Some sources include a surveyor's report and aerial photographs. He needs a wide background of knowledge. He must have a sense of proportion and drafting skill. Artistic ability is a help. All types of mechanical tools are available to him. See Fig. 2–12A, B, C and D.

There are four types of cartographers. A *geocartographer* is trained in geography. He is interested in small scale and special maps. A *topographer* has engineering preparation. He is interested in surveying and large scale topographic maps. An *aerocartographer* develops maps from aerial photographs. He specializes in photogrammetry. A *cartotechnician* has art training. He is prepared in engraving, printing, and photography. He receives layouts from the geocartographer and the topo-

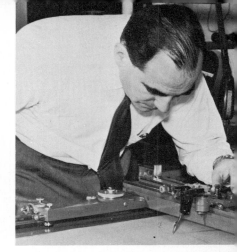

2–12A. *The cartographer uses a coordinate plotter in preparing a sheet showing latitude and longitude lines.*

cartographer and produces the finished maps.

Examples of cartographers' work are found in Chapter 21, Mapping.

TEACHING DRAFTING

There are a variety of drafting teaching positions. Junior high and senior high schools offer instruction in drafting as a part of the industrial arts program. Teachers offer their students a wide variety of experiences covering many areas of design and drafting. To qualify as teachers, they must complete a college program with a major in industrial arts.

Vocational programs prepare draftsmen for industry. If a student finds he enjoys drafting in his industrial arts experience, he can develop the extra skills and knowledge necessary in a vocational drafting program. The vocational drafting teacher must have industrial experience as a draftsman. He

2–11. *This is an art department where a wide variety of illustrations are developed.*

Aerospace and Defense Systems Operations
Philco-Ford Corporation

E. I. Du Pont

2–12B. *Water depths are plotted and then hand-drawn on nautical charts by the cartographer.*

is usually a graduate of a college level degree program.

Technical schools and institutes and junior colleges prepare design draftsmen. They emphasize mathematics and science. Before he can begin teaching, the technical drafting teacher must have industrial experience. For satisfactory employment a master's degree is often required.

Colleges and universities employ drafting teachers to prepare engineers and design and drafting technologists. Industrial experience is essential. These teachers are usually graduates of an engineering or degree level design and drafting curriculum. A master's degree is usually the minimum educational level that is acceptable. Many have the doctorate. See Fig. 2–13.

ENGINEERING AND SCIENCE TECHNICIANS

Technicians are the fastest growing occupational group in the United States. The term technician refers to technical workers whose jobs require both knowledge and use of scientific and mathematical theory, and specialized education in some aspect of technology or science. They generally work directly with scientists and engineers. There are a great number of technical positions. Job titles usually are descriptive of the technical level. Some examples are junior engineer, biological aid, or engineering technician. Some job titles are

2–12C. *Aerial photographs are used by cartographers.*

descriptive of work activity. Examples are quality control technician, tool designer, or materials tester.

Engineering and science technicians use complex electronic and mechanical instruments, laboratory apparatus, and drafting instruments. They conduct experiments and set up, calibrate, and operate instruments. They make calculations. They assist in developing experimental equipment

2–12D. *Skilled cartographers check and verify accuracy of charts by comparison with original information.*

2–13. *A drafting teacher works with students in a classroom situation. College degrees and industrial experience are necessary in order to teach.*

Dr. F. V. Sullivan
Kansas State College of Pittsburg

and models by making drawings and sketches.

The largest number of technicians are employed in the following general areas:

Aeronautical technicians work with engineers in the design and production of aircraft, rockets, missiles, and spacecraft.

Air conditioning, heating and refrigeration technicians generally specialize in some area. Examples are research and development, and design of layouts for heating, cooling and refrigeration systems. Some become involved in the manufacture of this equipment.

Chemical technicians work with chemists and chemical engineers in the development, production, sales, and use of chemical products and equipment.

Civil engineering technicians assist civil engineers in the planning and construction of highways, railroads, bridges, viaducts, dams, and other structures.

Electronics technicians work in all phases of industry utilizing electronic devices. This is a very large field. Common areas are communications, medical devices, computers, navigation equipment, and control instruments.

Industrial technicians work with production engineers on problems involving the efficient use of manpower, materials, and machines to produce goods and services.

	Approximate Number Employed
Draftsmen	327,000
Engineers	
Aerospace	60,000
Civil	180,000
Electrical	230,000
Mechanical	210,000
Chemical	50,000
Industrial	125,000
Industrial Designers	10,000
Architects	37,000
Commercial Artists and Illustrators	60,000
Engineering and Science Technicians	700,000
Teaching	
All secondary teachers	1,000,000
All college teachers	415,000

2–14. *Number of persons employed in various jobs using graphic communications skills.*

Mechanical technicians work in many areas of industry —automotive technology, diesel technology, tool design, machine design, and production technology.

There are many other rapidly developing technical areas.

These technicians receive their education in post-high school programs, found in junior colleges, area vocational-technical schools, and in-plant industrial training programs.

Students should have a good background in science and mathematics. They should enjoy working with tools, materials, and machines.

THE FUTURE

Fig 2–14 lists the approximate number of persons employed in occupations in which graphic communications play an important part. Before you decide to enter an occupation, you should study the latest information concerning the number of jobs available and the future needs. A good source of information is the *Occupational Outlook Handbook* published by the United States Department of Labor, Bureau of Labor Statistics, Washington, D.C. This is usually available in the guidance counselor's office.

The counselor can help you take inventory of your abilities. These can then be matched with the requirements for success in occupations in which you have an interest. Sound choices can be made when all this information is carefully considered.

30

Section Two

Skill Development

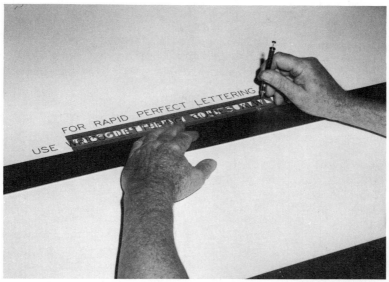

Wrico Lettering Guide

A lettering instrument used by draftsmen. The letters are cut out of the plastic template. A technical fountain pen is used to ink the letters.

Links in Learning

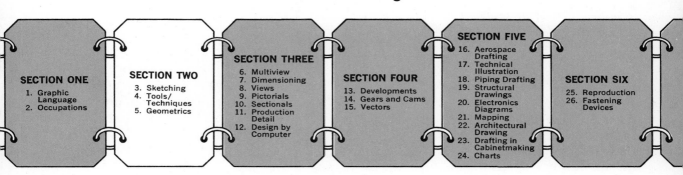

SECTION ONE
1. Graphic Language
2. Occupations

SECTION TWO
3. Sketching
4. Tools/ Techniques
5. Geometrics

SECTION THREE
6. Multiview
7. Dimensioning
8. Views
9. Pictorials
10. Sectionals
11. Production Detail
12. Design by Computer

SECTION FOUR
13. Developments
14. Gears and Cams
15. Vectors

SECTION FIVE
16. Aerospace Drafting
17. Technical Illustration
18. Piping Drafting
19. Structural Drawings
20. Electronics Diagrams
21. Mapping
22. Architectural Drawing
23. Drafting in Cabinetmaking
24. Charts

SECTION SIX
25. Reproduction
26. Fastening Devices

Chapter 3

Technical Sketching

Sketching is the oldest form man has used to record his thoughts. The cave man placed symbols on the walls of a cave and probably scratched a design in the dirt. Many of the drawings that have been found in the caves of southern France and near the Bay of Biscay show animals. Some of these drawings pointed out the spot where the animal could most easily be wounded.

Sketching has been used through the ages as a definite aid in recording shape descriptions. Its popularity has remained because it is a quick method of symbolizing one's ideas. Perhaps the greatest inventive genius in history to use sketching as a tool was Leonardo da Vinci (1452–1519). Records show that this talented man filled sketchbooks with illustrations which fully described his ideas. Numerous da Vinci sketches show alternate methods of construction as well as views of the device from different angles. See Fig. 3–1.

Thomas Edison (1847–1931), an outstanding contributor to technical progress, used sketching. He filled hundreds of laboratory notebooks with sketches.

Henry Ford, the man who "put America on wheels,"

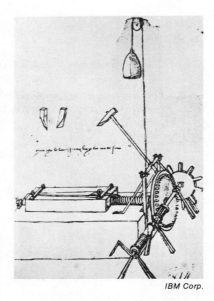

IBM Corp.

3–1. *Designers and engineers use sketches when planning new designs. This sketch was made by Leonardo da Vinci. It is a design for a machine to cut teeth on a file. Notice the use of a weight to provide power.*

used sketches in much of his work. Many of these sketches were used as the basis of the mass production system.

Sketching as it is used in industry today is the right hand of technology. No instruments are necessary. In the shop a technical sketch is used to help someone understand a blueprint. The sketch may be made on the back of an envelope, on a piece of scrap paper, on the corner of a print, or on an engineering change order. Sketching is a skill needed by the technically trained person.

Many people think that artistic talent is needed to make a technical sketch. *This is not true.* All that is required is an understanding of some basic step-by-step principles and practice to perfect these basic skills. Sketching plays a large role in all phases of industry. Almost every product that is marketed began as a crude sketch. Every engineer or designer must have some knowledge and a basic degree of skill in sketching. Many salesmen are required to sketch a proposed layout for a customer or for a sales report. Foremen in manufacturing industries frequently will make a hurried sketch to explain a drawing to a machine operator. Engineers, as well as teachers, must know how to sketch on the chalkboard. Many technical illustrators use a sketch to show the proposed method of illustration. Before beginning *any* type of drawing, the draftsman will make a sketch of the views that will be necessary to describe the object.

SKETCHING TOOLS

Pencil and Eraser. The equipment needed for sketching is not elaborate. Only two items are necessary—a pencil and an eraser. Any pencil will be sufficient. A medium soft pencil such as an F or HB drawing pencil or a common #2 writing pencil is preferred. A medium soft pencil is used because it can produce a soft, gray line for preliminary blocking-in of the sketch. The same pencil may be used for produc-

ing a bright, black line for the finished sketch. Remember the harder the pencil, the more difficult it is to produce a black line. The pencil should be sharpened to a long conical point. Approximately ⅜ inch of lead should be exposed. See Fig. 3–2.

The other piece of equipment that will be needed is an eraser. Usually a soft eraser such as a Pink Pearl or vinyl-type will do an adequate job. Both kinds of erasers are available either in block or stick

form. See Fig. 3–3A and B. Some draftsmen place a cap-type eraser on the end of their drawing pencil. See Fig. 3–3. Do not use an ink or ball-point pen eraser. These erasers contain coarse abrasive particles that will damage the surface of the paper.

Occasionally someone will use a small compass and a straight edge for making a technical sketch. These could be eliminated if the individual would learn the proper technique.

Paper. Two types of paper are used in sketching. Either plain or coordinate (paper with lines at right angles to each other) are used by draftsmen. Coordinate paper is used in many engineering offices and drafting rooms where a sketch must be made to an approximate scale. Fig. 3–4 shows some of the more common types of coordinate paper. The usual divisions are 4, 5, 6, 8, 10, 12, 16, and 20 spaces to the inch. The metric division available is 10 divisions to the centimetre.

Coordinate paper is usually sold in pads similar to a tablet. It is either printed on a heavy opaque or thin transparent paper. The lines are always printed in a light color such as orange, green, or blue. Blue grid lines are most commonly used on translucent paper because the grid will not reproduce when the sketch is printed (blue or white print). Some forms used

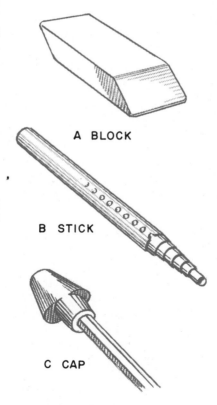

A - WOOD-CASED PENCIL

⅜

B - LEAD HOLDER

3–2. *Pencils sharpened for sketching.*

A BLOCK

B STICK

C CAP

3–3. *Commonly used erasers.*

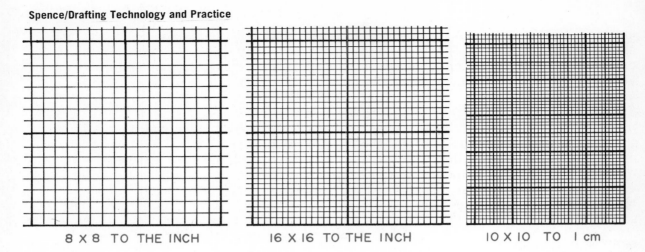

8 X 8 TO THE INCH   16 X 16 TO THE INCH   10 X 10 TO 1 cm

3–4. *Coordinate paper used in sketching.*

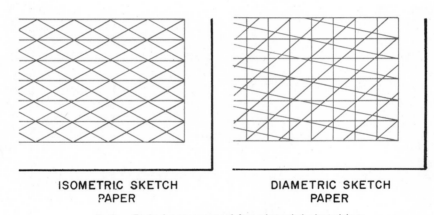

ISOMETRIC SKETCH
PAPER

DIAMETRIC SKETCH
PAPER

3–5. *Ruled paper used for pictorial sketching.*

by engineers in industry are printed on translucent coordinate paper. The engineer then sketches and has the message typed on the form to eliminate any misunderstanding. This form is then printed and distributed.

Special type rulings (divisions) are available for particular sketches. Ruled papers are available for pictorial type sketches. These are ruled with an isometric, oblique, diametric, or perspective grid and are available on either opaque or translucent paper. Fig. 3–5 shows two types of pictorial paper: isometric and diametric.

SKETCHING TECHNIQUES

Holding the Pencil. There is no absolute rule for holding the pencil. Hold the pencil in a natural position.

Most generally the pencil is held approximately 1½ to 2 inches from the point as is shown in Fig. 3–6. As a line is drawn, the pencil is rotated to keep the point from wearing flat. If the pencil is held closer to the point, it may be difficult to rotate. The pencil is held at an angle of 50 to 60 degrees

34

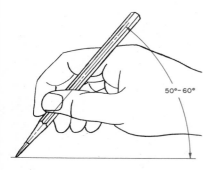

3–6. *How to hold the pencil when sketching.*

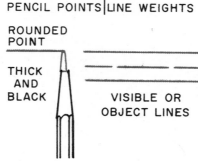

PENCIL POINTS|LINE WEIGHTS

ROUNDED POINT

THICK AND BLACK

VISIBLE OR OBJECT LINES

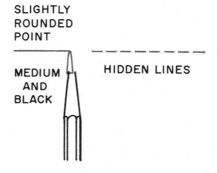

SLIGHTLY ROUNDED POINT

MEDIUM AND BLACK

HIDDEN LINES

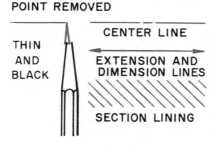

SHARP WITH POINT REMOVED

THIN AND BLACK

CENTER LINE

EXTENSION AND DIMENSION LINES

SECTION LINING

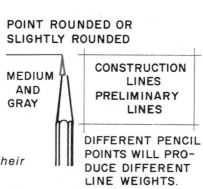

POINT ROUNDED OR SLIGHTLY ROUNDED

MEDIUM AND GRAY

CONSTRUCTION LINES
PRELIMINARY LINES

DIFFERENT PENCIL POINTS WILL PRODUCE DIFFERENT LINE WEIGHTS.

3–7. *Line symbols and their proper thickness.*

to the paper (parallel to the long bone of the arm). This angle may change depending upon the amount of pressure that is applied to the point. For example, the angle may be smaller, perhaps 40 to 50 degrees, for a preliminary or construction line and greater, 60 to 70 degrees, for a bright, black object or for hidden or center lines.

Line Thickness and Darkness. When beginning a sketch, use light, thin lines. These can be easily erased for changes. Use a medium soft pencil as 2H or H. See Fig. 3–7.

After the sketch is laid out, darken the lines. All lines should be very black. They will vary in thickness. See Fig. 3–7. Use the same line symbols as in instrument drawings. The different line thicknesses are used to call attention to the various parts of a drawing.

For example, the major outline is made thickest. It stands out above all other features.

Fig. 3–7 shows the types of pencil points used to make the different line thicknesses. Use a sharp point for thin lines. Dull the point to draw thicker lines.

Erasing. Be sure to clean the eraser before using it. Frequently the oil from one's face or hands on the erasing surface will cause a red or pink smudge. This smudge is difficult to remove. Any eraser may be cleaned by rubbing it on a piece of scrap paper.

Sketching Straight Lines. Methods of sketching straight lines vary with the individual. No line will ever be perfectly straight or uniform. In technical sketching a degree of freedom is desired in the character of the line, but try to make a line that is reasonably straight.

Horizontal Lines. Several methods are used to sketch horizontal lines. One used most frequently is the "point-to-point" method. See Fig. 3–8. One end of the line is located on the paper with a pencil point, and the other end spotted with a similar point. The pencil is held in a natural position and is moved back and forth from one point to the other. The pencil point is slightly above the surface of the paper. Be sure to keep your eye on the point toward which the pencil is to move.

35

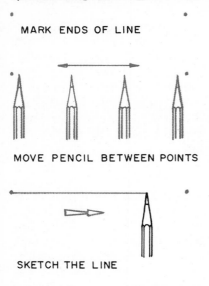

MARK ENDS OF LINE

MOVE PENCIL BETWEEN POINTS

SKETCH THE LINE

3–8. *Horizontal lines are sketched from left to right.*

Do not watch the pencil. After the pencil has been moved between the two points several times, lower the pencil and draw the line with a firm, free motion. Right-handed persons should sketch horizontal lines from left to right.

Some prefer to draw horizontal lines using a series of short strokes. See Fig. 3–9. These strokes can have a small

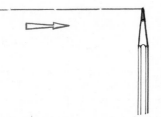

3–9. *Horizontal lines can be sketched as a series of short lines.*

36

gap between them. Another technique is to overlap the ends of each stroke. The hand is moved after each stroke is drawn. Either method is good. Use the one you can do the best.

The line sketched between points will have a tendency to be slightly curved, since your arm is pivoted from the elbow. To compensate for the arc, use a slight finger movement to pull the pencil toward the elbow. Some draftsmen prefer to compensate for this arc by moving the arm on the fleshy portion that rests on the table top.

Short horizontal lines are drawn with a finger and wrist movement. By combining a hand movement with the finger and wrist movement, the line will have a freer appearance. When the fingers and wrist are moved without moving the hand, the line will have a cramped appearance.

Vertical Lines. Vertical lines are drawn downward and toward the body. Once again begin practicing by spotting two points with your pencil on the paper. Move the pencil between the two points, keeping your eye on the spot closest to the bottom of the paper. Lower the pencil slightly until the point barely touches the paper. In this manner the direction of the line is established. Now draw the line.

If a series of parallel vertical lines are required, the "marking gage" method can be used to good advantage. See Fig.

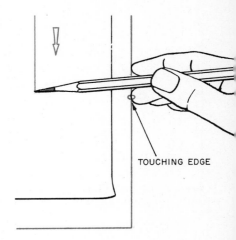

TOUCHING EDGE

3–10. *Vertical lines are sketched from top to bottom. The edge of the drawing can serve as a guide.*

3–10. Hold the pencil in its normal position with the tip of the middle or index finger touching the edge of the sketch pad. Place the pencil on the paper the desired distance the line is to be drawn from the edge. Then pull the pencil toward the body. If another line is to be drawn parallel to this line, simply move the pencil to that area and draw the line. If a single sheet of paper is being used, place the paper next to any straight edge, such as the working edge of a drawing board, desk top, or notebook. Horizontal lines may be drawn in the same manner by rotating the paper 90 degrees. Lines to be drawn near the center of the sheet pose some problems since the pencil must be extended almost to its full length.

Short vertical lines are

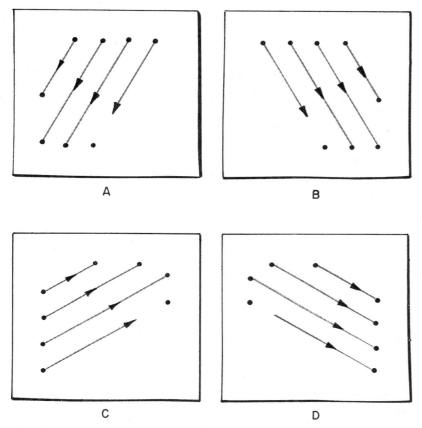

A   B

C   D

3–11.   *Directions to sketch inclined lines.*

analyze its basic geometric shapes. Is the object made up of squares, cubes, rectangles, prisms, triangles, pyramids, circles, spheres? All objects, with a few exceptions, may be broken down into basic geometric forms. In technical sketching it is important to learn how to describe graphically each of these shapes. Sketching a straight line is the foundation for sketching these geometric shapes.

Squares and Rectangles. To sketch a square, first lightly lay out two axes at right angles. See Fig. 3–12. Then using the pencil as a marking gage, make a mark on each axis representing half the width of the square. Sketch light lines through the marks. Then darken in the square with firm lines.

A rectangle is drawn in the same manner as the square. Use the pencil as an aid not only in laying out the same distance on either side of the axis, but in obtaining proportion.

If the rectangle is three times as long as it is high, place the pencil along the line representing the height. Move your hand along the pencil and position your thumb so the distance from the pencil point to the thumb corresponds with the height. Then move the pencil along the long axis, and three times that distance is the length.

Squares and rectangles may also be drawn by first sketching two parallel lines repre-

drawn by the same method as short horizontal lines. Move the pencil upward with a combined finger, wrist, and hand movement.

Remember all vertical lines may be sketched as horizontal lines if the paper is rotated.

Inclined Lines. Inclined lines which are nearly vertical should be drawn downward, Fig. 3–11A and B. Note in both instances the pencil is being pulled toward the body. For inclined lines that are nearly horizontal, sketch as

shown in Fig. 3–11C and D. With inclined lines, as with other straight lines, be sure to use two pencil "spots" to aid you.

Some students find they can sketch horizontal lines with more freedom and accuracy than inclined lines. For these students, it may be easier to rotate the sketch pad until the lines are in a horizontal position, and then sketch them.

Sketching Geometric Shapes. Pick any nearby object and

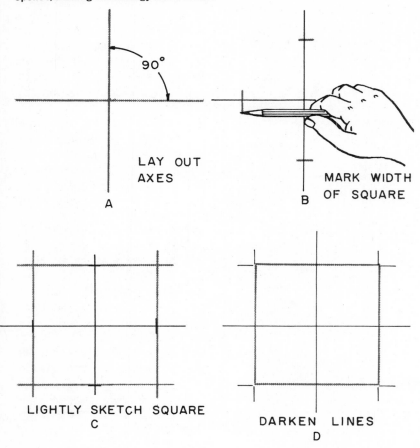

LAY OUT
AXES

A

MARK WIDTH
OF SQUARE

B

LIGHTLY SKETCH SQUARE
C

DARKEN LINES
D

3–12. *Squares and rectangles can be sketched by first laying out the axis. Then mark the width and length on the axis.*

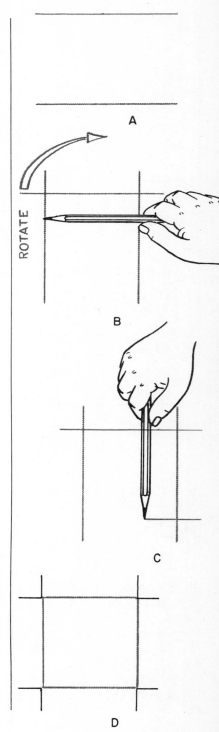

ROTATE

A

B

C

D

senting the width of the square. See Fig. 3–13A. Rotate the paper and sketch a light line representing the third side of the square (B). Use the pencil as a marking gage to transfer the width to one of the other sides with a mark. Lightly sketch in the fourth side and darken in the figure. A rectangle may be developed by the same method.

Sketching Circles and Arcs. Circles and arcs are easily sketched by first lightly drawing a square with sides equal to the circle's diameter. See Fig. 3–14A. Sketch two diagonals on the square. Then mark off the radii on the diag-

3–13. *Another way to sketch squares and rectangles.*

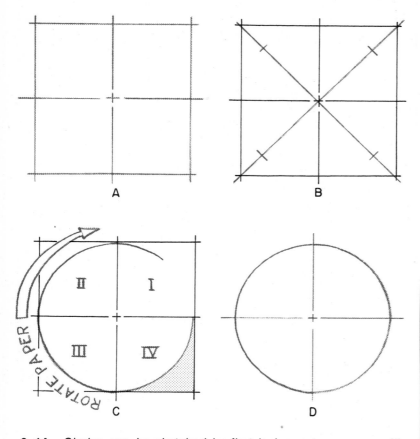

3–14. *Circles can be sketched by first laying out a square, with sides equal to the diameter of the circle.*

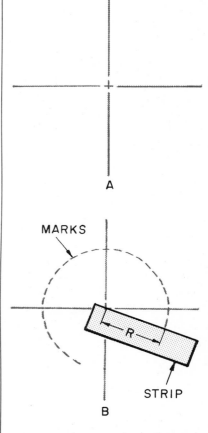

3–15. *Another way to sketch a circle.*

onals (B). Now begin to sketch in the circle (C). It is usually easier to draw an arc in one of the quadrants (I, II, III, or IV) than the other three. Where you start depends upon whether you are right or left-handed. After determining which area is easiest to draw an arc, simply move the sheet around so the next quadrant is in that position. Before darkening in the circle, check the shape of the area between the circle and square. The

areas between the circle and the square should be similar in shape.

Large circles and arcs may be drawn with a paper strip. First draw two axes as shown in Fig. 3–15A. Mark off the radius on a strip of scrap paper and move it around the center. Place as many marks as are necessary. Then sketch the circle through these points.

Another method of drawing large circles and arcs is to use

the pencil and hand as a compass, Fig. 3–16. Place the little finger as a pivot on the intersection of two axes you have drawn. Put the pencil point at the radius and rotate the paper. The hand must be held rigid in this position as the paper is rotated.

Small circles and arcs, no larger than $\frac{1}{2}$ or $\frac{5}{8}$ inch may be drawn without an enclosing box. These are made with one

39

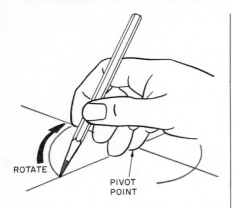

3–16. *To sketch a large circle, use the fingers as a compass.*

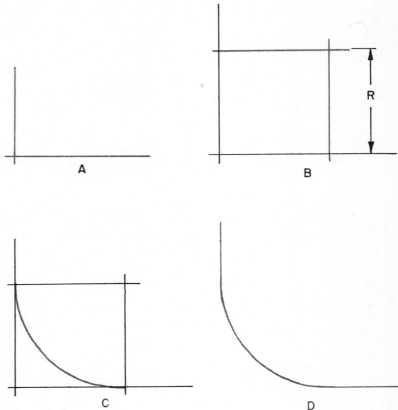

3–17. *Arcs are sketched by laying out the axis and forming a box. The side of the box is equal to the radius of the arc.*

complete movement, using the hand and finger.

Larger arcs are sketched in the same manner as circles. When sketching an arc, the usual method is to place it in an enclosing box. See Fig. 3–17. The side of the square is equal to the radius of the arc. Arcs may also be drawn by the paper strip or pencil and hand methods. Size will determine the method.

Sketching Ellipses. An ellipse has two axes of different lengths. The long axis is the major axis. The short axis is the minor axis. See Fig. 3–18.

The easiest method of sketching an ellipse is to first sketch a rectangle. The length and height are equal to the major and minor diameters. See Fig. 3–19A. Draw diagonals from the corners of the rectangle. Estimate one-third of half the diagonal and mark this distance from each corner (B). Then sketch in light tangent arcs at the mid-points of

the enclosing box (C). Now sketch in the ellipse with light lines. Check the areas between the ellipse and the rectangle. Be certain that they are the same size. When the ellipse has the proper shape— no flat sides or pointed ends, Fig. 3–20, darken it in.

Ellipses of *all sizes* should be placed in a rectangle. Use of a rectangle will insure the proper position of the ellipse. Small ellipses (the major axis no larger than ½ inch) can be sketched with a single hand and finger motion.

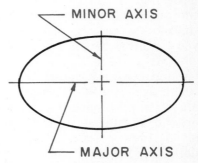

3–18. *An ellipse has a minor and a major axis.*

40

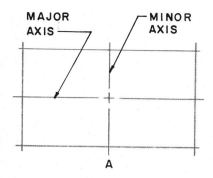

MAJOR AXIS — MINOR AXIS

A

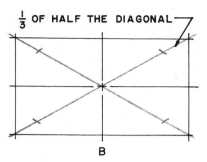

⅓ OF HALF THE DIAGONAL

B

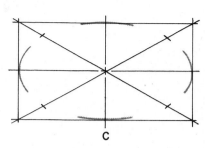

C

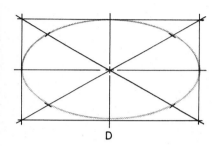

D

3–19. *An ellipse can be sketched by drawing a rectangle. One side is equal to the minor axis. The other is equal to the major axis.*

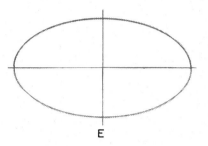

E

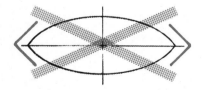

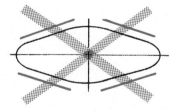

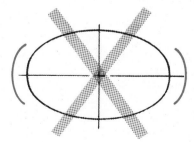

3–20. *An ellipse does not have pointed ends or flat sides.*

PROPORTION

Proportion is one of the key factors in producing a "good sketch." The term proportion refers to drawing parts of an object in the same size relationship as the object itself. It is not necessary to use a rule or scale in measuring the length of lines.

Proportion may be obtained by three different methods. They are (1) approximation by eye; (2) approximation by pencil and eye; and (3) actual measurement.

Approximation by Eye. This method is the most practical. It is the fastest since no measurements are made. First the object to be sketched is studied and the sizes of its

41

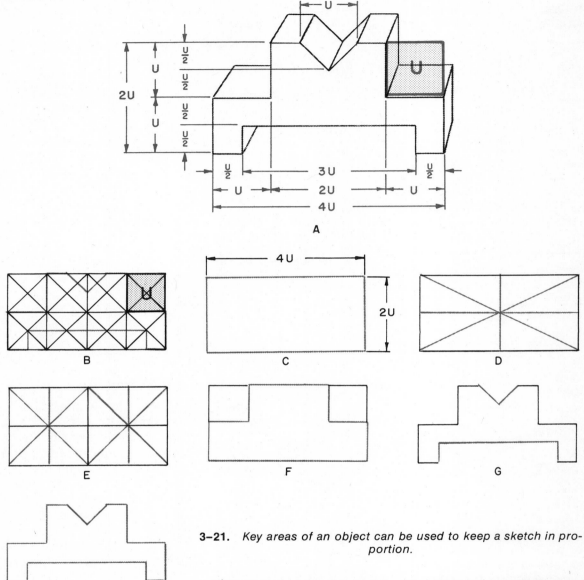

3–21. *Key areas of an object can be used to keep a sketch in proportion.*

various shapes are noted. Second, the size of each shape is compared with the other. Third, the size of each shape is compared with the overall width, height, and depth. This

relationship is gained by comparing general areas and distances visually.

After studying the object to be sketched, choose an area or shape which seems to relate to the entire object. In Fig. 3–21, Part A, the shaded area, U, has an approximate rela-

tionship of one-eighth of the total area. Upon further examination it is observed that the object is twice as long as it is high (4U by 2U). See Fig. 3–21, Part B. The shaded area, U, then becomes a key area in analyzing other parts of the object. The V-shaped notch and recess on the underside of the base each are a half unit high $\left(\dfrac{U}{2}\right)$.

Use the key area, U, as a basis for determining proportion. First lay out the base line and the general shape of the view. See Fig. 3–21, Part C. Next divide the basic shape into key areas (D and E) by diagonals. Then block in the basic shape (F). The height of the V-shaped slot and recess on the underside are each one-half unit. Divide one unit along the top and bottom rows by diagonals and lightly sketch in these features (G). Now erase the unneeded lines. Darken in the outline (H).

All features of objects do not neatly fall into units of quarters or one-eighths. The illustration used in Fig. 3–21 is designed to show the principle of using key areas.

Pencil and Eye. Another method of obtaining proportion is the pencil and eye method. Artists use this method frequently to obtain proportion and distance in their drawings and paintings. Fig. 3–22 illustrates the use of the pencil as an aid in obtaining proportion. The pencil is held comfortably in the hand with the arm outstretched at its full length. If the hand is moved closer to the eye, the dimension will decrease. Be sure to hold the hand the same distance from the eye for each measurement; otherwise, the

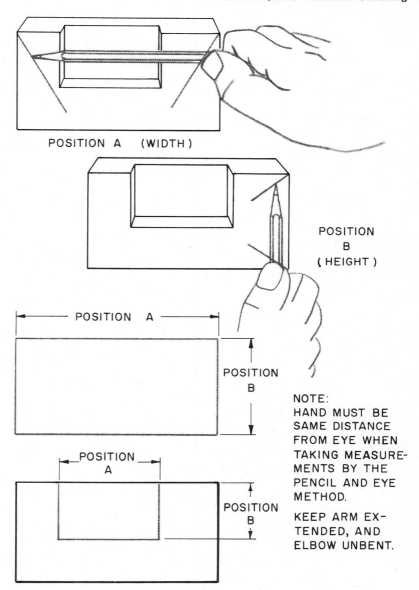

POSITION A   (WIDTH)

POSITION B (HEIGHT)

POSITION A

POSITION B

POSITION A

POSITION B

NOTE:
HAND MUST BE SAME DISTANCE FROM EYE WHEN TAKING MEASUREMENTS BY THE PENCIL AND EYE METHOD.

KEEP ARM EXTENDED, AND ELBOW UNBENT.

3–22.   *The pencil and eye method can be used to obtain proportion.*

dimensions will not be proportional. Each measurement is marked off on the sketch.

The pencil and eye method may also be used to obtain an approximate angle of an inclined surface or line. See Fig. 3–23. The pencil is moved until one side, or an imaginary line through the center of the pencil, is parallel to the line in question. Once the pencil is

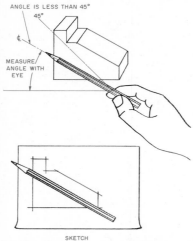

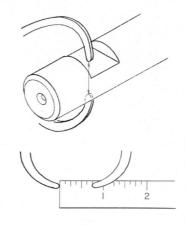

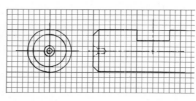

3–23. *Angles can be estimated using the pencil and eye method.*

parallel to the feature, estimate its position to a horizontal or vertical line. In Fig. 3–23, the line is slightly less than a 45-degree angle. It is best to use a 45-degree angle as your basis of judgment, since this angle is easiest to approximate. After gaining practice in sketching and becoming familiar with angles, you can also use 30- and 60-degree angles for estimating.

Actual Measurement. This method is most accurate because a scale, rule, inside or outside caliper is used to obtain measurements. These distances are then transferred to a coordinate-ruled sketch sheet. Fig. 3–24 shows a measurement taken with a pair of outside calipers and laid out on a ruled sheet (eight divisions per inch).

When several views compose a technical sketch, all

3–24. *The most accurate way to obtain size on a sketch is to lay out the actual measurements on coordinate paper.*

views are developed at the same time. Do not complete one view at a time. If all views are sketched at the same time, there is a greater probability that they will be in correct proportion.

PICTORIAL SKETCHES

Pictorial sketches are very useful in presenting ideas. The types of pictorial drawings are shown in Chapter 9, Pictorial Drawing. Study this chapter so these drawings are clearly understood. The terms isometric, oblique, and receding axis are defined and illustrated in Chapter 9. Suggestions to help make pictorial sketches follow.

Isometric Sketches. To make an isometric sketch, see Fig. 3–25.

1. Lay out the isometric axes, Fig. 3–25B.
2. Sketch an isometric box. The height, width, and depth of the box should be the same as those of the object, Fig. 3–25B and C.
3. Sketch on all details, Fig. 3–25D and E.
4. Erase unneeded lines.
5. Darken all lines.

Some uses of the isometric box are shown in Fig. 3–26.

To sketch circles in isometric, see Fig. 3–27. The procedure is much the same as sketching an ellipse.

1. Sketch the center lines of the circle in isometric.
2. Sketch an isometric box to enclose the circle. The sides of the box should equal the diameter of the circle.
3. Sketch the ellipse within the box.

Oblique Sketches. For making an oblique sketch, see Fig. 3–28.

1. Sketch the front view in the same manner as for a multiview drawing.
2. Sketch the receding axis. A 45-degree angle is most commonly used.
3. Estimate and mark the depth. Draw the back edges.
4. Sketch all details.
5. Erase unneeded lines.
6. Darken all lines.

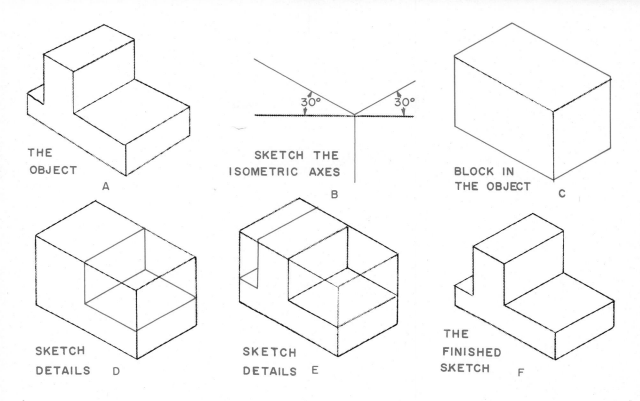

THE OBJECT

A

SKETCH THE ISOMETRIC AXES

B

30°   30°

BLOCK IN THE OBJECT

C

SKETCH DETAILS   D

SKETCH DETAILS   E

THE FINISHED SKETCH   F

3–25. *How to make an isometric sketch.*

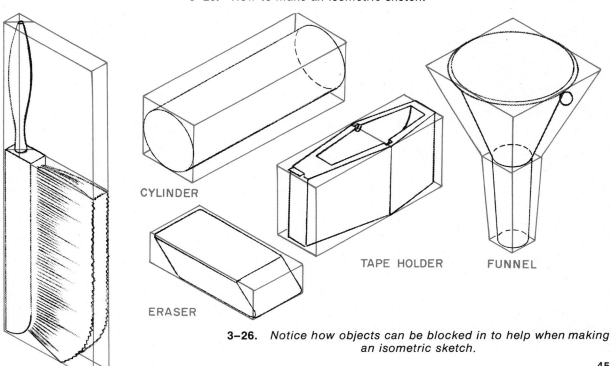

DUSTING BRUSH

CYLINDER

ERASER

TAPE HOLDER

FUNNEL

3–26. *Notice how objects can be blocked in to help when making an isometric sketch.*

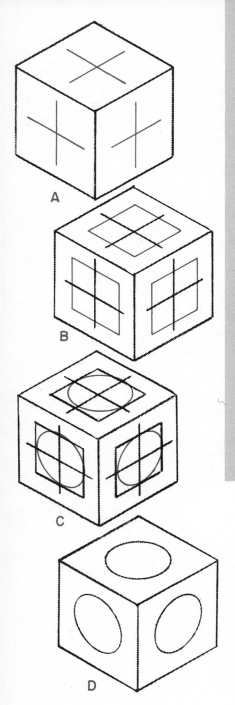

3-27. *Isometric circles are sketched by first drawing an isometric square. The sides of the square are equal to the diameter of the circle.*

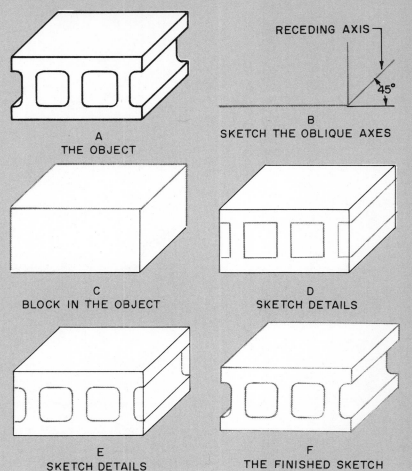

A
THE OBJECT

B
SKETCH THE OBLIQUE AXES

C
BLOCK IN THE OBJECT

D
SKETCH DETAILS

E
SKETCH DETAILS

F
THE FINISHED SKETCH

RECEDING AXIS
45°

3-28. *How to make an oblique sketch.*

Circles in oblique that are parallel with the front plane are drawn round. Those on the sides appear as ellipses. See Fig. 3-29.

To sketch an oblique ellipse, follow the same steps for sketching an isometric circle. See Fig. 3-27.

Sketching in Perspective. First study the principles behind perspective drawing found in Chapter 9, Pictorial Drawing. These steps are used to sketch in perspective. The major difference is accuracy. Since perspectives rely upon many projections, a sketch tends to become inaccurate. But, it can be very descriptive even though inaccurate.

Perspective grid paper is a big help for sketching. It has printed lines that will help with the projection of points. Fig. 3-30 shows a perspective sketch.

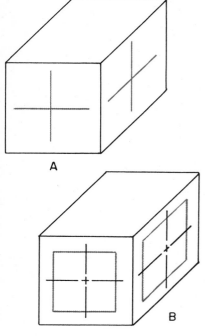

A

B

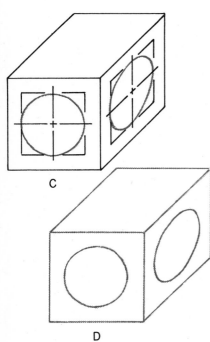

C

D

3-29. *How to sketch circles in oblique.*

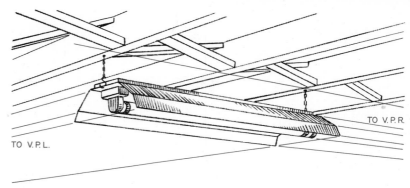

TO V.P.L.

TO V.P.R.

HORIZON LINE

3-30. *This is a two-point perspective sketch.*

Study Problems

The following problems will help develop skill in sketching. As sketches are made, pay close attention to proportion. Sketching skill and proportion are important to solving problems graphically. See P3-1.

1. Sketch the archery target twice as large as shown.
2. Sketch the hatbox three times as large as shown.
3. Make a sketch of the casement window. Enlarge it three times the size shown.
4. On graph paper make a full-size sketch of the irregular curve.
5. Enlarge the drawing of the explosive rivet three times.
6. Sketch the model car race track. Make the sketch twice as large as shown.
7. Study the section through the wall of a frame house. Sketch it twice as large as shown.
8. Make a sketch of the doghouse. Make it the same size shown.
9. Enlarge the drawing of the basketball backboard to twice the size shown.

The following problems can be found in the study problem section of Chapter 6, Multiview Drawing:

10. Make an isometric sketch of the rubber gasket, P6-4.
11. Make an isometric sketch of the asphalt paving block, P6-5.
12. Make an isometric sketch of the angle plate, P6-10.
13. Make an oblique sketch of the rapid release electrical connector, P6-15.
14. Make an oblique sketch of the V block, P6-35.

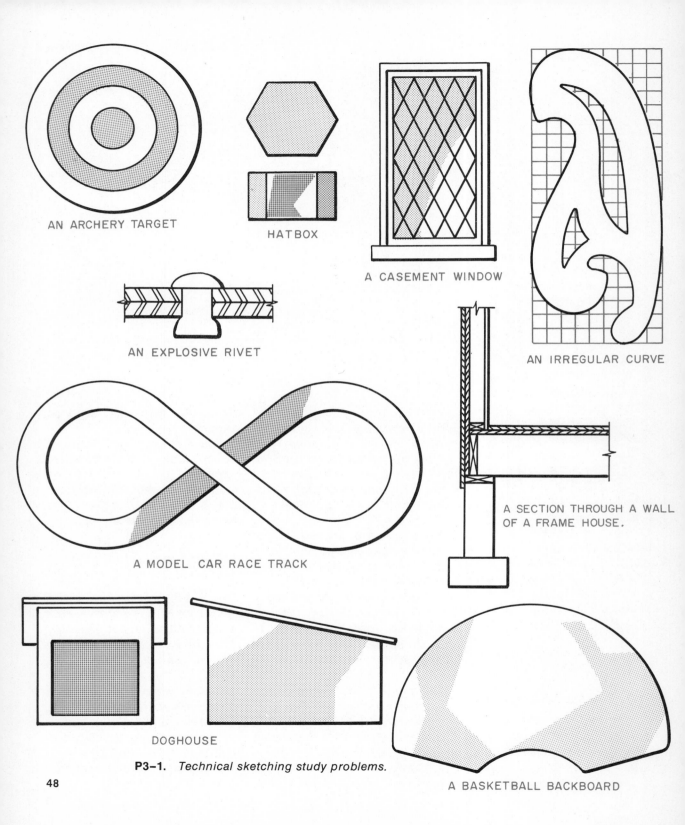

AN ARCHERY TARGET

HATBOX

A CASEMENT WINDOW

AN IRREGULAR CURVE

AN EXPLOSIVE RIVET

A MODEL CAR RACE TRACK

A SECTION THROUGH A WALL OF A FRAME HOUSE.

DOGHOUSE

P3–1. *Technical sketching study problems.*

A BASKETBALL BACKBOARD

Chapter 4

Tools and Techniques of Drafting

THE DRAWING SURFACE

The most commonly used material for the drawing surface is soft wood. It must be smooth and flat. Drawing boards and drafting table tops are made of this material.

Drawing boards are made from narrow strips of basswood. See Fig. 4–1. They are glued together and have a cleat on each end. The cleat can be wood or metal. The cleats prevent the board from warping. They also provide a working edge for the T-square.

Most drafting is done on tables, Fig. 4–2. They are made of wood or metal. The top is always soft wood. They have cleats on each end. The tops are adjustable to any angle desired.

Another type of drafting table is shown in Fig. 4–3. The top can be adjusted to any angle. The height of the top can also be easily changed. This permits the draftsman to use a regular chair instead of the high drafting stool.

Mayline Co.

4–3. *The top on this drafting table can be raised and lowered. It can be tilted to any angle desired.*

Mayline Co.

4–2. *A typical wood drafting table.*

Many draftsmen prefer to cover the top of the table with a linoleum or plastic material. It gives a surface that is a little softer than the wood top.

FASTENING PAPER TO THE BOARD

When placing the sheet to a drawing board or table top, it should be placed well up on the surface. If the paper is placed too close to the bottom, the hand and arm will not have adequate support. When fastening the paper to the board, begin by placing it near the left side of the board. See Fig. 4–4. Align the top edge of the sheet or border line with the top edge of the T-square. Now place your hand in the middle of the sheet and move the straight edge downward. Next take two pieces of drafting tape and place these in the upper two corners. Before taping the bottom corners, recheck the alignment of the sheet with the straightedge. If the sheet has slipped, correct the error by lifting one of the pieces of tape and realign.

Frederick Post Co.

4–1. *A wood drawing board. Notice the wood cleats on each edge.*

49

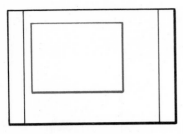

1. The paper is placed near the left and top of the drawing board.

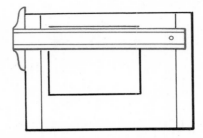

2. Line up the top of the paper with the top edge of the T-square.

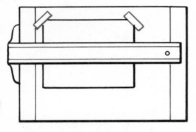

3. Slide the T-square down. Tape the top corners of the paper to the board.

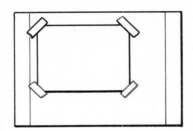

4. Smooth the paper and tape the lower two corners.

4-4. How to fasten paper to the drawing board.

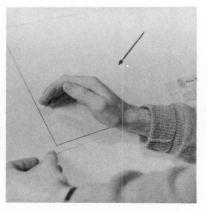

4-5. The paper is smoothed and pulled toward each bottom corner.

Now tape the two bottom corners. The sheet should be stretched by drawing the hand diagonally from an upper corner to the lower corner. Place a piece of tape in this corner. Do the same to the other corner. See Fig. 4-5.

Drafting Tape. Drafting tape is used to fasten the drafting paper to the board. See Fig. 4-6. Tape is preferred to staples or thumb tacks since it does not damage the drawing or board top. Do not use transparent or masking tape in place of drafting. The adhesive qualities of these tapes are such that they will damage the drawing paper when removed.

A very small piece of tape is all that is needed. It could be

3M Company

4-6. Drafting tape is used to hold drawings to the board.

¼ to ⅜ inches wide. Best procedure is to cut the tape from the roll with a scissors. This leaves a clean edge that is not as apt to stick to the straightedge.

THE T-SQUARE

The T-square is used to draw horizontal lines. It also serves as a base for triangles to draw vertical lines. The T-square has two parts—a head and a blade. See Fig. 4–7. They should be joined together firmly. No movement can be permitted.

A special type of T-square has an adjustable head. It is used to draw parallel inclined lines. See Fig. 4–8.

T-squares have clear plastic edges. This permits lines beneath the edge to be seen. Blade lengths are available from 18 to 60 inches.

Testing a T-Square for Straightness. The blade on a T-square must be perfectly straight if quality drawings are to be made. This can be tested by drawing a straight line the length of the blade through two points. Then turn the T-square over and try to draw a line through the same two points. Any space between these lines shows the blade is not straight. The amount of error is equal to half the space between the lines. See Fig. 4–9.

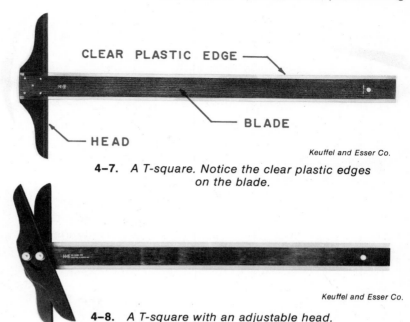

CLEAR PLASTIC EDGE

BLADE

HEAD

Keuffel and Esser Co.

4–7. *A T-square. Notice the clear plastic edges on the blade.*

4–8. *A T-square with an adjustable head.*

Keuffel and Esser Co.

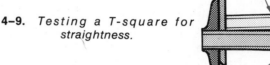

4–9. *Testing a T-square for straightness.*

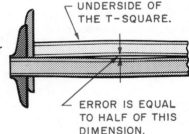

UNDERSIDE OF THE T-SQUARE.

ERROR IS EQUAL TO HALF OF THIS DIMENSION.

THE PARALLEL STRAIGHTEDGE

The parallel straightedge replaces the T-square. See Fig. 4–10. It moves up and down the drawing surface on a wire. The wire keeps it in a parallel position.

It has the advantage that it can be moved with one hand and will not lose its parallel position.

Mayline Co.

4–10. *A parallel straightedge mounted on a drawing board.*

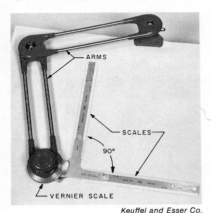

ARMS

SCALES

90°

VERNIER SCALE

Keuffel and Esser Co.

4–11. *An arm type drafting machine.*

DRAFTING MACHINES

There are two basic types of drafting machines. One is the arm type, Fig. 4–11. The other is the track type, Fig. 4–12. Drafting machines are a combination of several drafting tools. They replace the straight edge, triangles, scale, and protractor. The main advantage is that they reduce drafting time. The draftsman has all these tools at hand in one machine.

The two arms are fixed at a 90-degree angle to a round head. These arms serve as a straight edge, triangles, and scale. The head is held by the draftsman in his left hand. He moves the machine up and down and across the drawing at any angle. The round head has a pivot. A release button lets the draftsman rotate the arms to any angle desired. Angles in degrees are marked on the head. This replaces the protractor.

Drafting machines are made for right and left hand persons.

TRIANGLES

The two commonly used triangles are the 45 degree and the 30–60 degree. See Fig. 4–13. They are available in a variety of sizes. The size is the distance measured along the longest side of the right angle. They are made of clear and colored transparent plastic. Triangles are easily damaged and should be handled carefully.

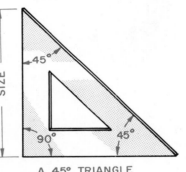

SIZE

45°

90°

45°

A 45° TRIANGLE

4–13. *Triangles used in drafting.*

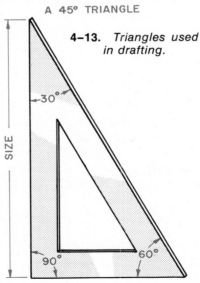

SIZE

30°

90°

60°

A 30°–60° TRIANGLE

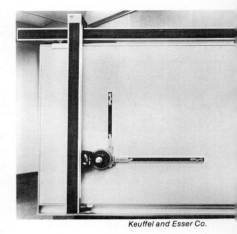

Keuffel and Esser Co.

4–12. *A track type drafting machine.*

Adjustable Triangles. Draftsmen are frequently required to draw angles at other than the usual 15-degree increment. To save time he will usually use an adjustable triangle. See Fig. 4–14. An advantage of the adjustable triangle is that it may be set at the proper angle, and lines may then be drawn parallel. Most adjustable triangles are graduated in half degrees. The movable portion of the triangle is held in position by a screw.

SCALES

A scale is used to measure or lay out a line on a drawing either in full size or larger or smaller than full size. The term *in scale* means the drawing can be larger, smaller, or the same size as the object. The size of the object and the sheet size will determine the scale of the drawing. Industrial

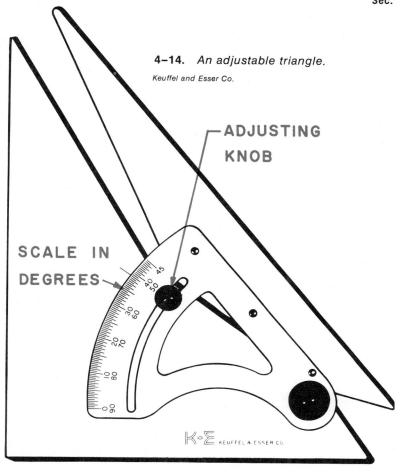

4–14. *An adjustable triangle.*

Keuffel and Esser Co.

— ADJUSTING
KNOB

SCALE IN
DEGREES →

45
40
50
30
60
20
70
10
80
0
90

K+E KEUFFEL & ESSER CO.

practice recommends that the object should be drawn full size whenever possible. It would be difficult to draw the balance wheel of a watch full size since it is so small. The scale should be increased greatly. A wheel for a freight car would be drawn smaller than full size.

Scales are made flat and triangular. A triangular scale is shown in Fig. 4–15. A flat scale is shown in Fig. 4–16. Fig. 4–17 shows all the available shapes. Scales are available that are open divided and fully divided. *Open divided* scales have a fully subdivided unit at the end. The remainder of the scale's length has only the main units of measure.

4–15. *A triangular scale. The measurement is marked with a short dash.*

4–16. *A flat scale. Notice the position of the pencil for marking the measurement.*

	TWO BEVEL	Wide base with complete visibility of both faces.
	OPPOSITE BEVEL	Easy to lift by tilting.
	FOUR BEVEL	Four faces for four scales.
	REGULAR TRIANGULAR	Permits full face contact with drawing.
	CONCAVE TRIANGULAR	Only edges of bottom scales are in contact with drawing.

Frederick Post Co.

4–17. *Shapes of scales available.*

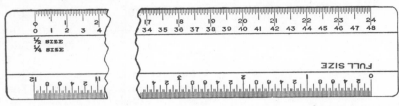

4-18. *A fully divided mechanical engineer's scale.*

T. A. Alteneder and Sons

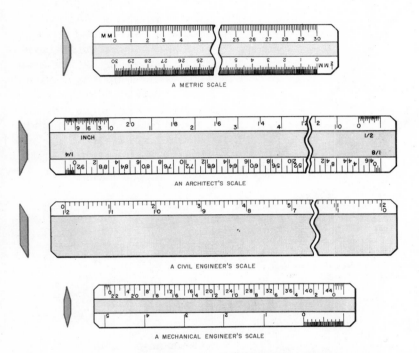

A METRIC SCALE

AN ARCHITECT'S SCALE

A CIVIL ENGINEER'S SCALE

A MECHANICAL ENGINEER'S SCALE

4-19. *Four types of scales used by draftsmen.*

The architect's scale, Fig. 4-19, is open divided. A fully divided scale has every unit along its length fully subdivided. See Fig. 4-18.

There are several types of scales used in drafting. These are the *mechanical* engineer's scale, the *architect's* scale, the *civil engineer's* scale, and the *metric* scale. These are shown in Fig. 4-19.

The *architect's scale* is used for working drawings of buildings. It is also used for drawing machine parts. All of the scales on this tool represent one foot. For example, the ¼ scale means that ¼ inch represents one foot on the drawing. The scales are ⅛, 3/32, 3/16, ¼, ⅜, ½, ¾, 1, 1½ and 3. At the end of each scale the unit used, as ½ inch, is divided like a 12-inch ruler. See Fig. 4-20. Here the ½-inch length is divided into 24 parts. Each part equals one-half inch. Following are the divisions found on the architect's scale.

Scale	Each Mark Represents
3	1/8"
1½	¼"
1	¼"
¾	½"
½	½"
⅜	1"
¼	1"
3/16	1"
3/32	2"
⅛	2"

The architect's scale is divided into feet along the entire length. Only the foot division at the end is divided into inches or fractions of an inch. To measure a length, both are used. To measure 3'5" start with the 0 mark. Count 3 feet to the left of the 0 and 5 inches to the right. See Fig. 4-20. These points are marked with a short line.

If the scale is not divided so a needed measurement is marked, estimate the measurement.

The 3-inch and 1½-inch scales are used to enlarge drawings. The 3-inch scale will enlarge a drawing 3 times. The 1½-inch scale will enlarge a drawing 1½ times.

The *mechanical engineer's* scale is divided to make drawings one-eighth size, quarter size, half size, or full size. See Fig. 4–19. If the scale is half size, this will reduce the drawing to half the true size. The 1/2-inch length is divided into fractions of an inch. The length of the scale is divided into inches.

The *civil engineer's* scale is divided into decimal parts. The divisions used are 10, 20, 30, 40, 50, and 60 parts to the inch. The scale marked 10 means that the inch is divided into 10 parts. See Fig. 4–19.

This tool is used for making maps. It is also used to make drawings that have decimal dimensions.

The metric system of measurement is finding wider use. This system is explained in detail in Chapter 7, Dimensioning. Scales are made with the metric increments, centimetres, and millimetres. A scale may have all metric scales or have some metric and some customary scales. A typical metric scale is shown in Fig. 4–19.

Generally a metric scale has a 30-centimetre length, fully divided into centimetres and millimetres. Common divisions include centimetres divided into millimetres, and centimetres divided into half-millimetres.

TEMPLATES

A template is a thin, flat, plastic tool with openings of different shapes cut into it.

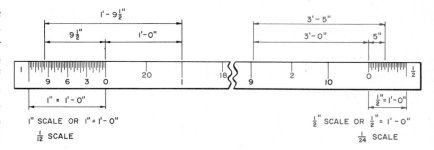

4–20. *These are the 1″ = 1′ − 0″ and 1/2″ = 1′ − 0″ scales on an architect's scale. Some scales are made with the measurements overlapping, as above. The colored scale is the 1/2″ = 1′ − 0″ scale. The black scale is the 1″ = 1′ − 0″ scale.*

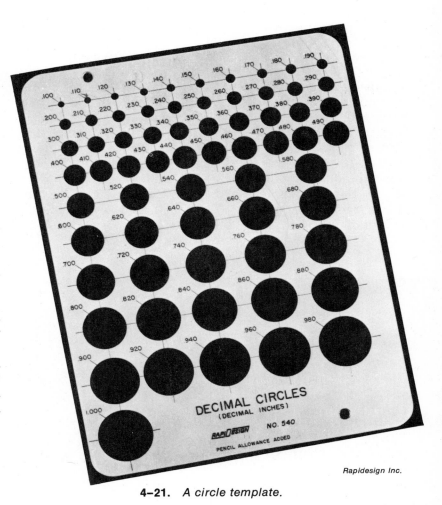

Rapidesign Inc.

4–21. *A circle template.*

See Fig. 4–21. The pencil is placed into the opening and moved along the outline. Most templates are made to allow for the thickness of the pencil lead.

There are many different kinds of templates. Some examples are circle, ellipse, triangles, squares, architectural symbols, and bolts.

DRAWING INSTRUMENT SETS

Fig. 4–22 shows a typical set of drawing instruments. There are many different kinds of sets. Some contain many more sizes of these tools. A basic set will contain a compass, dividers, and ruling pen.

The compass is used to draw circles and arcs. The compass shown in Fig. 4–22 is a bow compass. It has a center wheel adjustment. This wheel is turned to adjust the distance between the legs. Compasses without this adjustment are called friction compasses.

Compasses have a pin on one leg. The other leg has attachments to hold lead or an inking pen.

HOW TO USE A COMPASS

To use a compass, set the distance between the legs to the radius of the circle to be drawn. To get this setting, mark the distance with a scale on a paper. Place the pin leg on one mark. Adjust the other leg until it touches the second mark. See Fig. 4–23.

Hold the compass in one hand at the top. Turn clockwise. Tilt it a little in the direction the compass is moving. Keep enough pressure on the lead point to get a dark line. See Fig. 4–24.

The lead in a compass is sharpened on a file or sanding pad. It is formed to a wedge edge. A sharp, slanted surface is formed. The edges are then sanded lightly to produce the wedge point. See Fig. 4–25.

4–23. *To set the radius on a compass, mark the distance on the drawing. Adjust the compass to this mark.*

HOW TO USE THE DIVIDERS

Dividers look much like compasses. They have pin points on both legs. Dividers are made in center wheel and friction types.

The common uses for dividers are:

1. *To mark off equal spaces.* This is done by spacing the legs to the distance wanted.

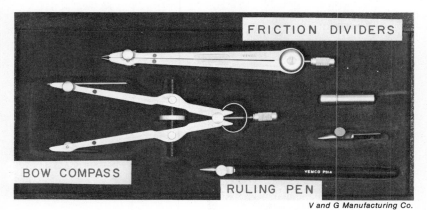

4–22. *A set of drawing instruments.*

FRICTION DIVIDERS

BOW COMPASS

RULING PEN

V and G Manufacturing Co.

4–24. *Swing the compass clockwise. Slant it in the direction it is moving.*

Place one point at the beginning of the line. Swing the other point until it touches the line. Mark a point here. Then swing the first leg around and mark the second point. See Fig. 4–26. Repeat this stepping process until the number of equal spaces wanted are marked.

2. *To divide a line into equal parts.* Draw the line to be divided. Mark the beginning and end of the line. Estimate the distance wanted. For example, a line is to be divided into five equal distances. Set the dividers at what you think is one-fifth of the distance. Place one point at the end of the line. Step along the line five times. If this does not reach the end of the line, open up the divider legs a little. If it reaches beyond the end, close them some. Step off five times again. Repeat until five steps reach exactly from one end of the line to the other.

3. *To transfer a distance from one place to another.* Set the dividers on the distance to be transferred from the drawing. Place the points on the drawing at the new position to which the distance is to be transferred. Mark where the divider legs touch the new line.

HOW TO USE A BEAM COMPASS

Another frequently used tool is the beam compass. See Fig. 4–27. It is used to draw large circles and arcs. One leg has a pin point. The other has lead and inking pen attachments. A pin point can be placed in the second leg to make a large divider. It takes two hands to swing an arc with this compass.

IRREGULAR CURVES

Irregular curves are instruments used to draw any noncircular curve. A noncircular curve is one consisting of

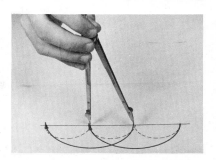

4–26. *Equal distances can be stepped off with dividers.*

tangent arcs of varying radii. They cannot be drawn with a compass. Noncircular curves are found on graphs, charts, involutes, spirals, and ellipses.

Curves are available in many different shapes and classifications. Some of these different types of curves are: rule curves, mechanical engineer's curves, ship's curves, and body sweeps. Each is used in a special field of drafting as their names imply.

Most commonly used in the field of drafting are the irregular curves. See Fig. 4–28.

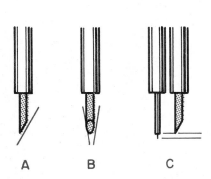

4–25A. *Sand a sharp, slanted surface.* **B.** *Sand edges slightly to form chisel point.* **C.** *Set pin slightly longer than lead.*

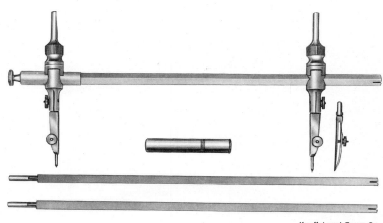

Keuffel and Esser Co.

4–27. *A beam compass.*

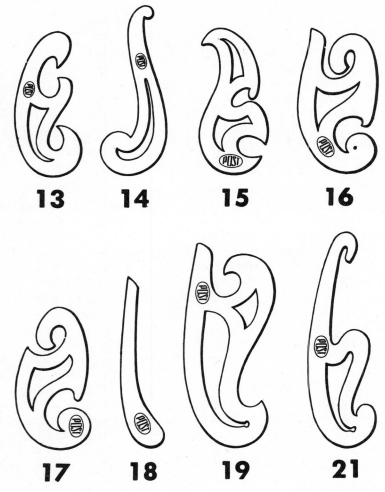

13 **14** **15** **16**

17 **18** **19** **21**

Frederick Post Co.

4-28. *A few of the many shapes available in irregular curves.*

Keuffel and Esser Co.

4-29. *Two types of flexible curves.*

Another type of curve is one that is called a flexible curve. See Fig. 4–29. These can be bent to any desired shape. Flexible curves may be made from metal, plastic, or rubber.

Drawing Irregular Curves. The curve is drawn by finding points that fall on the curve. The irregular curve is moved around until a section of it lines up through three or more points. The curved line is drawn through these points. The curve is then moved to a new position to join several more points. See Fig. 4–30.

DRAWING PENCILS

Drawing pencils are made with leads in 18 degrees of hardness. They run from 6B, very soft, to 9H, very hard. See Fig. 4–31. The most commonly used leads are graphite. The hardness is varied by the amount of clay added to the graphite.

Plastic leads are made for drawing on polyester drafting film. They are available in the same degrees of hardness as graphite leads. The range available is from 6H to 3B.

Two types of drawing pencils are used. One is a wood-cased pencil. See Fig. 4–32. The hardness of the lead is printed on one end of the wood. The other type is a mechanical pencil. See Fig. 4–33. Leads of various hardness are purchased and placed in the mechanical holder. The hardness is printed on the lead.

| 6B | 5B | 4B | 3B | 2B | B | HB | F | H | 2H | 3H | 4H | 5H | 6H | 7H | 8H | 9H |

| VERY SOFT | | RENDERING AND SHADING | | | | FREEHAND SKETCHING | FINISHED LINES | | CONSTRUCTION LINES | | | | | VERY HARD | | |

4-31. *The grades and uses for drawing pencils.*

HARDNESS

A. W. Faber-Castell

4-32. *Wood-cased drawing pencils. Notice the hardness indication on one end.*

T. A. Alteneder and Sons

4-33. *Mechanical drawing pencils.*

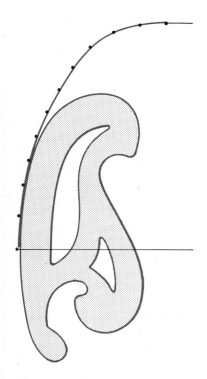

4-30. *The irregular curve is fitted to several of the points used to locate the line. This part of the line is drawn. The curve is turned until another part fits the next several points. This part of the curve is drawn.*

Sharpening the Pencil. Wood-cased pencils are sharpened with a draftsman's pencil sharpener. See Fig. 4-34. The end that does *not* show the hardness of the lead is sharpened. The pencil sharpener removes the wood only. It leaves about ³⁄₈ inch of lead exposed. See Fig. 4-35. Most drafting is done with a conical point. See Fig. 4-36.

The lead is pointed in a pencil pointer. See Fig. 4-37. The pencil is placed in the pointer and the top is rotated. The

4-34. *A pencil sharpener with draftsman's cutters cuts only the wood. The lead is not touched.*

4-35. *Drafting pencils are sharpened by cutting away the wood, leaving the lead untouched.*

59

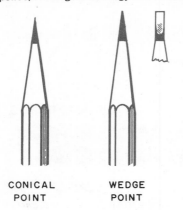

CONICAL
POINT

WEDGE
POINT

4–36. *The two types of pencil points used in drafting.*

lead rubs on an abrasive cylinder inside the pointer. Conical points can be formed by rubbing the lead on a fine file or sanding pad. See Fig. 4–38. As the pencil is drawn over the file, it is rotated between the fingers. This keeps the point from having flat spots.

For some jobs draftsmen prefer to use a wedge-shaped point. See Fig. 4–36. It is stronger than the conical point. It does not wear away as fast. Wedge points are formed with a file or sanding pad. The lead is drawn over the file several times on one side. Then this is repeated on the opposite side of the lead. The wedge should be long and gradual and run the entire ³/₈-inch length of the lead point.

After the pencil is pointed, wipe the point on a cloth. Special plastic cleaning pads are available. See Fig. 4–39. The point is stuck into the pad and rotated.

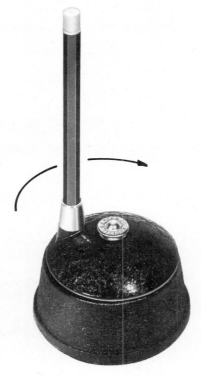

Keuffel and Esser Co.

4–37. *Lead is formed to a cone shape in a pencil pointer.*

PENCIL TECHNIQUE

Most drawings are made in pencil. They must be in condition to be reproduced by any of the methods of drawing reproduction. This requires that proper pencil technique be followed.

All pencil lines should be as dark as possible. The circles and arcs must be as dark as the straight lines. It may be necessary to use a softer lead in the compass to do this. The use of templates and the regular drafting pencil reduces this problem.

Line symbols should be properly used. The length of dashes in hidden lines should be uniform. Each dash should be dark and clear.

There is a tendency to draw extension and dimension lines light. They must be thin but as dark as all other lines.

Proper line width is essential. The more important lines are thicker. This calls attention to them. It helps when reading the drawing.

While it is best to draw a line the proper width and darkness with one stroke, it is sometimes necessary to go over a line several times. Be certain this does not change the width.

Proper pencil technique requires that pencils be sharpened properly. A visible line will require a thicker conical point than a center line. Pencils must be sharpened frequently.

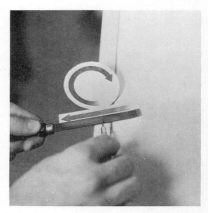

4–38. *Conical point is formed by pulling the pencil over a file and rotating it between the fingers.*

Keuffel and Esser Co.

4–39. *A pencil lead cleaner.*

The pencil should be kept close to the edge of the straightedge or triangle. If it leans in and out as the line is drawn, a wavy line will result.

Select the proper pencil for the job. Wide lines will require a softer pencil than thin lines. Experience will help decide which hardness is best. The 2H pencil is a good general purpose pencil for most lines.

The pencil should be soft enough to draw the line wanted without excess pressure. Too much pressure will form grooves in the paper. This will make it difficult to erase and make changes.

Always slant the pencil in the direction in which the line is to be drawn. A 60-degree angle is good.

INKING

Ink used for drawings is black waterproof drawing ink. This is sometimes called India ink. When inking a drawing, never allow the bottle to remain uncapped. If the stopper is not replaced, the ink will evaporate and thicken. When inking, always use a clean piece of cloth to wipe the pen.

The ruling pen, Fig. 4–40, is used to *rule* lines. It is never used freehand or for lettering. The technical fountain pen, Fig. 4–41, is used for ruling lines, freehand lines, and lettering. This type of pen has a large reservoir for ink. In addition, the point may be changed to produce different line widths.

The ruling pen is never dipped in the ink bottle. It is always filled by placing the dropper in the bottle between the nibs of the pen. When filling the ruling pen, open the nibs slightly. Then add the ink. Usually the ruling pen will hold a quarter inch of ink.

If the ruling pen is filled too full, it will have a tendency to drip from the pen. Be sure the

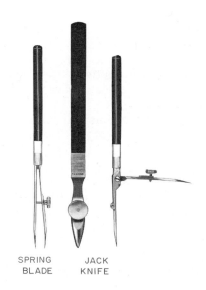

SPRING   JACK
BLADE   KNIFE

T. A. Alteneder and Sons

4–40. *Ruling pens for inking lines on drawings.*

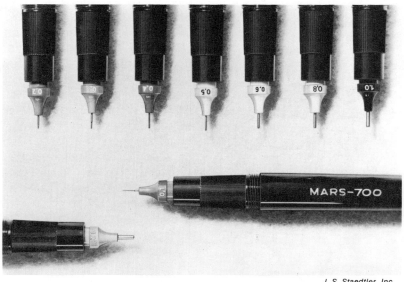

J. S. Staedtler, Inc.

4–41. *A technical fountain pen. Notice the variety of point sizes available.*

nibs of the ruling pen are clean, both inside and out. If dried ink remains on the outside of the nibs, the ink may flow to the outside of the nibs. This will increase the width of the line.

Hold the pen as you would a pencil. See Fig. 4–42. Test the line width on a scrap of paper. If the ink does not flow, draw the nibs across a piece of paper. If, after doing this, the ink is not flowing, squeeze the

nibs slightly. When inking horizontal lines, hold the pen perpendicular to the drafting medium and incline the pen to an angle of 60 degrees. Rule the horizontal lines from left to right. Vertical lines are ruled from bottom to top. Always remember to incline the pen in the direction of travel. The adjusting screw for the nibs is always held away from the ruling edge. Hold the pen with the thumb and index finger next to the adjusting screw. This will make it easier to adjust the line width.

The pen must be held perpendicular to the surface of the paper. See Fig. 4–43. If it is not held perpendicular to the surface of the paper, the nibs may be too close to the ruling edge and cause ink to run under it. If the nibs are too far away from the ruling edge, a ragged line will be produced. To help prevent ink from running under triangles and templates, glue a few pieces of cardboard on the bottom. This will lift them off the surface of the paper. Most straightedges have the plastic edge thinner than the wood section for inking purposes.

If the ink runs under the T-square or triangle, blot the ink as quickly as possible with a pen wiper. Allow the ink to dry thoroughly and then erase. Do not attempt to erase a wet, blotted ink area. Do not use an "ink eraser" or one which contains gritty particles. These pieces of abrasive will damage

the surface of the drafting medium. Never use a blotter to help dry the ink. Permit the ink to air-dry. To tell whether a line is dry, sight along the top of the drawing. If the line is shiny, it is wet; if it is dull, it is dry.

Prior to inking, sprinkle the drafting surface with "pounce" or "ink" powder. Rub the powder into the drafting medium with a clean cloth. Whisk any excess powder from the surface with a drafting brush. The powder will absorb any oil that may have been deposited on the drafting surface.

Before beginning to ink, test the line width on a piece of scrap paper—the same as that to be used for the drawing. Surface qualities of paper will make the line width vary.

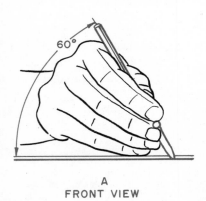

A
FRONT VIEW

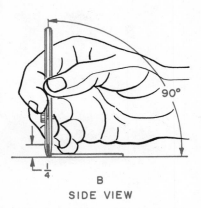

B
SIDE VIEW

4–42. *Hold the inking pen perpendicular to the paper. Slant 60 degrees in the direction the line is to be drawn.*

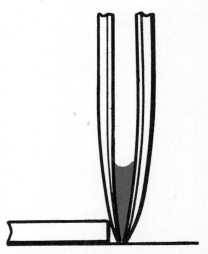

4–43. *The nibs of the ruling pen must be kept perpendicular to the paper.*

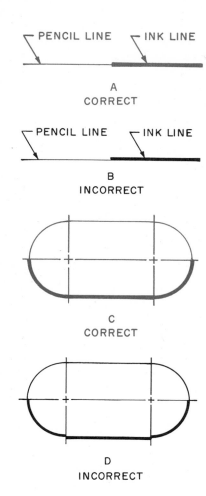

PENCIL LINE   INK LINE

A
CORRECT

PENCIL LINE   INK LINE

B
INCORRECT

C
CORRECT

D
INCORRECT

4–44. *Center the ink line on the pencil line.*

NON-ABRASIVE NON-SMUDGING
VINYL ERASER FOR DRAFTING FILM *Peel off Magic-Rub* · U.S.A. · A.W. FABER · 1960

101 U.S.A. *Pink Pearl* EBERHARD FABER

A. W. Faber

4–45. *Block and stick erasers.*

Check this line width against the alphabet of lines as found in your text.

Check the amount of ink in the pen before you rule a line. Make sure there is enough ink to complete the line.

Order of Inking. Always center the nibs of the pen over the line to be inked. See Fig. 4–44. This will insure that the inked lines will meet smoothly.

To obtain the best results when inking a drawing:

1. Ink all center lines, circular first, then horizontal, and vertical.

2. Ink all arcs and circles. Start with the largest arcs and circles first, then reduce the radius of the compass and ink the next largest.

3. Ink all horizontal visible lines.

4. Ink all vertical visible lines.

5. Ink all inclined visible lines.

6. Ink hidden lines.

7. Ink section lines, extension lines, dimension lines.

8. Ink arrowheads.

9. Ink all lettering for dimensions, notes, and titles.

ERASING

Three types of erasers are used in drafting. Each has a special purpose. A hard rubber eraser is used to remove dark heavy lines. A soft rubber eraser is used to remove light lines and to clean up smudges on a drawing. A vinyl eraser is used to remove plastic lead lines from polyester drafting film.

Erasers are available in block and sticks. The stick form is usually easier to use. See Fig. 4–45.

When erasing, use pressure with caution. Too much pressure can damage the paper. Stretch the paper in the area to be erased between two fingers. This holds it tight and reduces the danger of wrinkling the paper. Too much pressure can cause the eraser to overheat and leave a smudge.

Always keep the eraser clean. Rub it on scrap paper to remove any carbon on its surface.

Electric erasing machines are available. See Fig. 4–46. The machine is designed so it

Frederick Post Co.

4–46. *An electric eraser.*

can be operated with one hand. Since the eraser rotates at a high speed, very little pressure is needed. Hard and soft rubber and vinyl eraser points are made for the machine.

The *erasing shield,* Fig. 4–47, is an aid when removing a line near other lines that are to remain on the drawing. Select the opening in the shield that best fits the area to be erased. Place it on the drawing so the line to be erased appears in the opening. Rub the eraser over the opening. Hold the shield firmly to the paper so it will not slip.

Keuffel and Esser Co.

4–48. *A draftsman's dusting brush.*

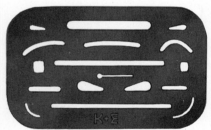

Keuffel and Esser Co.

4–47. *An erasing shield.*

KEEP DRAWINGS CLEAN

The finished drawing must be clean. Smudges spoil the appearance. They will reproduce when copies of the drawing are printed. Following are a number of things that will help.

1. Keep all tools clean. The straightedge, triangles, protractor, and template pick up carbon as they rub over the pencil lines. Wipe them frequently with a cloth. If the car-

bon will not wipe off, wash them with a damp cloth.

2. Use a dusting brush to remove eraser crumbs from the drawing. See Fig. 4–48. Do not brush away crumbs with the hand. This will cause pencil lines to smear.

3. To prevent the tools from smearing lines already drawn, place a clean paper over the lines.

4. Some draftsmen use a dry cleaning pad. See Fig. 4–49. It is a loosely woven sack filled with eraser crumbs. The pad is rubbed lightly over the surface of the drawing to remove graphite particles. If too much pressure is used, the lines are lightened. This makes it difficult to make prints from the drawing. Dry cleaning pads must be used with great care.

DRAWING HORIZONTAL LINES

Horizontal lines are drawn with the top edge of a T-square, parallel straightedge, or the horizontal blade on a drafting machine.

To use a T-square, right-handed persons place the head on the left edge of the drawing board. Left-handed persons usually use the right edge. A right-handed person

holds the head to the board with the left hand. Left-handed persons use the right hand. This hand is used to slide the T-square up and down the board. See Fig. 4–50.

Right-handed persons draw horizontal lines from left to right. Left-handed persons draw from right to left. See Fig. 4–51.

Slant the pencil at an angle of about 60 degrees in the direction the line is to be drawn. See Fig. 4–51. As the pencil is moved, it is rotated between the fingers. This keeps the point conical. The lead wears down evenly on all sides.

DRAWING VERTICAL LINES

Vertical lines are drawn with triangles. The triangle is

Frederick Post Co.

4–49. *A dry cleaning pad.*

placed on the top edge of the straightedge. A right-handed person holds the triangles with his left hand. A left-handed person holds them with his right hand.

Vertical lines are drawn from bottom to top by right-handed persons. See Fig. 4-52. Left-handed persons often find it easier to draw from top to bottom. Slant the pencil in the direction the line is to be drawn.

DRAWING INCLINED LINES

Inclined lines are drawn by placing the triangle against the top edge of the straightedge. Right-handed persons draw them in the directions shown in Figs. 4-53 and 4-54. Left-handed persons often draw in the opposite direction.

Several different angles can be drawn by combining the triangles. Some of these angles are shown in Fig. 4-55.

Angles can be laid out using a protractor. A *protractor* is a flat semicircular tool that is marked in degrees. See Fig. 4-56. There are two sets of degrees marked on the circular edge. These run from 0 degrees to 180 degrees. One set reads from right to left. The other reads from left to right.

To use the protractor, place the center point on the bottom at the corner of the angle. The base line is placed along one side of the angle. The degrees are then read on the circular scale. See Fig. 4-56. Place a mark at the center line and at

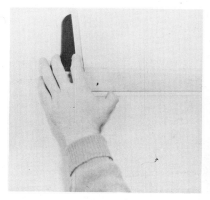

4-50. *The T-square is held to the left edge of the drawing board with the left hand.*

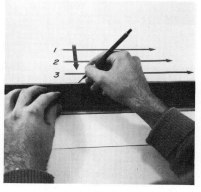

4-51. *Horizontal lines are drawn with the pencil slanted in the direction the line is to be drawn.*

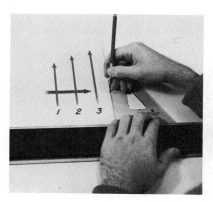

4-52. *Vertical lines are drawn from bottom to top.*

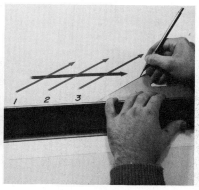

4-53. *Lines slanting to the right are drawn in this direction.*

the degree marking. Connect these with a straightedge. Inclined lines can also be drawn using adjustable triangles. See Fig. 4-14.

DRAWING LINES PERPENDICULAR TO EACH OTHER

To draw a perpendicular line to a horizontal line, use the 90-degree edge of a triangle. See Fig. 4-52.

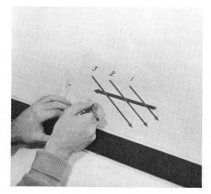

4-54. *Lines slanting to the left are drawn in this direction.*

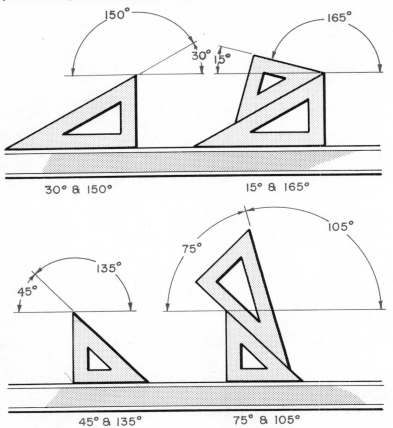

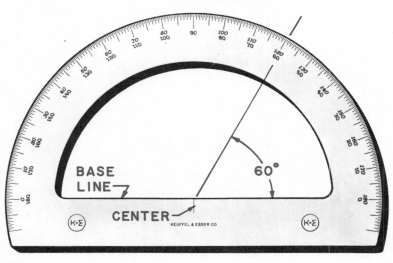

4-55. *Slanted lines can be drawn on commonly used angles by using triangles.*

4-56. *Angles can be measured with a protractor.*

Keuffel and Esser Co.

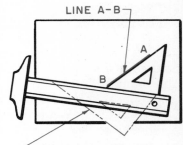

LINE A-B

A TRIANGLE MAY BE SUBSTITUTED FOR THE T-SQUARE.

A
LINE UP THE TRIANGLE AND T-SQUARE WITH LINE A-B.

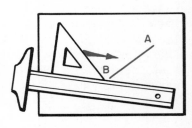

B
SLIDE THE TRIANGLE ALONG THE T-SQUARE.

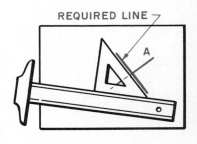

REQUIRED LINE

C
DRAW REQUIRED LINE PERPENDICULAR TO LINE A-B.

4-57. *How to draw lines perpendicular to each other.*

LINE A-B

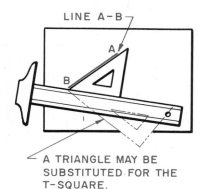

A TRIANGLE MAY BE
SUBSTITUTED FOR THE
T-SQUARE.

A
LINE UP THE TRIANGLE AND
T-SQUARE WITH LINE A-B.

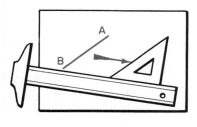

B
SLIDE THE TRIANGLE ALONG
THE T-SQUARE.

REQUIRED LINE

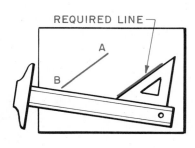

C
DRAW THE REQUIRED LINE
PARALLEL TO LINE A-B.

4–58. *How to draw parallel in-clined lines.*

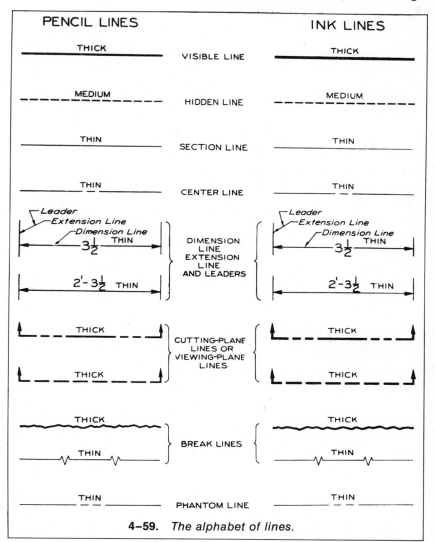

PENCIL LINES		INK LINES
THICK	VISIBLE LINE	THICK
MEDIUM	HIDDEN LINE	MEDIUM
THIN	SECTION LINE	THIN
THIN	CENTER LINE	THIN
Leader / Extension Line / Dimension Line THIN / $3\frac{1}{2}$ / $2'-3\frac{1}{2}$ THIN	DIMENSION LINE EXTENSION LINE AND LEADERS	Leader / Extension Line / Dimension Line THIN / $3\frac{1}{2}$ / $2'-3\frac{1}{2}$ THIN
THICK / THICK	CUTTING-PLANE LINES OR VIEWING-PLANE LINES	THICK / THICK
THICK / THIN	BREAK LINES	THICK / THIN
THIN	PHANTOM LINE	THIN

4–59. *The alphabet of lines.*

To draw lines perpendicular to inclined lines, triangles can be used. See Fig. 4–57. Place a triangle with one edge along the line. Place the T-square on the bottom of the triangle. Slide the triangle along the T-square until it reaches the point at which the perpendicular is wanted. Draw the required line.

DRAWING LINES PARALLEL WITH EACH OTHER

Place a triangle with one edge along the line. See Fig. 4–58. Place the T-square along the bottom of the triangle. Slide the triangle along the T-square away from the line. When it reaches the desired location, draw the desired parallel line.

ALPHABET OF LINES

A standard set of line symbols is used on drawings. See Fig. 4–59. Line symbols are drawn in three widths; thick, medium, and thin. The more important line symbols are thick. This makes them stand out on the drawing.

The thick lines include visible, cutting-plane, and break lines. The hidden line is medium width. Thin lines include hidden, section, extension, dimension, break, and phantom lines.

Study the symbols shown in Fig. 4–59. The thick ink line is a little wider than the thick pencil line. The medium and thin lines are about the same for pencil and ink drawings.

Following is a description of each line symbol illustrated in Fig. 4–60.

Visible lines are used to show all edges that are seen by the eye when looking at the object. They are shown by thick solid lines.

Hidden lines show surfaces that are not seen. They are shown by a dashed line. The dashes are about ⅛-inch long. The spaces are about 1/32 inch. These are often drawn larger if the drawing is large. These distances are estimated as the line is drawn. Do not take time to measure each dash.

Dimension lines are thin, solid lines. They are used to show the extent of the dimension. They contain the dimension number. A leader is one form of dimension line. It is

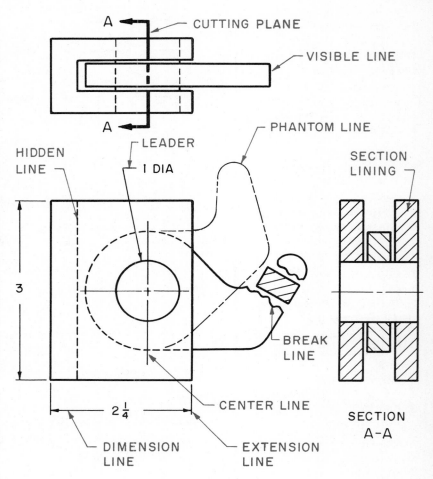

4–60. *Line symbols used on drawings.*

used to connect dimensions or notes to the drawing.

Extension lines are thin, solid lines used with dimension lines. They extend to the point to which the dimension line refers off of the view.

Center lines are used to locate the centers of holes or other parts of an object. They are thin lines made of long and short dashes. The long dashes

can be from ¾ to 1½ inches. The short dashes are about 1/16 inch. The length of the long dash will vary with the drawing and the length of center line needed.

Section lines are used to show a surface that has been cut in a section view. They are thin lines. Section lines are drawn parallel and spaced from 1/16 to ⅛ inch apart. The spacing varies with the area to be section lined. Larger areas have wider spacing between the lines. The usual angles for drawing section lines are 30, 60, and 45 degrees.

A *cutting-plane line* shows where a section has been taken. It is a thick line. Arrows on the end show the direction in which the section was taken. Two standard symbols are in use. One uses long and short dashes and the other uses long dashes of equal length.

Break lines are used to show that part of the object has been removed or broken away. Short breaks are made with a thick, solid, freehand line. Long breaks are drawn with a thin, solid line. This line has a Z symbol inserted in several places.

Phantom lines are used to show the position of part of an object that moves. If a handle is drawn with visible lines, its rotated position is shown with phantom lines. The symbol is a long dash followed by two short dashes. It is a thin line.

DRAWING PAPERS

Drawing paper is a heavy paper usually white, cream, or green. It is sold in rolls and sheet form. It is not used a great deal in industry. Drawings on paper cannot be reproduced by the blueprint or other reproduction processes.

Most types of drawing paper are smooth on one side and rough on the other. The smooth side is best for inking. The rough side is best for pencil drawings.

Tracing Paper and Vellum. Tracing paper is a thin, untreated, transparent paper. Treated paper has a transparentizing agent applied. It is called tracing vellum. Vellum is more transparent than untreated tracing paper. Vellums are made from 100 percent pure white rag stock. They withstand erasing without leaving marks. They are good for pencil and ink drawings. They do not discolor with age. They are tough and can stand a lot of handling.

Tracing Cloth. Tracing cloth is a fabric that has been transparentized. It is tough and strong. It will last years without depreciating. Cloths are used for important drawings that must be preserved for a long time.

Ink and pencil drawings can be made on tracing cloth. The dull side is used for drawing.

Polyester Drafting Film. Polyester film is a tough, transparent plastic drafting material. It is almost impossible to tear. It withstands much erasing. One big advantage is that it changes in size very little as humidity and temperature change. It can be used for drawings requiring a high degree of accuracy. Standard thicknesses are .003, .004, .005 and .007 inch.

The film can be used for ink and pencil drawings. It takes typing well. Special plastic-leaded pencils are used. These pencil lines do not smear as they do on paper. They are easily erased with a vinyl eraser.

Before inking, the surface should be wiped clean with a cloth dampened with a special cleaning solution. A special dry cleaning pad is used to maintain a clean, fingerprint-free surface.

Paper Sizes. Drafting papers, tracing papers, tracing vellums and polyester films are available in sheet and rolls. The standard sheet sizes are:

Type A   8½″ × 11″ and 9″ × 12″
Type B   11″ × 17″ and 12″ × 18″
Type C   17″ × 22″ and 18″ × 24″
Type D   22″ × 34″ and 24″ × 36″
Type E   34″ × 44″ and 36″ × 48″

The roll stock varies in width depending upon the material. The most common widths are 36 and 42 inches. The most common roll lengths are 20 and 50 yards. (Metric paper sizes are shown on p. 76.)

BORDERS

The borders used on drawings vary from one company to another. Commonly the border is ½ to 1 inch from the sheet edge. Larger drawings often have wider borders than small drawings. One system suggested by the American Society of Mechanical Engineers is:

A and B Sheets   ¼″ on all sides.

C Sheets   ⅜″ on all sides.

D and E Sheets   ½″ on all sides.

If sheets are to be bound, a one-inch border on the left is necessary. This is commonly found on sets of architectural drawings. Many companies buy paper with the border and title block printed.

TITLE BLOCKS

The title block frequently gives the following information: Part name and number, drawing date, scale, tolerances, material specifications, references, name of draftsman, tracer and checker, firm's name and address, drawing number and finishes. The size varies considerably from one company to another. It is usually located in the lower right corner of the drawing. Some are placed across the bottom of the drawing. See Fig. 4–61.

LETTERING

The most commonly used lettering style on engineering drawings is single stroke, capital, Gothic. This can be lettered vertically or inclined. See Figs. 4–62 and 4–63. This style of lettering is easily done and easy to read. Lower case letters are seldom used except on maps and some architectural drawings.

Usually a company will use all vertical or all inclined lettering. They are not mixed on a drawing. The vertical are easier to read but harder for some to letter. The slightest variance from the vertical is easily noticed. A draftsman should be able to letter well either way.

Clear lettering, with each letter carefully formed, is essential to an acceptable drawing. The information given in the dimensions and notes is just as important as any part of the drawing.

Forming the Letters. Study the letters shown in Figs. 4–62 and 4–63. Notice that some are wider than others. Keep this in mind as each letter is formed. Some of the letters are ⅚ths as wide as they are

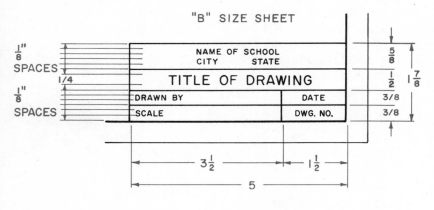

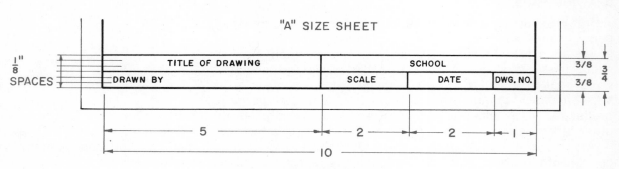

4–61. *Suggested title blocks.*

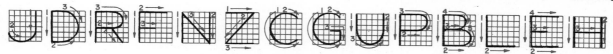

THESE LETTERS ARE $\frac{5}{6}$ AS WIDE AS THEY ARE HIGH.

THESE LETTERS ARE AS WIDE AS THEY ARE HIGH.

THIS W IS WIDER THAN IT IS HIGH.

NUMERALS ARE $\frac{5}{6}$ AS WIDE AS THEY ARE HIGH.

4–62. *These are vertical, upper case, Gothic letters and numbers.*

high. Others are as wide as they are high. One letter is wider than it is high. Which letter is this? The letters are kept in this same proportion. The higher a letter is made, the wider it becomes. Notice

that all numerals are $\frac{5}{6}$ths as wide as they are high.

Figs. 4–62 and 4–63 show the order in which each stroke is made. This order has been established to enable the letter to be formed in the easiest

way possible and retain its proportions.

Spacing Lettering. If the lettering is to be pleasing, the space between each letter must be considered. The space

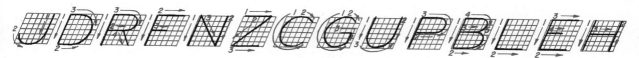

THESE LETTERS ARE $\frac{5}{6}$ AS WIDE AS THEY ARE HIGH.

THESE LETTERS ARE AS WIDE AS THEY ARE HIGH.

THIS W IS WIDER THAN IT IS HIGH.

NUMERALS ARE $\frac{5}{6}$ AS WIDE AS THEY ARE HIGH.

4–63. *These are slanted, upper case, Gothic letters and numbers.*

SPACES EQUAL

THE SPACE BETWEEN LETTERS

THE LETTERS APPEAR UNEQUALLY SPACED WHEN
THE SPACES BETWEEN THEM ARE EQUAL.

SPACES NOT EQUAL

THE SPACE BETWEEN LETTERS IN

LETTERS SHOULD BE SPACED SO THEY APPEAR
TO BE THE SAME DISTANCE APART.

4–64. *The spacing between letters in a word is important to the appearance of a drawing.*

ONE LETTER SPACE BETWEEN WORDS.

THE SPACE BETWEEN WORDS IS
EQUAL TO ONE LETTER.

4–65. *How to space between words on a drawing.*

VERTICAL GUIDE LINES

A GOOD DRAFTSMAN ALWAYS DRAWS
HORIZONTAL GUIDE LINES.

90°

HORIZONTAL GUIDE LINES

$67\frac{1}{2}°$

SLANTED GUIDE LINES

SLANT LETTERING IS ON AN ANGLE
OF $67\frac{1}{2}$ DEGREES

HORIZONTAL GUIDE LINES

4–66. *Guidelines are used when lettering.*

HEIGHT $67\frac{1}{2}$ | HEIGHT / HEIGHT

4–67. *How to letter fractions.*

between letters must appear to be equal. However, the space cannot be equal. See Fig. 4–64. Since the letters are different shapes, some must be closer together. The space between two letters with vertical parts, as H or I, should be larger than between letters with open space, as A or V. This spacing is done by eye.

The space between words is equal to one letter of the alphabet. Often the letter O is lightly sketched in to help with this spacing. See Fig. 4–65.

Guidelines for Lettering. A good draftsman always draws horizontal guidelines. These are spaced according to the height of the letter desired. See Fig. 4–66. The letters should touch each guideline, but should not cross guidelines.

Guidelines are drawn very lightly. They need only be dark enough to see. If made lightly enough, they need not be erased after the lettering is finished.

Vertical and inclined guidelines can be drawn. They are spaced at random. See Fig. 4–66.

Fractions are made as shown in Fig. 4–67. A fraction has the total height of two full numbers. The numbers do not touch the bar between them. This means they are drawn slightly smaller than normal. Guidelines are drawn for all fractions.

A lettering guide is used to draw guidelines. See Fig. 4–68. Notice the numbers on the bottom of the wheel. They are from two to ten. They show the height of the lettering in 32nds of an inch. The number eight is touching the vertical mark, Fig. 4–68. This means the guidelines will be 8/32- or 1/4-inch high.

The lettering guide slides along a straightedge. A pencil is placed in the holes. The guide is slid along the straightedge with the pencil. This draws very light horizontal guidelines. See Fig. 4–69.

Vertical guidelines can be drawn with the lettering guide. One side is on 90 degrees. The other is on 67½ degrees, required for inclined lettering. See Fig. 4–70.

Lettering Techniques. The height of lettering on most drawings is ⅛ inch. Fractions would be ¼ inch. Some of the important parts, as in the title block, are lettered larger.

A 2-H pencil is good for lettering. It should be kept sharp. Clear lettering cannot be done with a dull pencil.

When lettering, place your arm in a comfortable position. It should be supported on the drawing board. Turn the board on an angle if this helps. Do not hold the pencil too tightly. A relaxed grip will give a straighter line.

To prevent smearing a drawing while lettering, place a clean sheet of paper over it.

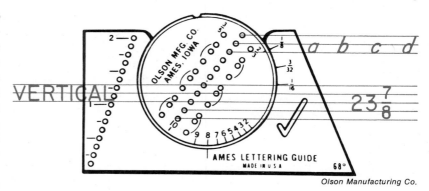

Olson Manufacturing Co.

4–68.   *Guidelines are drawn with this lettering guide.*

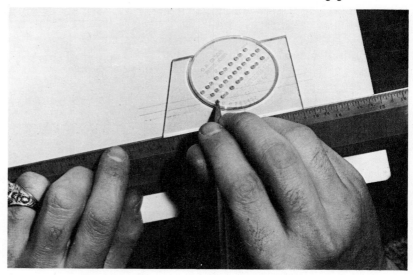

4–69.   *How to draw lines with the lettering guide.*

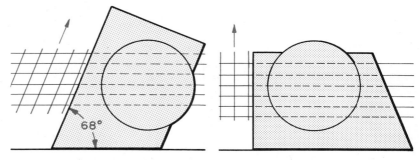

4–70.   *Drawing slanted and vertical guidelines with the Ames lettering instrument.*

73

This will keep the hand and arm from rubbing over the surface.

Lettering Instruments. Lettering instruments are a mechanical means of forming letters. There are several types available. One type has a plastic template with the letters sunk into it. See Fig. 4–71. A tool called a scriber is used with this. The scriber has one leg with a metal point. This slides into the letters sunk in the template. It has another leg that can hold a lead or ink pen. As the pin slides in the sunken letter, the other point forms the letter on the drawing. The template is often held in place on top of the straightedge. It is slid along the straightedge. Letters are formed one at a time.

This instrument has templates of many sizes. The width of the line drawn depends upon the height of the letter. Inclined letters can be drawn by adjusting the legs of the scriber.

Another lettering device is a plastic template with the letters cut out, shown at the beginning of Section Two. The pen traces along the edges of the letters. Notice the type of pen used. To form each word, the template is slid back and forth along the edge of the straightedge. Letters are formed one at a time.

Typewriters. Several kinds of typewriters are used in drafting rooms. The conventional platen typewriter is available with varying length carriages. The drawing is inserted as in a normal typewriter and the operator types the information. These typewriters are either manual or motor driven and have the standard keyboard. The advantage of the electric typewriter is a uniform impression for all letters. A typewriter is a timesaving tool when bills of materials, descriptions, specifications, and notes must be placed on the drawing.

When typing on a tracing, the operator may back up the tracing with carbon paper. The carbon paper is placed so that the coating is against the back of the tracing. In this manner, the carbon impression is on the back side of the tracing. Orange, yellow, and sepia colored carbon paper are most used. These three colors will block out more ultraviolet rays than other colors.

A different kind of typewriter can be used on drawings. This is called the Gritzner lettering typewriter, Fig. 4–72. Rather than taking the drawing to the typewriter, the typewriter is moved to the drawing. The typewriter moves on an indexing rail that replaces the horizontal scale of the drafting machine. The indexing rail may be placed against a parallel rule or T-square blade. As each key is struck, the machine will index one space to the left. The letters are rubber and are inked as the key is struck. Carbon backing cannot be used with this machine. The keyboard is ar-

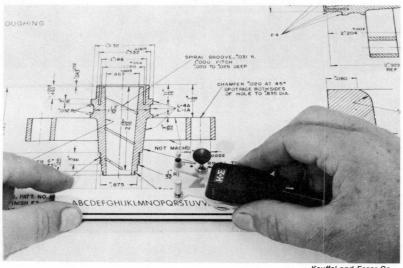

Keuffel and Esser Co.

4–71. *This lettering instrument uses a template to form the letters. It can make vertical and inclined letters.*

ranged in alphabetical order. The Gritzner weighs only 2 pounds and measures $6\frac{1}{8}''$ × $6\frac{5}{8}''$ × $3\frac{1}{8}''$. Three heights of type are available: .130 inch, .118 inch, and .078 inch. Most drawings are typed in capital letters.

Pressure-sensitive Letters. Pressure-sensitive letters are available in sheets. The letters have an adhesive on the back. They are removed from the sheet and pressed in place on the drawing. See Fig. 4–73. There are many different types of letters and several different systems for placing them on the drawing. Examples of different styles are shown in Fig. 4–74.

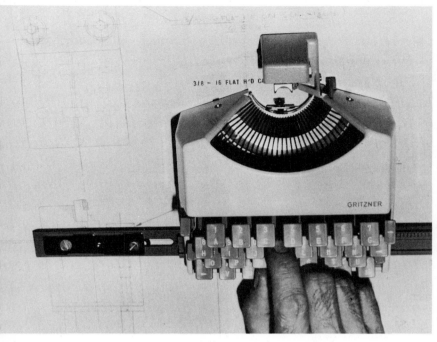

4–72. *A Gritzner lettering typewriter.*

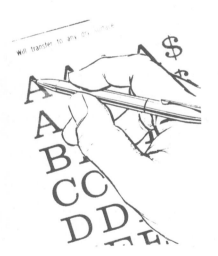

4–73A. *Position the letter; press letter to paper with a finger; rub it with a smooth tool, as a ball point pen.*

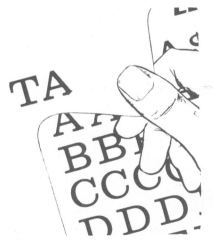

Craftint Manufacturing Co.

4–73B. *Lift sheet away. Letter is stuck to drawing. Place the next letter in position and repeat the operation.*

Craf-Type...The Best and

Craf-Type...The Bes

Craf-Type...Th

Craf-Type...The

Craf-Type...The Best and

Craf-Typ

CRAF-TYPE...T

crafty

Craftint Manufacturing Co.

4–74. *Quick transfer letters are sold in many styles and sizes.*

Build Your Vocabulary

Following are terms you should understand and use as a part of your working vocabulary. Write a brief explanation of what each means.

T-square	*Irregular curve*	*Phantom lines*
Parallel straightedge	*Protractor*	*Tracing paper*
Drafting machine	*Ruling pen*	*Vellum*
Triangle	*Visible lines*	*Tracing cloth*
Drawing to scale	*Hidden lines*	*Polyester drafting film*
Open divided scale	*Dimension lines*	*Title block*
Fully divided scale	*Center lines*	*Guidelines*
Template	*Section lines*	*Lettering guide*
Compass	*Cutting-plane lines*	*Lettering instrument*
Dividers	*Break lines*	*Pressure-sensitive letters*

METRIC DRAWING SHEET SIZES
(in millimetres)
Metric 'A' series
4A0 (2,378 x 1,682)
2A0 (1,682 x 1,189)
A0 (1,189 x 841)
A1 (841 x 594)
A2 (594 x 420)
A3 (420 x 297)
A4 (297 x 210)
A5 (210 x 148)
A6 (148 x 105)

Chapter 5

Geometric Figures and Constructions

The draftsman must work with angles and lines, solids, and geometric shapes, such as a hexagon. He must work with intersections of lines of many kinds. Constructions of this kind involve the principles of geometry. The draftsman must be able to use geometric constructions to make drawings properly. The following pages will present those used the most.

GEOMETRICAL SHAPES

The draftsman must be able to recognize geometrical shapes. They must be part of his working knowledge, since his drawings use these shapes.

Angles. An angle is a figure formed when two lines meet in a point, Fig. 5–1. An *acute angle* is one that is less than 90 degrees. A *right angle* is 90 degrees. An *obtuse angle* is more than 90 degrees. When

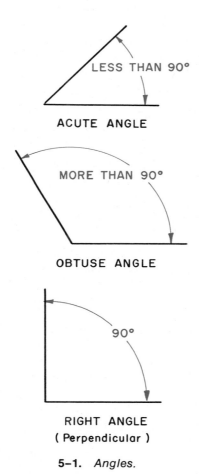

5–1. *Angles.*

two lines meet at a 90-degree angle, they are said to be *perpendicular.*

Triangles. A triangle is a figure formed by three lines meeting by twos at three points, Fig. 5–2. A *right triangle* has one

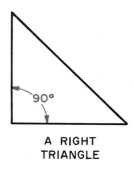

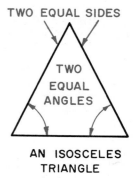

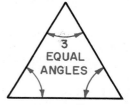

5–2. *Common triangles.*

77

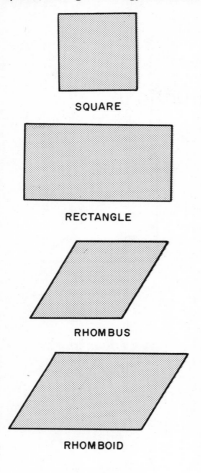

5-3. *Parallelograms.*

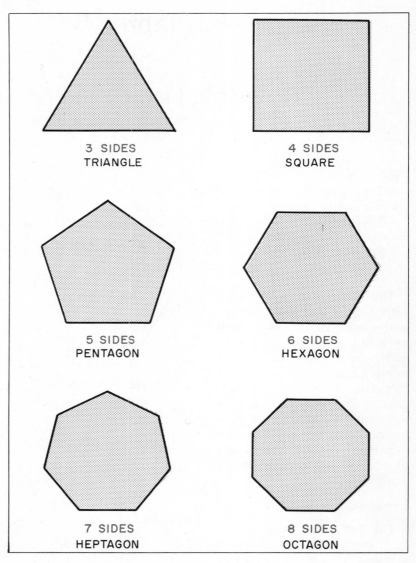

5-4. *Typical polygons.*

angle 90 degrees. An *isosceles triangle* has two sides and two angles equal. An *equilateral triangle* has all sides equal and all angles equal. All triangles contain 180 degrees. That is, the sum of the three angles always equals 180 degrees.

Parallelograms. Parallelograms are figures with the opposite sides parallel. Since they are parallel, they are equal in length. See Fig. 5-3. Par-

allel means to run in the same direction an equal distance apart but never meet. Two lines are parallel when they are drawn in the same direction an equal distance apart and would never meet if they were drawn to infinity. In-

finity means without a limit.

A *square* has four sides of equal length and four 90-degree angles. A *rectangle* has parallel sides of two different lengths. All angles are 90 degrees. A *rhombus* has parallel sides all the same length. No

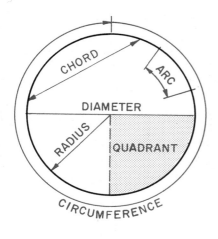

5-5. *Circles and arcs are described with these terms.*

angles are 90 degrees. A *rhomboid* has parallel sides of two different lengths. No angles are 90 degrees.

Polygons. A polygon is any figure with sides of straight lines. Figures with sides of equal length and equal angles are called regular polygons. See Fig. 5-4.

Circles and Arcs. Fig. 5-5 shows the terms used when describing circles and arcs. A *circle* is a closed curve. All points on the circle are the same distance from a point called the center. The *circumference* of a circle is the distance around the circle. It can be found by multiplying 3.1416 by the diameter of the circle. *Diameter* is a distance from one side of a circle through the center to the other side. *Radius* is the distance from the center to the outside. It is one-half the diameter. A *chord*

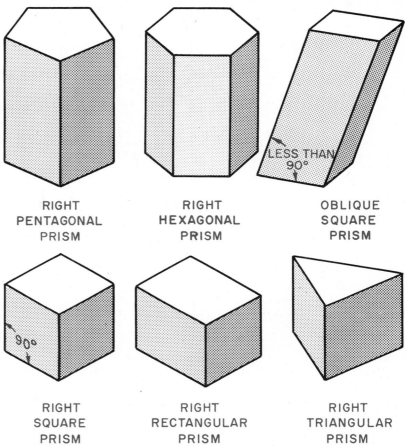

RIGHT PENTAGONAL PRISM

RIGHT HEXAGONAL PRISM

OBLIQUE SQUARE PRISM

RIGHT SQUARE PRISM

RIGHT RECTANGULAR PRISM

RIGHT TRIANGULAR PRISM

5-6. *Typical prisms.*

is any straight line crossing the circle in two places. An *arc* is a piece of the circumference of a circle. A *quadrant* is one-fourth of a circle.

GEOMETRIC SOLIDS

Solids have three dimensions. They have height, width, and depth. Some of the most used solids are explained in the following sections.

Prisms. A prism has two bases. The bases are parallel. They are the same size and are poly-

gons. They have three or more faces. See Fig. 5-6. A triangular prism has a triangular base. A rectangular prism has a rectangular base. A prism is called a right prism when the sides meet the base at an angle of 90 degrees. It is called an oblique prism when the sides meet the base at any other angle.

Cylinders. A cylinder is formed by moving a line in a circle around a central axis, Fig. 5-7.

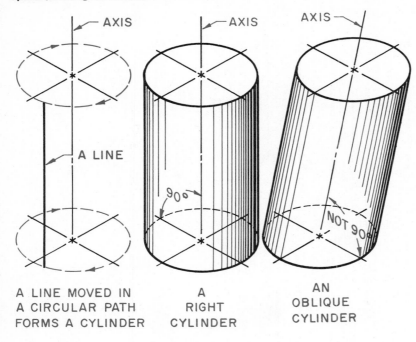

5-7. *Cylinders.*

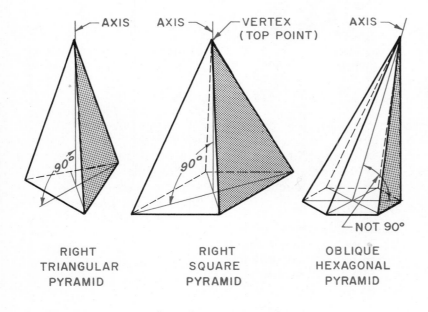

5-8. *Common pyramids.*

If the axis is perpendicular to the base, it is a right cylinder. If the axis is not perpendicular to the base, it is an oblique cylinder.

Pyramids. A pyramid has a polygon for a base. It can be any type of polygon, such as a triangle, square, or hexagon. Triangular faces rise from each edge of the base. They meet at a common point called the vertex. See Fig. 5-8. If the axis of the pyramid is perpendicular to the base, it is a right pyramid. If the axis is not perpendicular, it is an oblique pyramid.

Cones. A cone is formed by moving one end of a straight line in a circle and holding the other end fixed. See Fig. 5-9. A right cone has an axis perpendicular to the base. An oblique cone has an axis that is not perpendicular to the base.

Sphere. A sphere is formed by spinning a circle around a diameter. See Fig. 5-10. An example of a sphere is a basketball.

Torus. A torus is formed by moving a circle around a central axis. See Fig. 5-11. A doughnut is an example.

GEOMETRIC CONSTRUCTION

Bisect a Line. To bisect a line means to divide it into two equal parts. The steps are shown in Fig. 5-12.

1. Set a compass with a

radius greater than half the length of the line to be bisected. This is line AB.

2. Put the pin of the compass at one end of the line and swing an arc above and below the line.

3. Place the pin on the other end of the line and swing arcs crossing the first arc. This forms points C and D.

4. Connect C and D with a straight line. This line bisects line AB.

Bisect an Angle. To bisect an angle means to divide it into two equal angles, Fig. 5–13.

1. Set the compass at any length radius.

2. Put the pin at point A and swing an arc crossing the sides of the angle.

3. Put the pin on one of the intersections, as B. Swing an arc.

4. Put the pin on the other intersection, as C. Swing an arc that crosses the first arc.

5. Connect point A with the intersection of the arcs at D. This line bisects the angle.

Bisect an Arc. To bisect an arc means to divide an arc into two equal arcs, Fig. 5–14.

1. The arc to be bisected is AB. Connect AB with a straight line. This line is called a chord.

2. Bisect the chord in the same way you bisect any line. This bisects the arc. See Fig. 5–12.

When you bisect the chord of an arc, you bisect the arc.

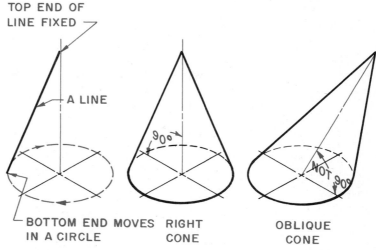

TOP END OF LINE FIXED

A LINE

90°

NOT 90°

BOTTOM END MOVES IN A CIRCLE

RIGHT CONE

OBLIQUE CONE

A CONE IS FORMED BY HOLDING ONE END OF A LINE FIXED AND MOVING THE OTHER IN A CIRCLE.

5–9. *Cones.*

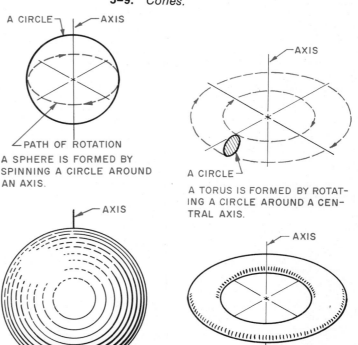

A CIRCLE

AXIS

PATH OF ROTATION

A SPHERE IS FORMED BY SPINNING A CIRCLE AROUND AN AXIS.

AXIS

A CIRCLE

A TORUS IS FORMED BY ROTATING A CIRCLE AROUND A CENTRAL AXIS.

AXIS

A TORUS

AXIS

SPHERE

5–10. *The sphere.*

5–11. *The torus.*

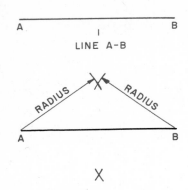

LINE A-B

RADIUS   RADIUS

A   B

1

X

2

SWING ARCS FROM THE END
OF THE LINE. THE RADIUS
MUST BE GREATER THAN HALF
THE LENGTH OF THE LINE.

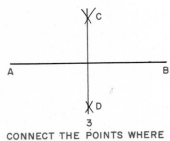

C

A   B

D

3

CONNECT THE POINTS WHERE
THE ARCS CROSS. THESE ARE
POINTS C AND D.

5–12. *To bisect a line with a compass.*

Divide a Line Into Equal Parts.

1. Line AB, Fig. 5–15, is the line to be divided. Suppose this line is to be divided into five equal parts.

2. Draw a line, as BC, at any angle to AB.

3. Mark off five equal lengths along this line. These lengths can be any convenient size. This locates point D.

4. Connect D with A. Draw lines parallel with DA through the other points on CD. Where they cross AB is one-fifth of that line.

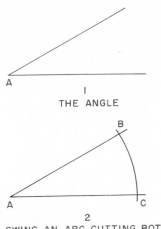

1

THE ANGLE

B

A   C

2

SWING AN ARC CUTTING BOTH
SIDES OF THE ANGLE.

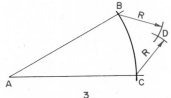

B   R
      D
   R

A   C

3

USING B AND C AS POINTS,
SWING ARCS THAT WILL CROSS.

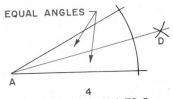

EQUAL ANGLES

D

A

4

DRAW A LINE FROM A TO D.
THIS IS THE BISECTOR.
THE ANGLES ARE EQUAL.

5–13. *To bisect an angle using a compass.*

This will work with any number of divisions.

Construct an Equilateral Triangle.
An equilateral triangle is one with all three sides the same length. See Fig. 5–16.

1. Draw one side, AB, to length.

1

THE ARC TO BE BISECTED.

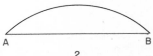

A   B

2

DRAW A LINE THROUGH THE
ENDS OF THE ARC.

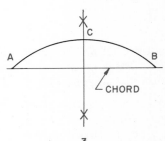

C

A   B

CHORD

3

BISECT THIS LINE. THE ARC
FROM A TO C IS THE SAME
LENGTH AS THAT FROM C TO B.

5–14. *To bisect an arc using a compass.*

2. Set a compass this same length.

3. Swing arc AC with A as a center.

4. Swing arc BD with B as a center.

5. Where the arcs cross is point E.

6. Connect E with A and B.

Draw a Square When the Length of the Diagonal Is Known.

1. Draw a circle with a diameter equal to the diagonal. See Fig. 5–17.

2. Draw the center lines at right angles to each other.

LINE AB TO BE EQUALLY DIVIDED INTO SEVERAL PARTS.

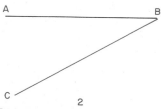

DRAW A LINE ON AN ANGLE FROM ONE END OF THE LINE TO BE DIVIDED. THE LINE MAY BE ANY LENGTH AND ANGLE.

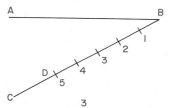

MARK OFF EQUAL SPACES, THE SAME NUMBER AS IS DESIRED TO DIVIDE LINE AB (5 IN THIS EXAMPLE.)

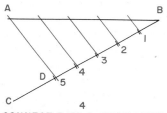

CONNECT D TO A. DRAW LINES PARALLEL TO AD THROUGH ALL OTHER POINTS ON BC.

5–15. To divide a line into equal parts.

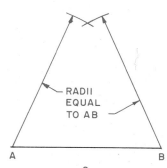

DRAW ONE SIDE OF TRIANGLE TO LENGTH.

SET A COMPASS TO LENGTH AB. SWING ARCS WITH A AND B AS CENTERS.

5–16. To draw an equilateral triangle using a compass.

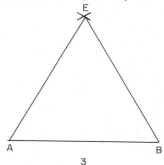

CONNECT POINT E, THE INTERSECTION OF ARCS, WITH A AND B. THIS FORMS THE TRIANGLE.

3. Connect the points where the center lines touch the circle.

Draw a Square When the Length of the Sides Is Known.

1. Draw a circle with a diameter equal to the length of the side of the square. See Fig. 5–18.

2. Draw the center lines of the circle at right angles to each other.

3. Draw 45-degree lines from each center line to the edge of the circle. These lines just touch the circle.

Draw a Hexagon When the "Across the Corners" Distance Is Known.

The term "across the corners" of a hexagon is the distance from one corner to the other. It is measured on a line that goes through the center of the hexagon. See Fig. 5–19.

1. Draw a circle with a diameter equal to the "across the corners" distance. See Fig. 5–20.

2. With A and B as centers, swing arcs until they cross the circle. The radius is the same as the radius of the circle.

3. Connect the points located to form a hexagon.

Draw a Hexagon When the "Across the Flats" Distance Is Known.

The term "across the flats" of a hexagon is the distance from one flat side to another. It is measured on a line that goes through the center of the hexagon. See Fig. 5–19.

1. Draw a circle with a diameter equal to the "across the flats" distance, Fig. 5–21.

2. Draw horizontal lines 90 degrees to vertical center line.

3. Draw lines on an angle of 60 degrees to the horizontal lines. They are drawn tangent to the circle.

Draw an Octagon

1. Draw a circle with a diameter equal to the "across the flats" distance. See Fig. 5–22.

2. Draw horizontal lines tangent to the top and bottom of the circle. They are 90 degrees to the center line.

83

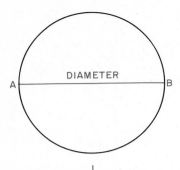

1
DRAW A CIRCLE WITH A
DIAMETER EQUAL TO THE
DIAGONAL OF THE SQUARE.

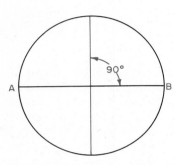

2
DRAW A CENTER LINE AT A RIGHT
ANGLE TO THE DIAMETER AB.

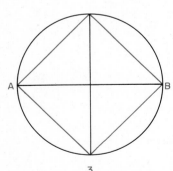

3
CONNECT THE POINTS WHERE
THE CENTER LINES TOUCH THE
CIRCLE.

5–17. *To draw a square when the
length of the diagonal is known.*

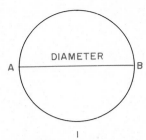

1
DRAW A CIRCLE WITH A
DIAMETER EQUAL TO THE
LENGTH OF THE SIDE OF THE
SQUARE.

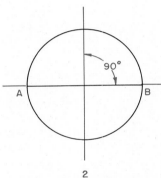

2
DRAW CENTER LINES AT RIGHT
ANGLES TO EACH OTHER.

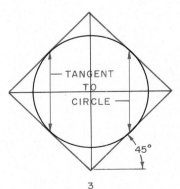

3
DRAW LINES AT 45° WITH THE
HORIZONTAL, THAT ARE
TANGENT TO THE CIRCLE.

5–18. *To draw a square when the
length of the sides is known.*

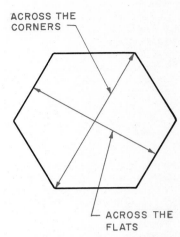

5–19. *Measurements used with
hexagons.*

3. Draw vertical lines tangent with the right and left sides. They are 90 degrees to the center line.

4. Draw 45-degree lines tangent to the circle. Where these lines cross the vertical and horizontal lines locates the corners of the octagon.

Construct an Ellipse. An ellipse is shown in Fig. 5–23. The long axis is called the major axis. The short axis is the minor axis. How to draw an approximate ellipse is shown in Fig. 5–24. An approximate ellipse is slightly inaccurate. However, it is accurate enough for most purposes.

To draw an approximate ellipse:

1. Draw the major and minor axes.

2. Draw a line connecting one end of each axis. This is line AC. Set a compass on radius OA. Swing an arc to locate point E.

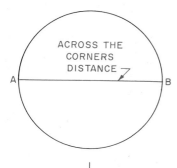

ACROSS THE
CORNERS
DISTANCE

A   B

1
DRAW A CIRCLE WITH A DIAME-
TER EQUAL TO THE ACROSS THE
CORNERS DISTANCE OF THE
HEXAGON.

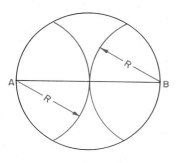

R

A   B

R

2
SWING ARCS FROM A AND B US-
ING THE RADIUS OF THE
CIRCLE.

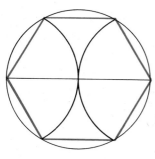

3
CONNECT THE POINTS LOCATED
ON THE CIRCLE.

5–20. *To draw a hexagon when
the "across the corners" distance is known.*

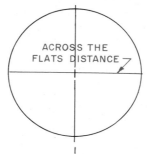

ACROSS THE
FLATS DISTANCE

1
DRAW A CIRCLE WITH A DIAME-
TER EQUAL TO THE ACROSS
THE FLATS DISTANCE. DRAW
THE CENTER LINES.

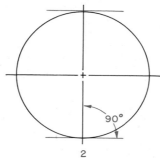

90°

2
DRAW HORIZONTAL LINES
TANGENT WITH THE CIRCLE.
THEY ARE PERPENDICULAR
TO THE VERTICAL CENTER
LINE.

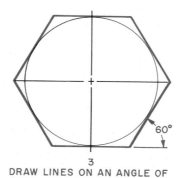

60°

3
DRAW LINES ON AN ANGLE OF
60° TO THE HORIZONTAL,
TANGENT TO THE CIRCLE.

5–21. *To draw a hexagon when
the "across the flats" distance is
known.*

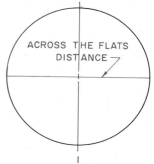

ACROSS THE FLATS
DISTANCE

1
DRAW A CIRCLE WITH A DIAME-
TER EQUAL TO THE ACROSS
THE FLATS DISTANCE OF THE
OCTAGON. DRAW THE CENTER
LINES.

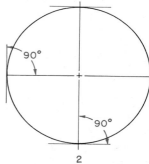

90°

90°

2
DRAW VERTICAL AND HORIZON-
TAL LINES TANGENT TO THE
CIRCLE. THEY ARE PERPENDIC-
ULAR TO THE CENTER LINES.

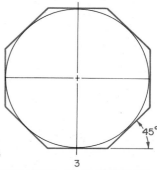

45°

3
DRAW LINES ON AN ANGLE OF
45° WITH THE HORIZONTAL AND
TANGENT TO THE CIRCLE.

5–22. *To draw an octagon when
the "across the flats" distance
is known.*

3. Set a compass on radius CE. Swing an arc to locate point F. Draw a line that is perpendicular to AF. It must bisect AF. This crosses AB at G, and CD at H. These are the centers of the arcs to form the ellipse. With dividers, locate points J and K. OG equals OJ. OH equals OK.

4. Draw lines KG, KJ, HJ.

5. Set a compass to radius HC. Swing an arc until it touches line GH and HJ. These are the tangent points. Repeat using K as the center. Set a compass to radius GA. With G as a center, swing the end of the ellipse. Repeat using J as a center.

Tangent Points. A tangent point is the exact point at which one of two joining lines stops and the other starts. See Fig. 5–25.

The use of tangent points is very important to good drafting. A draftsman always locates tangent points as he makes a drawing.

Tangent points must be found when a curved and straight line meet. To do this, first find the center of the curved line. Then draw a perpendicular from the center to the straight line. Where they cross is the tangent point.

When two curved lines meet, the tangent point must be found. This is done by drawing a line from one center to the other. Where it crosses the curved line is the tangent point.

Draw an Arc Tangent to Two Lines at 90 Degrees. This construction is shown in Fig. 5–26.

1. Draw the two lines meeting at 90 degrees. These are AB and BC.

2. Set a compass on the radius of the corner arc.

3. Put the pin on corner B. Swing an arc locating points D and E.

4. Set the pin on D and E and swing arcs. These cross at F. This is the center to be used to draw the arc wanted.

5. Set the pin at F and draw the corners from D to E. D and E are the tangent points. The straight lines must stop here. The curved line begins at D and must stop at E.

A line drawn from F to D and E is perpendicular to each side. Therefore, a line from a center point drawn perpendicular to a straight line locates the tangent point. This is shown in Fig. 5–27.

Draw an Arc Tangent to Two Lines Not at 90 Degrees. This is shown in Fig. 5–27.

1. Draw the two lines meeting at an angle other than 90 degrees.

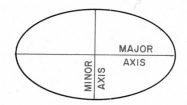

5-23. An ellipse.

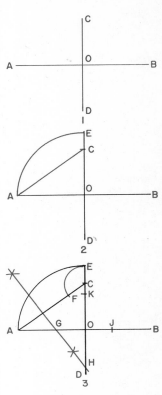

5-24. To draw an approximate ellipse.

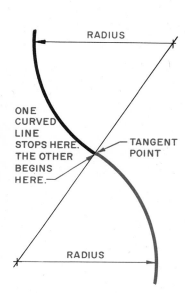

RADIUS

ONE CURVED LINE STOPS HERE. THE OTHER BEGINS HERE.

TANGENT POINT

RADIUS

THE TANGENT POINT BETWEEN TWO CURVED LINES.

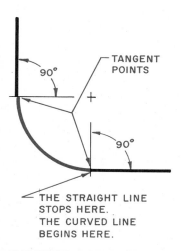

90°

TANGENT POINTS

90°

THE STRAIGHT LINE STOPS HERE. THE CURVED LINE BEGINS HERE.

THE TANGENT POINT BETWEEN A CURVED AND A STRAIGHT LINE.

5–25. *A tangent point is where one line stops and the one joining it begins.*

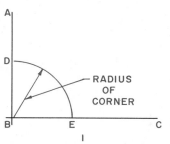

A

D

RADIUS OF CORNER

B   E   C

1

SWING AN ARC THAT IS THE RADIUS OF THE ROUND CORNER TO BE DRAWN. THIS LOCATES POINTS D AND E. THEY ARE THE TANGENT POINTS BETWEEN THE STRAIGHT LINES AND THE CORNER TO BE DRAWN.

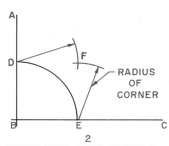

A

D   F

RADIUS OF CORNER

B   E   C

2

SWING ARCS FROM POINTS D AND E. USE THE RADIUS OF THE CORNER. THIS LOCATES POINT F. F IS THE CENTER USED TO DRAW THE CORNER.

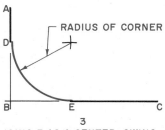

A

D

RADIUS OF CORNER

B   E   C

3

USING F AS A CENTER, SWING AN ARC FROM D TO E. D AND E ARE THE TANGENT POINTS. THE ARC ENDS AT THESE TWO POINTS.

5–26. *To draw an arc tangent to two straight lines meeting at a 90-degree angle.*

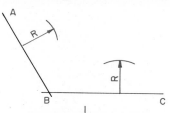

A

R

R

B   C

1

SWING ARCS HAVING A RADIUS EQUAL TO THE RADIUS WANTED AT THE CORNER.

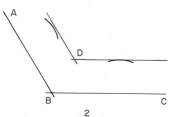

A

D

B   C

2

DRAW LINES TANGENT TO THESE ARCS AND PARALLEL TO THE LINES. THESE MEET AT D. D IS THE CENTER USED TO DRAW THE CURVED CORNER.

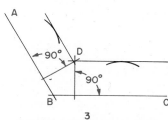

A

90°   D

90°

B   C

3

DRAW LINES FROM D, PERPENDICULAR TO AB AND BC. THESE ARE THE TANGENT POINTS FOR THE CURVED CORNER.

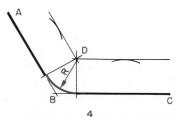

A

D

R

B   C

4

USING D AS A CENTER, DRAW AN ARC FROM ONE TANGENT POINT TO THE OTHER.

5–27. *To draw an arc tangent to two lines that are not 90 degrees to each other.*

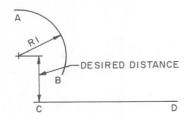

I
DRAW ARC AB (RI). MEASURE
THE DESIRED DISTANCE FROM
THE CENTER OF AB FOR LINE
CD. DRAW LINE CD.

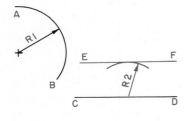

2
DRAW A LINE PARALLEL TO CD
AT A DISTANCE EQUAL TO THE
RADIUS OF THE ARC TO CON-
NECT AB AND CD. THE RADIUS
IS MARKED R2.

3
SWING AN ARC WITH A RADIUS
EQUAL TO THE RADIUS OF ARC
AB (RI) PLUS THE RADIUS OF
THE CORNER (R2). THIS
LOCATES POINT E. E IS THE
CENTER USED TO DRAW THE
ARC TO CONNECT AB AND CD.

5–28. *To draw an arc tangent to
a straight line and an arc.*

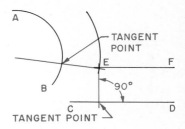

4
TANGENT POINT ON CD IS
FOUND BY DRAWING A PERPEN-
DICULAR FROM THE CENTER E
TO CD. THE TANGENT POINT
ON AB IS FOUND BY DRAWING
A LINE FROM THE CENTER E
TO THE CENTER OF AB.

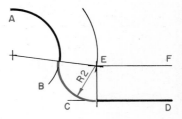

5
DRAW THE ARC WITH RADIUS
R2 FROM ONE TANGENT POINT
TO THE OTHER.

2. Set the compass at the radius of the arc wanted.

3. Select two points on lines AB and BC and swing arcs from them.

4. Draw lines tangent to these arcs. They must be parallel with AB and BC. They are the length of the radius away from AB and BC. The intersection of these new lines locates the center of the arc D.

5. Draw lines from the center D perpendicular to AB and BC. This locates the tangent points.

88

6. Draw the arc using D as the center. It goes from one tangent point to the other.

Draw an Arc Tangent to a Straight Line and an Arc. This is shown in Fig. 5–28.

Known are arc AB with radius R^1 and line CD. It is desired to connect these with an arc with radius R^2.

1. Draw a line, EF, parallel to CD. It is distance R^2 above CD.

2. Add together R^1 and R^2.

Set a compass this length.

3. Swing an arc with radius R^1 plus R^2 from the center of arc AB.

4. Where this arc crosses line EF is the center of R^2.

5. Find the tangent points on arc AB and line CD. Connect the centers of the two arcs. Where this crosses arc AB is one tangent point. Draw a perpendicular from center E to line CD. This is the tangent point.

6. Set a compass on R^2. Swing an arc from center E

connecting arc AB and line CD. Stop at each tangent point.

Draw an Arc Tangent to Two Arcs. This is shown in Fig. 5–29.

Known are arcs AB and CD. It is desired to connect these with an arc with radius R^3.

1. Add the radius of AB (R^1) and R^3. Set a compass this size.

2. Swing an arc from the center, E, of arc AB.

3. Add the radius of CD (R^2) and R^3. Set a compass this size.

4. Swing an arc from the center, F, of arc CD until it crosses the arc from AB. This is the center, G. It is the center used to draw the arc with radius R^3.

5. Connect centers EG and FG. The tangent point is where these lines cross arcs AB and CD.

6. Set a compass on R^3. Swing an arc using G as the center. It starts at the tangent point on AB and stops at the tangent point on CD.

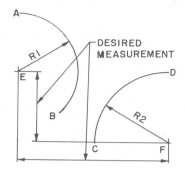

1
DRAW ARC AB (RI) FROM THE CENTER E. MAKE THE TWO DESIRED MEASUREMENTS TO LOCATE CENTER F. DRAW ARC CD (R2).

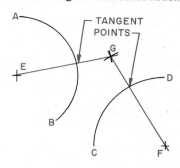

3
CONNECT THE CENTERS EG AND FG. WHERE THESE LINES CUT THE ARCS ARE THE TWO TANGENT POINTS.

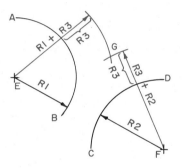

2
ADD THE RADIUS OF THE ARC AB (RI) AND THE RADIUS OF THE ARC TO CONNECT AB AND CD (R3). USING THIS RADIUS (RI PLUS R3), SWING AN ARC USING E AS THE CENTER. ADD THE RADIUS OF ARC CD (R2) AND R3. SWING AN ARC USING F AS THE CENTER. THIS LOCATES CENTER G. IT IS THE CENTER FOR THE ARC TO CONNECT ARCS AB AND CD.

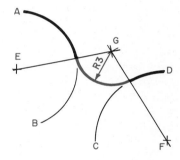

4
SET A COMPASS TO THE RADIUS OF THE ARC TO CONNECT AB AND CD (R3). SWING THE CONNECTING ARC FROM CENTER G. IT RUNS FROM ONE TANGENT POINT TO THE OTHER.

5–29.   *To draw an arc tangent to two other arcs.*

Build Your Vocabulary

Following are terms that you should understand and use as a part of your working vocabulary. Write a brief explanation of what each means.

Angle
Triangle
Parallelogram
Square
Polygon
Arc
Circle
Prism
Cylinder
Pyramid
Cone
Sphere
Torus
Bisect
Equilateral
Hexagon
Octagon
Ellipse
Tangent

Section Three
Communicating with Drawings

Many types of drawings are used to show design ideas. This drawing shows the artist's concept of a service module returning to earth.

Links in Learning

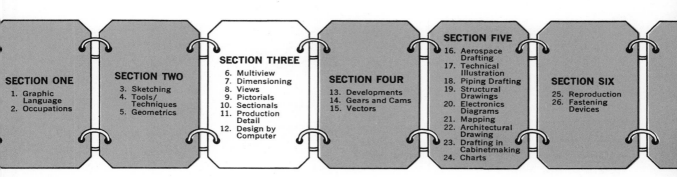

Chapter 6

Multiview Drawing

A designer or engineer uses drawings to present his ideas. A pictorial drawing of an object shows a picturelike view that is easy to understand, but it does not present the true shape of the parts. The view of the object as it actually appears to the eye also gives the viewer a description of the object as shown in Fig. 6–1. The most frequently used way to present all details of an object is by making a multiview drawing. A *multiview* drawing is a series of separate views of an object, arranged so each view is related to the other views. Fig. 6–2 shows a multiview drawing of a machinist's vise. Notice how each separate view on the multiview drawing shows one side of the object. These views are drawn on flat surfaces called *planes.*

PLANES

An examination of any object will show it has plane surfaces and curved surfaces.

A *plane* is a surface on which a straight line drawn between any two points on the surface lies on the surface. Any two lines in a plane are either parallel or must intersect. For example, the surface of a sheet of plywood that is perfectly smooth is a plane.

A plane in flat position, as the floor, is called a *horizontal* plane. A plane in an upright position, as the wall, is a *vertical* plane. The three planes

Columbian Vise Mfg. Co.

6–1. *This is a machinist's vise as it appears to the eye. It does not show the true shape of the parts of the vise.*

used in multiview drawings are *horizontal, profile,* and *frontal.* The *profile* and *frontal* planes are vertical. See Fig. 6–3. Multiview drawings are made by projecting lines from an object to these planes, which are called *planes of projection.*

Third Angle Projection. The three planes, Fig. 6–3, form four 90-degree angles called quadrants. These are identified by numbers 1, 2, 3, and 4. All multiview drawings used in North America place the object in the *third quadrant.* The views are projected to the planes of projection, as shown in Fig. 6–4. Since this object is in the third quadrant or angle, it is called *third angle projection.* In Europe the object is placed in the first quadrant; therefore, Europeans use first angle projection.

Plane Surfaces. The plane surfaces of an object can be of three types: normal, inclined, and oblique. See Fig. 6–5.

A *normal plane* is one that is perpendicular (meets at a 90-degree angle) to *two* regular planes of projection and parallel to the third plane. It

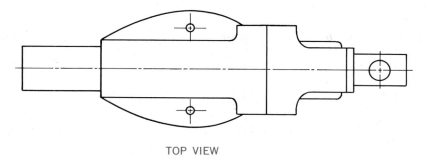

TOP VIEW

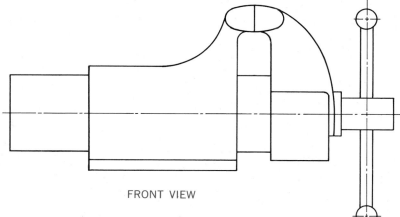

FRONT VIEW

6–2. *A three-view multiview drawing showing front, top, and side views of a machinist's vise.*

RIGHT SIDE
VIEW

appears *true size* on the plane to which it is parallel and as a *true length* line on the planes to which it is perpendicular. See Fig. 6–6.

An *inclined plane* is one that is perpendicular to *one* regular plane of projection and is at an acute angle (less than 90 degrees) to the other two planes. It appears *true length* on any regular plane of projection to which it is perpendicular and *foreshortened* (not true size) on any regular plane to which it is not perpendicular. See Fig. 6–7. Fig. 6–8 shows

an industrial product containing normal and inclined planes.

An *oblique plane* is one that is *not* perpendicular to any of the regular planes of projection. It appears *foreshortened* (shorter than true length) on all regular planes of projection. See Fig. 6–9. Fig. 6–10 shows a product containing normal, curved, inclined, and oblique planes.

LINES

The meeting of two planes forms a corner which appears as a straight line on a drawing.

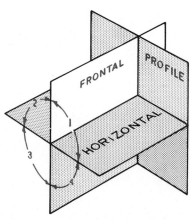

6–3. *Planes of projection form quadrants.*

93

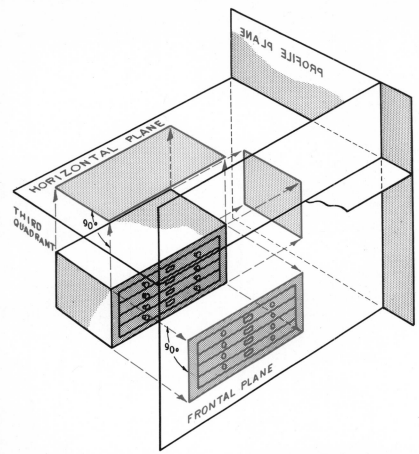

6-4. Third angle projection places the object in the third quadrant.

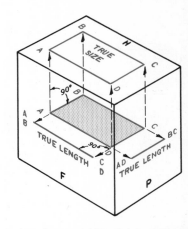

6-6. A normal plane appears true size on the plane to which it is parallel and true length on the plane to which it is perpendicular.

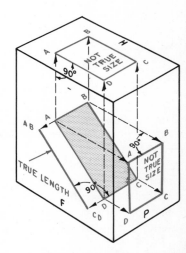

6-7. An inclined plane appears true length on the plane to which it is perpendicular and fore-shortened on any plane to which it is not parallel.

6-5. The Aluette II space craft was designed using normal, inclined, and oblique surfaces on the exterior. Can you find an example of each surface?

National Aeronautics and
Space Administration

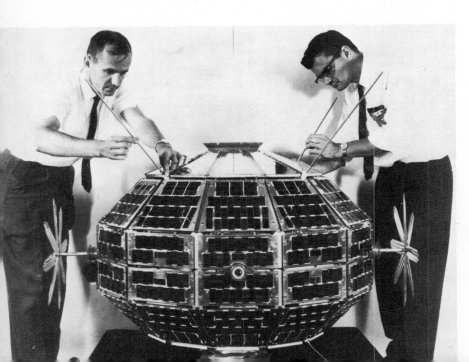

6–8. *Magnetic V blocks contain normal and inclined planes.*

Reid Tool Supply Co.

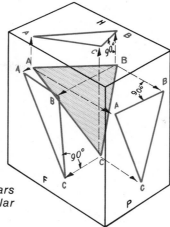

6–9. *An oblique plane appears foreshortened on all regular planes of projection.*

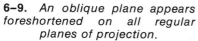

INCLINED SURFACE

CURVED SURFACE

OBLIQUE SURFACE

NORMAL

6–10. *A spade drill used in machine tool work. Notice the normal, inclined, oblique, and curved surfaces that make up the cutting edge.*

De Vlieg Microbore

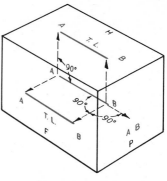

6–11. *A normal line appears true length on planes to which it is parallel and as a point on planes to which it is perpendicular.*

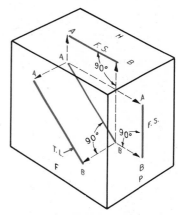

6–12. *An inclined line appears true length on any plane to which it is parallel.*

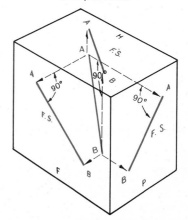

6–13. *An oblique line appears foreshortened on all regular planes of projection.*

Lines can be of three types: normal, inclined, and oblique.

A *normal line* is one that is parallel to *two* regular planes of projection and perpendicular to the other plane. It appears *true length* on the planes to which it is parallel and as a *point* on the plane to which it is perpendicular. See Fig. 6–11.

An *inclined line* is one that is parallel to *one* regular plane of projection and at an acute angle to the other two regular planes. It appears *true length* on the plane to which it is parallel and *foreshortened* on the other two planes to which it is not parallel. See Fig. 6–12.

An *oblique line* is one that is not parallel to any of the regular planes of projection. It appears *foreshortened* on all regular planes of projection. See Fig. 6–13.

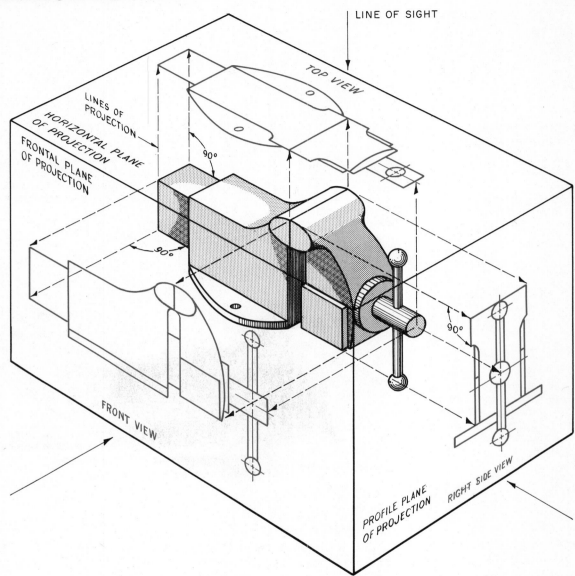

LINE OF SIGHT

TOP VIEW

LINES OF PROJECTION

HORIZONTAL PLANE OF PROJECTION

FRONTAL PLANE OF PROJECTION

90°

90°

FRONT VIEW

90°

PROFILE PLANE OF PROJECTION

RIGHT SIDE VIEW

6–14. *Glass box with views projected on the planes of projection.*

ORTHOGRAPHIC PROJECTION

The views of a multiview drawing are projected onto surfaces called planes of projection. Imagine that an object to be drawn is placed inside a hinged glass box with the sides of the object parallel with the sides of the box. The sides of the box are the *planes of projection.* The draftsman is on the outside looking through the glass at the object. See Fig. 6–14.

The surfaces of the object are projected onto the glass sides at right angles to the side. This *right angle projection* is called *orthographic projection.* All orthographic projections are at right angles to the plane of projection.

In order to show the views of the object on a flat surface

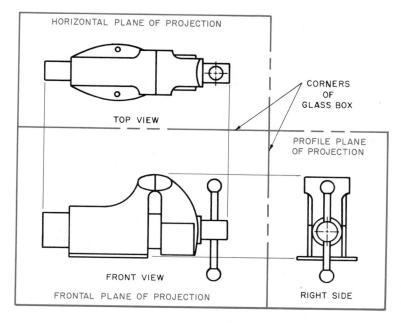

HORIZONTAL PLANE OF PROJECTION

TOP VIEW

CORNERS
OF
GLASS BOX

PROFILE PLANE
OF PROJECTION

FRONT VIEW

FRONTAL PLANE OF PROJECTION

RIGHT SIDE

6–15. *Planes of projection unfolded.*

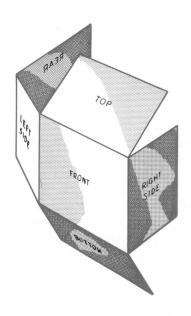

6–16. *Glass box partly "unfolded" to show the six planes of projection.*

bottom view is projected to a horizontal plane. The rear view is projected to a frontal plane.

It is possible to unfold the glass box so all six sides appear on a plane. This means that six views of the object can be drawn in this manner. See Fig. 6–17. It is good practice to draw only the views that are necessary to describe the object. This could take from one to six views.

Notice how the surfaces of the object project from one view to another. This projection is used in laying out and drawing the object.

Height, Width, and Depth. When referring to the overall size of an object, the terms height, width, and depth are used. *Height* is the perpendicular distance between horizontal planes. *Width* is the perpendicular distance between profile planes. *Depth* is the perpendicular distance between frontal planes. These are shown in Fig. 6–18.

In the front view, the width and height of the object are shown. See Fig. 6–17. The height projects to the side and rear views and the width to the top and bottom views. The side view shows the depth and height of the object. The depth is most easily transferred to the top, bottom, and left-side view using dividers or scale. Fig. 6–17 shows the depth measured on the side view with a scale and transferred to the other views with a divider.

(a sheet of paper), the glass box is unfolded and spread so it is flat or in one plane, Fig. 6–15. This is how the orthographic views appear on a multiview drawing.

The top section of the glass box forms the top view on a horizontal plane. It always folds up above the front view. The front view is projected on the frontal plane. The right-side view is projected on the profile plane. It always unfolds to the right of the front view.

The glass box has six sides. Views of the object can be projected orthographically (perpendicularly) from any of these sides. Each side has a name to describe the view projected from it. See Fig. 6–16.

The left-side view is projected to a profile plane. The

6-17. *The six regular orthographic views.*

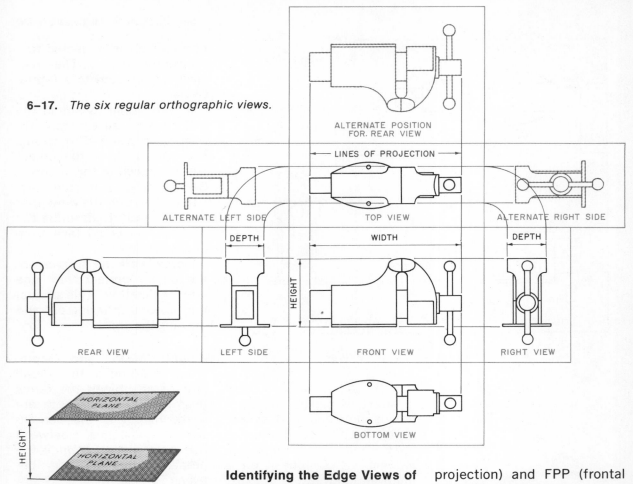

ALTERNATE POSITION
FOR. REAR VIEW

LINES OF PROJECTION

ALTERNATE LEFT SIDE TOP VIEW ALTERNATE RIGHT SIDE

DEPTH WIDTH DEPTH

HEIGHT

REAR VIEW LEFT SIDE FRONT VIEW RIGHT VIEW

BOTTOM VIEW

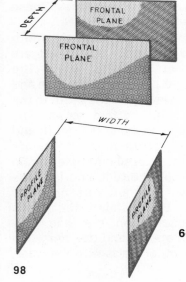

6-18. *Height, width, and depth.*

Identifying the Edge Views of the Planes of Projection.

Basic to understanding orthographic projection is an understanding of the planes of projection. In complex problems it is necessary to label these planes and their edge views properly so projection distances can be measured. See Fig. 6–19. When the line of sight is perpendicular to the *horizontal plane,* the profile plane and frontal plane appear in *edge view.* Fig. 6–19 shows these labeled PPP, (profile plane of projection) and FPP (frontal plane of projection).

When the line of sight is perpendicular to the *profile plane* the HPP (horizontal plane of projection) and the PPP (profile plane of projection) appear in *edge view.*

Fig. 6–20 shows the three planes placed in a normal position, forming the glass projection box. The edge views of the planes are labeled. The planes are unfolded into one plane, Fig. 6–21 showing how the *edge views* of planes are identified on a drawing. After this identification system is

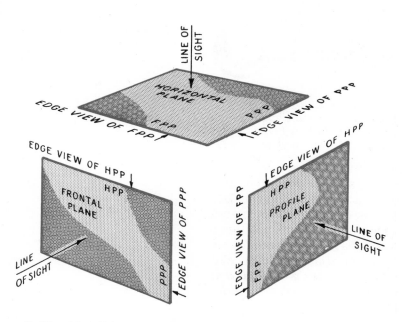

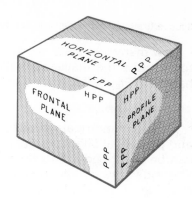

6-20. *The edge views of planes of projection as they relate to each other.*

6-19. *Identifying the edge views of the planes of projection.*

understood, it is not necessary to label the planes on every drawing. It is a good practice to follow until this is understood.

ONE-VIEW DRAWING

Some objects can be fully described by drawing one view and dimensioning. The draftsman must decide if the drawing gives all the information needed to make the article. Sometimes a simple note can remove the need for a view as shown in Fig. 6-22.

6-21. *Edge views of planes on the normal planes of projection.*

6-22. *One-view drawings.*

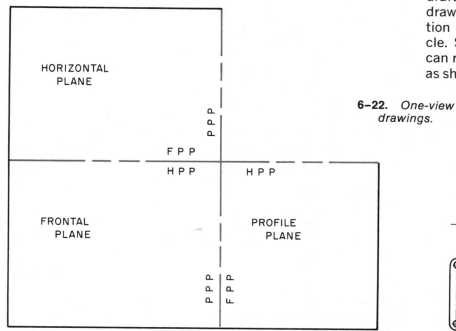

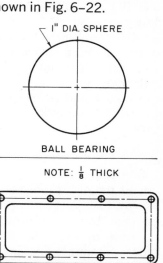

How many views are needed to describe the malleable ball handle in Fig. 6–23?

TWO-VIEW DRAWINGS

Most symmetrical, conical, and pyramidal-shaped objects can be described in two views. The side view of the bearing ring, Fig. 6–24, would be identical to the top view, and therefore is not needed. A top view of the adjustment knob, Fig. 6–24, would be identical to the side view, and therefore unnecessary.

THREE-VIEW DRAWING

Most objects require three views for complete shape description; therefore, the three-view drawing is most common. The front and right-side views give all the information needed except the shape of the vertical handle, Fig. 6–25. The top view shows it to be square.

Not all three-view drawings use the top, front, and right-side views. Any of the six regular orthographic views can be used. Fig. 6–26 shows a three-view drawing using right and left sides and a front view. A top view would be exactly like the front view and is unnecessary.

MORE THAN THREE VIEWS

Sometimes, an object will have a detail that will require a fourth, fifth, or sixth view. The decision to draw another view is made by the draftsman. It must present some information about the object that is

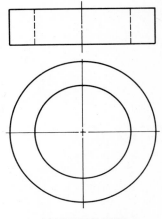

Reid Tool Supply Co.

6–23. *A malleable ball handle containing two spheres and a conical member.*

not clearly shown on the other views. In Fig. 6–27 the left-side view was needed to clearly show the shape of the hole. It could have been square, circular, or triangular. The clip, Fig. 6–28 uses a rear view to clarify details that are hidden in the front view.

PARTIAL VIEWS

It is accepted practice to draw views that are not complete, but which give enough information to describe the object. These are called *partial views*. They save space and time. They can involve: (A) a view of a portion of an object, (B) a half view of a symmetrical object, or (C) a partial view with a section. These are shown in Fig. 6–29.

If an object has features on the right and left sides that are different and require both side views, these views need not be complete. See Fig. 6–30. Notice that the hidden details on each end are not drawn. The same practice is shown in Fig. 6–26.

SELECTING THE VIEWS

Before drawing an object, the number of views needed must be decided. If two views

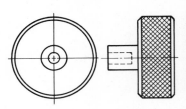

BEARING SLEEVE

ADJUSTMENT KNOB

6–24. *Two-view drawings.*

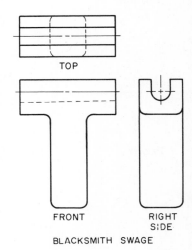

TOP

FRONT   RIGHT SIDE

BLACKSMITH SWAGE

6–25. *A three-view drawing.*

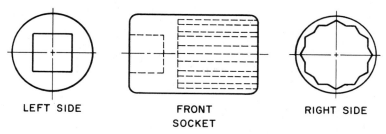

LEFT SIDE   FRONT   RIGHT SIDE
                   SOCKET

6-26. *A three-view drawing of a mechanic's socket. The left side view was necessary to show the shape of the hole in that end.*

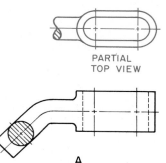

PARTIAL
TOP VIEW

A

PARTIAL TOP VIEW OF A GUIDE ROD.

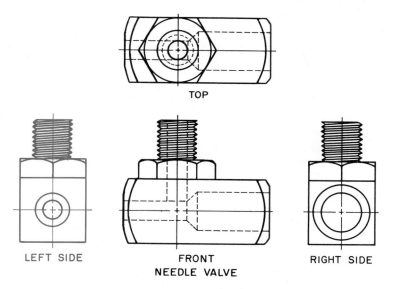

TOP

LEFT SIDE   FRONT   RIGHT SIDE
                 NEEDLE VALVE

6-27. *A fourth view, the left-side view, is needed to clarify details of the needle valve body and hub.*

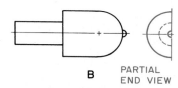

B

PARTIAL
END VIEW

PARTIAL END VIEW OF A SYMMETRICAL DIAMOND WHEEL DRESSER.

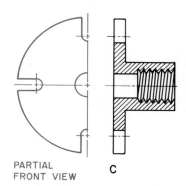

PARTIAL
FRONT VIEW

C

PARTIAL FRONT VIEW OF A WOOD LATHE FACEPLATE IN SECTION.

are needed, usually the top and front or front and right-side views are used. In a three-view drawing usually the front, top and right side are used.

The first view to be selected is the one that is to be the front view. It is the most important view of the object and usually the one that shows the shape most clearly. It is usually the longest view. See Fig. 6-31. The front view should be located so the least number of hidden lines appear. See

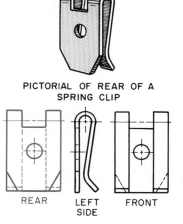

PICTORIAL OF REAR OF A SPRING CLIP

REAR   LEFT   FRONT
           SIDE

6-29. *Partial views simplify a drawing.*

6-28. *A spring clip requires a rear view to clarify details.*

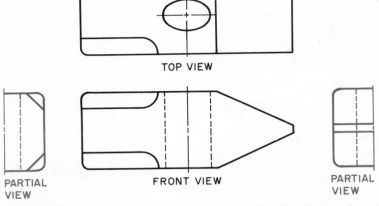

TOP VIEW

PARTIAL VIEW

FRONT VIEW

PARTIAL VIEW

6-30. *Partial views clarify the drawing by reducing the number of hidden lines.*

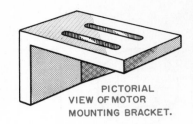

PICTORIAL VIEW OF MOTOR MOUNTING BRACKET.

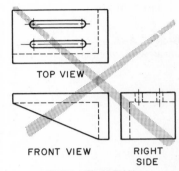

TOP VIEW

FRONT VIEW

RIGHT SIDE

POOR SELECTION

FRONT VIEW SHOWS SHAPE BUT DETAILS ARE HIDDEN.

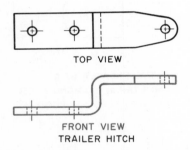

TOP VIEW

FRONT VIEW
TRAILER HITCH

6-31. *The front view shows the shape of the object. It is usually the longest view.*

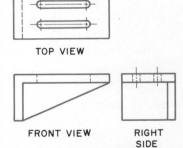

TOP VIEW

FRONT VIEW

RIGHT SIDE

GOOD SELECTION

FRONT VIEW SHOWS THE SHAPE AND THE INTERNAL SURFACES ARE VISIBLE.

6-32. *Motor mounting bracket.*

Fig. 6–32. When the motor mounting bracket was drawn with the open side to the front, the edges became visible.

The object should be drawn in a normal position. See Fig. 6–33. The cup drawn in any other position would be quite unnatural.

Alternate Positions of Views. The object should be drawn so it makes good use of the space on the drawing paper and the drawing presents a balanced appearance. See Fig. 6–34. In this case, an

alternate location for the right-side view was chosen to save space. It was projected off the top view. The same thing could be done with a left-side view. Fig. 6–17 shows an alternate location for a rear view. It can be drawn off the top view. A disadvantage is that it is upside down.

LAYING OUT THE DRAWING

Before a drawing can be started, several things must be decided. The size the object is to be drawn must be known. The number of views needed must be decided. The space needed for dimensions must be estimated. This is discussed in Chapter 7, Dimensioning. These space requirements are used to select the size paper needed. Drawing paper is sold in standard sizes. This is explained in Chapter 4, Tools and Techniques of Drafting.

The views should be spaced so they give the drawing a

balanced appearance. If the drawing occupies one full sheet, the right and left margins should be equal. Sometimes the bottom margin is slightly greater than the top margin.

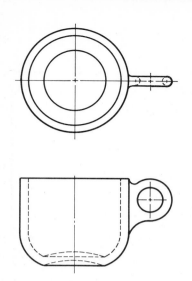

6–33. *An object is drawn in its normal position.*

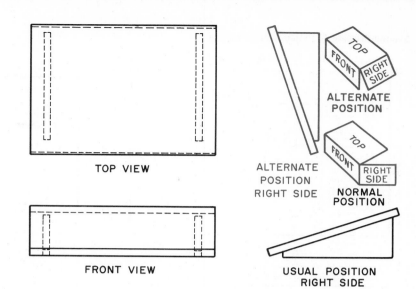

TOP VIEW

FRONT VIEW

6–34. *The alternate position of the side view of the drawing board balances the drawing and saves space.*

The overall dimensions of the object are usually "blocked in" first in all views. See Fig. 6–35C. Use a light, thin line. Center lines for cylindrical parts are drawn next.

Draw in the details. Start with the basic shape of the object. See Fig. 6–35D. Then draw circles and arcs. The minor details are drawn next. Carry each detail to all views before starting on a second detail. *Do not complete one view before starting on another.* Carry all views along together to completion.

After all the details are drawn, remove unnecessary lines. Light lines used to locate details are called construction lines. As you gain some experience, you will be able to draw the required lines the proper weight as you lay out the drawing. This is better than going back over the draw-

ing and darkening them. See the alphabet of lines in Chapter 4 for correct line weight.

If the object is cylindrical, locate the center lines in each view. Then draw the circles and arcs in each view. Next draw all horizontal lines in each view. Then draw all vertical lines. See Fig. 6–36.

Dimension the drawing and letter notes and other titles.

One method of spacing views on drawing paper is shown in Fig. 6–35. This method involves the following:

1. The sheet selected is size B, 11 inches by 17 inches. After ½-inch borders are drawn, a working space of 10 inches by 16 inches remains.

2. The object has overall dimensions of: width, 7½ inches; height, 4¼ inches; and depth, ¾ inches. See Fig. 6–35A.

3. Space needed between views for dimensions is: 2 inches between top and front view, 2¾ inches between front and side view.

4. The sum of the height, depth, and dimensioning space is therefore 7 inches, Fig. 6–35B.

5. The working space of 10 inches less the distance required by the object, 7 inches, leaves 3 inches to be divided equally above the top view and below the front view of the drawing.

6. The sum of the width, depth, and dimensioning space is 11 inches, Fig. 6–35C.

7. The working space of 16 inches less the distance required by the object leaves 5 inches to be divided equally for the right and left margins.

8. Details of the part, Fig. 6–35D, do not change the spacing.

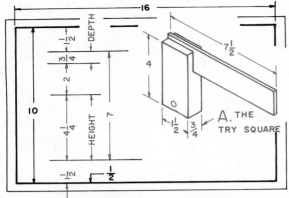

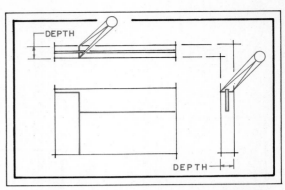

B. LAY OUT OVERALL HEIGHT AND DEPTH

D. ADD DETAILS STARTING WITH THE BASIC SHAPE OF THE OBJECT

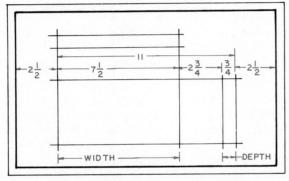

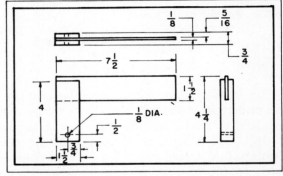

C. LAY OUT OVERALL WIDTH AND DEPTH. THIS "BLOCKS IN" EACH VIEW

E. REMOVE CONSTRUCTION LINES, ADD MINOR DETAILS, DIMENSIONS, AND LETTERING

6–35. *Spacing of views should leave room for dimensions and balance the drawing on the paper.*

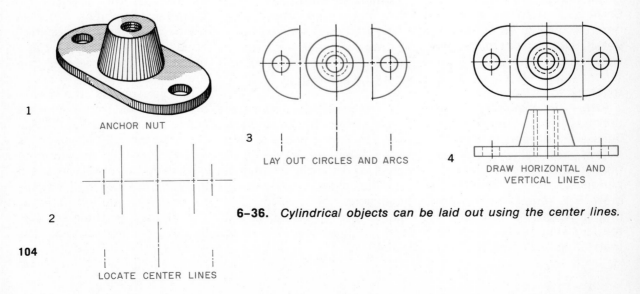

1

ANCHOR NUT

2

LOCATE CENTER LINES

3

LAY OUT CIRCLES AND ARCS

4

DRAW HORIZONTAL AND VERTICAL LINES

6–36. *Cylindrical objects can be laid out using the center lines.*

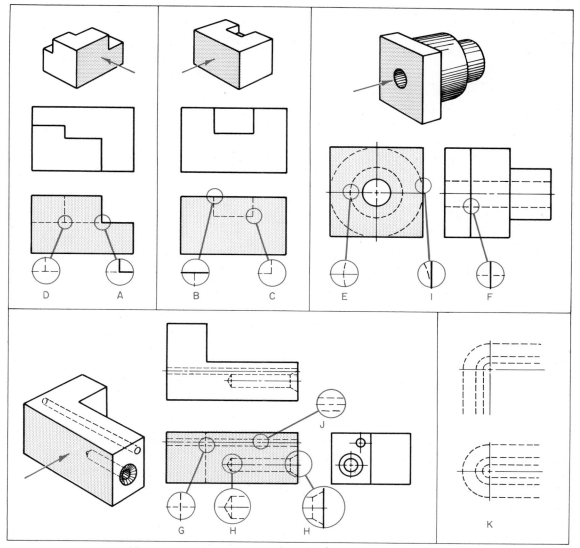

6-37. *Correct practices for drawing hidden lines.*

9. The completed view, properly spaced with dimensions between views, is shown in Fig. 6-35E.

HIDDEN LINES

Hidden lines are used to show details that are behind some part of the object. They are shown on a drawing by a line made of short dashes. The dashes are about $\frac{1}{8}$ inch in length and spaced $\frac{1}{32}$ inch apart. The top view, Fig. 6-34, has hidden lines to show the pieces below.

Fig. 6-37 shows the correct practices to use when drawing hidden lines.

Hidden lines:

• should leave a gap whenever a hidden dash would form a continuation of a visible line, Detail A.

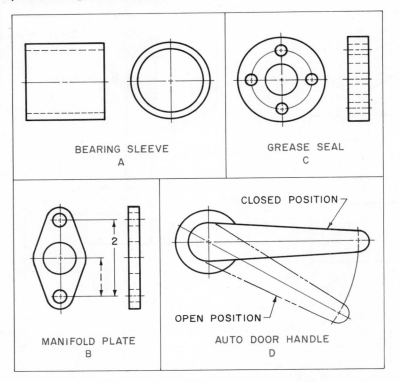

6–38. *Typical uses of center lines.*

and short dashes. The long dash is usually 3/4 inch to 1 1/2 inches and the short dash about 1/8 inch long. The center line extends about 1/4 inch beyond the outside edge of the object for which it was drawn, but is not used to connect the views.

In Fig. 6–38A the center of a symmetrical object is located. The center is usually located by crossing two short dashes. The center lines end with long dashes.

In Fig. 6–38B the center line continues across the entire object and serves to locate the holes and represent the center of the object. In Fig. 6–38C the center line is used to describe the circular path of the centers of a series of bolt holes. In Fig. 6–38D the path of motion made by the movement of a handle is recorded with a center line.

Center lines are useful in dimensioning. They assist in locating the position of circular parts in relation to the object as a whole. This is covered in Chapter 7, Dimensioning.

• should end by touching a visible line, Detail B.

• that form a corner should touch, Detail C.

• that meet should touch, Detail D.

• should cross center lines, Detail E.

• that cross a visible line should not touch the visible line, Detail F.

• that indicate planes which do not meet should not have dashes touching, Detail G.

• that indicate hidden holes should intersect at the corners, Detail H.

• should not touch visible lines because it would make

the visible lines appear to extend into the surface of the object, Detail I.

• that are close together and parallel should have the dashes staggered, Detail J.

• that form a corner should touch the center lines, Detail K.

CENTER LINES

Center lines are used to locate the centers of symmetrical objects and paths of motion. Symmetrical objects are those that are the same on both sides of the center lines. See Fig. 6–36. Center lines are drawn as a series of long

PRECEDENCE OF LINES

When two lines in a view fall together, the most important of the two is shown. The order of importance is:

1. Visible line
2. Hidden line
3. Center line

When a cutting-plane line falls on a center line, the cutting-plane line is shown.

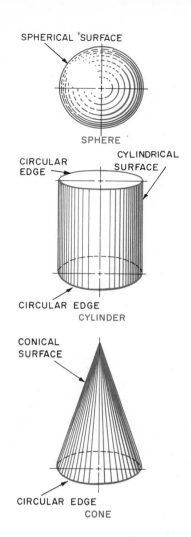

6–39. *Basic curved surfaces.*

PROJECTING CURVED SURFACE

The basic curved surfaces are cylindrical, spherical, and conical. See Fig. 6–39. Cylinders and cones appear as curves in views perpendicular with the axis of the object, Fig. 6–40. In other normal views, the cylinder projects as a rectangle and the cone as a triangle. The sphere projects as a circle in all normal views.

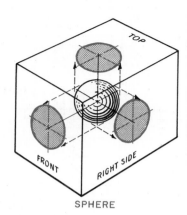

SPHERE

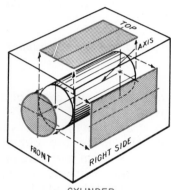

CYLINDER

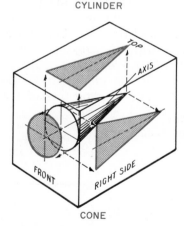

CONE

6–40. *Projections of basic curved surfaces.*

6–42. *Projection when all surfaces are curved.*

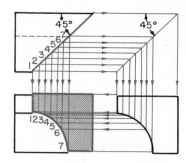

6–41. *Locate curved edges by projecting points along the edge.*

PROJECTING CURVED EDGES

Curved edges are projected by locating a series of points along the edge and projecting these to the other views. The points are then connected with an irregular curve to form the projected curve.

The number of points to use will vary with the curved edge. Usually the more points projected, the more accurate the projected curve will appear.

Fig. 6–41 shows an edge view of a piece of molding drawn actual size in the right-side view. The top view indicates the molding is cut on a 45-degree angle. Points 1, 2, 3, 4, 5 are located at random on the side view and projected to the top and front views and the curve is drawn where they intersect on the front view.

Fig. 6–42 shows projection when all surfaces are curved.

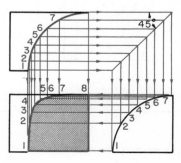

107

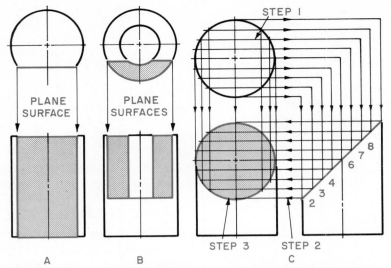

6-43. *Common intersections between plane and curved surfaces.*

INTERSECTIONS OF PLANES AND CURVED SURFACES

Fig. 6-43 shows typical intersections between a cylinder and a plane. At *A* a cylinder is cut by a vertical plane. At *B* a hollow tube is cut part way through by a vertical plane revealing an interior curved surface. At *C* the cylinder is cut by an inclined plane creating an elliptical surface. The elliptical surface can be projected by using the points system.

To draw the elliptical surface: (1) divide the circular view into several parts, (2) project these divisions to the other views, (3) locate the intersection of these projections with light dots. Connect these dots with an irregular curve. An ellipse template could be used or the ellipse could be constructed using one of the

6-45. *Surface intersections.*

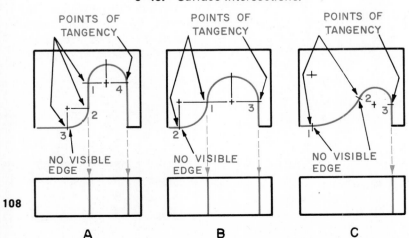

CIRCLE
ELLIPSE
PARABOLA
HYPERBOLA

CIRCULAR PLANE
A

ELLIPTICAL PLANE
C

PARABOLIC PLANE
D

HYPERBOLIC PLANE
B

6-44. *Conical sections generated by the intersection of a cone and a plane.*

methods described in Chapter 5, Geometric Figures and Constructions.

Fig. 6–44 illustrates typical intersections between a cone and a plane. At *A* the cone is cut by a horizontal plane, leaving a circular surface. At *B* it is cut by a vertical plane, creating a hyperbolic-shaped surface. At *C* the cone is cut by an inclined plane, leaving an elliptical surface. At *D* the cone is cut with an inclined plane, creating a parabolic-shaped surface.

The projection of cylindrical surfaces from view to view is shown in Fig. 6–45. At *A* notice a straight section, 1–2, occurs between the points of tangency of the two cylindrical surfaces. This appears as a straight, visible line in the front view. At point of tangency 3 no edge occurs; therefore, no visible line is projected to the front view.

At *B* the two cylindrical surfaces meet at the point of tangency. This is indicated on the front view as a straight vertical line.

At *C* the surface generated by the two intersecting curves is a continuous slope; therefore, in the front view no visible line appears.

Fig. 6–46 illustrates the intersection of other cylindrical surfaces and planes.

The intersection of cylinders is shown in Fig. 6–47. At *A* are

6–47. *Typical intersections between cylinders.*

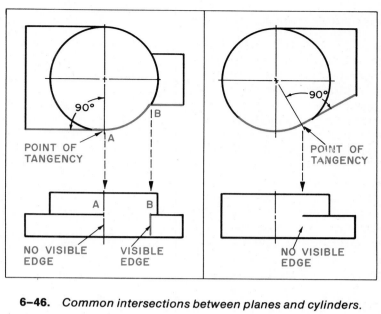

6–46. *Common intersections between planes and cylinders.*

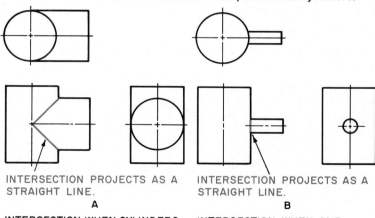

INTERSECTION PROJECTS AS A STRAIGHT LINE.
A
INTERSECTION WHEN CYLINDERS ARE THE SAME DIAMETER.

INTERSECTION PROJECTS AS A STRAIGHT LINE.
B
INTERSECTION WHEN ONE CYLINDER IS VERY SMALL.

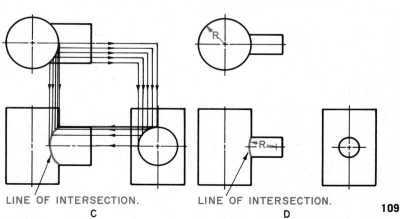

LINE OF INTERSECTION.
C
POINT METHOD FOR PLOTTING AN INTERSECTION.

LINE OF INTERSECTION.
D
APPROXIMATE METHOD FOR PLOTTING AN INTERSECTION.

109

two cylinders with the same diameter. They meet in a semi-elliptical curve that projects as a straight line to normal planes of projection. At *B* the intersecting cylinder is so small that the curvature of the intersection can be ignored. It is projected to the front view as a straight line. At *C* the intersection is plotted using the point method. This is done whenever the diameters of the intersecting cylinders are large enough to allow plotting the curve. The approximate method shown at *D* may be used when the diameter is small. Small intersecting curves are difficult to plot.

INDICATING ROUNDS AND FILLETS

External corners on cast and forged parts are usually rounded. Such a corner is called a round. See Fig. 6–48. The round makes it easier to cast and handle the part. A rounded corner can have any radius the designer decides is necessary.

A rounded interior corner is called a *fillet*. Fillets are necessary to prevent cast parts from fracturing at the corner. Fig. 6–48 illustrates the use of rounds and fillets. Notice that even though a corner is a curved surface, it appears as a line in the view to which it is perpendicular.

If fillets and rounds are dimensioned, it is usually with a note such as, "All fillets and rounds 1/8 inch radius unless otherwise specified." If they

110

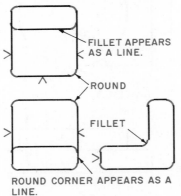

ALL FILLETS AND ROUNDS $\frac{1}{8}$ R. UNLESS OTHERWISE SPECIFIED.

FILLET APPEARS AS A LINE.

ROUND

FILLET

ROUND CORNER APPEARS AS A LINE.

A
ROUGH CAST ANGLE

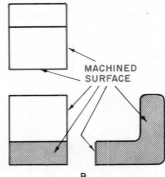

MACHINED SURFACE

B
ANGLE AFTER MACHINING

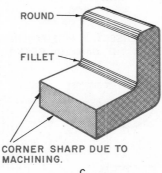

ROUND

FILLET

CORNER SHARP DUE TO MACHINING.

C
PICTORIAL OF ANGLE AFTER MACHINING

6–48. *Methods for indicating rounds and fillets before and after machining.*

are not dimensioned, the size is decided by the pattern-maker.

RUNOUTS

The intersection of fillets and rounds with other surfaces produces an extension of the curved surface. This extension is called a *runout*. See Fig. 6–49. These occur frequently on multiview drawings.

A runout arc may turn in or out, depending upon the shape and thickness of the intersecting areas.

Runouts can be drawn with a compass or an irregular curve. Very small intersections can be drawn freehand. It is common practice to draw runouts between small fillets and rounds by using the radius of the fillet. The length of the runout arc is usually one-eighth of a circle.

Fig. 6–49 illustrates how common filleted intersections are drawn.

CONVENTIONAL BREAKS

Sometimes on a multiview drawing it is necessary to break away part of an object to show inside details. The edges where the break occurs are marked with a *break symbol*. Long objects that have no distinctive features for most of their length can be drawn with some of the length removed. This shortens the drawing, and permits the object to be drawn to a larger scale. The break lines show that a piece has been removed.

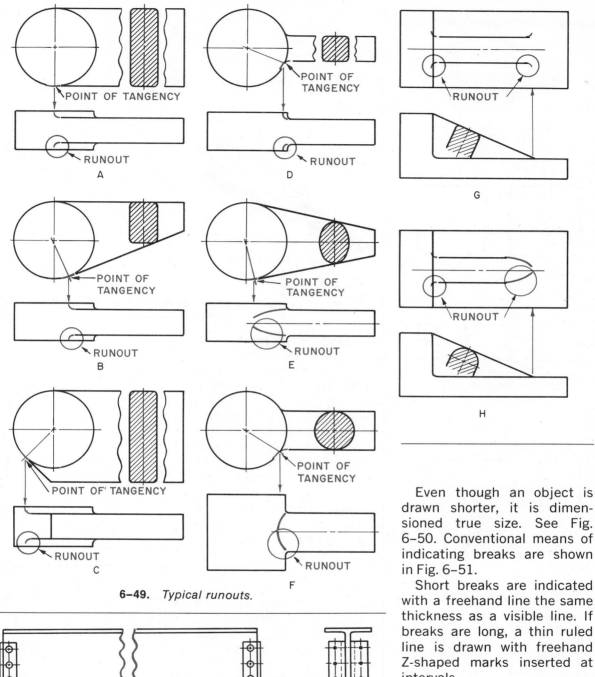

6–49. *Typical runouts.*

6–50. *A long object can be shortened when drawn by breaking out a section.*

Even though an object is drawn shorter, it is dimensioned true size. See Fig. 6–50. Conventional means of indicating breaks are shown in Fig. 6–51.

Short breaks are indicated with a freehand line the same thickness as a visible line. If breaks are long, a thin ruled line is drawn with freehand Z-shaped marks inserted at intervals.

Breaks in round shafts are indicated by an S-shaped symbol. This may be drawn freehand or with an irregular curve. Step A in Fig. 6–51

111

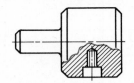

FREEHAND SHORT BREAK LINE
FOR METAL.

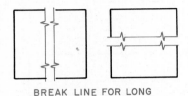

BREAK LINE FOR LONG
PIECES OF METAL.

FREEHAND BREAK LINE FOR
WOOD IN RECTANGULAR SHAPE.

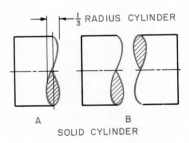

A   B
SOLID CYLINDER

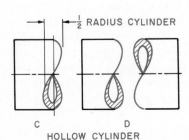

C   D
HOLLOW CYLINDER
CURVE IS DRAWN FREEHAND.

6-51. *Standard break symbols.*

112

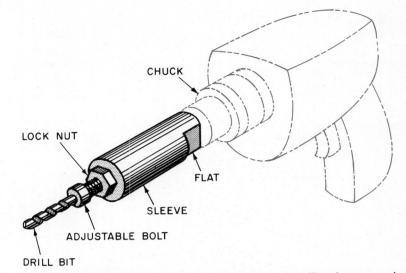

6-52. *Phantom view showing relationship of drill adapter and electric drill.*

indicates the proportions to use to draw the symbol for a solid shaft. Step B shows the symbol used to indicate a break in a rod.

Step C indicates the proportions for a hollow tube. Notice the symbol is wider than that for the solid shaft. The size of the small inside loop is estimated and frequently drawn in freehand. Step D shows the symbol used to indicate a break in a tube.

PHANTOM LINES

Phantom lines are used on multiview drawings to show the position of parts next to the object being drawn. See Fig. 6-52. The drill adapter is the object of importance. The electric drill is drawn in phantom lines to show the use of the adapter.

Phantom lines also show the possible positions of moving

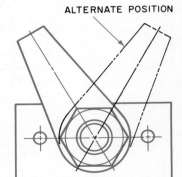

6-53. *Phantom view of a cabinet lock lever.*

parts. See Fig. 6-53. This shows the right and left positions of the cabinet lock.

Phantom lines are made of a series of one long and two short dashes. The long dash is ¾ inch to 1½ inches long. The short dash is ⅛ inch long. They are spaced 1/32 inch apart.

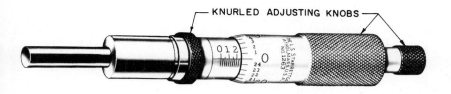

KNURLED ADJUSTING KNOBS

L. S. Starrett Co.

6–54. *Diamond knurling is used on the adjusting portions of this micrometer head.*

DIAMOND KNURL

STRAIGHT KNURL

6–55. *How to show knurling on a drawing.*

KNURLING

A knurl is a uniformly roughened surface. It is used on handles and knobs to provide a better grip. See Fig. 6–54. Shafts are also knurled so they will not slip when forced into a hole. The method for drawing knurls on multiview drawings is shown in Fig. 6–55. The dimensioning of knurls is discussed in Chapter 7.

Build Your Vocabulary

Following are terms that you should understand and use as a part of your working vocabulary. Write a brief explanation of what each means.

Multiview drawing
Plane
Frontal plane
Profile plane
Horizontal plane
Foreshortened
Inclined
Oblique
Orthographic projection
Partial views
Center line
Hidden line
Curved surface
Round
Fillet
Phantom line
Knurling

Study Problems

Surface Identification. Some surfaces are marked with letters of the alphabet, P6–1. Some edges are marked with numbers, as edge 1–2. Identify each surface and edge as normal, inclined, or oblique.

1. Surface identification

One-View Drawings. The products, P6–2 and P6–3 can be described with one view. Draw them full size.

2. Protractor
3. Plastic circle template

Two-View Drawings. The products, P6–4 through P6–9, can be described with two views. Select the two views that give complete details of the product. Make a detail drawing of each.

4. Rubber gasket
5. Asphalt paving block
6. Cutting edge for auto glass window tape trimmer
7. C Washer
8. Spherical radius locator button
9. Swing washer
10. Angle plate

Three-View Drawings. The products, P6–11 through P6–15, can be described with three views. Select the three views that give complete details of the product. Make a detail drawing of each.

11. Electric cable connection
12. Cold chisel

113

13. Hydraulic cylinder mounting bracket
14. Hydraulic hose connection
15. Rapid release electrical connector

Intersection Problems. The products, P6–16 through P6–24, involve the intersection of plane and curved surfaces. Select the views needed to describe each product completely. Make a detail drawing of each. Give special care to the drawing of intersecting surfaces.

16. Silo
17. Table leg
18. Speaker's stand
19. Mirror and stand
20. Christmas tree stand
21. Ball handle
22. Paperweight

23. Base for penholder
24. Tricycle frame

Phantom Line Problem. Draw the necessary views to describe the product, P6–25. Draw it with the lever in the highest position. Show the lever in the lowest position with phantom lines.

25. Pressure lever

Additional Problems. Make drawings to describe completely the following products. Select the side to be the front view. Decide which other views are necessary to describe the product.

26. Guide block
27. Blind flange
28. Angle bracket
29. Thumb screw blank before threading
30. Electric cable connector
31. Rivet set

32. Drawer pull
33. Curtain rod hanger
34. Chevron wood fastener
35. Machinist's V block
36. Universal right angle iron
37. Drill head mounting bracket
38. Spanner face wrench
39. Diamond pin
40. Yoke end
41. Reversing switch
42. Door stop
43. Mill fixture key
44. Roadway manhole casting
45. Friction plug
46. Water tube for rock drilling machine
47. Adjustable step block
48. Height gage base
49. Toolmaker's V block
50. Jig borer toolholder
51. Flanged turret mount
52. Precision boring spindle
53. Rocker arm
54. Pipe roller and base

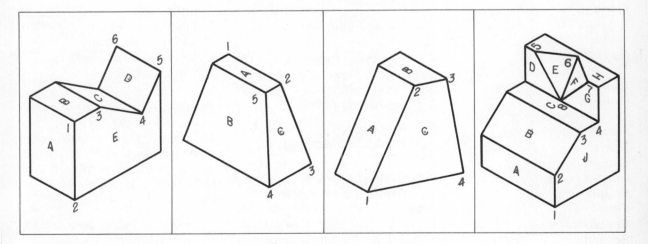

P6–1. *Surface identification figures.*

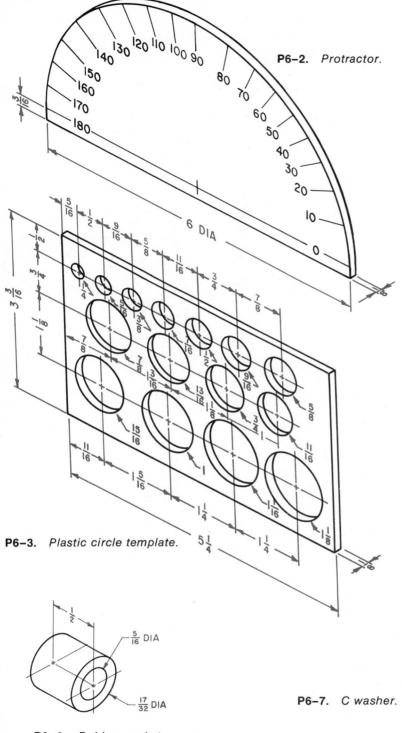

P6–2. *Protractor.*

P6–3. *Plastic circle template.*

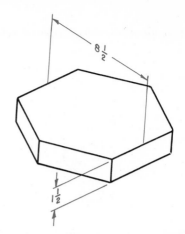

P6–5. *Asphalt paving block for industrial floors.*

FINISHED ALL OVER

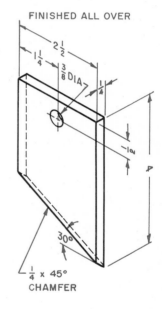

¼ x 45°
CHAMFER

P6–6. *Cutting edge for auto glass window tape trimmer.*

FINISHED ALL OVER

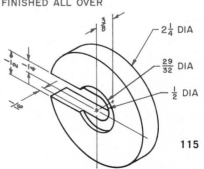

P6–4. *Rubber gasket.*

P6–7. *C washer.*

115

FINISHED ALL OVER

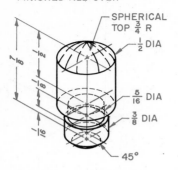

P6–8. *Spherical radius locator button.*

NOTE: SHADED SURFACES NOT FINISHED.
ALL FILLETS AND ROUNDS $\frac{1}{4}$ R.

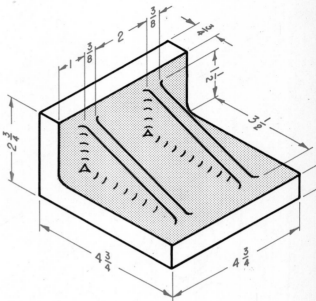

P6–10. *Angle plate.*

NOTE: THIS IS A CAST PRODUCT.

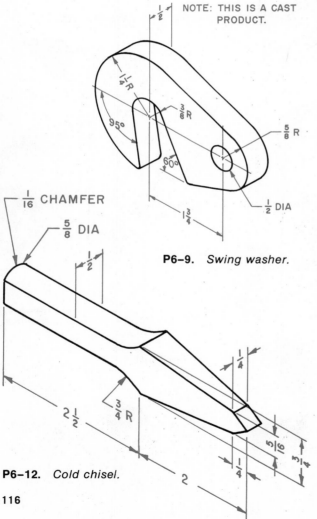

P6–9. *Swing washer.*

$\frac{1}{16}$ CHAMFER

$\frac{5}{8}$ DIA

P6–12. *Cold chisel.*

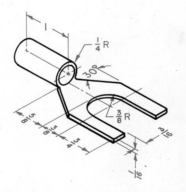

P6–11. *Electric cable connection.*

116

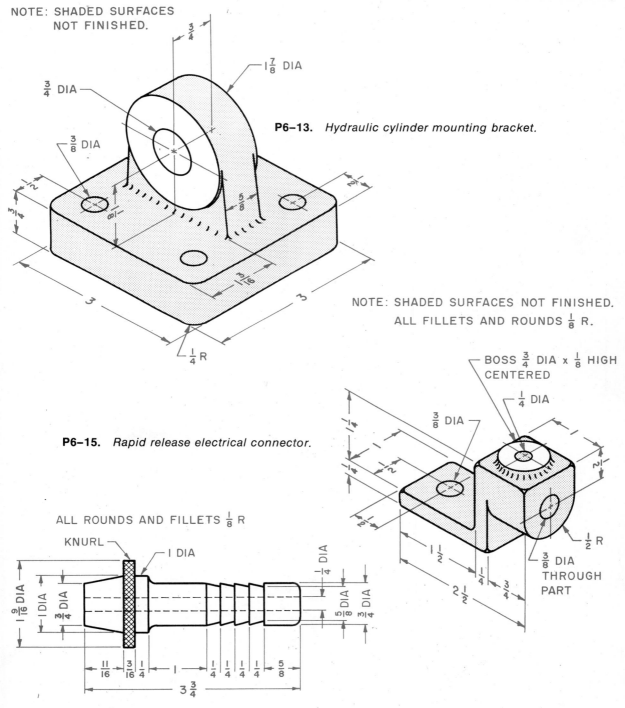

NOTE: SHADED SURFACES
NOT FINISHED.

$\frac{3}{4}$ DIA

$1\frac{7}{8}$ DIA

$\frac{3}{8}$ DIA

P6–13.   *Hydraulic cylinder mounting bracket.*

$\frac{1}{4}$ R

NOTE: SHADED SURFACES NOT FINISHED.
ALL FILLETS AND ROUNDS $\frac{1}{8}$ R.

BOSS $\frac{3}{4}$ DIA x $\frac{1}{8}$ HIGH
CENTERED

$\frac{1}{4}$ DIA

$\frac{3}{8}$ DIA

P6–15.   *Rapid release electrical connector.*

$\frac{1}{2}$ R

$\frac{3}{8}$ DIA
THROUGH
PART

$1\frac{1}{2}$

$2\frac{1}{2}$

ALL ROUNDS AND FILLETS $\frac{1}{8}$ R

KNURL

1 DIA

$\frac{1}{4}$ DIA

$1\frac{9}{16}$ DIA

1 DIA

$\frac{3}{4}$ DIA

$\frac{5}{8}$ DIA

$\frac{3}{8}$ DIA

$\frac{11}{16}$ $\frac{3}{16}$ $\frac{1}{4}$ 1 $\frac{1}{4}$ $\frac{1}{4}$ $\frac{1}{4}$ $\frac{1}{4}$ $\frac{5}{8}$

$3\frac{3}{4}$

P6–14.   *Hydraulic hose connection.*

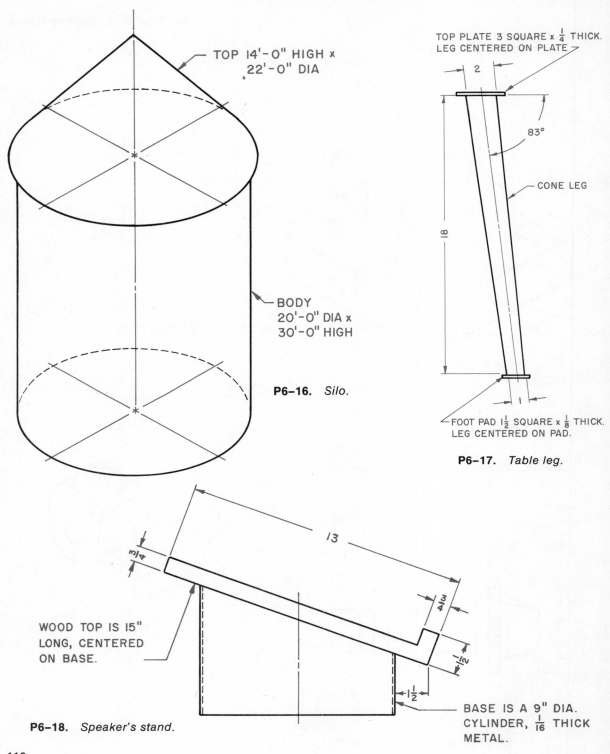

TOP 14'-0" HIGH x 22'-0" DIA

BODY
20'-0" DIA x
30'-0" HIGH

P6–16. *Silo.*

TOP PLATE 3 SQUARE x $\frac{1}{4}$ THICK.
LEG CENTERED ON PLATE

2

83°

CONE LEG

18

1

FOOT PAD $1\frac{1}{2}$ SQUARE x $\frac{1}{8}$ THICK.
LEG CENTERED ON PAD.

P6–17. *Table leg.*

13

$\frac{3}{4}$

$\frac{3}{4}$

$\frac{1}{2}$

$1\frac{1}{2}$

WOOD TOP IS 15"
LONG, CENTERED
ON BASE.

BASE IS A 9" DIA.
CYLINDER, $\frac{1}{16}$ THICK
METAL.

P6–18. *Speaker's stand.*

118

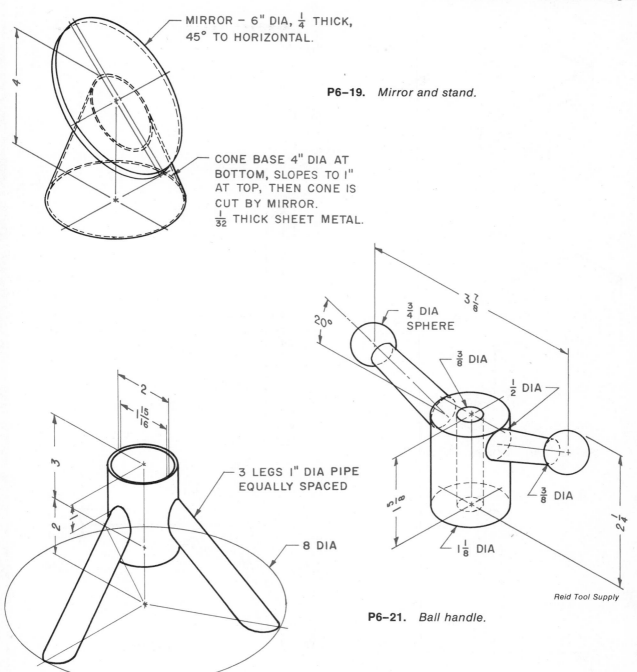

MIRROR – 6" DIA, $\frac{1}{4}$ THICK, 45° TO HORIZONTAL.

P6–19. *Mirror and stand.*

CONE BASE 4" DIA AT BOTTOM, SLOPES TO 1" AT TOP, THEN CONE IS CUT BY MIRROR. $\frac{1}{32}$ THICK SHEET METAL.

$\frac{3}{4}$ DIA SPHERE

$\frac{3}{8}$ DIA

$\frac{1}{2}$ DIA

20°

$3\frac{7}{8}$

$\frac{3}{8}$ DIA

$1\frac{5}{8}$

$2\frac{1}{4}$

$1\frac{1}{8}$ DIA

Reid Tool Supply

P6–21. *Ball handle.*

2

$1\frac{15}{16}$

3

2

1

3 LEGS 1" DIA PIPE EQUALLY SPACED

8 DIA

P6–20. *Christmas tree stand.*

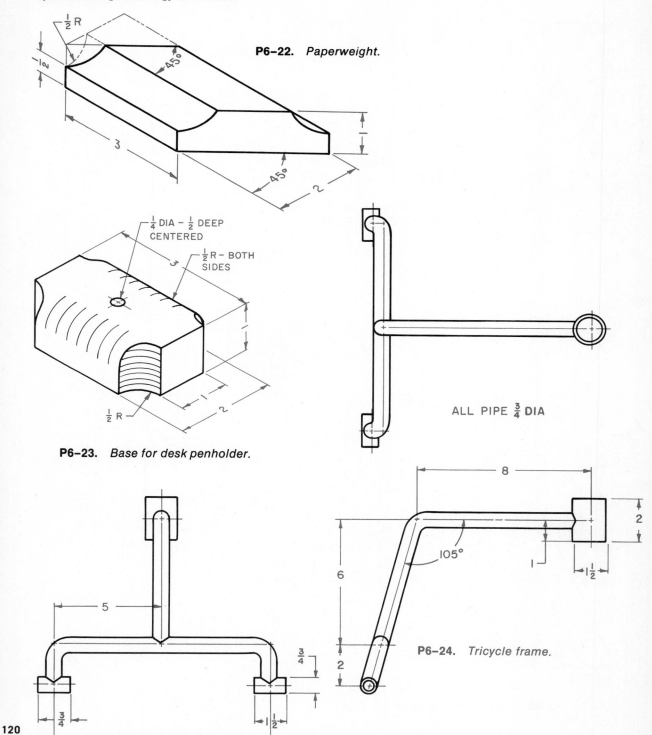

P6–22. *Paperweight.*

P6–23. *Base for desk penholder.*

ALL PIPE $\frac{3}{4}$ DIA

P6–24. *Tricycle frame.*

120

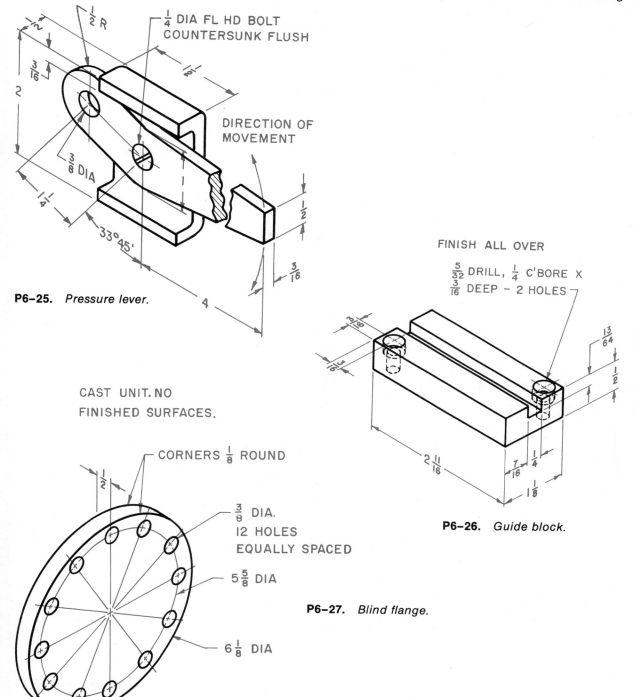

$\frac{1}{2}$ R

$\frac{1}{4}$ DIA FL HD BOLT
COUNTERSUNK FLUSH

$\frac{1}{2}$

$\frac{3}{16}$

2

$\frac{3}{8}$ DIA

$\frac{1}{4}$

33° 45'

4

DIRECTION OF
MOVEMENT

$\frac{1}{2}$

$\frac{3}{16}$

P6–25. *Pressure lever.*

FINISH ALL OVER

$\frac{5}{32}$ DRILL, $\frac{1}{4}$ C'BORE X
$\frac{3}{16}$ DEEP – 2 HOLES

$\frac{3}{16}$

$\frac{3}{16}$

$\frac{13}{64}$

$\frac{1}{2}$

2 $\frac{11}{16}$

$\frac{7}{16}$

$\frac{1}{4}$

1 $\frac{1}{8}$

P6–26. *Guide block.*

CAST UNIT. NO
FINISHED SURFACES.

CORNERS $\frac{1}{8}$ ROUND

$\frac{1}{2}$

$\frac{3}{8}$ DIA.
12 HOLES
EQUALLY SPACED

5 $\frac{5}{8}$ DIA

6 $\frac{1}{8}$ DIA

P6–27. *Blind flange.*

121

NOTE: SHADED SURFACES NOT FINISHED.
ALL FILLETS AND ROUNDS ¼ R.

CAST UNIT. NO
FINISHED SURFACES.

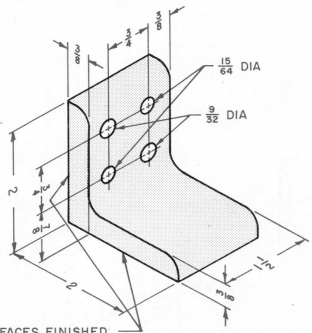

$\frac{15}{64}$ DIA

$\frac{9}{32}$ DIA

REAR SURFACES FINISHED

Carr Lane

P6–28. *Angle bracket.*

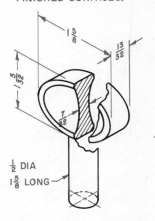

P6–29. *Thumb screw blank before threading.*

NOTE: SHADED SURFACES NOT FINISHED.
ALL ROUNDS ⅛ R.

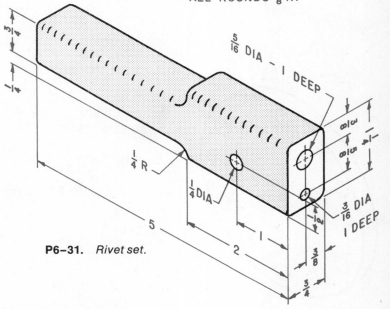

$\frac{5}{16}$ DIA – 1 DEEP

¼ R

¼ DIA

$\frac{3}{16}$ DIA
1 DEEP

P6–31. *Rivet set.*

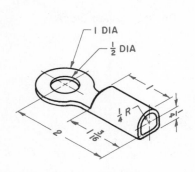

1 DIA

½ DIA

¼ R

P6–30. *Electric cable connector.*

NOTE: CAST UNIT.
       NO FINISHED SURFACES.
       ALL FILLETS AND ROUNDS $\frac{1}{8}$ R.

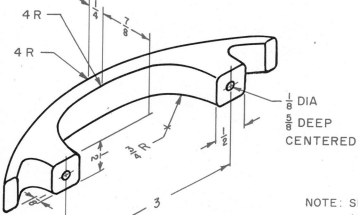

4 R
4 R

$\frac{1}{4}$
$\frac{7}{8}$

$\frac{1}{8}$ DIA
$\frac{5}{8}$ DEEP
CENTERED

$\frac{1}{2}$

$\frac{1}{4}$ R

$2\frac{1}{2}$

3

P6–32. *Drawer pull.*

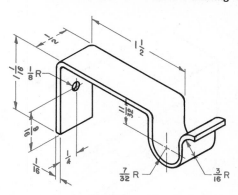

$\frac{1}{2}$

$1\frac{1}{2}$

$\frac{1}{16}$ $\frac{1}{8}$ R

$\frac{6}{16}$

$\frac{11}{32}$

$\frac{1}{16}$ $\frac{1}{4}$

$\frac{7}{32}$ R $\frac{3}{16}$ R

P6–33. *Curtain rod hanger.*

NOTE: SHADED SURFACES NOT FINISHED.
       ALL FILLETS AND ROUNDS $\frac{1}{8}$ R.

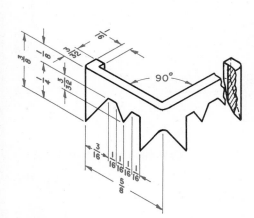

$\frac{1}{6}$
$\frac{1}{8}$
$\frac{3}{32}$
90°
$\frac{3}{8}$
$\frac{1}{4}$
$\frac{3}{32}$

$\frac{3}{16}$ $\frac{1}{16}$ $\frac{1}{16}$ $\frac{1}{16}$

$\frac{5}{8}$

P6–34. *Chevron wood fastener.*

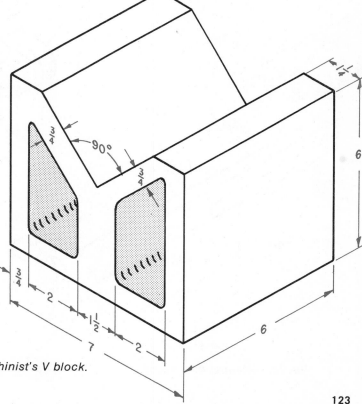

$1\frac{1}{4}$

$\frac{3}{4}$

90°

$\frac{3}{4}$

6

$\frac{3}{4}$

2

$1\frac{1}{2}$

7

2

6

6

P6–35. *Machinist's V block.*

123

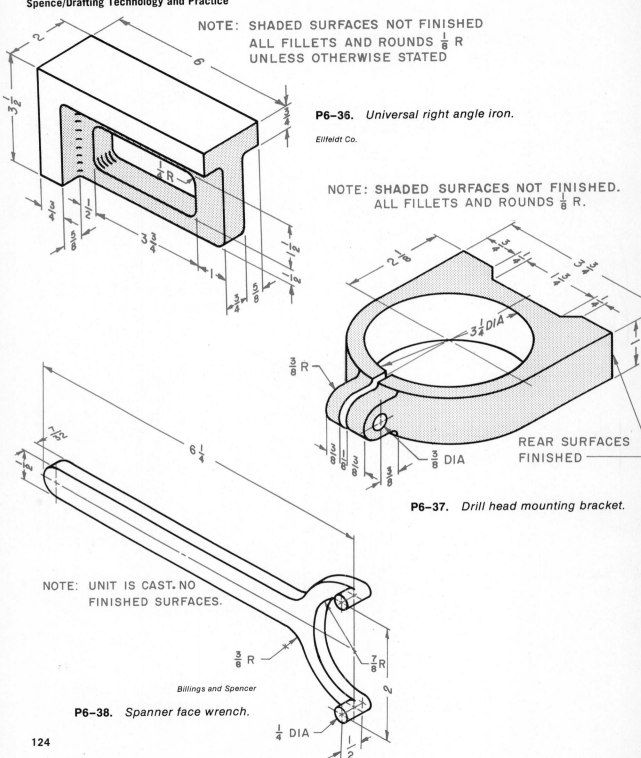

NOTE: SHADED SURFACES NOT FINISHED
ALL FILLETS AND ROUNDS $\frac{1}{8}$ R
UNLESS OTHERWISE STATED

P6–36. *Universal right angle iron.*

Ellfeldt Co.

NOTE: SHADED SURFACES NOT FINISHED.
ALL FILLETS AND ROUNDS $\frac{1}{8}$ R.

$3\frac{1}{4}$ DIA

$\frac{3}{8}$ R

$\frac{3}{8}$ DIA

REAR SURFACES
FINISHED

P6–37. *Drill head mounting bracket.*

NOTE: UNIT IS CAST. NO
FINISHED SURFACES.

$\frac{3}{8}$ R

$\frac{7}{8}$ R

Billings and Spencer

P6–38. *Spanner face wrench.*

$\frac{1}{4}$ DIA

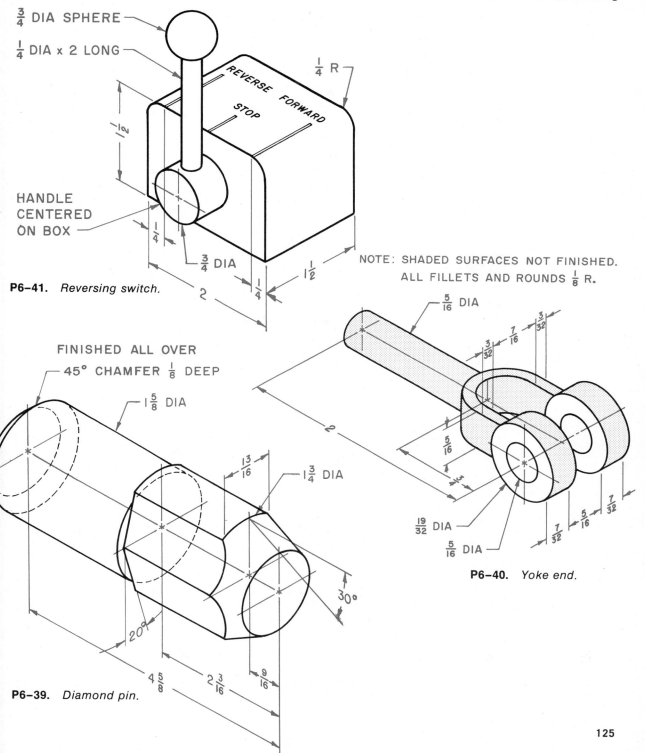

$\frac{3}{4}$ DIA SPHERE

$\frac{1}{4}$ DIA x 2 LONG

REVERSE FORWARD

STOP

$\frac{1}{4}$ R

$1\frac{1}{2}$

HANDLE
CENTERED
ON BOX

$\frac{1}{4}$

$\frac{3}{4}$ DIA

$1\frac{1}{2}$

$\frac{1}{4}$

2

P6–41. *Reversing switch.*

NOTE: SHADED SURFACES NOT FINISHED.
ALL FILLETS AND ROUNDS $\frac{1}{8}$ R.

$\frac{5}{16}$ DIA

$\frac{3}{32}$

$\frac{7}{16}$

$\frac{3}{32}$

$\frac{5}{16}$

2

$\frac{19}{32}$ DIA

$\frac{5}{16}$ DIA

$\frac{7}{32}$

$\frac{5}{16}$

$\frac{7}{32}$

P6–40. *Yoke end.*

FINISHED ALL OVER

45° CHAMFER $\frac{1}{8}$ DEEP

$1\frac{5}{8}$ DIA

$\frac{13}{16}$

$1\frac{3}{4}$ DIA

30°

20°

$4\frac{5}{8}$

$2\frac{3}{16}$

$\frac{9}{16}$

P6–39. *Diamond pin.*

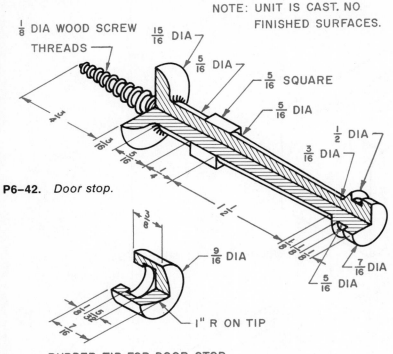

NOTE: UNIT IS CAST. NO FINISHED SURFACES.

$\frac{1}{8}$ DIA WOOD SCREW THREADS

$\frac{15}{16}$ DIA

$\frac{5}{16}$ DIA

$\frac{5}{16}$ SQUARE

$\frac{5}{16}$ DIA

$\frac{1}{2}$ DIA

$\frac{3}{16}$ DIA

P6–42. *Door stop.*

$\frac{9}{16}$ DIA

$\frac{7}{16}$ DIA

$\frac{5}{16}$ DIA

I" R ON TIP

RUBBER TIP FOR DOOR STOP

FINISHED ALL OVER

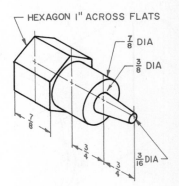

$45° \times \frac{1}{16}$

$\frac{7}{8}$ DIA

$\frac{1}{4}$ DRILL 2 HOLES

I DIA

ANGLE 30° FROM VERTICAL

$\frac{1}{4}$

$\frac{11}{16}$

P6–43. *Mill fixture key.*

NOTE: BOTH PARTS CAST. NO FINISHED SURFACES. ALL FILLETS AND ROUNDS $\frac{1}{4}$ R.

MANHOLE CASING

MANHOLE LID

13 SQUARE

12 SQUARE

LID $11\frac{3}{4}$ SQUARE

3 RIBS EQUALLY SPACED

15 SQUARE

10 SQUARE

P6–44. *Roadway manhole casting.*

FINISHED ALL OVER

HEXAGON I" ACROSS FLATS

$\frac{7}{8}$ DIA

$\frac{3}{8}$ DIA

$\frac{3}{16}$ DIA

$\frac{7}{8}$

$\frac{3}{4}$

$\frac{3}{4}$

P6–45. *Friction plug.*

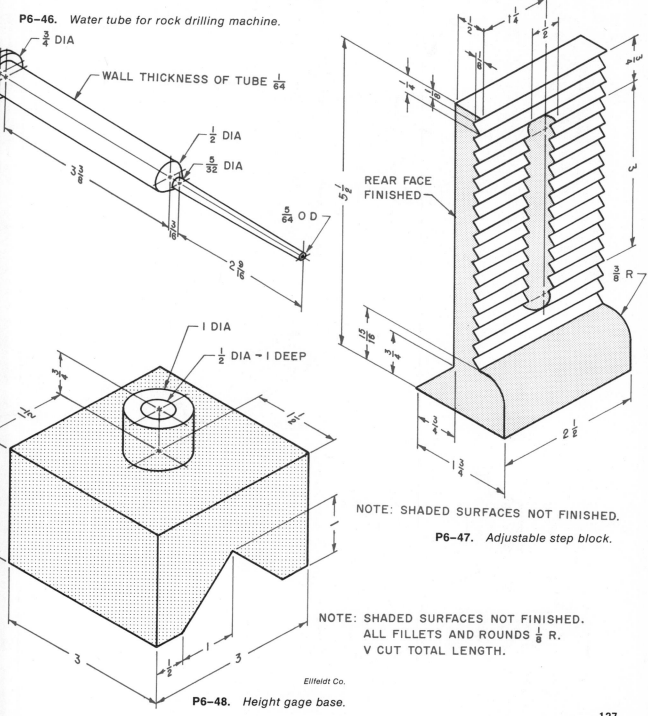

P6–46. *Water tube for rock drilling machine.*

$\frac{3}{4}$ DIA

WALL THICKNESS OF TUBE $\frac{1}{64}$

$\frac{1}{2}$ DIA

$\frac{5}{32}$ DIA

$\frac{5}{64}$ O D

$3\frac{3}{8}$

$\frac{3}{16}$

$2\frac{9}{16}$

REAR FACE FINISHED

NOTE: SHADED SURFACES NOT FINISHED.

P6–47. *Adjustable step block.*

I DIA

$\frac{1}{2}$ DIA – I DEEP

NOTE: SHADED SURFACES NOT FINISHED.
ALL FILLETS AND ROUNDS $\frac{1}{8}$ R.
V CUT TOTAL LENGTH.

Ellfeldt Co.

P6–48. *Height gage base.*

127

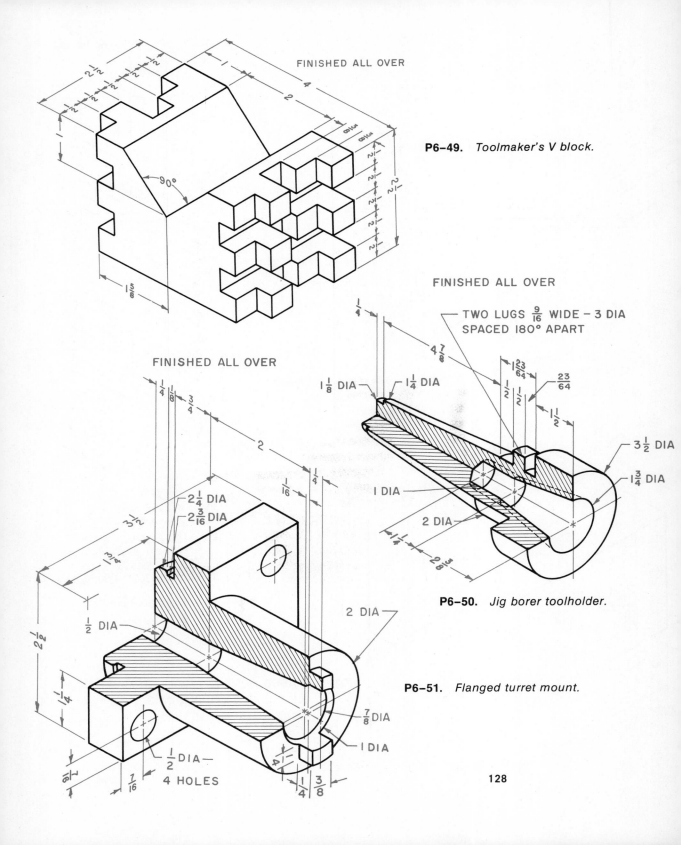

FINISHED ALL OVER

P6–49. *Toolmaker's V block.*

90°

FINISHED ALL OVER

TWO LUGS $\frac{9}{16}$ WIDE – 3 DIA
SPACED 180° APART

$1\frac{1}{8}$ DIA

$1\frac{1}{4}$ DIA

$3\frac{1}{2}$ DIA

$1\frac{3}{4}$ DIA

1 DIA

2 DIA

P6–50. *Jig borer toolholder.*

FINISHED ALL OVER

$2\frac{1}{4}$ DIA

$2\frac{3}{16}$ DIA

2 DIA

$\frac{1}{2}$ DIA

$\frac{7}{8}$ DIA

1 DIA

$\frac{1}{2}$ DIA
4 HOLES

P6–51. *Flanged turret mount.*

128

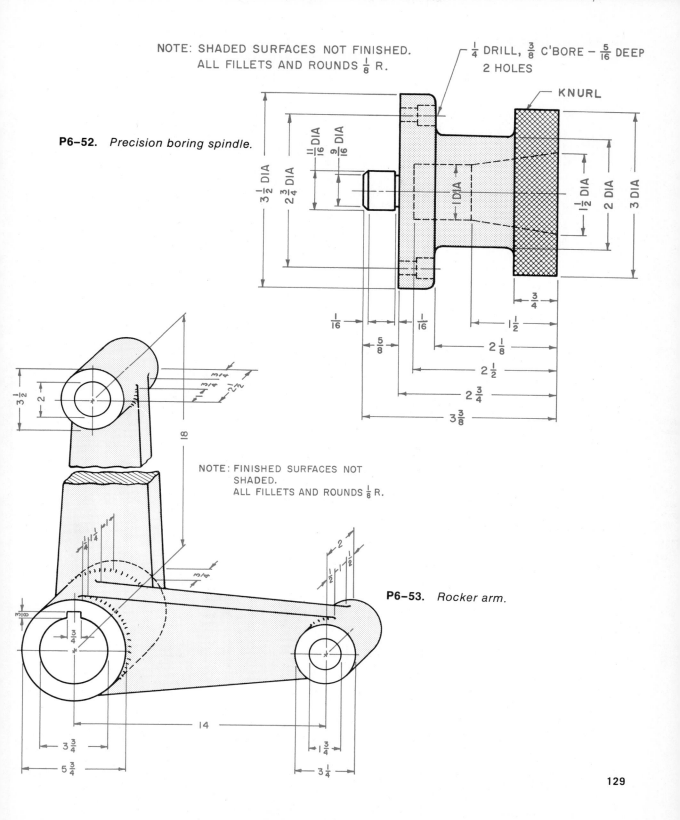

NOTE: SHADED SURFACES NOT FINISHED.
ALL FILLETS AND ROUNDS $\frac{1}{8}$ R.

$\frac{1}{4}$ DRILL, $\frac{3}{8}$ C'BORE $-\frac{5}{16}$ DEEP
2 HOLES

KNURL

P6–52. *Precision boring spindle.*

NOTE: FINISHED SURFACES NOT
SHADED.
ALL FILLETS AND ROUNDS $\frac{1}{8}$ R.

P6–53. *Rocker arm.*

NOTE: SHADED SURFACES NOT FINISHED.
ALL FILLETS AND ROUNDS $\frac{1}{8}$ R UNLESS NOTED.

PULLEY $4\frac{7}{8}$ LONG
OVERALL.

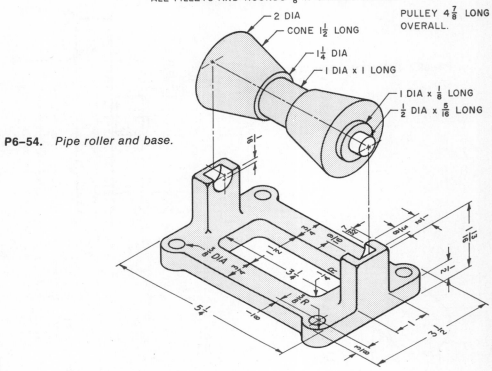

— 2 DIA
— CONE $1\frac{1}{2}$ LONG
— $1\frac{1}{4}$ DIA
— 1 DIA x 1 LONG
— 1 DIA x $\frac{1}{8}$ LONG
— $\frac{1}{2}$ DIA x $\frac{5}{16}$ LONG

P6–54. *Pipe roller and base.*

Chapter 7

Dimensioning

Any working detail drawing is made up of two elements— shape and size description. A shape description gives only the outline of the object. For example, Fig. 7–1 is a shape description of a cylinder. This could be a balance staff for a watch, a piece of shafting, a boiler for a model steam engine, or any cylindrical object. For the drawing to be meaningful, a *size* description must accompany the shape description. A size description will give all *sizes* or dimensions, general and specific *notes, material* designations, and *specifications* about the object. The size description must be complete enough so that it may be taken into the shop and produced without any additional explanation.

Dimensions placed on the drawing are the actual dimensions of the finished object. Actual dimensions are used even though the drawing may be smaller or larger than the object. To have a complete *working detail drawing*, two requirements must be met:

1. Sufficient views which describe the object; and,

2. Sufficient sizes and related information to produce the object.

Dimensioning is not difficult to learn if a few simple rules are followed. Perhaps most students encounter difficulty by neglecting to ask, "Could I make this object from this drawing without any additional information?"

DIMENSIONING STANDARDS

There are several sources for securing information on dimensioning standards. One is entitled, "Military Standards —Dimensioning and Tolerances." It is known as MIL STD-100. It sets forth practices for dimensioning and tolerances used in the departments of Army, Navy and the Air Force. It is published by the Standardization Division, Defense Supply Agency, Washington, D.C.

Another source is published by the American National Standards Institute, Inc., 1430 Broadway, New York, New York 10018. It is identified as Y14.5–1966 *DIMENSIONING AND TOLERANCING FOR ENGINEERING DRAWINGS.*

DIMENSION LINES

Dimension lines are thin, black, solid lines. They show the direction of most angular and linear dimensions. These lines show the extent of the dimension. They are drawn parallel to the surface or edge being described. The dimension line may be broken for the dimension. See Fig. 7–2. If the dimension has a tolerance, no break is provided in the line. A tolerance is an allowable variation from the size shown on the drawing. This is more fully explained later in this chapter. See Fig. 7–2. One part of the

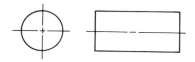

7–1. *Shape description gives only the outline of the object. Dimensions give the size of the object. Is this a pin for a watch or a boiler for a furnace?*

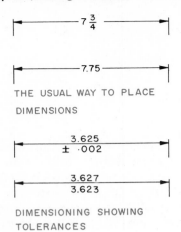

THE USUAL WAY TO PLACE
DIMENSIONS

DIMENSIONING SHOWING
TOLERANCES

7-2. *Dimension lines are usually broken for the dimension. An exception is made when showing tolerances.*

dimensional value is lettered above the line and the other is lettered below.

The dimension line nearest to the view is placed ³/₈ inch from the object line. All other dimension lines are spaced ¼ inch or more from the first dimension line. See Fig. 7-3. Object lines, center lines, or extension lines are not used as dimension lines.

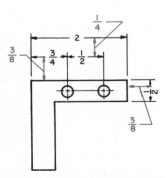

7-3. *It is important to space dimension lines properly.*

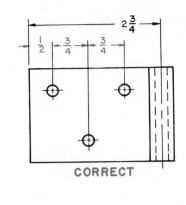

CORRECT

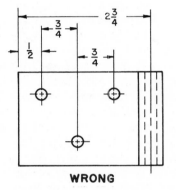

WRONG

7-4. *Line up dimension lines.*

Whenever possible, line up dimension lines to give the drawing an orderly appearance and make it easier to read. See Fig. 7-4.

ARROWHEADS

Arrowheads are used to show the beginning and ending of the dimension line. Arrowheads are drawn freehand and are made with two or three strokes. They may be open or closed. See Fig. 7-5. The length of the arrowhead on small drawings is approximately ⅛ inch and on large drawings the length may range up to ³/₁₆ inch. Regardless of the length, the arrowhead is

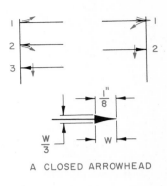

A CLOSED ARROWHEAD

AN OPEN ARROWHEAD

7-5. *Arrowheads may be open or closed.*

drawn one-third as high as it is long. All arrowheads on a drawing should be the same size.

EXTENSION LINES

Extension lines are used to extend lines on a view for dimensioning purposes. They are placed outside the view. They begin approximately ¹/₁₆ inch from the view and continue ⅛ inch beyond the last dimension line. See Fig. 7-6. They are thin, black lines. They

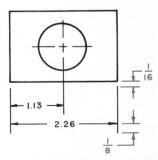

7-6. *Extension lines should not touch the object. They should extend beyond the dimension line.*

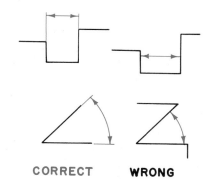

CORRECT   **WRONG**

7-7. *Dimension lines should terminate on extension lines.*

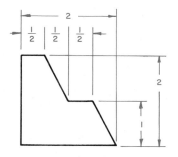

7-8. *The object line can be used as an extension line to avoid using very long extension lines.*

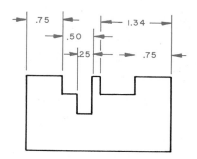

7-9. *Extension lines are broken if they cross arrowheads.*

have the same thickness as dimension lines.

It is considered poor practice to terminate a dimension line at an object line if the object line is in the same direction. See Fig. 7-7. If an extension line would cause confusion by being too long, an object line may be used as an extension line. See Fig. 7-8. However, this is not done unless absolutely necessary.

Extension lines may be broken when they are crossed by an arrowhead, Fig. 7-9. Extension lines are not broken when they cross each other or are crossed by a dimension line or object line. See Fig. 7-10.

When a point is located by extension lines, the extension line should pass through that point. See Fig. 7-11.

LEADERS

Leaders are thin lines used to direct dimensions, notes, and symbols to the intended place on the drawing. A leader

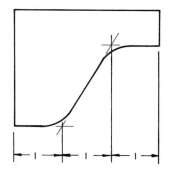

7-10. *Extension lines are not broken when they cross each other.*

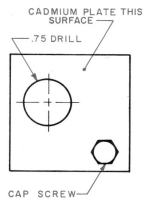

7-11. *Extension lines pass through points which they locate.*

is the same weight as a dimension line. An arrowhead is used on the end of a leader. See Fig. 7-12. A dot can be used instead of an arrowhead when the leader points to a surface. See Fig. 7-12. A leader is usually an oblique straight line with a short horizontal shoulder extending to the note. The shoulder is about 1/4 inch long. The shoulder extends to the *center* of the first or last letter or number of the note. The horizontal shoulder is never used to underline a note. A leader is most frequently

CADMIUM PLATE THIS
SURFACE

.75 DRILL

CAP SCREW

7-12. *Leaders are used to connect local notes to the drawing.*

133

drawn at an angle of 60 degrees to the horizontal. Other angles may be used. Two or more leaders to adjacent features should be drawn parallel. This makes it easier to read the drawing.

When a leader is directed to a circle or an arc, the leader should point to the center. See Fig. 7–12.

Some practices to avoid when using leaders are: (1) long leaders; (2) crossing other leaders; (3) leaders in a near horizontal or vertical position; and (4) small angles between leaders and object lines.

DIMENSION FIGURES

Numerals for almost all drawings are ⅛-inch high. The total height of a fraction is ¼ inch. The numerator (the value above the fraction line) and the denominator (the value below the fraction line) are slightly less than ⅛-inch high. See Fig. 7–13. The bottom of the numerator and top of the denominator should not touch the fraction line. The fraction line is heavier than the dimension line and is not slanted. Dimension figures in notes are the same size as regular dimension numerals.

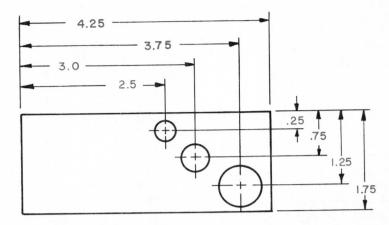

7–14. *Stagger dimensions to help make them easier to read.*

The heights of dimensional values expressed in decimal form are ⅛ inch. When a limit dimension is given, Fig. 7–2, the numerals above and below the dimension line are each ⅛-inch high. Note there is no fraction line and neither value touches the dimension line.

When several dimension lines are parallel the dimensions should be staggered. Staggering dimensions will make them easier to read. See Fig. 7–14.

When lettering dimensions or notes, always be sure to use guidelines.

CUSTOMARY AND METRIC SYSTEMS FOR LINEAR MEASUREMENT

Linear measurement means measurement along a straight line. In the United States and a few other countries the customary system is used. The standard unit of measurement for the customary system is the *inch.* Most countries use the International System of

Units. This is commonly called the metric system. The standard unit of measurement for the metric system is the *metre.* A metre is equal to 39.37 inches. A comparison of these systems is given in Fig. 7–15.

The Customary System. The basic unit, the inch, is divided into smaller parts providing a system of fine measurement.

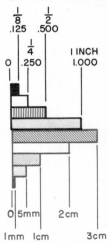

7–15. *A comparison of common fractions, decimal fractions, and metric measure.*

7–13. *Lettering is usually ⅛-inch high and fractions ¼-inch high.*

The inch can be divided into *common fractional* parts such as ³/₄, ¹/₂, ¹/₄, ¹/₈, ¹/₃₂, or ¹/₆₄ inch. It may also be divided into *decimal fractions* such as 0.1, 0.01, or 0.001 inch. The expression of decimal fractions is shown in Fig. 7–16.

Common fractional parts can also be expressed as decimal fractions. Following are some examples: ¹/₂″ = 0.500″, ¹/₄″ = 0.250″, or ¹/₈″ = 0.125″. A table of decimal equivalents of common fractions is found inside the front cover of this book.

The Metric System (SI). The international system of units is known as SI. This is an abbreviation of the French name for the system. SI base units are the *metre* for length, *kilogram* for mass, *second* for time, *ampere* for electric current, *kelvin* for temperature, *candela* for luminous intensity, and *mole* for amount of substance. Fig. 7–17. Note: Degree Celsius (Centigrade) will be commonly used for temperature, but it is not a base unit.

For most scientific and technical work the International System of Units is generally superior to other systems. This system is more widely accepted than any other as a common language in which scientific and technical data are expressed.

In addition to the six basic SI units, there are supplementary and derived units. Examples of these are electric resistance (ohm), force (Newton), or volume (cubic metre). The complete list can be found in the publication, *Standard Metric Practice Guide.*

For dimensioning purposes this text will consider suggestions for applying metric measure to length.

The metric system is a decimal system with the metre as the basic unit. The metre is divided into finer parts for precision measurement. These subdivisions, the decimetre (tenths of a metre), centimetre (hundredths of a metre) and millimetre (thousandths of a metre), are shown in Fig. 7–18.

Expressed in another way the metre can be divided into 10 decimetres, 100 centimetres, or 1,000 millimetres.

Inside the back cover of this book are tables used to convert common fractions and

---

One inch = 1″ = 1.00″
One-tenth inch = ¹/₁₀″ = 0.1″
One-hundredth inch = ¹/₁₀₀ = 0.01″
One-thousandth inch = ¹/₁,₀₀₀ = 0.001″
One ten-thousandth inch = ¹/₁₀,₀₀₀ = 0.0001″
One hundred-thousandth inch = ¹/₁₀₀,₀₀₀ = 0.00001″
One-millionth inch = ¹/₁,₀₀₀,₀₀₀ = 0.000001″

7–16.   *Customary system of linear measurement using decimal fractions.*

Quantity	Unit	SI Symbol
Length	metre	m
Mass	kilogram	kg
Time	second	s
Electric Current	ampere	A
Temperature	kelvin	K
Luminous Intensity	candela	cd
Amount of Substance	mole	mol

7–17.   *The basic units of SI.*

1 metre = 39.37 inches
1 metre = 10 decimetres (dm)
1 decimetre = 10 centimetres (cm)
1 centimetre = 10 millimetres (mm)

7–18.   *Metric system decimal divisions.*

decimal fractions in metric measure. Another table shows the metric equivalents of decimal fractions.

Following are examples to show how to use these tables. Suppose a drawing shows a part is to be machined to 55 millimetres. The machinist wants to know what this is in inches. The table for converting from millimetres to decimal inches shows that one millimetre equals 0.0394 inches. Therefore, 55 mm × 0.0394 inches = 2.1670 inches.

Suppose a drawing shows a part 1.50 inches long. To convert this to millimetres, use the table giving metric equivalents of decimal fractions. The table shows:

$$
\begin{aligned}
1 \text{ inch} &= 25.40 \text{ mm} \\
0.50 \text{ inch} &= \underline{12.70 \text{ mm}} \\
1.50 \text{ inches} &= 38.10 \text{ mm}
\end{aligned}
$$

This solution shows that 1.50 inches equals 38.10 millimetres. Stated in another way, it means that it equals 38 millimetres and $10/100$ of another millimetre. It should be emphasized that this is $10/100$ of a millimetre, not an inch. It should not be confused with decimal fractions used in the customary system.

The United States is rapidly moving to accept the metric system. Many major companies have already made the change. This change has been brought about by the rapidly expanding international markets. Products designed in one

CONVERSION CHART	
mm	inch
10.0	0.39
20.0	0.78
30.0	1.81
44.0	1.77
100.0	3.93

7–19. *A simple metric drawing with a conversion chart.*

country are often sold in another. Since most of the world has accepted the metric system, products must be designed and manufactured in that system in order to be marketable worldwide.

During the transition period from the customary to the metric system, several systems for dimensioning drawings can be used. Fig. 7–19 shows one system which is used widely. The drawing is totally metric, and the U.S. customary sizes are given in table form.

A metric scale such as might be used for this drawing is shown in Chapter 4, Tools and Techniques of Drafting.

Metric Dimensioning Practices. Following are suggested dimensioning practices when using the metric system. In many European countries the comma (,) is used instead of a decimal point (.). However, in much of the material published about the metric system, the decimal point is used instead of the comma. This practice will someday become standardized.

Any drawing using metric dimensions must have a note stating this. For example, the note "dimensions in millimetres" clearly shows this point. This is especially important when a decimal point is used.

Many things will continue to be expressed in nominal values. A nominal value is one used to name an object by giving approximate dimensions. For example, one-inch pipe is really 1.315 inches in its outside diameter. It is called one-inch pipe but the actual outside diameter is 33.40 mm. Screw threads will continue to use the nominal diameter. For example, a 1/2″–12 UNC screw thread will still be called this. Actually the dimensions used to manufacture it will be converted to metric values. Wire diameter and metal sheets will continue to be identified by established gage numbers, but the actual metric value will be given.

The symbols for SI units are capitalized if they were taken from a proper name, as K for kelvin. When the name is not abbreviated, it is not capitalized, as kelvin. Plural forms of unabbreviated SI units follow the usual rules of English. Abbreviated symbols are always written in singular form. For example, 50 millimetres or 50 mm.

Periods are not placed after SI unit symbols except at the end of a sentence.

Numbers having four or more digits are placed in groups of three, separated by a space instead of a comma. The spaces are counted from left to right. For example, 1 375 200 instead of 1,375,200 or 375 150.100 instead of 375,150.100. This avoids confusion since the Europeans use the comma as we use a decimal point.

DIMENSION UNITS

Drawings are dimensioned in any of the following units of measure:

1. Inches and common fraction of an inch: 2⅜″
2. Inches and decimal fractions of an inch: 4.750″
3. Feet, inches, and common fractions of an inch: 7′6½″
4. Feet and decimal fractions of a foot: 4.52′
5. Millimetre and decimals of a millimetre: 5.67 mm

If a drawing has both inches and millimetres, the mm symbol is used. If it is all in millimetres, this can be told with a note. The mm symbol is not needed. The symbol for an inch is ″, and for the foot ′. The symbol indicating inches is omitted if all the dimensions are given in inches. If a dimension may be misunderstood, the inch marks are used. Generally this is only true when the numeral one is involved. A hole one inch in diameter that is to be drilled would be noted 1″ drill. If the diameter would be over one inch, 1¼ for example, the inch marks would be omitted. No misunderstanding would occur when the dimension would be given as 1¼ drill.

When dimensions are expressed in feet and inches, both the symbols for feet and inches are indicated. A hyphen appears between the foot mark and the inch value. It is up to the company's drafting standards whether dimensions will be specified in feet and inches or all in inches. A practice followed by many industries is to give all dimensions below 72 inches in inches. Over 72 inches the value would be expressed in feet and inches.

Many industries that have manufacturing facilities in foreign countries use metric units for dimensioning. In this situation, if a piece is going to be manufactured both in the United States and a foreign country, it will be dimensioned in both inches and millimetres. The inch mark as well as the symbol for millimetres, mm, are used.

TYPES OF DIMENSIONS

Whether the drawing will be dimensioned in fractions, decimals, or in a combination of both is determined by the accuracy required. Industrial practice (the standards specified by a manufacturing concern) may dictate the choice.

Fractional dimensions are used for all phases of size description except when close tolerances are required. Generally, common fractions are used where accuracy must be no closer than ± 1/64 inch. Common fractions are used to indicate nominal sizes of materials or features such as holes, threads, and keyseats. For example: ¼–20 UNC-2A; 5/16 DIA; STOCK ⅝ × ⅞. The ways to show common fractional dimensions on a drawing are given in Fig. 7–20.

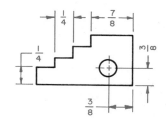

7-20. *Ways to show common fractional dimensions.*

Decimals are used when the accuracy must be closer than ± 1/64 inch. Decimals are used for all dimensions and designations. All decimals are carried to the same number of places as tolerances require.

Combinations of both fractional and decimal dimensions are used when all dimensions are given in decimals *except* nominal sizes. Features such as bolts, nuts, screws, threads, keyseats and keys are given by this standardized fractional designation.

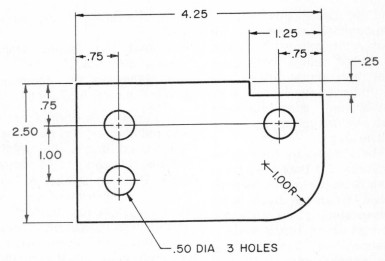

THE UNIDIRECTIONAL SYSTEM OF DIMENSIONING

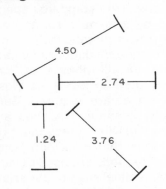

DECIMALS IN THE UNIDIRECTIONAL SYSTEM

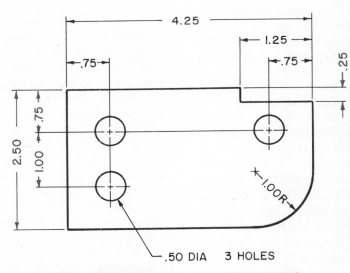

THE ALIGNED SYSTEM OF DIMENSIONING

7-22. *Two systems for placing dimensions.*

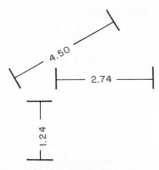

DECIMALS IN THE ALIGNED SYSTEM

7-21. *Ways to show decimal dimensions.*

DIMENSIONING SYSTEMS

Two systems are used for placing dimensions on a drawing—the *aligned system* and the *unidirectional system*. See Fig. 7-21. Both systems are acceptable and used by industry.

The *aligned system* is the older of the two methods. In this system all dimensions are placed along the dimen-

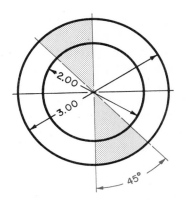

7–23. *Avoid placing aligned dimensions within the shaded area. They are hard to read.*

THEORY OF DIMENSIONING

Dimensioning any object is relatively easy if it is first broken down into its component geometric shapes. All objects are composed of one or more of the following geometric shapes: cylinder, prism, cone, pyramid, and sphere. See Fig. 7–24.

Fig. 7–25 illustrates a simple object broken into its component geometric shapes. In dimensioning, think in terms of positive and negative shapes. A positive feature is one which exists. A negative shape is one which is hollow or open and has the same general outlines as a positive feature. For example, a drilled hole is a negative cylinder. Dimensioning therefore consists of dimensioning *sizes* of these positive

sion lines. All dimensional values are lettered so they read from the bottom and right side of the drawing. See Fig. 7–22. All horizontal dimensions are lettered and read from the bottom of the sheet. Vertical dimensions are lettered and read from the right side of the sheet. For this reason, inclined dimension lines in the 45-degree zone should be avoided. See Fig. 7–23.

In the *unidirectional system* all dimensions are lettered and read from the bottom of the sheet, regardless of the direction of the dimension lines. See Fig. 7–21. The unidirectional system is easier to letter and read. The automotive and aircraft industries are pioneers with this type of dimensioning since many of their drawings are extremely large. Fractions are lettered with their fraction bar parallel to the bottom of the sheet.

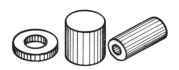

A – PRISMS

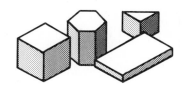

B – CYLINDERS

C – CONES

D – PYRAMIDS

E – SPHERICAL

7–24. *Basic geometric shapes.*

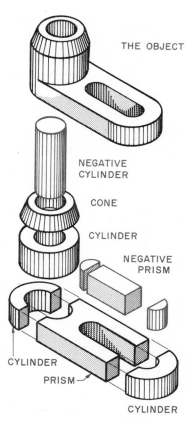

THE OBJECT

NEGATIVE CYLINDER

CONE

CYLINDER

NEGATIVE PRISM

CYLINDER

PRISM

CYLINDER

THE OBJECT BROKEN INTO GEOMETRIC FORMS

7–25. *Objects can be broken into various geometric forms.*

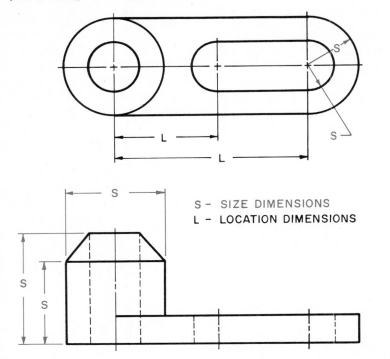

S - SIZE DIMENSIONS
L - LOCATION DIMENSIONS

7-26. *Size dimensions show dimensions of basic geometric shapes. Location dimensions show relative positions of these shapes.*

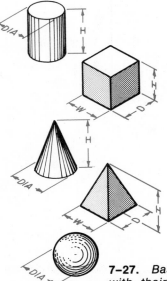

7-27. *Basic geometric shapes with their principal dimensions.*

and negative features *and* the *location* of these features.

Size dimensions are used to give the dimensions of basic geometric shapes. *Location* dimensions are used to position these shapes relative to each other. See Fig. 7-26.

As stated before, all geometric shapes are composed of three dimensions: *width, height,* and *depth.* Basic geometric shapes described previously are shown in Fig. 7-27, illustrating their three dimen-

sions. In the case of a cylinder or cone, only two dimensions would be necessary since two of the dimensions are replaced by a diameter. The sphere requires only one dimension, diameter, since it is completely round.

DIMENSIONING GEOMETRIC SHAPES

Prisms. The rectangular prism and variations of it are the most elementary. All rectangular prisms require three principal dimensions: *width, height,* and *depth.*

Three different methods of dimensioning a rectangular prism are shown in Fig. 7-28. Any views which are adjacent (next to each other) will have a common dimension. The dimension which is common to both views is usually placed between the views. For example, in Fig. 7-28A the height dimension is common to both the front and right-side views. In example B of that figure the width is the common dimension in the front and top view. When the piece has a uniform depth, it may be dimensioned as shown at C in Fig. 7-28. The depth dimension is given as a note. This practice is followed *only* when the piece has a *uniform depth.*

The correct method of dimensioning an irregular rectangular prismatic object is illustrated in Figs. 7-29, 7-30, and 7-31. Note in each of these figures the views are *not* connected by extension lines.

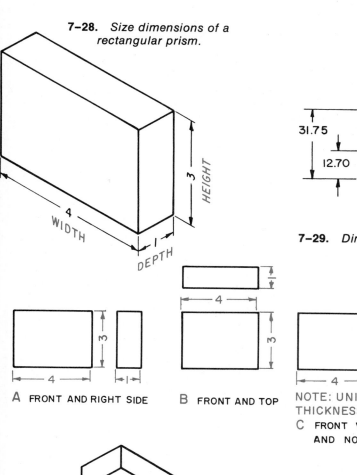

7-28. *Size dimensions of a rectangular prism.*

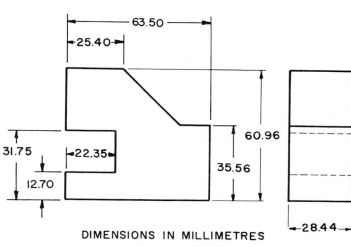

DIMENSIONS IN MILLIMETRES

7-29. *Dimensioning an irregular flat prism. Notice the use of metric measure.*

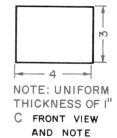

A FRONT AND RIGHT SIDE

B FRONT AND TOP

NOTE: UNIFORM
THICKNESS OF 1"
C FRONT VIEW
AND NOTE

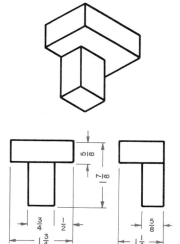

7-30. *Dimensioning a prismatic shaped object.*

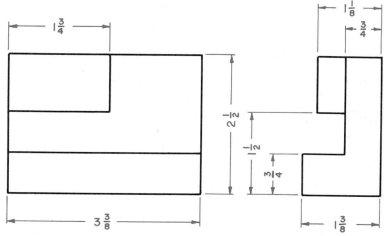

7-31. *Dimensioning a prismatic-shaped object.*

141

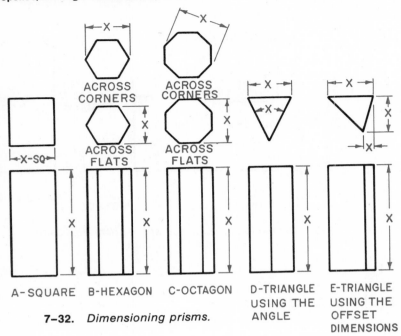

ACROSS CORNERS

ACROSS CORNERS

ACROSS FLATS

ACROSS FLATS

X-SQ

A-SQUARE   B-HEXAGON   C-OCTAGON   D-TRIANGLE USING THE ANGLE   E-TRIANGLE USING THE OFFSET DIMENSIONS

7-32. *Dimensioning prisms.*

A

B

C

7-33. *Dimensioning cylinders.*

added to the dimension. Both prisms may also be dimensioned by giving the distance "across corners" rather than "across flats." Never give both dimensions in the same view. A triangular prism may be dimensioned by giving an angle in degrees or by offset dimensions as in D and E of Fig. 7-32. In each case the dimensions have been placed where they appear with the most *clarity.* If you can follow this rule in dimensioning, you will encounter very few problems.

Cylinders. The diameter and length of a cylinder are given in the *rectangular view.* Remember that cylinders may be positive and negative. Fig. 7-33A shows how a positive cylinder is dimensioned. Fig. 7-33B and C show the application of this principle. Notice how some of the dimensions have been placed between the views and others beyond the view. Keep extension lines short and the diameter as close to the circular view as possible.

When the size dimension is given for a hole (a negative cylinder) the operation is specified, such as drill, bore, countersink, ream, spot-face, counterbore, or core. The dimension is given by note *in the circular view.* Both holes, Fig. 7-34, are dimensioned by note. The note in both cases gives the diameter and length of the hole. The radius of a cylinder is never given.

As a general rule any flat prism has the thickness given on the end view and all other dimensions on the outline view. This is usually the front view. See Fig. 7-29.

Other geometric-shaped prisms are dimensioned in a similar manner to flat rectangular prisms. A square prism, Fig. 7-32A can be dimensioned by adding the abbreviation SQ for square following the dimension. A hexagonal or octagonal prism, Fig. 7-32B and C, may be dimensioned by using only two dimensions. In some cases, as with the square, the abbreviation HEX or OCT may be

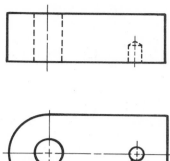

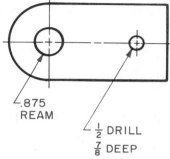

.875
REAM

½ DRILL
⅞ DEEP

7–34. *The diameter and length of the hole are given in the same view.*

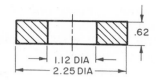

.62

1.12 DIA
2.25 DIA

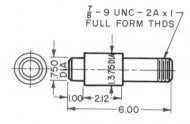

⅞ – 9 UNC – 2A x 1
FULL FORM THDS

.750 DIA

1.375 DIA

1.00 — 2.12

6.00

7–35. *The circular view may be omitted when the abbreviation, DIA, accompanies the dimension.*

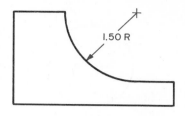

1.50 R

LARGE ARC

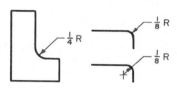

¼ R

⅛ R

⅛ R

SMALL ARCS

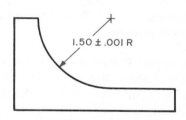

1.50 ± .001 R

ARC WITH TOLERANCE

The circular view of a cylinder may be omitted, if the abbreviation for diameter, DIA, appears with the diameter dimension. See Fig. 7–35. DIA may also be used with the dimension if the cylinder does not show in its circular shape.

Arcs. An arc is dimensioned by its radius. The radius is noted by the letter R. Where space permits, a radius dimension line is drawn from the radius center to the arc. An arrowhead is used to end the dimension line at the arc. See Fig. 7–36. When space is limited, a leader may continue from the arc and the dimension placed outside the arc.

When space is limited and the complete radius dimension line cannot be shown to scale,

the dimension line may be foreshortened. The portion of the dimension line ending with an arrowhead must be pointed toward the center. See Fig. 7–36.

If an arc does not show in its true shape, it may be dimensioned as a true radius. See Fig. 7–37. The term TRUE R *must* follow the dimension. This practice may save drawing an auxiliary view of a simple angular portion.

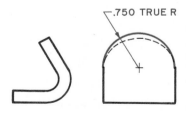

.750 TRUE R

7–37. *The true radius of foreshortened arcs can be dimensioned in this way. It eliminates the need to draw an auxiliary view.*

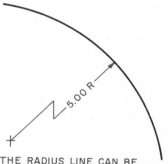

5.00 R

THE RADIUS LINE CAN BE FORESHORTENED FOR LONG ARCS.

7–36. *How to dimension arcs.*

143

1.75 SPHERE

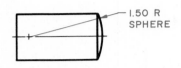

1.50 R
SPHERE

7-38. *Dimensioning spherical surfaces.*

Spherical surfaces may be dimensioned as shown in Fig. 7-38. The abbreviation for spherical, SPHER, is added to the R dimension.

Angular Dimensions. Angular dimensions are given in degrees, minutes and seconds. See Fig. 7-39. The symbols used on a drawing are degree°, minute', and second". If the angle is in degrees alone, it can be followed by the symbol ° or DEG. For example, 30° or 30 DEG. If the angle is in minutes alone, it should have 0° placed before it. For example, 0°-15'.

Angles can be given in decimal form. It is in degrees and decimal parts of a degree. For example, 30.5°.

If an angle has a tolerance, it should be in the same form as the angle. For example, 30° 15 ± 0° 5' or 30.25° ± .05°. Notice in the first example even the tolerance, if

in minutes alone, has 0° before it.

The usual ways to place angular dimensions on a drawing are given in Fig. 7-39.

Rounded End Shapes. The function of the piece will frequently determine the method of dimensioning rounded ends. One method is shown in Fig. 7-40. In this case the overall length along the center line is important, as well as the center to center distance. The radius is indicated but *not* dimensioned. By not dimensioning the radius, it is allowed to vary with the width of the piece.

If the overall length and width are not important, an object can be dimensioned as

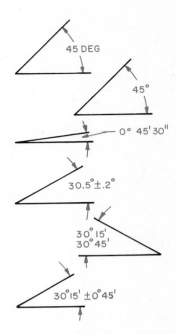

7-39. *Ways to dimension angular features.*

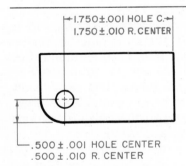

1.750±.001 HOLE C.
1.750±.010 R. CENTER

.500 ± .001 HOLE CENTER
.500 ± .010 R. CENTER

DIMENSION THIS WAY WHEN THE HOLE CENTER IS MORE CRITICAL THAN THE RADIUS CENTER.

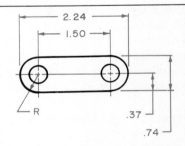

2.24
1.50
R
.37
.74

DIMENSION THIS WAY WHEN THE OVERALL LENGTH AND CENTER TO CENTER DISTANCES ARE CRITICAL.

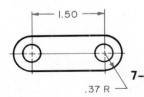

1.50
.37 R

DIMENSION THIS WAY IF THE OVERALL LENGTH AND WIDTH ARE NOT CRITICAL.

7-40. *How to dimension rounded ends.*

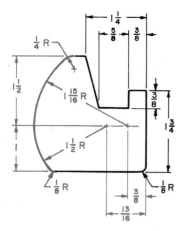

THE CURVED EDGE IS MADE
OF FOUR CURVES.

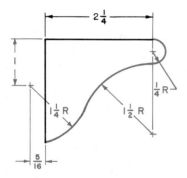

THE CURVED EDGE IS MADE
OF THREE CURVES.

7–41. *How to dimension curves made of several circular arcs.*

shown in Fig. 7–40. The centers are located. The radius establishes the width.

When the hole location is more critical than the location of the radius from the same center, the radius *is* specified. See Fig. 7–40. The overall dimension can be given by a reference dimension. Reference dimensions are discussed later in this chapter.

Curves Consisting of Circular Arcs. A curved line made up of two or more circular arcs should be dimensioned as shown in Fig. 7–41. The arcs are dimensioned by indicating the radii and locating the centers or points of tangency.

A noncircular or irregular curve may be dimensioned by giving the offset or coordinate dimensions. These dimensions are always given from a specific line, called a *datum*. See Fig. 7–42. Dimensioning from datums is discussed in more detail later in this chapter.

Cone, Pyramid, and Sphere. (Page 146). Cones are dimensioned by giving both the diameter and the altitude in the triangular view. See Fig. 7–43A. In some cases the diameter and angle formed by the sides are specified as in B. A *frustum* (a cut made perpendicular to the axis) of a cone may be dimensioned by indicating the altitude and *both* diameters. Observe that the abbreviation, DIA, is not used after the dimension since the circular view is shown. Another method of dimensioning frustums is by giving the altitude, one diameter, and the amount of taper per foot. The taper per foot is always specified by note and refers to the difference in diameter in one foot of length. Cones are usually dimensioned in one view.

Pyramids are dimensioned by giving the altitude dimension in the front view and the

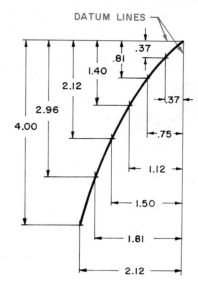

7–42. *A noncircular curve is dimensioned by locating coordinates from datum lines.*

dimension of the base in the top view. The dimensions are placed where they are the most meaningful. As with a square prism, the base of a square pyramid may be dimensioned by giving only one dimension followed by SQ, as in F. A sphere is dimensioned by giving its diameter, as in G.

LOCATION DIMENSIONS

Location dimensions are used to position geometric shapes relative to each other or to a particular line. Finished surfaces, center lines, or axis are used in locating geometric shapes. When you are going to dimension a particular view, first plan the size dimensions and then the location dimension. Usually location dimensions are given in three

145

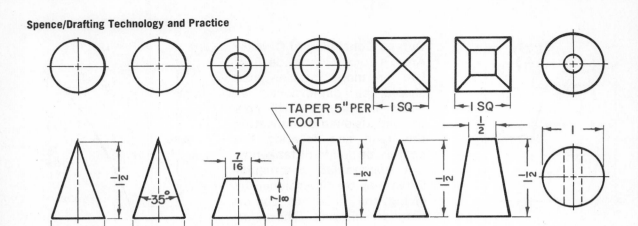

TAPER 5" PER FOOT

A-CONE   B-CONE   C-FRUSTUM OF A CONE   D-FRUSTUM OF A CONE   E-PYRAMID   F-PYRAMID   G-SPHERE

7–43. *Size dimensions of cones, pyramids, and spheres.*

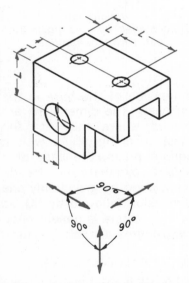

7–44. *Location dimensions are usually given in three mutually perpendicular directions.*

mutually perpendicular directions. These directions Fig. 7–44, *vertical*-bottom to top, and *horizontal*-front to back and side to side, will be adequate to locate all shapes.

Some knowledge of the function of the piece, its importance with mating pieces or how it fits with other parts of the object, and a basic knowledge of shop processes are all important in dimensioning practice. An understanding of these items is particularly valuable in determining how and what location dimension will be specified.

Prisms. Rectangular shapes are located from surface to surface or from a center line. See Fig. 7–45. Dimensions that locate surfaces may be from a datum line. Datums are used as a reference where dimensions may be critical. This is particularly true where one dimension must meet with another. See Fig. 7–52. *Point-to-point* distances are used for describing the location of prisms where the object is simple and when a high degree of accuracy is not required.

Cylinders. Cylinders are located from their center lines. If possible, always give the location dimension in the circular view of the cylinder. Fig. 7–46 shows how cylindrical holes may be located.

When a series of equally-spaced holes in a line are to be located, they may be indicated as shown in Fig. 7–47. The term, Equally Spaced, is placed in the note with the diameter of the holes.

Holes or position cylindrical features, when located on a circle or arc, may be dimensioned in several ways. Holes may be located by rectangular coordinates. See Fig. 7–48. They may be located by giving the radius of the arc or diameter of the circle and the note, Equally Spaced. See Fig. 7–49. Angular dimensioning may also be used to locate holes along a circular center line. See Fig. 7–50.

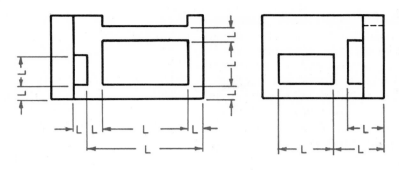

A - LOCATION USING SURFACES

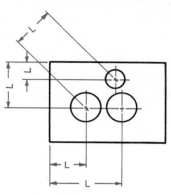

7-46. *Ways to locate cylindrical holes.*

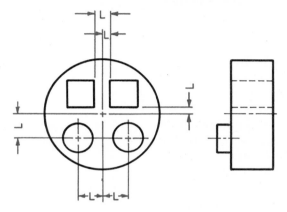

B - LOCATION FROM A CENTER LINE

7-45. *Location dimensions for prisms.*

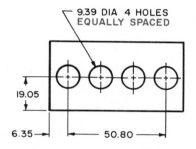

9.39 DIA 4 HOLES
EQUALLY SPACED

19.05

6.35

50.80

DIMENSIONS IN MILLIMETRES

7-47. *The term, equally spaced, can be used to locate holes.*

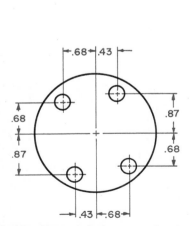

.68   .43

.68

.87

.87

.68

.43   .68

7-48. *Round holes can be located using coordinates.*

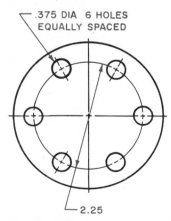

.375 DIA 6 HOLES
EQUALLY SPACED

2.25

7-49. *Equally spaced round holes can be located by giving the diameter of their center line and the term, equally spaced.*

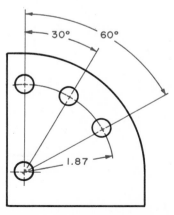

30°   60°

1.87

7-50. *Round holes can be located with a radius and angular dimensions.*

147

Distances Along Curved Surfaces. Linear distances along curved surfaces may be dimensioned as a chord or an arc. See Fig. 7–51. In either case, it should be clear that the dimension line indicates an arc or chord.

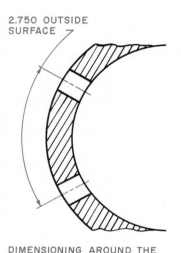

2.750 OUTSIDE SURFACE

DIMENSIONING AROUND THE ARC

2.531 ±.010

DIMENSIONING ACROSS THE CHORD

7–51. *Linear distances on curved surfaces can be dimensioned by giving the chord or length of the arc on the surface.*

Dimensions from Datums. Datums are points, lines, planes, and cylinders which are understood to be exact. These features are used for computation or reference as well as for giving location dimensions. When a datum is specified, Fig. 7–52, all features *must* be specified from this datum and with respect to one another. Two and often three datums must be established in locating a feature. Surfaces, center lines, or surface and a center line which serve as datums must be clearly identified or easily recognizable. Fig. 7–53 and 7–42 show two additional applications of datum dimensioning. To be most useful for measuring, the datums on the piece must be accessible during manufacture. Corresponding features on mating parts should be used as datums to insure proper assembly.

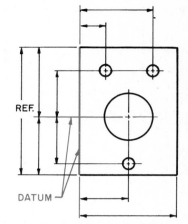

REF.

DATUM

7–52. *A center line or edge can be used as a datum.*

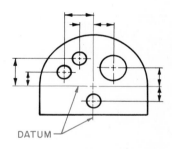

DATUM

7–53. *Datum dimensioning.*

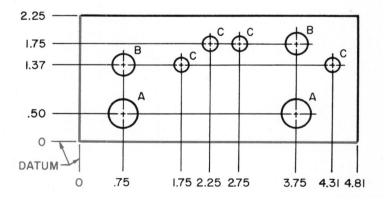

2.25
1.75
1.37

.50

0

DATUM

0   .75   1.75 2.25 2.75   3.75   4.31 4.81

SIZE SYMBOL	A	B	C
HOLE DIA	.500	.375	2.50

7–54. *Features can be located using ordinates.*

ORDINATE DIMENSIONING

Ordinate dimensioning is a form of rectangular datum dimensioning. All dimensions are measured from two or three mutually perpendicular datum places. The datum planes are indicated as *zero* coordinates and the dimensions from them are shown on extension lines. See Fig. 7–54. Note that the dimensions are shown at the ends of the extension lines. There are no dimension lines or arrowheads. Notice the system used to indicate hole diameters. Each hole size is given a letter designation. Sizes are shown by a table. This simplifies a drawing.

TABULAR DIMENSIONING

Dimensions from a mutually perpendicular datum should be listed in a table on the drawing instead of the view. Fig. 7–55 illustrates a drawing with tabular dimensioning. Tabular dimensioning may be used on drawings which require the location of a large number of similar shaped features. Note in this drawing the datums are indicated as X and Y surfaces and are not called datums. This method is used a great deal in drawing electronic printed circuit boards. An example can be found in Chapter 20, Electrical — Electronics Diagrams.

REFERENCE DIMENSIONS

A reference dimension is a dimension *without any specified tolerance*. This type of dimension is used for informational purposes only. See Fig. 7–56. Reference dimensions

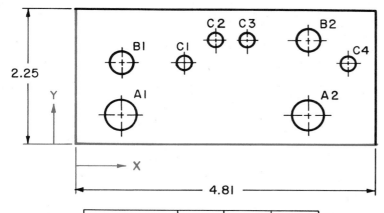

NO. REQ'D.		2	2	4
HOLE DIA		.500	.375	.250
		HOLE SYMBOL		
X →	Y ↑	A	B	C
.75	.50	AI		
3.75	.50	A2		
.75	1.37		BI	
3.75	1.75		B2	
1.75	1.37			CI
2.25	1.75			C2
2.75	1.75			C3
4.31	1.37			C4

7–55. *Features can be located using tabular dimensions.*

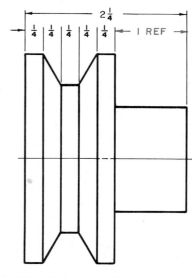

7–56. *Reference dimensions are used for informational purposes only.*

do not govern any machining or inspecting operations. All reference dimensions are indicated on drawings by the abbreviation REF directly following or under the dimension. Reference dimensions may be either size or location dimensions.

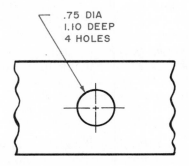

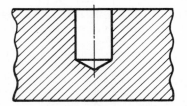

7–57. *Round holes are dimensioned by giving their diameter, depth, and the number of holes this size.*

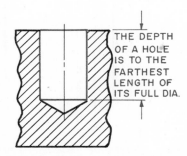

7–58. *A blind hole does not go all the way through.*

THE DEPTH OF A HOLE IS TO THE FARTHEST LENGTH OF ITS FULL DIA.

DIMENSIONING ROUND HOLES

Plain Round Holes. Round holes are dimensioned in various ways depending on design requirements and manufacturing methods. The diameter, depth and number of holes are always specified either by note or dimension. See Fig. 7–57. If it is not clear that the hole is to go completely through the piece, the abbreviation for through, THRU, should be added following the diameter of the hole. A hole which does not go completely through the piece is called a *blind* hole. The depth must be specified. See Fig. 7–58. The depth of a hole is identified by its shape rather than by the method of forming. Fig. 7–59 shows methods of dimensioning round holes.

Counterbored Holes. A counterbored hole is a cylindrical enlargment of a hole to a specified depth. The bottom of the counterbore is perpendicular to the axis of the hole. A typical application of a counterbored hole is to accommodate a fillister-head machine or capscrew. The diameter and depth of the counterbore should be indicated. Fig. 7–60 shows a counterbored hole dimensioned by a note. Notice how the data is given in the note. It includes the hole diameter, the counterbore diameter, and the depth of the counterbore. The abbreviation for counterbore is C'Bore. In some cases the thickness of

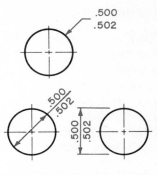

DIMENSIONING HOLES IN CIRCULAR VIEW

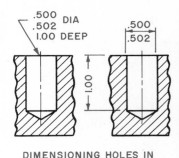

DIMENSIONING HOLES IN SECTION

7–59. *Dimensioning round holes.*

the remaining stock may be dimensioned rather than the depth of the counterbore. In this case the thickness is given as a dimension on the view and not in note form. See Fig. 7–60. The tool used to produce a counterbore is shown in Fig. 7–61.

Countersinking. Countersinking is the machining of a hole to a conical shape. A hole with a countersink is used to hold the head of a flathead screw. The diameter and angle of the

$\frac{7}{8}$ DIA $1\frac{1}{8}$ C'BORE
.125 DEEP 2 HOLES

DIMENSION TO CIRCULAR VIEW

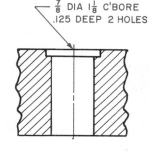

$\frac{7}{8}$ DIA $1\frac{1}{8}$ C'BORE
.125 DEEP 2 HOLES

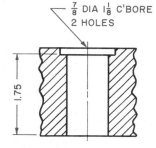

$\frac{7}{8}$ DIA $1\frac{1}{8}$ C'BORE
2 HOLES

1.75

DIMENSION TO SECTION VIEW

7–60. *Dimensioning counterbored holes.*

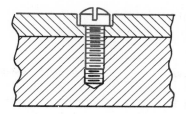

A MACHINE SCREW IN A
COUNTERBORE

THESE CUTTERS
FORM THE
COUNTERBORE

PILOT END LEADS
THE TOOL INTO
DRILLED HOLE

THIS TOOL IS USED TO
COUNTERBORE A ROUND
HOLE.

7–61. *Counterboring round
holes.*

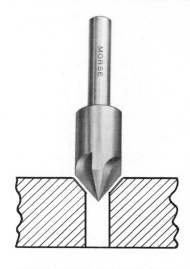

COUNTERSINK FORMING HOLE

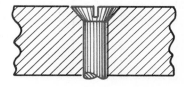

COUNTERSUNK HOLE
WITH FASTENER IN PLACE

7–62. *A countersink forms a
conical shape in the top of a
round hole.*

countersink used to produce
the cone-shaped enlargment
is shown in Fig. 7–62. Though
the countersink has an angle
of 82 degrees, it is actually
drawn at a 90-degree angle.
In Fig. 7–63 data given on the
note refers to the diameter of
the hole, the countersink
angle, and the diameter of the
countersink on the surface.

$\frac{3}{8}$ DIA 82° CSK
.75 DIA 4 HOLES

DIMENSION TO CIRCULAR VIEW

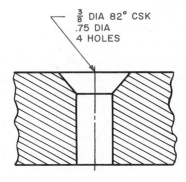

$\frac{3}{8}$ DIA 82° CSK
.75 DIA
4 HOLES

DIMENSION TO SECTION VIEW

7–63. *Dimensioning countersunk holes.*

151

$\frac{1}{2}$ DIA $\frac{3}{4}$ SPOT-FACE
4 HOLES

DIMENSION TO CIRCULAR VIEW

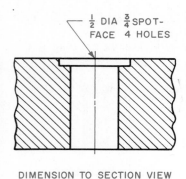

$\frac{1}{2}$ DIA $\frac{3}{4}$ SPOT-
FACE 4 HOLES

DIMENSION TO SECTION VIEW

7-64. *Dimensioning spot-faced holes.*

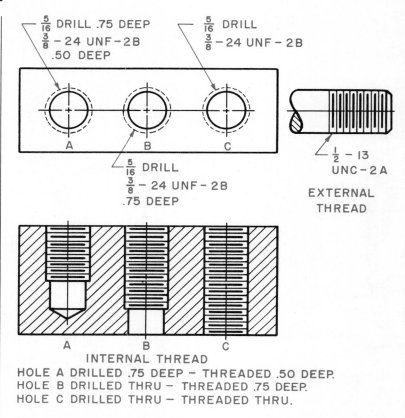

$\frac{5}{16}$ DRILL .75 DEEP
$\frac{3}{8}$ - 24 UNF - 2B
.50 DEEP

$\frac{5}{16}$ DRILL
$\frac{3}{8}$ - 24 UNF - 2B

A   B   C

$\frac{5}{16}$ DRILL
$\frac{3}{8}$ - 24 UNF - 2B
.75 DEEP

$\frac{1}{2}$ - 13
UNC - 2A

EXTERNAL
THREAD

A   B   C

INTERNAL THREAD
HOLE A DRILLED .75 DEEP — THREADED .50 DEEP.
HOLE B DRILLED THRU — THREADED .75 DEEP.
HOLE C DRILLED THRU — THREADED THRU.

7-65. *Typical thread notes. See Chapter 26, Fastening Devices, for detailed explanation.*

Spot-faced Holes. Spot-facing is the smoothing and squaring of the surface around a hole to seat a washer, nut, or head of a bolt. The same tool is used for spot-facing as for counterboring. The spot-facing is indicated by note, Fig. 7-64. The diameter of the spot-facing must be specified. The depth of the area to be spot-faced is understood and is not indicated. It is generally accepted that spot-facing is no deeper than $1/16$ inch; beyond this depth, it is classed as a counterbore.

THREADS

Threads are dimensioned with local notes. Tapped holes are dimensioned to the circular view whenever possible. External threads have the note lettered horizontally. See Fig. 7-65. The data in a thread note is explained in Chapter 26, Fastening Devices. Tools to cut threads are shown in Fig. 7-66.

SHAFT CENTERS

Shaft centers or countersunk center holes may be required in shafts or spindles during manufacturing or inspection. Primarily these are used for pieces that are to be turned on a lathe using one or two centers. The center drill is sharpened at an angle of 60 degrees. This angle is understood when it is dimensioned, Fig. 7-67. A center drill used for drilling a shaft center is shown in Fig. 7-68.

DIMENSIONING OTHER FEATURES

Chamfers. A narrow inclined flat surface along the intersection of two surfaces is

7–66A. *This tap cuts internal threads.*

7–66B. *This die cuts external threads.*

7–68. *A center drill. The small tip drills a straight hole. The larger cutting edge forms a 60-degree countersink.*

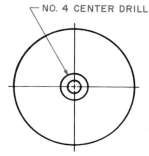

NO. 4 CENTER DRILL

DIMENSION IN CIRCULAR VIEW

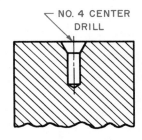

NO. 4 CENTER DRILL

DIMENSION IN SECTION VIEW

7–67. *Dimensioning a shaft center.*

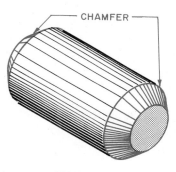

CHAMFER

CHAMFER ON
A CYLINDRICAL SURFACE

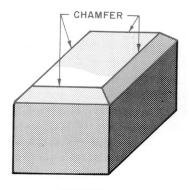

CHAMFER

CHAMFER ON
PLANE SURFACES

7–69. *A chamfer is a narrow, flat, inclined surface which removes the sharp corner where two surfaces meet.*

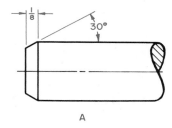

$\frac{1}{8}$ 30°

A

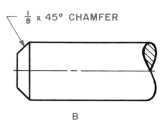

$\frac{1}{8}$ x 45° CHAMFER

B

7–70. *Dimensioning chamfers.*

called a chamfer. See Fig. 7–69. Chamfers are usually used on the ends of cylinders. The recommended method for dimensioning a chamfer is to give the angle and length, Fig. 7–70, Part A. If the piece is chamfered at an angle of 45 degrees, the size and angle (45°) is given. See Fig. 7–70B. The word chamfer may be omitted. This method is only used with angles of 45 degrees since the size dimension may apply to the radial or axial dimension. Chamfers are never measured along the angled surface.

Keys. Keys are used to prevent rotation of a shaft and hub. The key is sunk partly into the shaft and extends into the hub. Fig. 7–71 shows a shaft with a key being inserted into a hub. The recess in the shaft which holds the key is called the *keyseat.* Fig. 7–72 illustrates how a *keyseat* may be dimensioned when stock

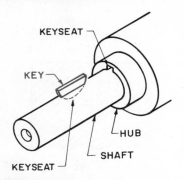

7-71. *Keys are used to prevent rotation between a shaft and a hub. See Chapter 26, Fastening Devices, for a detailed explanation.*

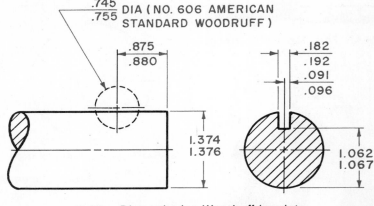

7-73. *Dimensioning Woodruff keyslots.*

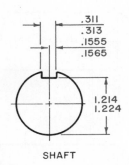

SHAFT

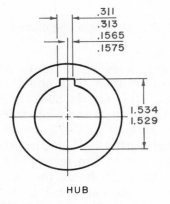

HUB

7-72. *Dimensioning keyseats for stock keys.*

keys are used. The dimensioning of a Woodruff key and keyseat are shown in Fig. 7-73. See Chapter 26, Fastening Devices, for more details.

Knurls. Knurling is the process of pressing a pattern into a metal surface. Fig. 7-74 shows a piece being knurled on a lathe. Knurls may be either diamond or straight, as shown in Fig. 7-75. Close tolerances are not necessary for knurls that provide a rough surface for gripping or that are used for decoration. For these purposes only the pitch of the knurl, type of knurl, and length of knurled area is specified. See Fig. 7-76. Knurls for a press fit between mating parts are dimensioned by a note that includes: diameter before knurling, minimum diameter after knurling, pitch, length of knurled area, and type of knurl. Fig. 7-76 shows an example of a straight knurl that is used for a press fit.

Tapers. A taper gradually decreases in diameter, thus forming a conical shape. The method in which a taper will be dimensioned is dependent upon the price and method of manufacture. Tapers are usually dimensioned in decimals.

Four dimensions control the taper:

1. Diameter at large end.
2. Diameter at small end.
3. Length of axis of the taper.
4. Amount or rate of taper or angle of taper.

Three of these four dimensions will be specified with a tolerance. The fourth may be given as a *reference dimension.* The three dimensions with tolerances are directly related to the function of the part. The reference dimension has the base influence on the function of the part.

Fig. 7-77 is an example of dimensioning a tapered section when the accuracy of the

154

7-74. *The knurling tool presses a pattern into the metal piece.*

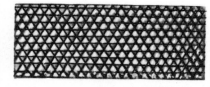

STRAIGHT KNURL

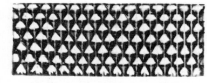
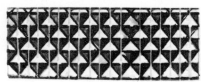

DIAMOND KNURLS

7-75. *Common types of knurls.*

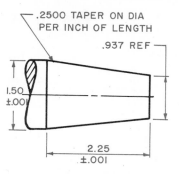

EXTERNAL TAPER

.2500 TAPER ON DIA
PER INCH OF LENGTH

.937 REF

1.50 ±.001

2.25 ±.001

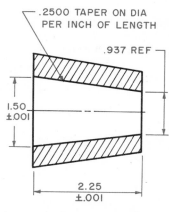

.2500 TAPER ON DIA
PER INCH OF LENGTH

.937 REF

1.50 ±.001

2.25 ±.001

INTERNAL TAPER

7-77. *The taper-per-inch method is used to dimension tapers when accuracy is important.*

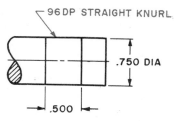

96 DP STRAIGHT KNURL

.750 DIA

.500

A KNURL FOR GRIPPING OR DECORATION

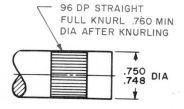

96 DP STRAIGHT
FULL KNURL .760 MIN
DIA AFTER KNURLING

.750
.748 DIA

A KNURL FOR A PRESS FIT

7-76. *Dimensioning knurls.*

taper is vital. The note on the drawing should always specify the taper or diameter per length. In this manner misinterpretation will be avoided since the taper is *not based on the radius.*

When the *length of the taper and diameter* are important and length of the part is not, it is dimensioned as shown in Fig. 7-78. The word BASIC in the note means the taper may vary from this value. Any variation, however, must fall within the zone established by the

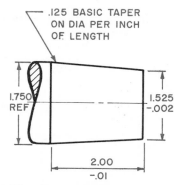

.125 BASIC TAPER
ON DIA PER INCH
OF LENGTH

1.750
REF

1.525
-.002

2.00
-.01

7-78. *Dimensioning a taper when taper diameter and length are important.*

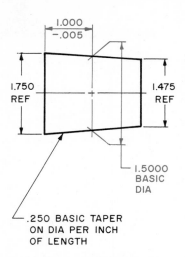

1.000
−.005

1.750
REF

1.475
REF

1.5000
BASIC
DIA

.250 BASIC TAPER
ON DIA PER INCH
OF LENGTH

7-79. *Basic diameter method for dimensioning a taper gives the exact location of the basic diameter.*

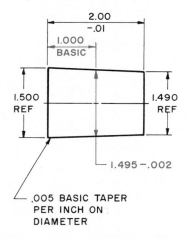

2.00
−.01

1.000
BASIC

1.500
REF

1.490
REF

1.495 −.002

.005 BASIC TAPER
PER INCH ON
DIAMETER

7-80. *The basic length method of dimensioning tapers is used when the taper is slight.*

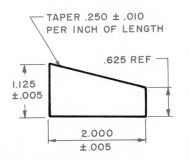

TAPER .250 ± .010
PER INCH OF LENGTH

.625 REF

1.125
±.005

2.000
±.005

7-81. *Dimensioning a flat taper.*

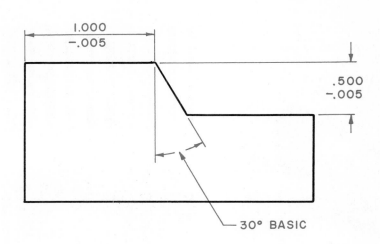

1.000
−.005

.500
−.005

30° BASIC

A
ANGLE DOES NOT HAVE A TOLERANCE

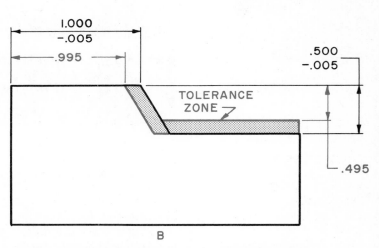

1.000
−.005

.995

.500
−.005

TOLERANCE
ZONE

.495

B
DIMENSIONS ESTABLISH A TOLERANCE ZONE

7-82. *Angular surfaces can be dimensioned by locating the linear distances and an angle.*

maximum and minimum end diameter. In Fig. 7–78 this is shown as 1.525–.002.

The *basic diameter method* of dimensioning tapers, Fig. 7–79, controls the size of the tapered section. The size of the tapered section is controlled as is its position in relation to some other surface. Both the basic taper diameter and its location are dimensioned. The basic diameter controls:

1. Axial position of the tapered section.
2. Sets up a tolerance zone for the taper.

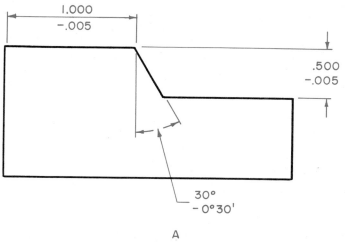

A
ANGLE WITH TOLERANCE

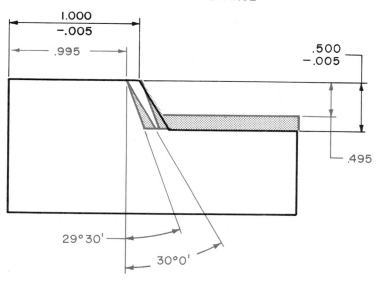

B
DIMENSIONS ESTABLISH THE TOLERANCE ZONE

7-83. *Angular surfaces can be dimensioned by locating the linear distances and determining the tolerance of the angle.*

controls the diameter as well as the position of the tapered section.

Flat Tapers. The methods recommended for dimensioning conical tapers can be adapted to flat tapers. Fig. 7-81 shows an example of dimensioning a flat taper. This is the same method that was used in Fig. 7-78.

Positioning Angular Surfaces. The location of angular surfaces may be achieved by giving linear dimensions and an angle, or by linear dimensions alone. Each arrangement of dimension and tolerances has the effect of specifying a particular tolerance zone. This zone will indicate the area in which the surface must lie. The shape and extent of the zone depends on the method of dimensioning and the method of indicating tolerances. Fig. 7-82 illustrates the use of linear dimensions and an angle to position an angular surface. In this example, the angular surface must lie between the planes, forming the tolerance zone. When the word BASIC follows the angular dimension, no specific tolerances can be applied to that angle. This means the angle may vary within the extent of the tolerance zone.

If the angular tolerance is expressed, the angle can be given without loss of precision. Fig. 7-83 illustrates this principle.

The tolerance zone that is created by the basic diameter will position the taper.

The *basic length* method, Fig. 7-80, is used when the taper on the part is extremely slight. You will note in this method the tolerance is applied directly to a diameter on the taper. The basic length

157

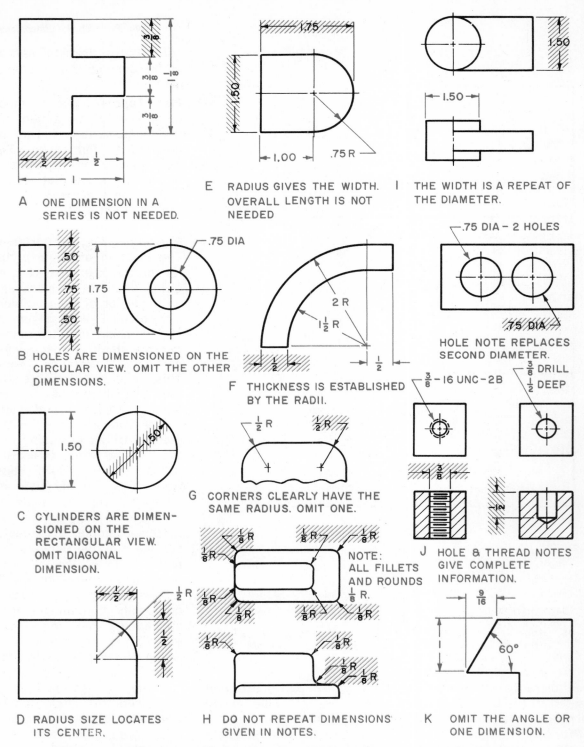

A ONE DIMENSION IN A SERIES IS NOT NEEDED.

B HOLES ARE DIMENSIONED ON THE CIRCULAR VIEW. OMIT THE OTHER DIMENSIONS.

C CYLINDERS ARE DIMENSIONED ON THE RECTANGULAR VIEW. OMIT DIAGONAL DIMENSION.

D RADIUS SIZE LOCATES ITS CENTER.

E RADIUS GIVES THE WIDTH. OVERALL LENGTH IS NOT NEEDED

F THICKNESS IS ESTABLISHED BY THE RADII.

G CORNERS CLEARLY HAVE THE SAME RADIUS. OMIT ONE.

H DO NOT REPEAT DIMENSIONS GIVEN IN NOTES.

I THE WIDTH IS A REPEAT OF THE DIAMETER.

HOLE NOTE REPLACES SECOND DIAMETER.

J HOLE & THREAD NOTES GIVE COMPLETE INFORMATION.

K OMIT THE ANGLE OR ONE DIMENSION.

7-84. *Examples of common errors committed by giving unneeded dimensions.*

158

END PRODUCT DRAWINGS

An *end product* drawing gives only the dimensions and notes for the final forms and sizes of the piece. No mention or reference is given to the method of manufacture. For example, only the diameter of a hole is given. No information is given whether it is to be drilled, reamed, punched, or formed by any other operation. End product dimensioning is usually used on drawings of parts that will be produced in large quantity for interchangeable assembly.

Many industrial corporations have a process engineering department. A portion of this group's responsibilities is to determine the manufacturing methods used to produce a piece. According to the accuracy specified on the drawing, the hole may be drilled, reamed, bored, or punched.

PROCESS DRAWINGS

In the manufacture of many pieces, preliminary operations are required to bring the piece to its final dimension(s). A *process drawing* will give the dimension for each preliminary operation. Proper allowances must be made for stock removal.

The choice between dimensioning the piece as an end product or process drawing depends on the purpose of the drawing. The type and extent of the industry will also play a part in determining how the drawing will be dimensioned.

EXCESS DIMENSIONS

A drawing should contain all the dimensions needed to make the object. However, dimensions should not be repeated anywhere on the drawing. Information contained in notes should not be shown again in dimensional form. Fig. 7–84 shows examples of common mistakes.

A. When there are a number of dimensions in a row, the last dimension is omitted. The overall dimension plus the other dimensions make this unnecessary.

B. Cylinders are dimensioned to their rectangular view. Holes are dimensioned to the circular view. The distance from the edge of a cylinder to the edge of a hole is not needed. The object is made by measuring from its center line.

C. The diameter is properly given on the rectangular view.

D. The center of the arc is located by the radius.

E. The overall length and width are not needed to make this object. They may be given as reference dimensions.

F. The center line is located from the end of the object. The thickness is established by the radii.

G. It is clear that several identical features have the same dimension. It is not necessary to dimension each one.

H. Information given in notes should not be repeated on a drawing.

I. The diameter gives the width.

J. Hole and thread notes give complete information.

K. The angle only needs the two outside dimensions or the angle and one dimension.

NOTES

Notes are used to give information that is not part of the normal dimensioning system. They should be stated as briefly as possible. The meaning must be clear.

Notes are lettered horizontally. Always use guidelines. Form letters carefully so they can be read easily. Notes that are to be related to a drawing with a leader should use leaders that are as short as possible. Avoid placing notes in crowded areas of the drawing. Do not place between views unless necessary.

There are two types of notes to place on a drawing, general and local.

General notes apply to the entire drawing. For example, when the note FINISH ALL OVER is used, it refers to all surfaces of the object. General notes are usually located in two places. Some are placed above or to the left of the title block. Materials, tolerances, and specifications are placed by the title block. Other notes are placed near the view to which they refer. They are lettered in an open area so they are easily seen. They should not be confused with a specific dimension.

7–85. *Standard abbreviations used on drawings.*

A		C (con't)		H (con't)	
Allowance	ALLOW	Countersink	CSK	Headless	HDLS
Alloy	ALY	Cross Section	XSECT	Heat	HT
Alternate	ALT	Cubic	CU	Heat Treat	HT TR
Alternating Current	AC	Cubic Foot	CU FT	Heavy	HVY
Aluminum	AL	Cubic Inch	CU IN	Hexagon	HEX
Ampere	AMP			Horizontal	HOR
Anneal	ANL	**D**		Horsepower	HP
Approximate	APPROX				
Area	A	Decimal	DEC		
Assemble	ASSEM	Degree	DEG	**I**	
Assembly	ASSY	Detail	DET		
Average	AVG	Diagonal	DIAG	Inch	IN
		Diameter	DIA	Inside Diameter	ID
		Dimension	DIM	Interior	INT
B		Drawing	DWG	Irregular	IRREG
		Drill	DR		
Back to Back	B TO B			**J**	
Balance	BAL	**E**			
Ball Bearing	BB			Joint	JT
Base Line	BL	Elevation	EL	Junction	JCT
Bevel	BEV	Estimate	EST		
Both Faces	BF	Exterior	EXT	**K**	
Both Sides	BS	Extra Heavy	X HVY		
Both Ways	BW	Extra Strong	X STR	Key	K
Bottom	BOT	Extrude	EXTR	Keyseat	KST
Bottom Face	BF				
Brass	BRS	**F**		**L**	
Break	BRK				
British Thermal Units	BTU	Fabricate	FAB	Left	L
Bronze	BRZ	Face to Face	F TO F	Left Hand	LH
Bushing	BUSH	Fahrenheit	F	Length	LG
Button	BUT	Feed	FD	Length Over All	LOA
		Feet	FT	Light	LT
		Figure	FIG	Line	L
C		Fillet	FIL	Locate	LOC
		Fillister	FIL	Long	LG
Cap Screw	CAP SCR	Finish	FIN		
Capacity	CAP	Finish All Over	FAO	**M**	
Case Harden	CH	Flange	FLG		
Cast Iron	CI	Flat	F	Machine	MACH
Cast Steel	CS	Flat Head	FH	Material	MATL
Casting	CSTG	Foot	FT	Maximum	MAX
Center	CTR	Front	FR	Metal	MET
Center Line	CL			Metre	M
Center to Center	C TO C	**G**		Miles	MI
Chamfer	CHAM			Miles Per Hour	MPH
Chord	CHD	Gage	GA	Millimetre	MM
Circle	CIR	Gallon	GAL	Minimum	MIN
Circumference	CIRC	Gasket	GSKT	Minute	MIN
Clear	CL	General	GEN		
Clockwise	CW	Glass	GL	**N**	
Cold Drawn	CD	Grade	GR		
Cold Drawn Steel	CDS	Grind	GRD	Negative	NEG
Cold Rolled	CR	Groove	GRV	Neutral	NEUT
Cold Rolled Steel	CRS			Nominal	NOM
Concentric	CONC			Normal	N
Connect	CONN	**H**		Not to Scale	NTS
Counterclockwise	CCW	Hard	H	Number	NO
Counterbore	CBORE	Head	HD		

O		R (con't)		T (con't)	
Obsolete	OBS	Reinforce	REIN	Thread	THD
Octagon	OCT	Revolution	REV	Threads per Inch	TPI
Outside Diameter	OD	Revolutions per Minute	RPM	Through	THRU
Overall	OA	Right Hand	RH	Tolerance	TOL
		Rivet	RIV	Tooth	T
		Rough	RGH		
P		Round	RD		
Part	PT			**U**	
Pattern	PATT	**S**		Unit	U
Perpendicular	PERP			Universal	UNIV
Pitch	P	Screw	SCR		
Pitch Circle	PC	Section	SECT		
Pitch Diameter	PD	Shaft	SFT	**V**	
Plate	PL	Sheet	SH		
Point	PT	Single	S	Valve	V
Pound	LB	Small	SM	Versus	VS
Pounds Per Square Inch	PSI	Spring	SPG	Vertical	VERT
		Square	SQ	Volt	V
		Standard	STD	Volume	VOL
Q		Steel	STL		
Quadrant	QUAD	Stock	STK	**W**	
Quality	QUAL	Surface	SUR	Washer	WASH
Quarter	QTR			Watt	W
				Weight	WT
		T		Width	W
R		Tangent	TAN		
Radial	RAD	Taper	TPR	**X Y Z**	
Radius	R	Template	TEMP	Yard	YD
Rectangle	RECT	Thick	THK	Year	YR
		Thousand	M		

ANSI Bulletin Y32.13–1950

Local notes apply to a specific part of the detail. For example, ⁵⁄₁₆–18UNC–2A is a thread note. It refers to a specific fastener. Local notes are connected to the detail with a leader. Place them as close to the detail as possible. Keep the leader as short as possible.

Abbreviations are commonly used in notes. Common standard abbreviations are shown in Fig. 7–85.

General notes are placed on the drawing after the size and location dimensions are in place.

DECIMAL DIMENSIONING

Since the dawn of interchangeable manufacture, the role of the designer-draftsman has become increasingly important. Before interchangeable pieces were made, the responsibility for accuracy rested largely with the workman in the shop. It was his skill and accuracy that determined the success of any project. In many cases the designer was also the producer in the shop. Each piece was made to fit the next piece as it was manufactured. Even though a hammer on a gunlock would appear identical with another produced by the same gunsmith, there was no indication it could be interchanged. If a breakdown occurred on a stream engine, no replacement part could be drawn from a stock supply.

Today the engineer, designer, and draftsman assume the responsibility of thinking through every detail of a product. Careful engineering, combined with precise dimensioning, are behind every product manufactured. Parts fabricated in various areas of our nation are brought together with assurance they will fit together and function properly.

Only through a system of precision dimensioning and proper expression of these values can interchangeable parts and mass production be accomplished. Few industrial companies manufacture all pieces which go into a product. Component parts are subcontracted to many companies throughout the country and world.

To meet the needs of mass production, it would be impossible to give dimensions as fractional values. Dimensions must be given to a fine degree with some *allowance over or under the specified size.* By indicating the location dimension of two holes with some allowable variance in their position, assembly is

possible. This allowable variance is called the *tolerance.*

To achieve these tolerance dimensions, precise measuring instruments are used. Measurements are then no longer given in fractional form but are expressed as *decimals.* Micrometers, Vernier calipers, height gages, and gage blocks are used for measuring. Some of these instruments are capable of measuring one ten-thousandth of an inch (0.0001). To work with and understand *tolerances,* some knowledge of decimals is necessary.

Decimal System. The draftsman may use a decimal rule or scale, Fig. 7–86, to lay out the drawing. Both the decimal rule and scale are graduated so an inch is divided into tenths (.10, .20, .30, .40) and even hundredths (.02, .04, .06, and .08). The smallest divi-

sion is .02. On both the scale and the rule, the .02 marks are longer and this makes it easier to read the scale.

Decimal dimensioning is based on the inch being divided into 50 parts. Each part of an inch ($1/50$) is equal to two hundredths (.02) of an inch. Decimal dimensioning consists of two digits following the decimal point—.02, .16, .38, .50, 1.14, 2.62.

All two-place decimals are expressed in even hundredth. —*not odd hundredths,* (.01, .03, .05). If an even hundredth were divided in two, as in obtaining a radius from a diameter, the result would remain a two-place figure. An even two-place decimal is easier to read on the scale than an odd two-place decimal.

Fractions to Decimals. When converting fractions to decimals, a table of decimal equivalents is used. The table is shown inside the front cover. The decimal equivalent for each fraction is given in two, three and four-place decimals. When dimensioning, the number of decimal places is determined by the accuracy required.

Two-place decimals are used where tolerance limits of ± .01 can be allowed.

Decimals to three or more places must be used for tolerance limits less than ± .010.

General Rules for Decimal Dimensioning. Dimensions are to be specified in two-place decimals except when con-

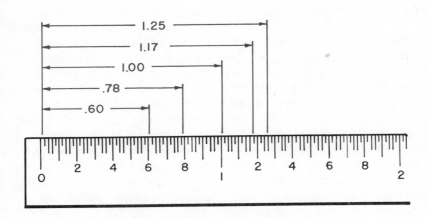

DECIMAL SCALE

7–86. *Graduations on the decimal scale and rule are identical. Both may be read to an even or odd hundredth of an inch. Each small division equals .02 inches.*

1.00		~~1.~~
1.40	X	~~1.4~~
1.16		~~1.1562~~
CORRECT		INCORRECT

7–87. *All dimensions should be expressed in two-place decimals.*

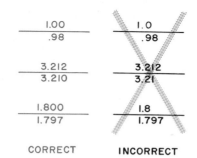

CORRECT		INCORRECT
1.00 / .98	X	~~1.0 / .98~~
3.212 / 3.210		~~3.212 / 3.21~~
1.800 / 1.797		~~1.8 / 1.797~~

7–88. *Upper and lower limits should contain the same number of decimal places.*

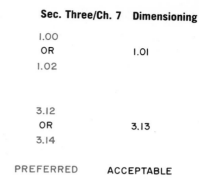

PREFERRED	ACCEPTABLE
1.00 OR 1.02	1.01
3.12 OR 3.14	3.13

7–89. *Dimensions should be expressed in even tenths or hundredths of an inch.*

verting from fractional dimensions. Another exception is also made when tolerances require more than two-place decimals. Fig. 7–87 illustrates the correct and incorrect method of expressing dimensions.

Tolerances have upper and lower limit dimensions in two or more decimal places. The upper and lower values contain the same number of decimal places. Fig. 7–88 illustrates this practice. The preferred method is to place the larger values above the smaller.

Try to work to the tenth and even hundredth increment of an inch. Odd hundredths may be used only when required for accuracy. See Fig. 7–89. This practice should be avoided as much as possible. When odd hundredths are divided by two, as from a center line to an edge or hole, they will result in more than two decimal places. 3.13 ÷ 2 = 1.565.

When ordinates are used to dimension curves or irregular lines, all dimensions are expressed in hundredths. These dimensions are scaled from the layout, as shown in Fig.

7–90. No effort is made to use even hundredths.

Rounding of Decimals. Rounding of decimals to a lesser number of places can be accomplished by the following method:

1. When the digit following the last digit to be retained is less than five, the last digit retained is not changed. See Fig. 7–91A.
2. If the digit following the last digit to be retained is more than five, the last digit retained is increased by one. See Fig. 7–91B.
3. When the digit following the last digit to be retained is exactly five, and the digit to be rounded off is even, that

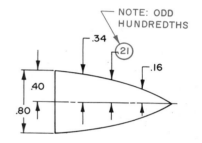

7–90. *When ordinates are used for dimensioning an irregular curve, dimensions may be expressed in odd hundredths.*

digit remains unchanged. See Fig. 7–91C.
4. If the digit following the last digit to be retained is exactly five, and the digit to be rounded off is odd, that digit is increased by one. See Fig. 7–91D.

7–91. *Rounding off dimensions.*

	4 PLACES	ROUNDED OFF TO 3 PLACES	2 PLACES
A	3.1262|4 = 3.1262	.786|2 = .786	5.13|3 = 5.13
B	1.1250|8 = 1.1251	6.187|6 = 6.188	.85|9 = .86
C	.3422|5 = .3422	4.276|5 = 4.276	1.78|5 = 1.78
D	10.7921|5 = 10.7922	2.125|5 = 2.126	4.87|5 = 4.88

163

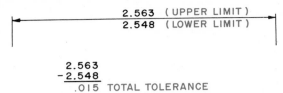

$$
\begin{aligned}
2.563 \quad &(\text{UPPER LIMIT}) \\
2.548 \quad &(\text{LOWER LIMIT})
\end{aligned}
$$

$$
\begin{array}{r}
2.563 \\
-\ 2.548 \\
\hline
.015 \quad \text{TOTAL TOLERANCE}
\end{array}
$$

7–92. *Tolerance is the difference between the upper and lower limits. It is the total amount of variation permitted in a piece.*

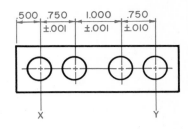

A
THE SUM OF THE TOLERANCES FROM X TO Y IS .012.

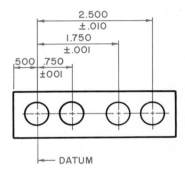

B
THE USE OF A DATUM PREVENTS THE ACCUMULATION OF TOLERANCES.

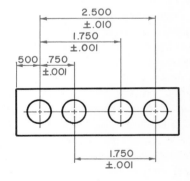

C
THE DISTANCE BETWEEN TWO POINTS CAN BE CONTROLLED BY A SPECIAL TOLERANCE DIMENSION.

7–93. *Ways to locate tolerance dimensions. The method to use depends upon the accuracy required.*

TOLERANCES AND LIMITS

Limits are the maximum and minimum values given for a specific dimension. A tolerance represents the total amount by which a specific dimension may vary. Thus the tolerance is the difference between the limits. See Fig. 7–92.

Tolerances are expressed in the same form as the dimension. If the dimension is in fractional form, the tolerance will then be expressed in fractions. Likewise, if the dimension is in decimals, the tolerance will be given in decimals to the same number of places.

In a "chain" of dimensions with tolerances, as shown in Fig. 7–93A, the overall variation in position is equal to the sums of the tolerances on the intermediate distances. By dimensioning from a datum, Fig. 7–93B, the overall accumulation is avoided. The tolerance given between the two features equals the sum of the tolerances on the dimensions from the datum. If the distance between the two points must be closely controlled, these distances should be dimensioned directly with a tolerance. See Fig. 7–93C.

Unilateral method of tolerances allows a variation from a design size in one direction only. The design size is one of the limits. The other limit is found by applying the tolerance to the design size. See Fig. 7–94A, where the design sizes are shown as 1.878 and 3.252. In the case of 1.878, the variation is minus, and in the 3.252 dimension, the variation is plus.

Bilateral method of tolerances permits a variation from a design size in *both* directions. See Fig. 7–94B. In the bilateral system, the design size is not one of the limits. The plus tolerance is added to the design size and the minus tolerance is subtracted to obtain the limits. The bilateral method is generally used for locational dimension or for any dimension that can vary in this direction.

Limit dimensions method gives only the largest and smallest permissible dimension, Fig. 7–95. This tolerance is the difference between limits.

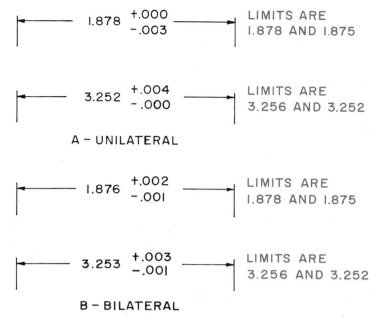

A – UNILATERAL

B – BILATERAL

7-94. *Giving tolerances by plus and minus figures.*

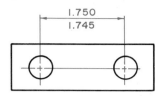

A
HIGH LIMIT IS PLACED ABOVE
THE LOW LIMIT ON LOCATION
DIMENSIONS.

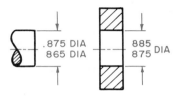

B
HIGH LIMIT IS PLACED ABOVE
THE LOW LIMIT ON SIZE
DIMENSIONS THAT SHOW SIZE
DIRECTLY.

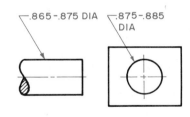

C
LOW LIMIT IS PLACED IN FRONT
OF THE HIGH LIMIT WHEN THE
DIMENSIONS ARE GIVEN IN
NOTE FORM.

7-95. *Accepted ways to place tolerance dimensions.*

Allowable variations may be expressed in one of the following ways:

1. By specifying the tolerance values, one plus and one minus, as shown in Fig. 7-94. This form is necessary if the plus variation differs from the minus variation. When the variation in one direction is zero, the tolerance method is unilateral. When neither variation is zero, the tolerance method is bilateral. Two variations in the same direction are not specified.

2. By specifying the tolerance values with a combined plus and minus sign ($\pm$), the single tolerance follows the $\pm$ sign. See Fig. 7-93. This method applies only if the plus and minus variation are equal.

3. By specifying the maximum and minimum limits of size and position, the numerical values may be expressed as shown in Fig. 7-95. The high limit is always placed above the low limit when the dimensions are given directly. The low limit always comes before the high limit when given in note form. In location dimensions which are given directly, not by note, the high limit is placed above.

4. By not stating both limits when specifying unilateral tolerances. This system should not be used if the unilateral system, Fig. 7-94 is used. Do not use both systems on one drawing.

a. A unilateral variation may be expressed without stating that the variation in the other direction is zero. See Fig. 7-96.

b. MIN (Minimum) or MAX (Maximum) is often placed after the numeral when

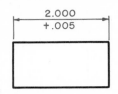

THESE LIMITS ARE 2.000 AND 2.005.

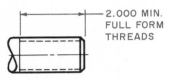

2.000 MIN. FULL FORM THREADS

MINIMUM LENGTH OF THREADS SHOULD BE 2.000.

.IO R MAX

CORNER RADIUS SHOULD NOT BE LARGER THAN .IOR.

7-96. *Ways to show single uni-lateral tolerances.*

other limits are not important. Depths of holes, thread lengths, chamfers, or radii are often limited in this manner. See. Fig. 7-96.

Placing Tolerance Dimension. Values indicating tolerances are placed:

1. To the right of and in line with the dimension numeral. See Fig. 7-97.

2. Below the dimension numeral with the dimension line between, Fig. 7-97.

3. To the right of the dimension numeral. The plus varia-

tion is above the level of the dimension line and the minus variation below the line. The dimension line is not drawn between the numerals indicating tolerance. See Fig. 7-97.

Placing Limit Dimensions. Numerals indicating limits are placed in one of the methods shown in Fig. 7-98.

1. One above the other with or without a dimension line between. This is used in *direct dimensioning* for size and location dimension.

2. In line with each other with a hyphen between. This is used for size dimension in note form.

DIMENSIONING FOR FITS

There are no complete sets of formulas or tables which the engineer, designer, or draftsmen can use to establish limits and tolerances. A few are found for each area of drafting (electrical, marine, mechanical). More specifically, the function and size of the piece and the method of machining will determine what limits and tolerances will be applied. The greater the limits of tolerance can be, the less costly the piece is to produce. Parts with large tolerance require less expensive tools to manufacture, lower labor costs, and reduced waste of parts. Tolerances should never be specified closer than necessary. In many instances it is desirable to use close limits to insure interchange-

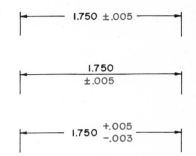

7-97. *Ways to place tolerance dimensions.*

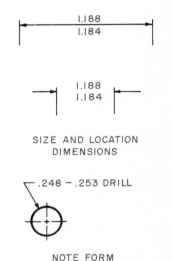

SIZE AND LOCATION DIMENSIONS

.248 - .253 DRILL

NOTE FORM

7-98. *Methods for placing limit dimensions.*

ability of parts. This avoids the extra cost of selective assembly. Fig. 7-99 shows the types of fits in which tolerances are involved. These are total tolerances and are related to machining processes that, under normal conditions, will produce work within the tolerances indicated.

Types of Fits. There are three general types of fits between parts—clearance, interference, and transitional.

Clearance fits are those in which an internal member fits in an external member with space between the parts. An example is shown in Fig. 7–99. Here the smallest hole is .751 and the largest shaft is .750. The space allowed between these is .001.

An *interference fit* has the internal member larger than the external member. In this case the shaft is larger than the hole. When they are brought together, the shaft is forced by pressure into the hole. See Fig. 7–99.

A *transistion fit* is one that could have either clearance or interference. When these parts are assembled, some will have clearance and some will have interference. See Fig. 7–99.

These fits have been standardized into five classes.

Standard Fits. Theoretically an infinite number of fits could be specified for any particular piece and its mate. Standard classes of fits have been designated to insure interchangeability. The type of fit is determined by the requirements of the piece. Standard classes of fits are:

RC Running or sliding fit
LC Locational clearance fit
LT Locational transitional fit
LN Locational interference fit
FN Force or shrink fit

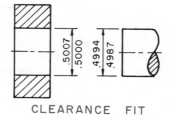

CLEARANCE FIT

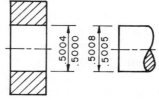

FORCE FIT

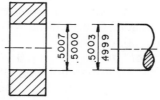

TRANSITIONAL FIT

7–99. *Clearance, force, and transitional fits.*

These letter symbols are used with numbers to represent the class of fit. A designation FN4 means a Class 4 force fit. The symbols and numbers do *not* appear on the drawing. The limits that are obtained from the tables are used to size the piece.

Description of Fits. The classes of fits are arranged in three general groups: running and sliding fits; locational fits; and force fits.

1. *Running and sliding fits* (RC) are intended to provide a running performance with a running performance with

suitable lubrication allowance in all range of sizes. The clearances of the first two classes, used chiefly as slide fits, increase more slowly with diameter than other classes.

a. RC1 *Close sliding fits* are intended for accurate location of parts which must be assembled without perceptible play (noticeable movement).

b. RC2 *Sliding fits* are intended for accurate location but with greater maximum clearance than RC1 parts that are made to fit, move, and turn easily but are not intended to run freely. In larger sizes, this fit may stick fast due to a temperature change.

c. RC3 *Precision running fits* are about the closest fits which can be expected to run freely. These are intended for precision work at slow speeds and light journal pressures. The RC3 fit is not suitable when appreciable temperature differences are likely to occur.

d. RC4 *Close running fits* are intended chiefly for running fits on accurate machinery with moderate surface speeds and journal pressures. These are used where accurate location and minimum play is desired.

e. RC5 and 6 *Medium running fits* are intended for higher running speeds and/or heavy journal pressures.

f. RC7 *Free running fits* are used where accuracy is not essential or where large temperature variations are likely to occur.

g. RC8 and 9 *Loose running fits* are used when materials such as cold rolled shafting and tubing made to commercial standards are involved.

2. *Locational fits* (LC, LT, and LN) are intended to determine only the location of mating parts. They may provide rigid or accurate location, as with interference fits or provide some freedom of location as with clearance fits. Locational fits are divided into three groups: clearance fits (LC); transitional fits (LT); and interference fits (LN).

Locational clearance fits (LC) are intended for parts which are normally stationary. These fits can be freely assembled or disassembled. They range from snug fits through medium clearance fits, to loose fastener fits when freedom of assembly is of prime importance.

Locational transitional fits (LT) are a compromise between clearance and interference fits. These are used when accuracy of location is important but either a small amount of clearance or interference is permissible.

Locational interference fits (LN) are used where accuracy of location is of prime importance. In addition these are used for parts requiring rigidity and alignment with

no requirements for bore pressure. These fits are not intended for parts designed to transmit frictional loads from one part to another. By virtue of tightness of fit, these latter conditions are covered by force fits.

3. *Force fits* (FN) or shrink fits are a special type of interference fit. These are characterized by maintenance of constant bore pressure throughout the range of sizes. The interference varies almost directly with the diameter. The difference between these values is small to maintain the resulting pressures within reasonable limits.

Light drive fits (FNI) are those requiring light assembly pressures and produce more or less permanent assemblies. These are suitable for thin sections, long fits, or in cast iron external members.

Medium drive fits (FN2) are suitable for ordinary steel parts or for shrink fits on light sections. These are about the tightest fits that can be used with high-grade cast iron external members.

Heavy drive fits (FN3) are suitable for heavier steel parts or for shorter fits in medium sections.

Force fits (FN 4 and 5) are suitable for parts which can be highly stressed or for shrink fits when heavy pressing forces required are impractical.

For interchangeable manufacture, the tolerances on dimensions are such that an acceptable fit will result from

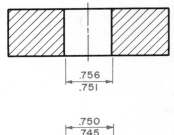

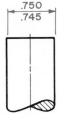

7–100. *Parts designed for interchangeability. The largest shaft is 0.750. The smallest hole is 0.751. This gives a minimum allowance of 0.001. The smallest shaft is 0.745 and the largest hole is 0.756. This gives a maximum clearance of 0.009. This is a clearance fit.*

an assembly of parts having any combination of actual sizes that are within tolerances. See Fig. 7–100.

When mating parts must fit together and do not need to be interchangeable, they may be dimensioned as shown in Fig. 7–101. For example, the $1^1/_{32}$-inch diameter need not be held to a close tolerance. This diameter will be machined to the size necessary for the desired fit based on the 1.000–.995 diameter.

Basic Hole and Basic Shaft System. To specify the dimension and tolerances of an internal and external cylindrical surface so they will fit together, it is necessary to begin calcu-

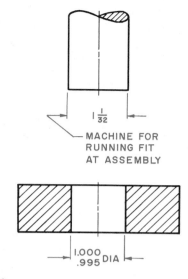

MACHINE FOR
RUNNING FIT
AT ASSEMBLY

$1\frac{1}{32}$

7–101. *Dimensioning for parts that are not interchangeable.*

1.000
.995 DIA

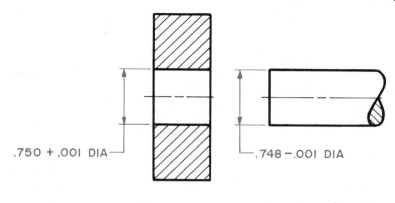

.750 + .001 DIA

.748 – .001 DIA

ALLOWANCE IS .002
CLEARANCE IS .004

7–102. *The basic hole system.*

lations with a minimum hole size or maximum shaft size. These are called the *basic hole system* or *basic shaft system.*

Selection of the size of the hole and shaft is usually in terms of a nominal size. *Nominal* size is used for general identification. It is expressed in fractions. For example, the nominal size of the hole and shaft, Fig. 7–102, is ¾ inch.

The *basic size* is the one to which limits are applied. It is a decimal. The base size, Fig. 7–102, is .750.

The *basic hole system* is for fits in which the design size of the hole is the basic size and *allowance* is applied to the shaft. See Fig. 7–102. *Allowance* is the difference between the maximum material conditions of mating parts. For example, it could

be described as the difference between the *smallest* possible hole and the *largest* shaft for a designated fit. *Clearance,* on the other hand, is the difference between minimum material conditions, such as

the *largest* hole and the *smallest* shaft for a prescribed fit.

Basic shaft system is for fits in which the design size of the shaft is the basic size and the allowance is applied to the hole. See Fig. 7–103.

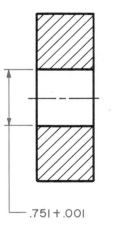

.751 + .001

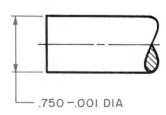

.750 – .001 DIA

ALLOWANCE IS .001
CLEARANCE IS .003

7–103. *The basic shaft system.*

Standardized Tables. The standardized data for running and sliding fits is given in Fig. 7–104. Data for locational fits is given in Figs. 7–105 through 7–107. Data for force fits is given in Fig. 7–108.

These tables are published by the American National Standards Institute. The publication is ANSI B4.1–1955, *Preferred Limits and Fits for Cylindrical Parts.* It sets forth the recommended preferred standard sizes, allowances, tolerances, and fits. They give the data for various sizes and types of fit. They set forth the standard limits for mating parts that will produce the fit specified. The tables are designed for the *basic hole system.*

How to Determine the Correct Fit. Most limit dimensions are figured on the basic hole system. The nominal size of the hole is converted to the basic theoretical hole size. The standard limits are then added or subtracted for the hole and shaft size. In the tables of standard fits, the nominal size ranges are expressed in inches along the left-hand column. To conserve space all other values (limits of tolerances and limits for the hole and shaft) are expressed in thousandths of an inch. For example 2.0 would be read as two-thousandths of an inch (.002); 1.4 would be read as fourteen ten-thousandths (.0014). To obtain the limits for a 1¾-inch shaft and hole, see Fig. 7–109.

1. Locate the RC3 fit column in the table of running and sliding fits, Fig. 7–109.

2. Find the nominal size range of the hole and shaft. 1.750 falls between 1.19–1.97. Read across to the RC3 column.

3. From the limits given for the hole (+0.6 and 0), and shaft (−1.0 and −1.6), add and subtract these to the basic size. Remember these figures are given in thousandths of an inch. These may be converted by multiplying by one-thousandth, therefore: 0.6 × .001 = .006.

Hole dimension:

$$\frac{1.7500 \text{ plus } .0006 = 1.7506}{1.7500 \text{ plus } .0000 = 1.7500}$$

Shaft dimension:

$$\frac{1.7500 \text{ minus } .0010 = 1.7490}{1.7500 \text{ minus } .0016 = 1.7484}$$

Under the heading, "Limits of Clearance," for a 1¾-inch RC3 fit, the value 1.0 and 2.2 are given. These are the minimum and maximum values of clearance between the hole and shaft. The minimum clearance (smallest hole and largest shaft) is .0010 inch. The maximum clearance (largest hole and smallest shaft) is .0022 inch.

FINISHED SURFACES

Since the early 1800's technical drawings have indicated when a surface is to be finished. A finished surface is one that is to be machined. To designate which surface is to be machined, a *finish mark* or surface symbol is used. Throughout the years a variety of marks have been used to designate finished surfaces. Finish marks have become standardized and today are universally accepted. They are used to indicate the degree of control over the surface.

There are two basic types of finish marks in use. These are noncontrol and control finish marks. They are discussed on page 179.

ANSI B4.1–1967 Preferred Limits and Fits for Cylindrical Parts

7–104. *American Standard running and sliding fits. Limits are in thousandths of an inch. Limits for hole and shaft are applied algebraically to the basic size to obtain the limits of size for the parts. The basic hole system is used. This also applies to Figs. 7–105 through 7–107.*

| Nominal Size Range Inches | Class RC 1 | | | Class RC 2 | | | Class RC 3 | | | Class RC 4 | | |
| | Limits of Clearance | Standard Limits | | Limits of Clearance | Standard Limits | | Limits of Clearance | Standard Limits | | Limits of Clearance | Standard Limits | |
Over To		Hole H5	Shaft g4		Hole H6	Shaft g5		Hole H6	Shaft f6		Hole H7	Shaft f7
0.04– 0.12	0.1 0.45	+0.2 0	−0.1 −0.25	0.1 0.55	+0.25 0	−0.1 −0.3	0.3 0.8	+0.25 0	−0.3 −0.55	0.3 1.1	+0.4 0	−0.3 −0.7
0.12– 0.24	0.15 0.5	+0.2 0	−0.15 −0.3	0.15 0.65	+0.3 0	−0.15 −0.35	0.4 1.0	+0.3 0	−0.4 −0.7	0.4 1.4	+0.5 0	−0.4 −0.9
0.24– 0.40	0.2 0.6	+0.25 0	−0.2 −0.35	0.2 0.85	+0.4 0	−0.2 −0.45	0.5 1.3	+0.4 0	−0.5 −0.9	0.5 1.7	+0.6 0	−0.5 −1.1
0.40– 0.71	0.25 0.75	+0.3 0	−0.25 −0.45	0.25 0.95	+0.4 0	−0.25 −0.55	0.6 1.4	+0.4 0	−0.6 −1.0	0.6 2.0	+0.7 0	−0.6 −1.3
0.71– 1.19	0.3 0.95	+0.4 0	−0.3 −0.55	0.3 1.2	+0.5 0	−0.3 −0.7	0.8 1.8	+0.5 0	−0.8 −1.3	0.8 2.4	+0.8 0	−0.8 −1.6
1.19– 1.97	0.4 1.1	+0.4 0	−0.4 −0.7	0.4 1.4	+0.6 0	−0.4 −0.8	1.0 2.2	+0.6 0	−1.0 −1.6	1.0 3.0	+1.0 0	−1.0 −2.0
1.97– 3.15	0.4 1.2	+0.5 0	−0.4 −0.7	0.4 1.6	+0.7 0	−0.4 −0.9	1.2 2.6	+0.7 0	−1.2 −1.9	1.2 3.6	+1.2 0	−1.2 −2.4
3.15– 4.73	0.5 1.5	+0.6 0	−0.5 −0.9	0.5 2.0	+0.9 0	−0.5 −1.1	1.4 3.2	+0.9 0	−1.4 −2.3	1.4 4.2	+1.4 0	−1.4 −2.8
4.73– 7.09	0.6 1.8	+0.7 0	−0.6 −1.1	0.6 2.3	+1.0 0	−0.6 −1.3	1.6 3.6	+1.0 0	−1.6 −2.6	1.6 4.8	+1.6 0	−1.6 −3.2
7.09– 9.85	0.6 2.0	+0.8 0	−0.6 −1.2	0.6 2.6	+1.2 0	−0.6 −1.4	2.0 4.4	+1.2 0	−2.0 −3.2	2.0 5.6	+1.8 0	−2.0 −3.8
9.85–12.41	0.8 2.3	+0.9 0	−0.8 −1.4	0.8 2.9	+1.2 0	−0.8 −1.7	2.5 4.9	+1.2 0	−2.5 −3.7	2.5 6.5	+2.0 0	−2.5 −4.5
12.41–15.75	1.0 2.7	+1.0 0	−1.0 −1.7	1.0 3.4	+1.4 0	−1.0 −2.0	3.0 5.8	+1.4 0	−3.0 −4.4	3.0 7.4	+2.2 0	−3.0 −5.2

Nominal Size Range Inches	Class RC 5			Class RC 6			Class RC 7			Class RC 8			Class RC 9		
	Limits of Clearance	Standard Limits		Limits of Clearance	Standard Limits		Limits of Clearance	Standard Limits		Limits of Clearance	Standard Limits		Limits of Clearance	Standard Limits	
Over To		Hole H7	Shaft e7		Hole H8	Shaft e8		Hole H9	Shaft d8		Hole H10	Shaft c9		Hole H11	Shaft
0.04– 0.12	0.6	+0.4	−0.6	0.6	+0.6	−0.6	1.0	+1.0	− 1.0	2.5	+1.6	− 2.5	4.0	+ 2.5	− 4.0
	1.4	0	−1.0	1.8	0	−1.2	2.6	0	− 1.6	5.1	0	− 3.5	8.1	0	− 5.6
0.12– 0.24	0.8	+0.5	−0.8	0.8	+0.7	−0.8	1.2	+1.2	− 1.2	2.8	+1.8	− 2.8	4.5	+ 3.0	− 4.5
	1.8	0	−1.3	2.2	0	−1.5	3.1	0	− 1.9	5.8	0	− 4.0	9.0	0	− 6.0
0.24– 0.40	1.0	+0.6	−1.0	1.0	+0.9	−1.0	1.6	+1.4	− 1.6	3.0	+2.2	− 3.0	5.0	+ 3.5	− 5.0
	2.2	0	−1.6	2.8	0	−1.9	3.9	0	− 2.5	6.6	0	− 4.4	10.7	0	− 7.2
0.40– 0.71	1.2	+0.7	−1.2	1.2	+1.0	−1.2	2.0	+1.6	− 2.0	3.5	+2.8	− 3.5	6.0	+ 4.0	− 6.0
	2.6	0	−1.9	3.2	0	−2.2	4.6	0	− 3.0	7.9	0	− 5.1	12.8	0	− 8.8
0.71– 1.19	1.6	+0.8	−1.6	1.6	+1.2	−1.6	2.5	+2.0	− 2.5	4.5	+3.5	− 4.5	7.0	+ 5.0	− 7.0
	3.2	0	−2.4	4.0	0	−2.8	5.7	0	− 3.7	10.0	0	− 6.5	15.5	0	−10.5
1.19– 1.97	2.0	+1.0	−2.0	2.0	+1.6	−2.0	3.0	+2.5	− 3.0	5.0	+4.0	− 5.0	8.0	+ 6.0	− 8.0
	4.0	0	−3.0	5.2	0	−3.6	7.1	0	− 4.6	11.5	0	− 7.5	18.0	0	−12.0
1.97– 3.15	2.5	+1.2	−2.5	2.5	+1.8	−2.5	4.0	+3.0	− 4.0	6.0	+4.5	− 6.0	9.0	+ 7.0	− 9.0
	4.9	0	−3.7	6.1	0	−4.3	8.8	0	− 5.8	13.5	0	− 9.0	20.5	0	−13.5
3.15– 4.73	3.0	+1.4	−3.0	3.0	+2.2	−3.0	5.0	+3.5	− 5.0	7.0	+5.0	− 7.0	10.0	+ 9.0	−10.0
	5.8	0	−4.4	7.4	0	−5.2	10.7	0	− 7.2	15.5	0	−10.5	24.0	0	−15.0
4.73– 7.09	3.5	+1.6	−3.5	3.5	+2.5	−3.5	6.0	+4.0	− 6.0	8.0	+6.0	− 8.0	12.0	+10.0	−12.0
	6.7	0	−5.1	8.5	0	−6.0	12.5	0	− 8.5	18.0	0	−12.0	28.0	0	−18.0
7.09– 9.85	4.0	+1.8	−4.0	4.0	+2.8	−4.0	7.0	+4.5	− 7.0	10.0	+7.0	−10.0	15.0	+12.0	−15.0
	7.6	0	−5.8	9.6	0	−6.8	14.3	0	− 9.8	21.5	0	−14.5	34.0	0	−22.0
9.85–12.41	5.0	+2.0	−5.0	5.0	+3.0	−5.0	8.0	+5.0	− 8.0	12.0	+8.0	−12.0	18.0	+12.0	−18.0
	9.0	0	−7.0	11.0	0	−8.0	16.0	0	−11.0	25.0	0	−17.0	38.0	0	−26.0
12.41–15.75	6.0	+2.2	−6.0	6.0	+3.5	−6.0	10.0	+6.0	−10.0	14.0	+9.0	−14.0	22.0	+14.0	−22.0
	10.4	0	−8.2	13.0	0	−9.5	19.5	0	−13.5	29.0	0	−20.0	45.0	0	−31.0

ANSI B4.1–1967 Preferred Limits and Fits for Cylindrical Parts

7–104. *American Standard running and sliding fits.*

Nominal Size Range Inches (Over–To)	Class LC 1 Limits of Clearance	Class LC 1 Standard Limits Hole H6	Class LC 1 Standard Limits Shaft h5	Class LC 2 Limits of Clearance	Class LC 2 Standard Limits Hole H7	Class LC 2 Standard Limits Shaft h6	Class LC 3 Limits of Clearance	Class LC 3 Standard Limits Hole H8	Class LC 3 Standard Limits Shaft h7	Class LC 4 Limits of Clearance	Class LC 4 Standard Limits Hole H9	Class LC 4 Standard Limits Shaft h9	Class LC 5 Limits of Clearance	Class LC 5 Standard Limits Hole H7	Class LC 5 Standard Limits Shaft g6
0.04– 0.12	0 / 0.45	+0.25 / − 0	+ 0 / −0.2	0 / 0.65	+0.4 / − 0	+ 0 / −0.25	0 / 1	+0.6 / − 0	+ 0 / −0.4	0 / 2.0	+1.0 / − 0	+ 0 / −1.0	0.1 / 0.75	+0.4 / − 0	−0.1 / −0.35
0.12– 0.24	0 / 0.5	+0.3 / − 0	+ 0 / −0.2	0 / 0.8	+0.5 / − 0	+ 0 / −0.3	0 / 1.2	+0.7 / − 0	+ 0 / −0.5	0 / 2.4	+1.2 / − 0	+ 0 / −1.2	0.15 / 0.95	+0.5 / − 0	−0.15 / −0.45
0.24– 0.40	0 / 0.65	+0.4 / − 0	+ 0 / −0.25	0 / 1.0	+0.6 / − 0	+ 0 / −0.4	0 / 1.5	+0.9 / − 0	+ 0 / −0.6	0 / 2.8	+1.4 / − 0	+ 0 / −1.4	0.2 / 1.2	+0.6 / − 0	−0.2 / −0.6
0.40– 0.71	0 / 0.7	+0.4 / − 0	+ 0 / −0.3	0 / 1.1	+0.7 / − 0	+ 0 / −0.4	0 / 1.7	+1.0 / − 0	+ 0 / −0.7	0 / 3.2	+1.6 / − 0	+ 0 / −1.6	0.25 / 1.35	+0.7 / − 0	−0.25 / −0.65
0.71– 1.19	0 / 0.9	+0.5 / − 0	+ 0 / −0.4	0 / 1.3	+0.8 / − 0	+ 0 / −0.5	0 / 2	+1.2 / − 0	+ 0 / −0.8	0 / 4	+2.0 / − 0	+ 0 / −2.0	0.3 / 1.6	+0.8 / − 0	−0.3 / −0.8
1.19– 1.97	0 / 1.0	+0.6 / − 0	+ 0 / −0.4	0 / 1.6	+1.0 / − 0	+ 0 / −0.6	0 / 2.6	+1.6 / − 0	+ 0 / −1	0 / 5	+2.5 / − 0	+ 0 / −2.5	0.4 / 2.0	+1.0 / − 0	−0.4 / −1.0
1.97– 3.15	0 / 1.2	+0.7 / − 0	+ 0 / −0.5	0 / 1.9	+1.2 / − 0	+ 0 / −0.7	0 / 3	+1.8 / − 0	+ 0 / −1.2	0 / 6	+ 3 / − 0	+ 0 / −3	0.4 / 2.3	+1.2 / − 0	−0.4 / −1.1
3.15– 4.73	0 / 1.5	+0.9 / − 0	+ 0 / −0.6	0 / 2.3	+1.4 / − 0	+ 0 / −0.9	0 / 3.6	+2.2 / − 0	+ 0 / −1.4	0 / 7	+3.5 / − 0	+ 0 / −3.5	0.5 / 2.8	+1.4 / − 0	−0.5 / −1.4
4.73– 7.09	0 / 1.7	+1.0 / − 0	+ 0 / −0.7	0 / 2.6	+1.6 / − 0	+ 0 / −1.0	0 / 4.1	+2.5 / − 0	+ 0 / −1.6	0 / 8	+4 / − 0	+ 0 / −4	0.6 / 3.2	+1.6 / − 0	−0.6 / −1.6
7.09– 9.85	0 / 2.0	+1.2 / − 0	+ 0 / −0.8	0 / 3.0	+1.8 / − 0	+ 0 / −1.2	0 / 4.6	+2.8 / − 0	+ 0 / −1.8	0 / 9	+4.5 / − 0	+ 0 / −4.5	0.6 / 3.6	+1.8 / − 0	−0.6 / −1.8
9.85–12.41	0 / 2.1	+1.2 / − 0	+ 0 / −0.9	0 / 3.2	+2.0 / − 0	+ 0 / −1.2	0 / 5	+3.0 / − 0	+ 0 / −2.0	0 / 10	+5 / − 0	+ 0 / −5	0.7 / 3.9	+2.0 / − 0	−0.7 / −1.9
12.41–15.75	0 / 2.4	+1.4 / − 0	+ 0 / −1.0	0 / 3.6	+2.2 / − 0	+ 0 / −1.4	0 / 5.7	+3.5 / − 0	+ 0 / −2.2	0 / 12	+6 / − 0	+ 0 / −6	0.7 / 4.3	+2.2 / − 0	−0.7 / −2.1

ANSI B4.1–1967 Preferred Limits and Fits for Cylindrical Parts

7–105. *American Standard clearance locational fits.*

Class LC 6 / LC 7 / LC 8 / LC 9 / LC 10 / LC 11

Nominal Size Range Inches Over To	Class LC 6 Limits of Clearance	Standard Limits Hole H8	Standard Limits Shaft f8	Class LC 7 Limits of Clearance	Standard Limits Hole H9	Standard Limits Shaft e9	Class LC 8 Limits of Clearance	Standard Limits Hole H10	Standard Limits Shaft d9	Class LC 9 Limits of Clearance	Standard Limits Hole H11	Standard Limits Shaft c11	Class LC 10 Limits of Clearance	Standard Limits Hole H12	Standard Limits Shaft	Class LC 11 Limits of Clearance	Standard Limits Hole H13	Standard Limits Shaft
0.04- 0.12	0.3	+0.6	−0.3	0.6	+1.0	−0.6	1.0	+1.6	− 1.0	2.5	+2.5	−2.5	4	+ 4	− 4	5	+ 6	− 5
	1.5	− 0	−0.9	2.6	− 0	−1.6	3.6	− 0	− 2.0	7.5	− 0	−5.0	12	− 0	− 8	17	− 0	−11
0.12- 0.24	0.4	+0.7	−0.4	0.8	+1.2	−0.8	1.2	+1.8	− 1.2	2.8	+3.0	−2.8	4.5	+ 5	−4.5	6	+ 7	− 6
	1.8	− 0	−1.1	3.2	− 0	−2.0	4.2	− 0	− 2.4	8.8	− 0	−5.8	14.5	− 0	−9.5	20	− 0	−13
0.24- 0.40	0.5	+0.9	−0.5	1.0	+1.4	−1.0	1.6	+2.2	− 1.6	3.0	+3.5	−3.0	5	+ 6	− 5	7	+ 9	− 7
	2.3	− 0	−1.4	3.8	− 0	−2.4	5.2	− 0	− 3.0	10.0	− 0	−6.5	17	− 0	−11	25	− 0	−16
0.40- 0.71	0.6	+1.0	−0.6	1.2	+1.6	−1.2	2.0	+2.8	− 2.0	3.5	+4.0	−3.5	6	+ 7	− 6	8	+10	− 8
	2.6	− 0	−1.6	4.4	− 0	−2.8	6.4	− 0	− 3.6	11.5	− 0	−7.5	20	− 0	−13	28	− 0	−18
0.71- 1.19	0.8	+1.2	−0.8	1.6	+2.0	−1.6	2.5	+3.5	− 2.5	4.5	+5.0	−4.5	7	+ 8	− 7	10	+12	−10
	3.2	− 0	−2.0	5.6	− 0	−3.6	8.0	− 0	− 4.5	14.5	− 0	−9.5	23	− 0	−15	34	− 0	−22
1.19- 1.97	1.0	+1.6	−1.0	2.0	+2.5	−2.0	3.0	+4.0	− 3.0	5	+6	− 5	8	+10	− 8	12	+16	−12
	4.2	− 0	−2.6	7.0	− 0	−4.5	9.5	− 0	− 5.5	17	− 0	− 11	28	− 0	−18	44	− 0	−28
1.97- 3.15	1.2	+1.8	−1.2	2.5	+3.0	−2.5	4.0	+4.5	− 4.0	6	+7	− 6	10	+12	−10	14	+18	−14
	4.8	− 0	−3.0	8.5	− 0	−5.5	11.5	− 0	− 7.0	20	− 0	− 13	34	− 0	−22	50	− 0	−32
3.15- 4.73	1.4	+2.2	−1.4	3.0	+3.5	−3.0	5.0	+5.0	− 5.0	7	+9	− 7	11	+14	−11	16	+22	−16
	5.8	− 0	−3.6	10.0	− 0	−6.5	13.5	− 0	− 8.5	25	− 0	− 16	39	− 0	−25	60	− 0	−38
4.73- 7.09	1.6	+2.5	−1.6	3.5	+4.0	−3.5	6	+6	− 6	8	+10	− 8	12	+16	−12	18	+25	−18
	6.6	− 0	−4.1	11.5	− 0	−7.5	16	− 0	−10	28	− 0	− 18	44	− 0	−28	68	− 0	−43
7.09- 9.85	2.0	+2.8	−2.0	4.0	+4.5	−4.0	7	+7	− 7	10	+12	−10	16	+18	−16	22	+28	−22
	7.6	− 0	−4.8	13.0	− 0	−8.5	18.5	− 0	−11.5	34	− 0	− 22	52	− 0	−34	78	− 0	−50
9.85-12.41	2.2	+3.0	−2.2	4.5	+5.0	−4.5	7	+8	− 7	12	+12	−12	20	+20	−20	28	+30	−28
	8.2	− 0	−5.2	14.5	− 0	−9.5	20	− 0	−12	36	− 0	− 24	60	− 0	−40	88	− 0	−58
12.41-15.75	2.5	+3.5	−2.5	5	+6	− 5	8	+9	− 8	14	+14	−14	22	+22	−22	30	+35	−30
	9.5	− 0	−6.0	17	− 0	−11	23	− 0	−14	42	− 0	− 28	66	− 0	−44	100	− 0	−65

ANSI B4.1–1967 Preferred Limits and Fits for Cylindrical Parts

7–105. *American Standard clearance locational fits.*

Nominal Size Range Inches		Class LT 1			Class LT 2			Class LT 3			Class LT 4			Class LT 6			Class LT 7		
			Standard Limits			Standard Limits			Standard Limits			Standard Limits			Standard Limits			Standard Limits	
Over	To	Fit	Hole H7	Shaft j6	Fit	Hole H8	Shaft j7	Fit	Hole H7	Shaft k6	Fit	Hole H8	Shaft k7	Fit	Hole H8	Shaft m7	Fit	Hole H7	Shaft n6
0.04	0.12	−0.15 / +0.5	+0.4 / −0	+0.15 / −0.1	−0.3 / +0.7	+0.6 / −0	+0.3 / −0.1							−0.55 / +0.45	+0.6 / −0	+0.55 / +0.15	−0.5 / +0.15	+0.4 / −0	+0.5 / +0.25
0.12	0.24	−0.2 / +0.6	+0.5 / −0	+0.2 / −0.1	−0.4 / +0.8	+0.7 / −0	+0.4 / −0.1							−0.7 / +0.5	+0.7 / −0	+0.7 / +0.2	−0.6 / +0.2	+0.5 / −0	+0.6 / +0.3
0.24	0.40	−0.3 / +0.7	+0.6 / −0	+0.3 / −0.1	−0.4 / +1.1	+0.9 / −0	+0.4 / −0.2	−0.5 / +0.5	+0.6 / −0	+0.5 / +0.1	−0.7 / +0.8	+0.9 / −0	+0.7 / +0.1	−0.8 / +0.7	+0.9 / −0	+0.8 / +0.2	−0.8 / +0.2	+0.6 / −0	+0.8 / +0.4
0.40	0.71	−0.3 / +0.8	+0.7 / −0	+0.3 / −0.1	−0.5 / +1.2	+1.0 / −0	+0.5 / −0.2	−0.5 / +0.6	+0.7 / −0	+0.5 / +0.1	−0.8 / +0.9	+1.0 / −0	+0.8 / +0.1	−1.0 / +0.7	+1.0 / −0	+1.0 / +0.3	−0.9 / +0.2	+0.7 / −0	+0.9 / +0.5
0.71	1.19	−0.3 / +1.0	+0.8 / −0	+0.3 / −0.2	−0.5 / +1.5	+1.2 / −0	+0.5 / −0.3	−0.6 / +0.7	+0.8 / −0	+0.6 / +0.1	−0.9 / +1.1	+1.2 / −0	+0.9 / +0.1	−1.1 / +0.9	+1.2 / −0	+1.1 / +0.3	−1.1 / +0.2	+0.8 / −0	+1.1 / +0.6
1.19	1.97	−0.4 / +1.2	+1.0 / −0	+0.4 / −0.2	−0.6 / +2.0	+1.6 / −0	+0.6 / −0.4	−0.7 / +0.9	+1.0 / −0	+0.7 / +0.1	−1.1 / +1.5	+1.6 / −0	+1.1 / +0.1	−1.4 / +1.2	+1.6 / −0	+1.4 / +0.4	−1.3 / +0.3	+1.0 / −0	+1.3 / +0.7
1.97	3.15	−0.4 / +1.5	+1.2 / −0	+0.4 / −0.3	−0.7 / +2.3	+1.8 / −0	+0.7 / −0.5	−0.8 / +1.1	+1.2 / −0	+0.8 / +0.1	−1.3 / +1.7	+1.8 / −0	+1.3 / +0.1	−1.7 / +1.3	+1.8 / −0	+1.7 / +0.5	−1.5 / +0.4	+1.2 / −0	+1.5 / +0.8
3.15	4.73	−0.5 / +1.8	+1.4 / −0	+0.5 / −0.4	−0.8 / +2.8	+2.2 / −0	+0.8 / −0.6	−1.0 / +1.3	+1.4 / −0	+1.0 / +0.1	−1.5 / +2.1	+2.2 / −0	+1.5 / +0.1	−1.9 / +1.7	+2.2 / −0	+1.9 / +0.5	−1.9 / +0.4	+1.4 / −0	+1.9 / +1.0
4.73	7.09	−0.6 / +2.0	+1.6 / −0	+0.6 / −0.4	−0.9 / +3.2	+2.5 / −0	+0.9 / −0.7	−1.1 / +1.5	+1.6 / −0	+1.1 / +0.1	−1.7 / +2.4	+2.5 / −0	+1.7 / +0.1	−2.2 / +1.9	+2.5 / −0	+2.2 / +0.6	−2.2 / +0.4	+1.6 / −0	+2.2 / +1.2
7.09	9.85	−0.7 / +2.3	+1.8 / −0	+0.7 / −0.5	−1.0 / +3.6	+2.8 / −0	+1.0 / −0.8	−1.4 / +1.6	+1.8 / −0	+1.4 / +0.2	−2.0 / +2.6	+2.8 / −0	+2.0 / +0.2	−2.4 / +2.2	+2.8 / −0	+2.4 / +0.6	−2.6 / +0.4	+1.8 / −0	+2.6 / +1.4
9.85	12.41	−0.7 / +2.6	+2.0 / −0	+0.7 / −0.6	−1.0 / +4.0	+3.0 / −0	+1.0 / −1.0	−1.4 / +1.8	+2.0 / −0	+1.4 / +0.2	−2.2 / +2.8	+3.0 / −0	+2.2 / +0.2	−2.8 / +2.2	+3.0 / −0	+2.8 / +0.8	−2.6 / +0.6	+2.0 / −0	+2.6 / +1.4
12.41	15.75	−0.7 / +2.9	+2.2 / −0	+0.7 / −0.7	−1.2 / +4.5	+3.5 / +0	+1.2 / −1.0	−1.6 / +2.0	+2.2 / −0	+1.6 / +0.2	−2.4 / +3.3	+3.5 / −0	+2.4 / +0.2	−3.0 / +2.7	+3.5 / −0	+3.0 / +0.8	−3.0 / +0.6	+2.2 / −0	+3.0 / +1.6

ANSI B4.1–1967 Preferred Limits and Fits for Cylindrical Parts

7–106. *American Standard transition locational fits.*

Nominal Size Range Inches Over To	Class LN 2 Limits of Interference	Class LN 2 Standard Limits Hole H7	Class LN 2 Standard Limits Shaft p6	Class LN 3 Limits of Interference	Class LN 3 Standard Limits Hole H7	Class LN 3 Standard Limits Shaft r6	Nominal Size Range Inches Over To	Class LN 2 Limits of Interference	Class LN 2 Standard Limits Hole H7	Class LN 2 Standard Limits Shaft p6	Class LN 3 Limits of Interference	Class LN 3 Standard Limits Hole H7	Class LN 3 Standard Limits Shaft r6
0.04–0.12	0 / 0.65	+0.4 / − 0	+0.65 / +0.4	0.1 / 0.75	+0.4 / − 0	+0.75 / +0.5	1.97– 3.15	0.2 / 2.1	+1.2 / − 0	+2.1 / +1.4	0.4 / 2.3	+1.2 / − 0	+2.3 / +1.6
0.12–0.24	0 / 0.8	+0.5 / − 0	+0.8 / +0.5	0.1 / 0.9	+0.5 / − 0	+0.9 / +0.6	3.15– 4.73	0.2 / 2.5	+1.4 / − 0	+2.5 / +1.6	0.6 / 2.9	+1.4 / − 0	+2.9 / +2.0
0.24–0.40	0 / 1.0	+0.6 / − 0	+1.0 / +0.6	0.2 / 1.2	+0.6 / − 0	+1.2 / +0.8	4.73– 7.09	0.2 / 2.8	+1.6 / − 0	+2.8 / +1.8	0.9 / 3.5	+1.6 / − 0	+3.5 / +2.5
0.40–0.71	0 / 1.1	+0.7 / − 0	+1.1 / +0.7	0.3 / 1.4	+0.7 / − 0	+1.4 / +1.0	7.09– 9.85	0.2 / 3.2	+1.8 / − 0	+3.2 / +2.0	1.2 / 4.2	+1.8 / − 0	+4.2 / +3.0
0.71–1.19	0 / 1.3	+0.8 / − 0	+1.3 / +0.8	0.4 / 1.7	+0.8 / − 0	+1.7 / +1.2	9.85–12.41	0.2 / 3.4	+2.0 / − 0	+3.4 / +2.2	1.5 / 4.7	+2.0 / − 0	+4.7 / +3.5
1.19–1.97	0 / 1.6	+1.0 / − 0	+1.6 / +1.0	0.4 / 2.0	+1.0 / − 0	+2.0 / +1.4	12.41–15.75	0.3 / 3.9	+2.2 / − 0	+3.9 / +2.5	2.3 / 5.9	+2.2 / − 0	+5.9 / +4.5

ANSI B4.1–1967 Preferred Limits and Fits for Cylindrical Parts

7–107. *American Standard interference locational fits.*

7–108. *American Standard force and shrink fits.*

ANSI B4.1–1967 Preferred Limits and Fits for Cylindrical Parts

Nominal Size Range Inches (Over–To)	Class FN 1			Class FN 2			Class FN 3			Class FN 4			Class FN 5		
	Limits of Interference	Standard Limits		Limits of Interference	Standard Limits		Limits of Interference	Standard Limits		Limits of Interference	Standard Limits		Limits of Interference	Standard Limits	
		Hole H6	Shaft		Hole H7	Shaft s6		Hole H7	Shaft t6		Hole H7	Shaft u6		Hole H7	Shaft x7
0.04– 0.12	0.05	+0.25	+0.5	0.2	+0.4	+0.85				0.3	+0.4	+ 0.95	0.5	+0.4	+ 1.3
	0.5	− 0	+0.3	0.85	− 0	+0.6				0.95	− 0	+ 0.7	1.3	− 0	+ 0.9
0.12– 0.24	0.1	+0.3	+0.6	0.2	+0.5	+1.0				0.4	+0.5	+ 1.2	0.7	+0.5	+ 1.7
	0.6	− 0	+0.4	1.0	− 0	+0.7				1.2	− 0	+ 0.9	1.7	− 0	+ 1.2
0.24– 0.40	0.1	+0.4	+0.75	0.4	+0.6	+1.4				0.6	+0.6	+ 1.6	0.8	+0.6	+ 2.0
	0.75	− 0	+0.5	1.4	− 0	+1.0				1.6	− 0	+ 1.2	2.0	− 0	+ 1.4
0.40– 0.56	0.1	+0.4	+0.8	0.5	+0.7	+1.6				0.7	+0.7	+ 1.8	0.9	+0.7	+ 2.3
	0.8	− 0	+0.5	1.6	− 0	+1.2				1.8	− 0	+ 1.4	2.3	− 0	+ 1.6
0.56– 0.71	0.2	+0.4	+0.9	0.5	+0.7	+1.6				0.7	+0.7	+ 1.8	1.1	+0.7	+ 2.5
	0.9	− 0	+0.6	1.6	− 0	+1.2				1.8	− 0	+ 1.4	2.5	− 0	+ 1.8
0.71– 0.95	0.2	+0.5	+1.1	0.6	+0.8	+1.9				0.8	+0.8	+ 2.1	1.4	+0.8	+ 3.0
	1.1	− 0	+0.7	1.9	− 0	+1.4				2.1	− 0	+ 1.6	3.0	− 0	+ 2.2
0.95– 1.19	0.3	+0.5	+1.2	0.6	+0.8	+1.9	0.8	+0.8	+ 2.1	1.0	+0.8	+ 2.3	1.7	+0.8	+ 3.3
	1.2	− 0	+0.8	1.9	− 0	+1.4	2.1	− 0	+ 1.6	2.3	− 0	+ 1.8	3.3	− 0	+ 2.5
1.19– 1.58	0.3	+0.6	+1.3	0.8	+1.0	+2.4	1.0	+1.0	+ 2.6	1.5	+1.0	+ 3.1	2.0	+1.0	+ 4.0
	1.3	− 0	+0.9	2.4	− 0	+1.8	2.6	− 0	+ 2.0	3.1	− 0	+ 2.5	4.0	− 0	+ 3.0
1.58– 1.97	0.4	+0.6	+1.4	0.8	+1.0	+2.4	1.2	+1.0	+ 2.8	1.8	+1.0	+ 3.4	3.0	+1.0	+ 5.0
	1.4	− 0	+1.0	2.4	− 0	+1.8	2.8	− 0	+ 2.2	3.4	− 0	+ 2.8	5.0	− 0	+ 4.0
1.97– 2.56	0.6	+0.7	+1.8	0.8	+1.2	+2.7	1.3	+1.2	+ 3.2	2.3	+1.2	+ 4.2	3.8	+1.2	+ 6.2
	1.8	− 0	+1.3	2.7	− 0	+2.0	3.2	− 0	+ 2.5	4.2	− 0	+ 3.5	6.2	− 0	+ 5.0
2.56– 3.15	0.7	+0.7	+1.9	1.0	+1.2	+2.9	1.8	+1.2	+ 3.7	2.8	+1.2	+ 4.7	4.8	+1.2	+ 7.2
	1.9	− 0	+1.4	2.9	− 0	+2.2	3.7	− 0	+ 3.0	4.7	− 0	+ 4.0	7.2	− 0	+ 6.0
3.15– 3.94	0.9	+0.9	+2.4	1.4	+1.4	+3.7	2.1	+1.4	+ 4.4	3.6	+1.4	+ 5.9	5.6	+1.4	+ 8.4
	2.4	− 0	+1.8	3.7	− 0	+2.8	4.4	− 0	+ 3.5	5.9	− 0	+ 5.0	8.4	− 0	+ 7.0
3.94– 4.73	1.1	+0.9	+2.6	1.6	+1.4	+3.9	2.6	+1.4	+ 4.9	4.6	+1.4	+ 6.9	6.6	+1.4	+ 9.4
	2.6	− 0	+2.0	3.9	− 0	+3.0	4.9	− 0	+ 4.0	6.9	− 0	+ 6.0	9.4	− 0	+ 8.0
4.73– 5.52	1.2	+1.0	+2.9	1.9	+1.6	+4.5	3.4	+1.6	+ 6.0	5.4	+1.6	+ 8.0	8.4	+1.6	+11.6
	2.9	− 0	+2.2	4.5	− 0	+3.5	6.0	− 0	+ 5.0	8.0	− 0	+ 7.0	11.6	− 0	+10.0
5.52– 6.30	1.5	+1.0	+3.2	2.4	+1.6	+5.0	3.4	+1.6	+ 6.0	5.4	+1.6	+ 8.0	10.4	+1.6	+13.6
	3.2	− 0	+2.5	5.0	− 0	+4.0	6.0	− 0	+ 5.0	8.0	− 0	+ 7.0	13.6	− 0	+12.0
6.30– 7.09	1.8	+1.0	+3.5	2.9	+1.6	+5.5	4.4	+1.6	+ 7.0	6.4	+1.6	+ 9.0	10.4	+1.6	+13.6
	3.5	− 0	+2.8	5.5	− 0	+4.5	7.0	− 0	+ 6.0	9.0	− 0	+ 8.0	13.6	− 0	+12.0
7.09– 7.88	1.8	+1.2	+3.8	3.2	+1.8	+6.2	5.2	+1.8	+ 8.2	7.2	+1.8	+10.2	12.2	+1.8	+15.8
	3.8	− 0	+3.0	6.2	− 0	+5.0	8.2	− 0	+ 7.0	10.2	− 0	+ 9.0	15.8	− 0	+14.0
7.88– 8.86	2.3	+1.2	+4.3	3.2	+1.8	+6.2	5.2	+1.8	+ 8.2	8.2	+1.8	+11.2	14.2	+1.8	+17.8
	4.3	− 0	+3.5	6.2	− 0	+5.0	8.2	− 0	+ 7.0	11.2	− 0	+10.0	17.8	− 0	+16.0
8.86– 9.85	2.3	+1.2	+4.3	4.2	+1.8	+7.2	6.2	+1.8	+ 9.2	10.2	+1.8	+13.2	14.2	+1.8	+17.8
	4.3	− 0	+3.5	7.2	− 0	+6.0	9.2	− 0	+ 8.0	13.2	− 0	+12.0	17.8	− 0	+16.0
9.85–11.03	2.8	+1.2	+4.9	4.0	+2.0	+7.2	7.0	+2.0	+10.2	10.2	+2.0	+13.2	16.0	+2.0	+20.0
	4.9	− 0	+4.0	7.2	− 0	+6.0	10.2	− 0	+ 9.0	13.2	− 0	+12.0	20.0	− 0	+18.0
11.03–12.41	2.8	+1.2	+4.9	5.0	+2.0	+8.2	7.0	+2.0	+10.2	12.0	+2.0	+15.2	18.0	+2.0	+22.0
	4.9	− 0	+4.0	8.2	− 0	+7.0	10.2	− 0	+ 9.0	15.2	− 0	+14.0	22.0	− 0	+20.0
12.41–13.98	3.1	+1.4	+5.5	5.8	+2.2	+9.4	7.8	+2.2	+11.4	13.8	+2.2	+17.4	19.8	+2.2	+24.2
	5.5	− 0	+4.5	9.4	− 0	+8.0	11.4	− 0	+10.0	17.4	− 0	+16.0	24.2	− 0	+22.0

Nominal Size Range Inches	Class RC 1			Class RC 2			Class RC 3		
	Limits of Clearance	Standard Limits		Limits of Clearance	Standard Limits		Limits of Clearance	Standard Limits	
Over To		Hole H5	Shaft g4		Hole H6	Shaft g5		Hole H6	Shaft f6
0.04–0.12	0.1 0.45	+0.2 0	−0.1 −0.25	0.1 0.55	+0.25 0	−0.1 −0.3	0.3 0.8	+0.25 0	−0.3 −0.55
0.12–0.24	0.15 0.5	+0.2 0	−0.15 −0.3	0.15 0.65	+0.3 0	−0.15 −0.35	0.4 1.0	+0.3 0	−0.4 −0.7
0.24–0.40	0.2 0.6	+0.25 0	−0.2 −0.35	0.2 0.85	+0.4 0	−0.2 −0.45	0.5 1.3	+0.4 0	−0.5 −0.9
0.40–0.71	0.25 0.75	+0.3 0	−0.25 −0.45	0.25 0.95	+0.4 0	−0.25 −0.55	0.6 1.4	+0.4 0	−0.6 −1.0
0.71–1.19	0.3 0.95	+0.4 0	−0.3 −0.55	0.3 1.2	+0.5 0	−0.3 −0.7	0.8 1.8	+0.5 0	−0.8 −1.3
1.19–1.97	0.4 1.1	+0.4 0	−0.4 −0.7	0.4 1.4	+0.6 0	−0.4 −0.6	1.0 2.2	+0.6 0	−1.0 −1.6

NOMINAL SIZE $1\frac{3}{4}$
BASIC SIZE 1.7500

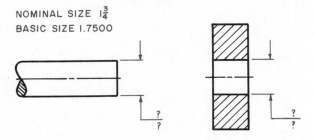

PROBLEM: FIND THE LIMITS IF AN RC 3 FIT IS DESIRED.

SHAFT

```
 1.7500      1.7500
- .0010    - .0016
 1.7490      1.7484
```

HOLE

```
 1.7500      1.7500
+ .0006      .0000
 1.7506      1.7500
```

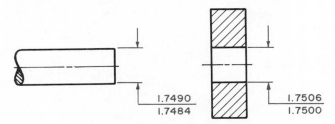

ANSI B4.1–1967 Preferred Limits and Fits for Cylindrical Parts

7–109. *How to find limits using the standard tables.*

Noncontrol Finish Marks.
These finish marks are used
when no control over the surface quality is indicated. The
noncontrol finish mark *does*
indicate a smooth finish. One
symbol used to designate that
a surface is to be machined
resembles the italic *f*. This
symbol is still used to some
extent, but is being replaced
by the newer symbol, V. The
symbols are placed on the *edge
view of the surface to be
finished,* and in each view
where that surface appears as
an edge. The finish mark is
always placed on the air side
of the surface. Fig. 7–110 illustrates the use of noncontrol finish marks. Both of these
finish marks are standardized.

Several classifications of
surfaces may be indicated by
a letter or a number near the
V or italic f. A note is usually
placed on the drawing to explain the meaning.

When the entire piece is to
have a finished surface, a note
is placed on the drawing to
this effect. It is worded FINISH
ALL OVER, or it may be abbreviated FAO. Once again,
no indication is given as to
the quality of smoothness.

Control Finish Marks. When
control of the quality of finish
is necessary, some exact
method must be used to give
this information. In general
the quality of finish depends
upon the height, width, and
direction of surface irregularities. The difference be-

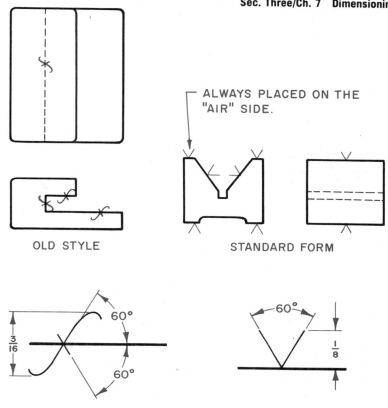

OLD STYLE   STANDARD FORM

NOTE:
THE OLD FORM FINISH MARK
LEANS FORWARD, AND IS
DRAWN FREEHAND.

NOTE:
THE STANDARD FORM IS
DRAWN MECHANICALLY
WITH A 30 – 60° TRI-
ANGLE.

7–110. *Noncontrol finish marks.*

tween a rough and smooth surface is determined by the
grade of abrasive material used
to produce the finish. A coarse
abrasive material will produce
a rough surface texture because of the size of the abrasive particles. See Fig. 7–111.
By comparison, a fine finish
is composed of more surface
irregularities that are not as
deep. The finer finish is obtained by using finer abrasive
particles.

ROUGH

FINE

7–111. *A comparison of rough
and fine finish greatly magnified.*

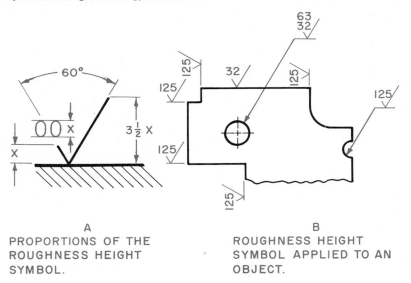

A
PROPORTIONS OF THE
ROUGHNESS HEIGHT
SYMBOL.

B
ROUGHNESS HEIGHT
SYMBOL APPLIED TO AN
OBJECT.

7–112. *The roughness symbol is used to specify the finish wanted on a surface.*

To indicate the degree of surface roughness, a standardized *roughness symbol* is used on the line indicating the surface. See Fig. 7–112. This standardized symbol is a check mark. The numerical value placed in the check mark refers to the roughness height of the finish. The long leg of the check mark is always on the right as the drawing is read. Symbols on horizontal lines or surfaces are read from the bottom of the drawing. Symbols on vertical or angular lines are read from the bottom or right side of the drawing. Surface control indications for any individual surface are in *one place only. They are not repeated in another view or section.*

When a uniform finish is to be produced over the entire piece, a note is placed on the drawing. Such a note might read: FINISH ALL OVER 32. No other finish marks or symbols indicating surface control should appear on the drawing.

When a part must be finished all over but to different degrees of roughness, a note should include the specification, such as: 63 ALL OVER EXCEPT AS NOTED. Surfaces to which the note does not apply should bear the individual symbols.

Additional values and markings are placed around the check mark to indicate type, direction, and amount of surface irregularities. This additional information includes roughness, waviness, and lay. See Fig. 7–113.

Roughness is the height rating of the relatively fine surface irregularities produced by the manufacturing process. Remember this process may be casting, forging, rolling, coating, molding, machining, grinding, buffing, burnishing, or extruding. The average roughness height is measured in *microinches.* A microinch is one millionth of an inch (0.000001). A table giving recommended roughness height values is shown in Fig. 7–114.

In addition to the roughness height, the width between these peaks is given. This distance is measured in thousandths of an inch (.001).

Waviness refers to the irregularities which are spaced farther apart than roughness peaks. Waviness may result from imperfections or vibrations of machine tools, deflections, or warping. Surface roughness may be thought of as being placed upon a wavy surface. Waviness height and width are measured in inches. The recommended waviness height values are shown in Fig. 7–115.

Lay is the direction of the predominate surface pattern produced by the tool marks. This may also be the result of the surface grain. Lay is usually dependent upon the production method used. Lay symbols are shown in Fig. 7–116.

Flaws are irregularities which occur at one place or at infrequent intervals on the surface. Flaws are such defects as cracks, scratches, holes, checks, and ridges.

BASIC ROUGHNESS SYMBOL

 ROUGHNESS HEIGHT RATING. Placed to the left of the long leg. Specification of one number indicates the maximum average. Any lesser rating is acceptable.

MAXIMUM AND MINIMUM RATINGS. Maximum and minimum average roughness ratings indicate a permissible range. Maximum rating is placed above the minimum as shown.

WAVINESS HEIGHT. Maximum waviness height valve is placed above an extension line. The value shown indicates maximum allowable waviness height.

WAVINESS WIDTH. Waviness width value, in inches, is placed above the extension line and to the right of the waviness height value. This is *not* used when contact area is specified.

CONTACT AREA. Minimum requirements for contact or bearing area are indicated by a percentage value placed above the extension line.

LAY. The lay designation symbol is indicated by the symbol placed to the right and slightly above the point of the surface roughness symbol.

SURFACE ROUGHNESS WIDTH. The surface roughness width value, in inches, is placed to the *right* of the lay symbol. The value is the maximum allowable roughness width.

ROUGHNESS WIDTH CUTOFF. The roughness width cutoff value is placed below the horizontal extension line. This number gives the maximum width in inches of surface irregularities to be included in the measurement of roughness height.

7–113. *Application of roughness, waviness, and lay symbols and their ratings.*

Roughness Height Values (in Microinches)				7–114. *Preferred series of roughness height values.*	Waviness Height Values (in Inches)		
	5	20	80	320	0.00002	0.0003	0.005
	6	25	100	400	0.00003	0.0005	0.008
1	8	32	125	500	0.00005	0.0008	0.010
2	10	40	160	600	0.00008	0.001	0.015
3	13	50	200	800	0.0001	0.002	0.020
4	16	63	250	1000	0.0002	0.003	0.030

7–115. *Preferred series of waviness height values.*

ANSI B46.1–1962 Surface Texture

Lay Symbols

Lay Symbol	Designation	Example
‖	Lay parallel to the line representing the surface to which the symbol is applied.	DIRECTION OF TOOL MARKS
⊥	Lay perpendicular to the line representing the surface to which the symbol is applied.	DIRECTION OF TOOL MARKS
X	Lay angular in both directions to line representing the surface to which symbol is applied.	DIRECTION OF TOOL MARKS
M	Lay multidirectional.	
C	Lay approximately circular relative to the center of the surface to which the symbol is applied.	
R	Lay approximately radial relative to the center of the surface to which the symbol is applied.	

ANSI B46.1–1962 Surface Texture

7–116. *Lay symbols.*

Contact area refers to the area in contact with its component surface. It is understood that the contact area is distributed equally over the surface with approximate uniformity.

Examples of Symbols and Ratings. Fig. 7–117 shows the use of control finish marks. It will be noted that in each case only those symbols and ratings which apply are used with the finish mark. The symbol is always made in a standard upright position.

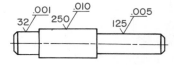

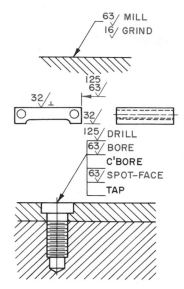

7–117. *Application of surface control finish symbols.*

Production Methods and Surface Roughness. The terms smooth and rough as used thus far in this section have been relative. What is smooth for a specific purpose may be rough for another application. It depends upon the function or purpose of the surface.

What determines the quality of a surface? Several factors influence the choice of surface roughness. The main consideration is what processes can be used to obtain the necessary dimensional characteristics. Another factor is the availability of machinery to perform that process. It is false economy to specify surfaces that are higher quality than are necessary.

Working Surfaces on bearings, pistons, and gears are typical of surfaces which require control. *Nonworking surfaces* seldom require any surface control. About the only exception is if the appearance demands a higher quality finish. Surface roughness should not be controlled on a drawing or specification sheet unless such control is essential to performance or appearance. Greater restriction of finish will increase production costs.

Roughness and smoothness are largely a result of the processing method. A machine or ground part may be rough or smooth for the purpose intended. The ideal surface characteristic for working surfaces may involve such operating conditions as:

1. Area in contact.
2. Load.
3. Speed.
4. Direction of motion.
5. Type and amount of lubricant.
6. Temperature.
7. Material and physical properties of parts.
8. Vibration.

Experimentation or experience with surfaces in similar functions are the best criteria to use in the selection procedure.

Varying machining processes will obviously produce a range of surface roughness. Fig. 7–118 shows some typical processes and their range of surface roughness values. It must be remembered that many factors can contribute to the quality of a finish produced by a specific process. For example, a drilled hole may vary between 63 and 250 microinches in roughness height. This range in roughness height may be due to:

1. Type of drill.
2. Speed of the drill.
3. Feed of the drill.
4. Condition of the drill.
5. Material.
6. Coolant.

Variation in any one of these factors can effect the surface quality. Each process listed in Fig. 7–118 has a range. Within the range of each process is an indication of the less frequent (rough), average, and more costly application. When specifying surface roughness, the recommended roughness

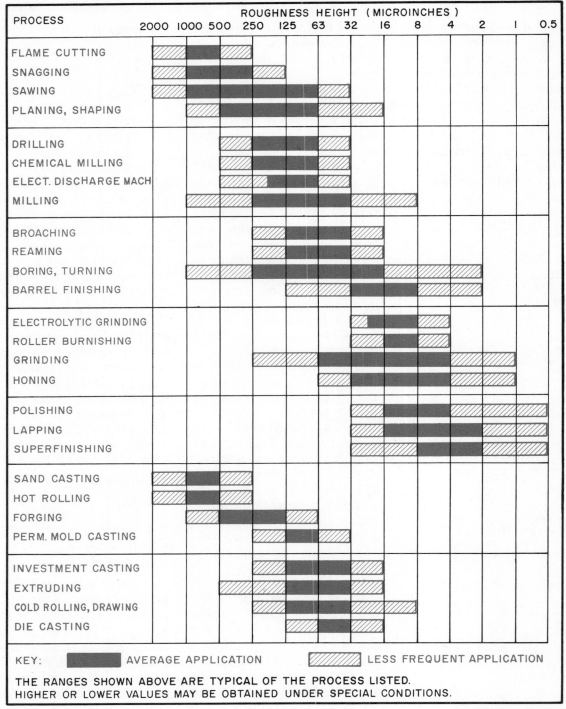

SURFACE ROUGHNESS PRODUCED BY COMMON PRODUCTION METHODS

KEY: ⬛ AVERAGE APPLICATION   ▨ LESS FREQUENT APPLICATION

THE RANGES SHOWN ABOVE ARE TYPICAL OF THE PROCESS LISTED.
HIGHER OR LOWER VALUES MAY BE OBTAINED UNDER SPECIAL CONDITIONS.

ANSI B46.1–1962 Surface Texture

7–118. *Typical machining processes and the range of surface roughness they produce.*

height values should be used. Do not specify a roughness height in between the recommended values. In specifying a roughness height for boring or turning, select 250, 125, 63, 32 or 16; do not select 200 or 75.

Additional Reading

American National Standards Institute, 1430 Broadway, New York. Publications issued by the ANSI as follows: *ISO Recommendation R 1,000* (International System of Units). *Measuring Systems and Standards Organizations. Orientation for Company Metric Studies.* American Society for Testing and Materials, 1916 Race Street, Philadelphia. *Standard Metric Practice Guide.*

Build Your Vocabulary

Following are some terms you should understand and use as a part of your working vocabulary. Write a brief explanation of what each means:

Dimension lines
Extension lines
Leaders
Metric system
Millimetre
Fractions
Decimals
Aligned system
Unidirectional system
Size dimensions
Location dimensions
Datum
Ordinate dimensioning
Tabular dimensioning
Counterbore
Countersink
Spot-face
Chamfer
Key
Taper
Knurls
Reference dimension
General notes
Local notes
Tolerance
Limit dimensions

Standard fits
Basic hole system
Basic shaft system
Noncontrol finish marks
Control finish marks
Roughness
Waviness
Lay
Flaws
Contact area

Study Problems

These problems give a variety of dimensioning situations. Their solutions will give experience in applying the techniques of dimensioning.

1–7. Solve Problems 1 through 7 by making freehand sketches on graph paper of the views given, P7–1 through P7–7. Then sketch on the dimensions. All of these objects can be dimensioned with fractions.

8. Draw the views for each part of the clamp, P7–8. Find
tolerances for the hole in the screw and the handle.
9. The marking gage head, P7–9, is formed with four radii. Draw the views and dimension completely.
10. Use a datum to locate the centers on the ball diameter gage, P7–10. Dimension in two-place decimals.
11. Dimension the lawn mower drive wheel, P7–11. Make the tolerance of the hole ± .01.
12. A plastic boat mooring buoy is shown in P7–12. Each square equals two inches. Draw half size and dimension in the metric system. Use millimetres as the basic unit.
13. The plumb bob, P7–13, can be dimensioned using millimetres. It has no critical size or location dimensions.
14. The lathe center, P7–14, has a tapered body. Draw and dimension in millimetres.
15. The cutting edge of the hollow punch must be drawn with a tolerance as shown in P7–15. Dimension the other parts with two-place dimensions. Notice the knurl.

16. The bed rail plate, P7–16, is a cast object. Dimension with fractions. Notice the countersunk screw holes.

17. A complex catch for a steel cabinet door lock is shown in P7–17. Dimension in two-place decimals. All dimensions are subject to a ± .01 tolerance.

18. The garden gate catch, P7–18, is stamped from a steel plate and bent to a 90-degree angle. Dimension completely in millimetres.

19. The drawer pull, P7–19, is cast in solid brass. Dimension completely in two-place decimals. Notice the threaded hole. Select from the thread tables found in Chapter 26 a suitable thread type.

20. The outside dimensions of the bearing journal, P7–20, can be dimensioned in two-place decimals. The bearing hole must have an RC4 fit. Locate the distance between the bolt holes with a tolerance of ± .001.

21. The clamp strap slot width, P7–21, must be accurate. Dimension with a tolerance of + .005. Dimension the other parts in two-place decimals.

22. The C washer, P7–22, does not need to be made very accurately. Use dimensions that permit low-cost manufacturing.

23. The microscope base, P7–23, is a cast metal object. Dimension it completely. The holes have a tolerance of ± .001. Notice the counterbore.

24. The pulley shown in P7–24 is used in the clutch on a motorcycle. The outside diameter must have a tolerance of ± .010. It is machined, so show the finish marks. The hole for the shaft has a tolerance of ± .001. Notice the keyseat for a square key.

25. The fixture jaw, P7–25, is machined on all surfaces. Show the finish marks. Dimension in two-place decimals.

26. All surfaces of the spherical washers shown in P7–26 are machined. Draw and dimension each separately. The spherical surfaces have a tolerance of ± .001. The other surfaces are ± .01.

27. The drill jig shown in P7–27 is used to locate holes accurately when the holes are to be drilled on a production job. Locate each hole from a datum. Dimension to three decimals. The tolerance is ± .003.

28. The center and the knurled ring of the adjustable center, P7–28, have an RC2 fit. Dimension completely with two-place decimals. Notice the threaded hole for a set screw.

29. Draw and dimension the machine handle, P7–29. It is cast with a tolerance of ± 1/64. The small end is machined to a tolerance of + .001. Notice the chamfer.

30. Locate the hole centers on the printed circuit board, P7–30, using tabular dimensions. Dimension to three decimals. All location dimensions have a ± .001 tolerance. Size dimensions have a ± .05 tolerance.

31. Plug A and B on the adjustable snap gage, P7–31, are machined to the tolerances shown. The two parts have an RC4 fit. All other dimensions have a tolerance of ± .001. Draw and dimension each part separately. The cylindrical surface of both plugs has roughness height value of 80 microinches.

32. Draw and dimension separately each part of the spring jack assembly, P7–32. Carefully dimension the tolerance of each part so they have the proper fit. The tolerance on the outer surfaces of the body and cap are ± 1/64. The tolerance on the inside hole is ± .010. All surfaces of the shaft are machined.

33. Draw and dimension separately each part of the revolving clamp assembly P7–33. The body is a cast unit. The pin and lock are machined on all surfaces. The washer is stamped. Notice the tolerance on the distance between the holes.

34. Solve the tolerance problems shown in P7–34.

35. Draw and dimension separately each part of the circle cutter, P7–35. Notice the tolerances on the machined bar.

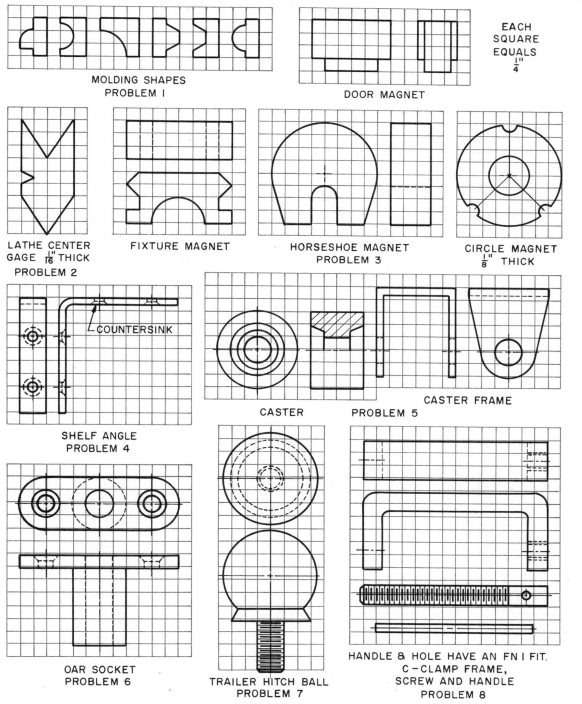

MOLDING SHAPES
PROBLEM 1

DOOR MAGNET

EACH
SQUARE
EQUALS
$\frac{1}{4}$"

LATHE CENTER
GAGE $\frac{1}{16}$" THICK
PROBLEM 2

FIXTURE MAGNET

HORSESHOE MAGNET
PROBLEM 3

CIRCLE MAGNET
$\frac{1}{8}$" THICK

COUNTERSINK

SHELF ANGLE
PROBLEM 4

CASTER

PROBLEM 5

CASTER FRAME

OAR SOCKET
PROBLEM 6

TRAILER HITCH BALL
PROBLEM 7

HANDLE & HOLE HAVE AN FN I FIT.
C-CLAMP FRAME,
SCREW AND HANDLE
PROBLEM 8

P7-1. *Dimensioning Study Problems.*

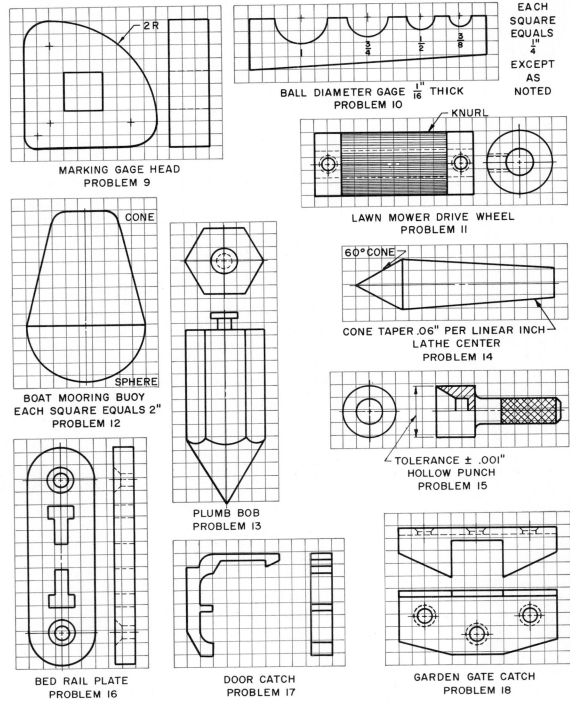

P7-2. *Dimensioning Study Problems.*

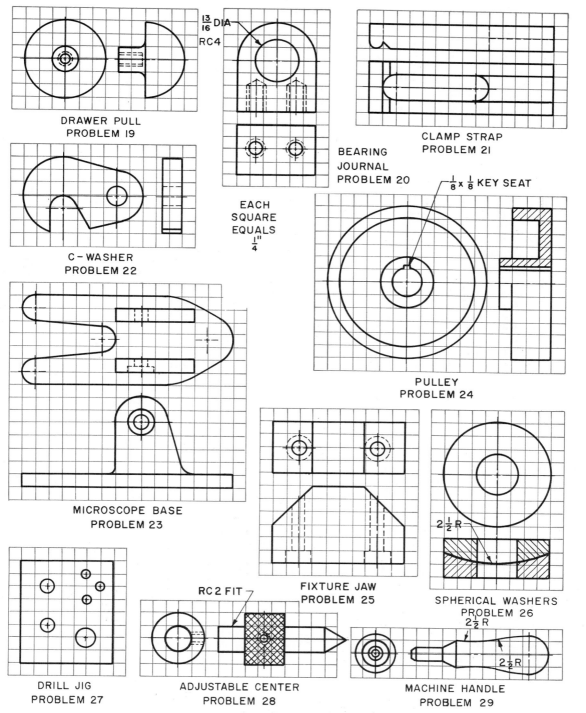

DRAWER PULL
PROBLEM 19

$\frac{13}{16}$ DIA
RC4

BEARING
JOURNAL
PROBLEM 20

EACH
SQUARE
EQUALS
$\frac{1}{4}$"

CLAMP STRAP
PROBLEM 21

C – WASHER
PROBLEM 22

$\frac{1}{8} \times \frac{1}{8}$ KEY SEAT

PULLEY
PROBLEM 24

MICROSCOPE BASE
PROBLEM 23

FIXTURE JAW
PROBLEM 25

$2\frac{1}{2}$ R

SPHERICAL WASHERS
PROBLEM 26

$2\frac{1}{2}$ R

RC 2 FIT

DRILL JIG
PROBLEM 27

ADJUSTABLE CENTER
PROBLEM 28

$2\frac{1}{2}$ R

MACHINE HANDLE
PROBLEM 29

P7–3. *Dimensioning Study Problems.*

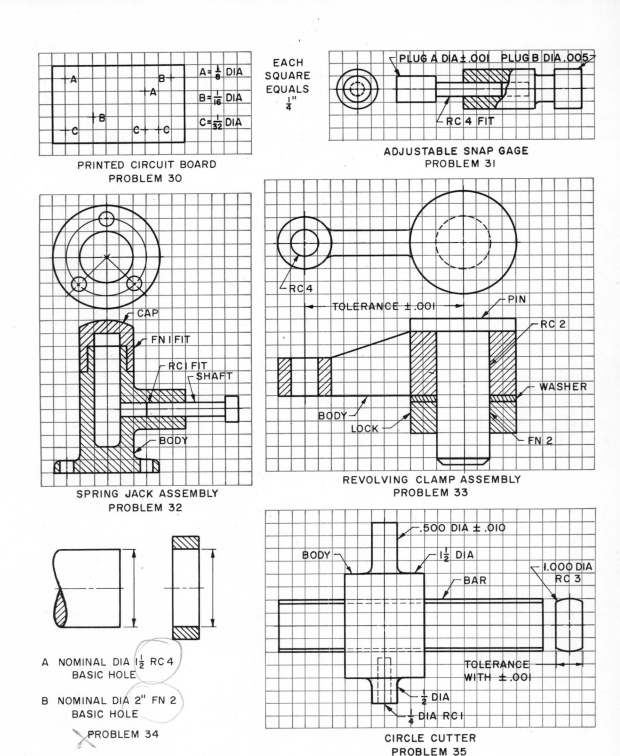

PRINTED CIRCUIT BOARD
PROBLEM 30

A = $\frac{1}{8}$ DIA

B = $\frac{1}{16}$ DIA

C = $\frac{1}{32}$ DIA

EACH
SQUARE
EQUALS
$\frac{1}{4}$"

PLUG A DIA ±.001   PLUG B DIA .005

RC 4 FIT

ADJUSTABLE SNAP GAGE
PROBLEM 31

CAP

FN 1 FIT

RC 1 FIT

SHAFT

BODY

SPRING JACK ASSEMBLY
PROBLEM 32

RC 4

TOLERANCE ±.001

PIN

RC 2

WASHER

BODY

LOCK

FN 2

REVOLVING CLAMP ASSEMBLY
PROBLEM 33

A   NOMINAL DIA 1$\frac{1}{2}$ RC 4
BASIC HOLE

B   NOMINAL DIA 2" FN 2
BASIC HOLE

PROBLEM 34

.500 DIA ±.010

BODY

1$\frac{1}{2}$ DIA

BAR

1.000 DIA
RC 3

TOLERANCE
WITH ±.001

$\frac{1}{2}$ DIA

$\frac{1}{4}$ DIA RC 1

CIRCLE CUTTER
PROBLEM 35

P7–4. *Dimensioning Study Problems.*

190

Chapter 8

Auxiliary Views, Descriptive Geometry, and Revolutions

AUXILIARY VIEWS

Some objects have surfaces that are *inclined* or *oblique* to regular planes of projection. An *inclined surface* is one that is perpendicular to *one* of the regular planes of projection. The voltmeter, Fig. 8–1, has the gage on an inclined surface. An *oblique surface* is not perpendicular to any of the regular planes of projection. The spade drill grinding fixture, Fig. 8–2, has several oblique surfaces. Study Figs. 6–6 through 6–10 in Chapter 6, Multiview Drawing, for a review of plane surfaces.

Inclined and oblique surfaces do not appear true size when projected to regular planes of projection. (Review the section in Chapter 6, Multiview Drawing, that discusses projection of planes and lines to regular planes of projection.) In order to show the true size and shape of these surfaces, auxiliary views must be drawn. An *auxiliary view* is an orthographic view with the line of sight not perpendicular to the regular planes of projection. The line of sight in an auxiliary view is perpendicular to the inclined or oblique surface. See Fig. 8–3.

An auxiliary view of an inclined surface is shown in Fig. 8–4. Notice the line of sight is perpendicular to the inclined plane. The inclined surface is shown true size in the auxiliary view.

Auxiliary views differ from regular views in one way. They are projected upon a plane that is not one of the regular frontal, horizontal, or profile planes. See Fig. 8–5. The plane of projection is called an auxiliary plane.

The principles of orthographic projection apply to *auxiliary views*. All lines of projection are perpendicular to the auxiliary plane. An auxiliary view is simply another orthographic view.

Fig. 8–5 shows a voltmeter with an inclined surface inside a glass box. An auxiliary plane is passed through the box parallel to the inclined surface. The surface is projected onto the auxiliary plane. Fig. 8–6 shows the glass box unfolded. The finished drawing is shown in Fig. 8–4.

INCLINED SURFACE

8–1. *A voltmeter with the gage mounted on an inclined surface.*

Triplett Electrical Instrument Co.

191

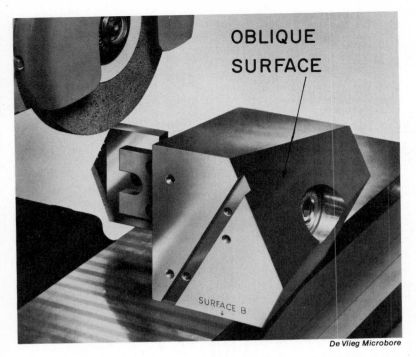

OBLIQUE
SURFACE

SURFACE B

De Vlieg Microbore

8-2. *A grinding fixture used to sharpen a spade drill. Can you identify the types of planes on the surfaces of the fixture?*

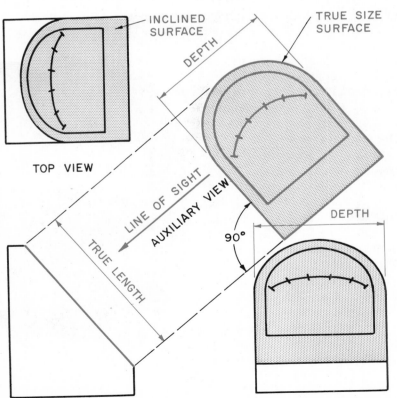

INCLINED SURFACE

TRUE SIZE SURFACE

DEPTH

TOP VIEW

LINE OF SIGHT

AUXILIARY VIEW

90°

TRUE LENGTH

DEPTH

FRONT VIEW

SIDE VIEW

HORIZONTAL PLANE

INCLINED PLANE

AUXILIARY VIEW LINE OF SIGHT

90°

FRONTAL PLANE

PROFILE PLANE

8-3. *An auxiliary view is one drawn with the line of sight perpendicular to the surface to be described.*

PRIMARY AUXILIARY VIEWS

Primary auxiliary views are those projected off a regular top, front, or side view. Primary auxiliary views are classified as elevation, right, left, front, and rear.

Elevation auxiliary views are those having the auxiliary plane perpendicular to the horizontal plane (top view). The auxiliary plane appears in edge view in the horizontal plane as shown in Fig. 8–7.

Right and left auxiliary views are those having the auxiliary plane perpendicular to the frontal reference plane (front view). The auxiliary plane appears in edge view in the frontal reference plane. This view can be a right or left auxiliary as shown in Fig. 8–8.

8-4. *An auxiliary view of the inclined surface of a voltmeter.*

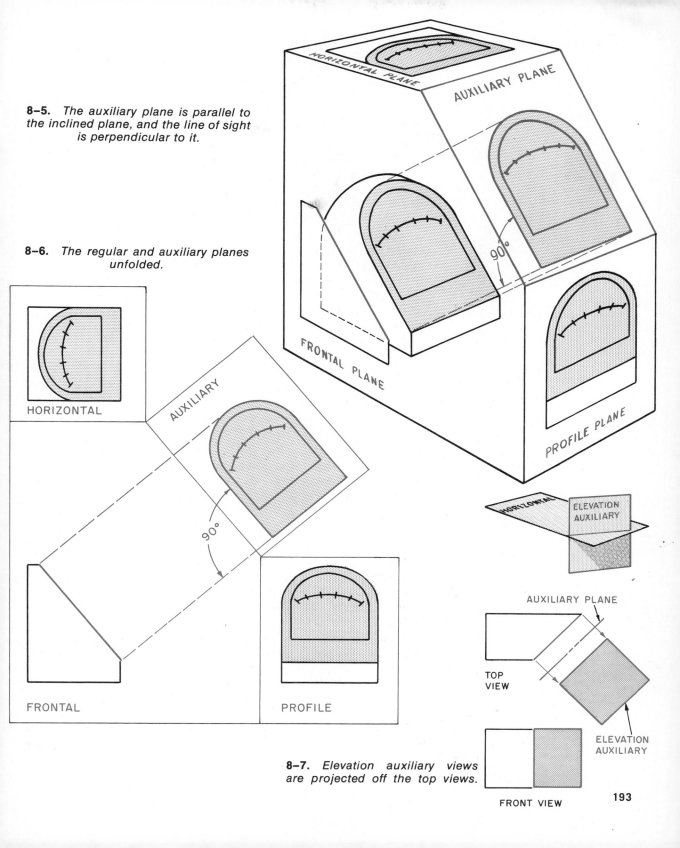

8–5. *The auxiliary plane is parallel to the inclined plane, and the line of sight is perpendicular to it.*

8–6. *The regular and auxiliary planes unfolded.*

HORIZONTAL PLANE

AUXILIARY PLANE

90°

FRONTAL PLANE

PROFILE PLANE

HORIZONTAL

AUXILIARY

90°

FRONTAL

PROFILE

HORIZONTAL

ELEVATION AUXILIARY

AUXILIARY PLANE

TOP VIEW

ELEVATION AUXILIARY

8–7. *Elevation auxiliary views are projected off the top views.*

FRONT VIEW

193

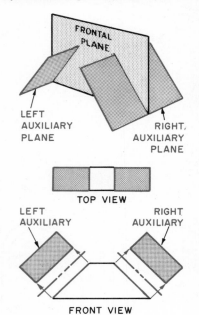

8-8. *Right and left auxiliary views are projected off the front view.*

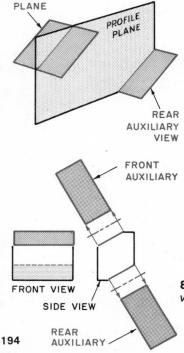

8-9. *Front and rear auxiliary views are projected off the side view.*

Front and rear auxiliary views are those having the auxiliary plane perpendicular to the profile reference plane (side view). The auxiliary plane appears in edge view in the profile reference plane as shown in Fig. 8-9.

Uses of Auxiliary Views. An auxiliary view or views helps to describe clearly the inclined or oblique surface. In Fig. 8-10 the top view is confused by hidden lines and a foreshortened surface. The auxiliary view helps clarify how the box-end wrench really appears.

Auxiliary views are used to show the true length of an edge, true shape of a surface, the true size of an angle, circular or irregularly shaped part, or the edge view of a plane.

Auxiliary views are essential to dimensioning. The inclined or oblique surface must appear true size to be properly dimensioned.

MAKING AN AUXILIARY VIEW

In making an auxiliary view, two things must be remembered. First, the auxiliary plane is drawn *parallel* to the surface to be drawn. Second, the line of sight from the auxiliary plane is *perpendicular* to the surface. These

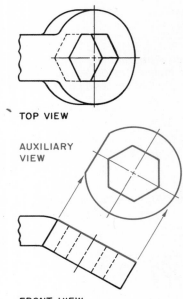

8-10. *An auxiliary view clarifies a drawing. This is a right auxiliary view.*

principles are illustrated in the problems that follow.

True Length of Oblique Lines. An oblique line is one that is not parallel to any of the regular planes of projection. Fig. 8-11 gives three regular views of an oblique line. The ends of the line are identified as C and D. These points project from view to view according to the principles of orthographic projection.

Since line CD is not parallel to any of the regular planes of projection, it is foreshortened in all views.

To find the true length of line CD, a *primary auxiliary plane of projection*, APP, must be established parallel to the line in one of the views.

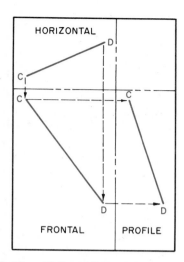

8-11. *Oblique lines appear fore-shortened in all regular views.*

When a line is parallel to a plane, it projects true length on that plane. In Fig. 8-12 an auxiliary plane, shown by line XY, was drawn in the top view parallel to line CD. Line XY represents the edge view of the HPP in the auxiliary view. Line UV represents the edge view of the HPP in the front view. Notice the location of the planes in the pictorial drawing, Fig. 8-12.

To find the true length of line CD, follow the steps illustrated in Fig. 8-12.

1. Establish a primary auxiliary plane of projection off one of the views. In Fig. 8-12 it was established off the top view. It was drawn parallel to line CD. It could have been taken off the front or side view as easily.

2. Project C and D from the top view perpendicular to the auxiliary plane.

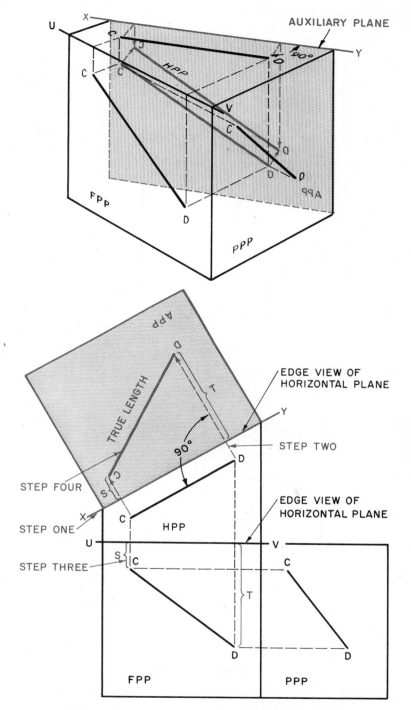

8-12. *Pictorial drawing showing an oblique line projected upon an auxiliary plane. A primary auxiliary is used to find the true length of an oblique line by projecting off the top view.*

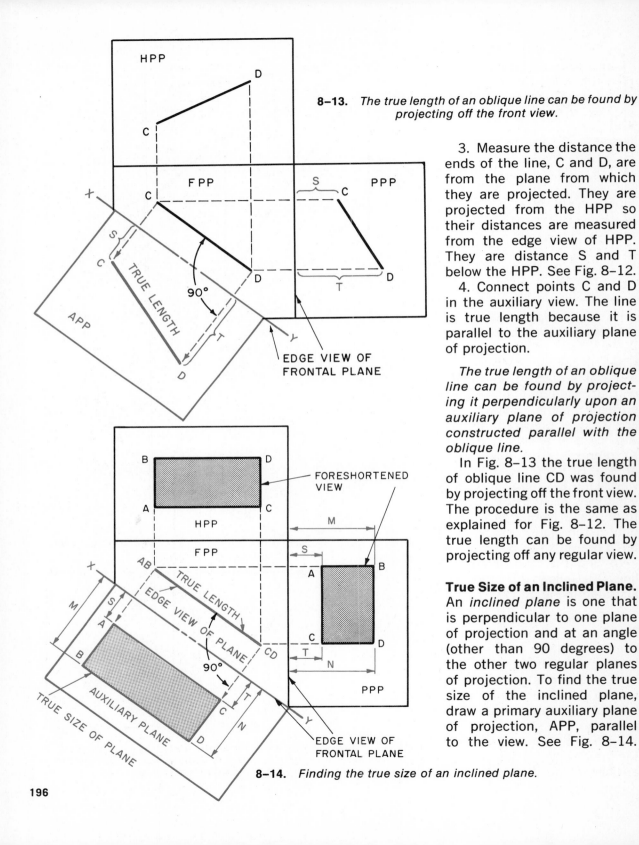

8–13. *The true length of an oblique line can be found by projecting off the front view.*

3. Measure the distance the ends of the line, C and D, are from the plane from which they are projected. They are projected from the HPP so their distances are measured from the edge view of HPP. They are distance S and T below the HPP. See Fig. 8–12.

4. Connect points C and D in the auxiliary view. The line is true length because it is parallel to the auxiliary plane of projection.

The true length of an oblique line can be found by projecting it perpendicularly upon an auxiliary plane of projection constructed parallel with the oblique line.

In Fig. 8–13 the true length of oblique line CD was found by projecting off the front view. The procedure is the same as explained for Fig. 8–12. The true length can be found by projecting off any regular view.

True Size of an Inclined Plane. An *inclined plane* is one that is perpendicular to one plane of projection and at an angle (other than 90 degrees) to the other two regular planes of projection. To find the true size of the inclined plane, draw a primary auxiliary plane of projection, APP, parallel to the view. See Fig. 8–14.

8–14. *Finding the true size of an inclined plane.*

Then project the inclined plane upon the auxiliary plane.

The auxiliary plane is drawn parallel to the edge view of the plane because the edge view is true length. Since it is true length, it will project true size on the plane parallel to it.

To find the true size of an inclined plane, follow the steps illustrated in Fig. 8–14.

1. Draw the primary auxiliary plane of projection, XY, parallel with the edge view of the inclined plane, A, B, C, D. It can be any convenient distance from true length edge view.

2. Project corners A, B, C, D from the edge view perpendicular to the auxiliary plane of projection.

3. Measure the distance corners A, B, C, D are from the edge view of the frontal plane. These distances are S, T, M, N. Locate the corners on the auxiliary plane.

4. Connect the corners A, B, C, D to form the true size of the inclined plane. This is shown pictorially in Fig. 8–15.

An inclined plane appears true size on any primary auxiliary plane to which it is parallel.

Drawing a Primary Auxiliary View. Fig. 8–16 shows a pictorial view of a bracket. This bracket has an inclined surface ABCD. To find the true size of this surface and the true shape of the bolt hole, an auxiliary plane must be drawn

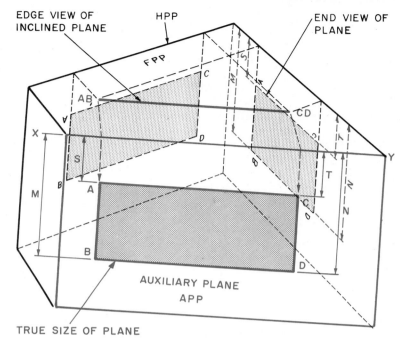

8–15. *An auxiliary plane is passed parallel to the edge view of the inclined plane.*

parallel to it. How this is done is shown in Figs. 8–17 and 8–18.

In Fig. 8–17 three regular views of the bracket are drawn. One surface, marked ABCD, is inclined and foreshortened in the top and side views. It appears as an edge in the front view. To find the true size of inclined surface ABCD the following procedure is used:

1. Draw an auxiliary plane, XY, parallel to the edge view of the inclined surface, as

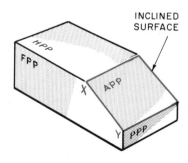

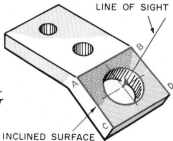

8–16. *The planes of projection. Pictorial view of the anchor bracket for an aircraft.*

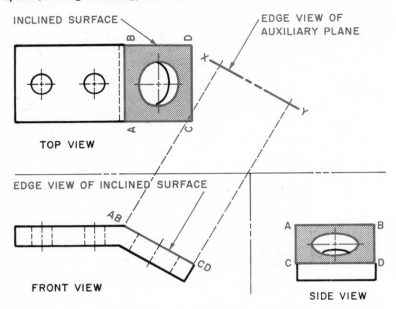

INCLINED SURFACE

EDGE VIEW OF AUXILIARY PLANE

TOP VIEW

EDGE VIEW OF INCLINED SURFACE

FRONT VIEW

SIDE VIEW

8–17. *The edge view of the auxiliary plane is drawn parallel to the edge view of the inclined surface.*

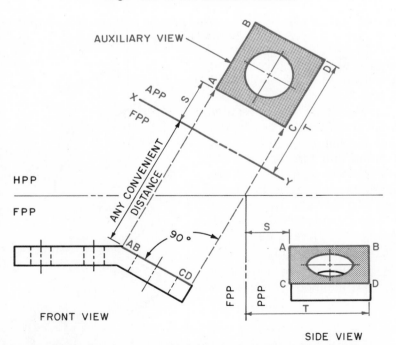

AUXILIARY VIEW

HPP

FPP

ANY CONVENIENT DISTANCE

90°

FRONT VIEW

SIDE VIEW

8–18. *The corners of the inclined surface are projected onto the auxiliary plane. They are located by measuring the distance from the frontal plane. This is a right auxiliary view.*

shown in Fig. 8–17. It can be located any convenient distance from the inclined plane. The auxiliary plane was drawn in the front view because it is here that the inclined surface appears as in edge view. Here it is true length.

2. Project the corners of the inclined surface perpendicular to the edge view of the auxiliary plane, XY. This is shown in Fig. 8–18.

3. Edge AC of the inclined surface is distance S from the frontal plane. (See the side view.) Since the edge view of the auxiliary plane is also a frontal plane, locate AC in the auxiliary view by measuring distance S from it.

4. Measure the distance T from the edge view of the frontal plane. Lay out this distance on the auxiliary view to locate corners B and D.

5. Connect corners ABCD in the auxiliary view. This gives the true size of the surface.

Full Auxiliary Views. In Fig. 8–18 the auxiliary view shows only the inclined surface. Since other parts of the object appear true size in the regular views, they are not projected to the auxiliary view. Fig. 8–19 illustrates an auxiliary of the entire object. Notice that all surfaces except the inclined surface are foreshortened. This does not make the drawing clearer or easier to understand. The purpose of the auxiliary view is to show the inclined surface true size.

Usually it is not necessary to draw full auxiliary views.

Partial Auxiliary Views. Usually an auxiliary view is used to show details of one part of an object. Such views are called partial auxiliary views. The true size surface, Fig. 8–18 is a partial auxiliary view.

If a surface to be drawn in an auxiliary view is symmetrical, it is common practice to use a reference plane. The reference plane is drawn through the center of the auxiliary view and the regular view, showing the foreshortened view of the surface. It is drawn parallel to the edge view of the surface. The view is laid out to the right and left of the reference plane auxiliary. See Fig. 8–20.

Fig. 8–20 shows a steel I-beam. Since it is symmetrical, the reference plane is placed through the center. The distances right and left of the center line are indicated by the letters X and Y. The distance X equals distance Y.

Occasionally only half of a symmetrical auxiliary view is drawn to save time and space on the paper. See Fig. 8–21. Since both halves of the flange are the same, it is only necessary to draw one half. Actually, the side view adds nothing to the clarity of the drawing and could be omitted.

Auxiliary views that are non-symmetrical can also be drawn using a reference plane.

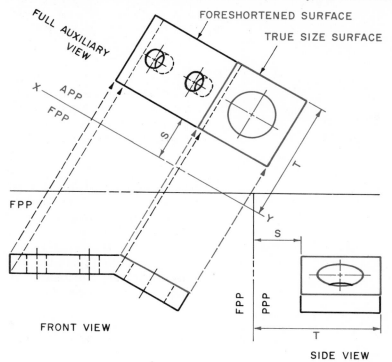

8–19. *An auxiliary view of the entire object.*

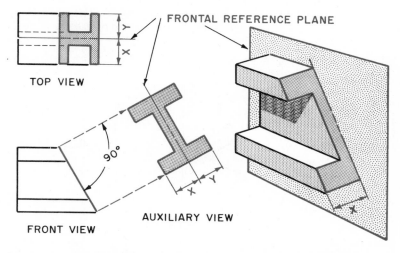

8–20. *A symmetrical auxiliary view can be constructed around a central reference plane. This is a right auxiliary view.*

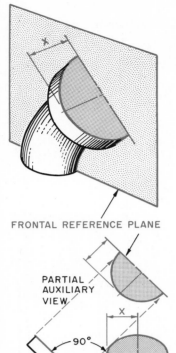

FRONTAL REFERENCE PLANE

PARTIAL
AUXILIARY
VIEW

90°

x

FRONTAL
REFERENCE
PLANE

FRONT VIEW   SIDE VIEW

8–21. *A half auxiliary view can be used to describe some symmetrical objects.*

If such a view is unilateral, it is drawn entirely on one side of the reference plane. A *unilateral object* is a one-sided part that has all projections falling in one direction. If it is bilateral, it is projected right and left of the reference plane. A *bilateral object* is one that has projections relating to two sides. See examples in Fig. 8–22.

Some objects can be clearly described by one regular view and partial views. Fig. 8–23 gives such a case. The full regular front view of the tack puller would add nothing to the clarity of the drawing. The top view is completely unnecessary and would take time to draw and occupy space on the paper. An alert draftsman will be quick to see the advantage of drawing only the side view, a partial front view, and the partial auxiliary.

Auxiliaries can be dimensioned. Normally dimensions are not given on inclined faces for location of holes or other features in any regular view.

Auxiliary Sections. Another use for auxiliary views is to show a section through an object. Fig. 8–24 shows the section through the hammer that is not parallel with a normal plane. The cutting-plane line is drawn through the place at which the section is desired. The auxiliary view is projected perpendicular to the cutting-plane line. Actually the cutting-plane line serves as the edge view of the auxiliary plane.

8–22. *An auxiliary on a unilateral object is drawn with the reference plane on one side of the object. An auxiliary of a bilateral object is drawn with projections on both sides of the reference plane.*

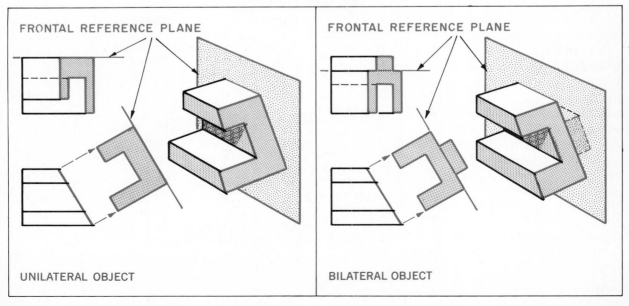

FRONTAL REFERENCE PLANE

FRONTAL REFERENCE PLANE

UNILATERAL OBJECT

BILATERAL OBJECT

TOP VIEW IS NOT NEEDED.

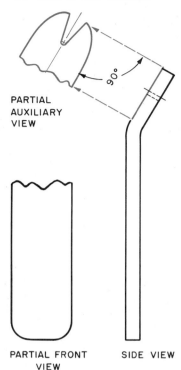

PARTIAL
AUXILIARY
VIEW

PARTIAL FRONT
VIEW

SIDE VIEW

8-23. *Drawing of a thumbtack puller uses auxiliary view to replace regular view.*

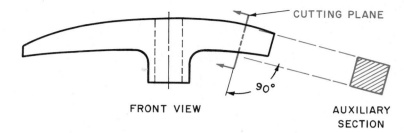

CUTTING PLANE

FRONT VIEW

90°

AUXILIARY
SECTION

8-24. *Upholsterer's hammer with an auxiliary section through one side.*

Curved Surfaces in Auxiliary Views. Curved surfaces present special problems. If they are perpendicular to the inclined surface, as the hole shown in Fig. 8-25, no special problem occurs because they appear as true shape in the auxiliary view.

If the curved surfaces meet the inclined plane on an angle or are not circular, they must be projected to the auxiliary view using the point method. This is discussed in the chapter on multiview drawings. The

curve to be projected should be divided into parts as shown in Fig. 8-26. Any number of points can be used. The more points used the more accurate the curve will be.

In Fig. 8-26, eleven numbered points are projected into the view in which the curve appears as an edge. This is the front view of the spotlight reflector shell.

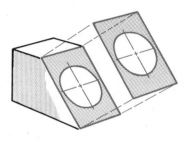

AUXILIARY VIEW

90°
90°

8-25. *Curved surfaces parallel with an inclined surface appear true size on the auxiliary view of the surface.*

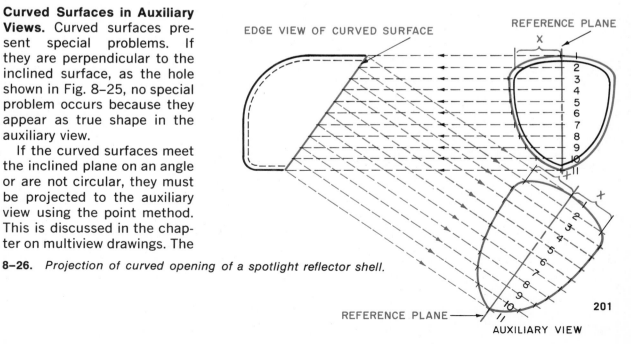

EDGE VIEW OF CURVED SURFACE

REFERENCE PLANE

X

REFERENCE PLANE

AUXILIARY VIEW

8-26. *Projection of curved opening of a spotlight reflector shell.*

201

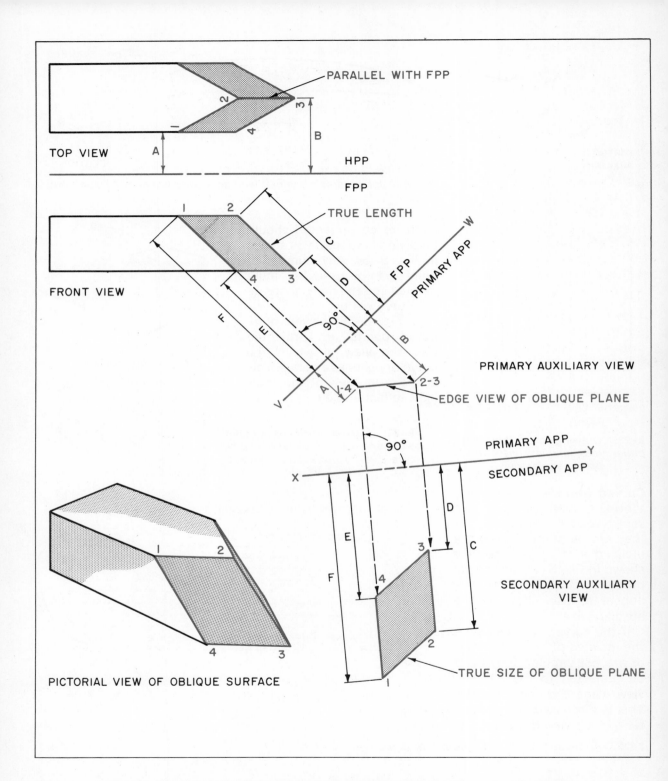

PARALLEL WITH FPP

TOP VIEW

HPP

FPP

TRUE LENGTH

FRONT VIEW

PRIMARY APP

PRIMARY AUXILIARY VIEW

EDGE VIEW OF OBLIQUE PLANE

PRIMARY APP

SECONDARY APP

PICTORIAL VIEW OF OBLIQUE SURFACE

SECONDARY AUXILIARY VIEW

TRUE SIZE OF OBLIQUE PLANE

8–27. *The true size of an oblique plane is found in the secondary auxiliary view.*

Next project the points from the edge view of the curved surface into the auxiliary view. Measure the distance each point is from the reference plane in the same manner as locating any part of an auxiliary view. This is the distance X, Fig. 8–26. Only half the surface needs to be measured because it is symmetrical. After the points are laid out, connect them with an irregular curve.

SECONDARY AUXILIARY VIEWS

Some surfaces are in such a position that they cannot be shown true size by projecting them upon normal planes or primary auxiliary planes. An oblique surface is a common example. Sometimes it is desired to view an object from a line of sight that is oblique to the normal planes. To do either of these, it is necessary to draw a secondary auxiliary. A *secondary auxiliary* is one that is projected off a primary auxiliary.

The procedure for drawing a secondary auxiliary view is the same as that for a primary auxiliary. *The only views needed are the two preceding views*. The views preceding the secondary auxiliary are the primary auxiliary and the front view, Fig. 8–27.

True Size of Oblique Planes. Fig. 8–27 shows a machine lathe tool. The end is ground so it contains two oblique surfaces. To find the true size of these surfaces, work with one of them. Surface 1, 2, 3,

4 was chosen. The general procedure is to find the edge view of the oblique surface by making a primary auxiliary. Then take a second auxiliary off the primary auxiliary. The line of sight is perpendicular to the edge view in the primary auxiliary. It is necessary to get the oblique plane in edge view because the only way to find the true size of a surface is to look perpendicular to it when it appears as an edge.

The steps to find the true size of an oblique plane are:

1. Find the oblique surface in its *edge view*. This is done by drawing a primary auxiliary plane, VW, perpendicular to a *true length line, 2, 3,* on the oblique plane. In plane 1, 2, 3, 4, edges 2, 3, and 1, 4 are true length in the front view because they are parallel with the frontal plane of projection. It is necessary to draw the auxiliary plane perpendicular to a true length line because in this way a plane will appear as an edge. In other words, when you look perpendicular to a true length line on a plane, you see the plane as an edge.

2. Corners 1, 4 are distance A from the frontal reference plane in the top view. Reference line VW is also an edge view of the frontal reference plane. Corners 1, 4 are therefore distance A from the FPP in the primary auxiliary view.

3. Corners 2, 3 are distance B from the FPP. See the top view.

4. Locate distances A and B in the primary auxiliary. This locates the plane in its edge view.

5. To see the plane true size, draw a *secondary auxiliary* parallel to the edge view of the plane. This is auxiliary plane XY.

6. In the front view the distances of the corners of the oblique plane from the primary auxiliary plane of projection can be measured. Corner 1 is distance F, 2 is distance C, 3 is distance D, and 4 is distance E from the primary APP.

7. Measure these distances from the primary auxiliary plane of projection in the secondary auxiliary view.

8. Connect the corners of the plane and it appears true size.

Fig. 8–28 shows a secondary auxiliary view of a control cable support. The purpose of making the secondary auxiliary is to see the oblique surface true size. In this way the *true size of the angle* can be found.

The steps to find the true size of the angle are:

1. Draw the top and front views.

2. Find a true length line in one of the views. Edge 6, 7 in the top view is chosen.

3. Construct an auxiliary plane of projection, WX, perpendicular to the true length line.

4. Draw the primary auxiliary view. The HPP appears

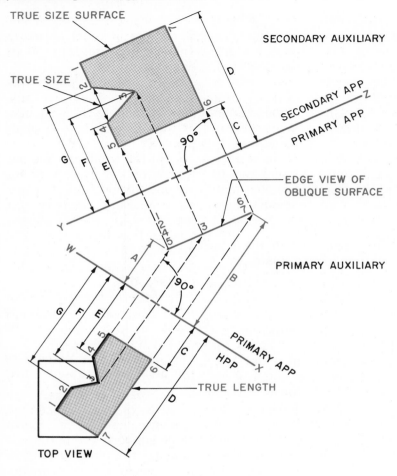

as an edge in the auxiliary view. It is the plane from which to measure. It also appears as an edge in the front view. Measure the distances each point of the plane are from the HPP. Locate these in the primary auxiliary.

5. Connect the points in the primary auxiliary. The oblique surface appears as a line.

6. Draw a secondary auxiliary plane of projection parallel with the edge view of the oblique surface.

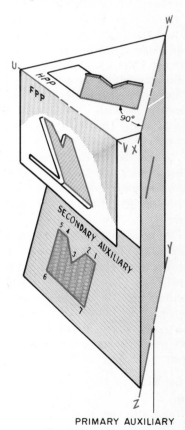

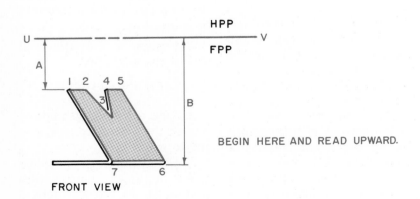

8-28. *The true size angle in the control cable support is found in the secondary auxiliary view.*

8-29. *Pictorial drawing of planes of projection used in Fig. 8-28.*

7. Project each point on the oblique surface perpendicular to the secondary APP.

8. The APP appears as an edge in the secondary auxiliary. It also appears as an edge in the second view back, the top view. Measure the distance each point is from the APP and locate in the secondary auxiliary.

9. Connect the points of the oblique surface. It is true size. The angle is true size.

The relationship between the planes is shown in Fig. 8–29.

DESCRIPTIVE GEOMETRY

Secondary auxiliary views are used to show true sizes, shapes, intersections, distances, piercing points, and angles of oblique surfaces and lines. Secondary auxiliary views are based on the theory of orthographic projection. This theory forms the basis of descriptive geometry. *Descriptive geometry* graphically shows the solution of space problems. Space problems involve the relationships of points, lines, and planes in space.

Actual Shortest Distance From a Point to a Line. Occasionally in solving design problems it becomes necessary to find the true distance between a point on an object and an edge of another object. A practical problem is presented in Fig. 8–30: how to transfer a fluid from the tank at point A to the pipe BC,

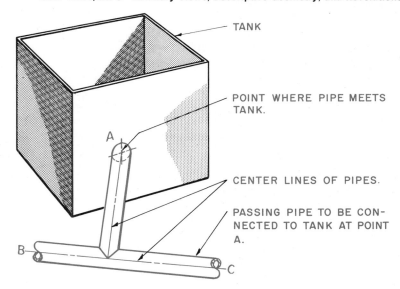

TANK

POINT WHERE PIPE MEETS TANK.

CENTER LINES OF PIPES.

PASSING PIPE TO BE CONNECTED TO TANK AT POINT A.

8–30. *Pictorial illustration of practical problem requiring the shortest distance from a point to a line.*

using the shortest connecting pipe possible.

The shortest distance between a point and a line is a line from the point perpendicular to the line. Find the line (in the problem, the pipe BC) in its point view. A straight line between the point (A) and the point view of the line (BC) is the shortest distance and can be measured on the drawing.

To determine shortest distance, see Figs. 8–31 and 8–32.

1. Find the true length of the section of pipe, BC, by constructing a primary auxiliary plane parallel to it. Use the center line of the pipe for measurements.

A line will appear true length on any plane to which it is

parallel. See Fig. 8–32. The connecting pipe is drawn perpendicular to pipe BC in this view because BC is true length. The connecting pipe is not true length in the primary auxiliary.

2. Construct a secondary auxiliary plane perpendicular to the true length of pipe BC. This will make the center line appear as a point. Whenever you look perpendicular to a true length line, it will appear as a point. See Fig. 8–32.

3. Carry point A into the secondary auxiliary view.

4. The distance between point A and the point view of pipe BC is the shortest distance. The connecting pipe is true length in this view; it can be measured on the drawing.

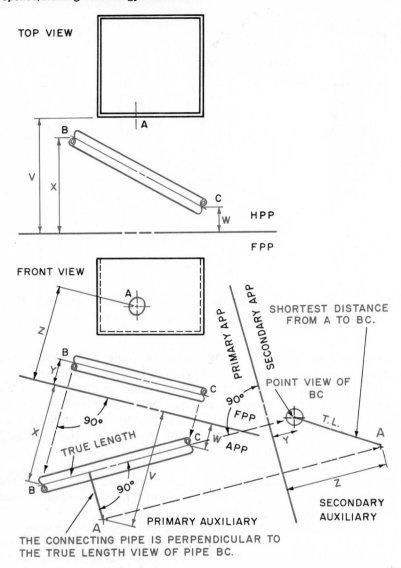

TOP VIEW

FRONT VIEW

HPP

FPP

V

X

W

B

C

A

Z

Y

X

B

C

A

90°

TRUE LENGTH

90°

C

W

V

90°

B

A

PRIMARY APP

SECONDARY APP

SHORTEST DISTANCE
FROM A TO BC.

POINT VIEW OF
BC

90°

FPP

T.L.

APP

Y

A

Z

SECONDARY
AUXILIARY

PRIMARY AUXILIARY

THE CONNECTING PIPE IS PERPENDICULAR TO
THE TRUE LENGTH VIEW OF PIPE BC.

8–31. *The shortest distance from a point to a line can be found in the secondary auxiliary view.*

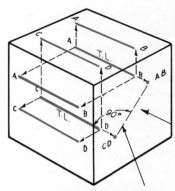

SHORTEST DISTANCE BETWEEN
LINES.

T.L. IS TRUE LENGTH.

8–32. *Lines that are parallel in one view are parallel in all regular and auxiliary views. The shortest distance between parallel lines is found by looking perpendicular to the true length of the lines.*

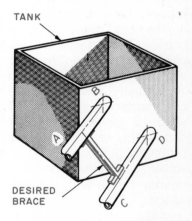

TANK

DESIRED
BRACE

8–33. *Pictorial illustration of a practical problem requiring the shortest distance between parallel lines.*

Shortest Distance Between Parallel Lines.

Lines or edges that are parallel appear parallel in all views. Sometimes a designer must ascertain the shortest distance between parallel elements of an object. Two pipes run from a tank, Fig. 8–33. It is necessary to find the distance between these pipes so a brace can be built.

The shortest distance between parallel lines is a line perpendicular to the lines.

The shortest distance between parallel lines can be found by measuring the dis-

206

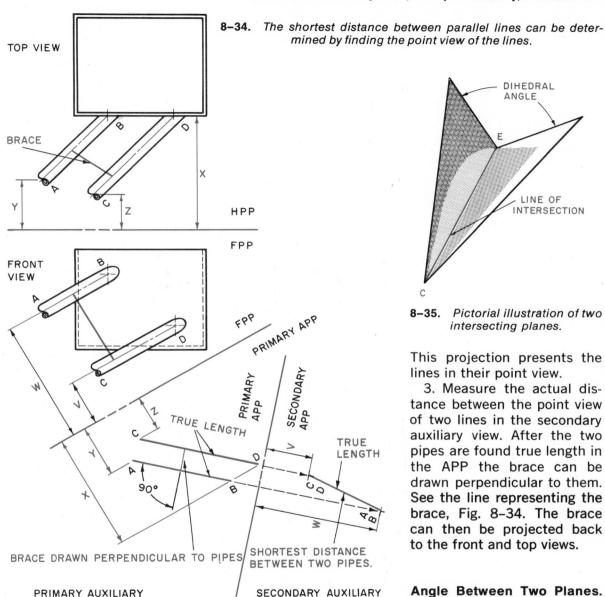

TOP VIEW

BRACE

HPP

FPP

FRONT VIEW

PRIMARY AUXILIARY   SECONDARY AUXILIARY

BRACE DRAWN PERPENDICULAR TO PIPES

SHORTEST DISTANCE BETWEEN TWO PIPES.

8-34. *The shortest distance between parallel lines can be determined by finding the point view of the lines.*

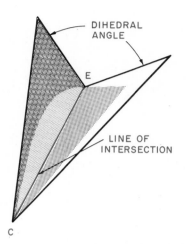

DIHEDRAL ANGLE

LINE OF INTERSECTION

8-35. *Pictorial illustration of two intersecting planes.*

This projection presents the lines in their point view.

3. Measure the actual distance between the point view of two lines in the secondary auxiliary view. After the two pipes are found true length in the APP the brace can be drawn perpendicular to them. See the line representing the brace, Fig. 8-34. The brace can then be projected back to the front and top views.

Angle Between Two Planes.
Some objects have two planes intersecting at an unknown angle. This angle is called a *dihedral angle.* See Fig. 8-35. The size of the dihedral angle is found by locating the line of intersection between the two planes in its point view.

tance between the lines when they appear in point view.

To determine the actual shortest distance between parallel lines:

1. Find the true length of the lines, Fig. 8-34, by con-

structing a primary auxiliary view parallel to the center lines. Use the center lines of the pipe for measurements.

2. Construct a second auxiliary perpendicular to the true length view of the lines.

207

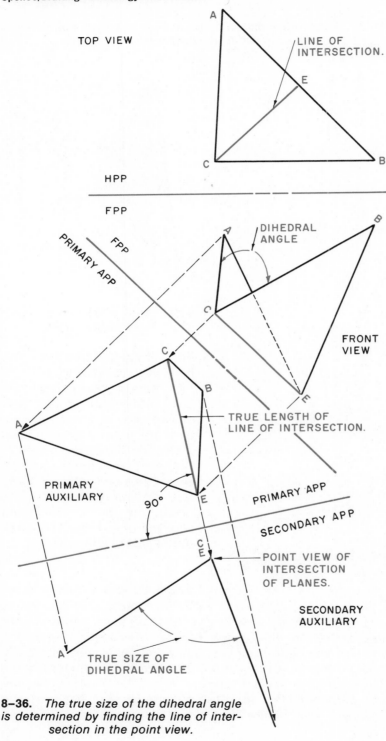

TOP VIEW

LINE OF INTERSECTION.

HPP

FPP

FPP

PRIMARY APP

DIHEDRAL ANGLE

FRONT VIEW

TRUE LENGTH OF LINE OF INTERSECTION.

PRIMARY AUXILIARY

90°

PRIMARY APP

SECONDARY APP

POINT VIEW OF INTERSECTION OF PLANES.

SECONDARY AUXILIARY

TRUE SIZE OF DIHEDRAL ANGLE

8–36. *The true size of the dihedral angle is determined by finding the line of intersection in the point view.*

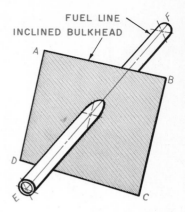

FUEL LINE

INCLINED BULKHEAD

8–37. *Pictorial illustration of fuel line piercing an inclined bulkhead.*

Then the angle can be measured on the drawing. See Fig. 8–36.

To find the dihedral angle:

1. The planes intersect at line CE. Draw a primary auxiliary to find the true length of this line. This is done by establishing an auxiliary plane parallel to CE and projecting the object into the primary auxiliary view.

2. Draw a secondary auxiliary line perpendicular to the true length of CE. This gives a point view of line CE.

3. When the planes BCE and ACE are projected into the secondary auxiliary, they appear as edge views and the true angle between them can be measured.

Location of the Point Where a Line Pierces a Plane. Sometimes a line, such as a fuel line in an aircraft, must pass through a bulkhead. In order to locate exactly where the line will pierce the bulkhead,

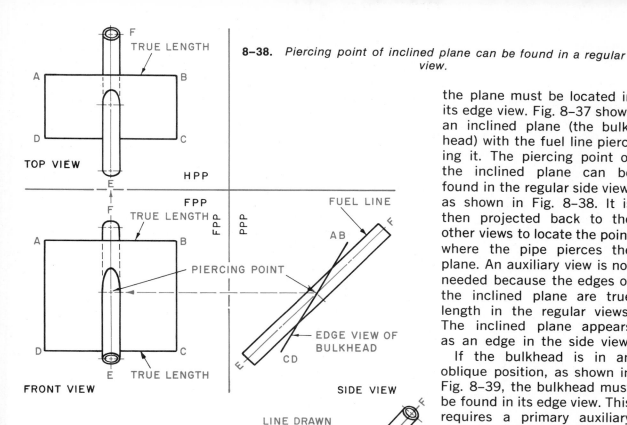

TOP VIEW

FRONT VIEW

F TRUE LENGTH

A B

D C

HPP

FPP

F TRUE LENGTH

A B

D C

E TRUE LENGTH

FPP

PPP

8-38. *Piercing point of inclined plane can be found in a regular view.*

FUEL LINE

F

AB

PIERCING POINT

EDGE VIEW OF BULKHEAD

CD

E

SIDE VIEW

the plane must be located in its edge view. Fig. 8–37 shows an inclined plane (the bulkhead) with the fuel line piercing it. The piercing point of the inclined plane can be found in the regular side view, as shown in Fig. 8–38. It is then projected back to the other views to locate the point where the pipe pierces the plane. An auxiliary view is not needed because the edges of the inclined plane are true length in the regular views. The inclined plane appears as an edge in the side view.

If the bulkhead is in an oblique position, as shown in Fig. 8–39, the bulkhead must be found in its edge view. This requires a primary auxiliary view. To find the piercing point of a line and an oblique plane:

1. Locate or construct a true length line on plane ABCD. In Fig. 8–40 the oblique plane shown has no true length lines in either the top or front

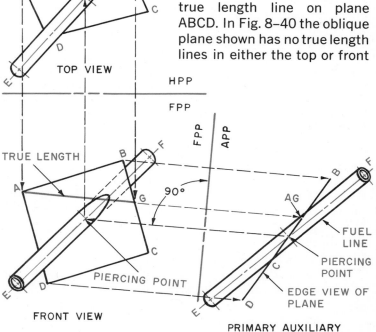

OBLIQUE BULKHEAD

FUEL LINE

F

A

B

D

C

E

8-39. *Pictorial illustration of a fuel line piercing an oblique bulkhead.*

8-40. *The piercing point of an oblique plane can be found in an auxiliary view.*

LINE DRAWN PARALLEL TO HPP

A B G

C

D

E

TOP VIEW

HPP

FPP

FPP

APP

TRUE LENGTH

B F

A

G

C

90°

D

FRONT VIEW

PIERCING POINT

E

B F

AG

FUEL LINE

PIERCING POINT

C

D

EDGE VIEW OF PLANE

E

PRIMARY AUXILIARY

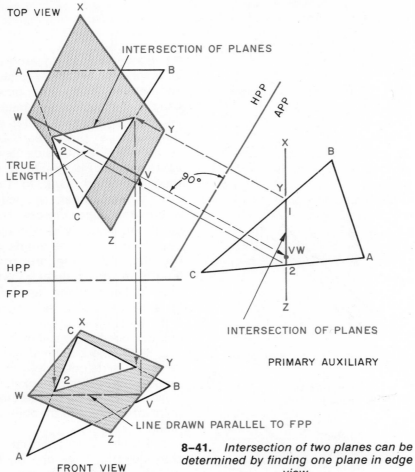

TOP VIEW

INTERSECTION OF PLANES

TRUE LENGTH

INTERSECTION OF PLANES

PRIMARY AUXILIARY

90°

HPP
APP

VW

HPP
FPP

LINE DRAWN PARALLEL TO FPP

FRONT VIEW

8–41. *Intersection of two planes can be determined by finding one plane in edge view.*

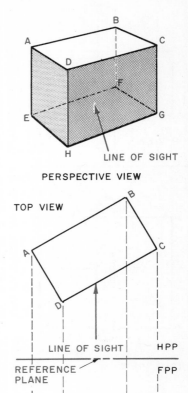

PERSPECTIVE VIEW

LINE OF SIGHT

TOP VIEW

LINE OF SIGHT

REFERENCE PLANE

HPP
FPP

FRONT VIEW

8–42. *Visibility may be found by inspection.*

view. It was necessary to construct such a line. In the top view line AG is drawn parallel to the FPP. When this is projected on to the front view, it appears true length.

2. *If the line of sight is parallel to a true length line on a plane, the plane appears in edge view.* To find the edge view, construct an auxiliary plane perpendicular to the true length line.

3. Project the plane ABCD on to the auxiliary plane.

4. Project the center line of the fuel line, EF, on to the auxiliary plane. Where it crosses the edge view of the plane is the piercing point.

5. To locate the piercing point in the regular views, project it back to them from the auxiliary view.

Intersection of Planes. It is a common task in design problems to have two planes intersect. To describe the object, it is necessary to locate the line of intersection between these planes.

The line of intersection of two planes is found by locating one plane in its edge view and noting where the second plane is cut by the edge view. To do this:

1. Select one of the planes and show it in edge view in a primary auxiliary. In Fig. 8–41 plane WXYZ is selected.

2. Locate or construct a true length line on plane WXYZ. Line WV is drawn in

210

the front view parallel to the HPP.

3. Project line WV to the top view. Here it is true length.

4. Draw an auxiliary plane perpendicular to the true length line, WV.

5. Project both planes to the auxiliary plane.

6. Where plane ABC crosses WXYZ is the line of intersection. It is labeled 1, 2.

7. Project points 1, 2 to the regular views. This locates the line of intersection. Notice that point 1 intersects side CB and point 2 intersects side AC.

8. Ascertain the visibility of the planes. This is explained in the next paragraph.

VISIBILITY

It is essential when drawing orthographic views to indicate which of several crossing lines is on top and therefore visible in the view under construction. The outline of a view will be visible, but it is sometimes difficult to tell whether lines within the outline are hidden or visible.

Many times visibility can be determined by inspecting the given views, Fig. 8–42. The outline is clearly visible in the front view. Edges DH and BF are in question and must have visibility established. The line of sight in the top view is directed toward the front view of the object. Since edge DH is nearest the reference line, it is visible. Corner BF is farthest away from the reference line and behind the face CDHG; therefore it is hidden.

Occasionally the visibility of nonintersecting lines cannot be determined by simple inspection, as illustrated in Fig. 8–43. Shown are two hydraulic lines controlling the blade of a bulldozer. To complete accurately a drawing of

8–43. *Visibility of nonintersecting lines is found by projection.*

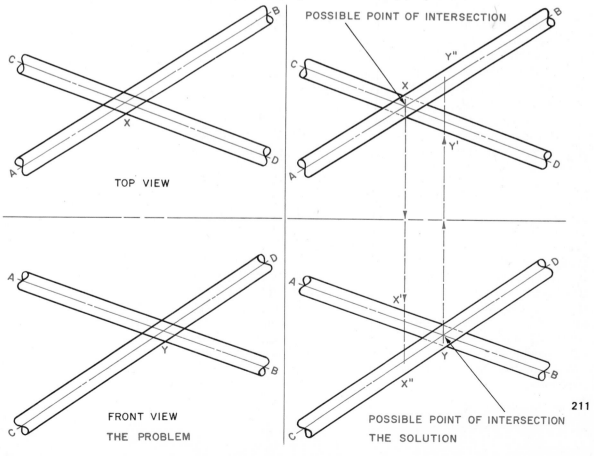

TOP VIEW

POSSIBLE POINT OF INTERSECTION

FRONT VIEW

THE PROBLEM

POSSIBLE POINT OF INTERSECTION

THE SOLUTION

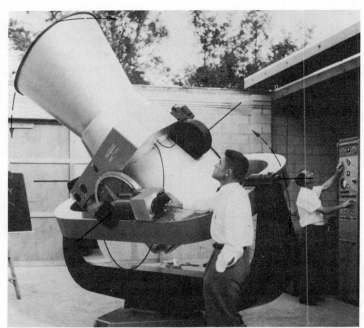

National Aeronautics and Space Administration

8-44. *The Baker-Nunn tracking satellite camera rotates on an axis of revolution.*

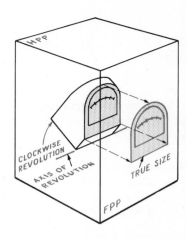

8-45. *The object is revolved until the inclined surface is parallel to a normal plane of projection. It appears true size when projected to this plane.*

the piping installation, it is necessary to show which pipe is visible (on top) in each view. The center lines of the pipes are used for drawing purposes. To ascertain visibility:

1. In the top and front views of the pipes appear two possible points of intersection. These are indicated by X and Y.

2. Project point X from the top to the front view. It intersects pipe AB at X' and CD at X". Since it intersects AB first in the front view, this pipe is visible in the *top view.*

3. Project point Y from the front view into the top view. It intersects pipe CD at Y' and AB at Y". Since it intersects CD first in the top view, CD is visible in the *front view.*

4. Draw the visible and hidden lines.

212

REVOLUTION

In the discussion on auxiliary views, the object is assumed to be in a fixed position. The draftsman moves to a different position (line of sight) if he wishes to view the object from other than the normal position. For example, to see an inclined plane true size, the draftsman moves until he is looking perpendicularly at the plane and draws on an auxiliary plane what he sees, as shown in Fig. 8-3.

Another method for solving such problems is by the revolution method. This allows the draftsman to remain in a fixed position. *The object viewed is revolved about an axis until it is parallel to a normal plane of projection,* as shown in Fig. 8-45. The inclined sur-

face appears true size when projected to the normal plane because it is parallel to the plane.

When an object is revolved, it turns about an axis. The tracking satellite camera, Fig. 8-44, has two axes of revolution.

Principles of Revolution. Following are principles that must be learned before revolution problems can be completely understood. They are illustrated in Figs. 8-46 and 8-47.

1. A point revolving about a line (an axis) forms a circular path. The axis is the center of the circle.

2. When the axis appears as a point, the path of the revolving point appears as a circle. See the top view, Fig. 8-42.

3. When the axis appears as a straight line perpendicular

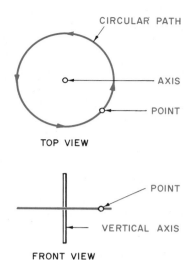

CIRCULAR PATH

AXIS

POINT

TOP VIEW

POINT

VERTICAL AXIS

FRONT VIEW

8–46. *The path of a point revolving about an axis is a circle when the axis appears as a point. The path of a point revolving about an axis is a straight line when the axis appears as a true length line.*

to the line of sight, the path of the revolving point appears as a straight line. The length of the line is equal to the diameter of the circular path of revolution and at right angles to the axis. See the front view, Fig. 8–46.

4. The path of revolution of a point about an axis appears to the eye to be an ellipse when the axis is inclined to the normal planes of projection. See Fig. 8–47.

The most commonly used types of revolutions involve rotating the object about a horizontal or vertical axis with the axis perpendicular to either the horizontal, frontal, or profile plane.

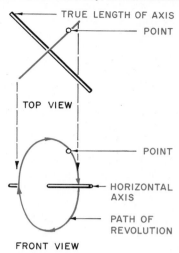

TRUE LENGTH OF AXIS

POINT

TOP VIEW

POINT

HORIZONTAL AXIS

PATH OF REVOLUTION

FRONT VIEW

8–47. *A revolving point appears to form an elliptical path when revolved about an axis inclined to a normal plane.*

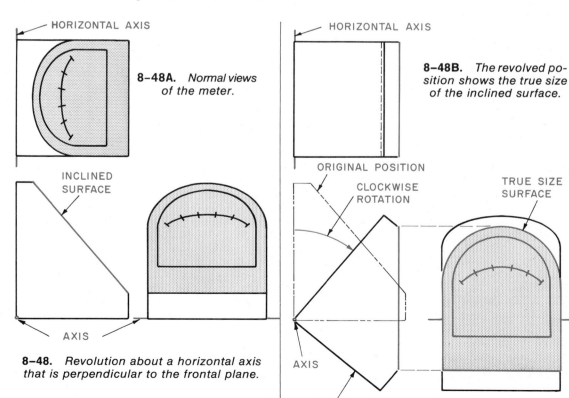

HORIZONTAL AXIS

8–48A. *Normal views of the meter.*

HORIZONTAL AXIS

8–48B. *The revolved position shows the true size of the inclined surface.*

INCLINED SURFACE

AXIS

ORIGINAL POSITION

CLOCKWISE ROTATION

TRUE SIZE SURFACE

AXIS

REVOLVED POSITION

8–48. *Revolution about a horizontal axis that is perpendicular to the frontal plane.*

213

Revolution with Horizontal Axis Perpendicular to Frontal Plane. Revolution about a horizontal axis is shown in Fig. 8–48. The surface of the meter is inclined and foreshortened in all normal views. If an axis is located at a lower edge and the object rotated clockwise until the inclined surface is parallel to the profile plane, it will project true size on that plane. The axis may be located at any convenient point. The object can be rotated clockwise or counterclockwise as seems to be best for the particular situation.

The size and shape of the object do not change. Since this is orthographic projection, all points project from view to view exactly the same as in any multiview problem.

Revolution with Horizontal Axis Perpendicular to Profile Plane. A revolution about an axis perpendicular to the profile plane is shown in Fig. 8–49. The axis is placed at a lower corner and the object is rotated in a counterclockwise direction until the inclined surface is parallel with the frontal plane where it appears true size. The revolved profile view is drawn first and the other views projected from it.

Revolution with Vertical Axis Perpendicular to Horizontal Plane. Revolution about a vertical axis is shown in Fig. 8–50. Notice the axis is placed through the center of the object in this illustration. It could have been located anywhere desired. The central location limits the space needed to revolve the object. The revolved top view is drawn first and the other views projected from it.

True Length of a Line by Revolution. While the true length of a line can be found by making an auxiliary view, using revolution is easier and

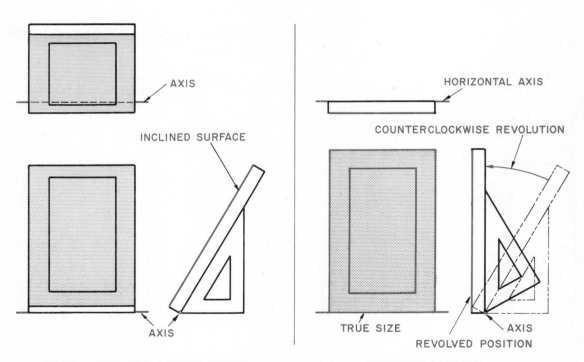

NORMAL VIEWS OF A PICTURE FRAME REVOLVED POSITION

8–49. *Revolution about a horizontal axis that is perpendicular to the profile plane.*

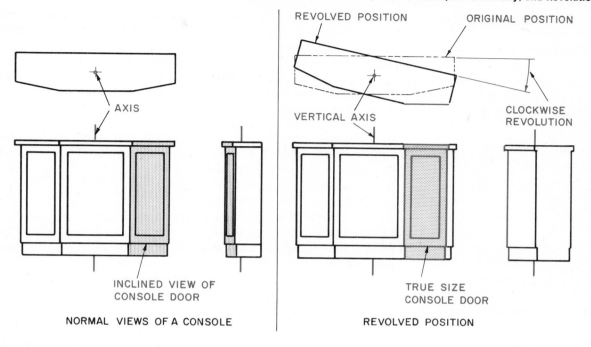

NORMAL VIEWS OF A CONSOLE

REVOLVED POSITION

8–50. *Revolution about a vertical axis that is perpendicular to the horizontal plane.*

faster since it does not require another view to be drawn.

Fig. 8–51 shows a table leg having an inclined plane. Edge AB is an oblique edge. To find the true length of that edge, assume a vertical axis at A in the top view. Revolve B in the top view clockwise until it is parallel to the frontal plane. An edge will appear true length in any plane to which it is parallel. Project B (revolved) into the front view. Project B' in the front view until it intersects B (revolved). Connect A' in the front view with B (revolved) in the front view. This is the true length of edge AB.

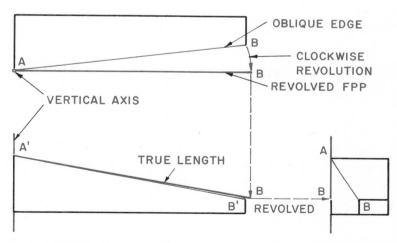

8–51. *The true length of an oblique line can be found by revolution.*

Build Your Vocabulary

Following are terms that you should understand and use as a part of your working vocabulary. Write a brief explanation of what each means.

Auxiliary view
Auxiliary plane
Secondary auxiliary views
Primary auxiliary views
Descriptive geometry
Dihedral angle
Visibility
Revolution
Axis of revolution

Study Problems

Primary Auxiliary Views. The problems shown in P8–1 through P8–6 can be solved by constructing a primary auxiliary view. Draw the necessary regular views plus the needed primary auxiliary views.

1. Tip for rock drilling machine, P8–1. Find the true size of the inclined cutting edge using a central reference plane.
2. End of a spade drill driver, P8–2. Show the true surface area of the inclined face.
3. Flag pole, P8–3. Find the true length of the guy wires. Use the auxiliary view method.
4. Grain chute, P8–4. How many degrees are angles ABC and BCD?

5. Eyepiece, P8–5. Draw the true shape of the inclined surface using the auxiliary view method.
6. Cable sling, P8–6. Find the true length of the cables making the sling and the angle they make with the surface of the crate.

Secondary Auxiliary Views. The problems shown in P8–7 through P8–9 can be solved by constructing a secondary auxiliary view. Draw the necessary regular views plus the needed auxiliary views.

7. Spade drill, P8–7. Draw true size the oblique cutting edge surface.
8. Spade drill grinding fixture, P8–8. Draw true size the two oblique surfaces.
9. Windshield, P8–9. Find the actual dimensions for each glass pane.

Curved Surfaces in Auxiliary Views

10. Space helmet shell, P8–10. Draw the true shape of the inclined front face. Draw the planes twice as large as shown.

Shortest Distance From a Point to a Line

11. TV antenna pole, P8–11. Construct the shortest guy wire possible from the top of the pole to roof ridge AB and rafter BC. Find the actual length of each guy wire.
12. Hopper to pipe connection, P8–12. Construct the shortest connecting pipe from the hopper at C to pipe AB. What is the length of the connecting pipe?

Shortest Distance Between Parallel Lines

13. Hopper to pipe connection, P8–12. Construct the shortest brace possible between pipes AB and DE. What is the true length of the brace?

True Size of a Dihedral Angle

14. Windshield, P8–9. Find the true size of the dihedral angles between the glass panes.
15. Feed hopper, P8–13. Find the true size of the angle between the inclined sides.

Locating Piercing Points

16. Pierced inclined block, P8–14. Locate the point where line EF pierces plane ABCD and GHIJ in the front and top views.
17. Pierced oblique surface, P8–15. Find the point where line EF pierces plane ABCD on the front and top views.

Intersections and Visibility

18. Intersecting planes, *P8–16. Find the line of intersection between planes ABCD and 1234. Also find which portions are hidden and visible by applying visibility techniques. Draw the planes twice as large as shown.*
19. Tower bracing, *P8–17. Which brace is visible in the top view? Front view? Use hidden lines to show this on your drawing.*

Revolutions

20. Engraving tool, *P8–18. Find the true size of the inclined surface. Use the revolution method.*
21. Pipe line, *P8–19. Find the true length and true angle of pipe B, D, F using the revolution method.*
22. Eyepiece, *P8–5. Draw the true shape of the inclined surface using the revolution method.*
23. Cable sling, *P8–6. Find the true length of the cables making the sling. Use the revolution method.*

Other Study Problems. Select the best and easiest method of solving the following problems:

24. Raised platform, *P8–20. Find the true length of the wire brace between the legs of the raised platform.*
25. Control cable, *P8–21. Find the true length of the control cable.*
26. Engraving tool, *P8–18. What are the actual dimensions of each side of the cutting edge and the number of degrees in each angle of the cutting surface?*
27. Wedge block, *P8–22. Line EF represents the center line of a hole to be drilled. Locate the point where the drill would pierce surface ABCD.*
28. Mine shaft, *P8–23. How long is the mine shaft? What angle does it make with the horizontal?*
29. Space craft bracket, *P8–24. Make an auxiliary drawing giving the true size of the inclined curved surface.*
30. Truck identification light, *P8–25. Show the true size of the lens using a one-half auxiliary view.*
31. Landing gear brace, *P8–26. Construct the shortest brace possible from point A on the fuselage to the edge BC of the landing gear. What is the true length of the brace?*
32. Hydraulic lines, *P8–27. Find the dimensions needed to make the brace indicated on the drawing. The brace should keep the three pipes equally spaced apart and be as short in length as possible. Allow 3/8-inch metal between the edge of each hole and the outside of the brace. Locate the brace on the pipe drawing. Make a dimensioned drawing of the brace.*
33. Geodetic satellite, *P8–28. Find the true size of the shielding panels, the laser reflector panels, and solar cell panels. Cut from cardboard and glue together to form a model.*

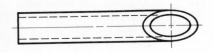

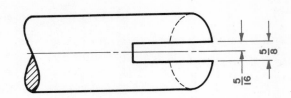

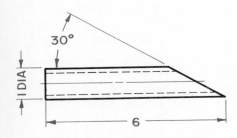

P8–1. *Tip for rock drilling machine.*

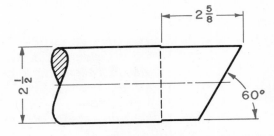

P8–2. *End of a spade drill driver.*

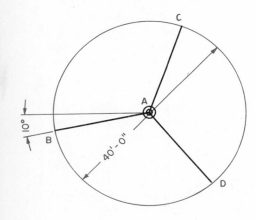

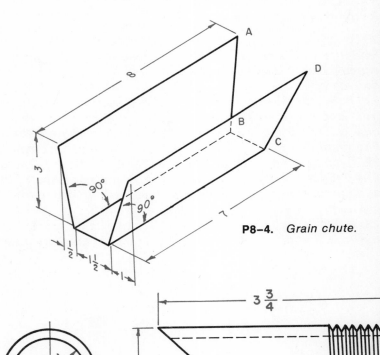

P8–4. *Grain chute.*

P8–3. *Flag pole.*

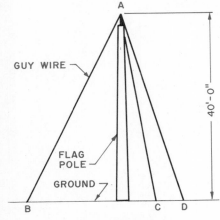

P8–5. *Eyepiece for transit.*

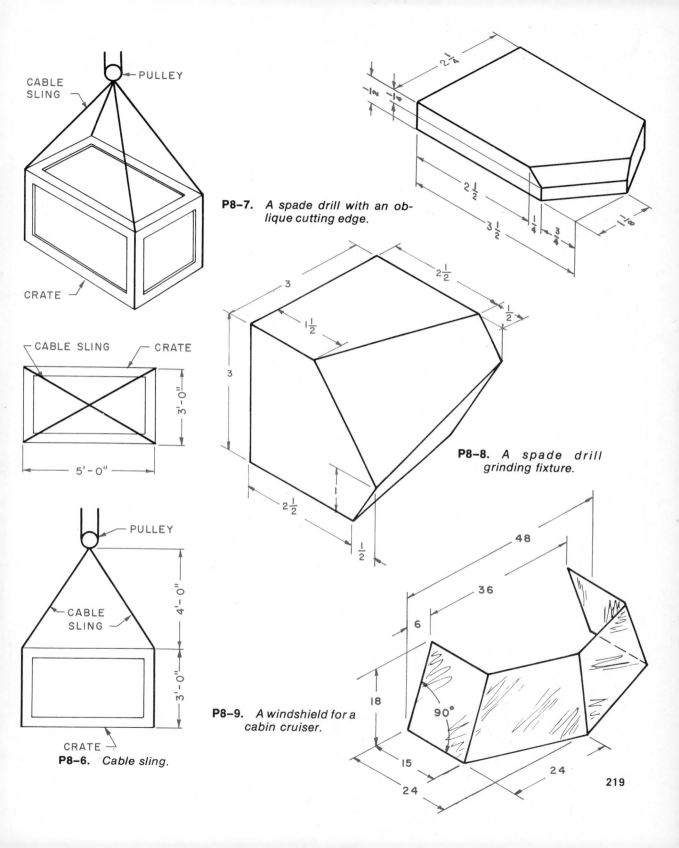

CABLE SLING

PULLEY

CRATE

CABLE SLING — CRATE

3'-0"

5'-0"

PULLEY

CABLE SLING

CRATE

4'-0"

3'-0"

P8-6. *Cable sling.*

P8-7. *A spade drill with an oblique cutting edge.*

$2\frac{1}{4}$

$\frac{1}{2}$ $\frac{1}{4}$

$2\frac{1}{2}$

$3\frac{1}{2}$ $\frac{1}{4}$ $\frac{3}{4}$ $\frac{1}{8}$

P8-8. *A spade drill grinding fixture.*

3

$2\frac{1}{2}$

$1\frac{1}{2}$ $\frac{1}{2}$

3

$2\frac{1}{2}$ $\frac{1}{2}$

P8-9. *A windshield for a cabin cruiser.*

48

36

6

18

90°

15

24

24

219

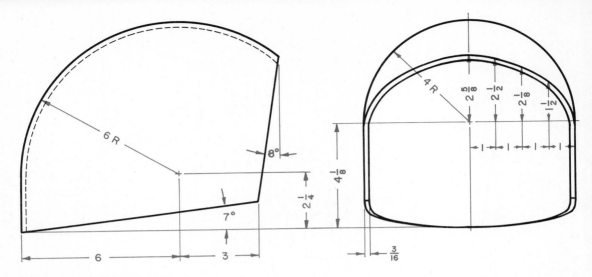

P8-10. *Space helmet shell.*

P8-12. *Hopper to pipe connection.*

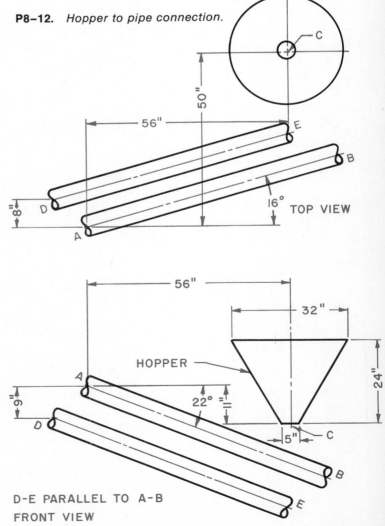

TOP VIEW

HOPPER

D-E PARALLEL TO A-B
FRONT VIEW

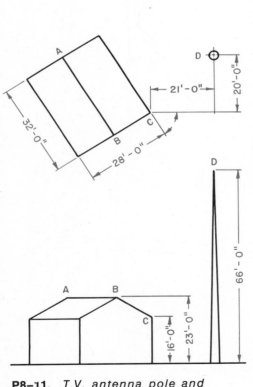

P8-11. *T.V. antenna pole and building.*

220

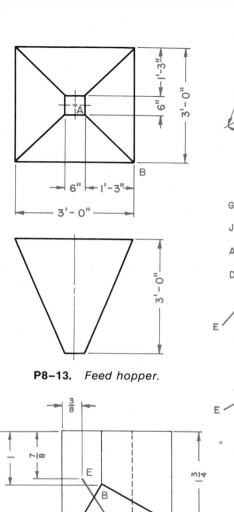

P8–13. Feed hopper.

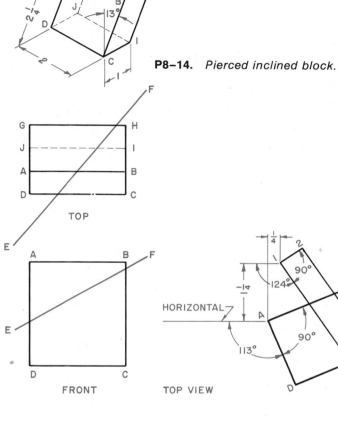

P8–14. Pierced inclined block.

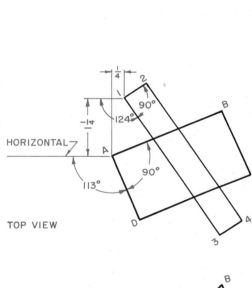

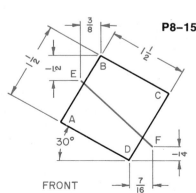

P8–15. Pierced oblique surface.

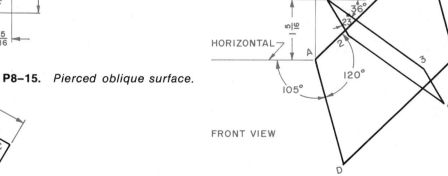

P8–16. Intersecting planes.

221

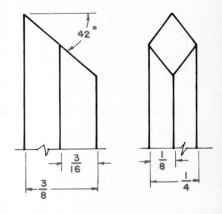

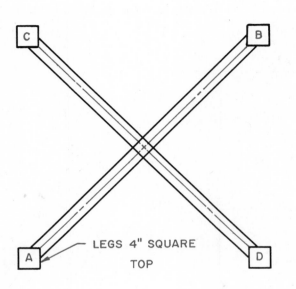

LEGS 4" SQUARE

TOP

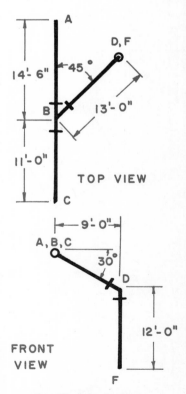

P8-18. *The cutting edge of an engraving tool.*

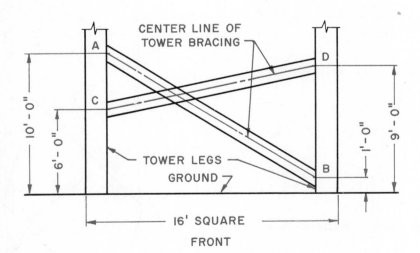

CENTER LINE OF TOWER BRACING

TOWER LEGS

GROUND

10'-0"

6'-0"

1'-0"

9'-0"

16' SQUARE

FRONT

P8-17. *Tower bracing.*

TOP VIEW

14'-6"

45°

13'-0"

11'-0"

9'-0"

30°

12'-0"

FRONT VIEW

P8-19. *A pipeline.*

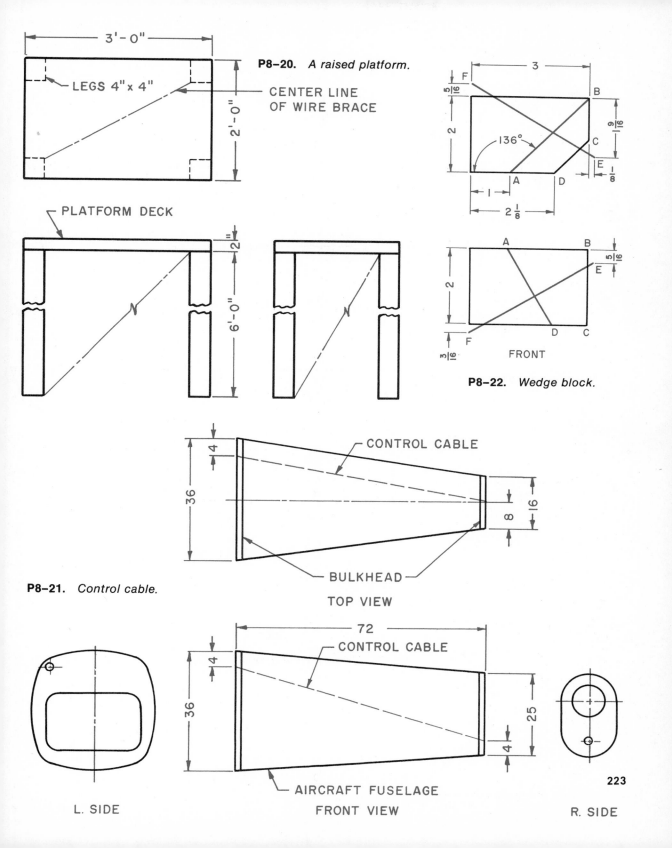

3'-0"

LEGS 4" x 4"

CENTER LINE
OF WIRE BRACE

P8-20. *A raised platform.*

2'-0"

PLATFORM DECK

2"

6'-0"

F

5/16

3

B

2

136°

9/16

C

E

1/8

1

A

D

2 1/8

A

B

5/16

E

2

D

C

3/16

F

FRONT

P8-22. *Wedge block.*

CONTROL CABLE

4

36

16

8

BULKHEAD

P8-21. *Control cable.*

TOP VIEW

72

CONTROL CABLE

4

36

25

4

AIRCRAFT FUSELAGE

L. SIDE

FRONT VIEW

R. SIDE

223

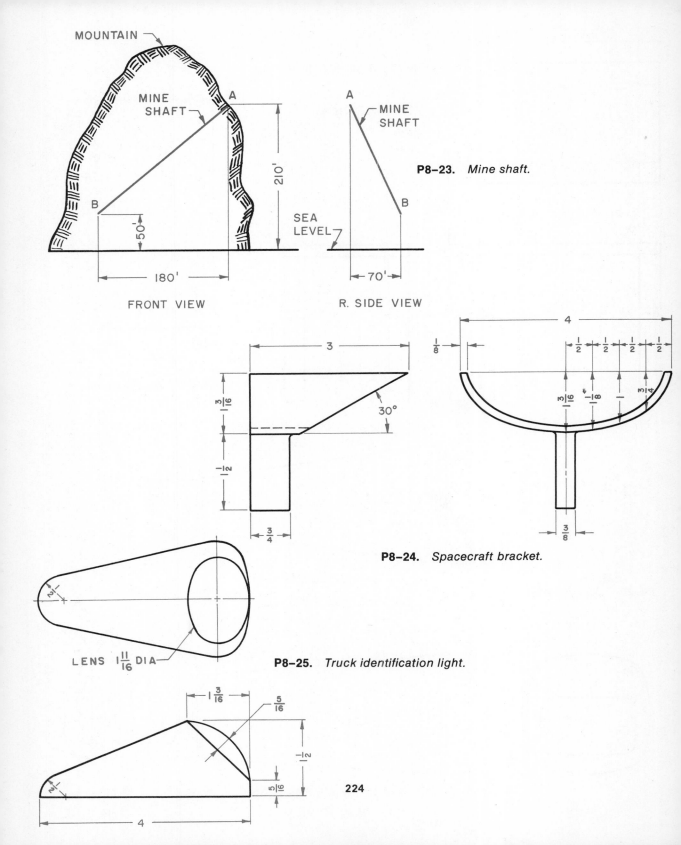

MOUNTAIN

MINE SHAFT

A

210'

SEA LEVEL

B

50'

180'

FRONT VIEW

A

MINE SHAFT

B

70'

R. SIDE VIEW

P8-23. *Mine shaft.*

3

$\frac{3}{16}$

$1\frac{1}{2}$

30°

$\frac{3}{4}$

4

$\frac{1}{8}$

$\frac{1}{2}$ $\frac{1}{2}$ $\frac{1}{2}$ $\frac{1}{2}$

$\frac{3}{16}$ $1\frac{1}{8}$ $3\frac{1}{4}$

$\frac{3}{8}$

P8-24. *Spacecraft bracket.*

$\frac{1}{2}$

LENS $1\frac{11}{16}$ DIA

P8-25. *Truck identification light.*

$1\frac{3}{16}$

$\frac{5}{16}$

$1\frac{1}{2}$

$\frac{5}{16}$

$\frac{1}{2}$

4

224

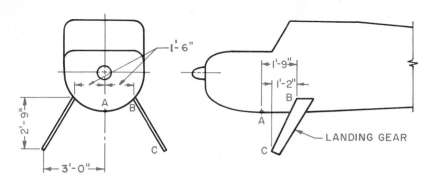

P8–26. *Landing gear brace.*

LANDING GEAR

BRACE TO BE DESIGNED

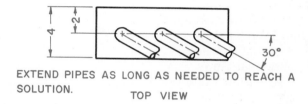

EXTEND PIPES AS LONG AS NEEDED TO REACH A SOLUTION.

TOP VIEW

P8–27. *Hydraulic lines.*

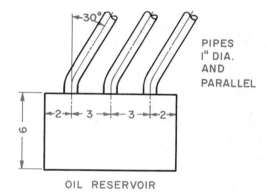

PIPES
1" DIA.
AND
PARALLEL

OIL RESERVOIR
FRONT VIEW

P8–28. *Geodetic satellite.*

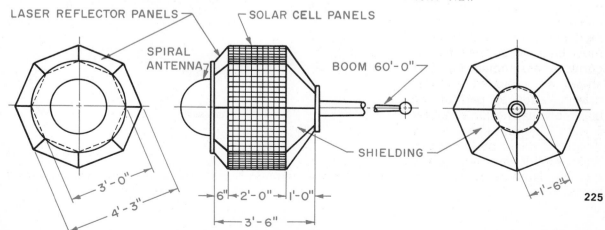

LASER REFLECTOR PANELS

SOLAR CELL PANELS

SPIRAL ANTENNA

BOOM 60'-0"

SHIELDING

Chapter 9

Pictorial Drawing

Seldom a day passes when you do not see some form of pictorial drawing. Books, newspapers, and magazines are filled with pictorial illustrations. Industry uses many types of pictorials to present information. Pictorial drawings help to explain a process or an assembly. Many purchased items come disassembled. Some form of instruction must be given so they may be correctly assembled. Fig. 9–1 shows an exploded pictorial drawing used as part of the assembly instruction sheet for a child's bicycle. Imagine how difficult it might be to put this together if it were not for the pictorial illustration. It would

226

be difficult to assemble if only an orthographic drawing were available. Many persons do not have the ability to understand an orthographic drawing. For those who have not had this type of training, pictorial drawings give the needed information.

Pictorial drawings have the advantage of showing three faces in a single view. In other words, it is a three-dimensional drawing. Orthographic projection shows only one face at a time. Each view can only show two dimensions. There is no suggestion of depth.

TYPES OF PICTORIAL DRAWINGS

Pictorial drawings can be divided into three main types: (1) axonometric, which is divided into isometric, dimetric, and trimetric; (2) oblique, which is divided into cabinet and cavalier, and (3) perspective. See Fig. 9–2. *Axonometric drawings* have the object

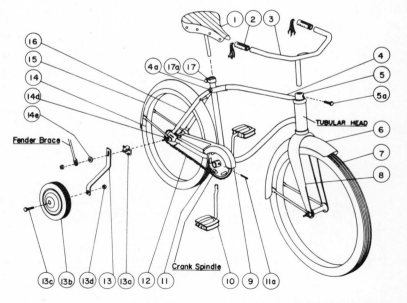

American Machine and Foundry Co., Inc.

9–1. *Pictorial drawings help show how things are to be assembled.*

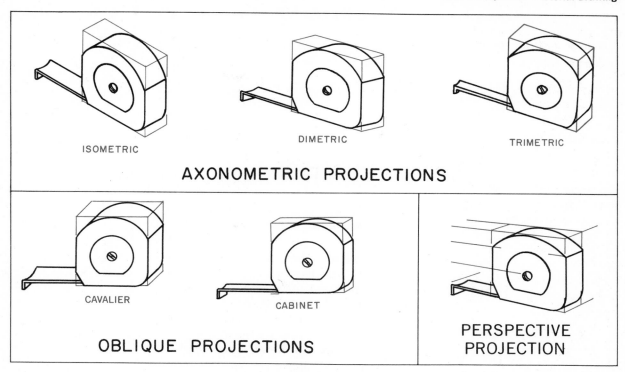

ISOMETRIC

DIMETRIC

TRIMETRIC

AXONOMETRIC PROJECTIONS

CAVALIER

CABINET

OBLIQUE PROJECTIONS

PERSPECTIVE PROJECTION

9-2. *Types of pictorial drawings.*

inclined to the plane of projection. *Oblique drawings* have the face of the object parallel with the plane of projection. *Perspective drawings* appear to converge toward some point in the distance.

ISOMETRIC DRAWING

Isometric drawing is based on the principle of rotation. Fig. 9-3 shows a V block being rotated. Fig. 9-4 shows the same idea orthographically. First the object is rotated 45 degrees. The axis of revolution is placed perpendicular to the horizontal plane. The object now shows two faces in the front and profile views. Next the axis of revolution is placed perpendicular to the

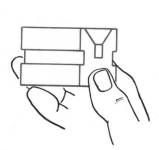

9-3. *The idea behind isometric drawing is rotation of the object.*

9-4. *To make an isometric drawing, the object is revolved around vertical and horizontal axes.*

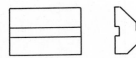

9-4A. *Three regular views of a V block.*

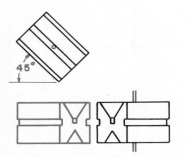

9-4B. *Place an axis of rotation perpendicular to the horizontal plane and rotate the object 45 degrees.*

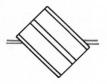

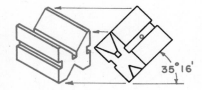

THE ISOMETRIC

9-4C. *Place the axis of rotation perpendicular to the profile plane and rotate the object 35° 16'.*

228

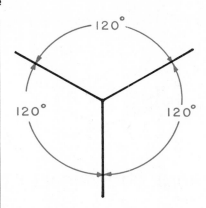

9-5. *Isometric axes are uniformly spaced at 120 degrees.*

profile plane. The object is then rotated downward. The angle the front edge makes with the vertical plane is 35°16'. Note that three faces of the object are now visible in the front and top views. The front view is shown in *isometric.*

All isometric drawings are based on three axes spaced 120 degrees apart. See Fig. 9-5. The axes may be placed in four different positions. Each position will show various faces of the object. Regardless of the position selected, each axis is still 120 degrees from the other. Fig. 9-6 shows the four positions of the isometric axes. All isometric drawings are based on these four positions. An alternate arrangement for the axes is shown in Fig. 9-7. One axis is drawn vertical. The other two axes are drawn 30

9-6. *The isometric axes can be placed in any of these four positions.*

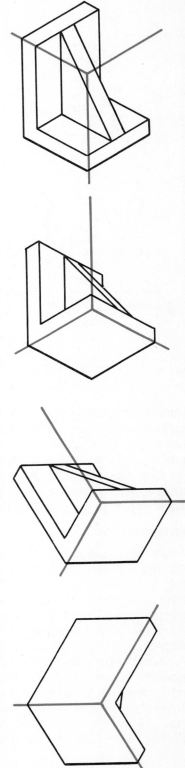

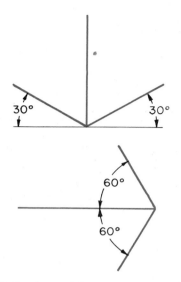

9–7. *Isometric axes can be drawn in these special positions.*

degrees to the right and left. When one of the axes is horizontal, the other two are drawn at 60 degrees to the horizontal. The width, depth, and height are measured on the isometric axes.

Making an Isometric Drawing. Any isometric drawing is built upon the isometric axes. Choose the position of the axes that will best illustrate the object. The following procedure may be used in making an isometric drawing. See Fig. 9–8.

1. Lay out the isometric axes.

2. Lay out the principal dimensions of the object on the axes. These are true length dimensions.

3. Draw a box which will enclose the object.

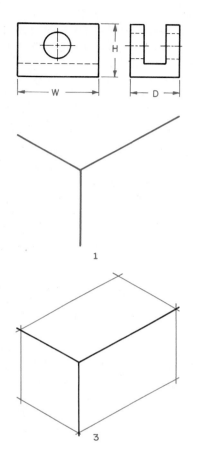

9–8. *How to make an isometric drawing.*

4. Locate other features by measuring along the edges of the enclosing box. All measurements must be made on isometric lines.

Hidden Lines. Hidden lines are usually not used in any form of pictorial drawing. To show all edges which are not visible would make the drawing difficult to read. In some cases, hidden lines may be used to show a feature of the

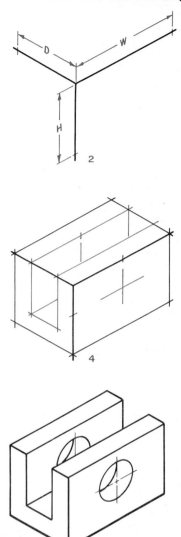

object which is not visible. Sometimes hidden lines can be eliminated by dimensioning the pictorial. For example, if two holes are visible with the third hidden, the note for the hole would read: $17/32$ DRILL, 3 HOLES EQUALLY SPACED ON $4\frac{1}{2}$ DIA B C.

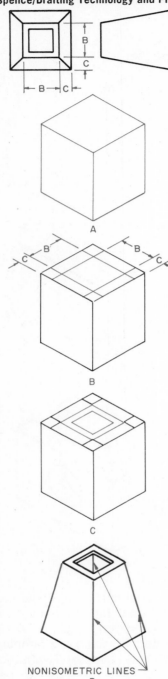

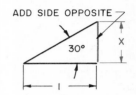

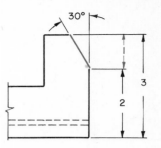

ADD SIDE OPPOSITE

Nonisometric Lines. All principal dimensions are laid out along the isometric axes. Some edges may be oblique or inclined. These do not show true length. In isometric drawings these lines are called *nonisometric lines.* These are lines that cannot be measured directly on the isometric axes. To draw a nonisometric line, both ends must first be located on isometric lines. The nonisometric line is then drawn between the two points. The best method of drawing a nonisometric line is to use the "box method" of construction. See Fig. 9–9. Construct an isometric box using the largest dimensions of the object. Then locate the edges of the various surfaces from the edges of the box.

For example, Fig. 9–9 shows two orthographic views of a base for an automotive safety stand. Using the three largest dimensions, a box is constructed. See Fig. 9–9, Part A. The top of the stand is located by using dimensions B and the offset dimension C. See Fig. 9–9, Parts B and C. The nonisometric corners are then drawn between the top and the bottom of the box.

Angles in Isometric. Occasionally it is necessary to lay out an angle in isometric. Angles in isometric, just as nonisometric lines, do not show in their true size. They cannot be measured in degrees with an ordinary protractor. Special isometric

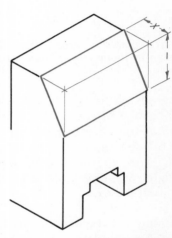

9–10. *The coordinate method for laying out an angle in isometric.*

protractors are available for laying out angles in isometric.

If an isometric protractor is not available, the angle may be laid out by the coordinate method. This method uses a series of coordinates parallel to the isometric axes. The size of the angle is stepped off on the coordinates. Fig. 9–10 shows a partial view of a drill press vise. First draw the angle in an orthographic view. Make

NONISOMETRIC LINES
D

9–9. *Nonisometric lines are drawn by locating their corners on isometric lines.*

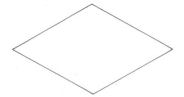

THE ISOMETRIC SQUARE

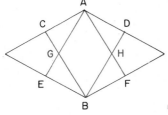

BISECT EACH SIDE

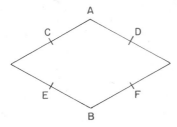

CONNECT BISECTORS
WITH VERTEX

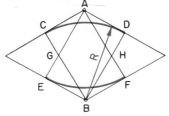

DRAW THE SIDES

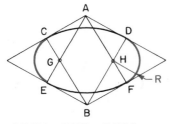

DRAW THE ENDS

9-11. *How to draw a four-center ellipse.*

this drawing the same scale as the isometric drawing. Add a third side to the angle, called the *side opposite,* thus forming a right triangle. It then may be stepped off on the edges of the isometric box.

Isometric Circles and Arcs. Circles in isometric will appear as ellipses. Arcs will appear as a part of an ellipse. Rather than plot ellipses by actual projection or coordinates, an *approximate method* is used. This method of drawing ellipses is called the *four-center approximate ellipse.* As the name of this ellipse indicates, it is approximate.

To draw a four-center ellipse see Fig. 9–11. First draw an isometric square. The length of the side of the square is equal to the diameter of the circle. Then bisect each of the four sides of the isometric square. Connect each bisector to the vertex of the opposite large angle. Where bisectors intersect will form the center points of the large and small arcs. With the compass point at B, draw arc CD. Using the same radius, draw arc EF. Now draw arc DF from point H, and in a similar manner, draw arc CE.

The sides of an isometric square, regardless of its position, may be easily bisected by using a 30–60 degree triangle and T-square. See Fig. 9–12. This method of finding the mid-point of each side can save considerable time.

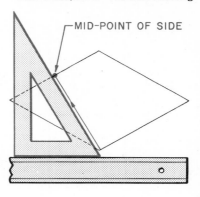

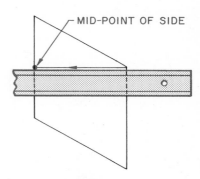

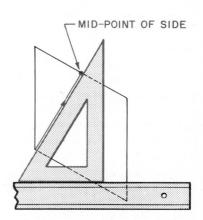

9-12. *The isometric square for a four-center ellipse can be easily bisected with a T-square and 30–60 degree triangle.*

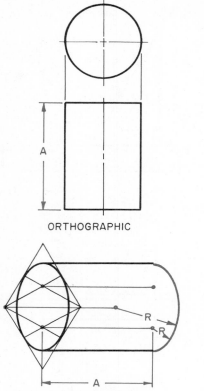

ORTHOGRAPHIC

ISOMETRIC

9-13. *Ellipses of the same diameter can be drawn by moving the centers of the first ellipse the length of the cylinder.*

To draw several isometric circles of the same diameter in parallel planes is not as difficult as it may seem. A cylinder of uniform diameter is an example of this problem. See Fig. 9–13. Locate the four centers for the approximate ellipse which is closest. These are located in the usual manner by first drawing the isometric square. The four centers for the farther ellipse are found by moving the centers

232

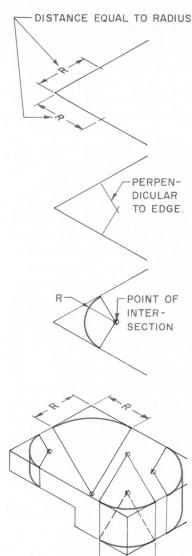

DISTANCE EQUAL TO RADIUS

R

R

PERPEN-
DICULAR
TO EDGE.

R

POINT OF
INTER-
SECTION

R   R

R   R

9-14. *How to lay out arcs in isometric.*

9-15. *The center for dimensioning an isometric arc follows the isometric axes.*

parallel to the proper axis. The length the centers are moved is equal to the length of the cylinder.

Arcs in isometric are drawn with the same ease as the approximate four-center ellipse. It is not necessary to draw the entire isometric square since an arc is only part of a circle. The same principle is used in drawing an arc as in drawing the circle. See Fig. 9–14. From the corner where the arc is to appear, lay off the radius of the arc on both edges. From each of these points, draw perpendiculars. Where the perpendiculars intersect forms the center of the arc in isometric. It should be pointed out that the center for the arc is not the center used for dimensioning. Fig. 9–15 illustrates this point.

Irregular Curves in Isometric. Irregular curves in isometric are drawn by plotting a

CENTER FOR ARC

R

CENTER FOR
DIMENSIONING

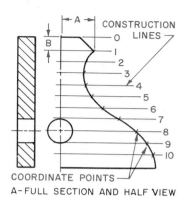

CONSTRUCTION LINES

COORDINATE POINTS

A-FULL SECTION AND HALF VIEW

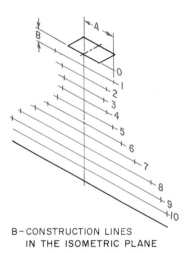

B-CONSTRUCTION LINES
IN THE ISOMETRIC PLANE

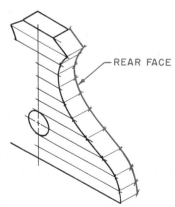

REAR FACE

9-16.   *Drawing an irregular curve in isometric.*

series of points, Fig. 9–16. First an orthographic view of the object must be drawn to the same scale as the isometric. Next a series of parallel construction lines are drawn on the orthographic view. These construction lines should be spaced equally. Where the curve intersects each construction line, a coordinate point is formed. See Fig. 9–16B. Now draw the construction lines in the correct isometric plane. Use the same spacing between the construction lines in the isometric as in the orthographic view. Points on the orthographic view are called coordinates. The coordinates are then transferred to the isometric view by dividers.

The back face is drawn similar to the method used for parallel isometric circles. Draw a series of parallel construction lines for the depth. Step off the thickness of the object along these construction lines. Now draw the curve by connecting the points. Use a tool called an irregular curve.

Centering an Isometric. There are several methods that can be used to center an isometric drawing on a sheet. The most widely used is the box method. See Fig. 9–17.

On a piece of scrap paper, draw the isometric block that encloses the entire object. Then draw an *enclosing box* around the isometric block. Equalize the distance from

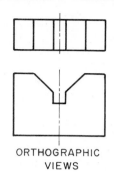

ORTHOGRAPHIC
VIEWS

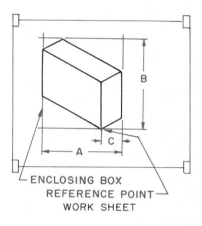

ENCLOSING BOX
REFERENCE POINT
WORK SHEET

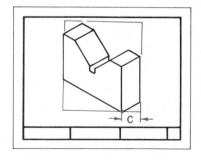

DRAWING SHEET

9-17.   *The box method is used to center an isometric drawing on the paper.*

233

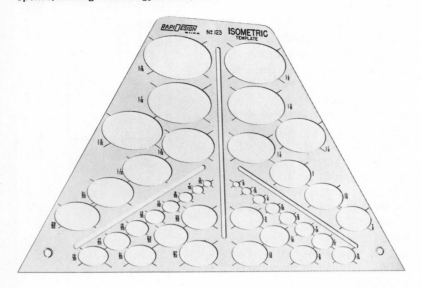

Rapidesign

9–18. *An isometric ellipse template can simplify the drawing of ellipses.*

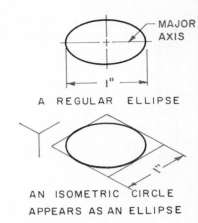

A REGULAR ELLIPSE

AN ISOMETRIC CIRCLE
APPEARS AS AN ELLIPSE

9–19. *The diameter of an isometric circle is measured along the isometric axis.*

the edges of the enclosing box to the borders of the paper upon which the isometric drawing is to be made. Mark the lower corner. The enclosing box is the *reference point* to use for the isometric drawing.

Isometric Drawing Aids. Templates are available which simplify the drawing of isometric circles. One type of isometric template is shown in Fig. 9–18. All isometric circles are based on the circle's *actual projection.* The circle is placed at an angle of 35°–16' to the plane of projection. The diameter of the isometric circle is measured along the isometric axis. A regular ellipse is measured along its major axis. See Fig. 9–19. Do

234

not confuse the regular ellipse and isometric circle templates.

Isometric grid paper is available to speed making an isometric drawing. See Fig. 9–20. The grids are printed on both tracing paper and opaque paper. Isometric grids printed on vellum are usually printed with nonreproductible ink. Any grid printed with nonreproductible ink will not show when reproduced. Only the pencil or ink drawing will print. Tracing paper pads are available with an opaque underlay grid sheet. The underlay sheet is placed beneath the tracing paper. It can be used over and over again.

Isometric Sections. Isometric sections are used to show interior detail. To show interior

construction by means of hidden lines would lead to confusion. To remedy this situation, any of the standard sections may be used. See Chapter 10, Sectional Views, for more information. In some instances the use of a section will eliminate an extra pictorial view.

In making an isometric section, lay out the entire object. See Fig. 9–21. Next locate the cutting plane. Then block in the section. Now erase the unneeded construction lines. Draw any minor details, such as holes or fillets. Add the section lines and darken all of the other lines.

The cutting plane used in an isometric *full section* is passed through the center of the object. The cutting plane is always parallel to one of the isometric axes. Features that are shown by the section are completed just as any other portion of the piece. Section

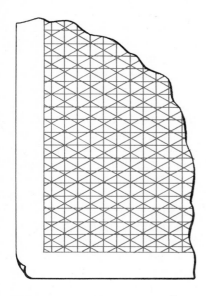

9–20. *Isometric grid paper simplifies making isometric drawings and sketches.*

lining may be drawn at any angle. It should not be parallel to any main edges of the cut surface. Section lines placed on an angle of 60 degrees will work in most instances.

In an isometric *half section,* the cutting plane is placed parallel to *both* receding axes. The cutting plane will intersect at the center line. See Fig. 9–22.

Broken-out sections are made by placing the cutting plane at *any* desired position. If possible, the cutting plane is placed parallel to an isometric axis. When broken-out sections are made of shafts, holes, or similar features, the

9–21. *When making an isometric section, the object is first drawn and then the part not needed is removed.*

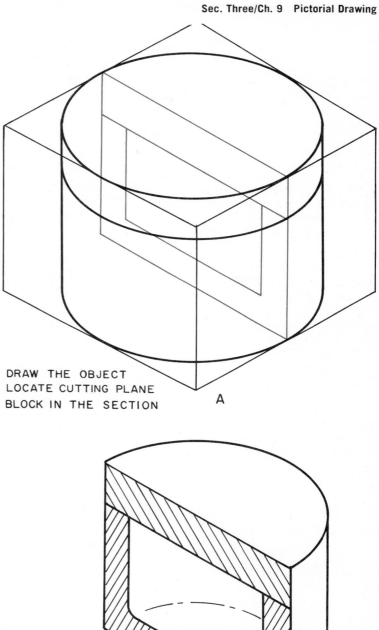

DRAW THE OBJECT
LOCATE CUTTING PLANE
BLOCK IN THE SECTION A

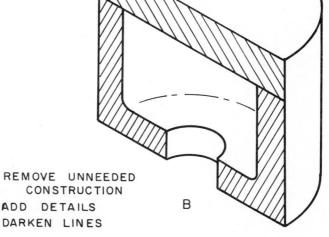

REMOVE UNNEEDED CONSTRUCTION
ADD DETAILS
DARKEN LINES B

235

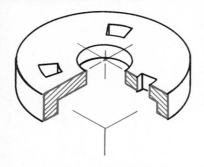

9-22. *In an isometric half section, the cutting planes follow the isometric axes.*

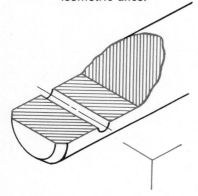

9-23. *The cutting plane in an isometric broken-out section will follow one of the isometric axes.*

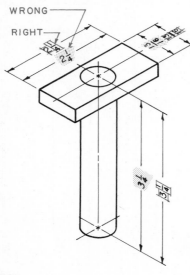

WRONG
RIGHT

9-24. *Dimensions using the aligned system are lettered parallel to the isometric axes.*

cutting plane passes through their axis. Fig. 9-23 is an example of this type of sectioning. Note the treatment of the broken area in section not cut by the plane.

PICTORIAL DIMENSIONING

The same basic rules are followed for dimensioning a pictorial as for multiview drawing. Study Chapter 7, Dimensioning. All dimensions must be clear and complete. Sufficient information must be given to describe the piece. There are two systems of dimensioning a pictorial drawing. These are the aligned and unidirectional systems.

When using the *aligned* system, all dimensions are positioned parallel with the isometric axes. See Fig. 9-24.

When using the *unidirectional* system, all dimensions are lettered vertically. See Fig. 9-25. The unidirectional system is faster and easier to use.

Some basic rules that will help in dimensioning are:

9-25. *Dimensions using the unilateral system are lettered vertically.*

A. Dimension lines and extension lines are parallel to the isometric axes.

B. Slope lines and guidelines for dimensions are always parallel to their dimension and extension lines.

C. Arrowheads lie in the same plane as the extension line and dimension value.

D. Place dimensions on visible features as much as possible.

E. Notes indicating a dimension *and* process are lettered parallel with the bottom edge of the paper. Figs. 9-24 and 9-25 show each of these principles.

OBLIQUE DRAWING

Perhaps the easiest type of pictorial drawing to make is the oblique. In oblique drawing, one of the faces of the object is parallel to the plane of projection. It shows in its true size and shape. Oblique drawing uses three axes just as isometric. Two of the oblique axes are placed at right angles to each other. The third axis

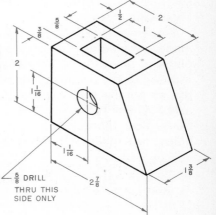

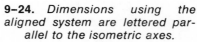

$\frac{5}{8}$ DRILL
THRU THIS
SIDE ONLY

(receding) is at any convenient angle to the horizontal. All receding edges are drawn parallel to this axis. There are three classifications of oblique: cavalier, cabinet, and general. Each type of oblique drawing has its principal face parallel to the plane of projection. See Fig. 9–26.

A *cavalier* drawing is one in which the receding axis makes any angle with the horizontal. Usually this angle is 30 or 45 degrees. The same scale is used on *all* axes. See the cube in Fig. 9–26. Measure the front face and the depth of the cube. The depth of the cube appears to be greater than the width and height. Actually they are equal. Some of this distortion can be eliminated by reducing the angle of the receding axis.

In *cabinet* drawing, the scale of the receding axis is reduced one half. See Fig. 9–26. The angle the receding axis makes with the horizontal can be any angle. By shortening the depth of the receding axis, a more natural appearance is obtained. Compare the cabinet drawing with the cavalier. The cabinet drawing gets its name from the early days in the furniture industry. Many cabinet drawings of furniture were prepared by craftsmen and designers.

The *general oblique* drawing is one in which the receding axis scale is variable. The scale may vary between full and one-half. On occasion the draftsman may even drop the

receding scale down to three-eighths. The scale is reduced until the object appears most natural. The angle between the receding axis and the horizontal usually is between 30 and 60 degrees.

Position of the Object. The general shape of the object will have some effect on its position. The most descriptive features of the object should be placed parallel to the plane of projection. This is very important when drawing a piece having irregular curves or circles. Placing the circular

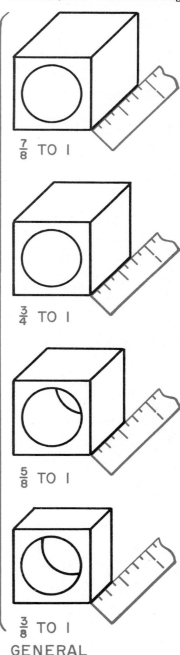

$\frac{7}{8}$ TO I

$\frac{3}{4}$ TO I

$\frac{5}{8}$ TO I

$\frac{3}{8}$ TO I

GENERAL

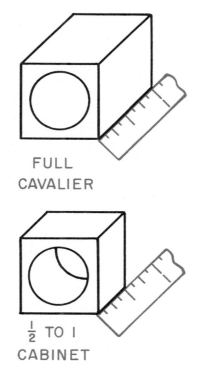

FULL CAVALIER

$\frac{1}{2}$ TO I

CABINET

9–26. *All forms of oblique drawing have one face parallel to a plane of projection.*

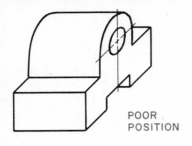

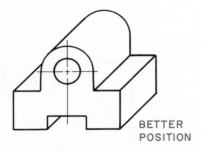

POOR POSITION

BETTER POSITION

9-27. *When possible, circular features should be placed parallel to the plane of projection.*

RECOMMENDED

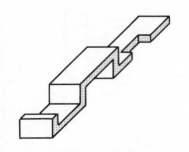

NOT RECOMMENDED

9-28. *The long dimension of a piece should be placed parallel to the plane of projection.*

238

features to the front will not only eliminate distortion but will make it easier to draw. All circles may be drawn with a compass or template. See Fig. 9–27. The largest dimensional face of the object is usually placed parallel to the plane of projection. This is true also of long objects. When a long object is oriented with its greatest dimension along the receding axis, too much distortion will result. See Fig. 9–28.

Angle of the Receding Axis. The angle of the receding axis may vary. For convenience a 30-, 45-, or 60-degree angle is used. The angle chosen should be one in which the details of the receding face are clear. If a great degree of detail appears on the top of the object, the angle of the receding axis would be increased. If this angle were small, 20 degrees for example, the detail could become too compressed. If the detail is too "flat" it may be difficult to read as well as draw. See Fig. 9–29. The draftsman must decide, based on his judgment, which angle is best for the object.

Circles and Arcs in Oblique. Circles and arcs not facing the plane of projection will appear as ellipses. The four-center ellipse is used for drawing ellipses in oblique. The principle used in drawing this ellipse is the same as was used in isometric. Most ellip-

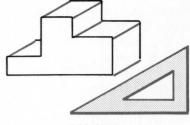

30° AXIS

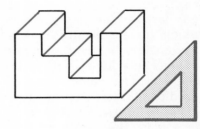

45° AXIS

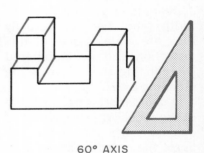

60° AXIS

9-29. *Usually a 30, 45, or 60 degree angle is used for the receding axis.*

ses drawn along the receding axis will appear distorted. This is due to the angle of the axis with the horizontal. The four-center approximate ellipse can *only* be used with *cavalier* drawings. Any other type of oblique drawing requires that the ellipse be plotted. The four-center cannot be used with cabinet or general type oblique drawings.

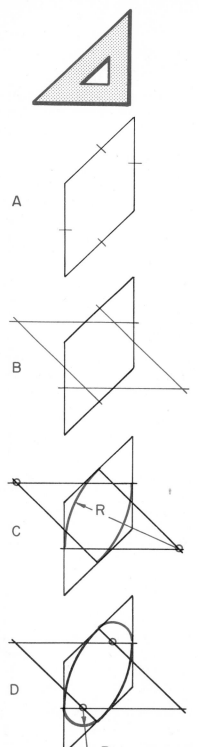

To draw a four-center approximate ellipse, first draw a square (parallelogram) on the receding face. See Fig. 9–30. The sides of the square must be equal to the diameter of the circle. Since this circle is lying in a receding plane, two of the sides of the square are parallel to the receding axis. Next bisect all four sides of the square. To find the midpoint of a side, use dividers of a scale. From these midpoints, draw perpendiculars to the sides of the squares. The centers for the arcs are formed where the perpendiculars intersect. Now draw arcs tangent to the sides of the square. If the angle of the receding axis is less than 30 degrees the perpendiculars will intersect *inside* the square. The perpendiculars will intersect *outside* the square if the angle of the receding axis is greater than 30 degrees.

To draw an ellipse in *cabinet* or any of the *general* types of oblique, the *offset method* must be used. An easy method of plotting an ellipse along the receding axis is shown in Fig. 9–31. As with any offset method of plotting, the circle must be drawn in its true size and shape. Enclose the circle in a square. Now draw two diagonals through the square. Where these diagonals intersect the circle, draw a horizontal line to the sides of the

9–30. *Constructing a four-center ellipse for a cavalier drawing.*

square. See Fig. 9–31A. These horizontals will then be used to locate the coordinates on the receding plane. Draw the square on the receding plane. Be sure to reduce the depth of the square according to the rest of the drawing. Draw two diagonals through the square. See Fig. 9–31B. At the point of intersection, draw two center lines parallel to two of the oblique axes. See Fig. 9–31C. Now transfer distance A from the orthographic to the pictorial. Step off this distance above and below the center lines. From this point draw lines parallel to the receding axis and intersecting the diagonals. The points thus formed will serve as coordinates in plotting the ellipse. Using eight points in plotting the ellipse is sufficient for most circles. Large circles appearing on the receding plane will require more coordinates. Connect the coordinates with an irregular curve.

Arcs in cavalier may be drawn by the same method as used in isometric. See Fig. 9–30. Set off the actual radius along the edges. From these points erect perpendiculars to form the center for the arc. Any other method of oblique drawing requires that arcs be set off with coordinates.

Dimensioning an Oblique. Any oblique drawing is dimensioned in the same manner as an isometric. Either the aligned or unidirectional method may be used. When

239

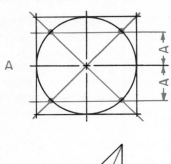

A

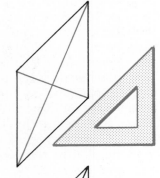

B

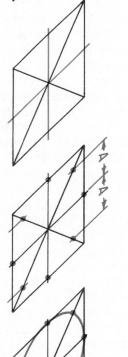

C

D

E

9–31. *An ellipse must be plotted by the offset method when used on a cabinet or general oblique drawing.*

dimensioning in the aligned system, those dimensions appearing in the oblique plane must lie in that plane. If the angle of the receding axis is small, it is difficult to letter dimensions in the oblique plane. Both types of guidelines, drawn parallel to the dimension and extension lines, are necessary. These will form parallelograms in which the figures are lettered. Dimensions for those features on the front face are lettered vertically. Frequently the unidirectional system proves to be best for obliques.

Angles in Oblique. Those angles appearing on the front face, or on a face parallel to the front, are laid out in their actual size. Any angle visible in the oblique plane must be drawn first in orthographic. The offset dimensions of the angle are laid out on the oblique plane. An angle to be drawn in cabinet or general type oblique means that one of the offset dimensions must be shortened. The amount of decrease is the same as the scale of the oblique axis.

DIMETRIC PROJECTION

A dimetric projection, as stated before, is a form of axonometric projection. When making a dimetric drawing, the object is turned so that two of the axes make the same angle with the plane of projection while the third is at a different angle. Edges which are parallel to the first two axes are drawn to the same scale. The edge parallel to the third axis is drawn to a different scale. Since two different scales are used, less distortion will be apparent. A dimetric drawing is laid out the same as the isometric drawing. A dimetric drawing is shown in Fig. 9–2.

In dimetric the angles formed by the receding axes and their scales are many and varied. Fig. 9–32 shows some of the more generally used positions of axes and scales. These scales and angles are approximate. If an adjustable triangle is not available, lay out the angles with a protractor. Use two triangles to draw lines parallel to the axes. Each of these axes may be reversed or inverted to show the piece to its best advantage.

Dimetric templates are available to aid in the construction of dimetric drawings. These templates not only have the elliptical openings but the appropriate scales as well. The templates shown are based on axes placed at 39.23 and 11.53 degrees. Scales are full size and .623 to 1. Grid paper is available to simplify sketching in dimetric. See Fig. 9–33.

TRIMETRIC PROJECTION

A trimetric projection is a form of axonometric projection. When making a trimetric

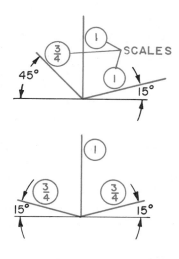

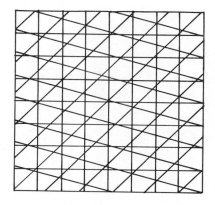

9-33. *Dimetric sketch paper with lines on 15 and 45 degrees.*

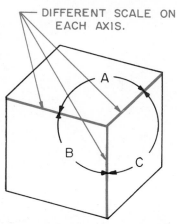

EACH AXIS IS ON A DIFFERENT ANGLE.

9-34. *An example of trimetric axes.*

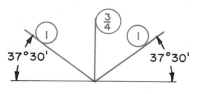

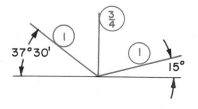

9-32. *These scales and axes are used most frequently in dimetric drawing.*

drawing, the object is turned so that each of the axes makes a different angle with the plane of projection. Each of the axes not only has a different angle but a different scale of reduction. See Fig. 9-34. Plastic trimetric angle and scale guides are available for drawing in a variety of positions.

The advantage of trimetric drawings is that they have less distortion than isometric or dimetric drawings. They are more difficult to make, especially if circular elements are present.

PERSPECTIVE DRAWING

A perspective drawing is a three-dimensional pictorial drawing. It shows the object as the eye sees it when looking from a particular point. A perspective drawing is much like a photograph. It makes the object appear lifelike.

When an object, such as a building, is viewed, the parts farthest from the eye appear smaller. See Fig. 9-35. If the edges of the building were extended, they would appear to meet on the horizon. When you are driving down a highway, the parallel sides of the road appear to come together in the distance on the horizon.

This chapter presents the conventional methods for laying out several types of perspectives. In industry these are used very little because they are difficult and time consuming. Industry uses mechanical aids to speed up perspective drawing. These will be discussed. It is important to study conventional practices so the theory of perspective is understood. This makes it possible to make proper use of perspective drawing aids.

Understanding Perspective. The forms of pictorial drawing that have been discussed are based on the principle that all projectors are parallel. In perspective drawing, the projectors are not parallel. They originate at a single point. See Fig. 9-36. This is called the *station point,* noted on drawings as SP. The station point is the eye of the observer. From his eye radiate visual rays to the visible corners of the object. In all forms of

241

Whitesitt Hall, School of Technology, Kansas State College of Pittsburg

9–35. *Objects appear to become smaller the farther they are from the eye. This is the principle used in perspective drawing.*

perspective drawing, the observer views the object through a plane of projection. This plane of projection is called a *picture plane,* noted on drawings as PP. It is on the picture plane that the object is drawn. This is much like standing in front of a window and drawing what you saw *on* the window. The window would act as the picture plane. The object usually rests on the ground plane. The ground plane intersects the picture plane at the *ground line.* It is noted on the drawing as GL. The *horizon line* is the line in the distance to which the visual rays meet. It is on eye level above the ground. It can be located above, below, or through the object. See Fig. 9–37. The *vanishing points* are located

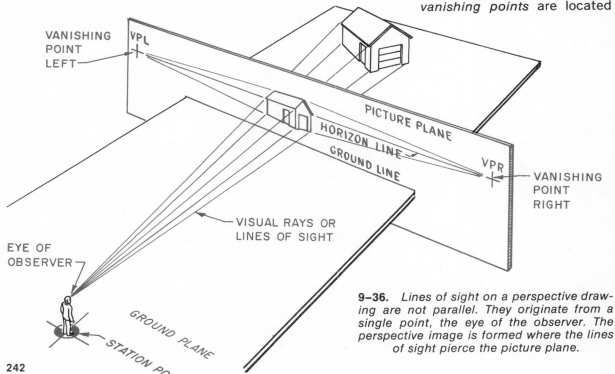

9–36. *Lines of sight on a perspective drawing are not parallel. They originate from a single point, the eye of the observer. The perspective image is formed where the lines of sight pierce the picture plane.*

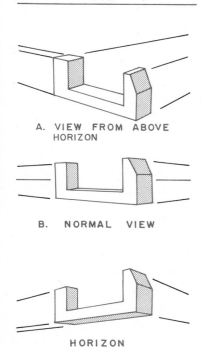

HORIZON

A.   VIEW   FROM   ABOVE

HORIZON

B.   NORMAL   VIEW

HORIZON

C.   VIEW   FROM   BELOW

on the horizon line where the lines of sight meet. These are the points at which the sides of the object appear to meet if they were extended to the horizon. It is noted on the drawing as VP.

Kinds of Perspectives. Two types of perspective drawings are parallel and angular. Fig. 9–38A illustrates *parallel* or one-point perspective, and B shows *angular* or two-point perspective. Angular perspective is the type of perspective with which we are most familiar. Each type of perspective

9–38.   *Each type of perspective has its own advantages.*

9–37. *An object can be placed above, on, or below the horizon.*

has its own advantages for a particular situation.

Parallel Perspective. One-point or *parallel perspective* gains its name from the fact that one of the faces of the object is parallel to the picture plane. Since one of the faces is parallel to the picture plane, only one vanishing point can exist. To prove this, hold an object so your line of sight is perpendicular to the front face and slightly below eye level. The edges that are parallel to your line of sight tend to converge at a single point. Some parallel perspectives have the appearance of being distorted. This distortion is

usually caused by poor placement of the station point.

To draw a parallel perspective, as shown in Fig. 9–39:

1. Draw the edge view of the picture plane, *PP*.

2. Draw the top view of the object with the front face touching the *PP*. In this position the front face of the object will be in its true size and shape. It can be drawn on a sheet of scrap paper since it is not needed after the perspective is finished.

3. Locate the station point, *SP*, in front and to the side of the object. *The SP should be at least twice the length of the object from the PP.* The closer the *SP* is to the *PP*, the more distortion will occur. In Fig. 9–39 the *SP* is placed more than three times the length from the *PP*.

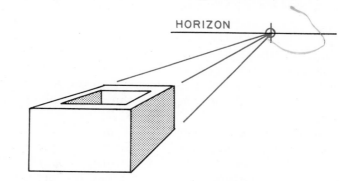

HORIZON

A.   PARALLEL   OR   ONE–POINT   PERSPECTIVE

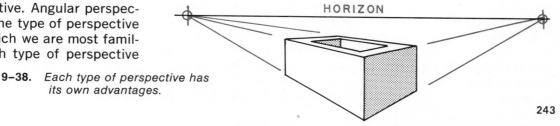

HORIZON

B.   ANGULAR   OR   TWO–POINT   PERSPECTIVE

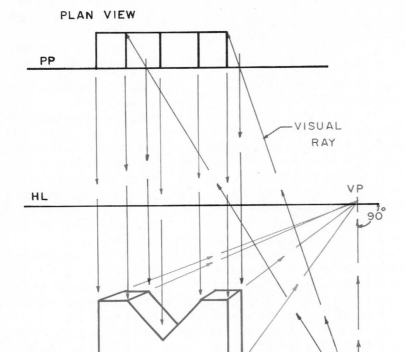

9–39. *A one-point or parallel perspective.*

corner is the location of that corner on the perspective. Repeat this for every corner.

10. Connect the corners to complete the view.

Angular Perspective. Two-point or *angular perspective* gets its name from the fact that the object is placed at an angle to the picture plane. The most frequently used angle is 30 degrees. Other angles commonly used are 15 and 45 degrees. See Fig. 9–40. Angular perspective is one of the most frequently used forms. It is readily adapted to almost any object or situation.

One of the most critical features in angular perspective is the location of the station point. The station point should be located far enough away from the object to prevent distortion. Usually this should be two or three times the width of the object. The closer the *SP* is to the object, the more distortion. The angle formed by the visual rays touching the outside edges of the object should not be any greater than 30 degrees. This angle is called the *angle of clear vision.* See Fig. 9–41.

Fig. 9–42 shows an angular perspective. To draw this type of perspective:

1. Draw the edge view of the picture plane, *PP.* Draw the top view of the object so that one corner, AB, touches the *PP.* This corner will be the only *true length edge* on the perspective.

4. Draw in the ground line, *GL.* It should be located where the bottom edge of the perspective is wanted on the drawing.

5. Lay out the front view of the object. It is projected from the top view to the ground line. Measure and draw the height using the same scale as for the plan view. (In the plan view, the object appears as seen from above.)

6. Draw the horizon line, *HL,* above the *GL.* It is located to give the type of view wanted. See Fig. 9–37 for examples.

7. Locate the vanishing point, *VP.* The *VP* is always located on the *HL* by projecting the *SP* perpendicular to the *HL.*

8. Draw visual rays from the *SP* to the corners of the object in the top view. Where they cross the picture plane is the point to project to the perspective view. Project these corners into the perspective area.

9. Project the corners from the front view to the vanishing point. The point of intersection between the vanishing line and the visual ray for the

244

PLAN
VIEW

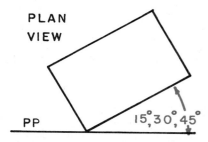

PP   15°,30°,45°

9-40. *The most commonly used angles on two-point or angular perspective.*

2. Locate the station point, *SP*, so the entire object is included in the angle of clear vision. Many times the *SP* is in line with the corner touching the *PP*. The *SP* can be located to the right or left of

PLAN
VIEW

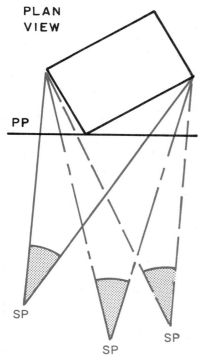

PP

SP
SP   SP

ANGLE OF CLEAR VISION 30° MAX.

9-41. *The location of the station point is governed by the face of the object to be emphasized and the angle of clear vision.*

this point, depending upon which face is to be emphasized.

3. Locate the vanishing point left, *VPL,* and vanishing point right, *VPR,* in the *PP.* To do this, draw lines from the *SP* to the *PP* parallel to the sides of the object in the plan view. This locates the vanishing points in the *PP.* They are marked SPC and SPD.

4. Draw in the ground line, *GL.*

5. Locate the horizon line, *HL,* at the desired distance above, on or below the *GL.* The normal observer's eye is 5'2" to 5'6" above the GL. See Fig. 9-37 for examples.

6. Locate the *VPL* and *VPR* in the elevation view by dropping points SPC and SPD from the *PP* to the *HL.* By dropping verticals from these points to the *HL,* we are now working in the elevation view of the perspective.

7. Project corner AB from the plan view to the ground line in the elevation view. Measure its true length on line AB.

8. Draw visual rays from the *SP* to the main features in the top view. See points E and F.

9. Where the visual rays have pierced the *PP,* drop perpendicular projectors into the perspective area.

10. Draw in the basic box shape of the object. Draw points A', B' to the VPL and VPR. Where the visual rays to E and F intersect the PP, drop projectors to the perspective area. Where they intersect the sides of the box locates the other corners. Corners E and F are now in perspective. Continue locating the other corners in the same manner to complete the perspective.

11. Special problems arise when locating surfaces that are not on the edges of the box. For example, in Fig. 9-43, the location of surface Y must be located. To do this, measure true length distance G on corner AB. AB is true length. The true length of measurement G can be measured here and projected to the proper vanishing point. The vertical edges are found by projecting them from the picture plane.

Circles and Arcs in Perspective. The arc or circle in perspective is seen as an ellipse. Ellipses vary greatly in appearance from those which resemble a circle to those looking like a cigar. Regardless of the ellipse's shape in perspective, it is more readily constructed in a square (trapezoid). By placing the ellipse in a square, it is easier to visualize its shape in perspective.

To draw the shape of a circle inclined to the picture plane, see Fig. 9-44.

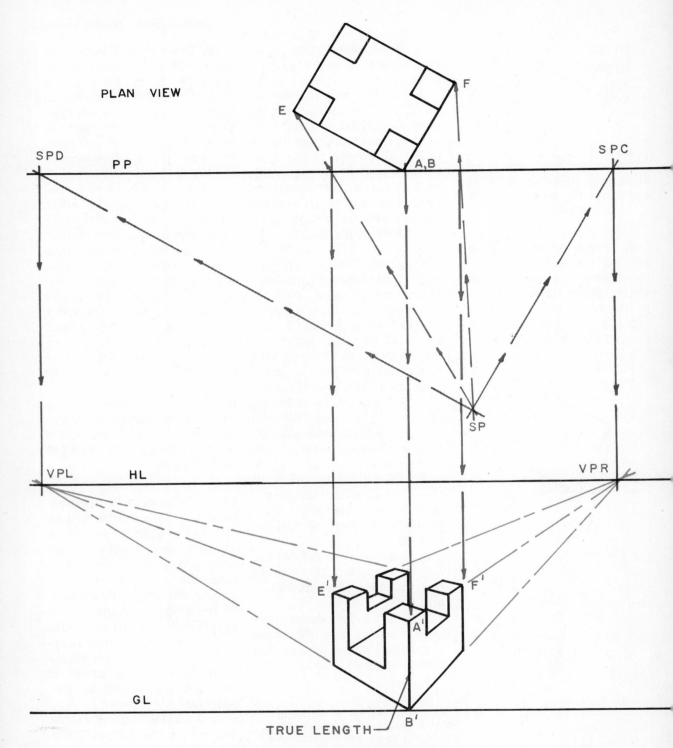

PLAN VIEW

SPD

PP

E

A,B

F

SPC

SP

VPL

HL

VPR

E'

F'

A'

GL

TRUE LENGTH

B'

9-42. *An angular or two-point perspective.*

PROJECTORS FROM
PICTURE PLANE

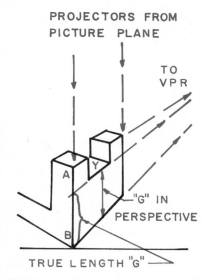

9-43. *Vertical distances are found by measuring their true length on the true length corner. Then project this to the proper vanishing point.*

1. Draw the elevation view of the circle at the side of the perspective area. Divide the circle into any number of equal divisions. It is suggested that the points be located 30 degrees apart. Number each of these divisions.

2. Draw an auxiliary of the circle in the plan view behind the picture plane, *PP*. Divide this circle into the same number of equal parts. Number each of these points identically with those in the elevation. Project each point and number to the plan view of the circle.

3. Draw visual rays from the station point, *SP,* to the numbered points in the plan view.

4. Drop verticals where the

visual rays have intersected the *PP*.

5. Draw the square in perspective by projecting across from the elevation view.

6. Draw horizontals from the points on the circle in the elevation view to the edge of the square. Carry each of these points from the edge of the square to the vanishing point, *VP*.

7. The coordinate points are located by the intersection of the verticals dropped from the *PP* and the vanishing lines drawn to the *VP*.

8. Connect the points with an irregular curve.

One of the least confusing methods used to draw a circle lying in the horizontal plane is by the *ground line method.* In this method the ground line, *GL,* is temporarily moved from its original position to the height of the circle. Fig. 9–45 shows how to plot a circle in two-point horizontal perspective.

1. Enclose the circle in a square, ABCD, in the plan view.

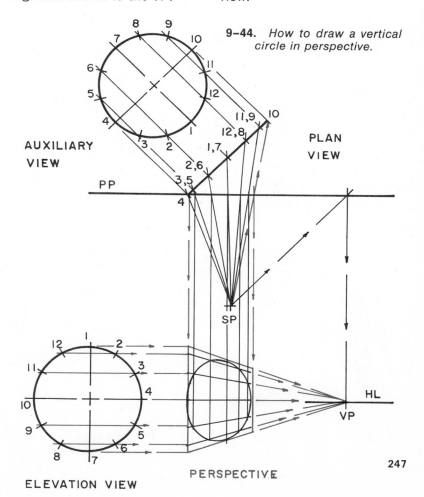

9-44. *How to draw a vertical circle in perspective.*

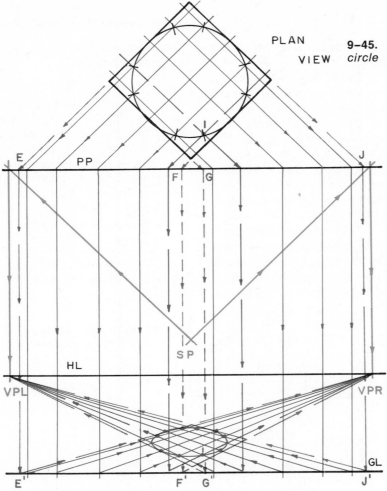

PLAN VIEW

9–45. *How to draw a horizontal circle in perspective using the ground line method.*

2. Divide the circle into equal parts.

3. Extend the sides of the square to the *PP* points E, F, G, and J. Where the sides of the square extended have touched the *PP,* drop verticals to the *GL.* From each of these points, carry lines to the vanishing point right and left, *VPR* and *VPL.* The square is thus formed in perspective.

4. Carry each of the points on the circle parallel to the sides of the square to the *PP.* Drop verticals from the *PP* to the *GL.* From these points of intersection on the *GL* extend lines to the vanishing points.

5. Where the corresponding lines to the vanishing points intersect, form the coordinates for the ellipse.

6. Draw in a smooth curve through each of the points.

True Perspective. True perspective foreshortens length,

248

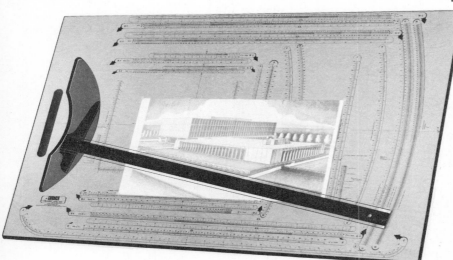

9–46. *A perspective drawing board. The scales permit drawing one, two, and three-point perspectives. The drawings are true perspective, eliminating the distortion found in graphic methods.*

John H. Klok, Inventor
The Bruning Division,
Addressograph Multigraph Corp.

width, and height. One- and two-point perspectives do not shorten the height. Fore-shortened height is especially valuable in making architectural perspective drawings of large buildings. The buildings are more lifelike and appear as they do to the eye.

There are a number of drafting devices available to use to draw true perspectives. These establish three vanishing points. One of these devices is shown in Fig. 9–46. The perspective drawing board contains a series of graduate scales and eight vanishing points. The extreme left vanishing point is represented by a concave curve. There are five vanishing points on the

9–47. *A true perspective drawn on the Bruning-Klok perspective board shown in Fig. 9–46.*

horizon line. There are two additional vanishing points below the horizon line.

This type of drafting device greatly speeds up perspective drawing. It increases the accuracy of the finished drawing. A true perspective is shown in Fig. 9–47.

Other Pictorial Techniques. Chapter 17, Technical Illustration, gives additional details on pictorial drawing. After you master the fundamentals of this chapter you are ready to try to draw some technical illustrations.

Build Your Vocabulary

Following are some terms you should understand and use as part of your working vocabulary. Write a brief explanation of what each means:

Isometric drawings
Dimetric drawings
Oblique drawings
Axes
Isometric lines
Nonisometric lines
Cavalier drawings
Cabinet drawings
Four-center ellipse
Parallel perspective
Angular perspective

Station point
Horizon
Picture plane
Ground line
Vanishing points

Study Problems

1. Make an isometric drawing of the try square, P9–1.
2. Make an isometric drawing of the carbide cutting tool inserts, P9–2. Draw twice actual size.
3. Make an isometric drawing of the diamond point cutting tool, P9–3.

4. Make an isometric drawing of the stop block, P9–4.
5. Make an isometric drawing of the machine tool cutter grinding wheel, P9–5.
6. Make an oblique drawing of the grinding wheel, P9–5. Compare it with the isometric drawing. Which was easier to draw? Which looks the most natural?
7. Make an isometric drawing of the jaw for an inside micrometer, P9–6. Notice it has a long irregular curve.
8. Make an isometric drawing of the crane trolley bracket, P9–7.

9. Make an oblique drawing of the crane trolley bracket, P9–7.

10. Make a cabinet oblique drawing of the cabinet shown in P9–8.

11. Make an oblique drawing of the packing fitting, P9–9. Draw it with the axis on 45 degrees. Then draw it with the axis on 30 degrees. Which appears more natural?

12. Study the objects shown in P9–10 through P9–15. Select the method of pictorial drawing that would be best to show each. Draw them with the method chosen. Compare your drawing with others who chose a different method or angle for the receding axis. Was your decision best?

13. Make a perspective drawing of the cabinet shown in P9–8. Draw the plan view to the scale 1/4" = 1'0". Locate the station point ten feet from the picture plane. Position the station point so both sides of the building are clearly visible. Place the horizon 6'0" above the ground line.

14. Make a perspective of the summer cottage in P9–16. Draw the plan view to the scale 1/4" = 1'0". Position the station point and horizon at the place you think gives the best perspective.

15. Make a dimetric drawing of the A-frame cottage. Select any of the recommended axes.

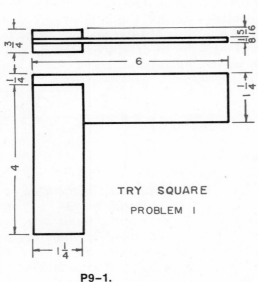

TRY SQUARE
PROBLEM 1

P9–1.

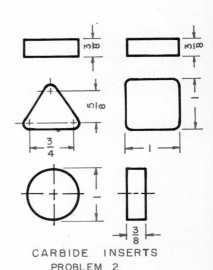

CARBIDE INSERTS
PROBLEM 2

P9–2.

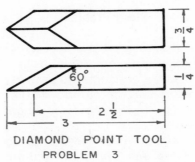

DIAMOND POINT TOOL
PROBLEM 3

P9–3.

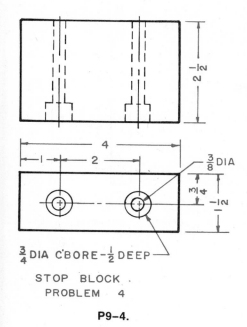

$\frac{3}{8}$ DIA

$\frac{3}{4}$ DIA C'BORE - $\frac{1}{2}$ DEEP

STOP BLOCK
PROBLEM 4

P9-4.

$\frac{1}{2}$ DIA

GRINDING   WHEEL
PROBLEM 5

P9-5.

P9-6.

$\frac{1}{8}$ R

$\frac{3}{8}$ DIA

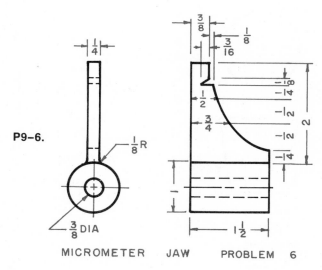

MICROMETER   JAW   PROBLEM 6

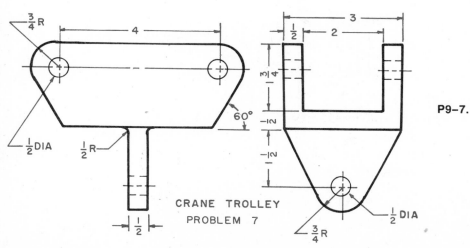

$\frac{3}{4}$ R

$\frac{1}{2}$ DIA

$\frac{1}{2}$ R

60°

P9-7.

CRANE TROLLEY
PROBLEM 7

$\frac{1}{2}$

$\frac{1}{2}$ DIA

$\frac{3}{4}$ R

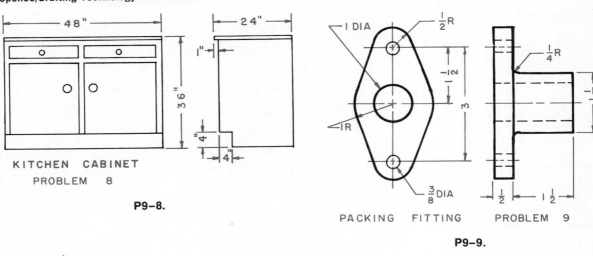

KITCHEN CABINET
PROBLEM 8

P9-8.

PACKING FITTING   PROBLEM 9

P9-9.

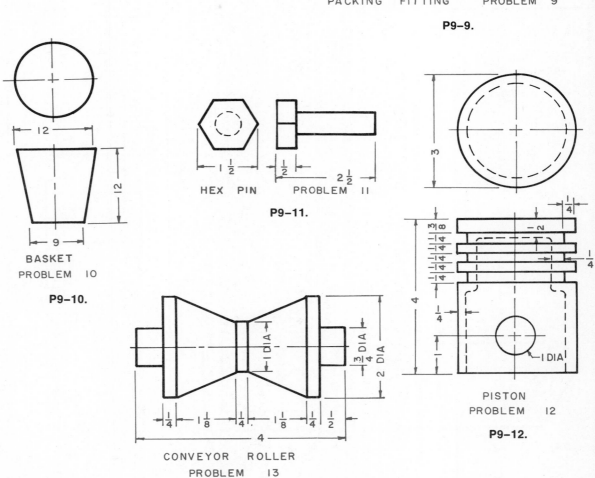

BASKET
PROBLEM 10

P9-10.

HEX PIN   PROBLEM 11

P9-11.

PISTON
PROBLEM 12

P9-12.

CONVEYOR ROLLER
PROBLEM 13

P9-13.

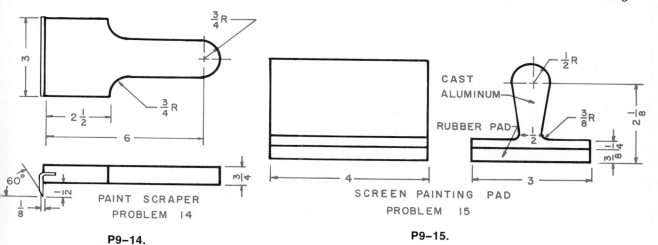

$\frac{3}{4}$R

$\frac{3}{4}$R

3

$2\frac{1}{2}$

6

60°

$\frac{3}{4}$

$\frac{1}{2}$

$\frac{1}{8}$

PAINT SCRAPER
PROBLEM 14

P9–14.

CAST
ALUMINUM

RUBBER PAD

$\frac{1}{2}$R

$\frac{3}{8}$R

$\frac{1}{2}$

4

3

$2\frac{1}{8}$

$\frac{1}{4}$

$\frac{3}{8}$

SCREEN PAINTING PAD
PROBLEM 15

P9–15.

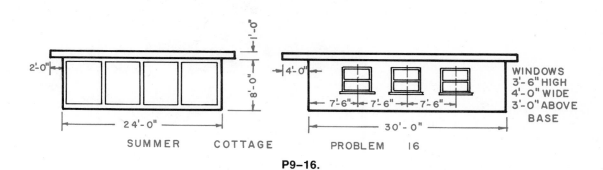

1'-0"

2'-0"

8'-0"

24'-0"

SUMMER   COTTAGE

4'-0"

7'-6"   7'-6"   7'-6"

30'-0"

WINDOWS
3'-6" HIGH
4'-0" WIDE
3'-0" ABOVE
BASE

PROBLEM   16

P9–16.

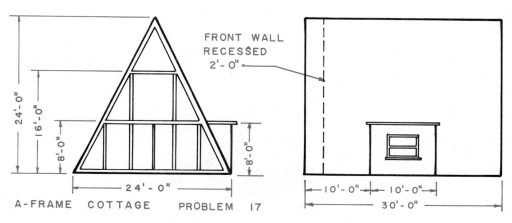

24'-0"

16'-0"

8'-0"

24'-0"

8'-0"

FRONT WALL
RECESSED
2'-0"

10'-0"   10'-0"

30'-0"

A-FRAME COTTAGE   PROBLEM   17

P9–17.

Chapter 10

Sectional Views

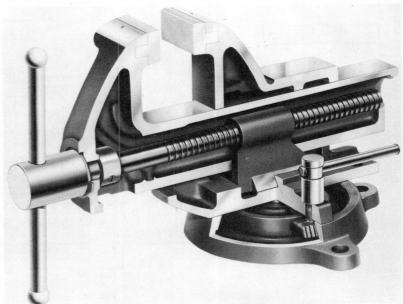

The Columbian Vise and Manufacturing Co.

10–1. *A sectional view is drawn as if part of the object is cut away, leaving the interior exposed. This is a technical illustration prepared for advertising purposes.*

Some objects have interior details that cannot be seen in the exterior views on a multiview drawing. If these interior details are simple, they are shown with hidden lines. If the hidden lines do not clearly show interior details, a sectional view is drawn. See Fig. 10–1.

A sectional view is made by passing an imaginary cutting plane through an object. One part is removed, leaving the interior details exposed. See Fig. 10–2A. It is much the same as sawing through the object and viewing the inside.

In Fig. 10–2 the interior details of the valve are not clear on the three-view drawing shown at C. The imaginary cutting plane (A) shows how

the object could be cut to show interior details. If the front part was actually removed, the valve would appear as shown at B. The multiview drawing with the sectional view is shown at D.

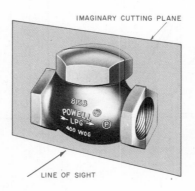

IMAGINARY CUTTING PLANE

LINE OF SIGHT

10–2A. *The imaginary cutting plane is passed through the lift valve parallel to the frontal plane.*

The William Powell Company

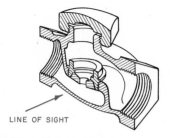

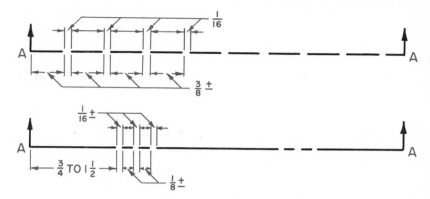

10–2B. *The surfaces cut by the cutting plane are exposed when the front part is removed.*

10–3. *Standard cutting plane symbols.*

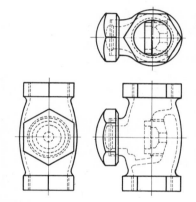

10–2C. *Three-view drawing of a lift check valve.*

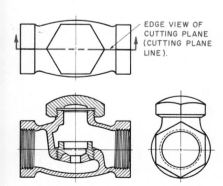

10–2D. *Surfaces cut by cutting plane.*

CUTTING-PLANE LINE

The imaginary cutting plane is shown on a drawing by symbols, Fig. 10–3. Both cutting plane symbols are used as standard practice. The cutting-plane line is placed on the view in which it appears in edge view. It is from this view that the section is projected. See Fig. 10–2, where the section is projected from the top view.

Usually the cutting-plane line has arrows at each end. These are at right angles to the line. The arrows point in the direction of the line of sight used to make the section.

The section is identified by placing capital letters by each arrowhead, as shown in Fig. 10–9. If it is obvious where the section has been taken, the letters are not used. If more than one section is made, letters are very useful to identify the sections.

If the section is made along the center line of a symmetrical object, the center line is not used. See Fig. 10–11.

Sectional views are projected in the same manner as regular orthographic views. The lines of sight are perpendicular to the cutting-plane line.

SECTION LINING

The surface exposed by the cutting plane is indicated by drawing very thin lines called section lines on the surface.

Symbols have been developed to indicate the type of material in the part sectioned. The commonly accepted symbols are shown in Fig. 10–4. These symbols indicate only the general type of material, such as steel, brass, and cast iron. These materials are made in many different types. For example, there are hundreds of different types of steel. The symbol does not give this information. It is given with a note.

It is standard practice to use the general symbol (cast iron) for all materials. If the specific material symbols are

255

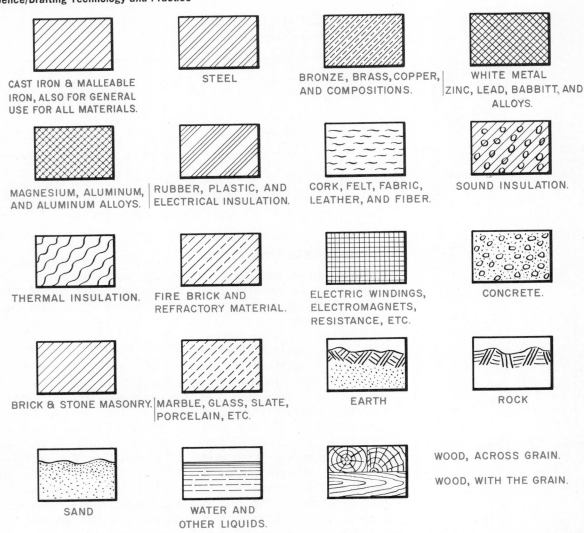

CAST IRON & MALLEABLE IRON, ALSO FOR GENERAL USE FOR ALL MATERIALS.

STEEL

BRONZE, BRASS, COPPER, AND COMPOSITIONS.

WHITE METAL ZINC, LEAD, BABBITT, AND ALLOYS.

MAGNESIUM, ALUMINUM, AND ALUMINUM ALLOYS.

RUBBER, PLASTIC, AND ELECTRICAL INSULATION.

CORK, FELT, FABRIC, LEATHER, AND FIBER.

SOUND INSULATION.

THERMAL INSULATION.

FIRE BRICK AND REFRACTORY MATERIAL.

ELECTRIC WINDINGS, ELECTROMAGNETS, RESISTANCE, ETC.

CONCRETE.

BRICK & STONE MASONRY.

MARBLE, GLASS, SLATE, PORCELAIN, ETC.

EARTH

ROCK

SAND

WATER AND OTHER LIQUIDS.

WOOD, ACROSS GRAIN.

WOOD, WITH THE GRAIN.

10–4. *Standard symbols for indicating materials in section.*

used, it is usually on a drawing where it is necessary to point out differences in materials. See Fig. 10–5. In this drawing the lathe foot is cast iron, the leveling pad steel, and the blocking wood.

The section lines in the general purpose symbol are spaced from 1/16 to 1/8 inch apart. The actual spacing used depends upon the drawing. On smaller areas the spacing is closer together than larger areas. See Fig. 10–6. Spacing on most drawings should be about 3/32 inch.

Very thin parts, such as sheet metal or gaskets, are sectioned solid black. If several thin pieces are touching, a small white space is left between the solid sections. See Fig. 10–7.

Section lines are usually spaced by eye. A spacing guide can be made by using graph paper. The graph paper is placed under the paper upon which the drawing is

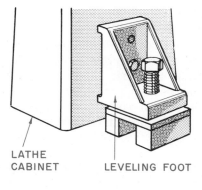

LATHE
CABINET

LEVELING FOOT

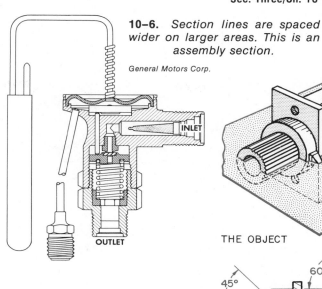

INLET

OUTLET

10-6. *Section lines are spaced wider on larger areas. This is an assembly section.*

General Motors Corp.

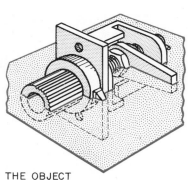

THE OBJECT

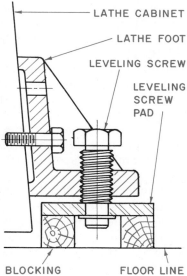

LATHE CABINET

LATHE FOOT

LEVELING SCREW

LEVELING SCREW PAD

BLOCKING   FLOOR LINE

Atlas Press Co.

10-5. *Leveling foot for a lathe. The use of material symbols helps to clarify the drawing.*

10-7. *Thin parts are sectioned solid.*

THE SECTION

45°   60°   30°

10-8. *A full section cuts the object fully through. Notice the different angles used on the section lining.*

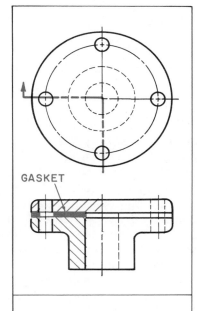

GASKET

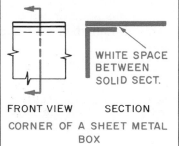

FRONT VIEW   SECTION

WHITE SPACE BETWEEN SOLID SECT.

CORNER OF A SHEET METAL BOX

being made. The printed lines show through and can be used for spacing.

The section lines are drawn on a 45-degree angle. Other angles can be used if the object has several parts. See Fig. 10–8. The angle can be changed if the 45-degree angle would make the section lines parallel with a side of the area sectioned. The use of different angles makes each part sectioned stand out clearly.

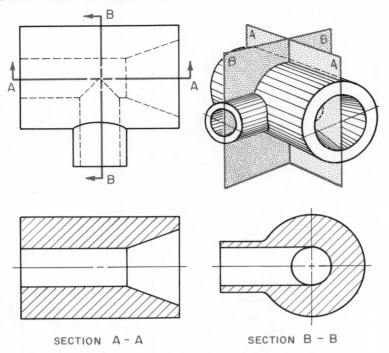

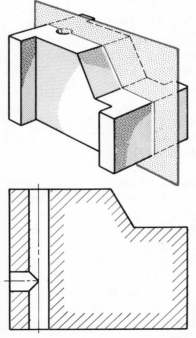

SECTION A - A

SECTION B - B

10-9. *When a part is sectioned in several views, the section lining is on the same angle and direction in all views.*

If a single piece is cut two or more times by the cutting plane, the section lining is in the same direction and at the same angle at each cut. See Fig. 10-8. Regardless of the number of views shown in section of the same piece, the section lining will be in the same direction. See Fig. 10-9.

Very large areas can be section lined as shown in Fig. 10-10. Only the outline of the area has lines drawn upon it. This is called *outline sectioning.*

10-10. *Section through a dial indicator base. Notice that outline sectioning is used on the large area.*

VISIBLE, HIDDEN, AND CENTER LINES

Sectional views are drawn to clarify a hidden interior detail. Since it would add unnecessary confusion to the sectional view to show parts of the object hidden behind the cut, surface *hidden lines* are usually omitted. Hidden lines are used only if they are essential to the clearness of a sectional view or if their use makes it practical to omit a regular external view. See Fig. 10-11.

Visible edges occurring behind the cut surface are usually drawn as shown in Fig. 10-11. In special cases, such as a removed or auxiliary section, visible features may be omitted. See Fig. 10-20.

Center lines should be placed on sectional views. The cutting-plane line has precedence over a center line. The edges of the cut surfaces are always drawn as visible lines.

FULL SECTIONS

A full section is one in which the cutting plane extends *fully* through the entire object, cutting it into two pieces. This is why it is called a *full section.* It can be parallel to the frontal, profile, or horizontal planes. Full sections are shown in Figs. 10-2, 10-5, 10-6, and 10-8.

OFFSET SECTIONS

The offset section is usually a full section. An offset section is one in which the cutting plane changes direction from the main axis of the object to show details not in a straight line. The cutting-plane line always changes direction at right angles. The change of direction is not shown in the sectional view. Notice in the drawing of the drill press base, Fig. 10–12, that more detail is revealed by the offset section than would be possible by a single full section.

HALF SECTIONS

If an object is symmetrical, interior detail can be fully described by cutting halfway through the object. Since the cutting plane passes through one half of the object, it is called a *half section*. See Fig. 10–13. The half of the object not removed shows the exterior of the object. One such view shows both internal details and external features.

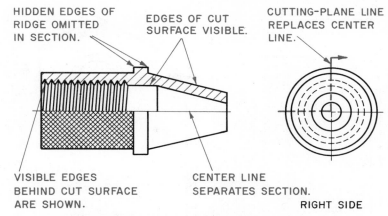

HIDDEN EDGES OF RIDGE OMITTED IN SECTION.

EDGES OF CUT SURFACE VISIBLE.

CUTTING-PLANE LINE REPLACES CENTER LINE.

VISIBLE EDGES BEHIND CUT SURFACE ARE SHOWN.

CENTER LINE SEPARATES SECTION.

RIGHT SIDE

FRONT VIEW OF A CHUCK HOUSING IN HALF SECTION.

10–11. *The use of hidden lines, visible lines, and center lines on section views.*

10–12. *The cutting plane for an offset section changes direction at 90-degree angles.*

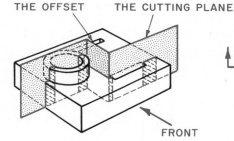

THE OFFSET

THE CUTTING PLANE

FRONT

DRILL PRESS BASE WITH OFFSET CUTTING PLANE.

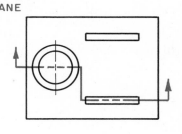

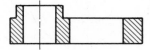

NOTICE THE OFFSET DOES NOT PROJECT TO THE SECTION VIEW.

DRILL PRESS BASE WITH PORTION IN FRONT OF CUTTING PLANE REMOVED.

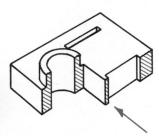

FRONT

10–13. *A half section is cut half way through the object. This is a turret tool post for a metal lathe.*

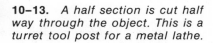

Enco Manufacturing Co.

259

CUTTING PLANE

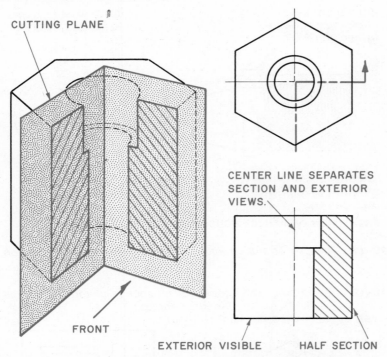

CENTER LINE SEPARATES
SECTION AND EXTERIOR
VIEWS.

FRONT

EXTERIOR VISIBLE   HALF SECTION

10–14. *A half section is made by passing the cutting plane half way.*

The half section is commonly separated from the exterior half by a center line. See Fig. 10–14.

Usually hidden lines are omitted on both sides of a half section. They are unnecessary since the section shows the hidden details. They can be used if needed for dimensioning. See Fig. 10–15.

BROKEN-OUT SECTIONS

Occasionally a section of one small interior detail is desired. Rather than draw an entire half or full section, a partial section, called a *broken-out* section, is drawn. It is assumed that the cutting-plane passes through the feature even though the cutting plane line is not drawn. A freehand break line limits the sectioned area. See the

ALL RADII TO BE $\frac{1}{8}''$
UNLESS OTHERWISE
SPECIFIED.

I. T. E. Circuit Breaker Co.

10–15. *A ceramic electrical spool insulator. The hidden lines aid in dimensioning.*

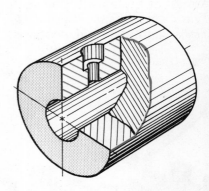

FREEHAND BREAK LINE

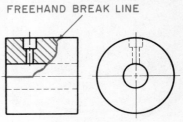

10–16. *Hidden details can be shown by a broken-out section.*

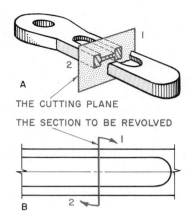

THE CUTTING PLANE

THE SECTION TO BE REVOLVED

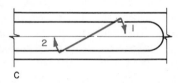

The section revolved parallel to the frontal plane.

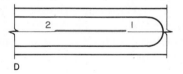

A revolved section is a "slice" across an object, turned until it is parallel to a plane of projection.

10–17. *The section to be revolved is perpendicular to the plane to which it is to be projected.*

broken-out section showing the counterbored hole in a bushing, Fig. 10–16.

REVOLVED SECTIONS

A revolved section is used to describe the cross section of an object, such as a spoke, rib, bar, or other details at one special point. This is done by cutting a section perpendicular to the length of the

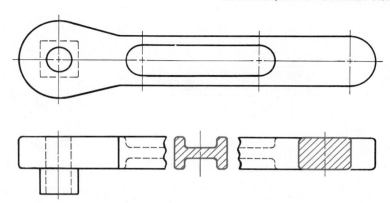

10–18. *The revolved section is drawn directly on a regular view.*

part to be described. This section is then revolved until it is parallel to a plane of projection. In Fig. 10–17 this section is shown as a "slice" taken through the handle of a ratchet wrench. At B the proposed section is located on the top view of the handle and the direction of rotation is indicated. At C the rotation is started. At D it is parallel to the frontal plane. Here it is in position to be projected to the front view and will appear true size in this view.

In Fig. 10–18 a typical two-view drawing of the ratchet wrench is shown. The section is revolved and drawn on the front view. Notice that the section is drawn directly on the front view. The center line which serves as the axis of revolution is always shown. Usually the revolved section is drawn with the object lines at the sides shown. See Fig. 10–18. The object lines can be broken if it makes the drawing easier to understand. See Fig. 10–18.

A frequent mistake made in projecting the revolved section is shown in Fig. 10–19. The section projects true size and shape regardless of the contour of the object.

TOP VIEW

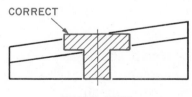

CORRECT

FRONT VIEW

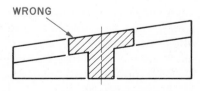

WRONG

FRONT VIEW

10–19. *The revolved section is true in size and shape.*

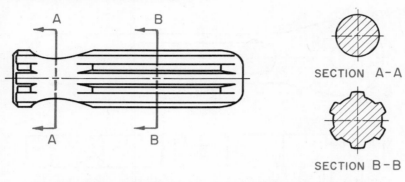

SECTION A-A

SECTION B-B

10-20. *Removed sections are drawn off the regular views.*

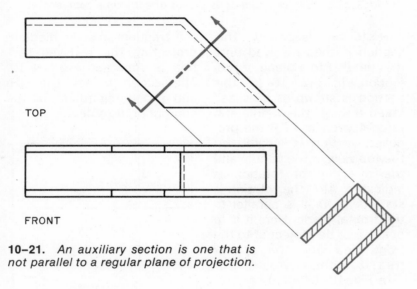

TOP

FRONT

10-21. *An auxiliary section is one that is not parallel to a regular plane of projection.*

drawing or in a vertical position with the first section on top.

Whenever possible the center line of the removed sections should be parallel to the corresponding lines in the normal projected position. Rotation of the view would make the drawing more difficult to read.

AUXILIARY SECTIONS

Another type of section is the auxiliary section. Some objects have angular elements. If a section through the object is to show the true size and shape at the point to be cut, the cutting plane must be perpendicular to the axis of the section. In this position the angular elements will not show true size on the normal reference planes and must be projected to an auxiliary reference plane. The projection is the same as a normal auxiliary view. See Fig. 10-21.

RIBS AND SPOKES IN SECTION

When sectioning circular objects, any element that is not solid around the axis of the object is not section lined. For example, if the cutting plane passes flatwise through a web, rib, gear tooth, spoke, or other such noncontinuous element, the element is *not section lined.* This is necessary to avoid giving a misleading appearance. The cutting plane A–A, Fig. 10–22, passes flatwise (parallel to the

REMOVED SECTIONS

Another means of clarifying the shape of an object with a sectional view is to remove it completely from the view. These are called *removed sections.* They are usually used when the space on the exterior view is not large enough to draw the required sections, as a revolved section, or if the sectional view is to be dimensioned. Frequently more than one section is needed.

The views are usually placed on the same sheet as the exterior views.

Notice on Fig. 10–20 how the cutting-plane line and section view are labeled so they can be identified. If the removed section is drawn to a larger scale than the exterior view, the scale of the sections must be indicated.

The section views should be arranged in alphabetical order from left to right on the

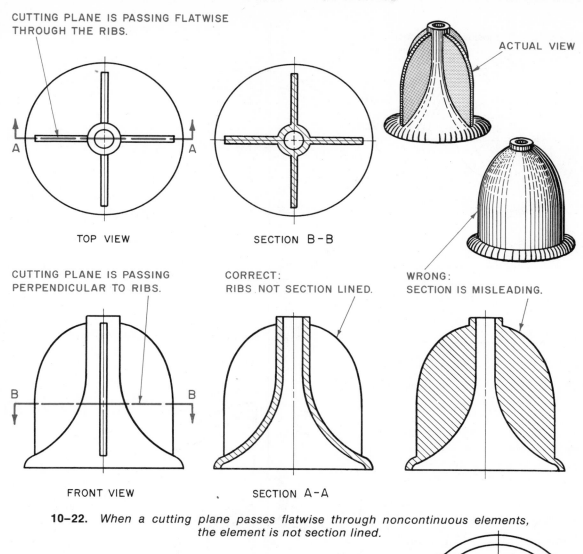

CUTTING PLANE IS PASSING FLATWISE THROUGH THE RIBS.

TOP VIEW

SECTION B-B

ACTUAL VIEW

CUTTING PLANE IS PASSING PERPENDICULAR TO RIBS.

CORRECT: RIBS NOT SECTION LINED.

WRONG: SECTION IS MISLEADING.

FRONT VIEW

SECTION A-A

10-22. *When a cutting plane passes flatwise through noncontinuous elements, the element is not section lined.*

face) through the ribs of a washing machine agitator. If the rib was section lined, it would give the impression that the agitator was a solid unit.

If the cutting plane passes perpendicular to a rib, it is sectioned in the usual manner. This is shown in section B–B of Fig. 10–22.

A steering wheel is shown in Fig. 10–23. It presents a sectioning problem similar to the washing machine agitator. The spokes are not section lined. If they were, the wheel would appear solid, rather than the open effect of spokes.

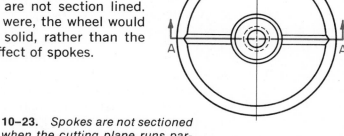

10-23. *Spokes are not sectioned when the cutting plane runs parallel through them.*

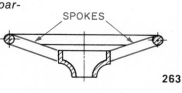

SPOKES

SECTION A-A

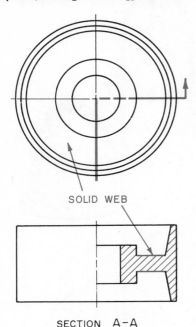

SOLID WEB

SECTION A-A

10–24. *A solid web is section lined.*

A section through a caster having a solid web is illustrated in Fig. 10–24. Since the web is solid, it is section lined.

REVOLVED FEATURES

Some objects needing a section have features that are shown more clearly if that feature is sectioned in a revolved position. Some typical features are ribs, slots and spokes. Fig. 10–25 shows a steel cap that is held in place with bolts placed in the three slots. If the lower slot was not revolved to a position parallel to a normal plane, the section would be confusing. In its revolved position, it appears true size in the section

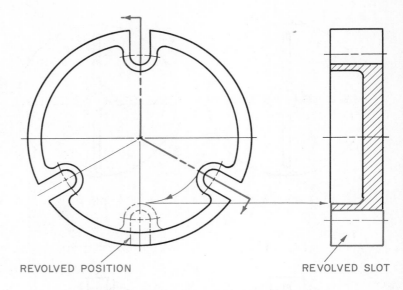

REVOLVED POSITION

REVOLVED SLOT

10–25. *The revolved slot clarifies the view.*

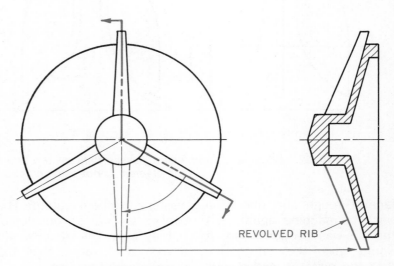

REVOLVED RIB

10–26. *A revolved rib makes the section appear normal.*

view. Fig. 10–26 shows a plastic dial from a television set. One of the lower ribs has been revolved. A band saw wheel is shown in Fig. 10–27. The section labeled "correct" shows the spoke in a revolved position. The section labeled "wrong" shows the spoke as it normally appears in an unrevolved position. Such a view is confusing and does not clearly reveal the desired information.

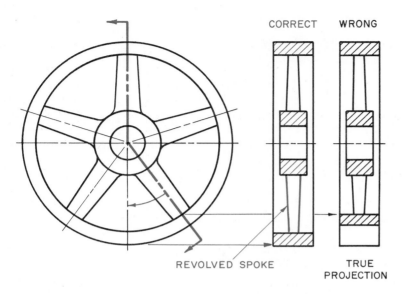

CORRECT   WRONG

REVOLVED SPOKE

TRUE PROJECTION

10-27. *Spokes are revolved until they are parallel to a regular plane.*

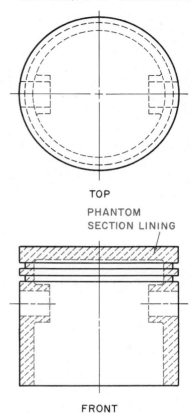

TOP

PHANTOM SECTION LINING

FRONT

10-29. *A phantom section can replace a regular section.*

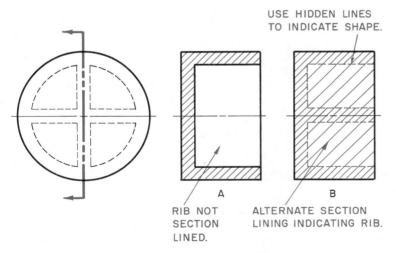

USE HIDDEN LINES TO INDICATE SHAPE.

A
RIB NOT SECTION LINED.

B
ALTERNATE SECTION LINING INDICATING RIB.

10-28. *Alternate section lining helps to identify a rib or other such part.*

ALTERNATE SECTION LINING

In some objects, if a rib or similar part is not section lined when a cutting plane passes through it parallel to its face, it is difficult to tell if the rib exists. The section, Fig. 10–28, shows a rib not section lined. Does a rib appear here or is the space open? The use of alternate section lining shows that it is a rib.

The spacing of the section lining over the rib is drawn using every other section line. The shape is indicated with hidden lines.

PHANTOM SECTIONS

A phantom section is drawn on top of a regular view. It shows interior shapes without drawing a separate sectional view. The section lining symbol used is like that for general materials except it is broken into short dashes. Fig. 10–29 shows a phantom section through a piston.

265

SHAFTS AND FASTENING DEVICES IN SECTION

If the cutting plane passes through a shaft, bearing, or a fastening device in its long direction, the device is not sectioned. See Fig. 10–30. Typical fastening devices are pins, keys, rivets, bolts, and set screws. Sectioning these things would serve no purpose because they have no interior

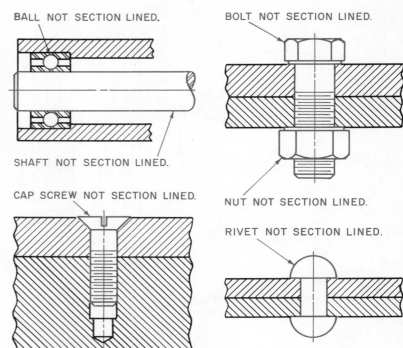

BALL NOT SECTION LINED.

SHAFT NOT SECTION LINED.

CAP SCREW NOT SECTION LINED.

BOLT NOT SECTION LINED.

NUT NOT SECTION LINED.

RIVET NOT SECTION LINED.

10–31.   Typical sections through fasteners.

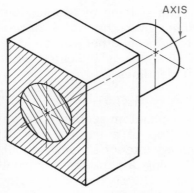

AXIS

When a cutting plane is along the long axis of a shaft, the shaft is not cut and section lined.

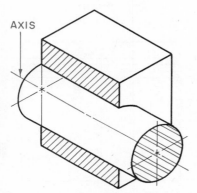

AXIS

When a cutting plane is perpendicular to the long axis of a shaft, the shaft is cut and section lined.

10–30.   *Sectioning shafts.*

details to show. Actually, the sectioning could prove confusing. If these parts are cut across their axis, they are section lined. Typical examples are shown in Fig. 10–31.

ASSEMBLIES IN SECTION

An important use for section drawings is showing how a device operates or how the parts are assembled. They also are used to show the relationship between parts. Sections made through an assembled device are called assembly sections. See Fig. 10–6.

Fig. 10–32 shows a pump. The interior details are complex. An assembly section can

be made to show these details. See Fig. 10–33.

Assemblies in section are discussed in greater detail in Chapter 11, Drawing for Production — Detail and Assembly.

PICTORIAL SECTIONS

Sectioning is used on pictorial drawings of single pieces and pictorial assemblies. Fig. 10–34 is an example of an assembly section drawn isometrically.

Fig. 10–35 shows a pictorial assembly made for use in a sales catalog. This type of drawing is discussed in Chapter 17, Technical Illustration.

10–32. *This pump has a complex interior structure.*

Buffalo Forge Co.

10–33. *An assembly section of the pump shown in Fig. 10–32. Notice that the general section lining symbol is used on all parts.*

Buffalo Forge Co.

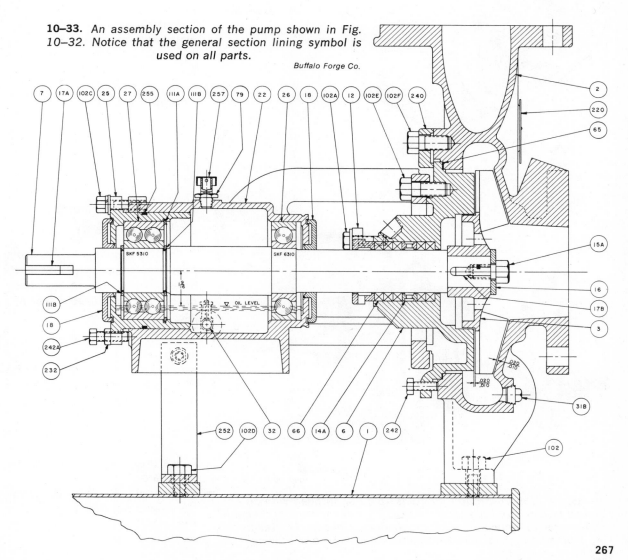

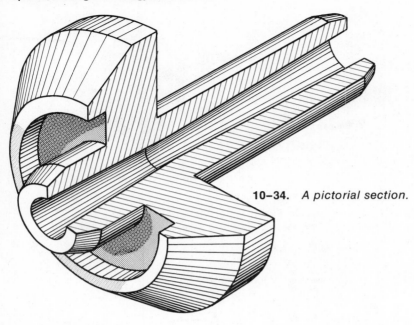

10-34. *A pictorial section.*

Build Your Vocabulary

Following are terms that you should understand and use as a part of your working vocabulary. Write a brief explanation of what each means.

Section view
Section lining
Cutting plane
Material symbols
Full section
Offset section
Half section
Broken-out section
Revolved section
Auxiliary section
Phantom section

De Vlieg Machine Co.

10-35. *A section through an adjustable boring head with an enclosed boring tool holding and adjusting mechanism. This section was made to appear lifelike by shading. Chapter 17, Technical Illustration, gives more information on this type of drawing.*

Study Problems

The following problems have the needed section indicated. Draw the regular views plus the required section.

Full Section

1. Laser shell, P10–1. Select and draw the views to describe the object. Use a full section to reveal interior details. Select your own scale.

2. Special purpose nut, P10–2. Draw the front and side views and a full section. Draw the scale 1″ = ½″.

3. Handwheel, P10–3. Draw the views needed to describe the object. Construct a full section. Draw full size.

Half Sections

4. Auto window handle knob, P10–4. Draw the necessary views and a half section. Select your own scale.

5. Candleholder, P10–5. Draw the required views and the section that best shows the details. Draw to the scale 1" = ½".

6. Hydraulic hose connection, P10–6. Draw the views required to describe the object. Use a section to show interior details. Draw full size.

7. Porcelain electrical bus support, P10–7. Draw the required views and a half section. Draw full size.

8. A steam pipe fitting, P10–8. Draw the front and top view. Show interior details with a half section. Draw to the scale ½" = 1".

Offset Section

9. Base for ball float trap, P10–9. Draw the front and top views. Draw an offset section. Draw full scale.

Broken-Out Section

10. Candlestick, P10–10. Do the front and top views and a partial section to show the hidden details. Draw full size.

Revolved Section

11. Nail set, P10–11. Draw the front and left side view and the indicated revolved sections. Draw full size.

12. Cast crank handle, P10–12. Draw the front and top views. Draw revolved sections one inch and three inches from the right end of the handle. Draw full size.

Removed Section

13. Gate valve handle, P10–13. Draw the necessary views and a removed auxiliary section taken two inches from the end of the handle. Draw full size.

Phantom Section

14. Jig borer tool holder, P10–14. Draw front and side views. Use phantom section lining to show clearly hidden details.

Assembly Section

16. Union for joining steel and plastic pipe, P10–15. Draw the necessary views to describe the object. Make a section to show the shape of the three parts and how they are assembled. Draw full size.

17. Bearing journal, P10–16. Draw front and side views. Draw a section showing interior details. Draw full size.

Study the following problems. Draw the regular views needed and use sections to clarify hidden details or replace a regular view.

18. Flange type gland, P10–17. Draw necessary views to describe the gland. Use a section to show interior details. Draw full size.

19. Machinist's center head, P10–18. Draw the necessary views. Construct an auxiliary section as indicated at A–A.

Draw a revolved section at B–B. Draw full size.

20. Jig borer tool holder, P10–14. Draw the front and right side views. Show interior details with a half section. Draw full size.

21. Sanding drum. Draw the front and side views, P10–19. Construct a section through the drum. Show details of the set screw in the shaft. Draw full size.

22. Plastic mounting bracket, P10–20. Draw the front and side views. Draw an offset section. Draw full size.

23. Cap for a quick-opening gate valve, P10–21. Draw the necessary views and a half section. Draw full size.

24. Cap and body of high pressure steam trap, P10–22A. Draw the needed views of the cap. Draw a section to show interior details. Draw full size.

25. Cap and body of high pressure steam trap, P10–22B. Draw the cap and body as assembled. Make a section to show interior details of these when assembled. Draw full size.

26. Bridge scupper. (A scupper is an opening that permits water to drain off a floor or roof.) Draw the views needed to describe the scupper, P10–23. Make sections showing hidden details on the front and side views. Select your own scale.

27. Bearing journal, P10–16. Draw the front and side views. Use phantom section lining to clearly show each part.

28. Air line filter, P10–24.

Draw the front and top views. Draw a full section. Draw full size.

29. Hydraulic accumulator, *P10–25. The accumulator is part of a hydraulic circuit. It fills with fluid when the circuit is under pressure. The fluid enters at the bottom and forces the piston up against the spring. If the fluid pressure drops, the spring pushes the piston down, forcing fluid into the circuit. This maintains an even pressure.*

Draw the necessary views to describe the accumulator. Make a full section to show interior details. Draw full size.

30. Band saw wheel, bearing, and hub assembly, *P10–26, A–D. Draw the needed views of each part. Section each to clearly show all details. Select your own scale.*

31. Band saw wheel, bearing, and hub assembly, *P10–26, A–D. Draw needed views of the unit assembled. Section the assembled drawing to show construction details and the relationship between parts. Select your own scale.*

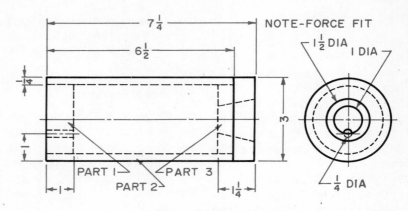

A SHELL FOR A LASER

P10–1.

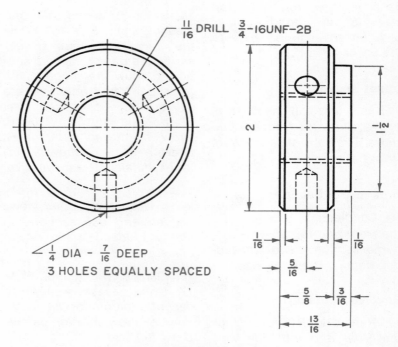

SPECIAL PURPOSE NUT

P10–2.

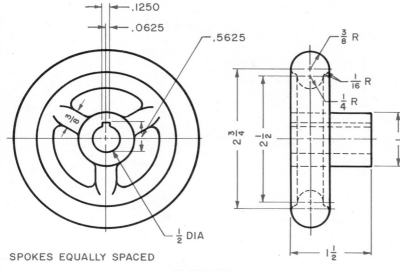

.1250

.0625

.5625

$\frac{3}{8}$ R

$\frac{1}{16}$ R

$\frac{1}{4}$ R

$2\frac{3}{4}$

$2\frac{1}{2}$

1

$1\frac{1}{2}$

$\frac{3}{8}$

$\frac{1}{2}$ DIA

SPOKES EQUALLY SPACED

HANDWHEEL

P10–3.

$\frac{3}{8}$

$\frac{7}{8}$

$\frac{5}{8}$

$1\frac{1}{8}$

$\frac{3}{4}$

$-1\frac{1}{4}$

$-1\frac{1}{4}$

$1\frac{1}{2}$

$\frac{15}{32}$ R

$\frac{1}{8}$ R

$\frac{3}{4}$

$-\frac{1}{8}$

$1\frac{1}{2}$

AUTO WINDOW HANDLE KNOB

P10–4.

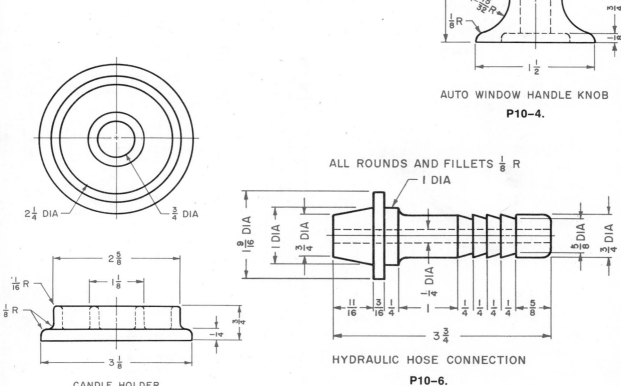

$2\frac{1}{4}$ DIA

$\frac{3}{4}$ DIA

$2\frac{5}{8}$

$1\frac{1}{8}$

$\frac{1}{16}$ R

$\frac{1}{8}$ R

$\frac{3}{4}$

$-1\frac{1}{4}$

$3\frac{1}{8}$

CANDLE HOLDER

P10–5.

ALL ROUNDS AND FILLETS $\frac{1}{8}$ R

1 DIA

$1\frac{9}{16}$ DIA

1 DIA

$\frac{3}{4}$ DIA

$\frac{1}{4}$ DIA

$\frac{5}{8}$ DIA

$\frac{3}{4}$ DIA

$\frac{11}{16}$

$\frac{3}{16}$

$\frac{1}{4}$

1

$\frac{1}{4}$

$\frac{1}{4}$

$\frac{1}{4}$

$\frac{1}{4}$

$\frac{5}{8}$

$3\frac{3}{4}$

HYDRAULIC HOSE CONNECTION

P10–6.

271

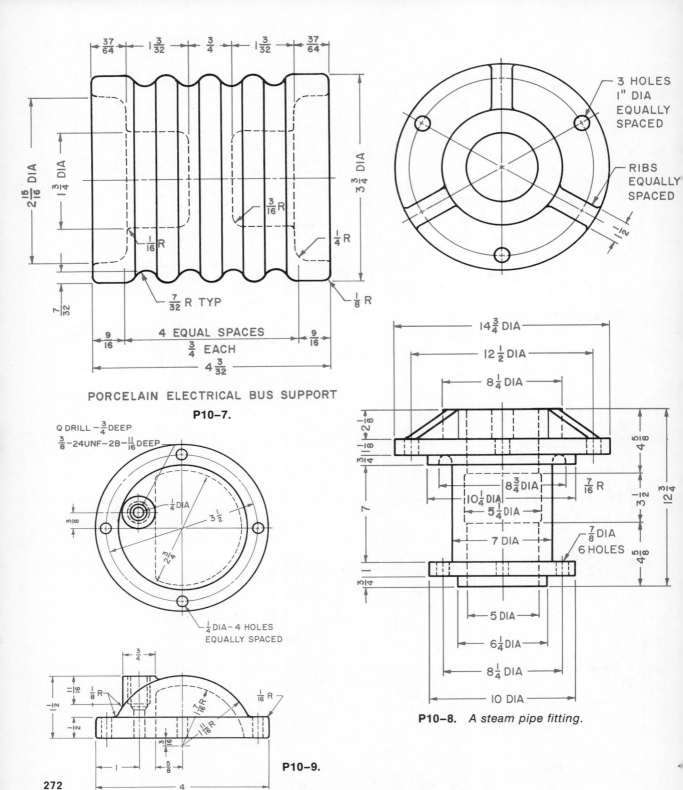

$\frac{37}{64}$ 1$\frac{3}{32}$ $\frac{3}{4}$ 1$\frac{3}{32}$ $\frac{37}{64}$

2$\frac{15}{16}$ DIA

1$\frac{3}{4}$ DIA

3$\frac{3}{4}$ DIA

$\frac{3}{16}$R

$\frac{1}{16}$R

$\frac{1}{4}$R

$\frac{7}{32}$

$\frac{7}{32}$ R TYP

$\frac{1}{8}$ R

$\frac{9}{16}$ 4 EQUAL SPACES $\frac{3}{4}$ EACH $\frac{9}{16}$

4$\frac{3}{32}$

PORCELAIN ELECTRICAL BUS SUPPORT

P10–7.

3 HOLES
1" DIA
EQUALLY
SPACED

RIBS
EQUALLY
SPACED

$\frac{1}{2}$

Q DRILL – $\frac{3}{4}$DEEP
$\frac{3}{8}$ –24UNF–2B– $\frac{11}{16}$DEEP

$\frac{1}{4}$DIA

$\frac{3}{8}$

3$\frac{1}{2}$

2$\frac{3}{4}$

$\frac{1}{4}$ DIA – 4 HOLES
EQUALLY SPACED

$\frac{3}{4}$

$\frac{11}{16}$

$\frac{1}{8}$ R

1$\frac{1}{2}$

$\frac{1}{2}$

$\frac{7}{16}$ R

$\frac{1}{16}$ R

$\frac{11}{16}$R

$\frac{3}{16}$

$\frac{5}{8}$

1

4

P10–9.

BASE FOR BALL FLOAT TRAP

14$\frac{3}{4}$ DIA

12$\frac{1}{2}$ DIA

8$\frac{1}{4}$ DIA

2$\frac{1}{8}$

1$\frac{1}{8}$

$\frac{3}{4}$

4$\frac{5}{8}$

8$\frac{3}{4}$ DIA

$\frac{7}{16}$ R

10$\frac{1}{4}$ DIA

5$\frac{1}{4}$ DIA

7

7 DIA

3$\frac{1}{2}$

12$\frac{3}{4}$

$\frac{7}{8}$ DIA
6 HOLES

1

$\frac{3}{4}$

4$\frac{5}{8}$

5 DIA

6$\frac{1}{4}$ DIA

8$\frac{1}{4}$ DIA

10 DIA

P10–8. *A steam pipe fitting.*

272

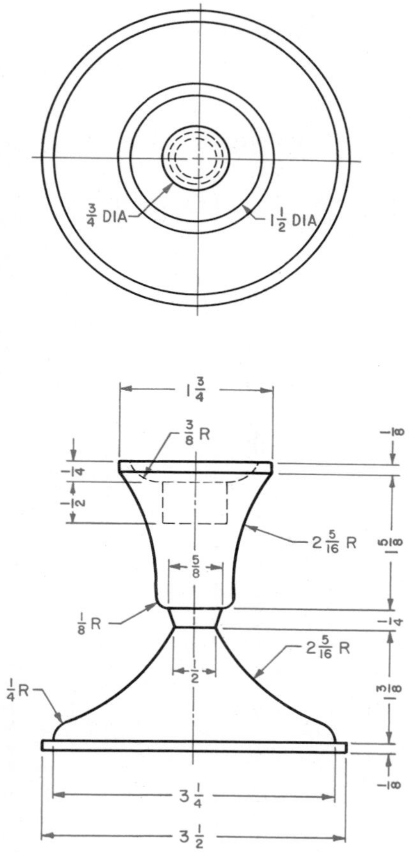

ALL FILLETS AND ROUNDS $\frac{1}{16}$ R

$\frac{3}{4}$ DIA — $1\frac{1}{2}$ DIA

$1\frac{3}{4}$

$\frac{3}{8}$ R

$\frac{1}{4}$

$\frac{1}{2}$

$2\frac{5}{16}$ R

$\frac{5}{8}$

$\frac{1}{8}$ R

$2\frac{5}{16}$ R

$\frac{1}{2}$

$\frac{1}{4}$ R

$3\frac{1}{4}$

$3\frac{1}{2}$

CANDLESTICK

P10–10.

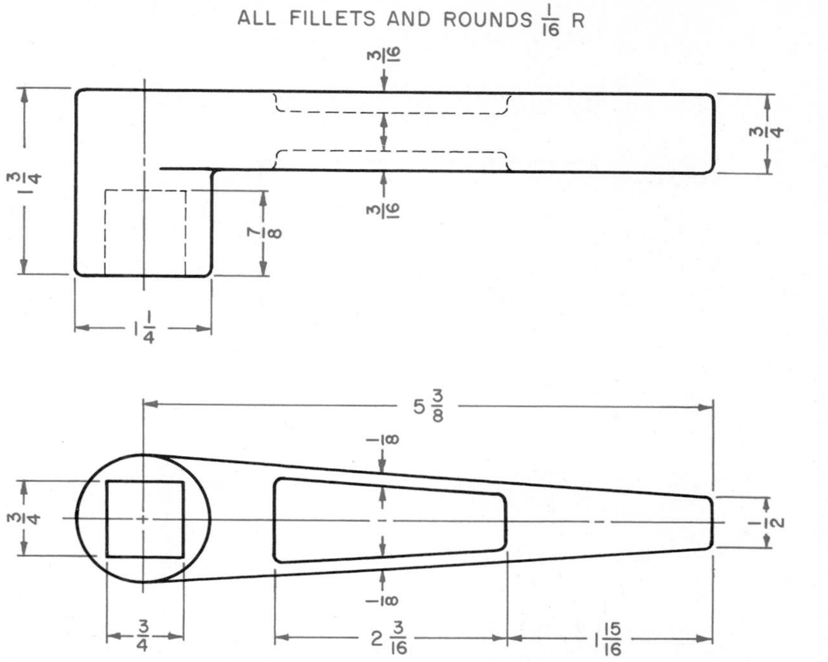

$\frac{3}{16}$

$\frac{3}{4}$

$1\frac{3}{4}$

$\frac{7}{8}$

$\frac{3}{16}$

$\frac{1}{8}$

$1\frac{1}{4}$

$5\frac{3}{8}$

$\frac{1}{8}$

$\frac{3}{4}$

$\frac{1}{2}$

$\frac{1}{8}$

$\frac{3}{4}$

$2\frac{3}{16}$

$1\frac{15}{16}$

CAST CRANK HANDLE

P10–12.

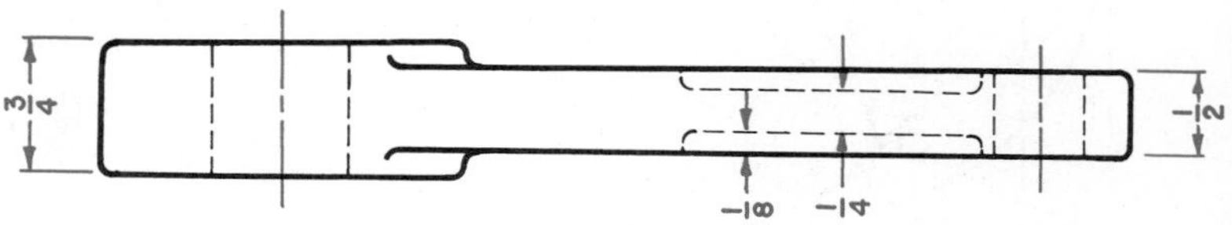

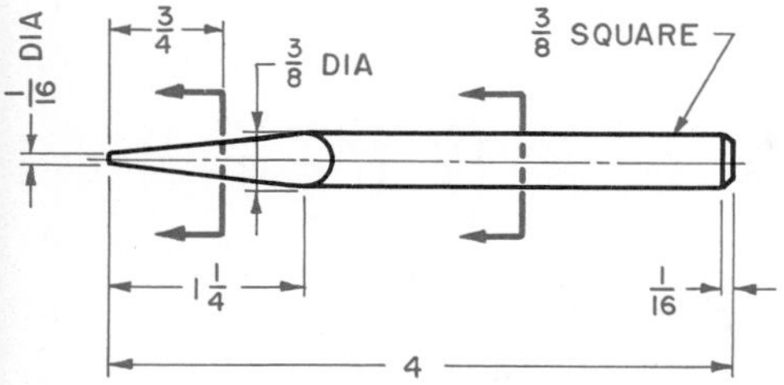

$\frac{3}{4}$

$\frac{1}{2}$

$\frac{1}{8}$

$\frac{1}{4}$

$\frac{1}{16}$ DIA

$\frac{3}{4}$

$\frac{3}{8}$ DIA

$\frac{3}{8}$ SQUARE

$1\frac{1}{4}$

$\frac{1}{16}$

4

NAIL SET

P10–11.

$\frac{1}{2}$ DIA

$\frac{1}{2}$ R

$\frac{1}{4}$ R

$\frac{1}{8}$ R

$\frac{1}{2}$ R

$1\frac{1}{2}$

$\frac{1}{2}$

$\frac{1}{4}$

$\frac{13}{16}$

$\frac{5}{8}$

$45°$

1

$\frac{3}{4}$ DIA

$2\frac{1}{8}$

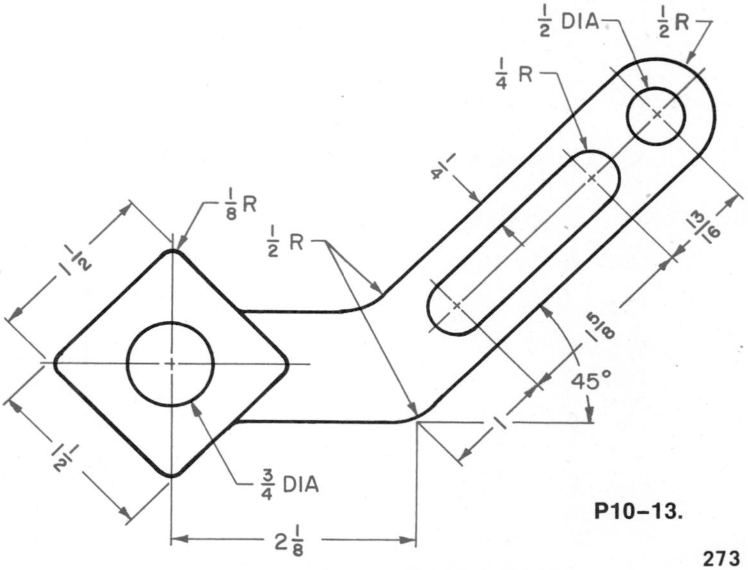

P10–13.

HANDLE FOR A GATE VALVE

273

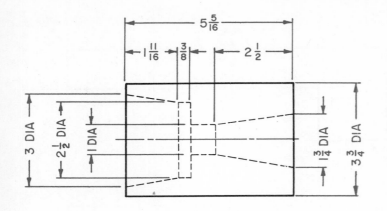

P10–14.

JIG BORER TOOL HOLDER

P10–15.

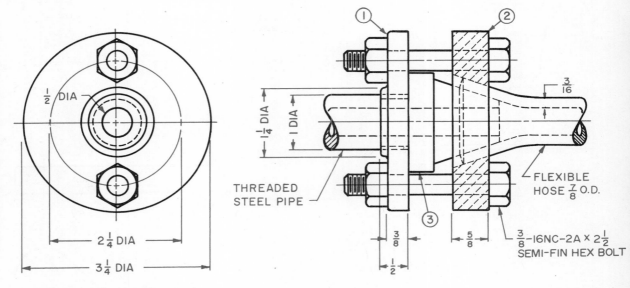

THREADED STEEL PIPE

FLEXIBLE HOSE $\frac{7}{8}$ O.D.

$\frac{3}{8}$-16NC-2A × 2$\frac{1}{2}$ SEMI-FIN HEX BOLT

ALL FILLETS AND ROUNDS $\frac{1}{16}$ R

NO.   PARTS LIST

1   CAP - CAST IRON

2   ANGLE PRESSURE PLATE - STEEL

3   PRESSURE CONE - PLASTIC

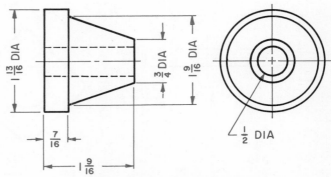

DETAIL OF PRESSURE CONE – PART NO. 3

UNION FOR JOINING STEEL AND PLASTIC PIPE

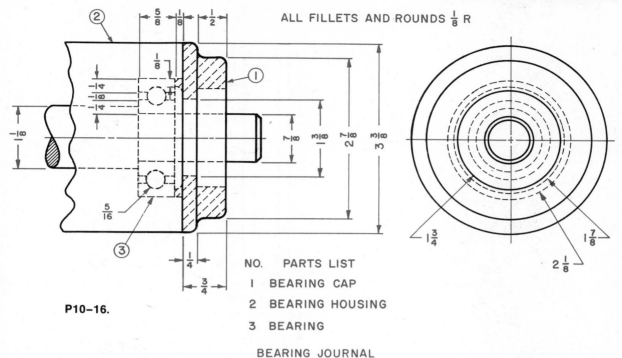

ALL FILLETS AND ROUNDS $\frac{1}{8}$ R

P10–16.

NO. PARTS LIST

1 BEARING CAP

2 BEARING HOUSING

3 BEARING

BEARING JOURNAL

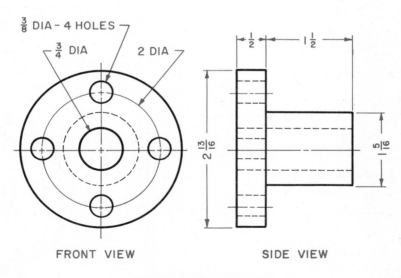

$\frac{3}{8}$ DIA – 4 HOLES

$\frac{3}{4}$ DIA

2 DIA

FRONT VIEW

SIDE VIEW

FLANGE TYPE GLAND

P10–17.

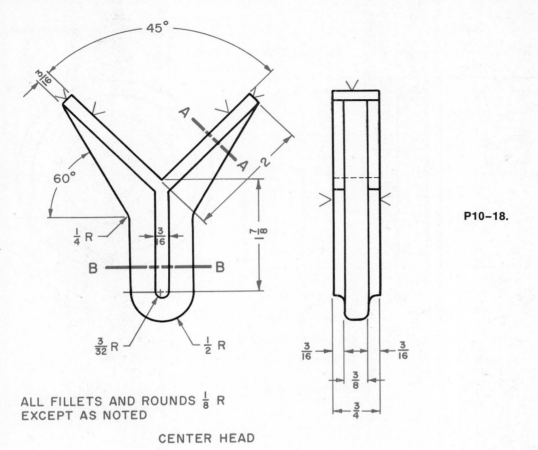

45°

$\frac{3}{16}$

60°

$\frac{1}{4}$ R

$\frac{3}{16}$

A

A

2

$\frac{7}{8}$

B

B

$\frac{3}{32}$ R

$\frac{1}{2}$ R

P10–18.

$\frac{3}{16}$

$\frac{3}{16}$

$\frac{3}{8}$

$\frac{3}{4}$

ALL FILLETS AND ROUNDS $\frac{1}{8}$ R
EXCEPT AS NOTED

CENTER HEAD

P10–19.

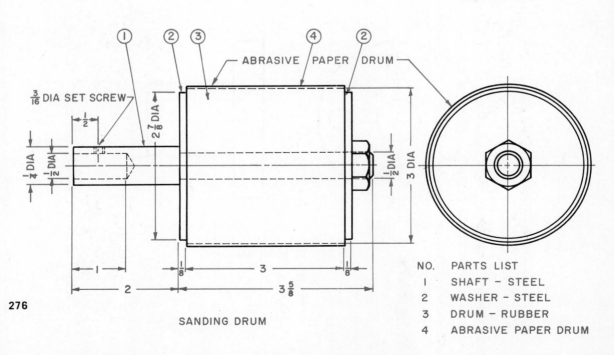

① ② ③ ④ ②

ABRASIVE PAPER DRUM

$\frac{3}{16}$ DIA SET SCREW

$\frac{1}{2}$

$2\frac{7}{8}$ DIA

$\frac{1}{4}$ DIA

$\frac{1}{2}$ DIA

$\frac{1}{2}$ DIA

3 DIA

$\frac{1}{8}$

3

$\frac{1}{8}$

1

2

$3\frac{5}{8}$

276

SANDING DRUM

NO.	PARTS LIST
1	SHAFT – STEEL
2	WASHER – STEEL
3	DRUM – RUBBER
4	ABRASIVE PAPER DRUM

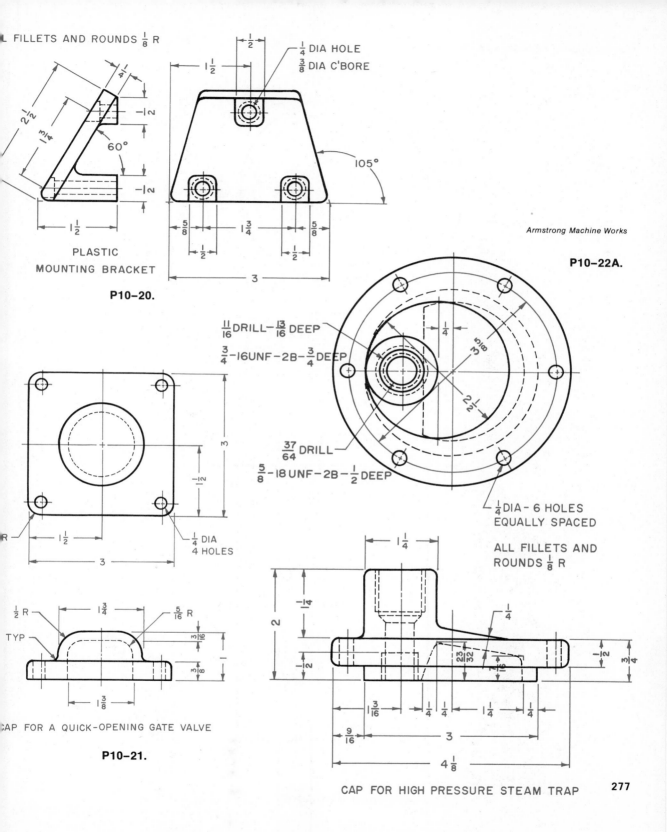

ALL FILLETS AND ROUNDS $\frac{1}{8}$ R

$\frac{1}{4}$ DIA HOLE
$\frac{3}{8}$ DIA C'BORE

2$\frac{1}{2}$

1$\frac{3}{4}$

$\frac{1}{4}$

$\frac{1}{2}$

60°

1$\frac{1}{2}$

1$\frac{1}{2}$

$\frac{1}{2}$

$\frac{1}{2}$

105°

$\frac{5}{8}$

1$\frac{3}{4}$

$\frac{5}{8}$

$\frac{1}{2}$

$\frac{1}{2}$

3

PLASTIC
MOUNTING BRACKET

P10–20.

Armstrong Machine Works

P10–22A.

$\frac{11}{16}$ DRILL – $\frac{13}{16}$ DEEP

$\frac{3}{4}$ – 16 UNF – 2B – $\frac{3}{4}$ DEEP

$\frac{1}{4}$

Ø3

2$\frac{1}{2}$

$\frac{37}{64}$ DRILL

$\frac{5}{8}$ – 18 UNF – 2B – $\frac{1}{2}$ DEEP

$\frac{1}{4}$ DIA – 6 HOLES
EQUALLY SPACED

3

1$\frac{1}{2}$

3

$\frac{1}{4}$ DIA
4 HOLES

1$\frac{1}{2}$

R

ALL FILLETS AND
ROUNDS $\frac{1}{8}$ R

$\frac{1}{2}$ R

1$\frac{3}{4}$

$\frac{5}{16}$ R

TYP

$\frac{3}{16}$

1

$\frac{3}{8}$

1$\frac{3}{8}$

CAP FOR A QUICK-OPENING GATE VALVE

P10–21.

1$\frac{1}{4}$

1$\frac{1}{4}$

$\frac{1}{4}$

2

$\frac{1}{2}$

$\frac{23}{32}$

1$\frac{1}{2}$

$\frac{3}{4}$

1$\frac{3}{16}$

$\frac{1}{4}$

$\frac{1}{4}$

1$\frac{1}{4}$

$\frac{1}{4}$

$\frac{9}{16}$

3

4$\frac{1}{8}$

CAP FOR HIGH PRESSURE STEAM TRAP

277

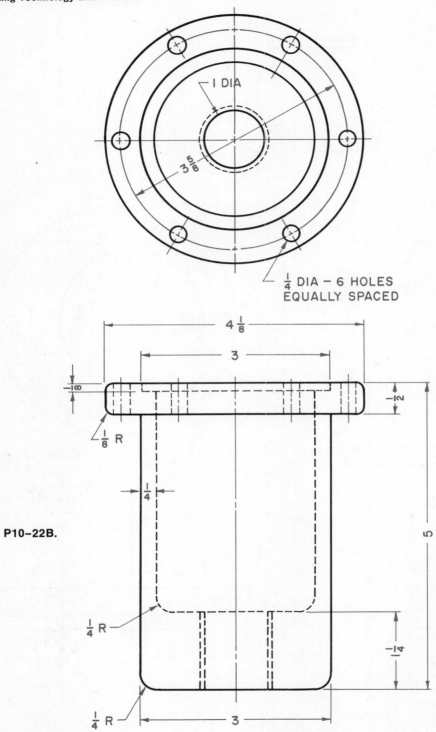

I DIA

$3\frac{5}{8}$

$\frac{1}{4}$ DIA – 6 HOLES
EQUALLY SPACED

$4\frac{1}{8}$

3

$\frac{1}{8}$

$\frac{1}{2}$

$\frac{1}{8}$ R

$\frac{1}{4}$

5

P10–22B.

$\frac{1}{4}$ R

$1\frac{1}{4}$

$\frac{1}{4}$ R

3

BODY FOR HIGH PRESSURE STEAM TRAP

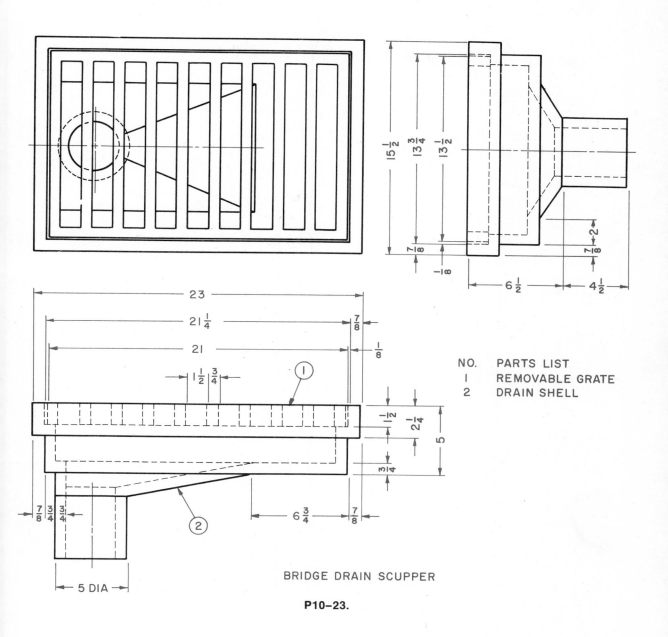

NO.	PARTS LIST
1	REMOVABLE GRATE
2	DRAIN SHELL

BRIDGE DRAIN SCUPPER

P10–23.

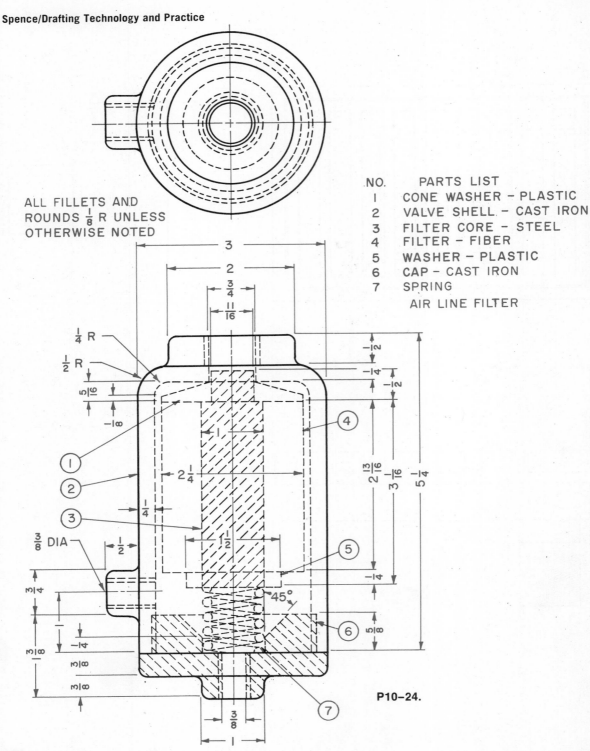

ALL FILLETS AND
ROUNDS $\frac{1}{8}$ R UNLESS
OTHERWISE NOTED

NO.	PARTS LIST
1	CONE WASHER – PLASTIC
2	VALVE SHELL – CAST IRON
3	FILTER CORE – STEEL
4	FILTER – FIBER
5	WASHER – PLASTIC
6	CAP – CAST IRON
7	SPRING
	AIR LINE FILTER

P10–24.

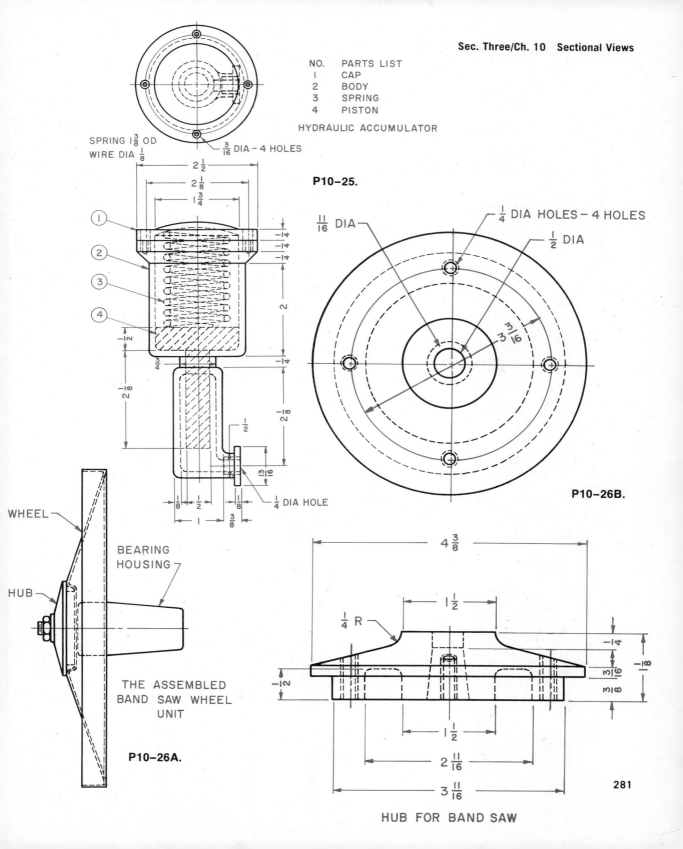

NO.   PARTS LIST
1   CAP
2   BODY
3   SPRING
4   PISTON

HYDRAULIC ACCUMULATOR

SPRING 1⅜ OD
WIRE DIA ⅛

$\frac{3}{16}$ DIA – 4 HOLES

2½
2⅛
1¾

–¼
–¼
–¼

2

–¼

2⅛

–½

¾

2⅛

½

13/16

⅛ ½ ⅛
1
⅜

¼ DIA HOLE

P10–25.

$\frac{11}{16}$ DIA

¼ DIA HOLES – 4 HOLES

½ DIA

3 3/16

P10–26B.

WHEEL

BEARING
HOUSING

HUB

THE ASSEMBLED
BAND SAW WHEEL
UNIT

P10–26A.

4⅜

1½

¼ R

–¼

–⅛

3/16

⅜

–½

1½

2 11/16

3 11/16

281

HUB FOR BAND SAW

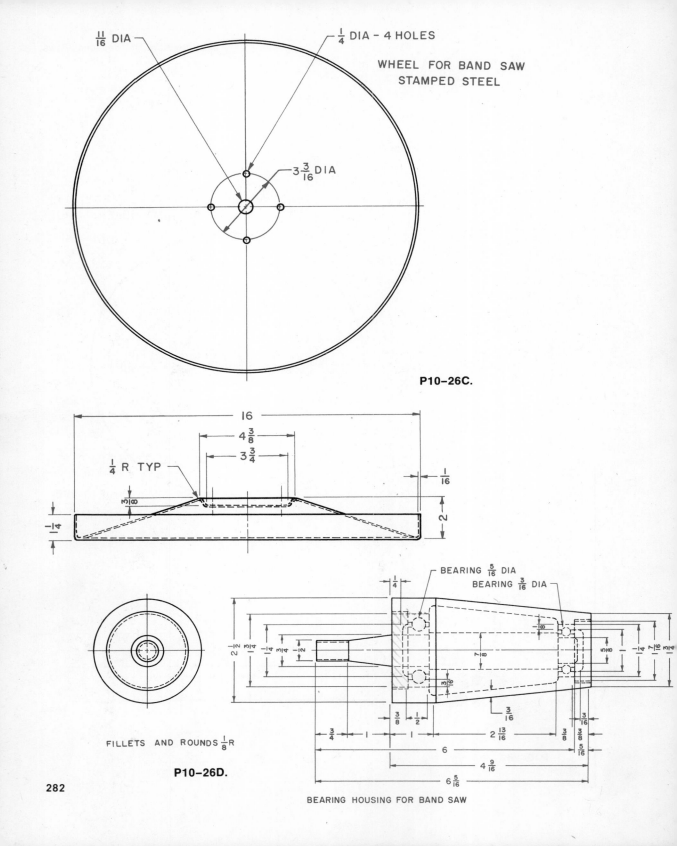

$\frac{11}{16}$ DIA

$\frac{1}{4}$ DIA – 4 HOLES

WHEEL FOR BAND SAW
STAMPED STEEL

$3\frac{3}{16}$ DIA

P10–26C.

16

$4\frac{3}{8}$

$3\frac{3}{4}$

$\frac{1}{4}$ R TYP

$\frac{3}{8}$

$\frac{1}{16}$

$\frac{1}{4}$

2

BEARING $\frac{5}{16}$ DIA

BEARING $\frac{3}{16}$ DIA

$\frac{1}{4}$

$2\frac{1}{2}$

$1\frac{3}{4}$

$1\frac{1}{4}$

$\frac{3}{4}$

$\frac{1}{2}$

$7\frac{1}{8}$

$\frac{1}{8}$

$\frac{5}{8}$

$\frac{1}{4}$

$7\frac{7}{16}$

$1\frac{3}{4}$

$\frac{3}{16}$

$\frac{3}{8}$

$\frac{1}{2}$

$\frac{3}{16}$

$\frac{3}{16}$

$\frac{3}{8}$

$\frac{3}{8}$

$\frac{5}{16}$

$\frac{3}{4}$

1

1

$2\frac{13}{16}$

6

$4\frac{9}{16}$

$6\frac{5}{16}$

FILLETS AND ROUNDS $\frac{1}{8}$ R

P10–26D.

BEARING HOUSING FOR BAND SAW

Chapter 11

Drawing for Production—Detail and Assembly

When designing a new product, the first drawings are usually freehand sketches. An engineer may make a large number of sketches before he is satisfied with the design. Usually the *final freehand sketch* is accompanied by other data, such as load or stress calculations and material specifications. From this information a draftsman draws a *preliminary design assembly*. This gives a more accurate picture of the object and aids in working out design details. This design assembly is an instrument drawing. It usually is drawn full size. Generally only *major* dimensions are given, while notes give details about material, finishes, desired clearances, and other needed data.

The draftsman uses the accurate preliminary design assembly as his source of information to develop *detail drawings* for each part. When size dimensions are needed, they are taken from the preliminary design assembly with a scale or dividers. Each part should be drawn and dimensioned so it will join properly with the other pieces of the device being designed. The draftsman must decide the exact sizes of the mating parts and set the tolerances needed so it will operate properly. He must make certain each part has sufficient clearance so the device can be assembled and will work. It is the job of the detail draftsman to draw each part so it can be manufactured easily. The draftsman must understand the shop processes used in manufacturing and how materials are worked.

The detailer must know what standard parts are available. These are parts, such as fasteners, washers, keys, and

Many complex production drawings were needed to design and make this track-type tractor.

Caterpillar Tractor Co.

283

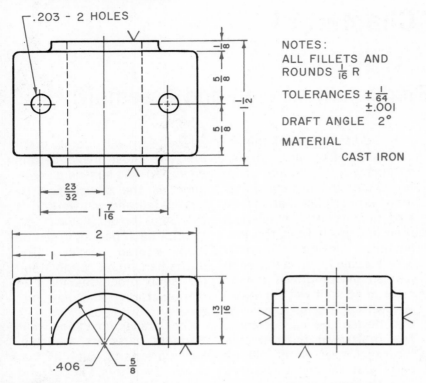

NOTES:

ALL FILLETS AND ROUNDS $\frac{1}{16}$ R

TOLERANCES $\pm \frac{1}{64}$
$\pm .001$

DRAFT ANGLE   2°

MATERIAL
                CAST IRON

.203 - 2 HOLES

11-1. *A detail drawing of the bearing cap from the bearing journal, Fig. 11-29. This drawing includes information needed by the pattern-maker and the machinist. It is called a casting detail drawing.*

bearings, that are manufactured in large quantities for general use. A standard part should be used whenever possible. These parts are usually purchased from another company. They are shown in the parts list for the device but are not detailed. Standard fastening devices are shown in Chapter 26. Others may be found in the catalogs of companies manufacturing these items.

DETAIL DRAWINGS

A *detail drawing* is a complete description of a single part of an object. The drawing may include auxiliary views, sections, or any other descriptive forms necessary. It includes dimensions, tolerances, materials, weight finish, scale of the drawing, and other information needed to describe the part completely. Fig. 11-1 illustrates a typical detail drawing.

The exact practice to follow when detailing varies with the type of industry and the shop practices used by the company. In general, each detail is placed on a separate sheet. Sometimes a small device with only a few parts is detailed on one sheet.

It becomes desirable in some industries to make separate drawings for each manufacturing process that a part must progress through. This is especially true if the part is large and complicated. For example, if a part is to be cast and then machined, a separate detail pattern drawing can be made from which the part will be cast. Then a separate detail drawing of the finished machined part will be prepared for use in the machine shop.

Typical types of detail drawings are pattern, machining, casting, forging, welding, and stamping detail drawings.

PATTERN DETAIL DRAWINGS

A *pattern detail* gives the information needed to make a pattern. A *pattern* is a duplicate of the part to be cast. It is made from wood, metal, or plastics. Three factors — *draft, shrinkage,* and *machining* — influence how this is to be drawn. *Draft* is a slight

L. S. Starrett Co.

11-2. *A shrink rule used by patternmakers.*

tapering of the casting. This helps remove the pattern from the sand mold. *Shrinkage* refers to making the pattern slightly larger than the desired size of the finished casting. This is necessary because the metal part shrinks when it cools. See Fig. 11–2. It shows a shrink rule.

Standard shrinkages for various metals in cast form have been established. These shrinkages are: cast iron and malleable iron, $1/8$ inch per foot; steel, $1/4$ inch per foot; brass, cooper aluminum, and bronze, $3/16$ inch per foot; and lead, $5/16$ inch per foot.

Shrink rules are made oversize for each of these metals. For example, the one-inch markings on a shrink rule for cast iron castings are actually $1/8$-inch longer than the actual inch. The pattern worker uses this rule when measuring a pattern for a cast iron casting. The pattern will then be enlarged the proper amount.

Machining refers to the removal of metal on a cast surface after the object has been cast. See Fig. 11–3 for one machining process.

If pattern detail dimensions include allowances for draft, shrinkage, and machining, the drawing shows the finished size of the rough casting, including the extra material allowed. However, this is not commonly done except on complex castings.

Pattern detail drawings, Fig. 11–4, show any holes to be formed by cores. A *core* is a

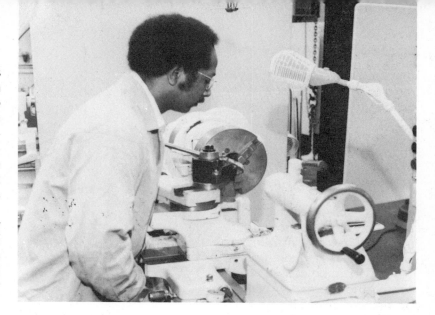

11–3. *This metal lathe is used to machine metal off round objects.*

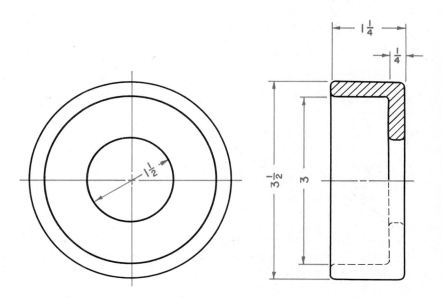

NOTES:
ALL FILLETS AND ROUNDS $\frac{1}{16}$ R
TOLERANCE $\pm\frac{1}{64}$
DRAFT ANGLE 2° IN ADDITION TO
DIMENSIONS SPECIFIED.

11–4. *A pattern detail drawing of a bearing cap. It gives the information needed by the patternmaker. It includes allowances for shrinkage and machining, and indicates the draft angle. The drawing does not tell where the part is to be machined or the finished size after machining.*

dry sand form inserted in the mold to form a hole or a cavity on the interior of the casting. Large holes are cored. Small holes are drilled and are not shown on a pattern detail drawing.

The edges of cast parts are rounded because it is difficult to cast sharp corners. The radius of fillets and rounds are indicated on the drawing as is the size of any cored parts.

If a part to be cast is rather simple, the information needed by the patternmaker and the machining information needed by the machinist are given together on one drawing. This is called a casting drawing. Casting drawings are discussed in a later section of this chapter.

MACHINING DETAIL DRAWINGS

A *machining drawing* includes only those details necessary for the machine shop to perform the machining operations. This generally includes the finished size, location of holes, finished surfaces, the quality of the finished surface, and tolerances on machined surfaces. Parts of the piece that are left rough (as cast) need no dimensions. Fig. 11–5 shows such a detail.

Notice in Fig. 11–5 how the machining drawing is dimensioned. The dimensions that locate finished surfaces or centers of holes that are machined are related directly from one another in order

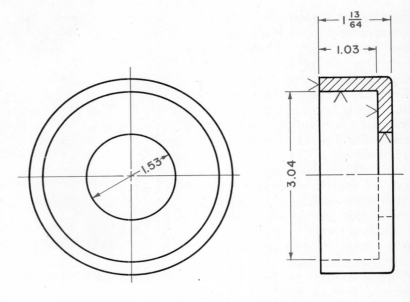

TOLERANCE ±.01

11–5. *A machining detail drawing of the bearing cap, Fig. 11–4. It gives the information needed by the machinist to produce the finished part. The overall dimensions are not needed since the machinist receives the part after it is cast.*

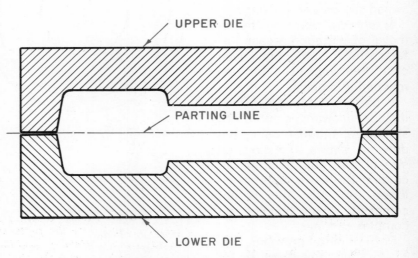

UPPER DIE

PARTING LINE

LOWER DIE

11–6. *A forging formed in a die.*

to maintain needed precision. Since cast holes and surfaces cannot be made to an accuracy greater than ± $\frac{1}{32}$, they must not be used to locate precision dimensions.

CASTING DETAIL DRAWINGS

If a part is not complex, a single drawing can be made that contains all the information the patternmaker and the machinist need. It is a combination of pattern and machining detail drawings and is called a *casting detail drawing*. See Fig. 11–1.

Notice that it contains the overall sizes of the rough casting as well as the precision dimensions needed for machining. The draft angle is given. The material is cast iron. This tells the patternmaker how much to allow for shrinkage.

FORGING DETAIL DRAWINGS

A forging drawing shows a part that is to be made by forging. Forging is a process in which the metal is heated and formed to shape with a power hammer. The piece is usually shaped in a *die*.

A *die* is a metal form with a cavity the shape of the part to be formed. Heated metal is placed between the dies. They are forced-closed, forcing the metal to the shape of the cavity. See Fig. 11–6.

Several practices used in drawing forgings are different from other types of drawings.

On the plan view, the *draft* is shown by lines drawn as if

the corners of the forging were sharp. See Fig. 11–7. *Draft* is the slope given to the surfaces to help remove the forging from the die. Standard draft for outside surfaces of all materials is 7 degrees. Inside draft angles for aluminum are 7 degrees and for steel 10 degrees. See Fig. 11–8.

A forging has a *parting line*. A *parting line* is the line where the two dies meet when forming the part. It is shown on the drawing using the center line symbol. It is labeled PL. See Fig. 11–8.

Forging drawings are dimensioned using the datum system. This avoids the accumulation of tolerances. See Fig. 11–8. The dimensions should refer to some important part of the object. A center line is most frequently used.

All dimensions parallel to the parting line should refer to the size at the bottom of the die impression. This is the narrowest part. The die cavity is widest at the parting line. See the cavity shown in Fig. 11–6.

All dimensions at right angles to the parting line should refer to the parting line. See Fig. 11–8.

Forging drawings should be drawn full scale whenever possible. Sectional views are used a great deal.

Tolerances. Forgings are designed with standard tolerances: thickness, shrinkage, die wear, mismatch, and machining tolerances.

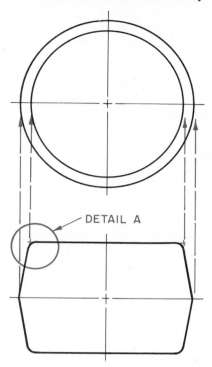

DETAIL A

PROJECTED SHARP CORNER

DETAIL A

11–7. *The rounded corners on a forging are projected and drawn as though they were sharp.*

Thickness tolerances refer to the overall height of the part. This dimension can vary depending upon how close the dies come when closed. See Fig. 11–9. Standard commercial tolerances and more accurate close tolerances are given in Fig. 11–10.

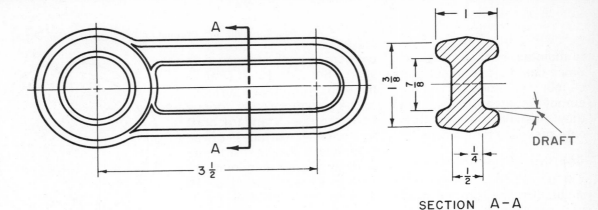

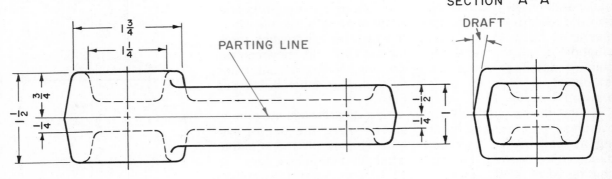

SECTION A-A

PARTING LINE

DRAFT

NOTES:
ALL DRAFT ANGLES 7°
UNLESS SPECIFIED.
FILLETS AND ROUNDS $\frac{1}{8}$ R.
EST. WEIGHT 15 LB.

TOLERANCES:
THICKNESS + 0.078 - 0.026
MISMATCH   + 0.024
DIE WEAR   + 0.053
SHRINKAGE + 0.045

11-8. *A forging drawing.*

Thickness Tolerances

Net Weight Up to Lb.	Commercial		Close	
	−	+	−	+
0.2	0.008	0.024	0.004	0.012
0.4	0.009	0.027	0.005	0.015
0.6	0.010	0.030	0.005	0.015
0.8	0.011	0.033	0.006	0.018
1	0.012	0.036	0.006	0.018
2	0.015	0.045	0.008	0.024
3	0.017	0.051	0.009	0.027
4	0.018	0.054	0.009	0.027
5	0.019	0.057	0.010	0.030
10	0.022	0.066	0.011	0.033
20	0.026	0.078	0.013	0.039
30	0.030	0.090	0.015	0.045
40	0.034	0.102	0.017	0.051
50	0.038	0.114	0.019	0.057
60	0.042	0.126	0.021	0.063
70	0.046	0.138	0.023	0.069
80	0.050	0.150	0.025	0.075
90	0.054	0.162	0.027	0.081
100	0.058	0.174	0.029	0.087

ANSI Y14.9–1958

Shrinkage tolerances relate to parts that become smaller when the heated metal cools. Standard tolerances are shown in Fig. 11–11. These are figured in the direction parallel to the parting line. See Fig. 11–12.

Die wear tolerances relate to the tendency for dies to wear and become larger. Standard tolerances are shown in Fig. 11–11. These are figured in the direction parallel to the parting line. See Fig. 11–12.

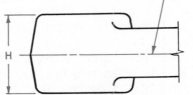

PARTING LINE

H

11-9. *The thickness tolerance is affected by the closing of the dies.*

Mismatch refers to slight size differences caused by the dies not lining up exactly when they close. See Fig. 11–13.

Notice in Fig. 11–12 that shrinkage, die wear, and mis-

11-10. *Thickness Tolerance for Forgings.*

Shrinkage and Die Wear Tolerances for Forgings

Shrinkage			Die Wear		
Length or Width Up to In.	Com-mercial + or −	Close + or −	Net Weight Up to — Lb.	Com-mercial + or −	Close + or −
1	0.003	0.002	1	0.032	0.016
2	0.006	0.003	3	0.035	0.018
3	0.009	0.005	5	0.038	0.019
4	0.012	0.006	7	0.041	0.021
5	0.015	0.008	9	0.044	0.022
6	0.018	0.009	11	0.047	0.024
Each Additional In. Add	0.003	0.0015	Each Additional 2 Lb. Add	0.003	0.0015

ANSI Y14.9–1958

11–11. *Shrinkage and die wear tolerances for forgings.*

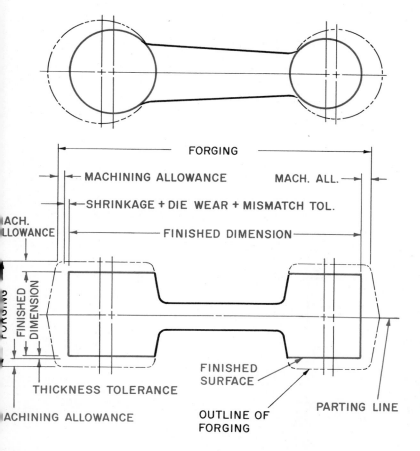

FORGING
— MACHINING ALLOWANCE — MACH. ALL. —
— SHRINKAGE + DIE WEAR + MISMATCH TOL. —
MACH. LLOWANCE
— FINISHED DIMENSION —
FORGING FINISHED DIMENSION
THICKNESS TOLERANCE
ACHINING ALLOWANCE
FINISHED SURFACE
OUTLINE OF FORGING
PARTING LINE

Mismatch Tolerances

Net Weight Up to Lb.	Commercial	Close
1	0.015	0.010
7	0.018	0.012
13	0.021	0.014
19	0.024	0.016
Each Additional 6 Lb. Add	0.003	0.002

ANSI Y14.9–1958

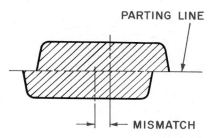

PARTING LINE

MISMATCH

11–13. *Mismatch is caused by the dies not lining up when they close to form a forging.*

match tolerances are added for a single total tolerance. *Machining tolerances* indicate the material needed to machine the forging to the finished size. See Fig. 11–12. This amount is in addition to shrinkage, thickness, and mismatch tolerances. Standard machining tolerances are shown in Fig. 11–14.

11–12. *The length of forged parts includes the finished dimension plus machining allowance, shrinkage, die wear allowance, and mismatch tolerance. The thickness of forged parts includes the finished dimension plus machining allowance and thickness tolerance.*

ANSI Y14.9–1958

Machining Allowance

	Under 1 Lb.	1 to 10 Lb.	11 to 40 Lb.	41 to 100 Lb.	101 to 200 Lb.
Compact Parts Gears, Discs	$1/32$	$3/64$	$1/16$	$3/32$	$1/8$
Thin Extended Parts	$1/16$	$1/16$	$3/32$	$1/8$	$5/32$
Long Parts Shafts	$1/16$	$1/16$	$3/32$	$1/8$	$3/16$

ANSI Y14.9–1958

11–14. *Machining allowance for forgings.*

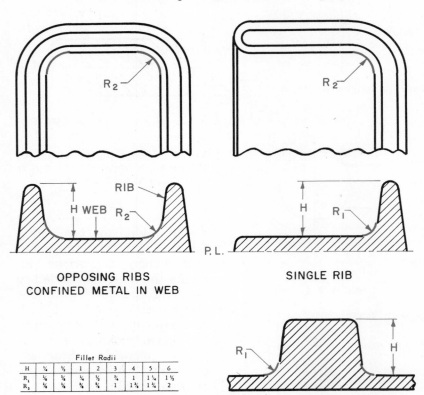

OPPOSING RIBS
CONFINED METAL IN WEB

SINGLE RIB

P. L.

Fillet Radii

H	$1/4$	$1/2$	1	2	3	4	5	6
R_1	$1/8$	$1/4$	$1/4$	$1/2$	$3/4$	1	$1\,1/4$	$1\,1/2$
R_2	$1/4$	$1/4$	$3/8$	$5/8$	1	$1\,3/8$	$1\,1/4$	2

ANSI Y14.9–1958

BOSS

11–15. *Minimum fillet radii for forgings.*

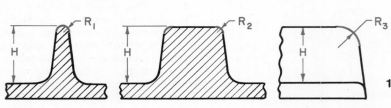

290 RIB   BOSS   END OF RIB

Corner Radii

H	$1/4$	$1/2$	1	2	3	4	5	6	7
R_1	$1/16$	$1/16$	$1/8$	$3/16$	$1/4$	$5/16$	$3/8$	$7/16$	$1/2$
R_2	$1/16$	$1/16$	$1/8$	$1/4$	$5/16$	$7/16$	$1/2$	$5/8$	$11/16$
R_3	$3/16$	$3/16$	$3/8$	$1/2$	$3/4$	1	$1\,1/8$	$1\,1/4$	$1\,1/2$

11–16. *Minimum corner radii for forgings.*

ANSI Y14.9–1958

The length of a forging is the sum of the finished length desired plus machining allowance, shrinkage, die wear, and mismatch required on each end.

The height of the forging is the sum of the finished height, thickness tolerance, and machining allowance required on each side.

The actual acceptable tolerances are based on the weight of the forging.

Corner and Fillet Radii. The size of the radii used for corners and fillets affects the quality of the forging and the life of the die. Small radii will cause die wear and produce seams or improperly filled sections in the forging. Figs. 11–15 and 11–16 give minimum radii for fillets and corners.

Making Forging Drawings. Forging drawings are made in two ways. The most frequently used method of making a forging drawing is to combine the forging and machining information on one drawing. The parts of the rough forging to be removed by machining are indicated by a dashed line. The finished size is indicated by a visible line. The difference between them is the tolerance. See Fig. 11–17.

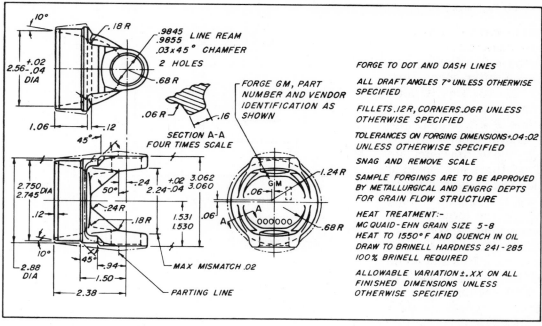

11–17. A single drawing showing both forging and machining details. The forging outline is shown by phantom lines, the finished machined surfaces by solid lines.

11–18. A forging drawing and a machining drawing. Notice the forging drawing has a finished machined surface indicated with phantom lines on the round shaft. It is related to the center line with dimensions.

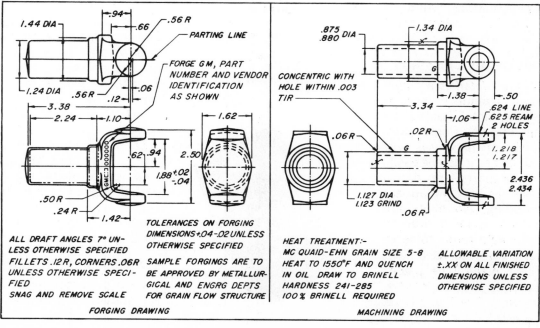

FORGING DRAWING

MACHINING DRAWING

A second way to detail forgings is to make separate drawings of the rough forging and the forging after machining. See Fig. 11–18. This is done if the part is so complex that the outline of the forging cannot be shown clearly on one drawing. These drawings are placed on the same sheet if possible. The forging drawing is to the left of the machining drawing. On the forging drawing, always indicate some surface of the finished part by phantom lines. This should be located to a forged surface with a dimension.

Forging drawings usually specify draft angles, parting lines, fillets and corner radii, tolerances, allowance for machining, material specifications, heat treatment, and part identification number. Usually all draft angles are the same and can be specified with a note such as, "All draft angles 7 degrees unless otherwise specified."

WELDING DETAIL DRAWINGS

A *welding detail* is a single part that is made of several pieces of metal joined by welding into a single unit. Each piece of the unit is fully dimensioned so it can be easily made. The parts are drawn assembled as they will appear after welding. No fillets are drawn on a welding drawing since the weld tends to serve the same purpose as a fillet on a cast piece. The pieces are drawn as they appear before welding. See Fig. 11–19.

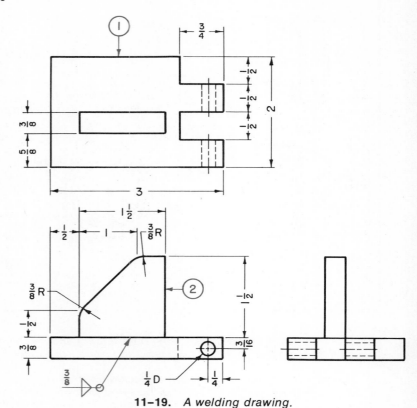

11–19. *A welding drawing.*

Parts List

No.	Req'd.	Description	Material
1	1	2 × ⅜ plate 3 long	SAE 1020
2	1	1½ × ⅜ plate 1½ long	SAE 1020

All joints between individual pieces making up the unit are drawn even though they will be covered by a weld on the finished piece. Each joint to be welded must have the proper symbol even though the actual welds are not drawn.

Each individual part of the unit is numbered. The specifications for each numbered part are given in a parts list.

The proper use of welding symbols is explained in Chapter 26, Fastening Devices.

STAMPING DETAIL DRAWINGS

Parts produced by stamping are made by pressing sheet

metal between dies under pressure. This makes the metal bend and in some cases stretch to fit the shape of the die. This process can involve simply forming a pre-cut piece to the desired shape. It also includes shearing or cutting the metal to shape (blanking) and punching holes in it. Some typical stamped products are shown in Fig. 11–20.

Some processes involved in making stampings are bending and forming, drawing, coining, blanking, punching, and trimming.

Bending and forming are the processes used to bend, flange, fold, and twist the metal to the desired shape. Usually the thickness of the metal is unchanged.

Drawing is a process in which the metal is stretched (drawn) over a form. The form has been made to the shape of the finished part. This usually causes the metal to become thinner in the section drawn. The rest of the part remains the same thickness.

Coining is the process by which metal is caused to flow under great pressure. The metal in the area coined becomes thinner. The metal from the thinned section flows to the areas not coined and they become thicker.

Blanking is the process of cutting flat stock to the size and shape of the finished piece.

Punching is a method of internal blanking. It cuts a hole or opening in the piece.

11–20.   *Examples of products formed using the stamping process.*

Trimming refers to operations such as cutting off excess material using dies.

Metal Gages. The thickness of metal sheets is indicated by a series of numbers called *gages*. Each gage number stands for a particular thickness. Fig. 11–21 gives the standard gage sizes with decimal equivalents for sheet and plate iron and steel.

The thickness of flat sheet stock under 1/4 inch is usually specified on the drawing in decimals. Thickness above this can be given in fractions. The thickness and type of steel is specified as, steel, 125–SAE 1010 Cr. Comm. Qual. Bright finish. This gives the type of material, steel, decimal thickness of .125, and type of steel.

Forming Holes. Holes in stampings are usually punched, extruded, or pierced. See Fig. 11–22. *Punched holes* are made by a punch which shears a clean hole to the finished size. *Extruded holes* are punched. The punch forming

U.S. Standard Gages for Sheet and Plate Iron and Steel

Number of Gage	Thickness		Weight	Number of Gage
	Approx. thickness in fractions of an inch	Approx. thickness in decimal parts of an inch	Weight per square foot in pounds avoirdupois	
10	9–64	.1406	5.625	10
11	1–8	.125	5.	11
12	7–64	.1094	4.375	12
13	3–32	.0938	3.75	13
14	5–64	.0781	3.125	14
15	9–128	.0703	2.8125	15
16	1–16	.0625	2.5	16
17	9–160	.0563	2.25	17
18	1–20	.05	2.	18
19	7–160	.0438	1.75	19
20	3–80	.0375	1.5	20
21	11–320	.0344	1.375	21
22	1–32	.0313	1.25	22
23	9–320	.0281	1.125	23
24	1–40	.025	1.	24
25	7–320	.0219	.875	25
26	3–160	.0188	.75	26
27	11–640	.0172	.6875	27
28	1–64	.0156	.625	28
29	9–640	.0141	.5625	29
30	1–80	.0125	.5	30

11–21. *U.S. standard gages for sheet and plate iron and steel.*

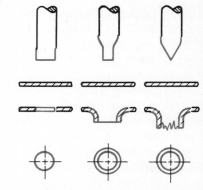

PUNCHED   EXTRUDED   PIERCED

11–22. *Methods for forming holes in stamped parts.*

MINIMUM RADIUS TWICE METAL THICKNESS FOR HIGHLY STRESSED PARTS.

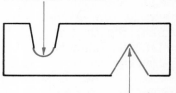

SHARP VERTEX PERMITTED FOR LOW STRESSED PARTS.

11–23. *Two types of notches used on stampings.*

the first opening is smaller than the final hole. The punch is enlarged beyond the tip. The enlarged portion is forced into the punched hole forming a flange. *Pierced holes* are made by a punch with a sharp point. It pierces the metal and forms a flange. It leaves a torn edge on the flange.

Notches. Two types of notches are used on stampings. One is a sharp V and the other has a rounded vertex. See Fig. 11–23. The sharp V is used only on parts with little stress, because the V notch will tend to crack if placed under stress. High-stressed parts must have a rounded vertex. The radius

of the round should be at least twice the thickness of the metal. It should be larger if the design permits.

Bend Allowances. When a stamping is formed, the blank must be made longer to allow for the material consumed in making the bend. This extra allowance is called *bend allowance.*

The procedure for computing bend allowance is explained in Chapter 16, Drafting in the Aerospace Industry. It also contains tables for figuring bend allowance in nonferrous metals and aluminum. If the finished length of the part is important, bend allowance

must be computed. If it is not, the part will be shorter than needed.

Corners. One system of forming corners of stampings, shown in Fig. 11–24, is explained in Chapter 16. This system is used primarily in the aircraft industry.

Fig. 11–25 shows other ways of forming corners. These are easier to make and are widely used in designing stampings.

294

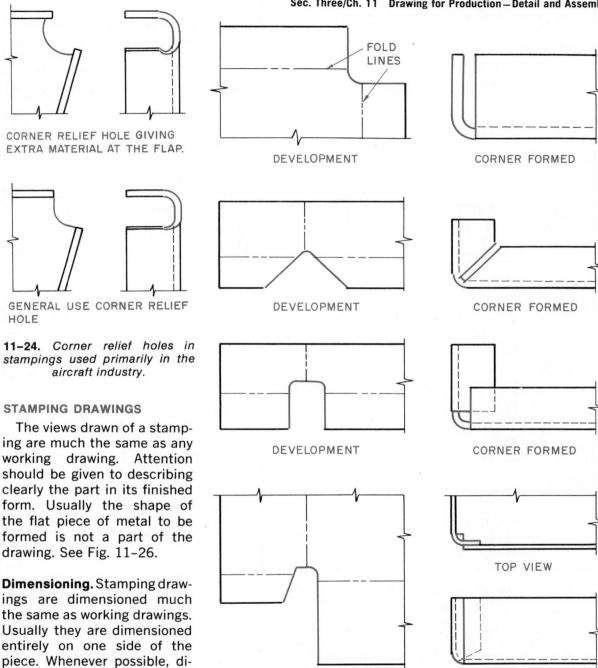

CORNER RELIEF HOLE GIVING
EXTRA MATERIAL AT THE FLAP.

GENERAL USE CORNER RELIEF
HOLE

11–24. *Corner relief holes in stampings used primarily in the aircraft industry.*

FOLD
LINES

DEVELOPMENT

DEVELOPMENT

DEVELOPMENT

DEVELOPMENT

CORNER FORMED

CORNER FORMED

CORNER FORMED

TOP VIEW

FRONT VIEW OF
FORMED CORNER

STAMPING DRAWINGS

The views drawn of a stamping are much the same as any working drawing. Attention should be given to describing clearly the part in its finished form. Usually the shape of the flat piece of metal to be formed is not a part of the drawing. See Fig. 11–26.

Dimensioning. Stamping drawings are dimensioned much the same as working drawings. Usually they are dimensioned entirely on one side of the piece. Whenever possible, dimensions should be given to intersections or tangent points.

Hole locations and other critical dimensions must be

11–25. *Common ways to prepare stampings to form corners.*

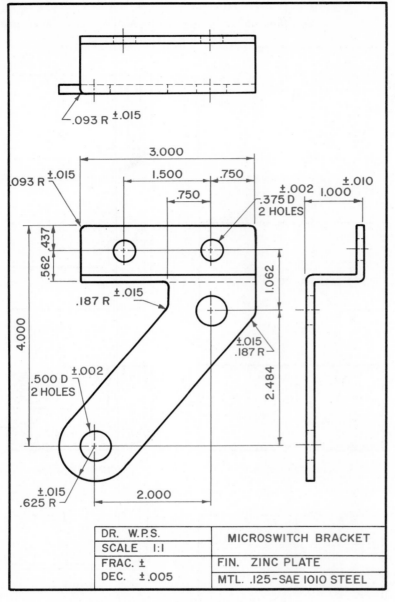

.093 R ±.015

3.000

1.500 .750

.750

.093 R ±.015

±.002 1.000 ±.010

.375 D
2 HOLES

.562 .437

1.062

.187 R ±.015

±.015
.187 R

4.000

2.484

.500 D ±.002
2 HOLES

±.015
.625 R

2.000

DR. W.P.S.	MICROSWITCH BRACKET
SCALE 1:1	
FRAC. ±	FIN. ZINC PLATE
DEC. ±.005	MTL. .125-SAE 1010 STEEL

11–26. *A stamping detail drawing.*

located from each other rather than from an outside edge. See Fig. 11–26.

ASSEMBLY DRAWINGS

It is often useful or necessary to show an object as it appears when all of its parts are assembled. This is called an *assembly drawing*. Assembly drawings are used to check detail drawings, to describe how a machine functions, to show how to install

a machine, give maintenance instructions, to show general design factors for sales purposes, to show subassemblies, or to simplify the assembly process during production. They can be in the form of an orthographic drawing or a pictorial drawing.

Planning an Orthographic Assembly Drawing. An assembly drawing shows all the parts of a device. It is much more complex than a detail drawing. The draftsman must understand how the device is supposed to operate and how it is to be assembled. He must decide which views are needed to show it best when it is assembled.

The usual procedure is to make a freehand sketch of the assembled device before making the finished drawing. This will make it possible to try several approaches. Since detailed drawings will be made, the assembly drawing does not need to include complete details for each part.

The handling of fastening devices, shafts, and bearings are the same as discussed in Chapter 10, Sectional Views. Dimensions are usually omitted though overall dimensions are sometimes used.

Selection of views depends upon the device. External views, sectional views, auxiliary views, and partial views can be used as required. The selection of the views follows the same principles as those used to make a multiview draw-

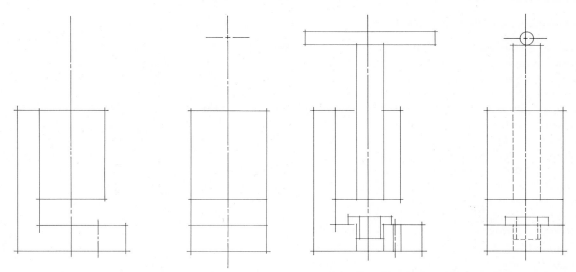

11–27A. *Locate the center lines and the main outline of the assembly.*

11–27B. *Block in details of the larger parts of the assembly.*

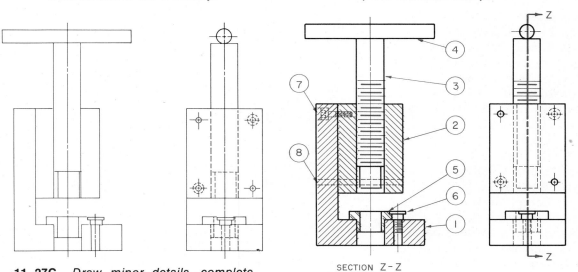

11–27C. *Draw minor details, complete details on major parts, and remove construction lines.*

SECTION Z–Z

11–27D. *Draw all lines to proper width, add section lining, threads, and other final line work. The parts list is finished.*

Ford Motor Company, Kansas City Assembly Plant

11–27. *How to make an assembly drawing. This device is a special tool designed to pierce holes in sheet metal.*

ing of a single part. The views must fully describe the entire device rather than one part.

As the views are drawn, it is best to block in the major features of the device in all views and add less important details next. Do not detail each piece one at a time. All preliminary construction must be drawn lightly and finished line work applied after all details are recorded. This procedure is the same as that for detail drawings.

Steps in Drawing an Orthographic Assembly Drawing. Assembly drawings are laid out much the same as detail drawings. Since the assembly is often used to check the correctness of the detail drawings from which it is made, accuracy is important. All views are laid out together. When one item is located on one view, it should be drawn in all views before going on to the next part of the layout.

A good plan to follow is:

1. Lay out the principal center lines and major outline of the assembled device. See Fig. 11–27A, page 297.

2. Block in the outlines of the important parts of the device. This is done with light construction lines but accuracy must be maintained. See Fig. 11–27B.

3. Locate and draw minor details. These are items such as fasteners, holes, ribs, and keys. See Fig. 11–27C.

4. Complete any minor details necessary to the shape description of various parts.

5. Remove the construction lines and darken lines of the drawing. See Fig. 11–27D.

6. Add section lines to surfaces cut by the cutting plane. See Fig. 11–27D.

7. Letter identification numbers and notes. See Fig. 11–27D.

Part Identification. Numbers are used to identify various parts of an assembly. Numbers are commonly assigned to parts in the order in which they are assembled. Another system gives the largest piece the smallest number; numbers get larger as parts get smaller.

The numbers are recorded on the drawing inside a circle usually ⅜ inch to ½ inch in diameter. The circle is connected to the part with a leader which may or may not have an arrowhead. If possible the circles should be arranged in vertical and horizontal rows, and should be off the view. They can appear on all sides of the view. The leaders can be at any convenient angle, but should not be in a vertical or horizontal position, cross each other, or obscure a detail on the drawing. The leaders are drawn so that if they were extended, they would cross the center of the numbered circle.

A parts list is included with the assembly drawing. Items on the list will vary with company policy, but a list usually includes part number, name of part, number of parts required for one machine, and material used to make the part. If the parts list is extensive, it can be placed on a separate sheet. See Fig. 11–28A and B.

Section Lining. Small cylindrical parts, such as shafts, bolts, and pins, are not usually sectioned in assembly drawings even though the cutting plane passes through them. Other items usually

11–28A. *The general assembly drawing of this pump is in fig. 11–28B.*

not section lined are gear teeth, keys, nuts, and bearings.

The proper section lining symbol for the material in each part is commonly used in assembly drawings. This helps separate the parts and makes reading the drawing easier.

The slope of the section lining should vary on adjacent parts to help clarify the drawing.

If two adjacent parts are spaced 1/32 inch or less apart, they are usually drawn with a single line—a shaft in a bushing is an example. If spaced wider than this, two lines are drawn with a clear space between.

The sections are indicated on exterior views with a cutting-plane line in the same manner as sections of details. The section drawing is identified with the title "Section" and the identifying letters, such as A–A. See Fig. 11–27D.

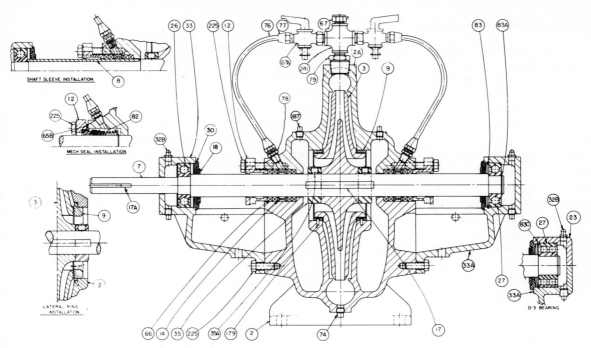

11–28B. *A general assembly drawing of a pump. This could also serve as a check assembly. Compare this assembly with the photo of the pump. Can you see how the assembly was drawn through the pump?*

Check Assembly Drawings.

After an engineering staff has designed a machine and draftsmen have made detailed drawings of each part, it is necessary to check the detail drawings for completeness and accuracy before the object is produced in quantity. This is frequently done by *inspection*. A checker goes over each drawing and checks it for completeness, mating fits, and to determine if it can be made by normal production methods. Inspection requires many calculations to check dimensions and tolerances to see if mating parts will fit and function properly.

Parts List			
Part Number	Name	Material	No. Required
2	Housing, Lower	Cast Iron	1
2A	Housing, Upper	Cast Iron	1
3	Impeller	Cast Iron	1
7	Shaft	Steel	1
8	Sleeve	Bronze	1
9	Bushing	Bronze	1
12	Flange	Cast Iron	1
14	Seal Spacer	Steel	1
17	Key	Steel	1
17A	Key	Steel	1
18	Bearing Ring	Steel	1
23	Bearing Housing, Upper	Cast Iron	1
26	Bearing		1
27	Bearing		1
30	Bearing Seal	Nylon	1
32B	Grease Fitting	Steel	2
33	Bearing Housing, Upper	Cast Iron	1
33A	Bearing Housing, Lower	Cast Iron	2
35	Impeller Collar	Steel	2
35A	Set Screw	Steel	2
65B	Seal	Nylon	2
66	Seal	Nylon	8
67	Lock Nut	Bronze	1
67A	Valve	Bronze	2
74	Plug	Bronze	1
76	Tubing	Copper	2
77	Fitting	Copper	4
78	Fitting	Copper	2
79	Fitting	Copper	1
82	Spring	Bronze	1
83	Seal	Nylon	1
83A	Ring	Bronze	1
83C	Lock Ring	Steel	1
179	Seal Lock	Steel	2
187	Plug	Bronze	2
225	Bolt	Steel	6
241	T-Fitting	Bronze	1

If time and expense are of little importance, a sample machine can be made to check the drawings. This frequently involves construction of dies, patterns, or special tools which usually make this means of checking too expensive. Actually, the checking process is intended to catch errors before great expense occurs.

Another frequently used method of checking involves making a *check assembly drawing* of the machine. This drawing shows the assembled machine with parts in their proper position. This provides a check on dimensions and assembly of the machine, and reduces the calculations necessary to complete a check. Inspection is required for a complete check with this type of drawing since tolerances, fit, and manufacturing processes are not evaluated. It is an inexpensive and effective way to check the details. See Fig. 11–28.

In check assembly drawing, accuracy is of extreme importance since the dimensions of the detail drawings are to be verified. Since the objective of this assembly is to check the details, it need not be as complete in detail as in a general assembly drawing.

GENERAL ASSEMBLY DRAWINGS

The general assembly drawing serves primarily to indicate how the various parts of a machine fit together. It usu-

11–29. *Working drawing assembly of a bearing journal. This gives complete dimensioned details of each part and shows the journal assembled.*

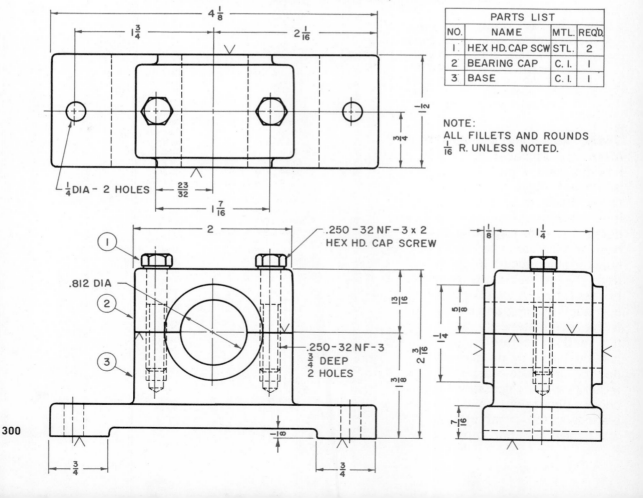

PARTS LIST			
NO.	NAME	MTL.	REQ'D.
1	HEX HD. CAP SCW	STL.	2
2	BEARING CAP	C. I.	1
3	BASE	C. I.	1

NOTE:
ALL FILLETS AND ROUNDS
$\frac{1}{16}$ R. UNLESS NOTED.

ally includes only the outlines of parts and indicates the movements intended. It is useful in the assembly phase of the production of a machine. The check assembly can often serve as a general assembly. See Fig. 11–28.

Additional examples of general assembly drawings are given in Chapter 10.

Subassembly Drawings. If a machine is too large to be shown with a single general assembly drawing, it is usually broken down into smaller functioning units called *subassemblies.* They serve the same purpose as the general assembly. A water pump is a subassembly of an automobile engine.

Working Drawing Assembly. If an assembled object is not too complex, a single drawing can be made to serve as detail drawing and assembly drawing, as shown in Fig. 11–29.

L. S. Starrett Co.

11–30. *A cutaway pictorial assembly drawing of a micrometer. This drawing is used to show construction details in sales materials.*

The necessary orthographic views are shown along with notes, dimensions, and sectioned portions when necessary. If one or two pieces cannot be clearly shown and dimensioned, they occasionally are drawn on the sheet as separate details.

Pictorial Assembly Drawings. There are two types of pictorial assemblies commonly used. One is the cutaway drawing, as shown in Fig. 11–30. The object is drawn with

a section removed to show interior detail. It is shaded to appear much like a photograph. This type of work is done by a technical illustrator and is commonly called production illustration.

A second means of presenting a pictorial assembly is with an exploded drawing, as shown in Fig. 11–31. Each part is drawn pictorially and arranged in the order it should be assembled. The parts are usually identified by a number and are occasionally shaded.

11–31. *An exploded pictorial assembly used in a plant to show how a device is to be assembled. It shows the location of each part and order in which the parts are assembled. They are not always shaded in this manner. This type of drawing is also useful in repair manuals.*

Chevrolet Motor Division, General Motors Corp.

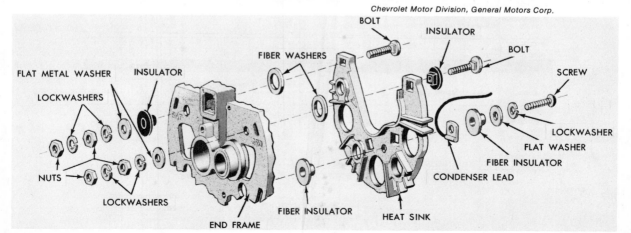

machine dimensions / plain — universal — vertical

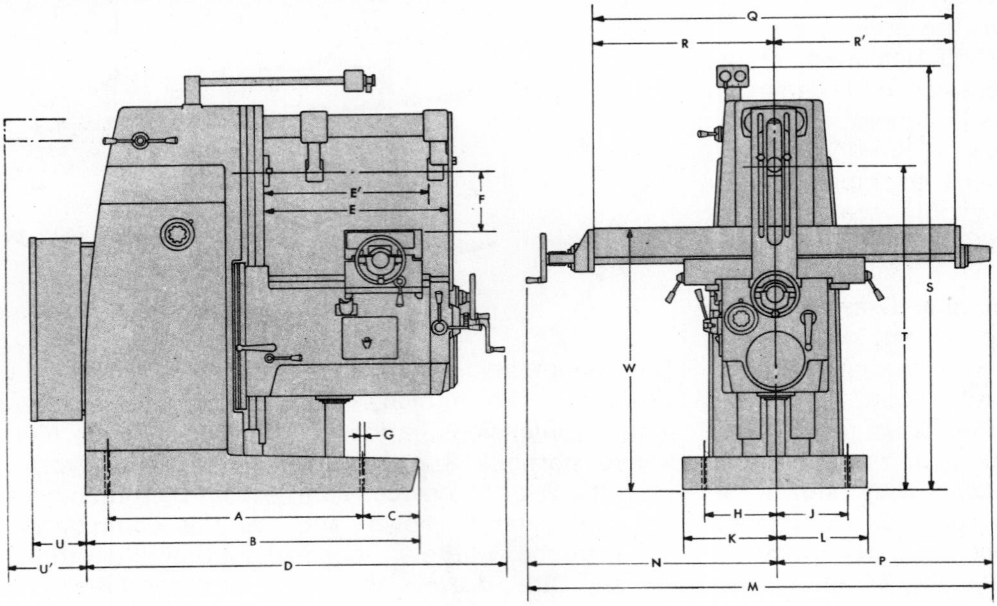

Machine	A	B	C	D	E Plain, Universal	E Vertical	E[1]	F Min.	F Max. Plain	F Max. Universal	F Max. Vertical	G	H Front
205 S-12	39½	52	9⅛	70½	29¾	14	26⅜	0	18	16⅜	19½	13/16	11
307 S-12	39½	52	9⅛	70½	29¾	14	26⅜	0	18	16⅜	19½	13/16	11

Machine	H Rear	J Front	J Rear	K	L	M Plain, Vertical	M Universal Conventional Lead	M Universal Low Lead	N Max.	P Max. Plain, Vertical	P Max. Universal Conventional Lead	P Max. Universal Low Lead	Q
205 S-12	11½	11	11½	14	14	73	73⅛	80	54	47	47⅛	56	56
307 S-12	11½	11	11½	14	14	79½	79⅝	86½	60½	53½	53⅝	60½	62½

Machine	R Max.	R[1] Max.	S Plain, Univ.	S Vert.	T Plain, Univ.	T Min. Vert.	T Max. Vert.	U	U[1]	V Vert.	W Min. Plain, Vert.	W Min. Univ.	W Max. Plain, Univ.
205 S-12	42	42	68	75	52⅜	50	54	8	22¾	½	34	35⅝	52⅜
307 S-12	48½	48½	68	75	52⅜	49	54	8	22¾	½	34	35⅝	52⅜

Machine	W Max. Vertical	X Vertical	AA	BB	CC	DD	EE	FF	GG	HH	JJ	KK	LL Min.	LL Max. Plain, Universal	LL Max. Vertical
205 S-12	49⅛	4	5⅝	3⅝	6 3/16	2¼	3¼	11/16	¾	1¼	½	12	7⅜	17⅝	19⅝
307 S-12	49⅛	5	5⅝	3⅝	6 3/16	2¼	3¼	11/16	¾	1¼	½	12	7⅜	17⅝	19⅝

Kearney and Trecker

11–32A. *An installation assembly drawing. It gives the overall dimensions for two machines of the same type but of different sizes.*

plan dimensions
plain — universal — vertical

Machine	MM	NN	PP	QQ	RR	SS	TT
205 S-12*	73	40	33	56	38¼	22	18¼
307 S-12	79½	46½	39½	62½	38¼	24	16¼

*Note: On vertical style SS = 24; TT = 16¼.

11–32B. *Plan dimensions for machine installation assembly.*

CODE

A — OIL DAILY with S.A.E. No. 20 oil

B — OIL WEEKLY with S.A.E. No. 20 oil

C — OIL MONTHLY with S.A.E. No. 20 oil

D — KEEP CLEAN and well oiled at all times

E — LUBRICATE gear teeth with Keystone No. 122 gear lubricant, or equivalent, to obtain smoother, more quiet operation. Remove oil and dirt before applying grease.

LUBRICATION CHART

12-INCH METALWORKING LATHES

IMPORTANT — LUBRICATE LATHE BEFORE OPERATING

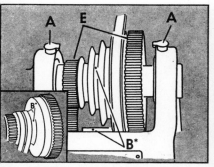

HEADSTOCK AND BACK GEARS
*Remove screw to oil bearings.

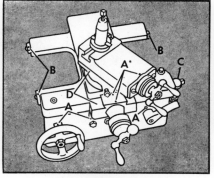

CARRIAGE
*Remove screw to oil bearings and cross feed gears.

TAILSTOCK — LEADSCREW — LEADSCREW BEARING — RACK
*About once a month clean with kerosene and a brush, then cover with oil.

Clausing, Kalamazoo, Mich.

11–33. *Pictorial assembly drawings used to give machine maintenance information.*

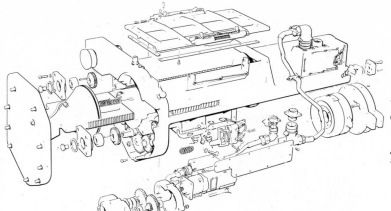

11–34. *A partially exploded assembly drawing. It is used by assembly workers and maintenance men.*

Joe Mickey, Graphic Arts Dept.
Sandia Corp., Albuquerque, N.M.

11–35. *An assembly drawing in section through an automobile engine showing the oil circulation system and partial assembly drawings showing details not clear on the main assembly. This is used in repair manuals prepared for mechanics.*

General Motors Corp.

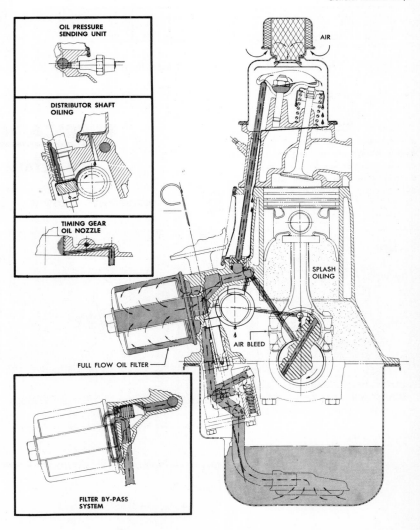

Such a drawing frequently accompanies a machine to assist in assembly and to identify parts for reordering when repairs are needed. They also are used for display purposes.

Additional drawings, examples, and problems on pictorial assemblies are given in Chapter 17, Technical Illustration.

Installation Assembly Drawings. An installation assembly drawing usually includes only the outline of the machine with overall dimensions necessary to indicate the amount of space needed to install it properly. Almost all details are omitted. Frequently, a machine will have a large number of dimensions or be manufactured in several sizes and dimensioning becomes complex. In this case, letters can be substituted for the dimensions and the actual dimensions recorded in table form on the drawing, as shown in Fig. 11–32. This is a form of

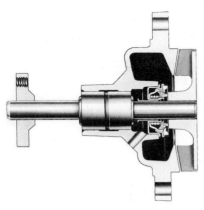

11–36. *A shaded assembly of an automobile water pump. It is drawn as a full section. This type of assembly drawing is useful in sales literature to show important features. It is also useful in technical manuals for auto repair.*

a "tabular drawing." See pages 302 and 303.

Maintenance Assembly Drawings. A maintenance assembly can be a general assembly or pictorial assembly. See Figs. 11–33 and 11–34. They are used to give lubrication, servicing, and operating information about a machine. Sometimes parts are identified by the manufacturer's part number used for ordering replacement parts.

Manufacturers supply manuals useful to those who must service their products. Assembly drawings are important to show clearly the operation of these units. Fig. 11–35 is an assembly drawing of an automobile engine. It shows the oil circulation system. The

drawing is kept as simple as possible. It only shows what is needed to serve its purpose.

Additional examples of maintenance assembly drawings are given in Chapter 17, Technical Illustration.

Catalog Assembly Drawings. Catalog assembly drawings are for inclusion in sales literature and parts catalogs. Their purpose is to explain the design and function of the object. Some are similar

Data for Parts List

Fig. No.	Part No.	Description
1	1260630–1	Nose Gear Steering Assembly
2	AN176–21A	Bolt
3	AN960–616L	Washer
4	NAS679A6	Nut
5	1243600–1	Link-Drag
6	NAS464P5A42	Bolt
7	Ang60–516L	Washer
8	NAS679A5	Nut
9	AN6–32A	Bolt
10	NAS679A6	Nut
11	0743624–2	Shimmy Dampner Attaching Parts
12	NAS464P3A32C	Bolt
13	NAS679A3	Nut
14	AN4H5A	Bolt
22	1241156–12	Wheel Assembly, Nose
56	1243610–2–6	Nose Gear Shock Strut Assembly
57	0743627–1	Cap, Wheel
58	AN936A8	Washer

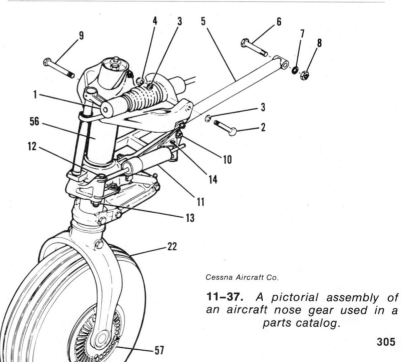

11–37. *A pictorial assembly of an aircraft nose gear used in a parts catalog.*

305

to general assemblies and have parts, as shafts, shaded to present a more lifelike resemblance. See Fig. 11–36.

Pictorial assemblies are also used. They are especially useful in parts catalogs. They usually show each part in an exploded pictorial view with the information for identify-ing each part by name and part number. See Fig. 11–31.

Some companies use pictorial drawings showing the exterior of the object or a subassembly in their parts catalogs. See Fig. 11–37. These can have parts in exploded positions as well as in final assembled positions.

These drawings include a parts list giving the manufacturer's part number and the name of the part.

Additional examples of catalog assembly drawings are given in Chapter 17, Technical Illustration.

Build Your Vocabulary

Following are terms that you should understand and use as a part of your working vocabulary. Write a brief explanation of what each means.

Preliminary design assembly
Detail drawings
Pattern detail
Draft
Shrinkage
Machining
Core
Machining drawing
Casting detail drawing
Forging detail drawing
Die
Parting line
Shrinkage tolerance
Die wear
Mismatch
Welding detail
Drawing metal
Stamping
Sheet metal bend allowance
Assembly drawing

Study Problems

Pattern Detail Drawings

1. Make a pattern detail drawing of the cap of the hoist ring, P11–1.
2. Make a pattern detail drawing of the optical mount, P11–2.
3. Make a pattern detail drawing of the wing nut on the bevel square, P11–3.

Casting Detail Drawings

4. Make a casting detail drawing for the cap of the hoist ring, P11–1. The material is steel.
5. Draw a casting detail drawing for the arm of the control handle, P11–4. The material is cast iron.
6. Draw a casting detail drawing for the base of the spring stop, P11–5. The material is cast iron.
7. Draw a casting detail for the hook of the fixture clamp, P11–6.

Machining Detail Drawings

8. Draw a machining detail drawing for the arm of the control handle, P11–4.
9. Draw machining detail drawings for the four parts of the spray gun nozzle, P11–7.
10. Draw a machining detail drawing of the optical cup, P11–2.
11. Draw a machining detail drawing of the yoke and the handle of the cam clamp, P11–8.
12. Draw machining detail drawings for the upper and lower plates of the adjustable angle, P11–9.
13. Draw a machining detail of the guide block on the wood shaper hold-down fixture, P11–10.

Forging

14. Draw a forging detail of the pin of the cabinet hinge, P11–11.

15. Draw a forging drawing and a separate machining drawing of the yoke of the foot pedal assembly, P11–12. Be certain to figure tolerances needed. The yoke weighs 0.4 pounds.

16. Draw a forging detail and a separate machining detail of the shaft of the T-handle socket wrench, P11–13. Be certain to figure tolerances needed. The shaft weighs three pounds.

17. Draw a forging detail drawing of the lathe power control link, P11–14. Show machining details on the same drawing. The inside of the bearing hub must be machined on both sides to receive a bearing with a $1\frac{1}{8}$-inch outside diameter. A $\frac{3}{8}$-inch diameter hole is to be drilled in the center of the bearing hub. All fillets and rounds should be held to the minimum permitted for forgings. The link weighs one-half pound. Be certain to add necessary tolerances to dimensions on the drawing.

18. Draw a forging detail of the foot brake arm, P11–15. Show machining details on the same drawing. All fillets and rounds should be held to the minimum permitted for forgings. The arm weighs three-fourths of a pound. Be certain to add all necessary tolerances to dimensions on the drawing.

Stamping Detail Drawings

19. Draw a stamping detail drawing for the cover for the optical cup and optical mount, P11–2.

20. Draw a stamping detail drawing for the brace of the foot pedal assembly, P11–12.

21. Draw a stamping detail of the blade of the bevel square, P11–3.

22. Draw a stamping detail drawing of the seat belt anchor, P11–16.

23. The unit shown in P11–17 is formed by stretching the metal during the stamping operation. Draw a stamping detail drawing of this unit.

24. The rearview mirror bracket, P11–18, has a rib stamped to give it stiffness. Draw a stamping detail drawing of the bracket.

25. Draw a stamping detail drawing of the motor bracket, P11–19.

26. Draw a stamping detail drawing of the flasher bracket, P11–20. Design the corners so they have enough clearance to be formed easily.

Welding Detail Drawings

27. Draw a welding detail drawing showing the union of the pedal and arm in the foot pedal assembly, P11–12.

28. Draw a welding detail drawing of the auto assembly drill fixture, P11–21.

29. Draw a welding detail drawing of the auto assembly fixture, P11–22.

30. An exploded view of a polishing head to be assembled by welding is shown in P11–23. Make a complete set of working drawings. Indicate the welds with the proper symbols.

Check Assembly Problem

31. Details of a spray gun nozzle are given in P11–7. Using the dimensions given, make a check assembly drawing.

General Assembly Problems

32. Make a general assembly of the control handle shown in P11–4. Include a parts list.

33. Make a general assembly of the spring stop shown in P11–5. Include a parts list.

34. Draw a general assembly of the fixture clamp shown in P11–6. Include a parts list.

35. Draw a general assembly of the optical cup, optical cup mount with the metal cover over the unit. Detail drawings are shown in P11–2. The optical cup screws into the cup mount. Include a parts list. A photo of the assembled unit with the cover removed is shown in Fig. 6–49 in Chapter 6, Multiview Drawing.

36. In P11–8 is shown an exploded pictorial view of a cam clamp. Draw this in assembled condition. Use sections where necessary to clarify details. Include a parts list.

Additional general assembly problems are given in the chapter on sections.

Working Drawing Assembly Problems

37. Make a working drawing

assembly of the hoist ring, P11–1.

38. An adjustable angle used in machine shops is shown in P11–9. Draw a working drawing assembly of this tool.

39. Make a working drawing assembly of the salt shaker, P11–24.

40. Make a working drawing assembly of the hinge shown in P11–11.

Pictorial Assembly Problems.

For additional information on pictorial assembly drawings, see Chapter 17, Technical Illustration.

41. Make a cutaway pictorial assembly drawing of the grease cup, P11–25.

42. Make a pictorial section of the navy pipe union, P11–26. Remove a quarter section to show assembly details.

43. Make an exploded assembly drawing of the tapered nose clamp, P11–27.

Installation Assembly Problem

44. Visit your industrial arts shop, select a machine, and make a simple installation assembly drawing of it, showing the overall size of the machine.

Maintenance Assembly Problems

45. Select a machine and make maintenance assembly drawings, showing where it should be lubricated. This could be a machine in your industrial arts shop.

Catalog Assembly

46. Prepare a catalog assembly drawing of the spring stop, P11–5. Use the principles of shading as explained in Chapter 17, Technical Illustration. Draw the spring stop as a full section.

Other Assembly Problems

47. Make an assembly drawing that would be most useful in a shop manual to show a machine operator how to assemble the wood shaper hold-down foot, P11–10.

48. Make an assembly drawing that would be most useful to a factory worker assembling the bevel square shown in P11–3.

Design Problems. Following are some problems requiring an original solution. There is more than one correct solution. Possible solutions are limited only by the imagination and ingenuity of the designer.

As solutions are sought, remember that simplicity, ease of manufacture, and cost are important. Use standard parts, as bolts, springs, and keys, whenever possible.

The complete solution will include freehand design sketches, and assembly, detail, and pictorial drawings needed to completely describe the solution.

Design Problem 1. P11–28 shows panels to serve as office partitions. Design a means of connecting these panels so that they can be fastened with only a screwdriver or pliers. The fastener must permit the panels to be taken apart rapidly without damage. This permits easy rearrangement of office space. The fastener should be as inconspicious as possible.

Design a leg to hold the panels 12 inches off the floor. The legs should be easily removed.

Design Problem 2. Design a hanger to secure the arm to the machine base, as shown in P11–29. The arm must be able to swing through an arc of 180 degrees and be removed from the machine base without the use of tools.

Design Problem 3. P11–30 shows a base for a machine. Shavings from the machine fill the base and have to be cleaned out frequently. Design a sheet metal door and the fastening devices to hold it in place. The door should be completely removable from the base without the use of tools. This design should allow for rapid replacement of the door.

Design Problem 4. Design a fitting that will hold bar A to bar B, as shown in P11–31. The fitting should permit bar B to slide up and down bar A. It should permit bar B to be tightened firmly to bar A any place along its length.

Bar B should be able to slide horizontally and lock in place.

Bar B should meet bar A at a 90 degree angle.

Design Problem 5. *Design a device to convert the hand grinder, P11—32, into a bench grinder. The holding device should permit the operator to have both hands free to hold the work to be ground. The center of the grinding wheel should be 3½ inches above the table when the grinder is in a horizontal position. The device should permit the grinder to swing above and below the horizontal at least 30 degrees and be locked in this position.*

Design Problem 6. *Design a fastening device which will enable the steel wall panels, P11—33 to be joined quickly yet permit disassembly so the panels can be moved or replaced. The fastening device should be easily installed with simple tools.*

Design Problem 7. *The rod shown in P11—34 moves in a vertical direction. Design a device or alter the rod or casting so a machine operator can easily raise or lower the rod but still lock it tightly in place. The device should provide positive stops in ½ inch increments. This enables the operator to lock the bar in ½ inch increments without measuring.*

Design Problem 8. *Design a gland to hold the bearing shown in P11—35 in place. A gland is a metal part that holds a bearing in place. The gland and the bearing should have a force fit of FN 1. The machine casting can have threaded holes added.*

Design Problem 9. *Design a pipe hanger that can be adjusted to carry pipes from 2 inches to 3 inches in diameter. The part of the hanger that is to be connected to the overhead must be adjustable so the pipe can be carried from 4 inches to 6 inches below the overhead. It should be designed so that it can be bolted to wood or metal overhead members. See P11—36.*

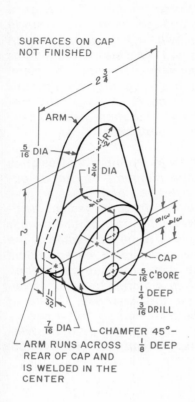

SURFACES ON CAP
NOT FINISHED

$2\frac{3}{4}$

ARM

$\frac{1}{2}$ R

$\frac{5}{16}$ DIA

$1\frac{3}{4}$ DIA

$\frac{3}{4}$

2

CAP

$\frac{5}{16}$ C'BORE
$\frac{1}{4}$ DEEP
$\frac{3}{16}$ DRILL

$\frac{11}{32}$

$\frac{7}{16}$ DIA

CHAMFER 45°-
$\frac{1}{8}$ DEEP

ARM RUNS ACROSS
REAR OF CAP AND
IS WELDED IN THE
CENTER

Carr Lane Manufacturing Co.

P11–1. *Hoist ring.*

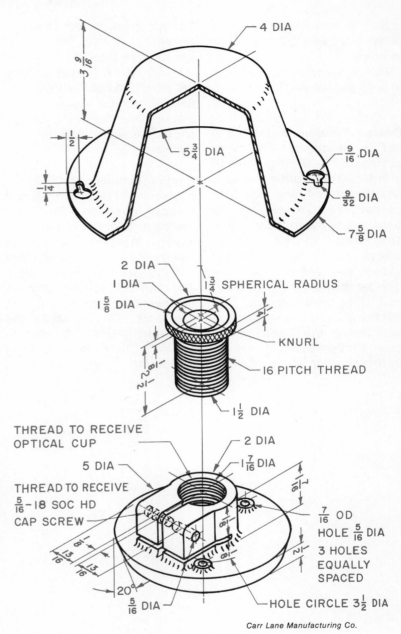

4 DIA

$3\frac{9}{16}$

$\frac{1}{2}$

$5\frac{3}{4}$ DIA

$\frac{1}{4}$

$\frac{9}{16}$ DIA

$\frac{9}{32}$ DIA

$7\frac{5}{8}$ DIA

2 DIA

1 DIA

$1\frac{5}{8}$ DIA

$1\frac{3}{4}$ SPHERICAL RADIUS

$\frac{1}{4}$

KNURL

16 PITCH THREAD

$\frac{1}{8}$

$2\frac{1}{2}$

$1\frac{1}{2}$ DIA

THREAD TO RECEIVE
OPTICAL CUP

5 DIA

THREAD TO RECEIVE
$\frac{5}{16}$–18 SOC HD
CAP SCREW

2 DIA

$1\frac{7}{16}$ DIA

$\frac{7}{16}$

$\frac{7}{16}$ OD
HOLE $\frac{5}{16}$ DIA

3 HOLES
EQUALLY
SPACED

$1\frac{1}{8}$

$\frac{13}{16}$ $\frac{13}{16}$

$2\frac{1}{2}$

20°

$\frac{5}{16}$ DIA

HOLE CIRCLE $3\frac{1}{2}$ DIA

Carr Lane Manufacturing Co.

P11–2. *Optical cup, mount, and cover.*

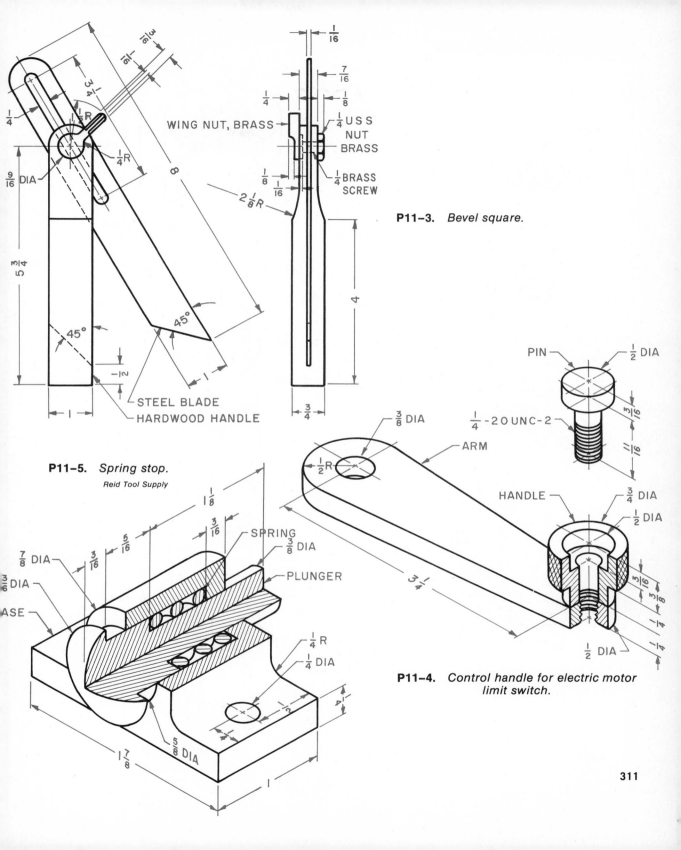

P11–3. Bevel square.

WING NUT, BRASS

$\frac{1}{16}$

$\frac{7}{16}$

$\frac{1}{4}$ · $\frac{1}{8}$

$\frac{1}{4}$ U S S NUT BRASS

$\frac{1}{8}$ · $\frac{1}{16}$

$2\frac{1}{8}R$

$\frac{1}{4}$ BRASS SCREW

4

$\frac{3}{4}$

$\frac{3}{16}$

$1\frac{1}{16}$

$3\frac{1}{4}$

$\frac{1}{4}$

$\frac{1}{2}R$

$\frac{1}{4}R$

8

$\frac{9}{16}$ DIA

$5\frac{3}{4}$

45°

45°

$-\frac{1}{2}$

1

1

STEEL BLADE

HARDWOOD HANDLE

PIN

$\frac{1}{2}$ DIA

$\frac{1}{4}$ -20UNC-2

$\frac{3}{16}$

$\frac{11}{16}$

ARM

$\frac{3}{8}$ DIA

$\frac{1}{2}R$

HANDLE

$\frac{3}{4}$ DIA

$\frac{1}{2}$ DIA

$3\frac{1}{4}$

$\frac{3}{16}$

$\frac{3}{8}$

$\frac{1}{4}$

$\frac{1}{2}$ DIA

P11–5. Spring stop.

Reid Tool Supply

$1\frac{1}{8}$

$\frac{5}{16}$

$\frac{3}{16}$

$\frac{3}{16}$

SPRING

$\frac{3}{8}$ DIA

PLUNGER

$\frac{7}{8}$ DIA

$\frac{3}{8}$ DIA

ASE

$\frac{1}{4}$ R

$\frac{1}{4}$ DIA

$\frac{1}{2}$

$\frac{1}{4}$

P11–4. Control handle for electric motor limit switch.

$\frac{5}{8}$ DIA

$1\frac{7}{8}$

1

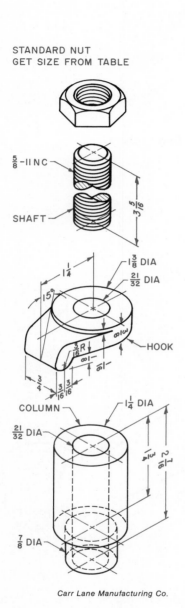

STANDARD NUT
GET SIZE FROM TABLE

$\frac{5}{8}$-11 NC

SHAFT

$3\frac{5}{16}$

$1\frac{3}{8}$ DIA

$\frac{21}{32}$ DIA

$1\frac{1}{4}$

15°

$\frac{3}{16}$ R

HOOK

$\frac{3}{4}$ $\frac{3}{16}$

COLUMN

$1\frac{1}{4}$ DIA

$\frac{21}{32}$ DIA

$2\frac{7}{16}$

$\frac{7}{8}$ DIA

Carr Lane Manufacturing Co.

P11–6. *Fixture clamp.*

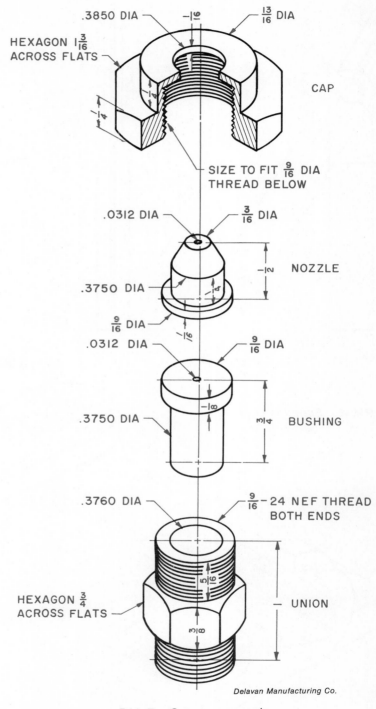

.3850 DIA $\frac{13}{16}$ DIA

HEXAGON $1\frac{3}{16}$
ACROSS FLATS

CAP

SIZE TO FIT $\frac{9}{16}$ DIA
THREAD BELOW

.0312 DIA $\frac{3}{16}$ DIA

NOZZLE

.3750 DIA

$\frac{9}{16}$ DIA

.0312 DIA $\frac{9}{16}$ DIA

.3750 DIA

BUSHING

.3760 DIA $\frac{9}{16}$-24 NEF THREAD
BOTH ENDS

HEXAGON $\frac{3}{4}$
ACROSS FLATS

UNION

Delavan Manufacturing Co.

P11–7. *Spray gun nozzle.*

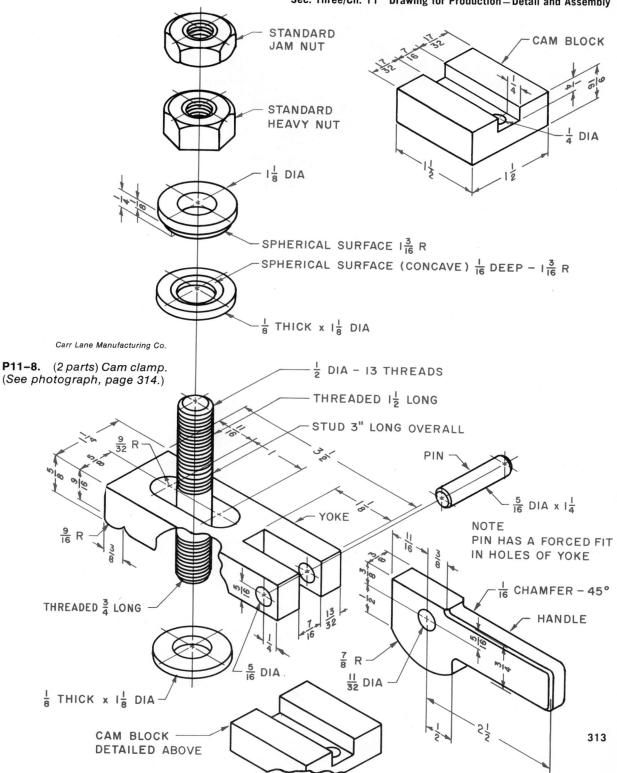

STANDARD
JAM NUT

STANDARD
HEAVY NUT

CAM BLOCK

$\frac{17}{32}$

$\frac{7}{16}$

$\frac{7}{32}$

$\frac{1}{4}$

$1\frac{1}{4}$

$\frac{9}{16}$

$\frac{1}{4}$ DIA

$1\frac{1}{2}$

$1\frac{1}{2}$

$1\frac{1}{8}$ DIA

$\frac{1}{4}$ - $\frac{1}{8}$

SPHERICAL SURFACE $1\frac{3}{16}$ R

SPHERICAL SURFACE (CONCAVE) $\frac{1}{16}$ DEEP – $1\frac{3}{16}$ R

$\frac{1}{8}$ THICK x $1\frac{1}{8}$ DIA

Carr Lane Manufacturing Co.

P11–8. (2 parts) Cam clamp.
(See photograph, page 314.)

$\frac{1}{2}$ DIA – 13 THREADS

THREADED $1\frac{1}{2}$ LONG

STUD 3" LONG OVERALL

$\frac{11}{16}$

1

$3\frac{1}{2}$

$\frac{1}{8}$

$\frac{1}{4}$

$\frac{9}{32}$ R

$\frac{5}{8}$

$\frac{9}{16}$

$\frac{3}{16}$

PIN

$\frac{5}{16}$ DIA x $1\frac{1}{4}$

NOTE
PIN HAS A FORCED FIT
IN HOLES OF YOKE

YOKE

$\frac{9}{16}$ R

$\frac{3}{8}$

$\frac{11}{16}$

$\frac{3}{8}$

$\frac{3}{8}$

$\frac{3}{16}$

$\frac{1}{2}$

$\frac{3}{8}$

$\frac{1}{16}$ CHAMFER – 45°

HANDLE

THREADED $\frac{3}{4}$ LONG

$\frac{5}{16}$

$\frac{1}{4}$

$\frac{7}{16}$

$\frac{13}{32}$

$\frac{5}{16}$ DIA

$\frac{7}{8}$ R

$\frac{11}{32}$ DIA

$\frac{5}{16}$

$3\frac{1}{4}$

$\frac{1}{8}$ THICK x $1\frac{1}{8}$ DIA

$\frac{1}{2}$

$2\frac{1}{2}$

CAM BLOCK
DETAILED ABOVE

313

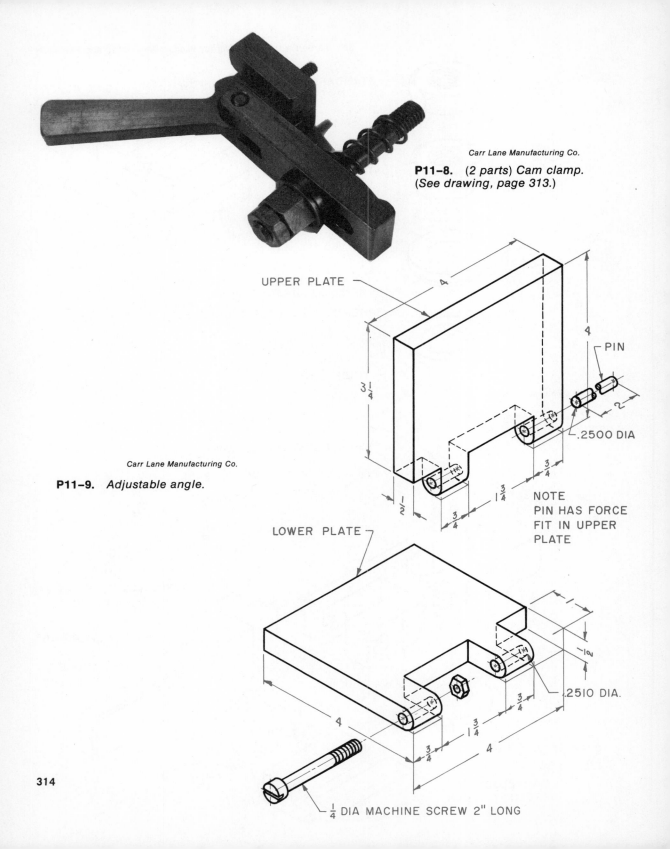

Carr Lane Manufacturing Co.

P11–8. (*2 parts*) *Cam clamp.*
(*See drawing, page 313.*)

UPPER PLATE

PIN

.2500 DIA

NOTE
PIN HAS FORCE
FIT IN UPPER
PLATE

Carr Lane Manufacturing Co.

P11–9. *Adjustable angle.*

LOWER PLATE

.2510 DIA.

$\frac{1}{4}$ DIA MACHINE SCREW 2" LONG

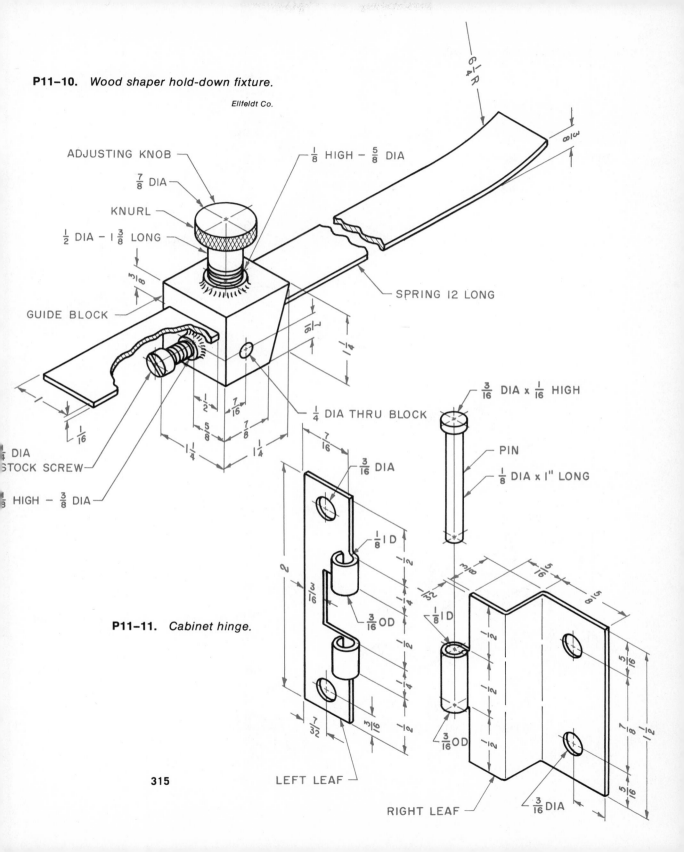

P11–10. *Wood shaper hold-down fixture.*

Ellfeldt Co.

$\frac{1}{4}$R

$\frac{3}{8}$

ADJUSTING KNOB

$\frac{1}{8}$ HIGH – $\frac{5}{8}$ DIA

$\frac{7}{8}$ DIA

KNURL

$\frac{1}{2}$ DIA – $1\frac{3}{8}$ LONG

$\frac{3}{8}$

GUIDE BLOCK

SPRING 12 LONG

$\frac{1}{2}$

$\frac{7}{16}$

$\frac{5}{8}$

$\frac{7}{8}$

$1\frac{1}{4}$

$1\frac{1}{4}$

$1\frac{1}{4}$

$\frac{7}{16}$

$1\frac{1}{4}$

$\frac{1}{16}$

$\frac{1}{4}$ DIA THRU BLOCK

DIA

STOCK SCREW

HIGH – $\frac{3}{8}$ DIA

$\frac{3}{16}$ DIA x $\frac{1}{16}$ HIGH

PIN

$\frac{1}{8}$ DIA x 1" LONG

P11–11. *Cabinet hinge.*

$\frac{7}{16}$

$\frac{3}{16}$ DIA

$\frac{1}{8}$ ID

$\frac{1}{2}$

$\frac{1}{4}$

2

$\frac{3}{16}$

$\frac{3}{16}$ OD

$\frac{1}{2}$

$\frac{1}{4}$

$\frac{7}{32}$

$\frac{3}{16}$

$\frac{1}{32}$

$\frac{1}{8}$ ID

$\frac{3}{8}$

$\frac{5}{16}$

$\frac{1}{2}$

$\frac{5}{16}$

$\frac{7}{8}$

$1\frac{1}{2}$

$\frac{5}{16}$

$\frac{3}{16}$ OD

$\frac{1}{2}$

$\frac{1}{2}$

$\frac{3}{16}$ DIA

LEFT LEAF

RIGHT LEAF

315

P11–12. *Foot pedal assembly.*

Ford Motor Co.
Kansas City Assembly Plant

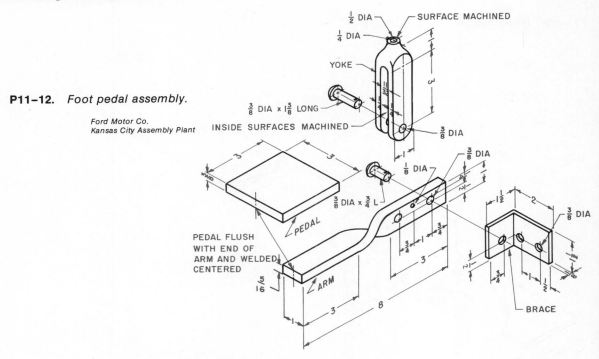

P11–13. *T handle socket wrench.*

Billings and Spencer

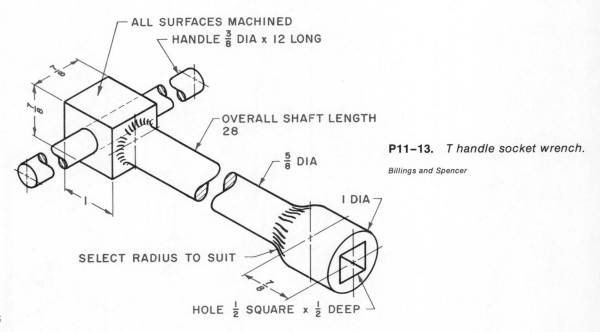

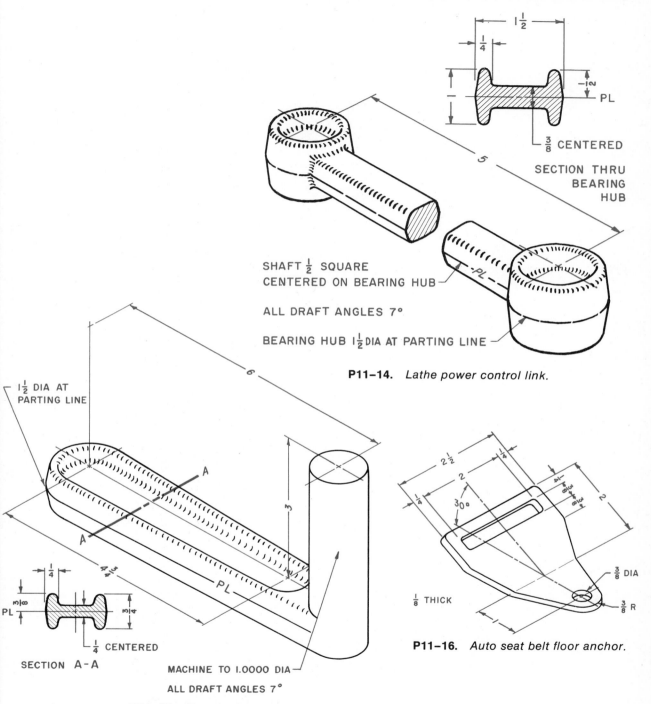

SHAFT $\frac{1}{2}$ SQUARE
CENTERED ON BEARING HUB

ALL DRAFT ANGLES 7°

BEARING HUB $1\frac{1}{2}$ DIA AT PARTING LINE

P11–14. *Lathe power control link.*

$1\frac{1}{2}$ DIA AT
PARTING LINE

SECTION THRU
BEARING
HUB

$\frac{3}{8}$ CENTERED

PL

SECTION A-A

$\frac{1}{4}$ CENTERED

MACHINE TO 1.0000 DIA
ALL DRAFT ANGLES 7°

P11–15. *Foot brake arm.*

$\frac{1}{8}$ THICK

$\frac{3}{8}$ DIA

$\frac{3}{8}$ R

30°

P11–16. *Auto seat belt floor anchor.*

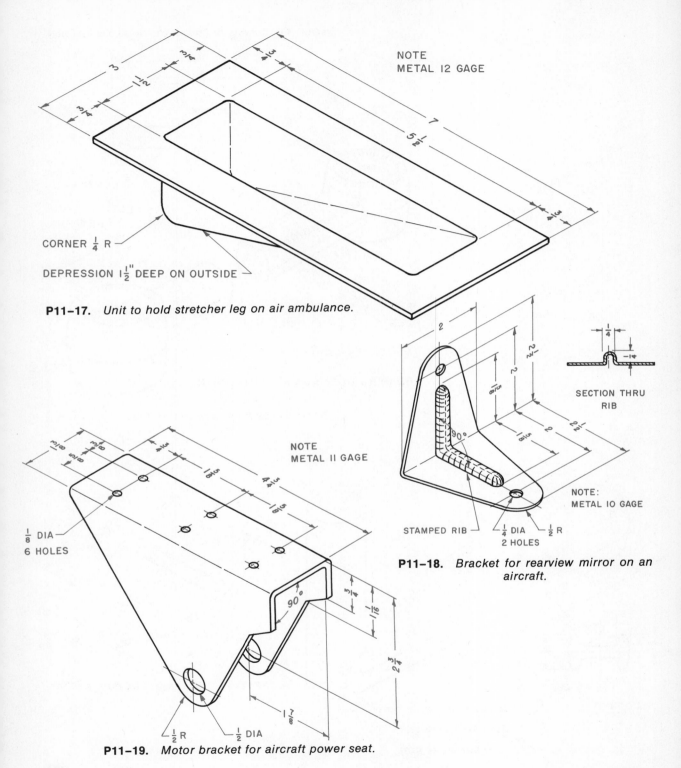

NOTE
METAL 12 GAGE

3

$\frac{3}{4}$

$1\frac{1}{2}$

$\frac{3}{4}$

7

$5\frac{1}{2}$

$4\frac{3}{4}$

CORNER $\frac{1}{4}$ R

DEPRESSION $1\frac{1}{2}''$ DEEP ON OUTSIDE

P11–17. *Unit to hold stretcher leg on air ambulance.*

NOTE
METAL 11 GAGE

$\frac{1}{8}$ DIA
6 HOLES

$\frac{1}{8}$ DIA

$\frac{3}{8}$

$\frac{1}{4}$

$\frac{3}{8}$

$4\frac{1}{4}$

$\frac{5}{8}$

$\frac{3}{4}$

$1\frac{1}{16}$

$2\frac{3}{4}$

90°

$\frac{1}{2}$ R

$\frac{1}{2}$ DIA

$1\frac{7}{8}$

P11–19. *Motor bracket for aircraft power seat.*

2

$2\frac{1}{2}$

2

$\emptyset\frac{1}{5}$

90°

$1\frac{1}{8}$

2

$2\frac{1}{2}$

$\frac{1}{4}$

$1\frac{1}{4}$

SECTION THRU
RIB

NOTE:
METAL 10 GAGE

STAMPED RIB

$\frac{1}{4}$ DIA
2 HOLES

$\frac{1}{2}$ R

P11–18. *Bracket for rearview mirror on an aircraft.*

318

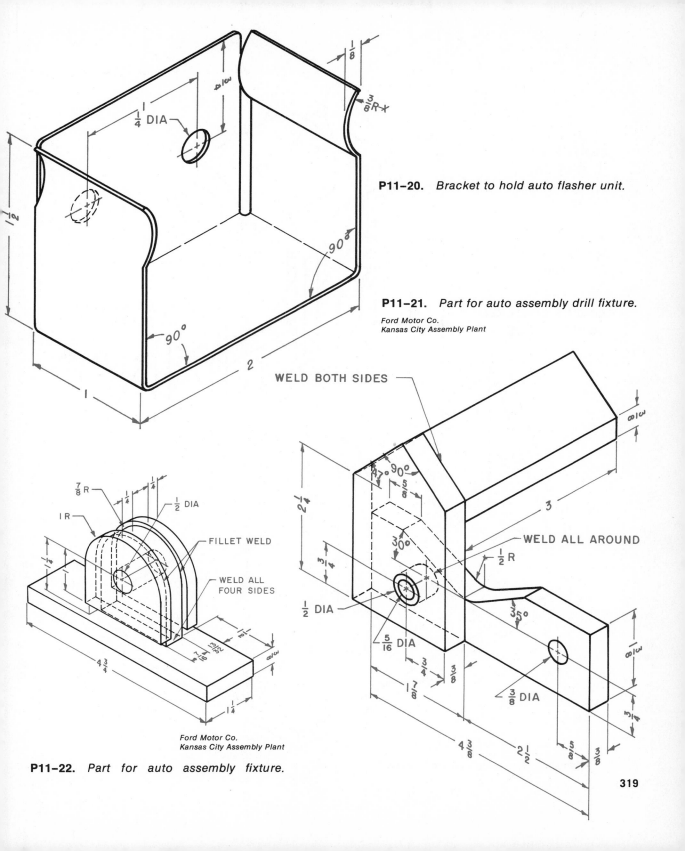

P11–20. *Bracket to hold auto flasher unit.*

$\frac{1}{8}$

$\frac{3}{8}$R

$\frac{3}{4}$

$\frac{1}{4}$ DIA

$90°$

$90°$

2

1

$1\frac{1}{2}$

1

P11–21. *Part for auto assembly drill fixture.*

Ford Motor Co.
Kansas City Assembly Plant

WELD BOTH SIDES

$\frac{3}{8}$

$45°$ $90°$

$\frac{5}{8}$

$2\frac{1}{4}$

3

WELD ALL AROUND

$30°$

$\frac{1}{2}$ R

$3\frac{1}{4}$

$\frac{1}{2}$ DIA

$\frac{5}{16}$ DIA

$35°$

$\frac{3}{8}$

$\frac{3}{8}$ DIA

$\frac{3}{4}$ $\frac{3}{8}$

$\frac{3}{8}$

$1\frac{7}{8}$

$4\frac{3}{8}$ $2\frac{1}{2}$ $\frac{5}{8}$ $\frac{3}{8}$

$\frac{3}{4}$

$\frac{7}{8}$ R

$\frac{1}{4}$ $\frac{1}{4}$

$\frac{1}{2}$ DIA

1 R

FILLET WELD

WELD ALL
FOUR SIDES

$\frac{1}{4}$

$\frac{7}{16}$

$\frac{13}{32}$

$\frac{1}{2}$

$\frac{3}{8}$

$4\frac{3}{4}$

$1\frac{1}{4}$

Ford Motor Co.
Kansas City Assembly Plant

P11–22. *Part for auto assembly fixture.*

319

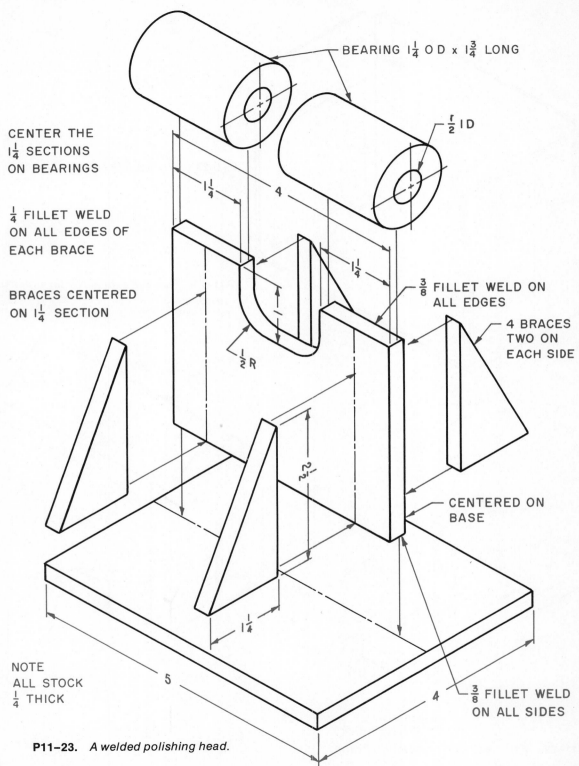

BEARING $1\frac{1}{4}$ O D x $1\frac{3}{4}$ LONG

$\frac{1}{2}$ ID

CENTER THE $1\frac{1}{4}$ SECTIONS ON BEARINGS

$\frac{1}{4}$ FILLET WELD ON ALL EDGES OF EACH BRACE

BRACES CENTERED ON $1\frac{1}{4}$ SECTION

$1\frac{1}{4}$

4

$1\frac{1}{4}$

$\frac{1}{2}$ R

$\frac{3}{8}$ FILLET WELD ON ALL EDGES

4 BRACES TWO ON EACH SIDE

1

$2\frac{1}{2}$

CENTERED ON BASE

$1\frac{1}{4}$

NOTE ALL STOCK $\frac{1}{4}$ THICK

5

4

$\frac{3}{8}$ FILLET WELD ON ALL SIDES

P11–23. *A welded polishing head.*

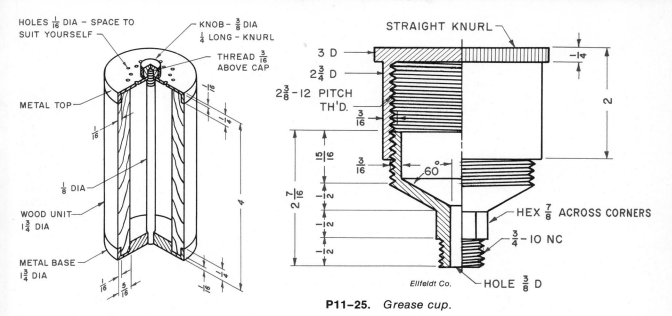

HOLES $\frac{1}{16}$ DIA – SPACE TO SUIT YOURSELF

KNOB – $\frac{3}{8}$ DIA $\frac{1}{4}$ LONG – KNURL

THREAD $\frac{3}{16}$ ABOVE CAP

METAL TOP

$\frac{1}{16}$

$\frac{1}{4}$

$\frac{1}{8}$ DIA

WOOD UNIT $1\frac{3}{4}$ DIA

4

METAL BASE $1\frac{3}{4}$ DIA

$\frac{1}{16}$ $\frac{5}{16}$

$\frac{1}{4}$

$\frac{1}{16}$

P11–24. *Salt shaker.*

STRAIGHT KNURL

3 D

$2\frac{3}{4}$ D

$2\frac{3}{8}$–12 PITCH TH'D.

$\frac{3}{16}$

$\frac{3}{16}$

$\frac{15}{16}$

$\frac{3}{16}$

$2\frac{7}{16}$

$1\frac{1}{2}$

$1\frac{1}{2}$

$1\frac{1}{2}$

60°

$-\frac{1}{4}$

2

HEX $\frac{7}{8}$ ACROSS CORNERS

$\frac{3}{4}$–10 NC

Ellfeldt Co. HOLE $\frac{3}{8}$ D

P11–25. *Grease cup.*

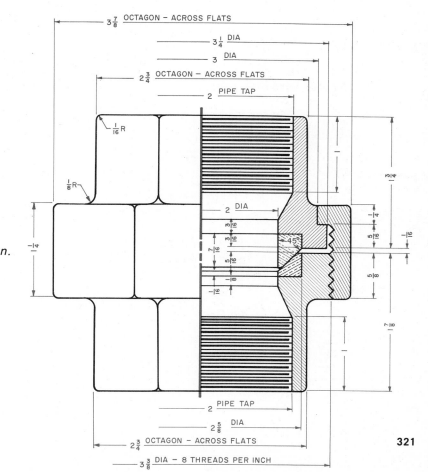

$3\frac{7}{8}$ OCTAGON – ACROSS FLATS

$3\frac{1}{4}$ DIA

3 DIA

$2\frac{3}{4}$ OCTAGON – ACROSS FLATS

2 PIPE TAP

$\frac{1}{16}$ R

$\frac{1}{8}$ R

$1\frac{1}{4}$

2 DIA

$\frac{3}{16}$

$\frac{3}{16}$

$\frac{7}{8}$

$\frac{5}{16}$

$\frac{1}{8}$

$\frac{1}{16}$

45°

$3\frac{1}{4}$

$\frac{1}{4}$

$\frac{5}{16}$

$\frac{1}{16}$

$\frac{5}{8}$

$1\frac{7}{8}$

P11–26. *A navy pipe union.*

2 PIPE TAP

$2\frac{5}{8}$ DIA

$2\frac{3}{4}$ OCTAGON – ACROSS FLATS

$3\frac{3}{8}$ DIA – 8 THREADS PER INCH

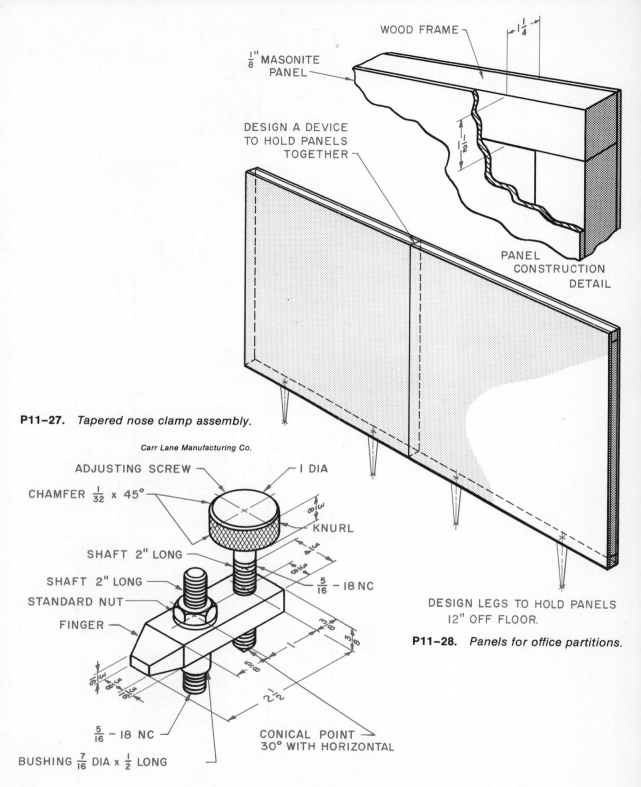

1⅛" MASONITE PANEL

WOOD FRAME

1¼

DESIGN A DEVICE TO HOLD PANELS TOGETHER

1½

PANEL CONSTRUCTION DETAIL

P11–27. *Tapered nose clamp assembly.*

Carr Lane Manufacturing Co.

ADJUSTING SCREW

CHAMFER $\frac{1}{32}$ x 45°

1 DIA

$\frac{3}{8}$

KNURL

SHAFT 2" LONG

$\frac{1}{4}$

$\frac{3}{8}$

$\frac{5}{16}$ – 18 NC

SHAFT 2" LONG

STANDARD NUT

FINGER

$\frac{3}{8}$

$\frac{3}{8}$

1

$\frac{5}{8}$

$\frac{3}{16}$

$\frac{3}{8}$

$\frac{3}{16}$

$2\frac{1}{2}$

$\frac{5}{16}$ – 18 NC

CONICAL POINT 30° WITH HORIZONTAL

BUSHING $\frac{7}{16}$ DIA x $\frac{1}{2}$ LONG

DESIGN LEGS TO HOLD PANELS 12" OFF FLOOR.

P11–28. *Panels for office partitions.*

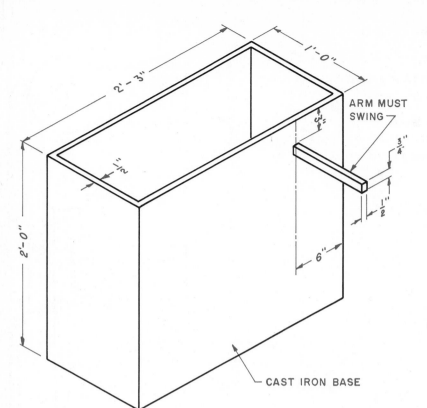

ARM MUST SWING

CAST IRON BASE

P11–29. *Movable arm design.*

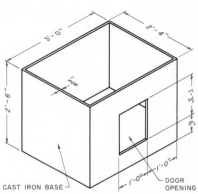

CAST IRON BASE

DOOR OPENING

P11–30. *Machine base.*

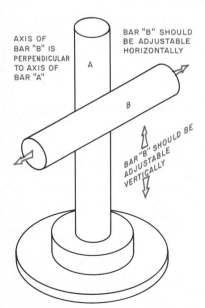

AXIS OF BAR "B" IS PERPENDICULAR TO AXIS OF BAR "A"

BAR "B" SHOULD BE ADJUSTABLE HORIZONTALLY

A

B

BAR "B" SHOULD BE ADJUSTABLE VERTICALLY

P11–31. *Adjustable bar design.*

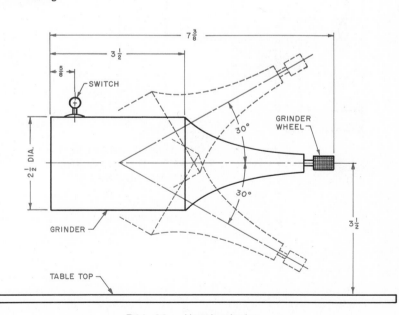

$7\frac{3}{8}$

$3\frac{1}{2}$

$\frac{5}{8}$

SWITCH

$2\frac{1}{2}$ DIA.

30°

GRINDER WHEEL

30°

$3\frac{1}{2}$

GRINDER

TABLE TOP

P11–32. *Hand grinder.*

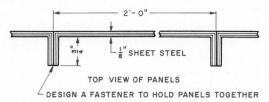

TOP VIEW OF PANELS

DESIGN A FASTENER TO HOLD PANELS TOGETHER

P11–33. *Steel wall panel.*

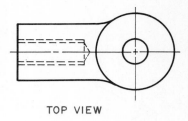

TOP VIEW

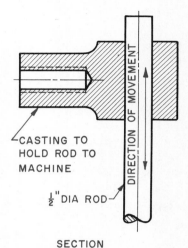

CASTING TO HOLD ROD TO MACHINE

½" DIA ROD

DIRECTION OF MOVEMENT

SECTION

P11–34. *Vertical bar stop lock.*

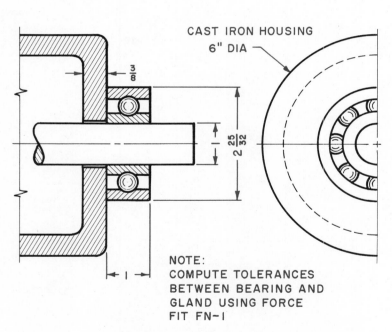

CAST IRON HOUSING 6" DIA

NOTE: COMPUTE TOLERANCES BETWEEN BEARING AND GLAND USING FORCE FIT FN–1

P11–35. *Gland design.*

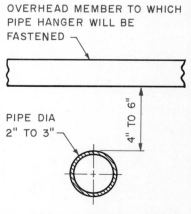

OVERHEAD MEMBER TO WHICH PIPE HANGER WILL BE FASTENED

PIPE DIA 2" TO 3"

4" TO 6"

P11–36. *Pipe hanger.*

Chapter 12

Computers in Design and Drafting

The design and drafting process is being influenced by many technical developments. Possibly the most significant changes in the years immediately ahead will be caused by the expanding use of numerically controlled machine tools and computers. While these developments will not replace the draftsman or designer, they will change his duties and the way he approaches a design situation.

DRAFTING FOR NUMERICAL CONTROL

Numerical control refers to the control of a machine tool by use of a punched or magnetic tape or punched card. The tape contains instructions that control the operations the machine is to perform. For example, the tape can program the machine to drill a hole in a particular place on the part being made. This operation is performed automatically in the same manner as when the machine is operated manually by a machinist. The difference is that the operations are performed without the machinist making adjustments and setting the drill or cutter in the place desired.

Fig. 12–1 shows a numerically controlled machine tool. It drills, counterbores, countersinks, reams, and mills. This machine uses a punched paper tape. The tape is placed

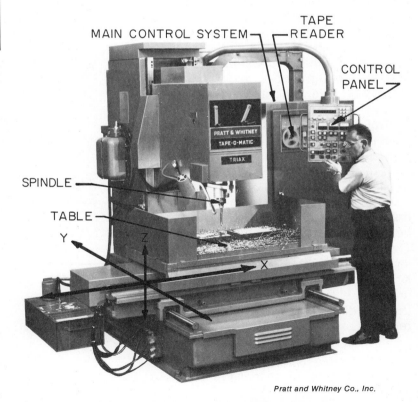

MAIN CONTROL SYSTEM — TAPE READER

CONTROL PANEL

SPINDLE

TABLE

Y Z X

Pratt and Whitney Co., Inc.

12–1. *A numerically controlled machine tool.*

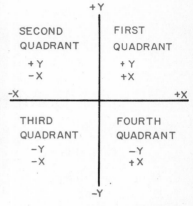

12-2. *A tape reader.*

Pratt and Whitney Co., Inc.

12-3. *Numerical control drawings using a coordinate dimensioning system can be placed in any one quadrant or a combination of the four quadrants.*

on the tape reader, Fig. 12–2. The tape reader transmits signals to the machine tool.

There are a number of numerical control systems available. One system uses coordinate dimensioning. Under this system, the engineering drawing staff produces a dimensioned drawing showing data needed for the numerical controlled operations. The object to be drawn is placed in a quadrant or a combination of the four quadrants. See Fig. 12–3.

In a typical dimensioned drawing, Fig. 12–4, the corners are located as points. Their location is noted by giving the coordinates from the X and Y axes. Holes to be drilled are located by giving the coordinates of their centers.

12-4. *To dimension drawings for numerical control, use coordinates for each corner and the center of each hole to be drilled. This drawing is in the first quadrant.*

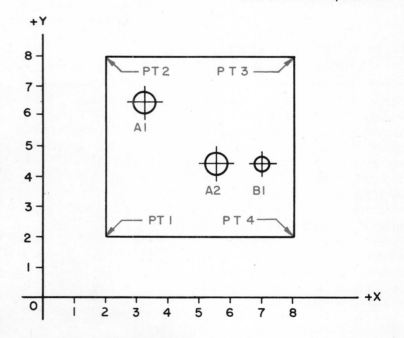

HOLE	COORDINATES	
	X	Y
A1	3.25	6.50
A2	5.50	4.50
B1	7.00	4.50

POINT	COORDINATES	
	X	Y
PT 1	2.00	2.00
PT 2	2.00	8.00
PT 3	8.00	8.00
PT 4	8.00	2.00

Study the numerically controlled machine tool, Fig. 12-1. The X axis controls movement of the table from left to right. The Y axis controls table movement forward and backward. These are both horizontal movements.

A third dimension to be considered is the Z axis. It controls the vertical movement of the spindle. The Z movement may be accomplished manually on some machines. On others it is controlled automatically by tape. One use of the Z axis is to control the depth to which a hole is drilled.

A process planner studies the drawing and decides the best sequence in which to perform the operations. This information is recorded on a process sheet. See Fig. 12-5. The process planner also establishes the setup point. This is the point from which the numerically controlled machine operator locates the center line of the spindle on the machine to the piece to be machined.

The tape is punched on a machine that looks much like a typewriter. See Fig. 12-6. The person doing this job is a tape punch operator. An example of the tape is shown in Fig. 12-7.

The punched tape is placed on the tape reader. It transmits signals to the unit that controls the table of the machine. The piece to be machined is fastened to the table. The setup point is

manually lined up with the center of the spindle. The tape then controls the movement of the table. It positions the piece beneath the spindle and starts the operation. After the first operation is performed, the tape moves the piece to the second position and initiates the process there. The number of consecutive operations that can be performed varies with the extent of the controls on the numerically controlled machine.

Numerically controlled machines are used to produce

PROCESS SHEET

M'S G	T A B	LONGITUDINAL X	T A B	TRANSVERSE Y	T A B	COMMENTS
	T	+ 00.000	T	+00.000	T	SET UP TO UPPER LEFT CORNER. (EOB)
%		% +00.213	T	− 02.127	T	DRILL 1" HOLES. (EOB)
	T	+03.614	T	−03500 (EOB)		
	T	&			T	TOOL CHANGE 3/8" DRILL (EOB)
&/	T	+04.350	T		T	DRILL 3/8" HOLES. (EOB)
	T	+05.000 (EOB)				
			T	− 04.791 (EOB)		
		&/	T		T	TOOL CHANGE 1" DRILL (EOB)

Pratt and Whitney Co., Inc.

12-5. *This is a process sheet upon which the process planner records instructions for the tape punch operator.*

Pratt and Whitney Co., Inc.

12-6. *A tape punch operator copies the process sheet by punching the corresponding keys on the tape punch machine. The numbers and instructions are converted into holes in the tape.*

12-7. *This is a 1-inch wide, 8-channel punched tape.*

Pratt and Whitney Co., Inc.

```
8
7
6
5
4
3
2
1
```

0 1 2 3 4 5 6 7 8 9 + −   TC   TAB
EOB REW
EOR   DELETE
RWST

STANDARD EIA 1 TAPE
showing channels at the right.

duplicate parts. They can do this more rapidly and accurately than a human operator. The punched tape program removed machining decisions from the judgment of individual machine operators. The accuracy of a part produced by an individual machine operator depends a lot upon his skill and judgment. When these functions are performed by a specialized group, the process planners, the operating instructions can be standardized and result in more efficient machine operations.

There are many different systems of numerical control available. It is a constantly changing field. The impact upon industrial production and drafting practice is becoming greater.

Numerical control does not eliminate the design and drafting process. The object to be made still needs to be designed through the drafting process. The route of an idea through the design process to the finished part is through engineering drawings. The numerical control process planner uses engineering drawings as the basis for developing the program to be put on the tape. The drawings are also used for the many other functions in manufacturing, such as inspection, quality control, and ordering materials needed.

The drafting process is being influenced some by numerical control. It appears that it

permits greater use of drawings combining the usual detail, assembly, and installation drawings. This combination drawing must still give complete information about the design, tolerances, materials, and other data found on single purpose drawings.

Many automatic drafting systems are designed so that they are usable for numerical control with minor changes. Automatic drafting systems are explained later in this chapter.

In some parts of the aerospace industry, undimensioned drawings are being used. Drawings for numerical control may permit some expansion of this practice. However, programming of numerical control tapes requires that specific dimensions be available. They must be obtained from a dimensioned drawing or by scaling an undimensioned drawing. Machines are being developed that will mechanize the digital pickup from the undimensioned drawings.

COMPUTER-AIDED DESIGN

The term, computer-aided design, means that the computer assists the designer in analyzing and changing a design within limits set by the designer. Most design problems have many possible solutions. In addition, many factors exist which have to be considered if the most efficient and economical design is to be found. The designer,

using the computer to help process the data, can test all possible solutions in a short time.

Available are several graphic data processing systems which use a large scale digital computer to analyze design situations and produce drawings of the proposed solution. The direct relationship between man and the computer enables the designer to get answers rapidly to his problems. It permits evaluation of his design ideas and accepts or rejects them. The computer can correctly interpret drawings and assist in making drawings showing the ideas of the human designer. Computers relieve the engineer and the draftsman of many calculations often needed. They also do much of the routine work connected with drawing layout and detail drawings.

The computer design system does not replace the man. The designer is the creative element. The computer can only respond to the design data it is given. The designer must have a working knowledge of engineering graphics. A sound understanding of descriptive geometry is important. The designer must not only understand the graphic language, but must know how to communicate with the computer.

DIGITAL COMPUTERS

A digital computer uses numerical values which represent a specific amount. They are not values which change continuously. The actual operations performed by a digital computer must be in sequence. A modern computer operates so fast it appears to operate continuously. The faster computers make more than one million computations in one second.

Digital computers have five parts: input, output, memory, arithmetic, and control. See Fig. 12–8.

The *input* is the means by which man is able to communicate with the computer. This is done with computer "language." The language is somewhat like a code or symbols. There are four methods of putting input into computer. One is a *keyboard* much like a typewriter. The keyboard translates the typed information into electronic symbols.

Punched cards are a second method of computer input. The computer reads rectangular shaped cards having holes in them. The cards go through a card reader. It translates the information punched in the card into electronic impulses.

A third input method is through *punched paper* tape. The computer uses a sensing device called a paper tape reader. The reader translates the holes in the tape into electronic pulses.

Magnetic tape input is a fourth method of communicating with a computer. It

uses a reading head much like those on tape recorders. The impulses on the magnetic tape are interpreted for the computer.

The *output* is the result of the work of a computer. A printer keyboard can be operated by electronic impulses within the computer. It will type information onto a readout sheet. It types across the page at a high rate of speed. Also in use is a *printer.* It is much like the printer keyboard except it prints an entire line at a time.

The output can be a *punched paper* tape. The punch equipment is generally part of the same equipment that punched the input paper tape.

Output is also available in the form of *punched cards* and *magnetic tape.*

Only the typewriter and printer produce output that

is immediately readable. The information on the other output forms is in symbolic language. It requires special devices to use the stored output. For example, this output could be used to drive an automatic drafting machine. It would have the data needed to produce a particular drawing.

The *memory* of a digital computer is the means provided to store information. The information in the memory is in two parts. Before a problem can be solved, all the information in both groups must be present. The first part of information is the *program*. It gives the order in which the computer is to perform the operations. The second group of information is the *data*. This is the material the computer is to process. All of this information must be

12–8.   *Major units of a computer system.*

stored in the computer in its "language."

The memory units differ in various computers. One type has a drum memory. The information is stored in numbered locations on a revolving drum. Another type of memory is stored in small magnetic cores. The polarity is reversed for changing from on to off position.

The *arithmetic* section of the computer processes the data according to the information given on the program.

The *control* part of the computer consists of automatic and operator-actuated controls. The sequence of events to take place is controlled by wiring special panels. Other parts of the computer system are controlled by switches and push buttons operated by a human.

The Digital Program. The digital program is a sequence of instructions written out in letters, numerals, and symbols. These instructions are written line after line in sequence. Each line is a single instruction to the computer.

A simple computer program is shown below. It describes machine computations for points on a straight line. The equation of the line is known. The program is for illustration only and not in any particular computer language.

Read X
V = E * X
Y = V + C
Print E, V, X, Y

The program shown reads as follows:

Insert value X
V = E multiplied by X (The * represents multiply.)
Y = V (found above) plus value C
Print values for E, C, X, Y

AUTOMATIC DRAFTING MACHINES

The output of a computer is tabular (numerical) in nature. The use of this output is often difficult for use in engineering and design situations. The tabular data is in the form of a punched tape, cards, magnetic tape, or direct output from a computer. Various devices have been developed to get a graphic picture of these results. This involves some type of plotting machine.

Automatic drafting machines produce highly accurate drawings from program language, crt (cathode ray tube) and light pen sketch, other drawings, or computer calculations. A programmed description is used by a control to move the drawing mechanism on a drafting table to produce a drawing. This description consists of a series of X and Y coordinates. They describe the paths of the lines to be drawn as well as instructions concerning the operation of the drafting table and drawing tools.

For complex drawings, the computer is used to prepare the program description. It can also digitize an existing drawing. *Digitizing* refers to the process by which graphics information is described with coordinate data. Digital drafting systems accept digital data inputs directly from computer or off-line storage media, as tapes or cards. The system then produces a permanent picture that can be viewed without magnification.

Complete automatic drafting systems are made of a digitizer, a drafting table, a control, and accessories. The drafting table changes command codes from a control unit into linework, thus producing a finished drawing. The pen on the drafting table is directed to move from an existing position to a new one along a path dictated by the control. It also draws or prints symbols and alphanumeric characters at precise locations.

Alphanumeric characters are either letters of the alphabet, numerals, or special symbols. They are digitized and stored in the control system under a code name. Each time the symbol is required, the control refers to the stored information and draws the symbol.

General purpose automatic drafting tables consist of a flat bed mounted on a base, Fig. 12–9. The bed is electrically tiltable to any angle from horizontal to vertical. The bed is usually a precision aluminum plate. High precision tables are made of steel plate. The drawing surface is

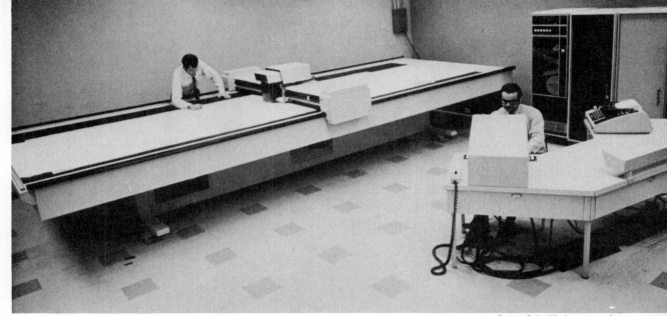

Gerber Scientific Instrument Co.

12-9. *A large area, high speed drafting table for general drafting, lofting, and digitizing. The top is electrically tilted. The drawing area is a rubber platen 5 feet wide, available in lengths from 8 to 24 feet. Standard accuracy is ±.004. Repeatability is ±.002. Speed is 400 inches per minute.*

covered by a precision ground neoprene rubber platen. It has small holes which are connected to a vacuum system. The vacuum holds the drawing material in place. Some plotters use a cylindrical plotting surface called a drum. See Fig. 12-10.

The carriages in digital automatic drafting systems are positioned by step motor drive mechanisms. They drive each carriage to a new position, using the digital drive impulses supplied by solid-state power amplifiers directly connected to the output pulse trains of the control. The standard drawing heads on the carriages hold ball-point pens, wet ink pens, or scribing tools.

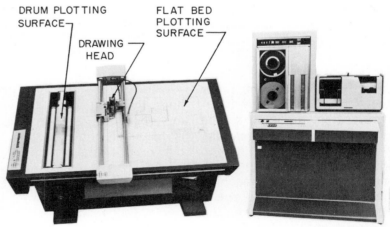

Auto-Trol Corp.

12-10. *A digital plotter having both flat bed and drum plotting surfaces. The flat bed size is available in 40" × 60" and 60" × 60". The drum handles paper 11, 24, 30, and 36 inches wide. It has input from punched cards, paper tape, and magnet tape. Resolution is 0.0005 inches on both the X and Y axes. An eight-pen tool holder is available. It draws at 10 inches per second.*

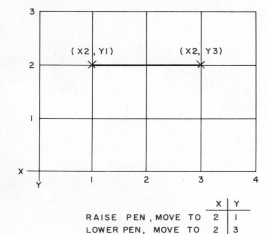

12-11. *How to draw a horizontal line.*

	X	Y
RAISE PEN, MOVE TO	2	1
LOWER PEN, MOVE TO	2	3
RAISE PEN		

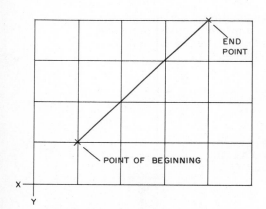

12-12. *An inclined line is drawn from a point of beginning to the end point located on the X-Y coordinates. The computer calculates the XY coordinates in very small increments. This is done so rapidly the pen moves in a straight line.*

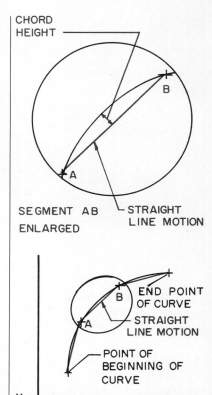

NOTE : CURVE GREATLY ENLARGED TO SHOW COORDINATES

12-13. *This curve is plotted using linear interpolation. The coordinates of the points on the curve are calculated using the chord height.*

The *control system* reads the coded coordinate information from the input source (a computer, tape, or cards). It converts the information into electrical signals which move the drafting machine head to the locations specified by the coordinates. The control computes the X and Y distances to be moved and sends pulses simultaneously to the X and Y step motors at such a rate that a straight line is drawn. See Fig. 12-11.

When drawing inclined lines, the head actually moves in steps. The size of each step is so small that the line is drawn as a straight line, Fig. 12-12. This is called *resolution*. Resolution varies with machines. It generally ranges from 0.002 to 0.0001. The X and Y step motors move the pen simultaneously in the X and Y directions. The resolution plus rapid calculations enable the pen to draw a straight line.

Many lines on a drawing are curved. They are drawn using interpolation.

Interpolation is the creation of intermediate points between the beginning and end points of the curve. These points are created mathematically inside the computer. The computer directs the pen on the automatic drafting

332

machine to draw the curve between these points.

There are three general types of interpolation methods in use today. These are linear, circular, and parabolic.

Linear interpolation means straight-line motion between closely spaced points on a curve or circle. See Fig. 12–13. This is the most generally used method. The chord height distance is used to create the points. The *chord height* is the distance between the arc and its chord. See Fig. 12–13. In this figure the distance between the points is greatly enlarged. A common chord height used is 0.001. In this case the distances between points on the curve would be

Circle Diameter	Curve Distance
1″	.0632″
2″	.0894″
4″	.1264″
6″	.1548″
8″	.1788″

These curve distances are frequently carried out to 20 or more decimal places. The computer calculates coordinates for the hundreds of points used to generate the curve at such a high rate of speed that they appear to be drawn continuously.

Circular interpolation requires as its input the beginning and ending points of a circle or portion of a circle along with coordinate points for the location of the center and the radius of the circle. By substituting the above in-formation into the general equation for a circle, it is possible to compute additional intermediate points on the circle.

Parabolic interpolation requires as its input the beginning and ending points of the curve to be drawn along with coordinate values of several intermediate points on the curve. The above information can be substituted in the general equation for a parabola and additional intermediate points calculated.

The three general methods of interpolation are employed by both automatic drafting machine controls and numerically controlled machines. In some cases, the calculation is performed by a *general purpose computer* and the information incorporated into the control tape that operates the numerically controlled machine or automatic drafting machine. In other cases, the calculations for intermediate points for each of the three interpolation methods are accomplished by the *control system* of the automatic drafting machine or the numerically controlled machine. In the latter case, the control system includes a small computer which provides the computing capability.

12–14. *A precision drawing of an integrated circuit.*

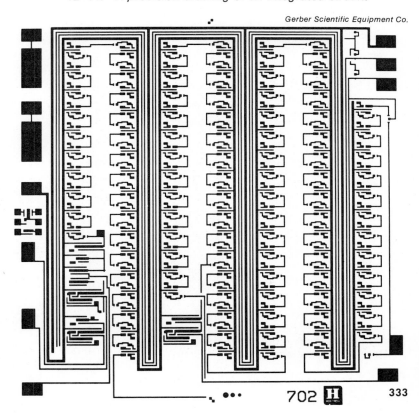

ELECTRICAL
DRAWING

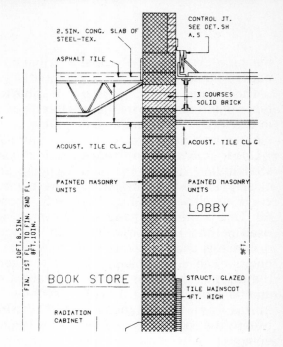

ARCHITECTURAL
SECTION

DETAIL
DRAWING

CHART

12-15. *Some typical drawings made on the Gerber 723 automatic drafting system.*

Gerber Scientific Instrument Co.

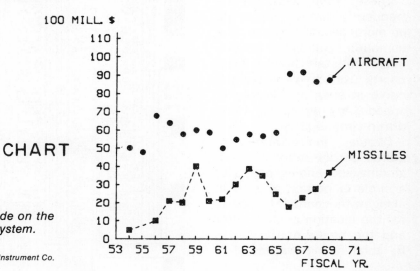

WEAPONS BUDGET

The *standard drafting head* is a tool holder. It holds a ball-point pen, wet ink pen, or scriber. It has a micrometer adjustment of height for various drafting material thicknesses. A ball-point pen drawing on paper is used for drawings where rapid display of information is needed. A wet pen produces quality drawings on frosted mylar or high quality drawing paper. A scribe is used with standard scribe coat. It is valuable when very fine lines and high accuracy are needed.

Some heads have as many as six or eight-position tool holders. This permits a choice of tools giving a variety of line widths or colors.

The most precise line work is obtained with an optical exposure head. It produces lines by exposing photographic emulsion on a base of material, such as mylar or glass, to a ray of light.

Automatic drafting systems are useful when precision needed is beyond that which the draftsman can produce. Automatic systems are also used to produce large volumes of noncreative drafting work or drawings with many repeated features. These systems can draw anything that can be described in terms of coordinates lying in a plane. Figs. 12–14 and 12–15 show a few drawings made on automatic drafting machines.

Another system used with the computer for design purposes is the cathode ray tube

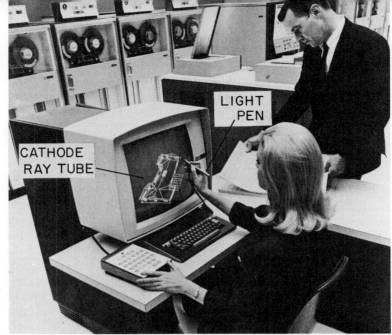

International Business Machines

12–16. *This display unit provides high speed visual communications between the user and the computer. The light pen and the keyboard are the means used to provide entry and to change the data in the computer.*

12–17. *This computerized drafting machine produces isometric, dimetric, and trimetric drawings from orthographic drawings. When a perspective module is added, it can draw one, two, and three-point perspective drawings.*

Perspective Systems, Inc.

(crt) and a light pen. See Fig. 12–16. The designer sketches on the face of the cathode ray tube with the light pen. This is a source of data input to the computer's memory units. He also uses a keyboard to introduce design data. The designer can store and recall drawings, change the shape, and erase and redraw corrections much the same as he can on paper. This process is extremely rapid and is of great assistance as the designer thinks through a solution. The final solution displayed on the cathode ray tube is then made into a permanent record photographically.

MACHINE-MADE PICTORIAL DRAWINGS

Pictorial drawings are used in industry to help overcome the problems many workers have in reading two-dimensional working drawings. The conversion of a two-dimensional drawing to a three-dimensional drawing is slow and costly. Fig. 12–17 shows a system which greatly reduces the time needed to make pictorial drawings from two-dimensional drawings.

The operator traces over two mutually perpendicular, flat orthographic views of the object. The machine produces a three-dimensional drawing. This unit has control modules that permit drawing isometric, dimetric, trimetric and perspective drawings. The views can be rotated from 0 degrees to 360 degrees about any one of the three axes. A steroview control module can be added which permits drawing both right-eye and left-eye perspective drawings. When viewed properly, three-dimensional illustrations are observed.

A device for digitizing, storing, and recreating two and three axes graphic data can be joined to this machine. It consists of a keyboard, data coupler, tape unit, and a digital interpolation computer. With this computer addition, the machine will digitize and record a three-dimensional drawing on magnetic tape. The tape can be used to produce as many illustrations as wanted on the plotter. The basic plotter is converted to a digital plotter. The plotter will search through a reel of magnetic tape to find a single drawing. This enables a company to create a storage bank of digital data. With this capability, the operator does not have to trace the entire orthographic drawing. He only needs to record two points to produce a straight line. He must record three points to generate a curve.

Build Your Vocabulary

Following are terms that you should understand and use as a part of your working vocabulary. Write a brief explanation of what each means.

Numerical control
X axis
Y axis
Z axis
Punched tape
Computer-aided design
Digital computer
Computer input
Punched cards
Magnetic tape
Computer output
Computer memory
Computer arithmetic section
Computer controls
Digital program
Automatic drafting machine
Digitizing
Resolution
Chord height distance

Section Four
Special Applications

Carlos B. Walker
Kansas City, Missouri

*A student solving a surface development problem. He is making a
paper model to check the solution.*

Links in Learning

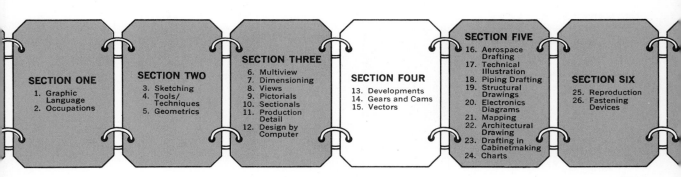

Chapter 13

Developments and Intersections

SURFACES

A surface is the exterior of an object. The outside of a sphere is a surface. The faces of a cube are surfaces. There are several types of surfaces — plane, single-curved, double-curved, and warped. See Fig. 13–1.

In a *plane surface,* if any two points are connected with a straight line, the line lies in the surface. The top of a table is a plane surface. A plane surface may have three or more sides.

A *single-curved surface* is one that can be unrolled to form a plane. A cylinder is an example.

A *double-curved surface* is formed by a curved line. It has no straight line elements. The surface of a sphere is an example. These surfaces cannot be developed exactly. Approximate developments can be made.

A *warped surface* is one that is neither plane nor curved. It cannot be exactly developed. Approximate developments can be made. The hoods on most automobiles are warped surfaces.

Surface Development. *Surface development* is the preparation of patterns or templates which, when properly formed, generate the surface desired. Examples of surface development of four basic geometric shapes are shown in Fig. 13–2.

Surface developments are used in many different industries to form many materials. Patterns are used to mark sheet metal for forming into pipes. Cardboard boxes are formed from develop-

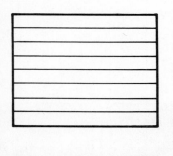

PLANE SURFACE

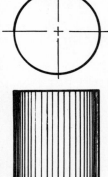

SINGLE–CURVED SURFACE

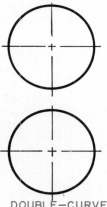

DOUBLE–CURVED SURFACE

WARPED SURFACE

13–1. *Types of surfaces.*

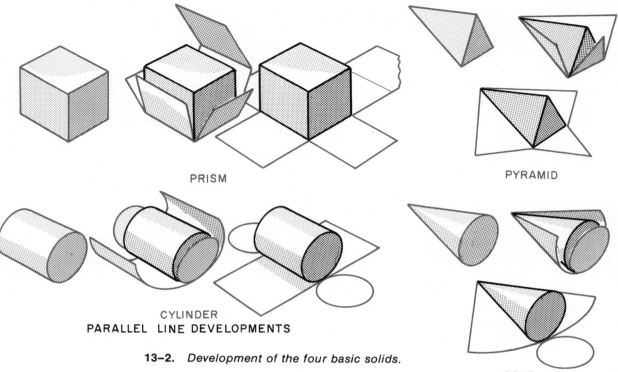

PRISM

PYRAMID

CYLINDER

PARALLEL LINE DEVELOPMENTS

13–2.   Development of the four basic solids.

CONE

RADIAL LINE DEVELOPMENTS

13–3A.   The box formed from a pattern before it is folded to finished shape.

13–3B.   The finished box.

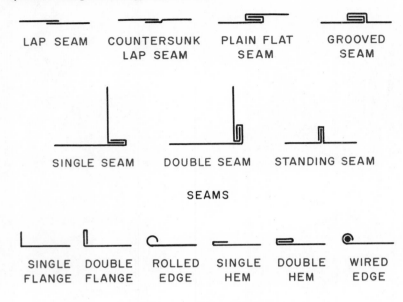

SEAMS

FLANGES AND HEMS

13-4. *Sheet metal joints.*

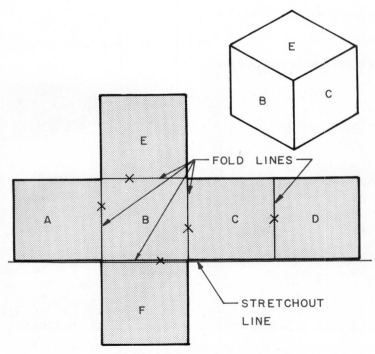

13-5. *Development of a cube.*

ments. Clothing is made from patterns.

The drawing of patterns therefore includes all surfaces of the object to be formed. It also includes parts used to hold the formed object together. Fig. 13–3 shows a box before it is formed. Notice the tabs used to glue the box together. Fig. 13–4 shows the common joints used to join sheet metal surfaces. The method of fastening will vary with the material used in the finished product.

Fig. 13–5 shows a pattern for a cube. Patterns are built around a long line, called *stretchout line.*

Folds are shown with thin, solid lines. They are sometimes marked with an "X" to help identify them. The faces of the pattern upon which the fold lines are drawn are the inside faces. When the pattern is folded, these are inside the object.

Before drawing a pattern, decide where the seams will be located. This will depend on the object and the desired appearance. Fig. 13–6 shows a pattern for a mailbox. The vertical seam was placed at the back edge so it would be somewhat hidden and improve the appearance of the product. Notice the metal tabs for joining the corners. The corners of tabs are usually cut on a 30-, 45-, or 60-degree angle. The angle used is the one that makes it easiest to clear the corner for a fold. The tabs on the top edges are used to cover the sharp metal edges.

Most objects fall into one of three types of development. These are *parallel line developments,* *radial line developments,* and *triangulation.* *Parallel line developments* are used for objects having parallel elements. Prisms and cylinders use parallel line developments. *Radial line developments* are those which have the fold lines coming from one point. Pyramids and cones are examples. *Triangulation* is a means of developing a surface by breaking it into a series of triangles. Objects made of a combination of curved and plane surfaces are developed by triangulation.

DEVELOPMENT OF RIGHT PRISMS

A *right prism* has its base perpendicular to its axis. All prisms develop into rectangular shaped patterns. They use the parallel line method. The development of right prisms is shown in Figs. 13–6 and 13–7.

To develop the mailbox shown in Fig. 13–6:

1. Draw the stretchout line. Its length is the sum of the width of the four sides. This is 14 inches. Mark each distance on the stretchout line. These are points 1, 2, 3 and 4.

2. At each of these points, draw vertical fold lines. Measure the length of each corner on these lines. These are points 5, 6, 7 and 8.

3. Connect these points to complete the pattern of the faces.

4. Draw needed tabs.

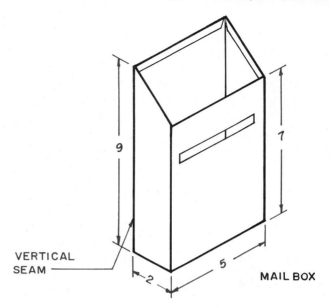

MAIL BOX

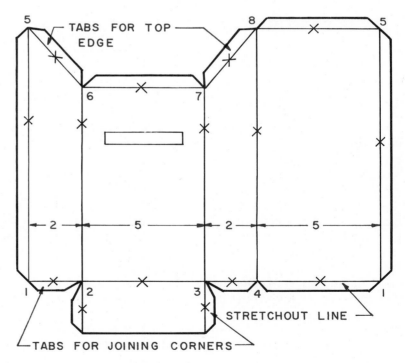

THE   PATTERN

13–6.   *A pattern drawing. This mailbox is a truncated right prism.*

341

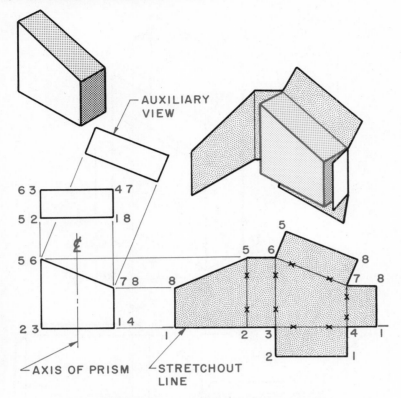

AUXILIARY
VIEW

6 3   4 7

5 2   I 8

5 6

7 8   8

2 3   I 4

₵

AXIS OF PRISM   STRETCHOUT
LINE

5 6   8

5 6   7 8

I   2 3   4 I

2   I

13–7. *Development of a truncated right prism.*

In Fig. 13–7 is shown the development of a truncated prism. A *truncated prism* has one of its surfaces cut on an angle. The procedure for drawing the development is the same as explained for the mailbox. The only difference is the inclined surface. This surface is not true size in the top view. On the development, all lines must be true length. One way to find the true size is to make an auxiliary view. It should be noted that the true length of the inclined surface appears on the front view. The true depth is on the top view.

Measurements can be made in these views.

DEVELOPMENT OF AN OBLIQUE PRISM

An *oblique prism* has its axis at angle to the base. This angle is other than 90 degrees. See Fig. 13–8.

Oblique prism development uses the same method as for right prisms. One difference is that the oblique prism development does not unfold in a straight line, so the stretchout line is placed through the center of the prism.

To develop an oblique prism:

1. Draw the top and front views.

2. Draw an auxiliary right section. This is perpendicular to the sides of the prism. The auxiliary section gives the true length of the sides of the prism.

3. Project the stretchout line from the location of the right section. This was projected from the front view, Fig. 13–4.

4. Measure the true length widths of the faces along the stretchout line. These are shown by points 1, 2, 3 and 4.

5. Draw lines through these points perpendicular to the stretchout line, forming the corners of the prism.

6. Project the end points of the corners to the development. This locates the top and bottom ends of these corners.

7. Connect the corners to finish the development.

CYLINDERS

Cylinders may be thought of as prisms. See Fig. 13–9. The number of sides of the imaginary prism are infinite. This means there is no limit to the number of sides. The more sides the prism has, the more it resembles a cylinder. Each of the sides of the prism forms an edge. These edges are called elements. An *element* is an imaginary straight line on the surface of a cylinder. All elements on a cylinder are parallel to the axis. Any cylinder whose bases are perpendicular to its axis will develop into a rectangular pattern. All cylinders are developed by the parallel line method.

The length of the stretch-out line of a cylinder is equal to its circumference. The *circumference* is the distance around the cylinder. The stretchout line is always perpendicular to the axis of the cylinder. If the cylinder is oblique, a right-section view must be made to obtain the circumference. The circumference can be found by multiplying the diameter by 3.14.

Development of a Right Cylinder. To develop a right cylinder, two opposite views are required. See Fig. 13–10. One of these views must show the diameter. The other must show the height.

To make an approximate development:

1. Divide the circular view into a number of equal parts. The more divisions, the more accurate the pattern.

2. Number each of the divisions on the circular view.

3. Draw the stretchout line. On the left end, draw a perpendicular line. Make it the same length as the height of the cylinder.

4. Measure the straight line distance between two divisions on the circular view.

5. Step off this distance on the stretchout line. Be certain to step off as many divi-

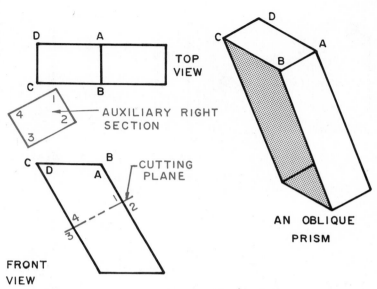

TOP VIEW

AUXILIARY RIGHT SECTION

CUTTING PLANE

FRONT VIEW

AN OBLIQUE PRISM

I. DRAW AN AUXILIARY RIGHT SECTION.

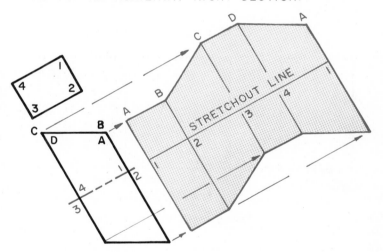

STRETCHOUT LINE

2. DRAW THE STRETCHOUT LINE.
3. LOCATE THE TRUE LENGTH WIDTHS.
4. LOCATE THE ENDS OF THE CORNERS.

13–8. *Development of a pattern for an oblique prism.*

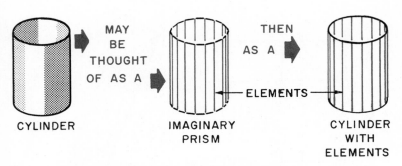

13–9. *Cylinders may be thought of as imaginary prisms.*

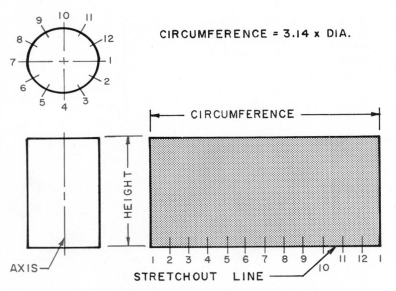

CIRCUMFERENCE = 3.14 x DIA.

13–10. *Development of a right cylinder.*

ference, multiply the diameter by pi. Pi is equal to 3.1416. For example, a cylinder with a diameter of 2 inches has a circumference of 6.28 inches. The length of the stretchout line would be 6.28 inches.

Development of a Truncated Right Cylinder. A truncated right cylinder is one that is cut at an angle to its axis. This angle is one that is other than 90 degrees. The end that is cut will result in a curved line in the development. To make the development, two adjacent views are necessary, commonly the top and front views. See Fig. 13–11.

To make this development:

1. Divide the circular view into a number of equal parts.

2. Project these points into the front view. The seam has been placed on the shortest element, Fig. 13–11.

3. Draw the stretchout line perpendicular to the axis of the cylinder.

4. Measure the straight-line distance between two divisions on the circular view.

5. Step off this distance on the stretchout line. Step off as many divisions as are on the circular view. This gives the circumference.

6. Draw perpendicular lines through each point on the stretchout line.

7. Project the points on the inclined edge in the front view to corresponding lines

sions as are on the circular view. This gives the circumference.

6. On the last division, draw a perpendicular to the stretchout line. Draw the top edge of the cylinder parallel with the stretchout line.

This method for obtaining circumference is approximate. The distances used are the chordal distances. A chord is shorter than its arc. A more accurate method of obtaining circumference is by mathematics. To obtain the circumin the stretchout. Connect these points with an irregular curve.

DEVELOPMENT OF ELBOWS

Elbows are used to change the direction of a pipe. They are made of two, three, four or more pieces. The number of pieces depends upon the pipe diameter. Elbows are used on round and square pipe. A round two-piece elbow is shown in Fig. 13–11. Three-piece elbows are shown in Fig. 13–12. A round four-piece elbow is shown in Fig. 13–13. To develop an elbow, draw the front view. This shows the number of pieces, the angle of bend, and the true length of the elements of each piece.

To develop the four-piece round elbow shown in Fig. 13–13:

1. Draw the heel and throat radii. See Fig. 13–13A.

2. Divide the angle of bend into the proper number of pieces. See Fig. 13–13B. The middle pieces of the elbow are twice as large as the end pieces. The angles are figured by using the following formula: Number of spaces = number of pieces × 2 − 2. Since Fig. 13–13 shows a four-piece elbow, the angles are 4 × 2 − 2 = 6. The angle of bend, 90 degrees, is divided by 6. Each angle is 15 degrees.

3. Draw the 15-degree angles on the front view. See Fig. 13–13B.

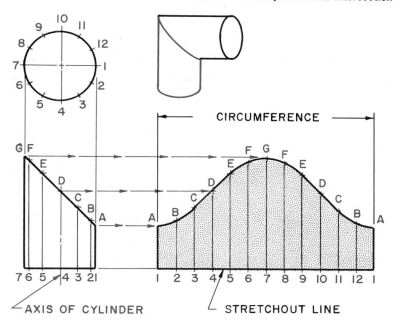

13–11.   *Development of a truncated right cylinder.*

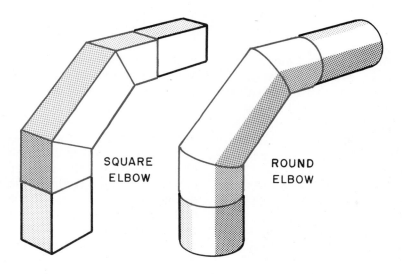

13–12.   *Rectangular and round three-piece elbows.*

345

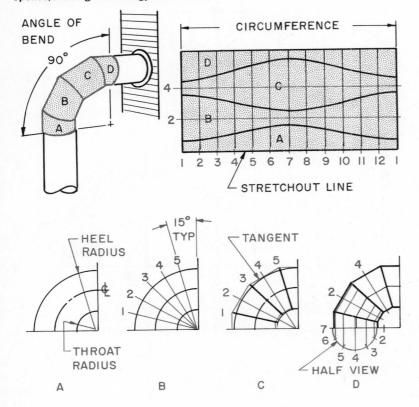

13-13. *Pattern for a round four-piece elbow.*

pieces are laid out together, each curve developed serves two pieces of the elbow. This also staggers the seams. The pattern A and C, Fig. 13-13, will have seams at the throat. Parts D and B will have seams at the heel.

RADIAL LINE DEVELOPMENT

Some objects are developed by the radial line method. Radial lines are used to develop forms such as pyramids and cones. Their edges come together in a common point.

Development of a Right Rectangular Pyramid. A right rectangular pyramid is shown in Fig. 13-14. The axis is perpendicular to the base. To draw a development of this figure:

1. Draw the top and front views.
2. Find the true length of one edge by the rotation method. The steps are shown in Fig. 13-14.
3. Mark point A on the paper. Set a compass with a radius equal to the true length edge A-1. Swing an arc.
4. Draw edge A-1 from A to one end of the arc. Step off the chords of the four base edges of the pyramid. These are 1-2, 2-3, 3-4, and 4-1. Connect these points with point A. This gives the fold lines. Connect the base points 1, 2, 3, and 4. This completes

4. Draw the miter lines. These are the places the pieces of pipe are to be joined. These are lines 1, 3, and 5, Fig. 13-13C. Since the center pieces are twice as large as the ends, they are each 30 degrees.
5. Draw lines tangent to the arcs of each piece. See Fig. 13-13C.
6. Draw a half view of the end of the pipe. See Fig. 13-13D. Divide it into an equal number of parts.
7. Project these divisions to the front view. They are drawn parallel with the surface of each part of the elbow.
8. Lay out the stretchout line. The length is equal to the circumference of the pipe.
9. Lay out the distances between the points on the half circular view on the stretchout line. Proceed to finish the development, using the same technique as explained for a truncated cylinder. Each piece of pipe is a truncated cylinder.

Lay out the pattern as shown in Fig. 13-13. When all four

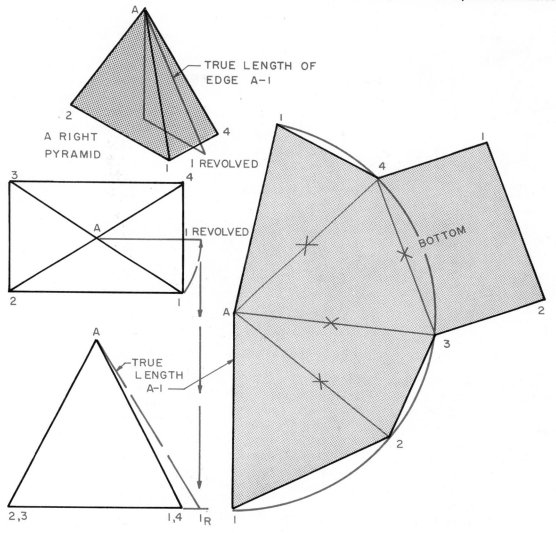

13-14. *Development of a right rectangular pyramid.*

the development of the sides.

5. The bottom can be drawn off any of the based edges. Draw lines perpendicular to the base edge. In Fig. 13-14, edge 3-4 was used. Measure the true length of width. This is 1-4. It is found on the top view.

Development of a Truncated Right Pyramid. A truncated pyramid is one that is cut on an angle to its axis. This angle is one other than 90 degrees.

Fig. 13-15 shows the steps for making the development. It uses the same procedure as explained for a right pyramid

shown in Fig. 13-14. The only difference is that the true length of several edges must be found. These are ÓA and OB, Fig. 13-15.

It is best to draw the object as a normal right pyramid. Then locate the true length of each side that is shorter.

347

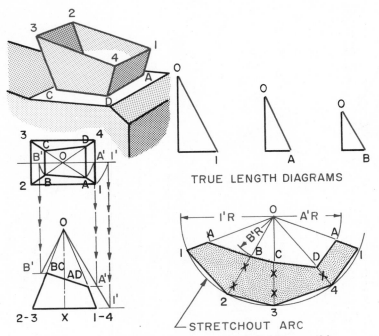

TRUE LENGTH DIAGRAMS

STRETCHOUT ARC

13–15. *Development of a truncated right pyramid.*

Developing Cones. Developing a cone is similar to developing a pyramid. A cone may be thought of as a many-sided pyramid. See Fig. 13–16. Each edge of the imaginary pyramid may be thought of as an element. The development is made by placing each of the elements next to each other.

Developing a Right Cone. Fig. 13–17 shows how to develop a right cone and a frustum of a right cone. The developments are identical except the frustum of a right cone has the top cut on an angle of 90 degrees to the axis.

To develop a right cone:

1. Draw the top and front views of the cone.

2. Divide the circular view into a number of equal parts.

3. Project these points to the base of the cone in the front view.

4. Connect the points from the base to the top of the cone. This is point 0. These form elements of the cone. The only true length elements are 0–1 and 0–7. They are true length because they are parallel to the frontal plane.

5. Using 0–1 as the radius, draw an arc. This is the stretchout line.

6. Measure the straight line distance between two of the points located on the top view. For example, use the distance from 1 to 2. Step off this distance along the arc until it has the same number of points as are on the top view. This is the circumference of the base of the cone.

7. Connect the ends of the arc with point 0. This completes the development of the right cone.

A frustum of a cone follows these same steps. In addition, measure the true length distance from the top of the cone, point 0, to the flat top surface. This is distance A, Fig. 13–17. Draw an arc on the development with this radius. This locates the top edge of the cone.

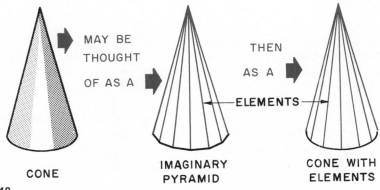

CONE

IMAGINARY PYRAMID

CONE WITH ELEMENTS

13–16. *A cone may be thought of as an imaginary pyramid.*

Development of a Truncated Right Cone.

A truncated right cone is one which is cut on an angle to its axis. This angle is one other than 90 degrees.

The development is similar to that used for developing a cone. See Fig. 13–18. First develop the object as a cone. Then find the true length of each of the elements from the top, point 0, to the truncated surface. Use the rotation method. Then locate this true length on the elements on the development. Connect these points with an irregular curve. This forms the top edge of the truncated right cone.

TRIANGULATION

Triangulation is used to develop surfaces which cannot be developed by the parallel or radial line methods. Surfaces developed by triangulation are approximate. It is done by dividing the surface into a series of triangles. The true length of the sides of each triangle must be found. Then the triangles are drawn next to each other on a flat surface. An example of this method is the development of an oblique cone.

Development of an Oblique Cone.

Oblique cones can have a circular or elliptical base. See Fig. 13–19. Either type can be developed approximately by triangulation.

To develop an oblique cone, see Fig. 13–20.

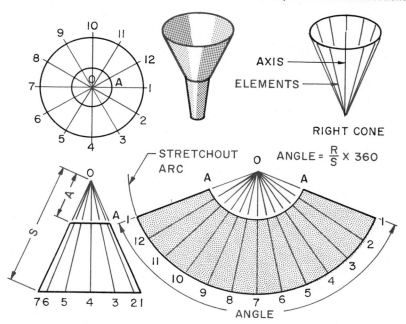

$$\text{ANGLE} = \frac{R}{S} \times 360$$

13–17. *Development of a right cone and a frustum of a right cone.*

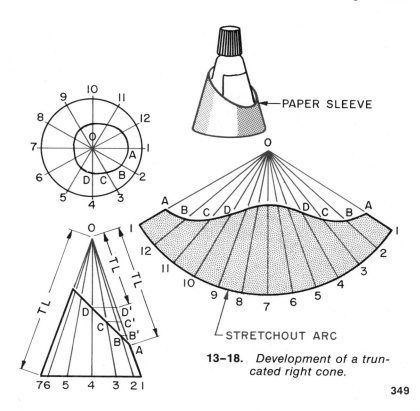

13–18. *Development of a truncated right cone.*

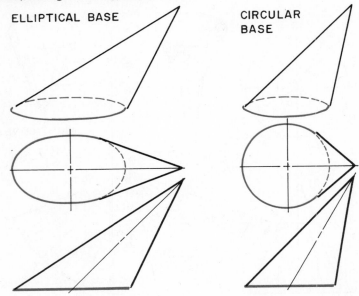

ELLIPTICAL BASE   CIRCULAR BASE

13–19. *Oblique cones can have elliptical or circular bases.*

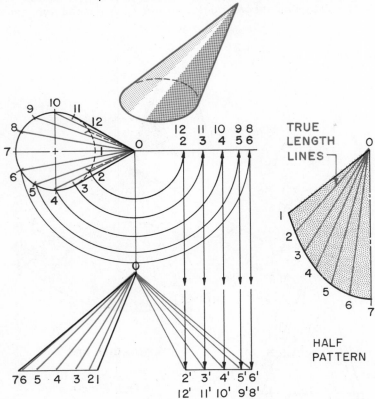

TRUE LENGTH LINES

HALF PATTERN

TRUE LENGTH DIAGRAM

1. Draw the top and front views.

2. Divide the circular view into a number of equal parts.

3. Project these to the base in the front view.

4. Draw the elements from each division on the base to the top of the cone. This divides the surface into a series of triangles.

5. Find the true length of each element. Only 0–7 and 0–1 are true length, Fig. 13–20. Find the true length of the others by rotation.

6. On the development, lay out each triangle in the order in which they are found on the views of the cone. The first triangle was 0–1–2, Fig. 13–20. First lay out true length side 0–1. Then measure the straight line distance between 1–2 on the top view. Using 1 on the development as a center, swing an arc with this radius. Next swing an arc with a radius equal to the true length of element 0–2 using 0 as the center. Where the arcs cross is one point on the pattern. This is point 2. Repeat these steps for each element. Connect the points with an irregular curve.

TRANSITION PIECES

Transition pieces are used to connect pipes of different sizes or shapes. Fig. 13–21

13–20. *Development of an oblique cone.*

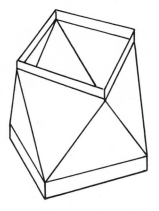

A – SQUARE TO SQUARE

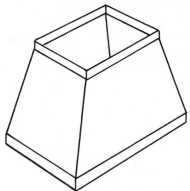

B – SQUARE TO SQUARE

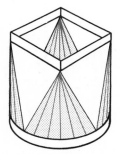

D – SQUARE TO ROUND

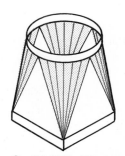

C – ROUND TO SQUARE

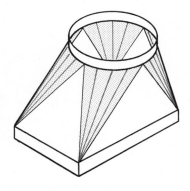

E – OFFSET ROUND TO SQUARE

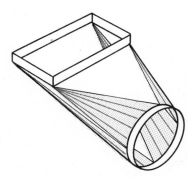

F – ELBOW SQUARE TO ROUND

13–21. *Some typical transition pieces.*

shows some examples. Notice that they are made of plane and curved surfaces. Sometimes both are found in one transition piece.

Development of a Round to Square Transition Piece. This development is similar to developing an oblique cone. However, it is made of plane triangular surfaces and curved corners. The development is shown in Fig. 13–22.

1. Draw the top and front views of the piece.

2. Divide the circular view into a number of equal divisions.

3. Project these to the front view.

4. Draw the elements for the curved corners on the front view.

5. Rotate these elements to find their true length.

6. To begin the pattern, lay out one true length line for one of the plane surfaces. This is edge 1–A, Fig. 13–22. This would be the seam. It is at the back of the hood, where it is less likely to be seen.

7. Measure the distance from A to B on the top view. Draw it perpendicular to line 1–A. Connect B with 1 on the development. Line 1–B is the true length.

8. Lay out the triangles forming the corner. These are marked B–1′–2′, B–2′–3′ and B–3′–4′.

9. Lay out the next large triangle. This is B–4′–C. The true length of BC is in the top

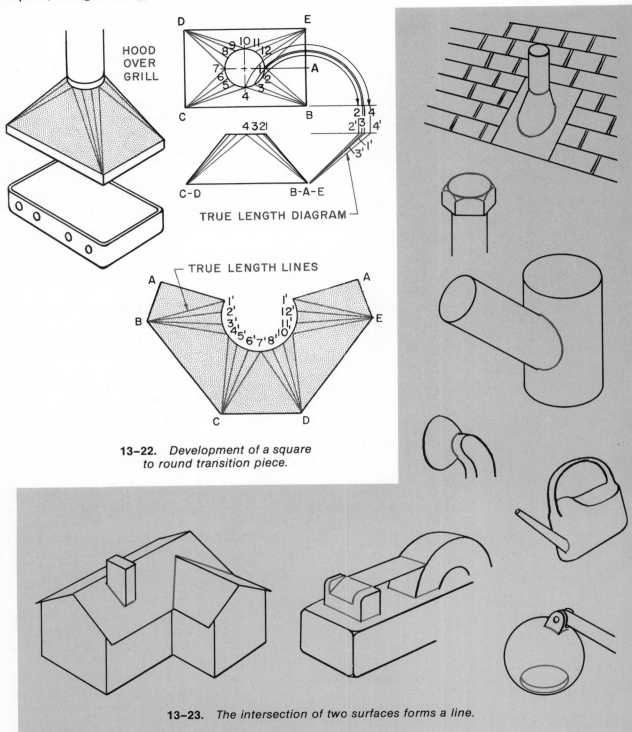

HOOD
OVER
GRILL

TRUE LENGTH DIAGRAM

TRUE LENGTH LINES

13-22. *Development of a square to round transition piece.*

13-23. *The intersection of two surfaces forms a line.*

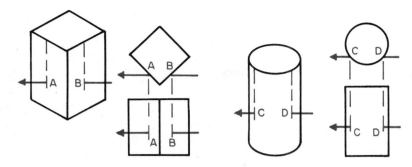

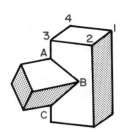

13–24. *The intersection of a line and a plane or curved surface is a point.*

view. The true length of C–4′ is found by revolution.

10. Continue these steps until all surfaces of the pattern are drawn.

11. Draw a curve through the points forming the circular opening.

INTERSECTIONS OF SOLIDS

Intersections are formed when two surfaces meet. The surfaces can be plane, curved, or spherical. The intersection of two plane surfaces is a straight line. The intersection of a plane and a curved or spherical surface is a curved line. The intersection of curved or spherical surfaces is a curved line. See Fig. 13–23.

When a line intersects a plane or curved surface, a

point is formed. This is called a *piercing point*. See Fig. 13–24.

Intersection of Two Prisms at Right Angles to Each Other. The intersection of two prisms can be found by locating the piercing points of the edges of one solid on the surface of the other solid. The piercing point is the point at which the edge touches the surface.

To find the intersection, see Fig. 13–25.

1. Draw the orthographic views of the two prisms.

2. Locate the piercing points. These are marked A, B, C, and D. In this problem they are found in the top view.

3. Project the piercing points to the view on which the intersection is seen. Where the

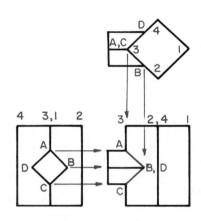

13–25. *Intersection of two prisms at right angles to each other.*

projection crosses the piercing line is the piercing point.

4. Connect the points with straight lines.

These same piercing points are used to make a development of the two prisms. Development for the intersecting prisms, Fig. 13–25, is shown in Fig. 13–26.

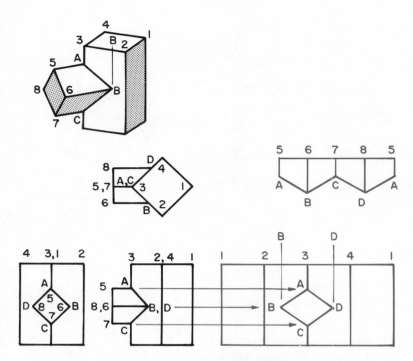

13–26. *The development of two intersecting prisms.*

Intersection of Two Prisms Oblique to Each Other. The steps to find this intersection are the same as those for prisms at right angles. See Fig. 13–27.

1. Draw the orthographic views.

2. Locate the piercing points of the edges of the intersecting planes. These are points A, B, C, and D.

3. Project these to the view on which the intersection is visible. Locate them on the proper edges in this view.

4. Connect the points with straight lines.

USING CUTTING PLANES

Cutting planes can be used to help locate points on a line of intersection. These are imaginary planes like those used in making section drawings. They can be used to help find intersections on prisms and cylindrical objects. They are especially useful on cylindrical objects.

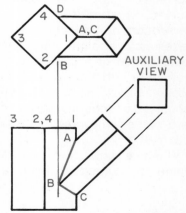

13–27. *Intersection of two prisms oblique to each other.*

To develop the large prism:

1. Draw the stretchout line. Locate the true length sizes. The true length is found in the top view in this example. These are edges 1–2, 2–3, 3–4, 3–1.

2. Measure the true height of the large prism. This is found on the front view.

3. Locate the piercing points. These are marked A, B, C, and D. The true length between the points is found in the top view. Points A and C are located on edge 3. Points B and D are found by measuring their true distance from edge 3 on the top view.

4. Connect the piercing points with straight lines.

The same steps are used to develop the small prism.

1. Draw the stretchout line. True lengths are found on the side view in this problem. These are points 5, 6, 7, and 8.

2. Measure the true length of each edge. This is from the end of the prism to the piercing point. These are on the top and front views. Lay out these distances on the development.

3. Connect the piercing points.

Intersection of Two Cylinders at Right Angles. The line of intersection of two cylinders is found by using a series of cutting planes. See Fig. 13–28. To use cutting planes:

1. Draw the necessary orthographic views. The circular and side views of both cylinders are needed.

2. Pass several cutting planes through both cylinders. In Fig. 13–28 this is shown on the top view. They are marked A, B, C, and D.

3. Locate these cutting planes on the circular view of both cylinders. To do this on the front view, a half view of the cylinder is drawn. Since the cylinder is symmetrical, this is all that is needed.

4. Project the points of intersection between the cylinders and the cutting planes to the view on which the intersection will be drawn. In Fig. 13–28 this is shown on the front view. Where these cross is a point on the line of inter-

section. Connect these points in a smooth curve.

Development of Intersecting Cylinders. The points of intersection used to locate the line of intersection are also used when drawing a development. This is shown in Fig. 13–28.

1. Draw the development of the large cylinder, described in Fig. 13–10.

2. Draw the center line. Lay out the lines representing the cutting planes. These are numbered 1 through 7.

3. From the front view, project the points of intersection located by cutting planes to the development. This gives a series of points on the development. Connect these with a smooth curve. This is the shape of the opening to which the small cylinder fits.

The development of the small cylinder is shown in Fig. 13–11.

Intersection of Two Cylinders Oblique to Each Other. The location of the line of intersection of two cylinders oblique to each other is shown in Fig. 13–29. Notice that an auxiliary view of the small cylinder is drawn. The cutting-plane lines from the top view are located on the auxiliary. The projection of points of intersection is the same as explained for Fig. 13–28.

The development of these cylinders is the same as explained in Fig. 13–28.

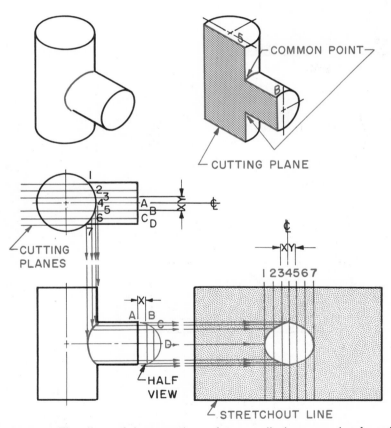

13–28. *The line of intersection of two cylinders can be found using cutting planes. This figure also shows how to find the development of the intersection.*

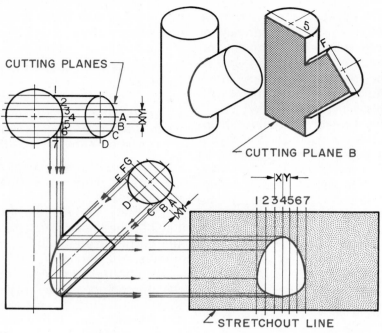

CUTTING PLANES

13–29. *The intersection of two cylinders oblique to each other can be found using cutting planes. This figure also shows how to find the development of the intersection.*

CUTTING PLANE B

STRETCHOUT LINE

Intersection of a Prism and a Cylinder. The cutting plane method is used to find the intersection of a prism and cylinder. See Fig. 13–30. Cutting planes A, B, C, and D are used to locate the points of intersection. The steps are the same as explained in Fig. 13–28. Notice that the line of intersection is two curves.

Cutting Planes on Cones. There are three possible methods of cutting cones. These are shown in Fig. 13–31. The cutting plane can be parallel with the axis of the cone. The surface produced is a parabola.

13–30. *The intersection of a prism and a cylinder can be found using cutting planes.*

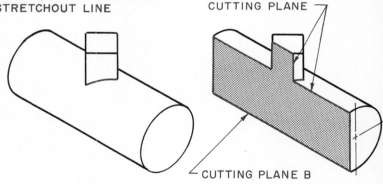

ELEMENTS FORMED BY
CUTTING PLANE

CUTTING PLANE B

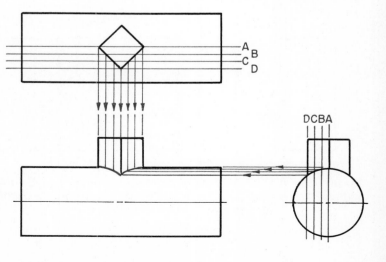

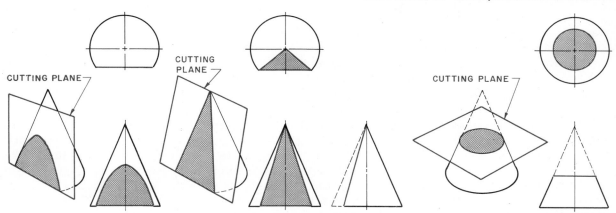

WHEN A CONE IS CUT PARALLEL TO ITS AXIS, A HYPERBOLA IS FORMED.

WHEN A CONE IS CUT FROM ITS APEX, A TRIANGLE IS FORMED.

WHEN A CONE IS CUT PERPENDICULAR TO ITS AXIS, A CIRCLE IS FORMED.

13-31. *There are three possible ways to cut a cone.*

A second method is to pass the cutting plane through the apex of the cone. It cuts the base at two points. This produces a triangular surface.

A third method is to pass a cutting plane perpendicular to the axis. This produces a circular surface.

Intersection of a Cone and Cylinder Using Radial Lines. Fig. 13-32 shows how to find the intersection of a cone and cylinder. This method uses radial lines. The cutting plane passes through the apex to the base of the cone.

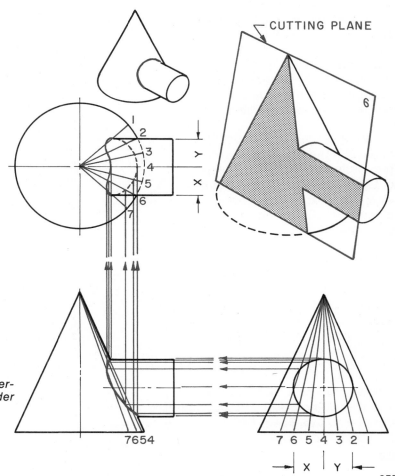

13-32. *Finding the line of intersection of a cone and a cylinder using radial lines.*

1. Draw the orthographic views.

2. Locate the cutting planes on the view in which the cylinder appears as a circle. In Fig. 13–32 this is shown in the end view. The cutting planes are marked 1 through 7. The outside planes, 1 and 7, pass through the point where the center line touches the outside of the cylinder. This is the extreme width. Any number of planes can be used.

3. Locate these cutting planes on the top view.

4. Next locate the cutting planes on the front view.

5. Now move back to the side view. Project the points where the cutting planes cross the circle to the front view. Where they intersect the same cutting plane is a point on the line of intersection.

For example, find cutting plane 6 in the side view. It cuts the circle in two places. These cuts are projected to the front view. Where they cross cutting plane 6 is a point on the line of intersection.

6. Project the points of intersection in the front view to the top view. The same method for locating points is used.

7. Connect the points with a smooth curve.

Intersection of a Cone and a Cylinder Using Cutting Planes Parallel to the Base. Fig. 13–33 shows how to find the in-

tersection of a cone and a cylinder using cutting planes parallel to the base of the cone. As the cone is cut a circle is formed.

1. Draw the orthographic views. The top, front, and side views are needed.

2. Draw the cutting planes on the side view. Any number of planes can be used. The more planes used, the more accurate will be the line of intersection.

3. Project the planes to the front and top views. In the top

view they will appear as circles.

4. On the side view, measure the distance from the center line to the point where each plane crosses the cylinder. For example, find plane E. The points of intersection are marked 1 and 2. They are distance Y from the center line.

5. Find plane E in the top view. Measure distance Y from the center line. Project it until it crosses plane E. This locates two points on the line of intersection. Do this for each plane.

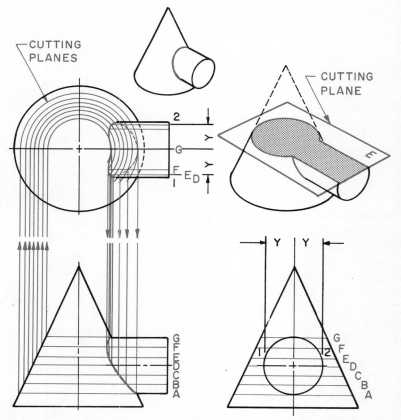

13–33. *Intersection of a cylinder and a cone using cutting planes parallel to the base of the cone.*

6. Project these points from the top view to the front view. Where they cross the proper cutting planes is a point on the line of intersection.

7. Draw a smooth curve between these points.

BEND ALLOWANCE

When sheet metal is bent to form a corner, extra metal is needed. If an accurate pattern is required, the extra metal needed for the corner must be calculated. This is called *bend allowance.* Bend allowance is usually not necessary on very thin metals. The outer surface will stretch, while the inner surface will compress enough to form the bend. Thicker metals require a bend allowance.

Fig. 13–34 shows a metal corner. The shaded area shows the bend allowance. Detailed instructions for finding bend allowance are found in Chapter 16, Drafting in the Aerospace Industry.

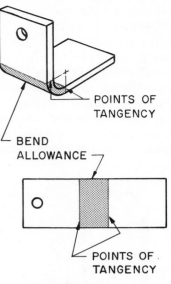

13–34. *Bends in thicker sheet metal require extra metal to form the bend. This is called bend allowance.*

Build Your Vocabulary

Following are some terms you should understand and use as part of your working vocabulary. Write a brief explanation of what each means:

Plane surface
Single curved surface
Double curved surface
Warped surface
Surface development
Parallel line developments
Radial line developments
Triangulation
Intersection
Piercing point
Bend allowance

Study Problems

1. Design a cardboard box to hold the bottle shown in P13–1. Make a full-size pattern. Remember to include tabs to hold the box together. Cut out the pattern and fold it into the finished box.

2. Develop the pattern for a rural mailbox. See P13–2. Design the back so it can be spot-welded in place. Design the front so it can be opened with a hinge.

3. Draw full-size patterns for the furnace and piping, P13–3. Cut out the patterns and assemble the installation. Draw to the scale 1" = 6".

4. Develop the pattern for the flower pot, P13–4. Put tabs on the sides. They fold in. Draw the bottom. The bottom is fastened to the tabs.

5. P13–5 shows a sheet metal flashing unit. It is used to keep rain water from leaking into a house where the sewer vent pipe comes through the roof. Make patterns for both pieces.

6. Develop patterns for the three parts of the ventilation head, P13–6. It is used to keep rain from entering the ventilator pipe.

7. P13–7 shows an overhead projector. Develop the pattern for the base and brace. Draw the head and lens cylinder together to show the line of intersection. Develop the pattern for the head. Show the

opening for the lens cylinder.
8. Develop the pattern for the vacuum cleaner nozzle, P13–8.
9. P13–9 shows several fittings for a dust collection system. Fitting A reduces the size of the pipe and changes its direction 45 degrees. Reducer fitting B changes direction 90 degrees. Fitting C gives a 60-degree change. Develop the patterns for these fittings.
10. Develop the pattern for the metal base of the lectern shown in P13–10.
11. Draw the top and front views of the pedestal base, P13–11. Then draw the line of intersection between the column and the base. Next draw the pattern for the leg.
12. P13–12 shows a part of a post for scaffolding used by painters. It has two brackets welded to it for horizontal supports. Draw the top and front views of the unit. Find the line of intersection. Then develop the pattern for the bracket.

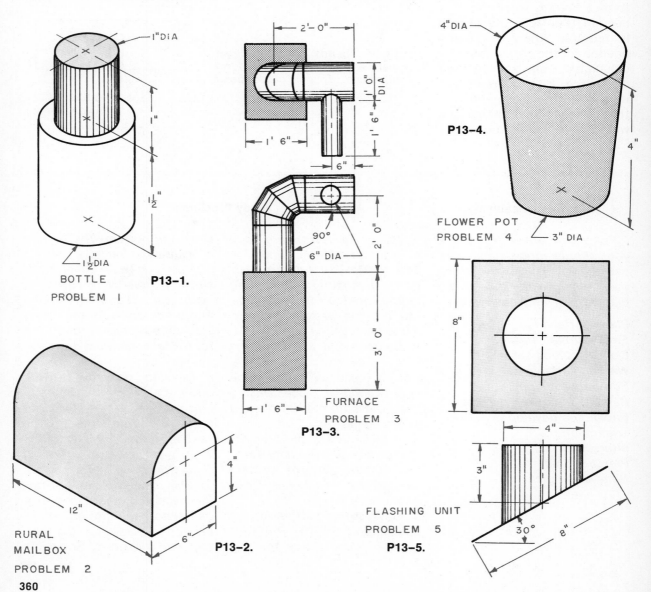

1"DIA

1"

1½"

1½"DIA

BOTTLE
PROBLEM 1

P13–1.

12"

6"

4"

RURAL
MAILBOX
PROBLEM 2

P13–2.

2'-0"

1'-6"

1'0" DIA

1'6"

6"

90°

6" DIA

2'0"

3'0"

1'6"

FURNACE
PROBLEM 3

P13–3.

4"DIA

P13–4.

4"

3" DIA

FLOWER POT
PROBLEM 4

8"

4"

3"

FLASHING UNIT
PROBLEM 5

30°

8"

P13–5.

360

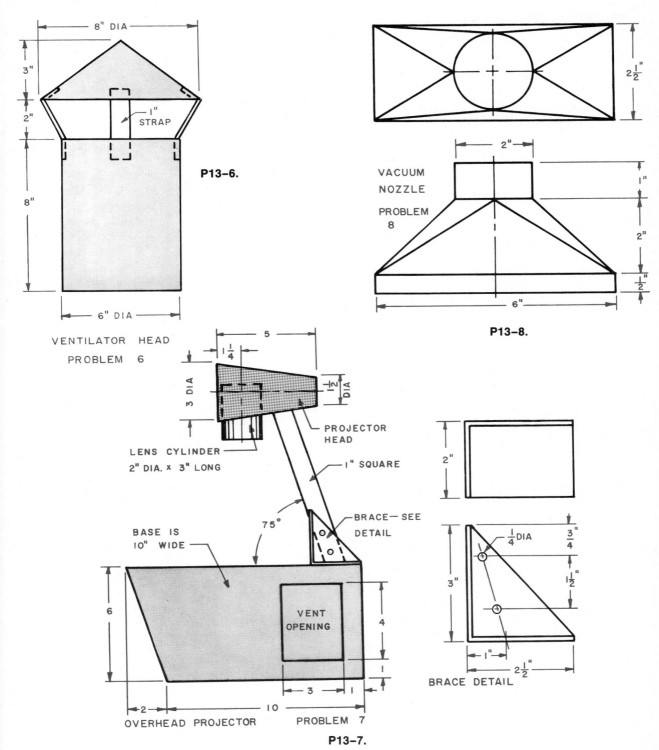

8" DIA

3"

2"

1" STRAP

8"

6" DIA

P13-6.

VENTILATOR HEAD

PROBLEM 6

$2\frac{1}{2}$"

2"

VACUUM
NOZZLE

PROBLEM
8

1"

2"

$\frac{1}{2}$"

6"

P13-8.

5

$1\frac{1}{4}$

3 DIA

$\frac{1}{2}$ DIA

PROJECTOR
HEAD

LENS CYLINDER
2" DIA. × 3" LONG

1" SQUARE

BRACE— SEE
DETAIL

75°

BASE IS
10" WIDE

2"

6

VENT
OPENING

4

1

2

3

1

10

$\frac{1}{4}$ DIA

$\frac{3}{4}$"

3"

$1\frac{1}{2}$"

1"

$2\frac{1}{2}$"

BRACE DETAIL

OVERHEAD PROJECTOR   PROBLEM 7

P13-7.

361

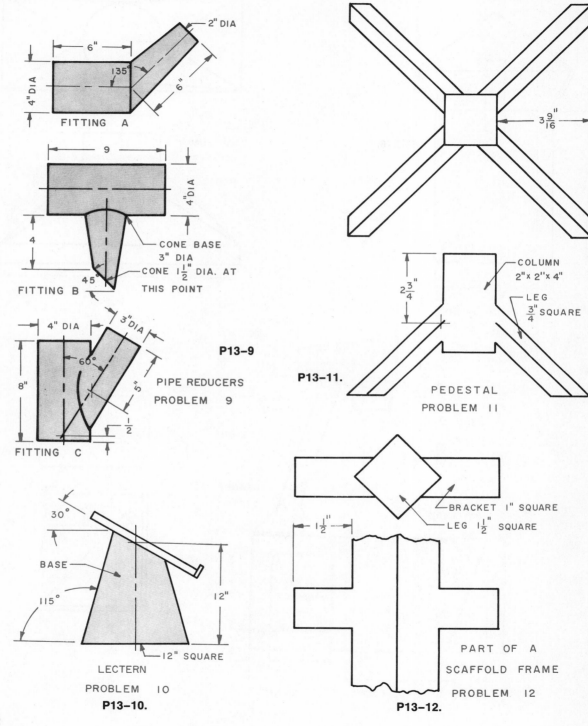

2" DIA

6"

135°

4" DIA

6"

FITTING A

9

4" DIA

4

CONE BASE
3" DIA

CONE 1½" DIA. AT

45°

FITTING B

THIS POINT

P13–9

4" DIA

3" DIA

60°

8"

5"

½

PIPE REDUCERS
PROBLEM 9

FITTING C

3 9/16"

COLUMN
2" x 2" x 4"

LEG
¾" SQUARE

P13–11.

2¾"

PEDESTAL
PROBLEM 11

30°

BASE

115°

12"

12" SQUARE

LECTERN
PROBLEM 10

P13–10.

BRACKET 1" SQUARE

LEG 1½" SQUARE

1½"

PART OF A
SCAFFOLD FRAME
PROBLEM 12

P13–12.

Chapter 14

Gears and Cams

GEARS

Gears are used to transfer rotary motion from one moving shaft to another. They also can be used to change the direction of rotation. By using gears of different sizes, the speed of rotation can be made slower or faster.

Fig. 14–1 shows two gears of the same size and number of teeth. When the driving gear makes one revolution, the second gear does also. Notice the second gear rotates in the opposite direction of the drive gear.

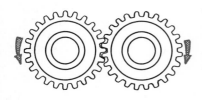

14–1. *Gears the same size rotate at the same speed.*

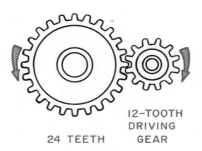

24 TEETH 12–TOOTH DRIVING GEAR

14–2. *The speed of rotation is governed by the size of the gear.*

Fig. 14–2 shows two gears of different sizes. The small gear has half the number of teeth found on the larger gear. When the smaller drive gear makes one revolution, the larger gear turns one-half a revolution. This slows down the larger gear. This has a 2:1 gearing ratio.

Study the series of gears shown in Fig. 14–3. Which direction does gear D rotate?

The design of gears is a difficult engineering problem. The engineer uses involute geometry. He must consider the strength required. The problem of wear must be considered. Since gears are complex to make, the designer must consider how they are to be made and the degree of accuracy needed. This chapter will show the details necessary to draw several commonly used gears.

GENERAL PRACTICES AND TERMS FOR MAKING GEAR DRAWINGS

Materials. Drawings should specify the required material.

Heat Treatment. When required, heat treatment should be specified and the hardness range shown. Reference can be made to the process standard or a specification.

Marking. Include all required markings. These can include part number, serial number, inspection symbol, or any other identifying number.

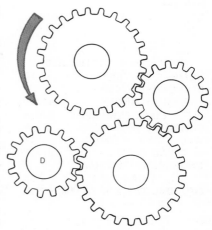

14–3. *Can you tell which direction gear D rotates?*

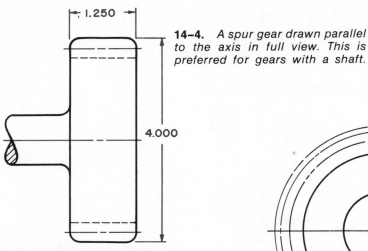

14-4. *A spur gear drawn parallel to the axis in full view. This is preferred for gears with a shaft.*

COMPLETE DRAWING WOULD
SHOW GEAR TOOTH DATA AS
IN THE NEXT FIGURE.

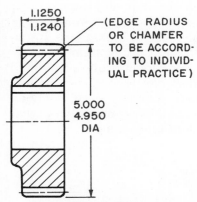

(EDGE RADIUS
OR CHAMFER
TO BE ACCORD-
ING TO INDIVID-
UAL PRACTICE)

SPUR-GEAR TOOTH DATA

NUMBER OF TEETH	18
DIAMETRAL PITCH	4
PRESSURE ANGLE	14°30'
PITCH DIAMETER (REF)	4.500
CIRCULAR THICKNESS	.3927
WHOLE DEPTH (MIN)	.5393
WORKING DEPTH	.5000
CHORDAL ADDENDUM	
(MAX OD)	.255
CHORDAL THICKNESS	.392-.395

ANSI Y14.7–1958 Gears, Splines, and Serrations

14-5. *Gears with holes or hubs are drawn in section. This drawing contains the minimum data for dimensioning the teeth.*

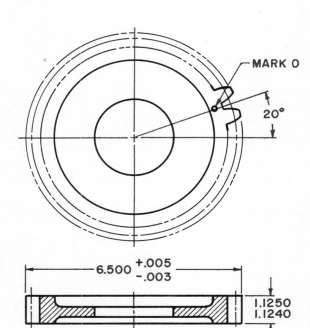

MARK 0

20°

SPUR-GEAR TOOTH DATA

NUMBER OF TEETH	24
DIAMETRAL PITCH	4
PRESSURE ANGLE	20°0'
PITCH DIAMETER	6.000
CIRCULAR THICKNESS	.3927
CIRCULAR PITCH	.7854
ADDENDUM	.2500
BASE DIAMETER	5.3382
WHOLE DEPTH (MIN.)	.5393
WORKING DEPTH	.5000
MEASURING PIN DIAMETER	.0100
MEASUREMENT OVER PINS	6.314 – 6.312

ANSI Y14.7–1958 Gears, Splines, and Serrations

14-6. *Sometimes a circular view of the gear is drawn. A few teeth can be drawn if it helps clarify the drawing. This drawing contains data needed when the pin measurement system of checking tooth thickness is to be used.*

Manufacturing Process. Whenever possible, this is left up to the company making the gear.

Quality of Finish. The method of finishing can be specified, such as ground, lapped, or shaved.

Type of Drawing. The type of drawing used depends upon the engineering of the gear and its use. Examples are shown on the following pages.

Choice of Views. Preferred is a view parallel with the axis. See Fig. 14–4. This is generally true with gears, pinions, or worms that are made on a shaft. Gears with holes or hubs are usually drawn in section. The section is passed through the axis. See Fig. 14–5. Sometimes a full or partial circular view is shown. See Fig. 14–6. Often a single view is all that is needed.

Gear Tooth Data. This varies with the type of gear. Examples are shown in the drawings on the following pages.

Showing Gear Teeth. The individual teeth are not usually drawn. In circular views, the addendum circle (outside), root, and pitch lines are shown by short and long dashes.

Dimensioning. Two types of dimensions are needed—those related to the teeth and those showing the dimensions for the gear blank. Gear dimensions that must be accurate are in decimals. The drawing shows the dimensions of the finished gear.

Dimensions that are measured during the manufacture of the gear are given tolerances. An example is the outside dimension (addendum circle). Dimensions used to determine other values but not

measured during manufacture are reference dimensions. They are labeled REF. An example is the pitch diameter.

Before drawing a gear, it is necessary to understand the terminology used to identify the parts. See Fig. 14–7.

Pitch Circle. An imaginary circle running through the point on the teeth at which the teeth from mating gears are tangent. The size of the gear is indicated by the pitch circle. For example, a six-inch gear has a pitch circle of six inches. Most of the gear dimensions are taken from the pitch circle.

Pitch Diameter. The diameter of the pitch circle.

Addendum. The distance from the pitch circle to the top of the teeth.

Addendum Circle. A circle formed by the outside surface

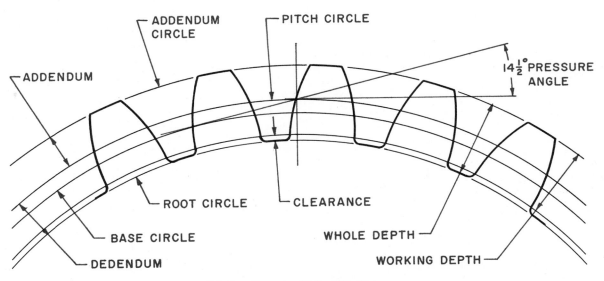

14–7. *Gear tooth terminology.*

of the teeth. The diameter is found by adding twice the addendum to the pitch circle. This is sometimes called the outside diameter.

Dedendum. The distance from the pitch circle to the bottom of the tooth.

Root Circle. A circle passing through the bottom of the teeth.

Root Diameter. The diameter of the root circle. It is found by subtracting twice the dedendum distance from the pitch circle.

Base Circle. A circle from which the tooth profile is generated. It is found by drawing a 14½ or 20-degree line through the pitch point. The angle used depends upon the gear design.

Pressure Angle. The angle used to determine the tooth shape. It is the angle at which pressure from the tooth of one gear is passed to the tooth of another gear. Two angles are in general use. These are 14½ and 20 degrees. The 20-degree angle is gradually replacing the 14½-degree angle. It has smoother and quieter running characteristics. It has greater load-carrying ability.

Circular Pitch. The distance along the pitch circle from a point on one tooth to the same point on the next tooth. See Fig. 14–6.

Whole Depth. The total height of a tooth. It can be found by adding the addendum and the dedendum.

Working Depth. The distance a tooth extends into the space between the teeth on another gear.

Clearance. The difference between the working depth and the whole depth.

Boston Gear Division
North American Rockwell

14–9. *Spur gears.*

Circular Thickness. The thickness of a tooth measured along the pitch circle. This is measured along the arc of the pitch circle. See Fig. 14–6.

Chordal Thickness. The thickness of a tooth measured along a chord of the pitch circle. See Fig. 14–6.

Chordal Addendum. The distance from the top of the tooth to the chordal thickness. See Fig. 14–8.

Face Width. The width of the tooth measured from one face to the other.

Number of Teeth. The number of teeth on the gear.

Diametral Pitch. A ratio equal to the number of teeth per inch of pitch diameter. For example, a four-pitch gear has four teeth for every inch of pitch diameter.

Spur Gears. A spur gear is shown in Fig. 14–9. It is used to transfer power and motion from one rotating shaft to another parallel with it.

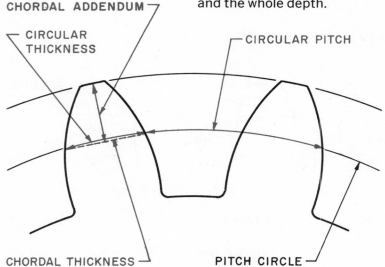

CHORDAL ADDENDUM

CIRCULAR THICKNESS

CIRCULAR PITCH

CHORDAL THICKNESS

PITCH CIRCLE

14–8. *Measurements for circular thickness, circular pitch, chordal thickness, and chordal addendum.*

A spur-gear drawing for a gear with a shaft is shown in Fig. 14–4. Generally gears with shafts are drawn in full view rather than in section.

A typical spur gear drawing for a gear with a hole is shown in Fig. 14–5. Gears with holes are usually drawn in section. Gear tooth data are recorded below the drawing. If the entire gear is to be dimensioned, standard dimensioning prac-

tices are used. A completely dimensioned gear is shown in Fig. 14–10.

An accurate way to gage tooth thickness is to measure with micrometers over or between cylindrical gage pins. The pins are placed between diametrically opposite tooth spaces. See Fig. 14–11. If this is to be done, the pin measurement figures are placed on the drawing as shown in Figs. 14–6

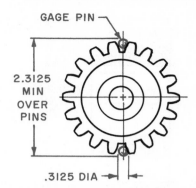

14–11. *How to dimension tooth thickness when gage pins are to be used.*

or 14–11. Since the pin measurements are used for checking the size, the addendum diameter is used as a reference dimension.

Another way to measure tooth thickness is to mesh it with a master gear. A master gear is one carefully made and checked to the tolerances wanted. It is used only for checking other gears as they are made. If this way of measuring is to be used, the gear drawing should have the information shown in Fig. 14–12. A master gear is generally used for checking fine pitch gears.

The testing radius is the radius of the circle that is tangent to the pitch circle of the master gear or the pitch line of the master rack at which the tooth thickness equals half the circular pitch.

Bevel Gears. Bevel gears are used to transfer power between shafts whose axes intersect. See Fig. 14–13. The

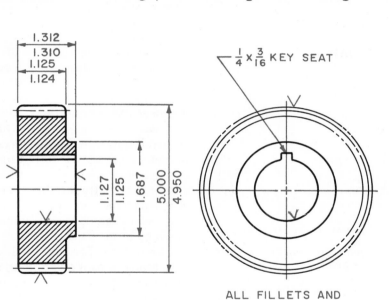

ALL FILLETS AND ROUNDS $\frac{1}{8}$ R

SPUR – GEAR TOOTH DATA

NUMBER OF TEETH	18
DIAMETRAL PITCH	4
PRESSURE ANGLE	14°30'
PITCH DIAMETER (REF)	4.500
CIRCULAR THICKNESS	.3927
WHOLE DEPTH (MIN)	.5393
WORKING DEPTH	.5000
CHORDAL ADDENDUM (MAX O D)	.255
CHORDAL THICKNESS	.392 – .395

14–10. *A completely dimensioned spur gear drawing.*

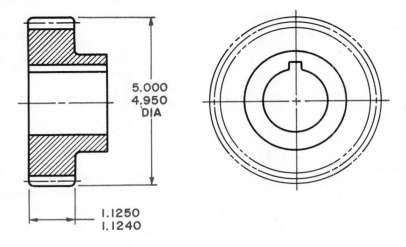

5.000
4.950
DIA

1.1250
1.1240

SPUR-GEAR TOOTH DATA

NUMBER OF TEETH	18
DIAMETRAL PITCH	4
PRESSURE ANGLE	14° 30'
PITCH DIAMETER	4.5000
ADDENDUM	.2500
WHOLE DEPTH (MIN)	.5393
WORKING DEPTH	.5000
CIRCULAR THICKNESS	.3927 – .3924
TESTING RADIUS WHEN IN TIGHT MESH WITH	
STANDARD MASTER GEAR (OR RACK)	4.750 – 4.745
AS A QUALITY CLASSIFICATION	

ANSI Y14.7–1958 Gears, Splines, and Serrations

14-12. *How to dimension a spur gear when tooth thickness is to be measured with a master gear.*

Boston Gear Division
North American Rockwell

14-13. *Bevel Gears*

axes may intersect at any angle. However, the most common intersection is at 90 degrees. When bevel gears are the same size and meet at a right angle, they are called *miter gears.* When a small bevel gear meets a large bevel gear, the small gear is called a *pinion.* Bevel gears are designed in pairs. They use the same involute tooth form as spur gears except the tooth is tapered.

Bevel gear terminology is shown in Fig. 14–14. Most of the terms used for spur gears apply to bevel gears. The general dimensions shown on a bevel gear drawing are diameter, pitch angle, cone distance, and face width. These are shown in the view that shows the gear's axis.

Fig. 14–15 shows a drawing with data for the gear teeth. If the entire gear blank is dimensioned, standard dimensioning practices are used. Fig. 14–16 shows a drawing of mating bevel gears. Each gear is dimensioned in the same way as shown in Fig. 14–15.

Worm and Worm Gears. Worms and worm gears are used to transmit power between shafts at right angles that do not intersect. A worm is a form of a screw. It has a thread the same shape as a tooth rack.

A worm gear looks much like a spur gear. It has a different tooth form. See Fig. 14–17. The tooth is formed to fit the curvature of the screw thread on the worm.

Fig. 14–18 lists the terms used to describe the parts of a worm and worm gear. The worm is described in the same manner as screw threads. The lead angle of the worm must be specified. The lead angle is the angle between the thread helix and the plane of rotation. The pitch diameter of the worm is the mean of the working depth. Fig. 14–19 gives the information needed on a worm drawing.

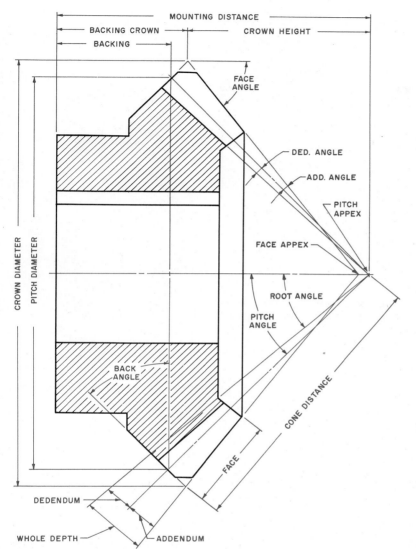

14-14. *Terms used to identify the parts of a bevel gear.*

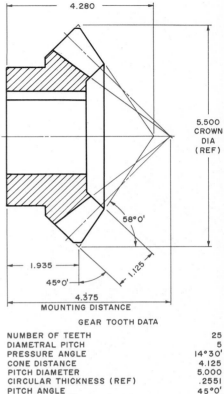

GEAR TOOTH DATA

NUMBER OF TEETH	25
DIAMETRAL PITCH	5
PRESSURE ANGLE	14°30'
CONE DISTANCE	4.125
PITCH DIAMETER	5.000
CIRCULAR THICKNESS (REF)	.2551
PITCH ANGLE	45°0'
ROOT ANGLE	38°0'
ADDENDUM	.125
WHOLE DEPTH (APPROX)	.370
CHORDAL ADDENDUM	.128
CHORDAL THICKNESS	.250
PART NUMBER OF MATING GEAR	1056
TEETH IN MATING GEAR	25
SHAFT ANGLE	90°0'
BACKLASH (ASSEMBLED)	.010
TOOTH ANGLE (APPROX)	31°30'

ANSI Y14.7-1958 Gears, Splines, and Serrations

14-15. *How to draw and dimension a straight bevel gear.*

A worm gear drawing is shown in Fig. 14–20. The data shown is needed to describe the teeth. The other parts of the gear blank are dimensioned following regular dimensioning practices.

Gear Rack. A rack is a straight piece of material with teeth cut on one side. See Fig. 14–21. Racks are usually rectangular. Round stock can be used. The rack changes rotary motion to reciprocating mo-

tion. Fig. 14–22 gives the terms used to identify the parts. They are the same as used for gears.

Fig. 14–23 shows how to dimension the teeth on a rack. The other size and location dimensions are placed in a normal manner.

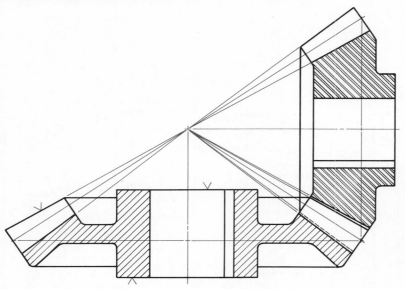

WORM

WORM
GEAR

14–16. *Mating bevel gears are sometimes drawn together. They can be dimensioned using standard dimensioning practices shown in this chapter.*

*Boston Gear Division
North American Rockwell*

14–17. *A worm and worm gear.*

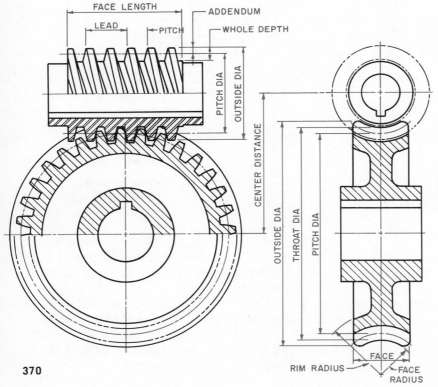

14–18. *Terms used to describe teeth on worms and worm gears.*

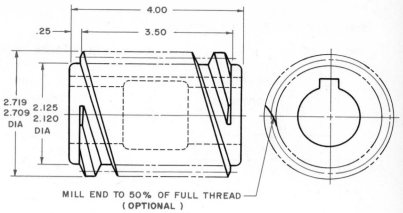

14–19. *A worm drawing.*

ANSI Y14.7–1958 Gears,
Splines, and Serrations

MILL END TO 50% OF FULL THREAD
(OPTIONAL)

4.00
.25
3.50
2.719 / 2.709 DIA
2.125 / 2.120 DIA

WORM TOOTH DATA

NUMBER OF THREADS	6
PITCH DIAMETER (NOMINAL)	2.212
AXIAL PITCH	.750
LEAD RIGHT (OR LEFT) HAND	4.500
LEAD ANGLE	32°56'
NORMAL PRESSURE ANGLE (NOMINAL)	14°30'
ADDENDUM	.480
WHOLE DEPTH (APPROX)	1.000
NORMAL CHORDAL ADDENDUM	.238
NORMAL CHORDAL THICKNESS	.375 –.370
WORM GEAR PART NUMBER	1013

CENTER LINE OF WORM

ROTATION

1.00 / .98
1.430
6.000 REF CENTER DISTANCE
10.50 / 10.48 DIA
.875 / .874
3.563 / 3.561
3.126 / 3.124 DIA
4.88 / 4.86 DIA
10.000 / 9.996 DIA
$\frac{3}{32}$ R
.50 / .48
2.00 / 1.96
3.504 / 3.496

ALL FILLETS AND ROUNDS
$\frac{1}{8}$ R EXCEPT AS NOTED

14–20. *A worm-gear drawing.*

ANSI Y14.7–1958 Gears,
Splines, and Serrations

WORM - GEAR TOOTH DATA

NUMBER OF TEETH	41
PITCH DIAMETER	9.000
ADDENDUM	4.80
WHOLE DEPTH (APPROX)	1.000
WORM PART NUMBER	2103
BACKLASH ASSEMBLED	.010 – .015
HOB NUMBER	82

WORM DATA (REFERENCE)

NUMBER OF THREADS	6
AXIAL PITCH	.750
LEAD RIGHT (OR LEFT) HAND	4.500
PITCH DIAMETER (NOMINAL)	2.212
LEAD ANGLE	32°56'
NORMAL PRESSURE ANGLE (NOMINAL)	14°30'

PINION

RACK

14–21. *A rack with a spur-gear pinion.*

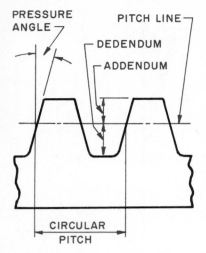

14–22. *Terms used to identify the parts of rack teeth.*

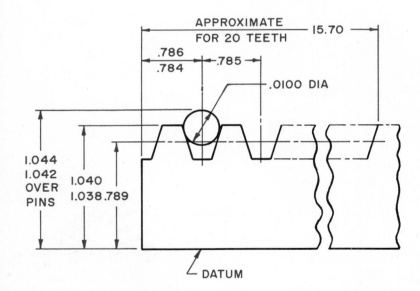

RACK TOOTH DATA

PRESSURE ANGLE	20°0'
TOOTH THICKNESS, PITCH LINE	.3927
WHOLE DEPTH	.5393
MAXIMUM ALLOWABLE PITCH VARIATION	.001
ACCUMULATED PITCH ERROR, MAXIMUM	.020

14–23. *Dimensioning rack teeth.*

Drafting Standards. Standards relating to gears can be found in the following publications of the American National Standards Institute:

ANSI B6.1, Standard Tooth Forms for Involute Spur Gears

ANSI B6.6, Gear Tolerances and Inspection

ANSI B6.7, 20 Degree Involute Fine-Pitch System for Spur Gears

ANSI B6.9, Design for Fine-Pitch Worm Gearing

ANSI B6.10, American Standard Gear Nomenclature

ANSI B6.11, Inspection of Fine-Pitch Gears

ANSI B6.12, Nomenclature of Gear-Tooth Wear and Failure

ANSI B6.13, System for Straight Bevel Gears

ANSI Y14.7, Gears, Splines, and Serrations

CAMS

A cam is an irregular-shaped disc used to change rotary motion into an irregular reciprocating motion. The cam is fastened to a revolving shaft. A shaft, called a *follower,* touches the curved surface of the cam. As the cam rotates, the follower moves along the curved surface of the cam. This causes the follower to move. See Fig. 14–24.

There are two general types of cams, disc and cylindrical. See Figs. 14–25 and 14–26. With the disc cam, the follower moves in a plane perpendicular to the axis of the shaft. The follower moves up and down as the cam rotates. See Fig. 14–24. The cylindrical cam moves the follower back and forth in a plane parallel with the axis of the shaft. See Fig. 14–27.

There are several common types of cam followers. See Fig. 14–28. The type used depends upon design features needed. Speed of rotation and stresses are factors considered when selecting a cam.

CAM MOTIONS

Cams produce three kinds of motion. These are uniform, harmonic, and uniformly accelerated and decelerated.

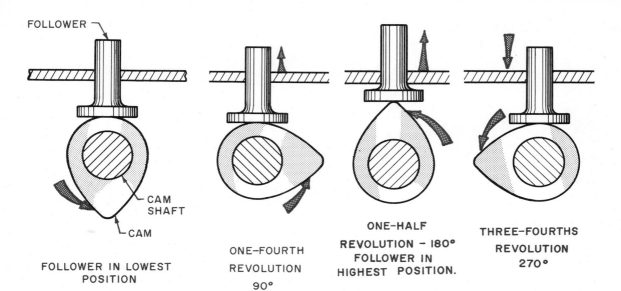

FOLLOWER

CAM SHAFT

CAM

FOLLOWER IN LOWEST POSITION

ONE-FOURTH REVOLUTION 90°

ONE-HALF REVOLUTION – 180° FOLLOWER IN HIGHEST POSITION.

THREE-FOURTHS REVOLUTION 270°

14-24. *The cam changes rotary movement to reciprocating movement.*

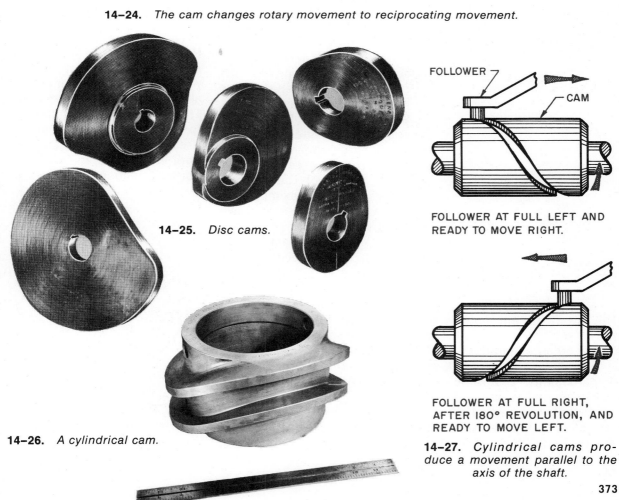

14-25. *Disc cams.*

14-26. *A cylindrical cam.*

FOLLOWER

CAM

FOLLOWER AT FULL LEFT AND READY TO MOVE RIGHT.

FOLLOWER AT FULL RIGHT, AFTER 180° REVOLUTION, AND READY TO MOVE LEFT.

14-27. *Cylindrical cams produce a movement parallel to the axis of the shaft.*

373

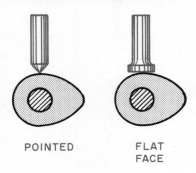

POINTED

FLAT FACE

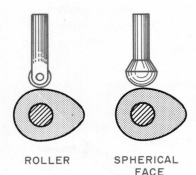

ROLLER

SPHERICAL FACE

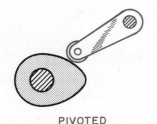

PIVOTED

14-28. *Common types of followers.*

The layout of cam motions is shown on a diagram called a displacement diagram.

Displacement Diagrams. A displacement diagram is a curve that represents the motion of the follower through successive units of time of one cam rotation. See Fig. 14–29.

374

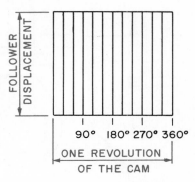

90° 180° 270° 360°

ONE REVOLUTION
OF THE CAM

14-29. *Grid for displacement diagram.*

The *length* of the diagram represents one revolution of the cam. Generally the length is drawn to equal the circumference of the work circle. The *working circle* has a radius equal to the distance from the center of the cam shaft to the highest point in the cam rise. See Fig. 14–30. This length equals 360 degrees. It is divided into segments. Often these are 10 degrees.

The *height* of the diagram represents the maximum rise of the cam follower. It is drawn to scale. This is called *follower displacement*. This height can be divided into time intervals or angle of cam rotation. The *time interval* is the time it takes the cam to move the follower to that height. The angle of cam rotation is the height the follower moves when the cam rotates a certain number of degrees.

Uniform Motion. Uniform motion means the follower rises and falls at a constant speed. It is a straight-line motion. It

is diagrammed in Fig. 14–31. The follower is to rise uniformly two inches through 180 degrees. Connect the beginning point A with the ending point B with a straight line. This is a diagram of uniform motion. It is common practice to draw an arc at each end of a uniform motion diagram. This eliminates an abrupt starting and stopping of the follower at the beginning and ending of the interval. The curve has a radius of one-third the rise. See Fig. 14–31. The straight line is drawn tangent to these arcs.

Harmonic Motion. A cam having harmonic motion is one which lifts the follower gradually from its starting position. The speed of rise increases to a point halfway in the full rise. The speed then decreases until it reaches its full rise. Such a motion relieves the

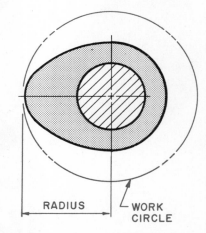

RADIUS WORK CIRCLE

14-30. *The work circle can be used as the length of the displacement diagram.*

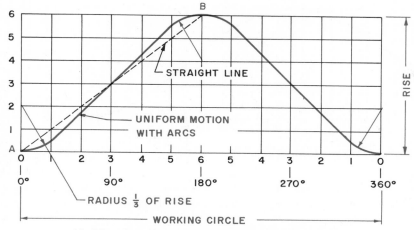

14–31. *A uniform motion displacement diagram.*

6. Project the points on the semicircle across the diagram. Where they intersect the line with the same number is a point on the harmonic curve.

7. Connect the points in a smooth curve.

Uniform Accelerated and Decelerated Motion. Uniform acceleration and deceleration is similar to harmonic motion. The difference is that uniform acceleration and deceleration occur at a constant speed. Harmonic motion increases and decreases at a constantly changing speed. Uniform accelerated and decelerated motion is smoother than harmonic motion.

A displacement diagram for uniform accelerated and decelerated motion is shown in Fig. 14–33. To draw this diagram:

1. Draw the length and rise of the diagram.

2. Divide the part of the

shock of starting and stopping as the follower moves through a rise and fall.

A displacement diagram for harmonic motion is shown in Fig. 14–32. To draw the diagram:

1. Draw the length of the working circle.

2. Draw the rise line.

3. Construct a semicircle on the rise line. The radius equals half the rise.

4. Divide the semicircle into a number of equal parts. Number the points.

5. Divide the part of the diagram's length to be used for the harmonic rise into the same number of parts. Number the points. In Fig. 14–32 this is 180 degrees.

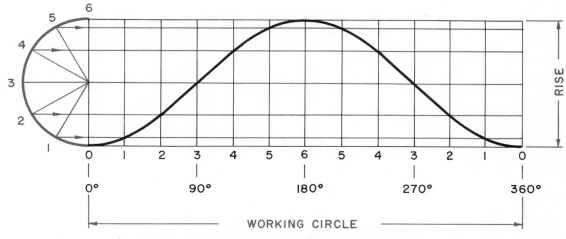

14–32. *A harmonic motion displacement diagram.*

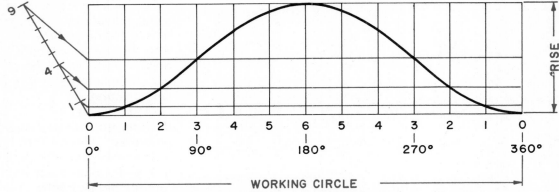

14–33. *A uniform accelerated and decelerated displacement diagram.*

length to be used for uniformily accelerated and decelerated motion into a number of equal parts.

3. Now divide half the rise into distances proportional to the square of the distances on half the length. Each point on the length is squared as 1^2, 2^2, and 3^2. This gives proportional parts as 1, 4, 9, and so forth, for marking the rise. Draw a line on any angle from Point A. Mark off equal dis-

tances to equal the largest squared number on the rise. Draw a line from the largest number to the center of the rise. Draw lines parallel to this to the rise. Project these points on the rise across the diagram parallel to the length line. This gives one half the curve.

The other half of the curve is the reverse of the half plotted. Points can be located using dividers.

4. Connect the points to complete the curve.

Dwell. Dwell means a period of time during which the follower does not move. This is shown on a displacement diagram as a horizontal line. See Fig. 14–34.

Displacement Diagrams with a Combination of Motions. Cams can contain a combination of the motions just discussed. Each motion is dia-

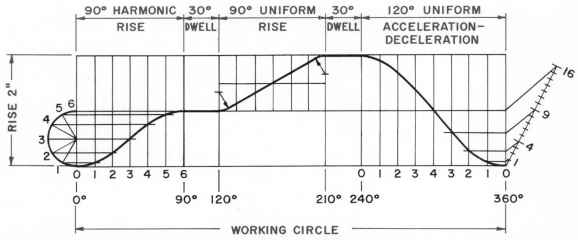

14–34. *A displacement diagram for a cam having a combination of motions.*

gramed as explained in Figs. 14–28 through 14–30. An example is shown in Fig. 14–34. Here the follower rises one inch in 90 degrees using harmonic motion. The follower dwells for 30 degrees. It then rises one inch in uniform motion as the cam turns 90 degrees. At the top of the rise it dwells for 30 degrees. It falls using uniform accelerated and decelerated motion through 120 degrees.

Drawing a Disc Cam Profile with an Offset Follower. To draw the cam profile shown in Fig. 14–35:

1. Draw the displacement diagram to the same scale as the cam drawing.

2. Locate the center of the shaft on which the cam will operate.

3. Draw the base circle from the center of the shaft. Its radius is equal to the distance from the center of the shaft to the center of the follower wheel at its lowest position. Where it crosses the center line of the follower is the lowest position. This is point 0 on the displacement diagram.

4. Draw the offset circle. It is drawn using the center of the shaft. The radius is equal to the offset.

5. Draw the center line of the follower.

6. Divide the offset circle into the same equal divisions used on the displacement diagram.

7. Number the divisions on the offset circle, beginning

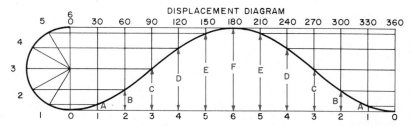

DISPLACEMENT DIAGRAM

CAM DATA: HARMONIC MOTION CAM, FOLLOWER OFFSET 7/16" WITH RISE OF 1", BASE CIRCLE 1/2" R, HUB 5/8" DIA, SHAFT 3/8" DIA, KEY 1/8" SQ., FOLLOWER ROLLER 3/8" DIA.

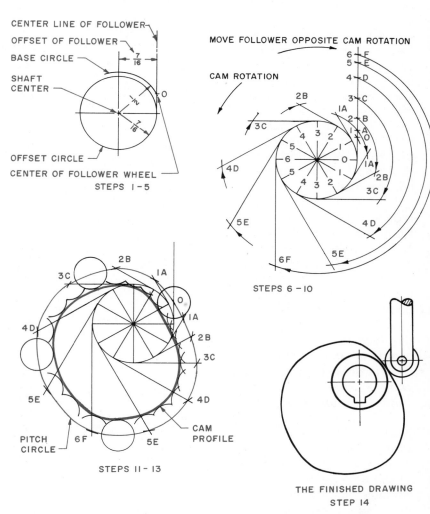

14–35. *How to draw a disc cam profile with an offset follower.*

with zero and in a direction opposite the direction of the cam rotation. The 0 point is where the center line of the follower is tangent with the offset circle. Number each division the same as on the displacement diagram.

8. Draw lines through each point on the offset circle tangent to the circle. Number these the same as the points on the circle.

9. Lay out the rise distances for each point on the center line of the follower. These begin at the center of the follower wheel. These distances are taken from the displacement diagram. Number each point as on the displacement diagram. These positions are the rise of the follower as the cam rotates.

10. Set a compass with a radius equal to the distance from the center of the cam shaft to point 1 on the center line of the follower. Swing an arc until this crosses tangent line 1. Repeat this for each point on the follower center line. This locates the center of the follower wheel in various positions around the cam.

11. Connect these points with a smooth curve. This forms the *pitch curve.*

12. Set a compass to the radius of the follower wheel. Using the pitch curve as a center, swing a series of arcs. These arcs are tangent to the working face of the cam. The more arcs used, the more accurate the cam profile drawn.

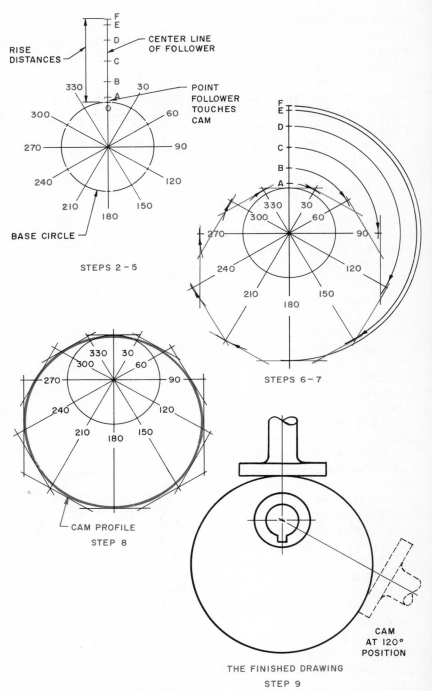

STEPS 2 − 5

STEPS 6 − 7

CAM PROFILE
STEP 8

THE FINISHED DRAWING
STEP 9

CAM AT 120° POSITION

14–36. *How to draw a disc cam profile with a flat face follower on the center line of the cam.*

13. Draw a smooth curve tangent to the cam wheel arcs. This forms the finished cam profile.

14. Draw hub, keyseat, and follower.

Drawing a Disc Cam Profile with a Flat Face Follower. To draw the cam profile shown in Fig. 14–36:

1. Draw the displacement diagram. The drawing shown in Fig. 14–36 uses the displacement diagram shown in Fig. 14–35.

2. Locate the center of the cam shaft.

3. Draw the base circle. Its radius is equal to the distance from the center of the cam shaft to the face of the follower in its lowest position. This point is where the base circle crosses the center line of the follower.

4. Lay out the follower rise distances on the center line of the follower. These distances are found on the displacement diagram. Number in the same order as they are on the displacement diagram.

5. From the center of the cam shaft, draw lines at the same degrees as used on the displacement diagram. Label these. They are the center lines of the followers at each of these locations.

6. Using the center of the cam shaft, swing arcs from each rise distance until they cross the degree line on which they were measured on the displacement diagram.

7. At each of these points of intersection, draw a line perpendicular to the degree line. These represent the position of the follower at these points.

8. The cam profile is drawn by constructing a smooth curve tangent to these perpendicular lines. Notice that the point of contact between the cam and the follower is not always along the center line. This is illustrated by the phantom followers drawn in a number of positions.

9. Draw the shaft, hub, keyseat, and follower.

Build Your Vocabulary

Following are some terms you should understand and use as part of your working vocabulary. Write a brief explanation of what each means:
Gear
Spur gear
Bevel gear
Pitch circle
Addendum circle
Circular pitch
Worm gear
Worm
Rack
Cam
Cam follower
Displacement diagram
Uniform motion
Harmonic motion
Accelerated motion
Decelerated motion

Study Problems

1. Make working drawings of spur gears using the data shown in P14–1. They use a standard square key. Calculate the working depth, whole depth, and the diameter of the addendum circle.
2. Make a drawing of the bevel gear described in P14–2. Fully dimension the gear. Develop any dimensions that may be missing by measuring your drawing. Use a stock square key.
3. Make a drawing of the worm described in P14–3. It uses a stock square key.
4. Study the data for the worm gear, P14–4. Make a dimensional drawing. A stock square key is used.
5. Data for a rack are shown in P14–5. Make a dimensioned drawing of the rack.
6. Draw a displacement diagram for a cam having uniform motion with arcs and a working circle 7 inches long. The rise is 2½ inches.
7. Draw a displacement diagram for a cam having harmonic motion. The working circle is 9 inches and the rise 2 inches.
8. Draw a displacement diagram for a cam having uniform accelerated and decelerated motion. The working circle is 6 inches and the rise 3 inches.
9. Draw the displacement diagram and the cam profile for the following: Harmonic

motion cam with an offset roller follower: Follower offset, $7/8$"; rise, $1^1/2$"; base circle, 1" R; hub, $1^1/4$" dia.; shaft, $5/8$" dia.; stock square key; follower roller, $3/4$" dia.

10. Draw the displacement diagram and the cam profile for the following: Uniform motion cam with flat face follower on the center line of the cam; rise, 2"; base circle, 1" R; hub, $1^3/8$" dia.; shaft, $3/4$" dia.; stock square key.

11. Draw the displacement diagram and the cam profile for the following: Cam with flat face follower on center line; up 1" in 90° harmonic rise; dwell, 60°; up 1" in 90° harmonic rise; dwell, 30°; down 2" in 90° uniform accelerated and decelerated motion; base circle, $1^1/4$" R; hub, $1^3/8$" dia.; shaft, $3/4$" dia.; stock square key.

P14–1.
Spur Gear Data

Problem Number	No. of Teeth	Dimetral Pitch	Face	Addendum	Dedendum	Clearance	Pitch Circle Diameter	Hole	Style	Pressure Angle	Hub Dia.	Hub Proj.
1	12	8	$1^1/2$	.125	.115	.010	1.500	$3/4$	Plain	20°	—	—
2	12	4	$3^1/2$	.250	.235	.015	3.000	$1^1/16$	Plain	20°	$2^1/4$	1
3	36	12	1	.092	.080	.010	3.000	$5/8$	Web	$14^1/2$°	$1^1/2$	$5/8$
4	24	24	$1/4$	.047	.042	.005	1.000	—	Plain with Shaft	$14^1/2$°	—	—
5	66	6	2	.188	.178	.010	11.000	$1^1/4$	Spoke	20°	$3^1/2$	$1^1/2$

P14–2.
Bevel Gear Data

Number of Teeth	21	Face	.40
Pitch Dia.	1.81	Backing	.50
Crown Dia.	1.87	Backing Crown	.56
Back Angle	46°	Mounting Distance	1.44
Dedendum Angle	5°-30′	Hub Projection	$11/16$
Addendum Angle	5°–0′	Hub Diameter	$1^3/8$
Pitch Angle	46°–30′	Hole	$5/8$

P14–3.
Worm Data

Dimetral Pitch	8
Hub Projection	$5/8$ each end
Hub Dia.	$1^3/16$
Length of Threads (Face)	$1^3/4$
Hole	$3/4$
Pitch Dia.	1.500
Lead	.3927 right
Pressure Angle	$14^1/2$°
Lead Angle	4°–46′
Addendum	.125
Whole Depth	.240

P14–4.

Worm Gear Data

No. of Teeth	20
Diametral Pitch	8
Pitch Dia.	2.50
Outside Dia.	2.94
Throat Dia.	2.68
Style	Plain
Hole	¾
Face	¾
Hub Dia.	1¾
Hub Projection	¾
Rim Radius	1.37
Face Radius	0.75

P14–5.

Rack Data

No. of Teeth (Pitch)	8
Face Width	1½
Overall Thickness	1½
Back to Pitch Line	1.375 +.000 −.008
Addendum	.125
Whole Depth	.240
Pressure Angle	20°
Length	12

Chapter 15

Vectors

Problems of a technical nature can be solved by graphical means. This gives a practical, visual means for finding solutions. These solutions are developed by using vectors.

A *vector* is a line that is given *direction* and represents a force, velocity, or some other measure of magnitude. The direction is measured in degrees. The force is indicated by drawing the vector to scale. For example, the vector, Fig. 15-1, has a direction of 30 degrees with the horizontal and a magnitude of 500 pounds.

Vectors are used in engineering in an area called mechanics. Mechanics is that branch of engineering that deals with the state of rest or motion of bodies under the action of a force. The forces can be pulling or pushing on the body. Mechanics is used in studying technical problems, such as the strength of roof structures, trusses, beams, engine performance, and fluid flow. These problems can be solved by mathematical or graphical means. This chapter explains the use of vectors to develop graphical solutions.

DEFINITION OF TERMS

Before any discussion of vectors can take place, it is necessary to understand certain terms used.

A *force* is something which tends to produce a pushing or pulling motion on an object. It is necessary to know the direction and magnitude of the force.

Direction of a force is the line in which the force acts. It is shown graphically in degrees from the horizontal. See Fig. 15-1.

The *magnitude* is the amount of force. It is expressed in some standard unit. Examples are pounds and miles per hour.

A force has a *point of application*. This is the point at which the force is applied. It is either pulling or pushing at this point. See Fig. 15-1.

Coplanar forces are those acting in a single plane. This could include two or more forces. See Fig. 15-2.

Concurrent forces are those all acting upon a single point. They may or may not be in the same plane. Those shown in Fig. 15-2 are concurrent and coplanar.

Noncoplaner forces are those that are not all in the same plane. See Fig. 15-2.

The *resultant* is a single vector that can replace the other vectors. See Fig. 15-2.

DRAWING A VECTOR DIAGRAM

How to draw a vector diagram is shown in Fig. 15-4. The point of application is established. The information about the forces must be

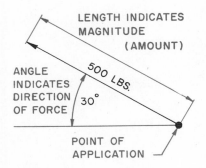

LENGTH INDICATES MAGNITUDE (AMOUNT)

500 LBS.

ANGLE INDICATES DIRECTION OF FORCE

30°

POINT OF APPLICATION

SCALE 1"=250 LBS

15-1. *A vector represents a force. It shows the direction and magnitude of the force.*

known. Fig. 15–3 shows a point having two forces pulling on it. Each is represented by a vector. The vectors are drawn from the point of application in the direction given. They are drawn to scale. The selection of scale is a decision of the draftsman. Any suitable scale can be used. The scale selected must be large enough to permit the solution to be measured carefully. In Fig. 15–1, the scale used was 1″ = 250 pounds. The vector represents 500 pounds. It was drawn two inches long.

The architect's scale and the civil engineer's scale are used. If the scale is based on a fraction of an inch, use the architect's scale. For example, 1/2″ = 1,000 pounds. The civil engineer's scale is divided on the basis of 10, 20, 30, 40, 50, and 60 parts to the inch. Sometimes this is easier to use.

The direction of the force is shown with an arrow on the end of the vector.

Accuracy in drawing is vital. Use a sharp-pointed, hard-leaded pencil.

ADDITION OF TWO CONCURRENT COPLANAR VECTORS

The following is a simple example of the addition of coplanar forces. See Fig. 15–5. Suppose two men find it necessary to move a filing cabinet. One man pushes with a force of 120 pounds. The other man pushes with a force of 90

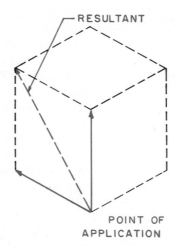

POINT OF APPLICATION

CONCURRENT COPLANAR FORCES

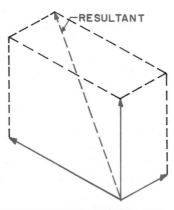

RESULTANT

CONCURRENT NONCOPLANAR FORCES

15–2. *Coplanar forces lie in one plane. Noncoplanar forces are not all in the same plane.*

pounds. The total force being used to move the cabinet is 210 pounds. Both forces were applied in the same direction. Since this was true, both forces were added.

During a picnic two groups of men decide to have a "tug

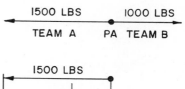

15–3. *Forces acting in opposite directions can be added graphically.*

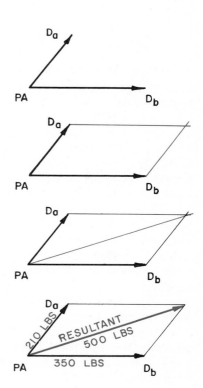

15–4. *A simple concurrent coplanar vector problem may be solved by closing the parallelogram and drawing a diagonal.*

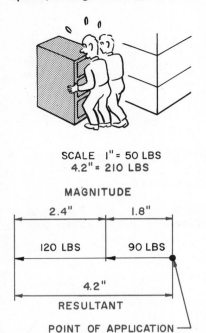

SCALE 1" = 50 LBS
4.2" = 210 LBS

MAGNITUDE

2.4"   1.8"

120 LBS   90 LBS

4.2"

RESULTANT

POINT OF APPLICATION

15-5. *Forces may be added graphically to determine the total force.*

of war." The total force exerted by Team B was 1,000 pounds. Team A put forth a total force of 1,500 pounds. Fig. 15–3 shows the graphical addition of these two forces. The resultant would be 500 pounds. These two forces were applied in opposite directions. In both examples, the resultants were simple addition of the quantities.

How can two forces be added when the total is less than the sum? Changes in temperature are an example of this condition. Assume the temperature at 1:00 p.m. was 70 degrees. In the next two hours the temperature became 15 degrees warmer. The

reading at 3:00 p.m. was 85 degrees. Or suppose the temperature was −10 degrees and four hours later it warmed to 22 degrees. The temperature would be 12 degrees. These temperatures were added: $-10° + (22°) = +12°$. This type of addition is called *adding algebraically.* Whenever two or more quantities of like sign are added, simple addition is used. The quantities may be *all* positive or *all* negative. The total will then be positive or negative, depending on the sign. When both men were pushing the cabinet, the forces were in the same direction. These can be said to be both positive. When two quantities are added and have *unlike* signs, their sum is the *difference* between the two. In other words, the two quantities are subtracted. The sign of the greater quantity will be the sign of the result. If +25 and −10 are added, the result would be $+25 + (-10) = +15$. The smaller quantity has been subtracted from the larger. The sign of the larger quantity is the sign of the result. In solving coplanar vector problems graphically, forces having the same sign act in the same direction. In the case of the "tug of war," the forces were acting in opposite directions. One force is considered positive and the other negative.

If the forces are not in the same line, they still may be added. Fig. 15–4 shows two vectors representing two

forces. They act on point PA. To find the force and direction of pull on PA, add the vectors graphically. To do this, draw lines parallel with the vectors. They go through the ends of the vectors. This forms a parallelogram. Then draw a diagonal from the point of application. The diagonal is the resultant force. The magnitude of the resultant is found by measuring the length of the diagonal. The resultant is a force that, if applied in the direction shown, would replace the other two forces.

Another example of this type of graphical solution is shown in Fig. 15–6. A plane is flying due west at an air speed of 250 mph. The wind is blowing 75 mph. from a southeasterly direction (315 degrees from north). Considering these two factors, what will be the air speed and direction of the plane?

By plotting the direction and magnitude (speed) of the plane and the wind, a parallelogram can be drawn. A diagonal from the PA becomes the resultant. In the illustration, the re-

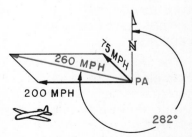

260 MPH
75 MPH
N
PA
200 MPH
282°

15-6. *Vectors can be used to solve speed and direction problems.*

sultant and speed would be 260 mph. at 282 degrees or N 48° W.

A triangle can be used to save time in solving concurrent coplanar problems. A diagonal (resultant) of a parallelogram cuts it into two equal triangles. See Fig. 15–7. The opposite sides of a parallelogram are equal. Since this is the case, a triangle can be used to solve the problem. Note the sides of the triangle are the same length and direction as in one half of the parallelogram. When plotting forces in a triangle, they are

placed *tip* to *tail*. The resultant is drawn from PA with its tip touching the tip of vector D_a.

The vectors are moved keeping the same magnitude and direction until they form two sides of a triangle.

ADDITION OF THREE OR MORE CONCURRENT COPLANAR VECTORS

More than two vectors may be added in a manner similar to that shown in Fig. 15–4. Three, four, or more vectors

may be added as long as they are in the same plane. One method that is used develops a series of parallelograms. See Fig. 15–8. This figure shows how to find the resultant of three vectors in one plane.

First find the resultant of two of the vectors. In Fig. 15–8, Step 2, vectors D1 and D2 are used. Their resultant is 180 pounds. Then this resultant is made into a parallelogram with vector D3. See Fig. 15–8, Step 3. The resultant is 270 pounds. This is the resultant for the three vectors, D1, D2, and D3.

If more than three vectors are involved, the same system is used. It requires the use of additional parallelograms.

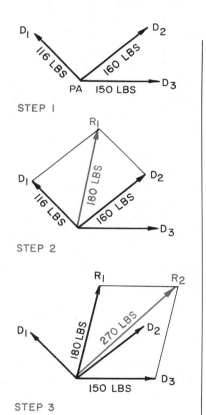

15–7. *A force triangle can be used to solve simple concurrent coplanar vector problems.*

STEP 1

STEP 2

STEP 3

15–8. *More than two concurrent coplanar forces can be added by drawing a series of parallelograms.*

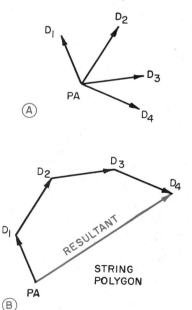

15–9. *Three or more concurrent coplanar forces can be added by using a string polygon.*

A second method of adding vectors is by the polygon method. This is only used for three or more vectors. A polygon is a geometric figure bounded by straight lines. The polygon method is similar to the triangle method of addition. This method is shown in Fig. 15–9 using four vectors. The resultant can be found by moving the vectors so the tip of one touches the tail of the next. See Fig. 15–9B. As the vectors are moved and connected, keep each in the original direction. The magnitude of each must be kept accurate. The resultant is found by connecting the tail of one, PA, to the tip of the last vector, D4. This resultant is the magnitude and direction of the four vectors on point PA.

COMPONENTS

Thus far, separate vector forces and their resultants have been explained. These produced a single vector that would give the effect of several individual vectors.

A *component* is a single force that needs to be divided into several individual vectors. It is the opposite of the situations explained thus far.

Fig. 15–10 shows a component with a magnitude of 400 pounds. It is desired to split this into two vectors with the directions shown. The direction of the desired vectors must be known.

The parallelogram method can be used. The component is the diagonal. Connect lines representing the vectors wanted with the ends of the component. These must be in the desired direction. Draw a parallelogram using these as sides. This marks the vectors to length. Measure them using the same scale used to draw the component. This gives the magnitude of each vector.

EQUILIBRIUM

Equilibrium is a condition of balance between opposing forces. Fig. 15–11 shows two equal forces pulling in opposite directions. Since they are equal, they are balanced. Neither force changes the position of point P. These forces are in equilibrium.

Fig. 15–12 shows a space diagram with three vectors drawn to scale. Is this diagram in equilibrium? This can be found by constructing a vector

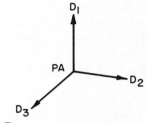

(A) SPACE DIAGRAM

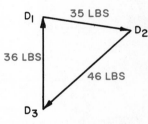

(B) VECTOR TRIANGLE

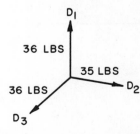

(C) SPACE DIAGRAM WITH NOTATIONS

15–12. *Forces in equilibrium are in balance with each other.*

15–10. *A force can be resolved into its components.*

15–11. *These forces are in equilibrium.*

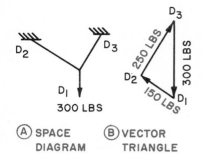

A SPACE
DIAGRAM

B VECTOR
TRIANGLE

15–13. *A vector triangle is used to find the forces necessary to hold the 300-pound sign in equilibrium.*

triangle, shown in Fig. 15–12B. Each side is drawn with the original magnitude and direction. An attempt is made to form a triangle. If a triangle can be formed, the diagram is in equilibrium.

An equilibrant force is one that will balance one or more forces. An equilibrant force acts in the opposite direction from a resultant force.

Fig. 15–13 shows a 300-pound weight supported by two cables. It is desired to find the forces in the two cables that will support the weight. These forces may be resolved graphically. To do this, draw a **vector triangle.** Begin by laying off the 300-pound force in its correct magnitude and direction. Next the two cables are laid off in their respective directions. When the triangle closes, scale the distances. These are the forces that are necessary to hold the weight in equilibrium.

Another example of the same type of problem is shown in

Fig. 15–14. In this situation, a sign hangs from a support above a building entrance. To determine the correct size cable, it is necessary to know the forces acting on the cable. To resolve the forces, first draw the vertical component in its correct magnitude. Now draw the support and the cable in their proper direction. These are the unknown forces in the diagram. The force polygon

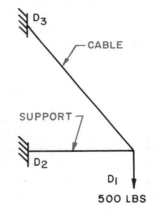

A SPACE DIAGRAM

B VECTOR TRIANGLE

15–14. *The forces in the cable and support are found by drawing a vector triangle.*

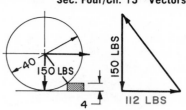

15–15. *How much force is required to roll the wheel over the four-inch block?*

will now close since the forces are in equilibrium. Scaling the lines will indicate the forces acting on the cable and support.

The same principles may be used in finding the force required to move a wheel over an obstruction. This problem may be solved just as the previous two have been. See Fig. 15–15. A force is applied level with the center of the wheel. The wheel has a 40-inch diameter and weighs 150 pounds. The forces acting on the wheel are unknown. How much force will be required to make the wheel rise over the block? First, draw the known force. This is a vertical force of 150 pounds. Next draw the other two forces parallel to the forces they represent. The horizontal member of the force triangle is the amount needed to move the wheel over the block. Suppose the horizontal force were raised above the center. Would the amount be smaller or larger?

BOW'S NOTATION

Bow's notation is a system of designating coplanar forces.

387

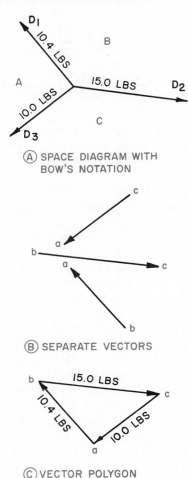

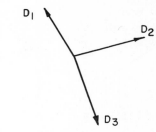

(A) SPACE DIAGRAM WITH
BOW'S NOTATION

(B) SEPARATE VECTORS

(C) VECTOR POLYGON

15–16. *These three forces are in equilibrium with Bow's notation.*

This method of noting forces acting in the same plane is widely used in engineering practice. Bow's notation simply uses capital letters between lines of action. See Fig. 15–16. Any line of action or force can be designated by the letters in the adjacent spaces. In the illustration, force D_1 is referred to as AB. Force D_2 is called BC. Force D_3

is identified as CA. The letters are usually placed between the forces in a clockwise manner. The first letter notes the tail end (A). The arrow end is noted by the second letter (B). The tip end of force D_2 is C and the tail end is B. Thus the force is called BC. It is standard practice to read the forces in a clockwise manner.

When using Bow's notation in finding the resultant of coplanar forces, an assumed resultant must be added. The letters, AR, are commonly used to note the assumed resultant. See Fig. 15–17. The AR may be placed between *any* of the forces. It has been placed at random between forces D_3 and D_1. When using an assumed resultant, begin lettering the forces between the AR and the first known force. In this example, the letter A was placed between AR and D_1. Now proceed clockwise, noting the spaces between the forces. The force polygon may now be drawn, noted, and closed. See portion C of this figure. Once the polygon has been closed, the resultant can be drawn on the vector diagram in its proper magnitude and direction.

Fig. 15–18 shows the application of Bow's notation on a series of parallel forces. Observe that the forces are lettered in a clockwise manner. Force D_1 is identified as AB, D_2 as BC, and D_3 as CD. Reaction R_2 is noted as DE and R_1 as EA. Note each of the forces and reactions have been read clockwise.

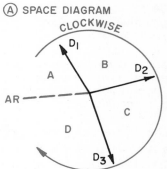

(A) SPACE DIAGRAM

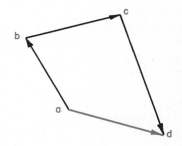

(B) SPACE DIAGRAM WITH BOW'S NOTATION

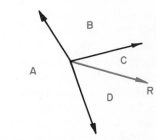

(C) VECTOR POLYGON

(D) SPACE DIAGRAM WITH RESULTANT

15–17. *Bow's notation applied when finding a resultant or equilibrant.*

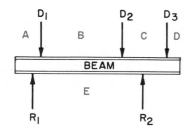

15–18. *An application of Bow's notation in nonconcurrent coplanar parallel forces.*

CONCURRENT NONCOPLANAR FORCES

Any series of forces that have the same point of application and are not acting in the same plane are called *concurrent noncoplanar* forces. See Fig. 15–2. Fig. 15–19 shows three forces acting from a single point but not in the same plane. It is desired to find the resultant of forces OA, OB, and OC. One of the easiest methods is to use a series of parallelograms. First, draw the top and front views of the vectors. Then find the resultant (R₁) of forces OA and OB in both views. See Fig. 15–19, Parts 1 and 2. Next

15–19. *The resultant of three concurrent noncoplanar forces may be found by using a series of parallelograms.*

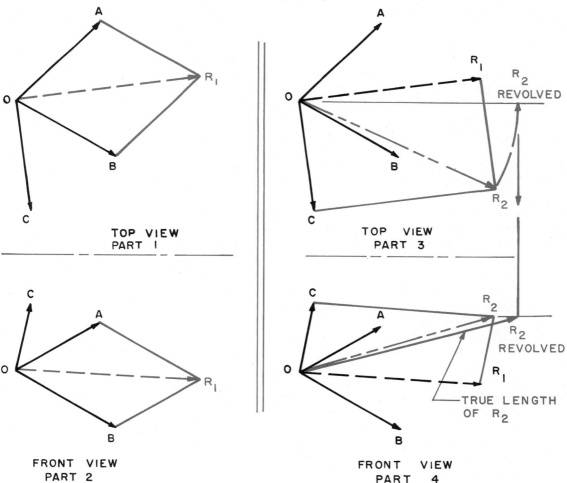

STEP I. FIRST FIND RESULTANT R_1 OF TWO OF THE FORCES.

STEP 2. NOW FIND THE RESULTANT OF R_1 AND THE THIRD FORCE. THIS IS R_2. REVOLVE R_2 TO FIND ITS TRUE LENGTH.

combine R_1 with force OC giving the resultant R_2. See Fig. 15–19, Parts 3 and 4. To find the magnitude of the final resultant, revolve R_2 parallel to a plane of projection. In the example, R_2 was revolved parallel to the vertical plane. The true length is then scaled in the front view.

An alternate method of finding the resultant is by using a parallelepiped. A *parallelepiped* is a prism with a square or rectangular base. See Fig. 15–20. The resultant of three concurrent noncoplanar forces is the body diagonal of a parallelepiped. Fig. 15–20 shows a parallelepiped with a body diagonal representing a resultant. Fig. 15–21 uses the same three forces as shown in Fig. 15–19. From the terminal points of each force, construct the sides of a parallelepiped. For example, draw edge AX parallel to OB in both views. Now side BX is drawn parallel to OA in both views. Then edge CY is drawn parallel to OA. The same is done with YZ and XZ. When the figure is completed, the body diagonal is drawn. The body diagonal, OR, is the resultant. To find its magnitude, it must be revolved parallel to one of the planes of projection. In this case, R was revolved parallel to the horizontal plane.

In some instances it is necessary to resolve a force into its components. If the direction and sense of the components are known, their magnitude can be determined. A

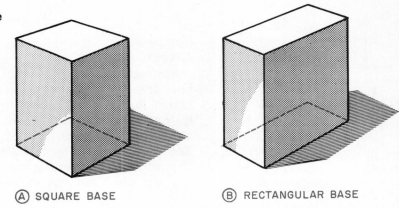

(A) SQUARE BASE   (B) RECTANGULAR BASE

15–20. *A parallelepiped can be either a square or rectangular-based prism.*

parallelepiped may be used to determine the components. Fig. 15–22 shows three unknown forces and their resultant. It shows the top and front views. In order to find the components, begin by drawing the opposite end of the parallelepiped. See Part B of Fig. 15–22. This is begun at the end of the resultant. In both views, draw two sides of the base, RE and RD, parallel to OB and OA. The third step is to determine where the edge, OC, pierces the base of the parallelepiped. Part C of the figure shows the application of determining where a line pierces a plane. Where line OC pierces the base is the length of one of the edges of the parallelepiped. Next the base can be completed from point C_p, shown in Fig. 15–22D. The magnitudes of forces OA, OB, and OC can be scaled at this time; however, the parallelepiped is usually completed, Part E, and visibility determined, Part F. To obtain the

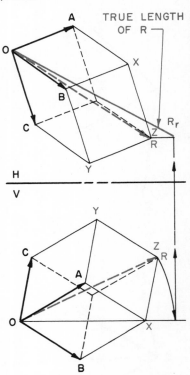

15–21. *The resultant of three concurrent noncoplanar forces may be found by using a parallelepiped.*

magnitudes of each force, OA, OB, and OC must be revolved parallel to a plane of projection.

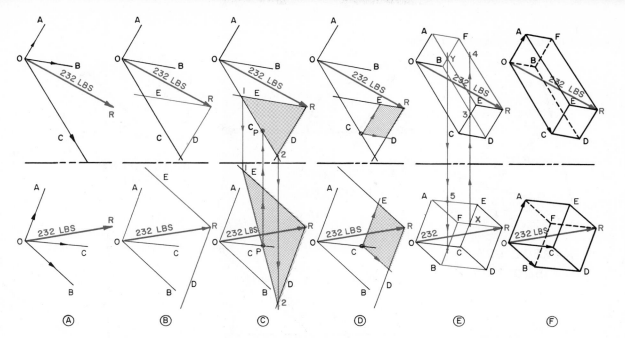

15–22. *The components of a resultant can be found by drawing a parallelepiped. The resultant is the body diagonal.*

Study Problems

Build Your Vocabulary

Following are some terms you should understand and use as part of your working vocabulary. Write a brief explanation of what each means:

Vector
Force
Direction
Magnitude
Coplanar forces (Co·planer)
Concurrent forces
Noncoplanar forces
Resultant
Parallelogram (Par'al·lel'o·gram)
Polygon
Component
Equilibrium
Parallelepiped (Par'al·lel'e·pi'ped)

Following are a variety of force problems. Work carefully and accurately.

1. Two cars are tied together. One pulls with a force of 5,000 pounds. The other pulls with a force of 3,500 pounds. Draw a diagram of these forces. If each car is trying to pull the other under full power, which car would pull the other?

2. Find the resultant of the concurrent coplanar vectors shown in P15–1. Use the parallelogram method.

3. Find the resultant of the concurrent coplanar vectors shown in P15–2. Use the triangle method.

4. Find the resultant of the three concurrent coplanar vectors shown in P15–3. First, solve it with the parallelogram method. Then use the string

polygon method. How close are the results?

5. Find the actual speed and direction of flight of the airplane shown in P15–4.

6. Find the coplanar components of the force shown in P15–5.

7. Check the vector diagram, P15–6, to see if it is in equilibrium.

8. Find the force on cable AB and AC, P15–7.

9. Find the true magnitude of the resultant of the three concurrent noncoplanar forces A, B and C, P15–8. The lengths shown on the drawing are the lengths to use to lay out the problem. These vectors do not appear true length on the drawing. Measure the force of the resultant, using the scale 1" = 200 pounds.

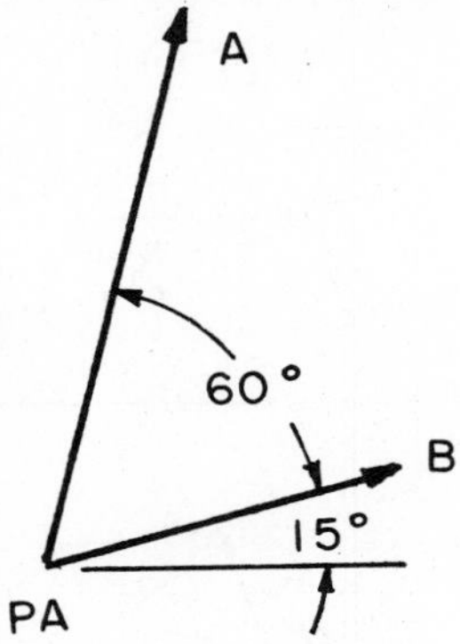

A

60°

B

15°

PA

A = 2500 LBS
B = 1500 LBS

P15-1

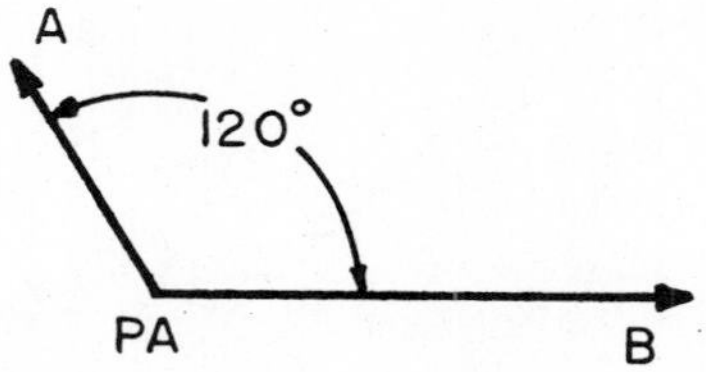

A

120°

PA

B

A = 500 LBS
B = 1000 LBS

P15-2

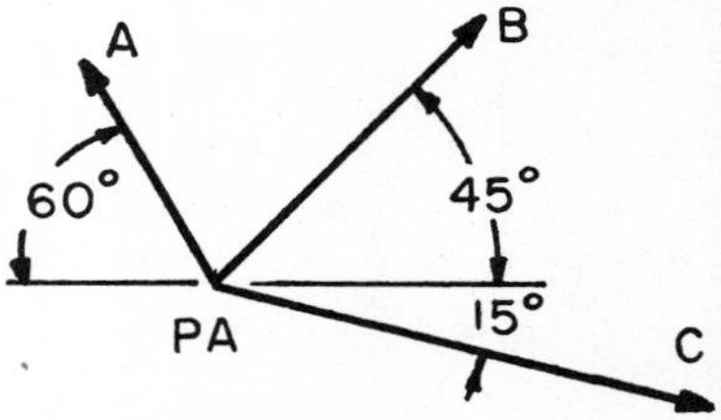

A

B

60°

45°

PA

15°

C

A = 1000 LBS
B = 1500 LBS
C = 2000 LBS

P15-3

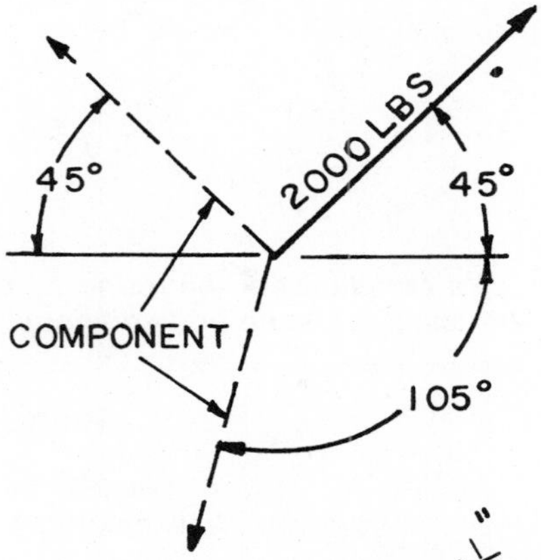

45°

2000 LBS

45°

COMPONENT

105°

P15-5

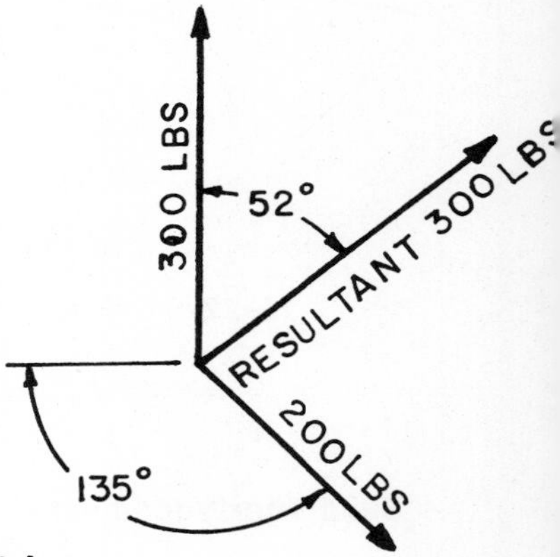

N

135°

225°

PLANE

75 MPH

350 MPH

WIND

P15-4

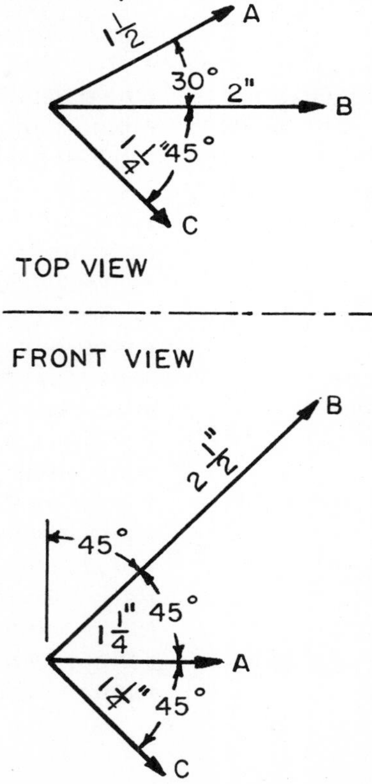

300 LBS

52°

RESULTANT 300 LBS

135°

200 LBS

P15-6

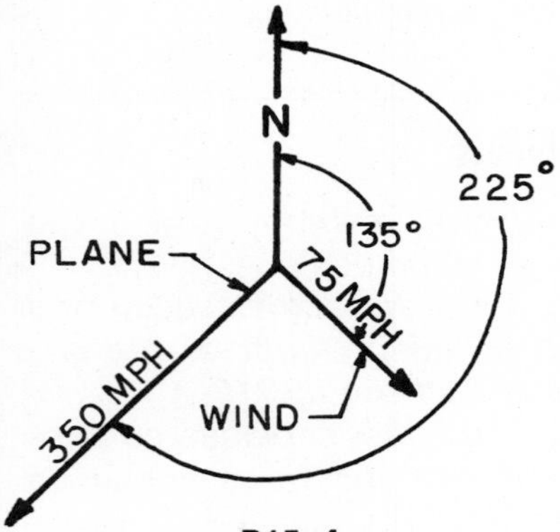

$1\frac{1}{2}$"

A

30°   2"   B

$1\frac{1}{4}$"   45°

C

TOP VIEW

FRONT VIEW

B

$2\frac{1}{2}$"

45°

$1\frac{1}{4}$"   45°

A

$1\frac{1}{4}$"   45°

C

P15-8

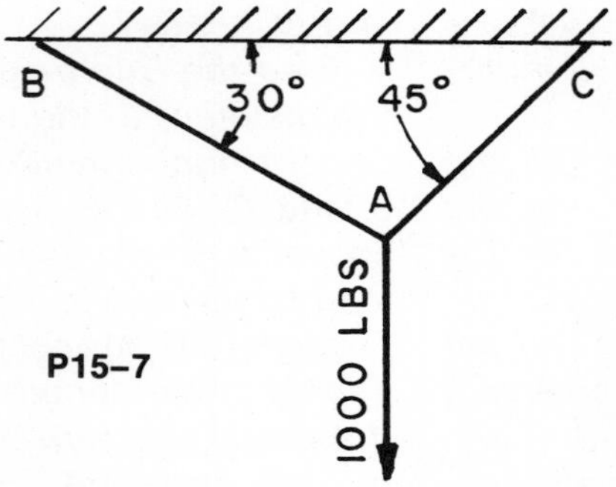

B

30°   45°

C

A

1000 LBS

P15-7

Section Five
Industrial Applications

Phillips Petroleum Co.

This refinery is used in the production of high octane motor fuels and distillates. Designing the maze of pipes required thousands of drawings.

Links in Learning

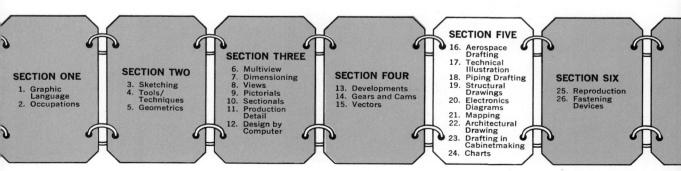

SECTION ONE

1. Graphic Language
2. Occupations

SECTION TWO

3. Sketching
4. Tools/Techniques
5. Geometrics

SECTION THREE

6. Multiview
7. Dimensioning
8. Views
9. Pictorials
10. Sectionals
11. Production Detail
12. Design by Computer

SECTION FOUR

13. Developments
14. Gears and Cams
15. Vectors

SECTION FIVE

16. Aerospace Drafting
17. Technical Illustration
18. Piping Drafting
19. Structural Drawings
20. Electronics Diagrams
21. Mapping
22. Architectural Drawing
23. Drafting in Cabinetmaking
24. Charts

SECTION SIX

25. Reproduction
26. Fastening Devices

Chapter 16

Drafting in the Aerospace Industry

The designing of aircraft and space vehicles is a difficult and complex process. It requires the services of a variety of engineers and technicians. These men must understand the theory of flight, aircraft design, and materials used in building aircraft. They must understand structural design and the forces an aircraft must withstand while in flight. See Fig. 16–1.

WHAT MAKES AN AIRCRAFT FLY?

The earth is surrounded by a sea of air about seven miles thick. Air has measurable weight, density, resistance, and temperature much the same as water. This must be remembered when applying the principles of aerodynamics to aircraft design. *Aerodynamics* is the study of air movements and the forces acting on an aircraft in flight.

At a low altitude the air is thicker and heavier than that at higher altitudes. A plane can fly slowly at low altitudes because of the greater density of the air. A plane can fly much faster at high altitudes because the resistance is less in the thinner air. However, an aircraft can reach an altitude so high that the air will be too thin to support it.

16–1. *The design of an aircraft is a complex process. The aircraft must withstand many forces. In addition to the forces of flight, this float plane must withstand the shock of a water landing.*

Cessna Aircraft Co.

394

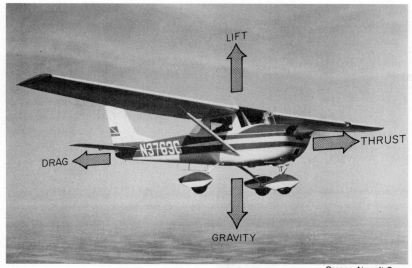

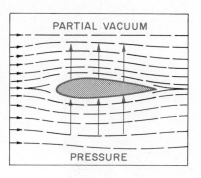

16–3. *Air flowing over the wing gives the aircraft lift.*

Cessna Aircraft Co.

16–2. *An aircraft is subjected to four forces—thrust, drag, lift, and gravity.*

Aerodynamic Forces. An aircraft is acted upon by four aerodynamic forces—thrust, lift, drag, and gravity. See Fig. 16–2.

Thrust. *Thrust* is a force that moves the aircraft through the air. The propeller makes a vacuum in front of itself. The vacuum sucks the aircraft along from the front. Thrust is also developed by jet engines.

Lift. *Lift* is the force that takes the aircraft off the ground and sustains it in flight. Lift is obtained by the wing. This is why wing design is so important.

As the aircraft moves, air flows over and under the wing. A partial vacuum is developed below and above the wing. This gives an upward pressure (lift). See Fig. 16–3.

Drag. *Drag* is a force which hinders the flight of the aircraft. Air resists and slows down objects moving against it. This resistance is called drag.

The shape of the object passing through the air influences how much drag it will have. Objects that are not streamlined, such as a round or flat surface, not only have drag from air hitting the surface but form a vacuum behind them. This serves as suction to retard forward movement. See Fig. 16–4.

A streamlined shape offers less drag. The air flows smoothly over it and no vacuum is formed at the rear.

Many things other than wing shape can cause drag. The fuselage, landing gear, and rivet heads are examples. The

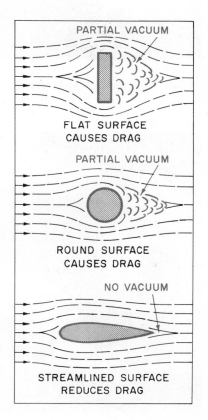

16–4. *The shape of the surface determines the amount of drag.*

National Aeronautics and Space Administration

16–5. *A one-ninth scale model of the XC-142 Tri-Service V/STOL tilt wing aircraft.*

National Aeronautics and Space Administration

16–6. *A wind tunnel with the XC-142 Tri-Service V/STOL tilt wing aircraft in position for testing.*

designer must consider the drag effect of every exterior part of an aircraft.

Gravity. *Gravity* is a force which tends to pull the aircraft to the earth. It must be overcome by lift. Aircraft designers attempt to reduce the force of gravity by reducing the weight of the aircraft.

KINDS OF DRAWINGS IN AEROSPACE INDUSTRIES

The aerospace industry uses drawings of all kinds. It takes thousands of drawings to completely describe an aircraft. These drawings are much the same as those in other industries. In common use are pictorial, detail, assembly, casting, forging, electrical, installation, machining, sheet metal, lofting, and many other kinds of drawings.

Major aircraft manufacturers usually have their own drafting standards. General industry standards are set by the Aerospace Standards, Society of Automotive Engineers, Inc.

This chapter deals with drafting problems and techniques that are somewhat unique to the aerospace industry. They are presented to give a sample of drafting experiences in the industry.

AIRCRAFT DESIGN

As stated before, the design of an aircraft is very complex and difficult, requiring the services of many engineers and draftsmen. Thousands of drawings and tests are made before the first experimental model is ready to be manufactured.

Decisions must be made concerning the performance expected. The size and load the aircraft must carry must be decided. Once the basic needs are decided, the design staff begins work. The aerodynamics group must provide the designers with the basic information they need to make preliminary scale drawings. These are usually $1/12$ or $1/16$ of actual size. From these drawings a scale mock-up and wind tunnel model are built. See Figs. 16–5 and 16–6.

The dimensions for the mock-up fuselage and the wing ordinates are furnished by the loft. The loft is the section of the design staff that ascertains the design dimensions. The ordinates are the dimensions above and below the chord of the wings. See Fig. 16–50.

After the wind tunnel model proves satisfactory, a full size mock-up is built. See Fig. 16–7. A *mock-up* is an exact model of the aircraft. It is made of a variety of materials. Much of it is wood since wood can be easily shaped. The mock-up is used to help with the design and installation of the systems such as electrical, hydraulic, and communications.

After the mock-up is approved by the engineering staff, the loft staff begins to make drawings for the actual parts. The parts are then produced and assembled into flying experimental aircraft. These parts are usually hand-made since changes will occur. The experimental aircraft is flown and design changes made until the end result has proven satisfactory. The loft is constantly involved in the redesign processes.

After the experimental model is approved and all changes made, the aircraft can go into production. The loft is involved in making certain that all templates for the production of parts are correct and accurate. The loft is responsible for keeping the manufacturing staff supplied with the latest drawings and templates. The loft notifies them of all changes decided upon after manufacturing begins.

Sometimes technical illustrations are made to help decide the locations of components. They also help show the relationship between the parts. See Fig. 16–8.

NOMENCLATURE

The aerospace industry has developed names for the parts of an aircraft. The draftsman must know these since they are the language of the industry. The more common names are shown in Figs. 16–9 through 16–13.

It is of special importance to notice that the word *aft* is used when referring to the *rear* of the aircraft. The *front* is *forward*.

The Boeing Company

16–7.   *A mock-up of a supersonic transport. It is full size and built of wood, steel, and aluminum. It took fifty tons of steel, 42,800 lineal feet of lumber, and 3,500 sheets of plywood to build. The tail is five stories high, and the overall length is about the size of a football field.*

M-2 LIFTING BODY RESEARCH VEHICLE

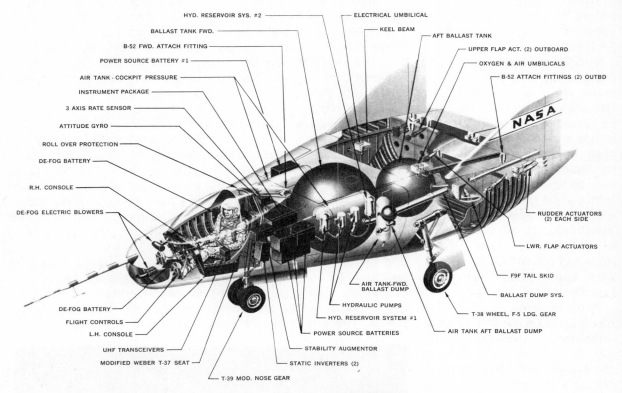

HYD. RESERVOIR SYS. #2 —
BALLAST TANK FWD. —
B-52 FWD. ATTACH FITTING —
POWER SOURCE BATTERY #1 —
AIR TANK - COCKPIT PRESSURE —
INSTRUMENT PACKAGE —
3 AXIS RATE SENSOR —
ATTITUDE GYRO —
ROLL OVER PROTECTION —
DE-FOG BATTERY —
R.H. CONSOLE —
DE-FOG ELECTRIC BLOWERS —

ELECTRICAL UMBILICAL
KEEL BEAM
AFT BALLAST TANK
UPPER FLAP ACT. (2) OUTBOARD
OXYGEN & AIR UMBILICALS
B-52 ATTACH FITTINGS (2) OUTBD

NASA

RUDDER ACTUATORS (2) EACH SIDE
LWR. FLAP ACTUATORS
F9F TAIL SKID
BALLAST DUMP SYS.
T-38 WHEEL, F-5 LDG. GEAR
AIR TANK AFT BALLAST DUMP

DE-FOG BATTERY —
FLIGHT CONTROLS —
L.H. CONSOLE —
UHF TRANSCEIVERS —
MODIFIED WEBER T-37 SEAT —
T-39 MOD. NOSE GEAR —

AIR TANK-FWD. BALLAST DUMP
HYDRAULIC PUMPS
HYD. RESERVOIR SYSTEM #1
POWER SOURCE BATTERIES
STABILITY AUGMENTOR
STATIC INVERTERS (2)

National Aeronautics and Space Administration

16-8. *A pictorial drawing of an experimental aircraft. This is useful in the design of the aircraft. It shows the location of the internal components.*

BASIC FORMING OPERATIONS FOR SHEET METAL PARTS

The draftsman must be familiar with the processes used to form sheet metal parts. The process to be used will sometimes influence his drawing.

Sheet metal parts are laid out on flat stock by following a template. A *template* is a pattern made to the exact size and shape of a part. It is drawn by the loftsman. It is very accurate.

The blank part is shaped by cutting, sawing, shearing, routing, or stamping it out on a press. It is cut to the exact size of the template.

The blank must then be formed to the contours provided by loft drawings when the part helps form an exterior contour of the aircraft. This could be a simple straight-line bend. See Fig. 16–14. These can be formed with a brake, press, or flanging machine.

More complex straight-line bends can be formed by rolling and drawing the blank.

A more difficult situation occurs when a blank must be flanged or formed along a curved line. Presses and flanging machines can be used. Sometimes the part is formed by hand, using blocks shaped to exact finished contour of the part.

Parts of cylindrical or conical shapes are formed on

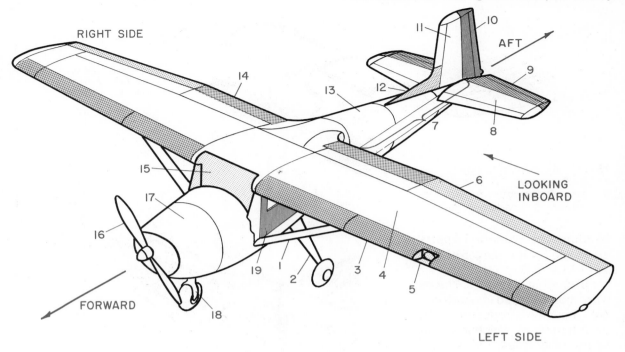

RIGHT SIDE

AFT

LOOKING INBOARD

FORWARD

LEFT SIDE

Cessna Aircraft Co.

16-9.   *Identification of major parts of an aircraft.*
1. Wing strut 2. Main landing gear 3. Leading edge 4. Wing 5. Landing light 6. Aileron 7. Fuselage tail cone 8. Stabilizer 9. Elevator 10. Rudder 11. Vertical fin 12. Dorsal 13. Fuselage 14. Flap 15. Windshield 16. Propeller 17. Engine cowl 18. Nose gear.

16-10.   *Major parts of vertical fin, rudder, and dorsal assembly.*
1. Dorsal, vertical fin 2. Skin, leading edge 3. Skin, vertical fin 4. Reinforcement assembly 5. Rib 6. Butt, rudder 7. Skin, rudder 8. Skin, trailing edge 9. Tip, rudder 10. Spar, rudder 11. Tip, vertical fin 12. Spar, assembly, vertical fin 13. Taillight.

Cessna Aircraft Co.

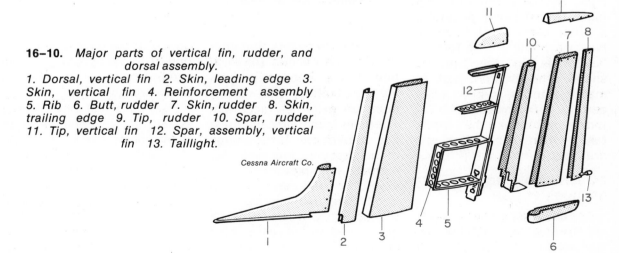

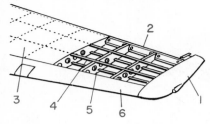

16–11. *Wing terminology.*
1. Tip assembly 2. Wing root
3. Skin 4. Stringer 5. Rib or
airfoil 6. Leading edge.

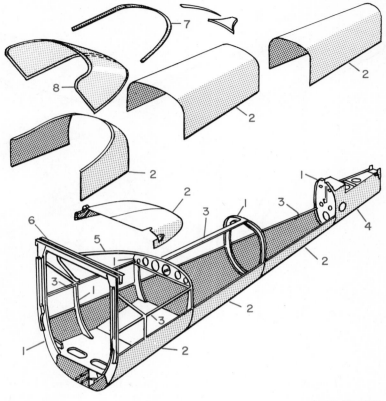

Cessna Aircraft Co.

16–12. *Fuselage Tail Cone Assembly.*
1. Bulkhead 2. Skin 3. Stringer 4. Aft tail cone assembly
5. Channel—rear window 6. Spar, rear door post 7. Retainer,
window 8. Rear window.

bending rolls. Sometimes parts are made by spinning. See Fig. 16–15A and B.

Parts that have surfaces with complicated curvatures are usually produced with drop hammers, mechanic and hydraulic presses, and stretching presses. The parts are usually formed with dies. *Dies* are metal forms machined to the shape of the finished part they are to form. See Fig. 16–16. The die is made to match the contours on the loft drawings. The dies and drawings must be extremely accurate.

Other parts are formed by twisting. Sometimes a form is handmade to the finished shape. The twisted part is matched to the form to make certain it is correct. The information to make the form is taken from the loftsman's drawings.

FASTENING METHODS IN AIRCRAFT DESIGN

The usual fastening methods are riveting, screwing, bolting, spot welding, and joining with adhesives. In some companies the loftsman shows the location of holes needed for riveting and bolting. This usually must be extremely accurate. The holes in mating parts must line up perfectly. If the degree of accuracy is too great for normal manufacturing processes, the holes may be placed in one part and omitted from the mating part. The two parts are placed in the assembled position and the mating holes drilled in the assembled position. If this procedure is to be used, it must be indicated on the drawings.

Fig. 16–17 illustrates one system for locating holes in sheet metal parts. When dimensioning the hole patterns of mating parts on separate drawings, fully dimension the hole pattern on one drawing. On the drawing with the mating part, show the hole pattern but dimension only one hole,

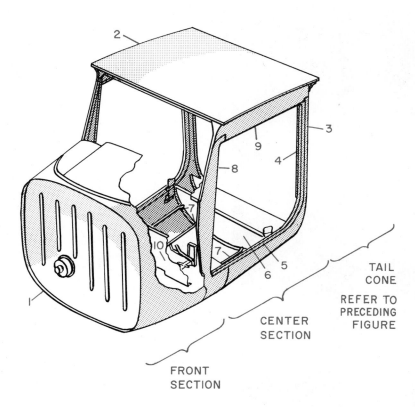

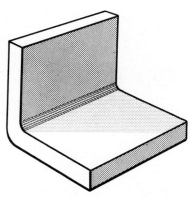

16–14. *Simple straight-line bend.*

16–13.

1. Firewall 2. Cabin, top skin 3. Rear door post assembly 4. Jamb, door rear 5. Sill, door lower 6. Bulkhead assembly 7. Stiffener, belly skin 8. Door post 9. Rib, fuselage center section 10. Bulkhead.

Cessna Aircraft Co.

giving the distances to the edge of the part. Then give the number of the drawing that shows the fully dimensioned hole pattern.

Usually the distance between each hole is not dimensioned. The overall distance from the center of the first hole to the center of the last hole is given. Following this dimension, in parentheses, the number of equal spaces is given. In Fig. 16–17 the distance is

shown as 2.50 (2S). This means there are three holes equally spaced.

Fig. 16–18 gives recommended spacing standards for distances between rivets.

Notice in Fig. 16–17 how the hole size and rivet are specified with a note.

A standard minimum safe distance for rivet holes from the edge of sheet metal parts is 1½ times the hole diameter measured from the center of

the hole. More detailed data is shown in Fig. 16–19. To read this table, use the following example. The center of a $3/32$-diameter rivet with a protruding head should be .19 inches from the edge.

The diameter of rivets used in aircraft assembly has been determined by testing and experience. Fig. 16–20 gives recommended rivet diameters for various sheet material thicknesses and casting thicknesses.

While a standard system of welding symbols is in general use, some aircraft companies have developed their own. One system is shown in Figs. 16–21 and 16–22.

The weld symbol is given, plus a note giving the company specifications for the weld. This is a simpler system than that in use in other industries.

Another thing that must be indicated on a drawing is *dimpling*. This is a conical dent in the metal. It is much like a countersunk hole. Metal is machined away to make the hole.

16–16. *Forming sheet metal with dies.*

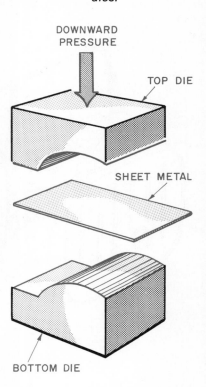

Cessna Aircraft Co.

16–15A. *The metal blank is being spun against a form the shape of the finished part.*

Die open with sheet metal in position.

Cessna Aircraft Co.

16–15B. *The part after spinning is completed.*

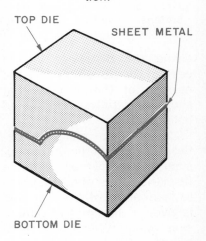

Die closed with sheet metal formed.

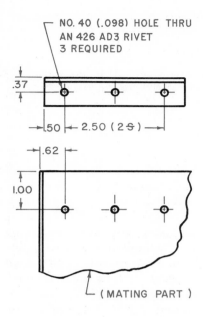

NO. 40 (.098) HOLE THRU
AN 426 AD3 RIVET
3 REQUIRED

.37

.50 |← 2.50 (2 S) →|

.62

1.00

(MATING PART)

Cessna Aircraft Co.

16–17. *Locating holes in sheet metal parts.*

16–19.
Rivet Edge Distance*

Rivet Dia.	Protruding Head**	Flush Head**
3/32	.19	.25
1/8	.25	.31
5/32	.31	.34
3/16	.38	.41
1/4	.50	.50

*Stress requirements may supersede these values.

**Minimum distance from edge of metal in inches.

Cessna Aircraft Co.

16–18.
Rivet Spacing

Rivet Dia.	Desirable* for Production	Minimum				
		Machine Riveting		Hand Riveting		Rivets Through Castings
		Protruding Head	Flush	Protruding Head	Flush	
3/32	.75 or more	.49	.49	.32	.35	.38
1/8	1.00 or more	.53	.53	.38	.41	.50
5/32	1.25 or more	.56	.56	.45	.50	.62
3/16	1.50 or more	—	—	.54	.59	.75
1/4	1.50 or more	—	—	.67	.78	1.00

*Stress requirements may supersede these values.

Cessna Aircraft Co.

Recommended Rivet Diameters for Various Material Thicknesses

Rivet Dia.	Sheet Thickness*	Casting Thickness
$3/32$	.020 or more	The rivet diameter, unless type "A" (1100 aluminum) rivets are used, shall not be greater than the thickness of that part of the casting being riveted.
$1/8$	.025 or more	
$5/32$	.032 or more	
$3/16$	.040 or more	
$1/4$	.063 or more	

*When materials of varying gages are joined by riveting, the outside sheets should equal or exceed these values.

Cessna Aircraft Co.

See Fig. 16–23. The purpose is to get the head of a rivet or bolt flush with the surface of the part.

A large number of standard, mass-produced fastening devices are made especially for the aerospace industry. The designer and draftsman should use these whenever possible. It is expensive to manufacture a special fastening device that is used for a single purpose.

DRAFTING TECHNIQUES

Drafting in the aerospace industry is much the same as in other industries. Following are a few drafting techniques of special importance.

Since the loft drawings and templates must be extremely accurate, any pencil drawings are usually made with a 9H pencil. A chisel-shaped lead is used for long lines. A sharp cone-shaped point is used for details. All lines are very thin. The alphabet of lines is not used on loft drawings and templates.

Long straight lines are drawn using a metal straight edge. Straight edges are made as long as 12 feet. Points along

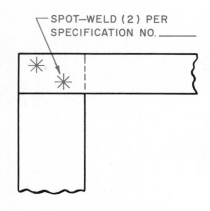

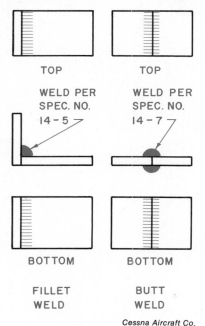

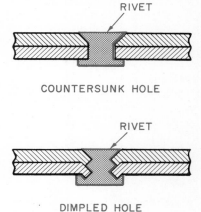

Cessna Aircraft Co.

16–21. *System for indicating spot-welds.*

Cessna Aircraft Co.

16–22. *System for indicating fillet and butt welds.*

16–23. *Dimpled holes are used to set rivet heads flush with the metal surface.*

straight lines are very carefully located. The loftsman must work to tolerances of at least ±0.03. Some jobs require tolerances of ±0.005.

Curved lines are drawn by connecting the points laid out with a spline and ships curve. The spline is a rectangular flexible strip. It is held in place with spline weights. See Fig. 16–24.

In some companies the spline and metal straight edge have been replaced by the computer plotter. Straight and curved edges are drawn by a tape-controlled drafting machine. The data is punched on tape, which then controls the plotter. See Chapter 12, Computers in Design and Drafting, for additional information.

If a layout is to be drawn on a metal sheet, the lines are marked with a flat scribe. A scribe is a very sharp, metal tool. The side touching the straight edge is flat. The point is usually kept about .006 wide.

Viewing the Aircraft. Whenever possible, all parts or assemblies are drawn as if the observer were at one of the following positions: (1) the rear of the aircraft looking forward, (2) the left side of the aircraft looking inboard, or (3) the top of the aircraft looking downward. See Fig. 16–25.

All drawings are assumed to be from one of these directions. If views are made from

Cessna Aircraft Co.

16–24. *A loftsman using a spline and weights to lay out a curve on a loft.*

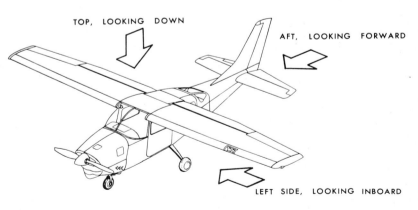

Cessna Aircraft Co.

16–25. *Aircraft drawings are made looking from the rear of the aircraft forward, the left side looking inboard, or the top looking down.*

16–26.

Drawing and Form Numbering

Digits 1 & 2 (Model)	Digit 3 (Major Group)	Digit 4 (Subdivision of Group)	Digit 5, 6 & 7 (Unassigned)	Dash No.'s
03–190 SERIES 04–120, 140 & 150 05–170 SERIES 06–305 07–180 SERIES 08–310 & 320 09–325 10–321 11–620 12–200 SERIES 13–OMITTED 14–336 15–337 16–188 17–172J 18–327	0–COMPLETE AIRCRAFT	0–Airplane Assy. 1–Opt. Equip. List 2–Jig Point 3–Three-View 4–Exterior Styling 5–Nameplates-Placards	Not assigned to any particular group or sub-assembly. Individual sets may be used to distinguish major model changes within the same model group. 0700000–182 0700600–182E	Assigned in normal sequence.
	1–FUSELAGE	0–Entire Fuselage 1–Center Section 2–Aft Section 3–Fwd. Section 4–Seats 5–Upholstery 6–Fuel System 7–Doors		
	2–WING	0–Entire Wing 1–Strut 2–Structure Assy. 3– 4–Ailerons 5–Flaps 6–Fuel 7–Booms		
	3–EMPENNAGE	0–Entire Empennage 1–Fin 2–Stabilizer 3–Rudder 4–Elevator		
	4–ALIGHTING GEAR	0–Entire Alighting Gear 1–Main Gear 2–Auxiliary Gear 3–Nose Gear		
	5–ENGINE COMPARTMENT	0–Engine Instl. 6–Fuel & Oil 1–Engine Mount Systems 2–Cowling 3–Firewall 4–Exhaust System 5–Baffles		
	6–CONTROLS	0–Controls Instl. 1– 2–Wing Flap Controls		
	7–ELECTRICAL & ELECTRONICS	0–Complete Instl.		1–99–Complete Instl. 200–Cable Assy. 300–Transmitter 400–Power Supply 500–Cable Instl. 600–Antenna 700–Mic. & Spkr.
	8–HYDRAULIC	0–Complete Instl. 1–Main Gear & Doors 2–Brakes 3–Nose Gear & Doors		
	9–SPARES			

Cessna Aircraft Co.

another direction, the drawing should be clearly labeled. An example would be "view looking at right side."

Numbering of Drawings. Aircraft design and production requires hundreds of drawings. The problem of keeping a record of these drawings is great. It is important to be able to file and find a drawing rapidly. Various numbering systems are in use. One system of numbering uses a seven-digit number to identify each drawing. The *first two digits* indicate the model number assigned to the aircraft being designed. The *third digit* indicates a major assembly, as a wing. The *fourth digit* indicates a minor assembly, as a flap on the wing. The last *three digits* are assigned by the drafting supervisor to specific detail drawings. Fig. 16–26 gives the details for this system.

Examine the drawing number shown in Fig. 16–27. It is 0412112. This number indicates it is for aircraft models 120, 140, and 150. It refers to the aft section of the fuselage. It is detail drawing number 112. Sometimes following the seven-digit identification number is a dash number.

Dash Numbers. A dash number following the seven-digit production drawing number is used when a drawing has more than one piece shown. Each piece is numbered so it can be

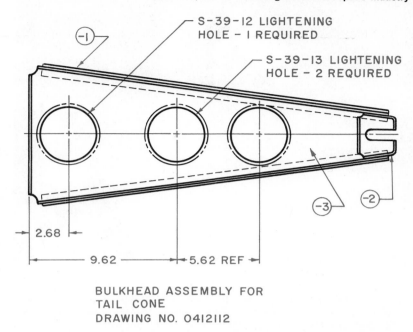

BULKHEAD ASSEMBLY FOR
TAIL CONE
DRAWING NO. 0412112

16–27.   *A layout drawing for a bulkhead assembly for a tailcone.*

easily identified. The number is usually placed on the drawing inside a circle. It is connected by a leader to the part it indicates. It can be placed on the drawing as a note. See Fig. 16–27. This is an assembly of three parts.

Each part is identified by the drawing number plus the dash number. In Fig. 16–27 the first part would be called 0412112–1. The second part is 0412112–2. The third part is 0412112–3.

Dash numbers are used on production drawings and lofts.

LAYOUT DRAWINGS

The individual parts of an aircraft are made from working drawings. Before these can be made, the complete unit involved must be designed. This design work is done on a layout drawing. A *layout drawing* is an accurate drawing of the parts used to solve, evaluate, or check the solution of a design problem. An example would be the mechanism to operate the wing flaps. Some layouts are drawn to scale. Others are full size and can be 20 or 30 feet long. They are used to give essential information about basic contours, design requirements, peculiar design conditions, weight limitations, and maintenance and production requirements. They relate these to overall configuration, operation, adjustment, clearance, tolerances, materials, assembly, and installation.

Cessna Aircraft Co.

16–28. *The aluminum sheet to form a template is being coated with a light-sensitive material. The vacuum frame for exposing the loft to this surface can be seen in the left rear with the lid open.*

Layouts are used to define the original design ideas. They provide the basis for drawing detail, assembly, and installation drawings. The layout is a record of the latest design information. It frequently requires revision.

Types of Layouts. There are two basic types of layouts, design layouts and checking layouts.

Design layouts are prepared to determine the detail, assembly, or installation arrangement. They define design requirements and the best solution of the design problem. Design layouts are used to develop or check paths of motion and motion analysis. They are used to check or develop the geometry of alighting gears, control surfaces, controls, doors, and other moving components.

Design layouts are also used to determine the possible interferences, clearances, and tolerances the design will allow under normal and abnormal operating conditions.

Checking layouts are made to check on the accuracy of the design. They are also used to check the proper relationship of mating parts, clearances, sizes, developed lengths, and relation of parts as dimensioned on detail, assembly, or installation layouts. They show the parts in an assembled position. They are usually full size.

Checking layouts are made after the detail, assembly, and installation drawings are made.

This is done so design and drawing errors can be discovered before the parts are manufactured.

Drawing the Layout

1. All the factors involved in the unit under design must be carefully recorded.

2. Freehand sketches are made of the proposed design solution. These can be pictorial or orthographic sketches.

3. The amount of space necessary to make the drawing must be determined. This includes the main layout, any special views needed, and design data.

4. Layouts are given drawing numbers according to the standard practice in the industry.

5. Layouts should be full size whenever possible. If parts are very small, they can be drawn larger than full size.

6. No parts list is placed on layout drawings.

7. Callouts must be complete. Callout is a frequently used shop term that refers to a note containing information such as material, finish, and processes. This information is also shown on the parts list of the production drawings.

8. A layout can be on several sheets.

9. Standard parts and materials should be used whenever possible. Examples are screws, washers, bolts, and rivets.

10. The layout need only be complete enough to serve its intended purpose. Each part

shown need not be complete in every detail. When the layout is used as a guide for making the production drawings, minor details can be designed by the draftsman.

11. A layout should include all details and notes basic to the design and operation of each part shown. Special study should be made of possible interference points, production and assembly operations required, and the correctness of the design.

12. Unusual design considerations should be clearly noted on the layout. This could include data such as special clearances, unusual tolerances, or stress requirements.

13. If the order of assembly of parts is not perfectly clear, this should be noted.

14. All critical dimensions and tolerances should be given. Each part is not dimensioned. It is accurately drawn.

15. If the layout is of only a portion of a unit, it should include some major item to be used as a reference base to indicate the proper location of the partial unit.

16. Placement of views is the same as standard practice for production drawings.

17. Layouts must be extremely accurate. Usually a sharp 9H pencil is used.

18. Layout revisions follow the same practice the industry uses for production drawings.

Fig. 16–27 shows an example of a layout drawing for a

bulkhead assembly for a tail cone.

TEMPLATE

As stated earlier in this chapter, a template is a pattern used as a guide for the manufacture of parts. It is usually metal. One type used has a thin, lead coating. The surface is painted. When the template is laid out, the scriber cuts through the paint to the lead. This leaves a bright line showing.

Another method for making templates is a photographic process. A sheet of aluminum is coated with a material sensitive to light. See Fig. 16–28. The original loft drawing of the part is placed next to the sensitive surface. They are placed in a vacuum frame and exposed to a light. The metal sheet is then treated with a developer.

The lines from the loft become visible on the sensitive metal surface. The aluminum sheet is sprayed to fix permanently the image on the surface.

The sheet is then cut to the shape of the lines on the surface. It is filed to the exact size. The line printed on the aluminum is half removed. The center of the line is actually the finished edge of the template.

The process of using a sensitized aluminum sheet is much faster than scribing the loft on a lead-coated sheet and usually more accurate.

Fig. 16–29 shows a finished template.

MASTER PLASTERS

As an aid to developing the dies for forming of compound curves in the sheet metal outside surfaces of the aircraft, master plasters are sometimes made. Compound curves are found in the wing tips, stabilizer tips, wheel farings, and fuselage surfaces.

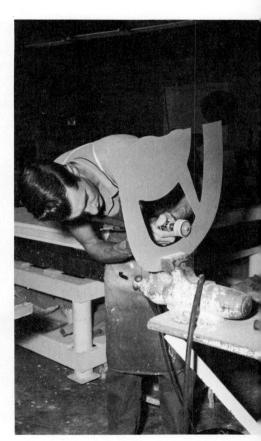

Cessna Aircraft Co.

16–29. *A metal template being ground to final shape.*

1. The core of the plaster model is made from threaded steel rods. Templates are made of the contour and are fastened to the rods every six inches. In the photo above, notice that one template has been fastened to the rods.

2. Wire is formed around steel core rods to give the model shape and serve as a base for the plaster.

3. The model is built by adding layers of plaster. The plaster is smoothed so the edges of the templates show. The final shape is obtained this way.

4. This is the finished plaster model of the forward fuselage structure assembly. It is accurate to +0.005.

Cessna Aircraft Co.

16–30. How a full-size plaster model is built. This is the forward fuselage structure that houses the engine.

Plasters are made after the exterior of the aircraft has been designed. After the plasters are built, they are examined visually for contour faults.

Fig. 16–30 shows a master plaster being made. It is built with steel rods as the core. Metal templates of the contour are secured in place about every six inches. Wire mesh is formed around these to hold the plaster. The final form is made by adding layers of plaster. These are worked to the final shape.

The master plaster gives the die makers an exact example to use as they make the dies to stamp the sheet metal parts.

PROBLEMS IN AIRCRAFT SHEET METAL LAYOUT

The techniques of sheet metal layout and development presented in Chapter 13, Developments and Intersections, are used a great deal in aircraft drafting. Following are selected techniques that are of special importance in this industry.

Straight-Bend Flat Pattern.
One type of drawing made by aircraft draftsmen is the development of a flat pattern for a sheet metal part that is to be formed with straight bends. Fig. 16–31 gives a detail of a part with a 120-degree straight bend.

The size of the flat pattern used to make this part is the sum of each side from the end

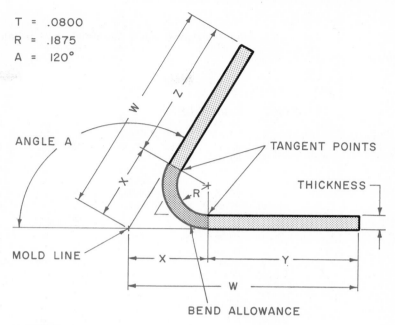

$T = .0800$
$R = .1875$
$A = 120°$

ANGLE A

TANGENT POINTS

THICKNESS

MOLD LINE

BEND ALLOWANCE

Y AND $Z = W - X$

WIDTH OF FLAT PATTERN $= Y + Z +$ BEND ALLOWANCE.

16–31. *How the size of a template with a straight bend is found. This example is a 120-degree bend.*

to the tangent point (distances Y and Z, Fig. 16–31) plus the material needed to form the bend, called *bend allowance*.

Locating the Tangent Point Dimension. In order to find the size of a flat pattern, it is necessary to know the distance from the mold line to the tangent point. This is distance X, Fig. 16–31. The *mold line* is the point at which the sides of the part would meet if extended.

The distance from the mold line to the tangent point is found using the formula $(T + R)$ times the tangent of

one-half Angle A. T is the thickness of the metal. R is the radius of the curve. Angle A is the angle of bend.

The following is based on Fig. 16–31.

$X = (T + R)$ tan ½ angle A
$X = (.080 + .1875)$ tan 60°
$X = .2675 \times 1.7320$
$X = .4633$

Distances Y and Z are found by subtracting distance X from the overall length, distance W.

FIGURING BEND ALLOWANCE

Bend allowance is found by using a bend-allowance chart.

16–32.

Bend Allowance* For 1 Degree of Bend in Nonferrous Metal

Radius	Thickness of Material											
	.016	.020	.025	.032	.040	.051	.064	.072	.081	.091	.102	.125
1/32	.00067	.00070	.00074	.00079								
1/16	.00121	.00125	.00129	.00135	.00140	.00145	.00159	.00165				
3/32	.00176	.00179	.00183	.00188	.00195	.00203	.00213	.00220	.00226	.00234	.00243	
1/8	.00230	.00234	.00238	.00243	.00249	.00258	.00268	.00274	.00281	.00289	.00297	.00315
5/32	.00285	.00288	.00292	.00297	.00304	.00312	.00322	.00328	.00335	.00343	.00352	.00370
3/16	.00339	.00342	.00347	.00352	.00358	.00367	.00377	.00383	.00390	.00398	.00406	.00424
7/32	.00394	.00397	.00401	.00406	.00412	.00421	.00431	.00437	.00444	.00452	.00461	.00479
1/4	.00448	.00451	.00456	.00461	.00467	.00476	.00486	.00492	.00499	.00507	.00515	.00533
9/32	.00501	.00506	.00510	.00515	.00521	.00530	.00540	.00546	.00553	.00561	.00570	.00588
5/16	.00557	.00560	.00564	.00570	.00576	.00584	.00595	.00601	.00608	.00616	.00624	.00642
11/32	.00612	.00615	.00619	.00624	.00630	.00639	.00649	.00655	.00662	.00670	.00679	.00697
3/8	.00666	.00669	.00673	.00679	.00685	.00693	.00704	.00710	.00717	.00725	.00733	.00751
13/32	.00721	.00724	.00728	.00733	.00739	.00748	.00758	.00764	.00771	.00779	.00787	.00806
7/16	.00775	.00778	.00782	.00787	.00794	.00802	.00812	.00819	.00826	.00834	.00842	.00860
15/32	.00829	.00833	.00837	.00842	.00848	.00857	.00867	.00873	.00880	.00888	.00896	.00915
1/2	.00884	.00887	.00891	.00896	.00903	.00911	.00921	.00928	.00935	.00943	.00951	.00969
17/32	.00938	.00942	.00946	.00951	.00957	.00966	.00976	.00982	.00989	.00997	.01005	.01023
9/16	.00993	.00996	.01000	.01005	.01012	.01020	.01030	.01037	.01043	.01051	.01058	.01078
19/32	.01047	.01051	.01055	.01060	.01065	.01073	.01083	.01091	.01098	.01105	.01114	.01132
5/8	.01102	.01105	.01109	.01114	.01121	.01129	.01139	.01146	.01152	.01160	.01170	.01187
21/32	.01156	.01160	.01164	.01170	.01175	.01183	.01193	.01200	.01207	.01214	.01223	.01241
11/16	.01211	.01214	.01218	.01223	.01230	.01238	.01248	.01254	.01261	.01269	.01276	.01296
23/32	.01265	.01268	.01273	.01276	.01283	.01291	.01301	.01309	.01316	.01322	.01332	.01350
3/4	.01320	.01323	.01327	.01332	.01338	.01347	.01357	.01363	.01370	.01378	.01386	.01405
25/32	.01374	.01378	.01381	.01386	.01392	.01401	.01411	.01418	.01425	.01432	.01441	.01459
13/16	.01429	.01432	.01436	.01441	.01447	.01456	.01466	.01472	.01479	.01487	.01494	.01514
27/32	.01483	.01486	.01490	.01494	.01501	.01509	.01519	.01527	.01534	.01540	.01550	.01568
7/8	.01538	.01541	.01545	.01550	.01556	.01565	.01575	.01581	.01588	.01596	.01604	.01623
29/32	.01592	.01595	.01599	.01604	.01611	.01619	.01629	.01636	.01643	.01650	.01659	.01677
15/16	.01646	.01650	.01654	.01659	.01665	.01674	.01684	.01690	.01697	.01705	.01712	.01732
31/32	.01701	.01704	.01708	.01712	.01718	.01727	.01737	.01745	.01752	.01758	.01768	.01786
1	.01755	.01759	.01763	.01768	.01774	.01783	.01793	.01799	.01806	.01814	.01823	.01841

*In inches.

Cessna Aircraft Co.

See Fig. 16–32. This chart gives the amount of material needed to bend a sheet metal part through an angle of one degree. The left column gives the radius of the bend. Across the top is the thickness of the metal.

The following problems will show how the actual size of a pattern is found using bend allowance.

Bend Allowance for 90-Degree Bends. Fig. 16–33 shows a sheet metal part 0.125 inch thick, formed with a radius of 1/4 inch through a 90-degree straight bend. The flat sides are each 1 1/2 inches long to the tangent point. These plus the length of the metal in the curved corner give the total width of the flat pattern. The curved corner runs from tangent point A to tangent point B.

To find the amount of material needed to form the curved section, look at the bend-allowance chart shown in Fig. 16–32. The bend allowance for metal 0.125-inch thick with a radius of 1/4 inch is 0.00533 inches per one degree bend. Since the bend is through 90 degrees, the length of the curved section is 90 times 0.00533 inches or 0.47970 inches.

The total width of the pattern is 1½ inches + 0.47970 inches + 1½ inches or 3.47970. See the flat pattern shown in Fig. 16–33.

Bend Allowance for Open-Bevel Bends. An *open-bevel bend* is one forming an obtuse angle. See Fig. 16–34. The bend allowance and total size of the pattern is found in the same way as the 90-degree bend. In Fig. 16–32, the bend allowance is 0.00195 inches per degree of bend. The amount of material needed is 0.00195 inches times 60 or 0.11700 inches. The total width of the pattern is 1.00000 + 0.11700 + 1.50000 or 2.61700 inches.

Bend Allowance for Closed-Bevel Bends. A *closed-bevel bend* is one forming an acute angle. See Fig. 16–35. The bend allowance in this problem is 0.00390. The amount of material needed is 0.00390 inches times 120 or 0.46800 inches. The total width of the pattern is 2.25000 + 0.46800 + 1.50000 or 4.21800 inches.

Bend Radii. The radii to be used when forming sheet metal parts must be carefully chosen. Tests have shown the most practical radii for the type of material being formed. The draftsman must always refer to a table of bend radii. See

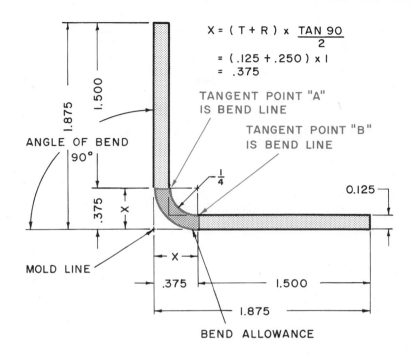

$$X = (T + R) \times \frac{TAN\ 90}{2}$$
$$= (.125 + .250) \times 1$$
$$= .375$$

TANGENT POINT "A" IS BEND LINE

TANGENT POINT "B" IS BEND LINE

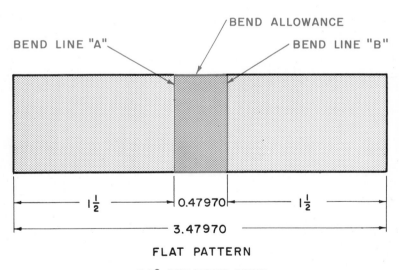

FLAT PATTERN

90° STRAIGHT BEND

16–33. *How to find the bend allowance for a 90-degree straight bend.*

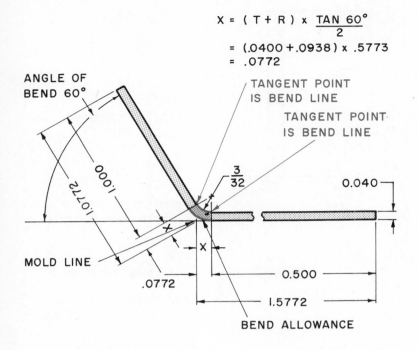

$$X = (T + R) \times \frac{TAN\ 60°}{2}$$
$$= (.0400 + .0938) \times .5773$$
$$= .0772$$

ANGLE OF
BEND 60°

TANGENT POINT
IS BEND LINE

TANGENT POINT
IS BEND LINE

1.0772

1.000

$\frac{3}{32}$

0.040

MOLD LINE

X

X

.0772

0.500

1.5772

BEND ALLOWANCE

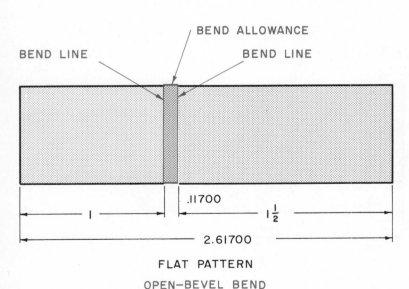

BEND ALLOWANCE

BEND LINE

BEND LINE

.11700

1

1$\frac{1}{2}$

2.61700

FLAT PATTERN

OPEN—BEVEL BEND

16-34. *How to find bend allowance for an open-bevel bend.*

Fig. 16–36. The largest permissible radii should be used whenever possible. This is easier to form and reduces the chances of the metal cracking.

RELIEF HOLES

When sheet metal is formed, it has a tendency to kink or tear at the corners where the bends come together. It is standard practice to relieve the strain at the corner by cutting holes. This prevents the metal from cracking or tearing at this point.

Various companies have different ways of deciding the size of the relief hole. A common plan is shown in Fig. 16–37.

Details for drawing a general use relief hole are shown in Fig. 16–38. If more material is needed on the flange, the layout shown in Fig. 16–39 is used. Sometimes extra material is necessary for riveting purposes.

How to Draw General Use Relief Holes

1. Lay out the template, including inner and outer bend lines. See Fig. 16–38.

2. Draw a line connecting the points where the inner and outer bend lines intersect. The center of the relief hole is on this line.

3. Measure 0.06 inches along the center line from the intersection of the inner bend lines. This is the center of the relief hole.

4. Draw the relief hole. Swing the arc until it reaches the tangent point run from the center of the hole.

5. Draw a perpendicular from the edge of the template to this tangent point.

How to Draw Relief Holes Giving More Flange Material. Follow steps 1–3 as given for general relief holes.

4. Draw relief hole. See Fig. 16–39.

5. Draw lines perpendicular to the sides of the template running to the point of intersection of the inner bend lines. The arc of the hole ends on these lines.

LIGHTENING HOLES

Some parts of the aircraft have sections of metal removed to reduce the weight. These are usually round sections. They are called lightening holes. Since the metal removed was not performing a useful purpose, the lightening holes do not reduce the structural strength of the part. They have flanges around the edges to add stiffness to the part.

Fig. 16–40 shows how lightening holes are drawn in elevation and side views.

The size and location of lightening holes are usually decided by the engineering staff. The draftsman must be familiar with company standards and notes so he can locate and draw the holes.

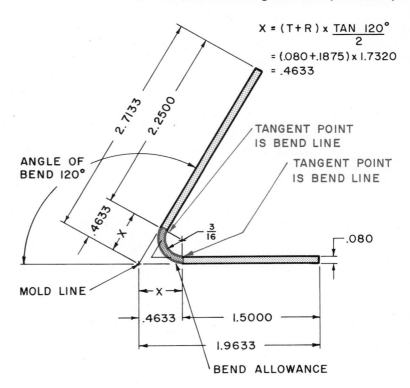

$$X = (T+R) \times \frac{\text{TAN } 120°}{2}$$
$$= (.080 + .1875) \times 1.7320$$
$$= .4633$$

ANGLE OF BEND 120°

2.7133

2.2500

.4633

X

MOLD LINE

TANGENT POINT IS BEND LINE

TANGENT POINT IS BEND LINE

$\frac{3}{16}$

.080

X

.4633   1.5000

1.9633

BEND ALLOWANCE

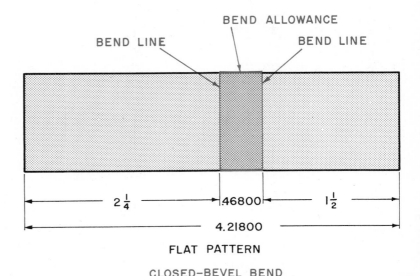

BEND ALLOWANCE

BEND LINE   BEND LINE

$2\frac{1}{4}$ 46800 $1\frac{1}{2}$

4.21800

FLAT PATTERN

CLOSED–BEVEL BEND

16–35.   *How to find bend allowance for a closed-bevel bend.*

16-36.
Standard and Minimum Bend Radii for Sheet Stock

Material	.012 S	.012 M	.016 S	.016 M	.018 S	.018 M	.020 S	.020 M	.025 S	.025 M	.028 S	.028 M	.032 S	.032 M	.036 S	.036 M	.040 S	.040 M	.050 S	.050 M	.063 S	.063 M	.071 S	.071 M	.080 S	.080 M	.090 S	.090 M	.100 S	.100 M	.125 S	.125 M	.160 S	.160 M	.190 S	.190 M	.200 S	.200 M	.224 S	.224 M
ALUMINUM 2024-0, 1100½ H, 6061 SW or ST	.06	.03	.06	.03			.06	.03	.06	.03			.06	.06			.09	.06	.12	.06	.12	.09	.16	.12	.19	.12	.19	.12	.22	.16	.25	.19	.38	.25	.50	.34	.50	.34		.47
ALUMINUM 1100-0, 3003-0, 6151-0, 6061-0	.06	.06	.06	.03			.06	.06	.06	.03			.06	.06			.06	.06	.06	.06	.09	.06	.12	.09	.12	.09	.12	.09	.16	.12	.16	.12	.19	.16	.25	.19	.28	.19		.19
ALUMINUM 2024-T3 or T4	.06	.06	.06	.06					.09	.06			.12	.09			.16	.09	.19	.12	.22	.16	.28	.22	.34	.25	.38	.28	.44	.34	.53	.44	.75	.66	1.00	.84		1.25		1.25
ALUMINUM 7075-0 or SW			.03				.03		.06				.06				.06		.06		.12	.12			.19	.19			.22	.22	.28	.28	.44	.44	.75	.75	.56		.75	.75
ALUMINUM 7075-T6			.09				.12		.12				.16				.19		.25				.34		.38		.50				.90		1.2				1.3		1.5	1.25
STEEL					.03	.03					.06	.03			.06	.06			.06	.06	.12	.09	.16	.12	.16	.12			.25	.16			.31	.25	.38	.28	.44	.34	.50	.38
MAGNESIUM FS-1 (AZ31) SOFT HOT			.06				.06		.06				.06				.09		.09		.16		.16		.19		.19		.25		.25		.38				.38			
MAGNESIUM FS-1 (AZ31) SOFT COLD																	.22		.25		.31		.38		.45		.45		.50		.62		.75				1.00			
MAGNESIUM FS-1 (AZ31) HARD HOT			.09				.09		.12				.16				.19		.25		.31		.38		.45		.45		.50		.62		.75				1.00			
MAGNESIUM FS-1 (AZ31) HARD COLD			.19				.19		.25				.31				.38		.50		.62		.85		.85		1.00		1.00		1.25		1.62				1.88			

S = Standard radius.
M = Minimum radius.

M = Minimum radius.
All bend radii apply to inside radius.

Cessna Aircraft Co.

Relief Hole Radii

Metal Thickness	Minimum Radius
.025 – .040	5/32
.040 – .064	3/16
.064 – .091	1/4

16-37. *One way to size relief holes.*

Lightening holes are indicated on a drawing by a note. The following example is based on the S–39 lightening hole standard of Cessna Aircraft Company. In Fig. 16–40 the note reads S–39–8 lightening hole. This means Cessna standard S–39 dash number 8 lightening hole. The data for this standard is shown in Fig. 16–41.

The left column of the table is the dash number. The "D" column is the diameter of the hole at the mold line. The "H" dimension is the depth of the bend on a 120-degree angle from the mold line to the edge of the flange. The "clear diameter" column is the diameter to the point of tangency.

The note, S–39–8, refers to a hole with a D of 3.25 inches, H of .19 inches, and a clear diameter of 3.50 inches. This is found in Fig. 16–41 by referring to dash number 8.

LOFTING

Lofting is laying out the shapes and contours of an aircraft at full size. It requires great accuracy. The templates used in the manufacture of aircraft parts are made from loft drawings. The person making loft drawings is called a loftsman. Fig. 16–42 shows a greatly reduced loft.

Surfaces developed in lofting vary from the very simple to the complex. The simplest example is a plane (flat) surface. Other surfaces may have a simple bend. They are rather easy to draw.

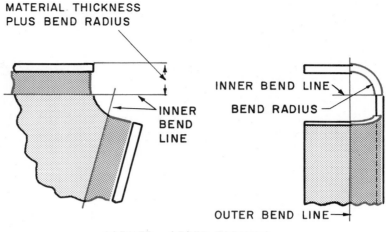

CORNER AFTER FORMING

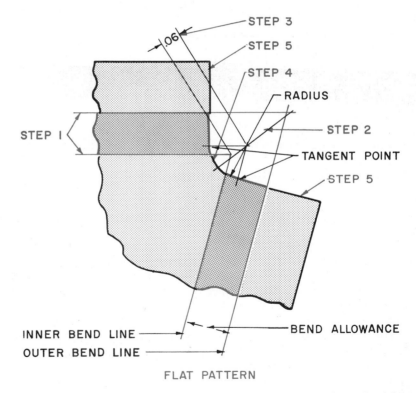

FLAT PATTERN

Cessna Aircraft Co.

16–38. *Relief hole for general use.*

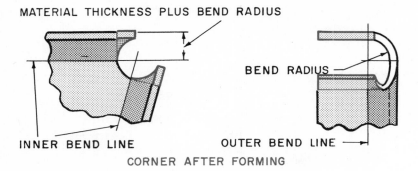

MATERIAL THICKNESS PLUS BEND RADIUS

BEND RADIUS

INNER BEND LINE

OUTER BEND LINE

CORNER AFTER FORMING

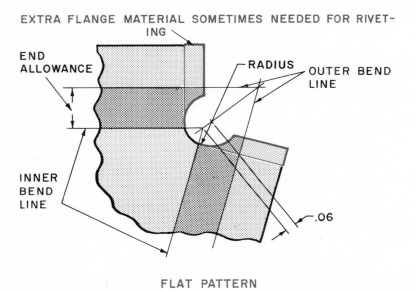

EXTRA FLANGE MATERIAL SOMETIMES NEEDED FOR RIVET-ING

END ALLOWANCE

RADIUS

OUTER BEND LINE

INNER BEND LINE

.06

FLAT PATTERN

Cessna Aircraft Co.

16–39. *Relief hole layout when extra flange material is needed.*

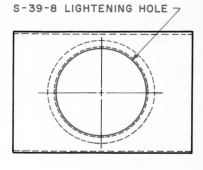

S-39-8 LIGHTENING HOLE

LIGHTENING HOLE WITH FLANGE DOWN

S-39-8 LIGHTENING HOLE

LIGHTENING HOLE WITH FLANGE UP

16–40. *How to draw lightening holes.*

A more difficult situation arises when a bend occurs that develops a warped surface. See Fig. 16–43.

The loftsman gives the lines for the finished size. A master template layout man makes full-size developments (patterns) of these surfaces, which are drawn on a flat plane. He must make allowances for the shortening effect metal has when it is formed. This means he must draw the template slightly larger than the actual finished size. This requires him to know how metal thickness and bends influence the size of the part after it is formed.

The loftsman must work with great care and accuracy. He must understand lines and surfaces in three dimensions.

Also he must understand the metal-forming processes used to shape rods, bars, tubes, and flat sheets of metal to the desired contours.

Since lofts are true size, many are too large to be drawn on a typical drafting table. These can be drawn on the

418

loft floor. The loftsman works by sitting or lying on the floor on top of the drawing. Some companies use low benches for drawing lofts. Some lofts are made on vertical drafting boards.

Fuselage Loft. The loftsman determines the shape or contour of the aircraft using the basic dimensions developed by the engineering department. The engineers select the length, height, and width requirements. They develop the other dimensions needed along the entire length of the fuselage. The loftsman uses these to make his overall size layout. He then establishes the *reference lines* he will use to develop the actual contour. The *reference lines* used are station lines, water lines, and buttock lines.

The *water lines* are edge views of horizontal planes passed through the aircraft when viewed from the side. One water line is selected as Water Line O.

Other water lines are drawn parallel to this and above and below it. Those above are positive and those below negative. The abbreviation used is WL. A WL+1 is the first water line above the center. A WL–1 is the first water line below the center. See Fig. 16–44.

Water lines are usually three or six inches apart.

Station lines are the edge views of vertical planes drawn through the aircraft in its side

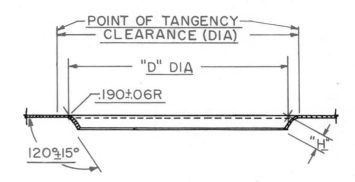

Dash No.	"D" Dia. ±.06	"H" ±.06	Clear Dia.	Dash No.	"D" Dia. ±.06	"H" ±.06	Clear Dia.
00	1.00	.09	1.25	13	4.50	.25	4.75
0	1.25	.09	1.50	14	4.75	.25	5.00
1	1.50	.12	1.75	15	5.00	.25	5.25
2	1.75	.12	2.00	16	5.25	.25	5.50
3	2.00	.12	2.25	17	5.50	.25	5.75
4	2.25	.19	2.50	18	5.75	.25	6.00
5	2.50	.19	2.75	19	6.00	.25	6.25
6	2.75	.19	3.00	20	6.25	.28	6.50
7	3.00	.19	3.25	21	6.50	.28	6.75
8	3.25	.19	3.50	22	6.75	.28	7.00
9	3.50	.19	3.75	23	7.00	.28	7.25
10	3.75	.19	4.00	24	7.25	.28	7.50
11	4.00	.25	4.25	25	7.50	.28	8.00
12	4.25	.25	4.50	—	—	—	—

16–41. *Lightening hole standards.*

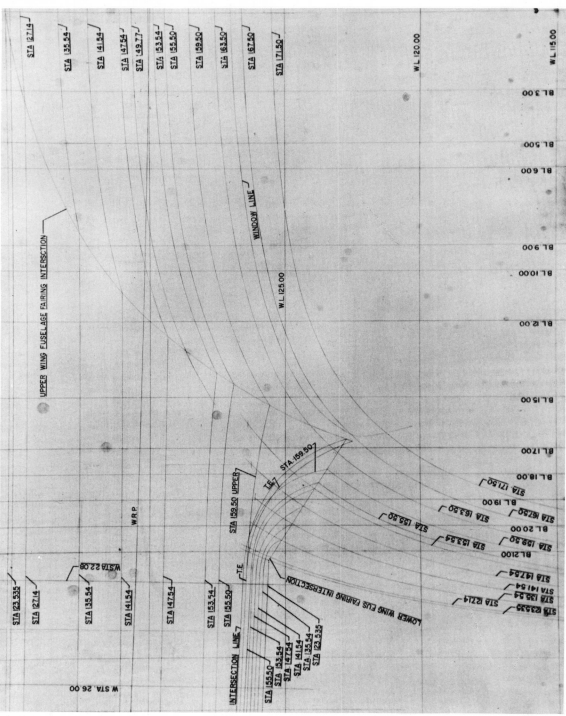

16–42. *A typical loft. It gives the exact shape of the aircraft at each of the stations indicated on the drawing. Notice the locations of the buttock lines, station lines, and water lines. After studying the section on fuselage lofts, refer back to this drawing.*

Cessna Aircraft Co.

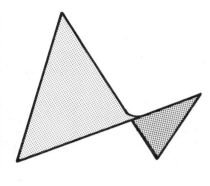

16–43.   *A warped surface.*

and plan views. They are perpendicular to the water and buttock lines. They begin at the nose of the fuselage, which is usually always drawn at the left. See Fig. 16–44. They are placed at 2½, 5, 7½, 10, 15, 20, 30, 40, 50, 60, 70, 80, 90, and 100 per cent of the length of the aircraft. The water lines plus the station lines divide the aircraft into horizontal and vertical segments.

The plan view of the aircraft is divided by buttock lines and station lines. A *buttock line* is the edge view of a plane running the length of the aircraft. Buttock line 0 is a center line through the ship. Buttock lines are designated as right or left of the center line, when viewing the aircraft from the rear. See Fig. 16–44. BLlL refers to the first buttock line to the left of the center line. BLlR refers to the first line to

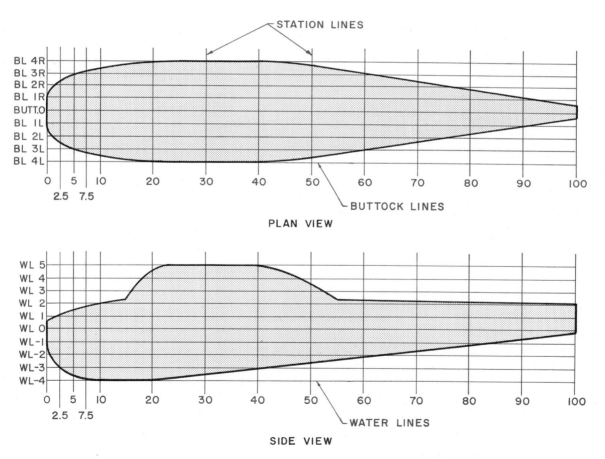

PLAN VIEW

SIDE VIEW

16–44.   *Water, buttock, and station lines as located for original full-size lofts.*

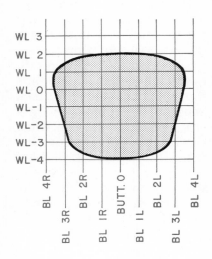

16–45. *Station lines drawn by using water and buttock lines.*

the right of the center line. Buttock lines are usually spaced three or six inches apart.

The plan view is divided by station lines exactly the same as the side view.

Stations. A station is a section through the fuselage cut perpendicular to the other reference lines.

Each station can be drawn by using the water lines and buttock lines. See Fig. 16–45. These are drawn as viewed from the front of the aircraft. The points of intersection between the various reference lines are connected by *faired*

lines. A *faired line* is a smooth, curved line. This gives a smooth, clean line to the aircraft. All dimensions within the aircraft are taken from these faired stations.

Engineering Layout of Fuselage. After the loftsman has made the full-size lofts using the reference line layout, engineers locate a new set of stations. These stations are located in inches from the nose of the aircraft. They are placed wherever it is necessary to know the shape, rather than at equally spaced distances. All stations are then in relation to the nose of the aircraft. For

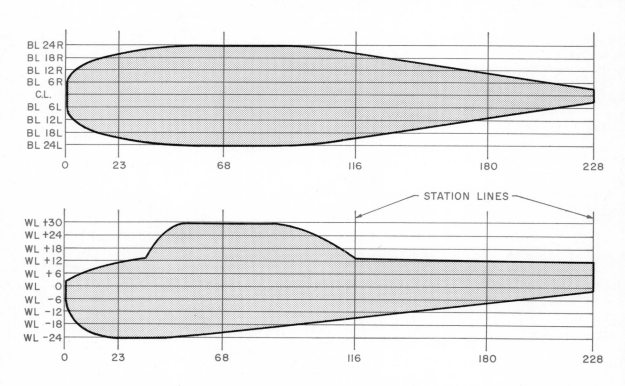

16–46. *Water, buttock, and several station lines located in inches on an engineering layout.*

example, station 68 is 68 inches from the nose.

This new engineering layout calls buttock line 0 the center line. All reference lines right and left of this are now in inches. The water line 0 is still used as a reference line, but all lines above and below it are in inches. See Fig. 16–46.

The Fuselage Body Plan. Fig. 16–47 shows a fuselage body plan. It is a common type of drawing used to show the contour and location of the stations. It can also show the stringers, window and door frames, pilots enclosure, and other openings in the fuselage.

Notice that the left contours show the aft section of the fuselage and the right contours the forward section. Since the fuselage is symmetrical, only one half need be drawn.

Laying Out A Bulkhead Contour. A bulkhead is an upright partition inside the fuselage. One is pictured in Fig. 16–12. It is usually made to fall on a station line.

The bulkhead contour is plotted by locating points on the contour. One way this is done is for the engineering staff to decide on the maximum breadth (width) of the fuselage, the height at the center line (buttock line 0), and a shoulder point. The shoulder point is a point on the contour between the maximum breadth point and the point indicating maximum

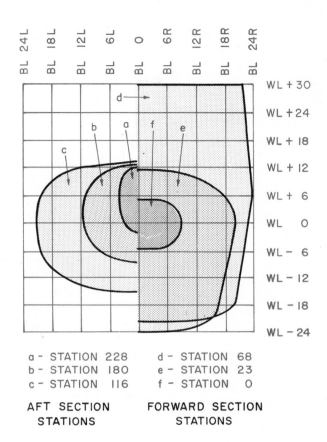

a – STATION 228 d – STATION 68
b – STATION 180 e – STATION 23
c – STATION 116 f – STATION 0

AFT SECTION   FORWARD SECTION
STATIONS   STATIONS

16–47. *End view of fuselage body plan. Stations are located as shown in the previous figure.*

height. These are located on the bulkhead shown in Fig. 16–48. The points are located according to the data in a table of ordinates. A line is faired between the points. This gives the contour.

Fig. 16–49 gives the ordinates for the three bulkheads in the tail cone shown in Fig. 16–12. Station 95 is the largest. Station 173.41 is the smallest. Point 1 is the maximum breadth. Point 2 is the

height above and below the water line 0. Point 3 is a point on the contour.

Fig. 16–48 shows how the bulkhead for station 95 was plotted.

The water line and buttock line ordinates, Fig. 16–49, are given for three points. Some are labeled "lower." These form the lower part of the bulkhead. Some are labeled "upper." These form the upper part of the bulkhead.

Point 1 is located where waterline +2.30 and buttock line 14.50 meet. See Fig. 16–48. Each point is located in this manner. They are then connected with a faired line. Fairing the contour requires the services of an experienced loftsman. He can usually fair the contour by eye using irregular curves. There is a method for plotting this type of curve, but it is too complex to present in this text.

Wing, Stabilizer, Rudder, and Control Surface Lofts. The contour of each rib or airfoil is lofted in a manner similar to the fuselage. They are designed around a chord line. A *chord line* is a straight line drawn from the front of the airfoil to the extreme rear edge. See Fig. 16–50. An *airfoil* is the contour above and below the chord line.

The engineering staff determines the length of the chord line. They give the dimensions above and below the chord line at various percentages of the chord line length. These dimensions are called *ordinates*. The chord line is divided into percentages as shown in Fig. 16–50. The airfoil is then completed for each percentage line by locating the points (ordinates) above and below the chord line. This gives points on the airfoil contour. A faired line is drawn through the points to develop the final shape. This gives the shape of one rib.

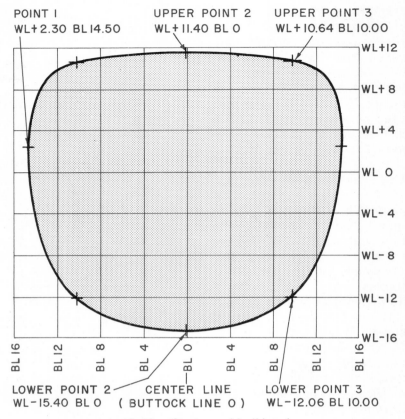

POINT I
WL+2.30 BL 14.50

UPPER POINT 2
WL+11.40 BL 0

UPPER POINT 3
WL+10.64 BL 10.00

LOWER POINT 2
WL-15.40 BL 0

CENTER LINE
(BUTTOCK LINE 0)

LOWER POINT 3
WL-12.06 BL 10.00

16–48. *Contour of bulkhead.*

The total wing is designed by laying out stations. See Fig. 16–51. Each station is an airfoil location. They are located in inches from the center line of the aircraft.

Wing Ordinate Tables. Standard tables of wing ordinates have been developed by the National Advisory Committee for Aeronautics (NACA). These tables give stations and ordinates for the upper surface and lower surface of a wide variety of wings. See Fig. 16–52.

The National Advisory Committee for Aeronautics has been discontinued and replaced by the National Aeronautics and Space Administration.

How to Draw Airfoils in Tapered Section of a Wing. In Fig. 16–51 airfoils 107, 127, 147, and 168 are shown in the tapered section of the wing. Each airfoil was plotted using the NACA 2412 table of ordinates. The ordinates are found in the same way as explained for airfoil 22.

Bulkhead	Ordinates for Three Bulkheads in a Tail Cone					
Station in Tail Cone	Point 1*		Point 2*		Point 3*	
	WL	BL	WL	BL	WL	BL
95.00 Lower	+2.30	14.50	−15.40	0	−12.06	10.00
95.00 Upper	+2.30	14.50	+11.40	0	+10.64	10.00
133.31 Lower	+2.64	9.76	−10.76	0	− 9.16	6.00
133.31 Upper	+2.64	9.76	+11.40	0	+11.02	6.00
173.41 Lower	+3.34	4.82	− 6.02	0	− 3.70	4.00
173.41 Upper	+3.34	4.82	+11.42	0	+10.12	4.00

*Ordinates are in inches.

Cessna Aircraft Co.

16–49. *Ordinates for bulkheads in a tail cone.*

The problem is to find the length of the chord for each airfoil.

One method that is satisfactory for a rough layout is to make an accurate drawing of the wing and measure the lengths on the drawing. This is not accurate enough for actual aircraft design.

The length of the chords in the airfoils in the tapered section of the wing can be found using simple trigonometry. Examine the wing drawing, Fig. 16–53. The first thing that must be found is the angle of the taper. This is found by solving the following formula:

Tangent of unknown angle
$$= \frac{\text{side opposite angle}}{\text{side adjacent to angle}}$$

$$\text{Tan } A = \frac{18}{102} = .1764$$

$$A = 10°$$

A table of tangents shows that .1764 is the tangent for 10 degrees. Therefore, Angle A is 10 degrees.

Now that the angle of the wing taper is known, the length of the chords in the taper can be found by simple trigonometry.

The length of the airfoil at station 107 will be solved. See Fig. 16–54.

The distance from A to B is known to be 42 inches. The distance from B to C must be found. The length of the chord of airfoil 107 is 42 inches plus the distance from B to C.

Side BC is found using the following formula:

$$\text{Tangent of } 10° = \frac{\text{Side BC}}{\text{Side BD}}$$

$$.1764 = \frac{BC}{82}$$

$$82 \,(.1764) = BC$$

$$14.46 \text{ inches} = BC$$

The chord of the airfoil at station 107 is 42 inches plus 14.46 inches or 56.46 inches.

The same formula and procedure can be used to find the other airfoil chords. They are drawn in Fig. 16–55.

The wing plan just explained is extremely simple. Usually a wing is designed using several different NACA ordinate tables. For example, the tip frequently has a different airfoil than the airfoil at the fuselage.

425

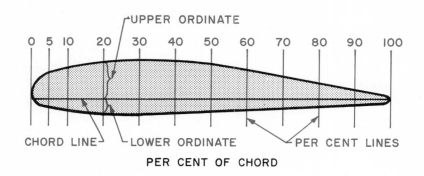

UPPER ORDINATE

0 5 10   20   30   40   50   60   70   80   90   100

CHORD LINE   LOWER ORDINATE   PER CENT LINES

PER CENT OF CHORD

16–50. *An airfoil.*

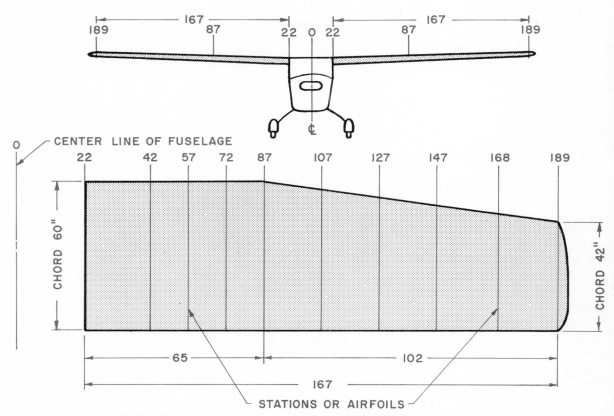

THE DIMENSIONS ARE IN INCHES FROM THE CENTER LINE OF THE AIRCRAFT.

16–51. *Stations on a wing locating airfoils.*

Airfoil Ordinate Tables

NACA 2412*
(Stations and ordinates given in per cent of airfoil chord)

Upper surface		Lower surface	
Station	Ordinate	Station	Ordinate
0	....	0	0
1.25	2.15	1.25	−1.65
2.5	2.99	2.5	−2.27
5.0	4.13	5.0	−3.01
7.5	4.96	7.5	−3.46
10	5.63	10	−3.75
15	6.61	15	−4.10
20	7.26	20	−4.23
25	7.67	25	−4.22
30	7.88	30	−4.12
40	7.80	40	−3.80
50	7.24	50	−3.34
60	6.36	60	−2.76
70	5.18	70	−2.14
80	3.75	80	−1.50
90	2.08	90	−0.82
95	1.14	95	−0.48
100	(0.13)	100	(−0.13)
100	....	100	0

L.E. radius: 1.58.
Slope of radius through L.E.: 0.10.

NACA 2415
(Stations and ordinates given in per cent of airfoil chord)

Upper surface		Lower surface	
Station	Ordinate	Station	Ordinate
0	....	0	0
1.25	2.71	1.25	−2.06
2.5	3.71	2.5	−2.86
5.0	5.07	5.0	−3.84
7.5	6.06	7.5	−4.47
10	6.83	10	−4.90
15	7.97	15	−5.42
20	8.70	20	−5.66
25	9.17	25	−5.70
30	9.38	30	−5.62
40	9.25	40	−5.25
50	8.57	50	−4.67
60	7.50	60	−3.90
70	6.10	70	−3.05
80	4.41	80	−2.15
90	2.45	90	−1.17
95	1.34	95	−0.68
100	(0.16)	100	(−0.16)
100	....	100	0

L.E. radius: 2.48.
Slope of radius through L.E.: 0.10.

NACA 2418
(Stations and ordinates given in per cent of airfoil chord)

Upper surface		Lower surface	
Station	Ordinate	Station	Ordinate
0	....	0	0
1.25	3.28	1.25	−2.45
2.5	4.45	2.5	−3.44
5.0	6.03	5.0	−4.68
7.5	7.17	7.5	−5.48
10	8.05	10	−6.03
15	9.34	15	−6.74
20	10.15	20	−7.09
25	10.65	25	−7.18
30	10.88	30	−7.12
40	10.71	40	−6.71
50	9.89	50	−5.99
60	8.65	60	−5.04
70	7.02	70	−3.97
80	5.08	80	−2.80
90	2.81	90	−1.53
95	1.55	95	−0.87
100	(0.19)	100	(−0.19)
100	....	100	0

L.E. radius: 3.56.
Slope of radius through L.E.: 0.10.

*National Advisory Committee for Aeronautics.

16-52.

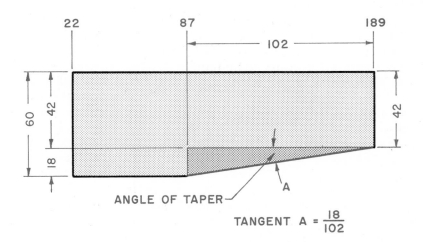

$$\text{TANGENT A} = \frac{18}{102}$$

16-53. *Solving for angle of taper.*

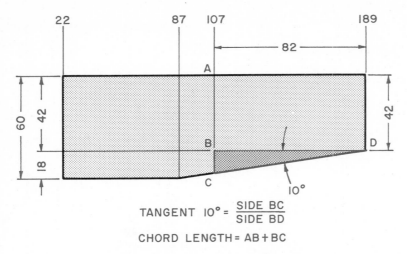

$$\text{TANGENT } 10° = \frac{\text{SIDE BC}}{\text{SIDE BD}}$$

$$\text{CHORD LENGTH} = AB + BC$$

16–54. *Solving for length of chord.*

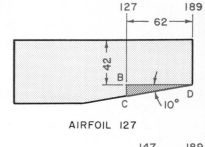

AIRFOIL 127

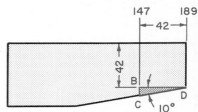

AIRFOIL 147

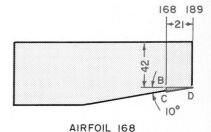

AIRFOIL 168

16–55. *Layout for figuring chord length of airfoils in tapered wing sections.*

The airfoils between have to be plotted using proportions. Some wings taper on both the leading and trailing edges. Others are planned with twist in them. These calculations are too complex to be presented in this text.

How to Lay Out Airfoils. The following procedure is based on the NACA airfoil ordinate tables. These tables have been engineered to provide design data for a wide variety of airfoils. The actual airfoil that is to be used for a wing is selected by the engineering staff. They also decide on the length of the chord.

Selected as an example was the NACA 2412 airfoil table. See Fig. 16–52. It is commonly used for small aircraft. The columns labeled "station" are the locations of each ordinate in per cent from the leading edge. The columns labeled "ordinate" give the distance in inches the contour of the

airfoil is above or below the chord. The data under the heading "upper surface" is for plotting the contour of the top surface of the airfoil. Under "lower surface" is the data for plotting the lower surface of the airfoil.

The airfoils to be designed are for the wing, Fig. 16–51. The airfoils at stations 22, 42, 57, 72, and 87 have the same chord length, 60 inches, and are identical. The airfoil for station 189 has a chord of 42 inches. The chord length for stations 107, 127, 147, and 168 has to be found.

Drawing the Airfoil at Station 22

Given: NACA table 2412
Fig. 16–52
Chord 60 inches

Procedure:

1. Multiply each figure in the NACA 2412 table by the chord length, 60. This produces a table of stations and ordinates for laying out this

airfoil. This new table is shown in Fig. 16–56. When the figures in the station column, Fig. 16–52, are multiplied by 60 inches, the answer is the distance of that station in inches from the leading edge of the airfoil. Notice the radius of the leading edge given at the bottom of the chord length. This is multiplied by the chord length. The slope of the radius through the leading edge is *not* changed.

2. Draw the chord to the exact length. See Fig. 16–57, Step One. If the airfoil is drawn

Station Distance*	Upper Ordinates	Lower Ordinates
0	—	0
0.75	1.29	0.99
1.50	1.79	1.36
3.00	2.48	1.81
4.50	2.98	2.08
6.00	3.38	2.25
9.00	3.97	2.46
12.00	4.36	2.54
15.00	4.60	2.53
18.00	4.73	2.47
24.00	4.68	2.28
30.00	4.34	2.00
36.00	3.82	1.66
42.00	3.11	1.28
48.00	2.25	0.90
54.00	1.25	0.49
57.00	0.68	0.29
60.00	0.08	0.08

Leading edge radius 0.948.
Slope of radius through leading edge 0.10.

*Distance in inches from leading edge.

16-56. *Ordinates for airfoil at station 22 on wing, Fig. 16-51.*

to scale, select the scale before proceeding.

3. Notice at the bottom of the data, Fig. 16-56, that the radius of the leading edge is 0.948 inches. The center for this radius is located on a line through the leading edge having a slope of 0.10. Measure one inch horizontally from the leading edge and vertically 0.10 inches. This gives the line used for drawing the leading edge radius. See Fig. 16-57, Step Two.

4. Draw the curve of the leading edge. Radius is 0.948 inches. See Fig. 16-57, Step Three.

5. Examine wing ordinates, Fig. 16-56. The distances above chord (upper surface) and below chord (lower surface) are given in inches.

The stations locating these are in inches. They are measured from the leading edge. It is 0 inches. Measure and draw the location of each ordinate. They are perpendicular to the chord.

Measure the ordinates above and below the chord for each station. Locate this with a short dash crossing the station line. Locate the upper and lower ordinates without moving the scale if possible. See Fig. 16-57, Step Four.

6. Connect the plotted points with an irregular curve. Draw tangent to the curve forming the leading edge of the airfoil. See Fig. 16-57, Step Five.

The airfoil for the wing tip has a 42-inch chord. It can be plotted and drawn using the same steps just presented. The only difference is that a new set of ordinates must be computed. The NACA 2412 table data should be multiplied by 42 to find the needed ordinates.

Additional Reading

Abbott, I. H., and A. E. Von Doehhoff. *Theory of Wing Sections.* New York: Dover Publications, Inc., 1959.

Anderson, N. H. *Aircraft Layout and Design.* New York: McGraw-Hill Book Co., 1946.

Leavell, Stuart, and Stanley Bungay. *Standard Aircraft Handbook.* Fallbrook, Calif.: Aero Publishers, Inc., 1958.

Build Your Vocabulary

Following are terms that you should understand and use as a part of your working vocabulary. Write a brief explanation of what each means.

Aerodynamics
Thrust
Lift
Drag
Gravity
Ordinates
Mock-up
Template
Layout drawing
Master plasters
Relief holes
Lightening holes
Lofting
Water lines
Station lines
Buttock lines
Chord lines
Airfoil

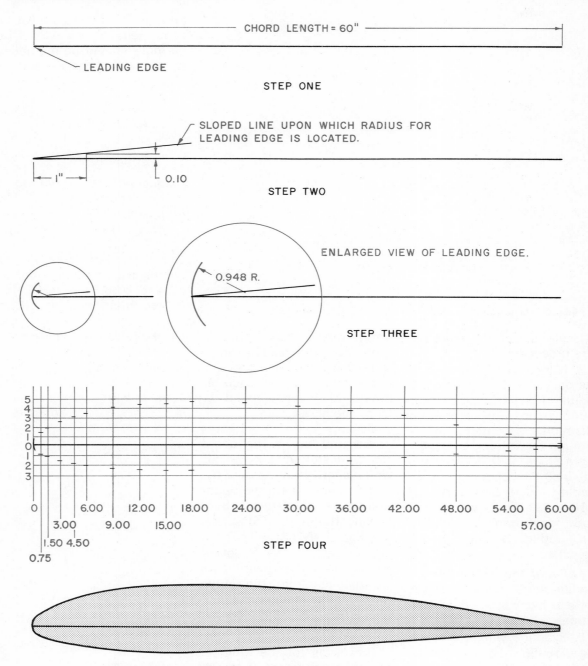

CHORD LENGTH = 60"

LEADING EDGE

STEP ONE

SLOPED LINE UPON WHICH RADIUS FOR
LEADING EDGE IS LOCATED.

1"   0.10

STEP TWO

ENLARGED VIEW OF LEADING EDGE.

0.948 R.

STEP THREE

0   3.00   6.00   9.00   12.00   15.00   18.00   24.00   30.00   36.00   42.00   48.00   54.00   60.00

1.50 4.50   57.00

0.75

STEP FOUR

AIRFOIL FOR STATIONS 22, 42, 57, 72 AND 87.

STEP FIVE

16–57. *How to draw an airfoil.*

Study Problems

1. Make a freehand sketch of the aircraft pictured in Fig. 16–9. Identify each major part. Letter the name on the drawing and connect to the part with a leader. Sketch it so it fills an 8½ × 11 inch sheet of paper.

2. Make a drawing of the flat sheet metal part, P16–1. Draw full size. It is to have rivet holes along the 3.25-inch and 5.95-inch edges. Select the proper size rivet. Determine the distance from the edges of the part for the rivet holes and the spacing between rivets. Draw the rivet locations on the drawing using standard practice. Identify the solution on the drawing with a callout.

3. Following are descriptions of several drawings. Develop the drawing number for each.

A. A drawing contains the bracket wing used to hold a strut to the wing. The numbers assigned by the drafting supervisor to the drawing are 734. The part is for the 170 series aircraft.
B. A drawing contains the details of a bulkhead for the tail cone of a 150 series aircraft. The drawing number is 325.
C. A drawing contains the assembly of three parts of the elevator for the 190 series aircraft. The drawing number is 605. The parts have dash numbers 1, 2, and 3. Develop the drawing numbers used to identify each part.

4. Draw a pattern for making the push-pull channel, P16–2. Draw full size. Record the overall dimensions on the drawing. Locate all holes and bend lines by dimensioning.

5. Draw a pattern for making the filter mounting bracket assembly shown in P16–3. Draw full size. Record the overall dimensions on the drawing. Locate bend lines by dimensioning.

6. Develop the patterns for the parts of the vertical tail and bumper horn shown in P16–4. Draw full size. Record the overall dimensions on the drawing. Locate bend lines by dimensioning.

7. Develop a pattern for the bracket of the rudder hinge shown in P16–5. Draw full size. Record the overall dimensions on the drawing. Locate bend lines by dimensioning.

8. Develop a pattern for the bracket of the aileron and elevator pulley shown in P16–6. Draw full size. Record the overall dimensions on the drawing. Locate bend lines by dimensioning.

9. Develop a full-size pattern for the vertical tail bumper shown in P16–7. Completely dimension the pattern. Locate the bend lines by dimensioning.

10. Design a cabin door handle and cavity so the handle is flush with the inside surface of the door. See P16–8. Keep distances X and Y as small as possible. Design the handle so it fits the hand comfortably. Design the cavity so it permits easy access to the handle. The cavity opening can be any shape that permits good access to the handle.

Make a design layout to check for interferences and clearances.

Make a full-size model of the cavity opening from cardboard. Make a full-size model of the handle from clay. Assemble them and see if the design is good. Can the handle be easily operated? Is there enough clearance in the cavity so the handle can be easily grasped?

11. Make a design layout of the propeller and spinner shown in P16–9. P16–10 shows an assembly view. Part 16 is an assembly of two pieces. It is made up of a round disc through which the bolts holding Part 18 extend. Part 17 is a small sheet metal part that is fastened to the disc. It is used to secure the screws, Part 15, from the dome assembly, Part 14. Part 17 has four screw holes.

Part 18, a bulkhead assembly, has six holes, Number 19, equally spaced.

Design the contour of the dome assembly, Part 14, to suit yourself. The section cut out of the dome assembly to go over the propeller should be large enough so it does not touch the propeller.

12. Draw the sheet metal pattern for the ceiling light channel shown in P16–11. Draw full size. Locate the bend lines.

13. Draw the pattern for the landing light frame shown in P16–12. Draw full size. Locate the bend lines and the bend allowance.

14. Solve the following problems: What are the minimum and standard bend radii for .063 gage 2024 aluminum; .160 gage 6061–0 aluminum; .028 gage steel sheet?

15. Develop a full-size pattern for the junction box shown in P16–13. Relieve the corners. Locate the bend lines on the drawing. Label the bend allowance.

16. Develop a full-size pattern for the cable clip shown in P16–14. Relieve the corners using relief holes at the corners. Since these will be riveted, use the relief hole layout giving extra flange material. Locate the bend lines on the drawing. Label the bend allowance.

17. Make a full-size drawing of the door jamb shown in P16–15. Draw the lightening holes as indicated. Draw a section through one of the lightening holes. Dimension the section drawing.

18. Plot the bulkhead contours for stations 133.31 and 173.41 from data given in Fig. 16–49. If possible, draw full size.

19. Plot a main wing airfoil using the NACA 2415 table of ordinates. See P16–9. The chord length is 55 inches. If possible, draw full size.

20. Plot a wing tip airfoil using the NACA 2412 table of ordinates. The chord length is 36 inches. If possible, draw full size.

21. Compute the chord length of the airfoils in the taper section of the wing, P16–16.

22. Draw the wing, P16–16 to scale. Carefully locate each airfoil in the tapered section. Measure the chord lengths. Compare the measured chord length with the calculated length in Problem 21.

23. P16–17A, B, and C give the basic dimensions for an aircraft.

Using this data, develop the following lofts:

A. Draw the top and side views of the fuselage. Lay out the water lines and buttock lines. Space them six inches apart. Locate the stations on the drawing. These are engineering stations. Draw to the scale $1/4'' = 3''$.

B. Lay out the bulkhead contour for each station. Use the scale $1/4'' = 3''$.

C. Draw the wing station layout as shown in Fig. 16–51. Use the scale $1/4'' = 3''$.

D. Lay out the airfoil for each wing station. Use the NACA 2412 airfoil ordinate table. Use the scale $1/4'' = 3''$. Since this is a rather complex prob-

lem, various class members could be assigned to solve parts of it. For example, several students could be assigned to develop the wings. Each could develop one or two airfoils. These could be cut out of heavy cardboard or balsa wood and assembled into a wing. Balsa stringers could be used between the airfoils. The wing skeleton could be covered with thin paper to produce a finished wing. The fuselage could be developed in the same manner.

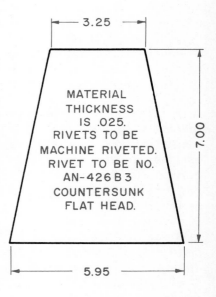

P16–1. Sheet metal part with rivet holes.

MATERIAL .063 THICK

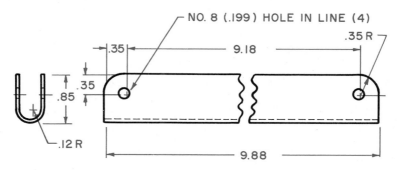

NO. 8 (.199) HOLE IN LINE (4)

.35 R

.35

9.18

.35

.85

.12 R

9.88

PUSH PULL CHANNEL

P16-2.

Cessna Aircraft Co.

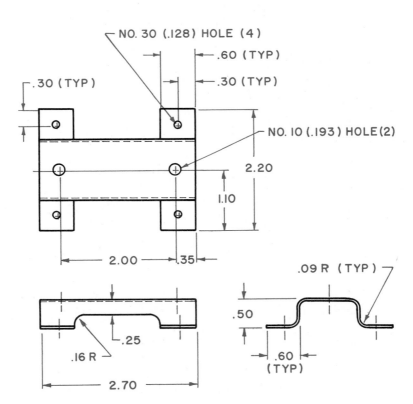

NO. 30 (.128) HOLE (4)

.60 (TYP)

.30 (TYP)

.30 (TYP)

NO. 10 (.193) HOLE (2)

2.20

1.10

2.00

.35

.25

.16 R

2.70

.09 R (TYP)

.50

.60
(TYP)

FILTER MOUNTING BRACKET

P16-3.

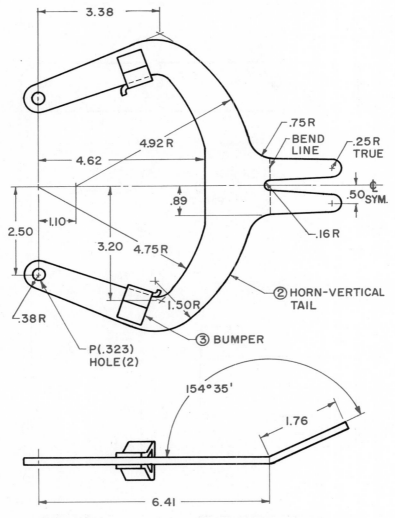

① ASSEMBLY OF VERTICAL TAIL AND BUMPER HORN

Cessna Aircraft Co.

P16–4.

434

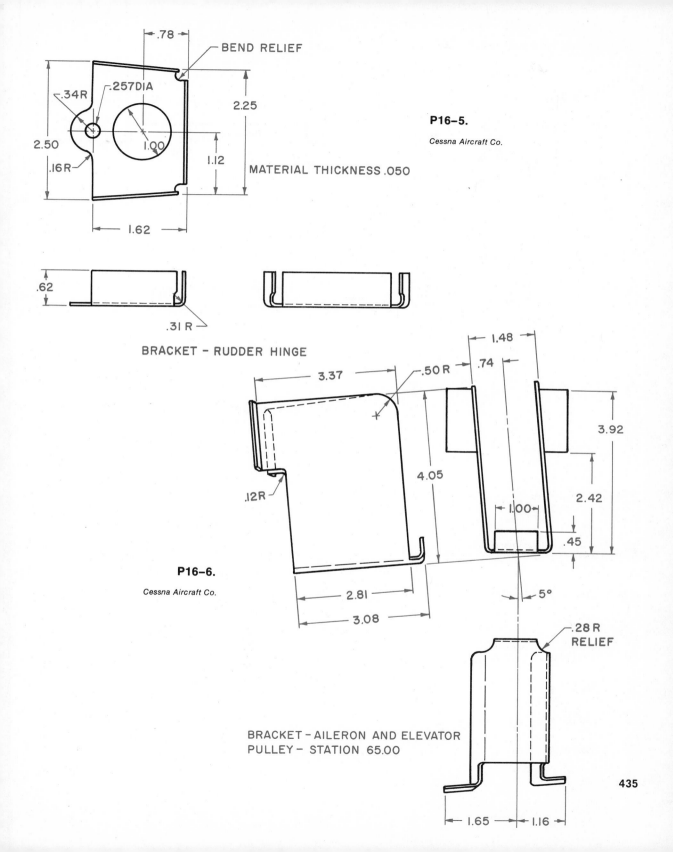

.78

BEND RELIEF

.34R   .257DIA

2.50

.16R

1.00

1.62

2.25

1.12

P16–5.

Cessna Aircraft Co.

MATERIAL THICKNESS .050

.62

.31 R

BRACKET - RUDDER HINGE

3.37   .50 R   .74   1.48

.12R

4.05   3.92

2.42

1.00   .45

P16–6.

Cessna Aircraft Co.

2.81   5°

3.08

.28 R
RELIEF

BRACKET - AILERON AND ELEVATOR
PULLEY - STATION 65.00

1.65   1.16

435

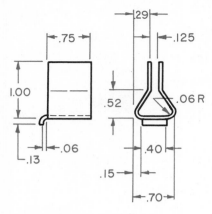

③ BUMPER-VERTICAL TAIL

Cessna Aircraft Co.

P16-7.

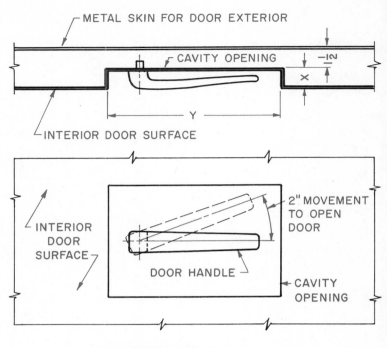

DOOR HANDLE AND CAVITY
P16-8.

P16-9. *Propeller and spinner assembly.*

Cessna Aircraft Co.

<u>PART</u> <u>I</u>DENTIFICATION

14. Dome Assembly

15. Screw

16. Bulkhead Assembly

17. Nutplate

18. Bulkhead Assembly

19. Screw Hole

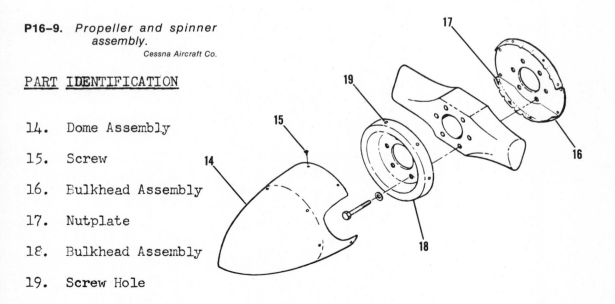

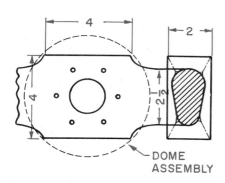

BASIC PROPELLER
DIMENSIONS

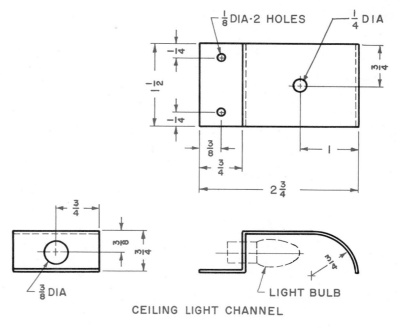

CEILING LIGHT CHANNEL

P16–11.

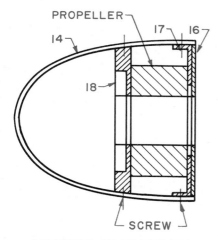

ASSEMBLY OF PROPELLER
AND SPINNER
P16–10.

P16–12. *Landing light frame.*

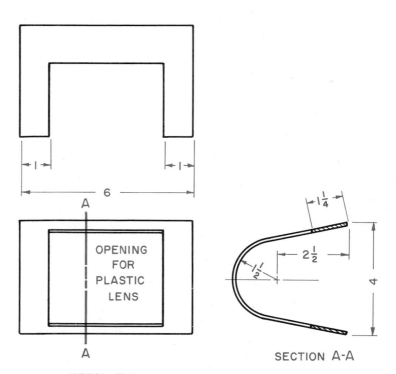

METAL FRAME FOR LANDING LIGHT IN
LEADING EDGE OF WING

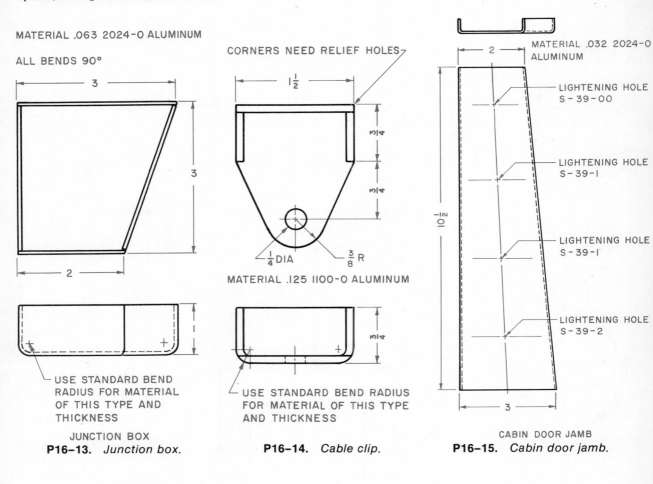

MATERIAL .063 2024-O ALUMINUM

ALL BENDS 90°

CORNERS NEED RELIEF HOLES

MATERIAL .032 2024-O
ALUMINUM

LIGHTENING HOLE
S – 39 – OO

LIGHTENING HOLE
S – 39 – I

LIGHTENING HOLE
S – 39 – I

LIGHTENING HOLE
S – 39 – 2

$\frac{1}{4}$ DIA $\frac{3}{8}$ R

MATERIAL .125 1100-O ALUMINUM

USE STANDARD BEND
RADIUS FOR MATERIAL
OF THIS TYPE AND
THICKNESS

JUNCTION BOX
P16–13. *Junction box.*

USE STANDARD BEND RADIUS
FOR MATERIAL OF THIS TYPE
AND THICKNESS

P16–14. *Cable clip.*

CABIN DOOR JAMB
P16–15. *Cabin door jamb.*

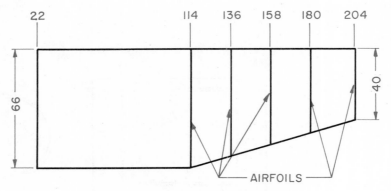

AIRFOILS

P16–16. *Airfoil stations in a simple wing.*

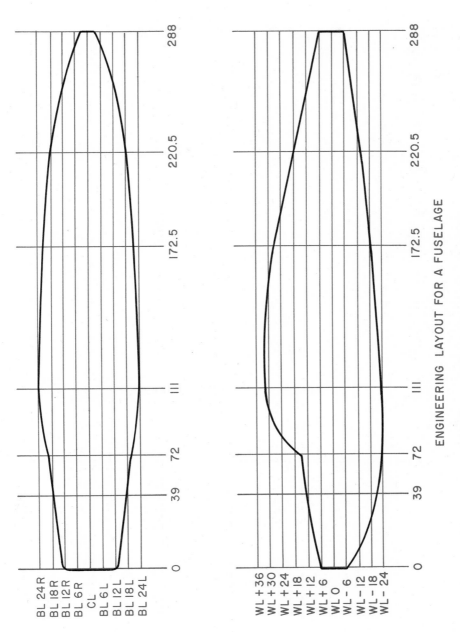

ENGINEERING LAYOUT FOR A FUSELAGE

P16-17A. *Engineering layout for a fuselage.*

Bulkhead		Bulkhead Ordinates					
Station in Fuselage		Point 1		Point 2		Point 3	
		WL	BL	WL	BL	WL	BL
0	Lower	+ 3.00	15.0	− 6	0	− 2.5	12.0
	Upper	+ 3.00	15.0	+ 6	0	+ 5.25	12.0
39	Lower	+ 4.0	18.50	−19.5	0	−13.25	15.0
	Upper	+ 4.0	18.50	+12.5	0	+ 9.75	15.0
72	Lower	+ 4.5	21.00	−23.75	0	−14.75	16.5
	Upper	+ 4.5	21.00	+15.00	0	+13.25	16.5
111	Lower	+12.0	24.00	−23.50	0	−15.5	18.50
	Upper	+12.0	24.00	+33.00	0	+33.00	18.50
172.5	Lower	+ 6.0	22.50	−17.75	0	− 9.25	17.50
	Upper	+ 6.0	22.50	+27.50	0	+24.0	17.50
220.5	Lower	+ 6.75	19.50	−13.25	0	− 6.5	15.0
	Upper	+ 6.75	19.50	+18.50	0	+14.25	15.0
288	Lower	+ 1.25	3.00	− 6.0	0	− 5.5	1.5
	Upper	+ 1.25	3.00	+ 6.0	0	+ 3.75	1.5

P16–17B. *Bulkhead dimensions.*

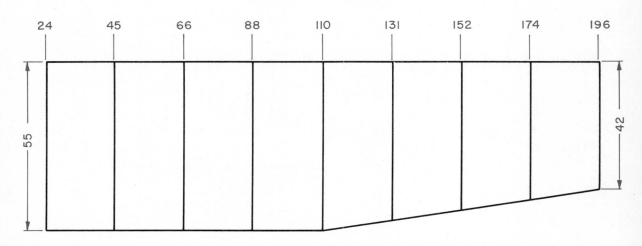

P16–17C. *Wing stations.*

Chapter 17

Technical Illustration

R. K. Lebedeff, Lockheed Aircraft Corp. Missiles and Space Co.

A technical illustration.

Technical illustrations are pictorial drawings made from orthographic drawings, mock-ups, models, or photographs. Their purpose is to present technical information in the form of easily understood pictures. For example, they may show how a device is to be assembled. While some may contain a few dimensions, the individual parts are not dimensioned. This information is found on the working drawings.

The common types of pictorial drawings used in technical illustration are isometric, dimetric, trimetric, oblique, and perspective. Study Chapter 9, Pictorial Drawing, for details on how to make each of these. Once these funda-mentals are mastered, you can begin to make technical illustrations.

USES AND TYPES OF ILLUSTRATIONS

Two kinds of illustrations are common: engineering and publication illustrations. *Engineering illustrations* are made for use in manufacturing, assembly, and installation. These drawings are accurately and carefully drawn. They show as simply as possible the information needed. Shading and other ways of making them attractive are not used. Workers having little ability to read working drawings can use engineering illustrations. A simple illustration is shown in Fig. 17–1A. It clearly shows how the adjustable angle fits together.

Publication illustrations are used in catalogs, maintenance and repair manuals, and sales and advertising copy. They are sometimes shaded to present a more lifelike appearance. They do not require the accuracy of engineering illustrations. The parts must be

441

kept in the proper proportion and relationship, but the illustrator frequently estimates the size and shape. It is important to produce a clear and attractive drawing. See Fig. 17–16.

SKILLS REQUIRED OF THE ILLUSTRATOR

The illustrator must be able to read working drawings. He must be able to use isometric, dimetric, trimetric, and perspective drawing techniques. These are explained in detail in Chapter 9, Pictorial Drawing. The illustrator must be able to use an airbrush and work with color. Skill in draft-ing and freehand sketching as well as a knowledge of art is needed. In Fig. 17–1B an illustrator is shown at work.

PROCEDURE FOR MAKING A TECHNICAL ILLUSTRATION

It is good practice to follow a definite plan when making a technical illustration. While experienced illustrators tend to develop their own procedures, the following plan has proven useful.

1. Understand the purpose of the illustration. Is it to show assembly, installation, maintenance, or repair information?
2. Study available material on the device to be illustrated. This could be a working drawing or the device itself.
3. Make several freehand sketches until a good layout has been developed.
4. Draw with instruments an accurate illustration in pencil.
5. Make the final tracing of the illustration.
6. Add lettering and notes.
7. Mount on cardboard if desired.

Purpose of the Illustration. It is important to know if the illustration is to be used for assembly, advertising, or some other purpose. This will help decide how much time should be spent on accuracy, shading,

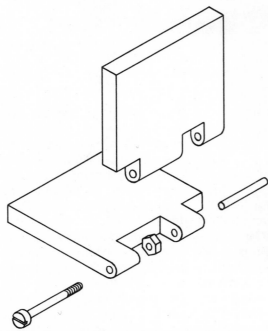

17–1A. *A simple technical illustration of an adjustable angle.*

Harlan Krug, Northrop Norair, Hawthorne, California, and Industrial Art Methods Magazine

17–1B. *This artist is making a full-color illustration. He is working from a photograph and a model. This work requires a great deal of skill and artistic talent.*

color, and other factors. If it is made for use under a military contract, military specifications must be followed. If it is to be reproduced by printing, the size it is to be printed must be decided. Since drawings are reduced photographically for printing, the amount of reduction must be known. This is usually 2 to 1 or 1½ to 1. A 2 to 1 reduction means the drawing will be reduced one-half the size it is drawn.

For example a 4 × 6 inch illustration will be printed 2 × 3 inches.

Study Available Material. Usually illustrations are made from working drawings. See Fig. 17–2, Part A. Study these so each part is clearly understood and the relationship between parts is known. Sometimes the illustration is made from the device itself. It can be disassembled for study.

Make Freehand Sketches. Try several freehand sketches until the object is in a position in which it can most clearly serve its purpose. Usually the object is placed so it appears in its normal position. Accuracy is not vital in this step but parts should be in proportion. Decide if an isometric, perspective, or other pictorial is best. See Fig. 17–2, Parts B and C. Make certain no parts have been omitted.

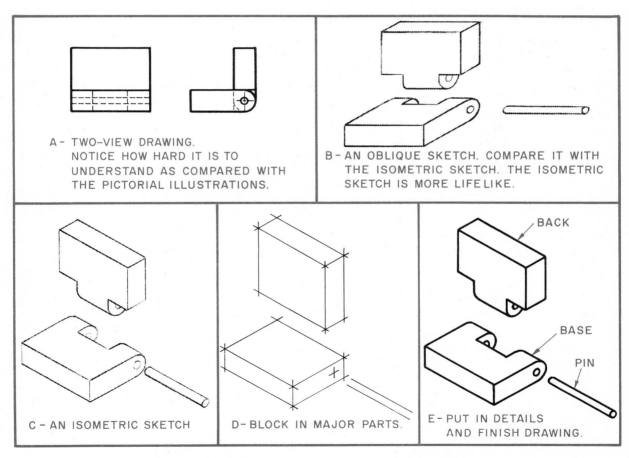

A - TWO-VIEW DRAWING. NOTICE HOW HARD IT IS TO UNDERSTAND AS COMPARED WITH THE PICTORIAL ILLUSTRATIONS.

B - AN OBLIQUE SKETCH. COMPARE IT WITH THE ISOMETRIC SKETCH. THE ISOMETRIC SKETCH IS MORE LIFELIKE.

C - AN ISOMETRIC SKETCH

D - BLOCK IN MAJOR PARTS.

E - PUT IN DETAILS AND FINISH DRAWING.

17–2. *A technical illustration made from a working drawing.*

Draw the Instrument Illustration. An instrument illustration must be carefully made since the final tracing will be made from it. Locate and draw center lines. Block in parts with construction lines. See Fig. 17–2, Part D. Complete the drawing by developing each part to its finished size and shape. See Fig. 17–2, Part E.

If an object has duplicate parts, draw one and use it to trace the others. This is fast and insures that all the parts will be the same size.

Construction lines should be made very light. If they tend to become confusing, erase them after they have served their purpose.

Since many objects have rounded or cylindrical parts, the task of drawing them pictorially is time consuming. Templates should be used to speed up drafting time. If they are not available, they can be made by drawing a variety of circle sizes pictorially and tracing them where needed.

If the illustration is to be reproduced by the blueprint or diazo process, the object lines on the instrument drawing can be darkened in pencil, construction lines removed, and it can be used for the final copy. If it is to be used for photographic reproduction, the lines are usually inked. While the instrument illustration can be inked, it usually is rather dirty and should be traced in ink on a separate sheet.

Make the Final Tracing. If the instrument illustration is to be traced in ink, it can be done on vellum, tracing cloth, drafting film, or two-ply Strathmore paper. Before tracing, make certain the instrument illustration is correct. Make any changes before inking.

If the inking is done on a transparent material, as vellum, the vellum can be placed over the instrument illustration and traced. If it is to be an opaque material, as Strathmore paper, it will have to be redrawn or traced on with carbon paper.

Several companies supply transparent sheets with standard fasteners, as bolts and nuts, printed on them. These are cut from the sheet and pressed into place on the drawing. Their adhesive back makes them stick in place. See Fig. 17–3. This speeds up the drafting time and produces high quality, clear copy.

The width of line on the ink illustration must be wide enough to be reduced for photographic reproduction. Remember that when a drawing is reduced, the line width is also reduced the same

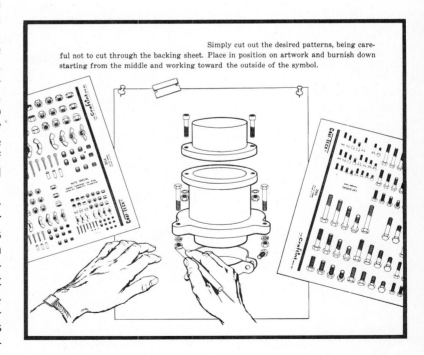

17–3. *Standard fasteners are available on adhesive-backed acetate sheets.*

amount. A line drawn 0.06, when reduced 2 to 1, will only be 0.03 wide.

Lettering and Notes. The parts of an object in a technical illustration sometimes require identification. This can be done by numbering each part or lettering the name of the part on the drawing. See Figs. 17–19 and 17–21. The numbers may be enclosed in a circle. See Figs. 17–19 and 17–27. If some details about the parts must be given, the parts are usually numbered and a parts list is used to give the details. See Figs. 17–19 and 17–24. The placement of the identifying numbers influences the appearance of the drawing and the ease with which it is read. The identifying name or number is connected to the part with a leader. The leader can be straight or angled.

Identifying numbers may be hand-lettered, machine-lettered, or placed on the drawing by cutting them from sheets having the numbers

printed on a transparent, adhesive-backed material. See Fig. 17–4 for samples. The method for using these letters is shown in Fig. 17–5.

If the name of the part is to be lettered on the drawing, the name should be placed to enhance the appearance. The lettering is usually placed horizontally and tied to the part with a leader. See Fig. 17–25.

Mounting the Illustration. If the illustration is to be displayed, it can be mounted on a heavy cardboard, called mountboard. It can be framed

with a matt in the same manner as a painting. Frequently it is covered with a transparent material such as acetate.

SHADING TECHNIQUES

Shading is a technique of varying light intensity on surfaces by lines or tones. It is used frequently in technical illustrations.

Engineering illustrations are usually not shaded. The only time shading is used is when it makes the drawing easier to understand. Shading takes time and therefore costs a company money. Publication

Letter-Press...The Most 123
Letter-Press... The Most Vers 123
Letter-Press...The Most Ve 123
Letter-Press...The Most Ve 12
Letter-Press...The Most 1234
Letter-Press...The Most Versatile Trans 12
Letter-Press..12

Craftint Manufacturing Company

17–4. *A few of the styles available in adhesive-backed lettering.*

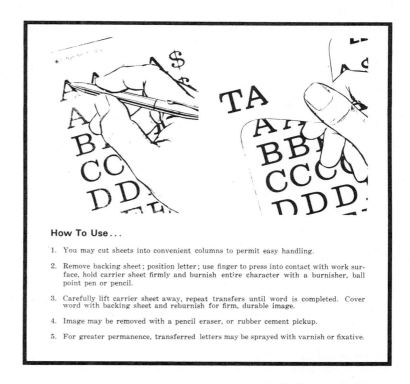

How To Use...

1. You may cut sheets into convenient columns to permit easy handling.

2. Remove backing sheet; position letter; use finger to press into contact with work surface, hold carrier sheet firmly and burnish entire character with a burnisher, ball point pen or pencil.

3. Carefully lift carrier sheet away, repeat transfers until word is completed. Cover word with backing sheet and reburnish for firm, durable image.

4. Image may be removed with a pencil eraser, or rubber cement pickup.

5. For greater permanence, transferred letters may be sprayed with varnish or fixative.

Craftint Manufacturing Company

17–5. *How to use adhesive-backed lettering.*

illustrations usually have some simple form of shading to improve the appearance. Shading provides the appearance of depth.

Study the shading techniques used on the illustrations in this chapter. You will notice some differences in technique, depending upon the skills and desire of the illustrator. All variations are satisfactory if they contribute to the primary purposes of shading—pleasing in appearance and clarity.

Source of Light. The direction of the source of light assumed to be shining on the object must be decided. While this is commonly taken on a 45-degree angle from the upper left corner of the drawing board, any direction can be used. The angle chosen is the one which will best show the planes of the object.

The sides directly facing the light source will have no shading. Surfaces with smaller angles receive less light and therefore have some shade. Those surfaces directly opposite the light source are in deep shadow and have the darkest shading. See Figs. 17–7, 17–9, and 17–10.

Exterior Line Shading. The simplist form of line shading is done by drawing the exterior lines of the object considerably heavier than those within the object. The interior lines are frequently broken. See Fig. 17–6. It is common practice to break the heavy object line if it is going to intersect another line from a different plane. This helps to show the separation of the planes in the illustration.

Solid Line Shading. Another method of line shading is shown in Fig. 17–7. A shadow effect is created by drawing a wide, solid, black band parallel to the edges of the object farthest from the light source. One or two single lines are drawn beyond the black band. They are spaced by eye.

Surface Line Shading. Still another line shadow technique is shown in Fig. 17–8. The lines are spaced by eye. The darker the area, the closer the lines are drawn. Usually the lines in the areas with darkest shadows are drawn thicker than those in the lighter areas.

The lines should usually parallel the edge of the surface to receive the shadow. In other words, vertical surfaces should have vertical shadow lines. Illustrators usually shade lightly one-fifth of the surface nearest the light source, leave two-fifths clear,

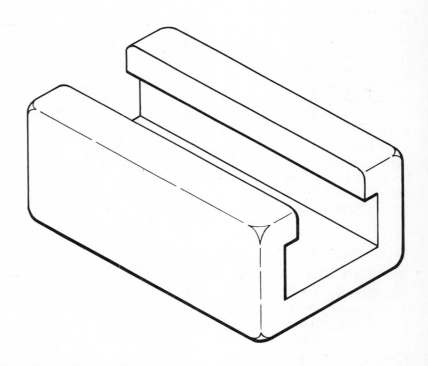

17–6. *This illustration shows the simplest form of shading. It uses a heavy line to indicate a cast shadow, a medium width line to indicate visible edges, and a light, broken line to indicate a rounded surface. The light comes from the upper left.*

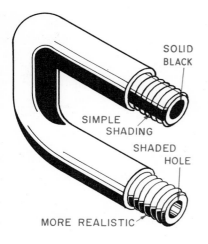

SOLID
BLACK

SIMPLE
SHADING

SHADED
HOLE

MORE REALISTIC

and shade heavily the two-fifths farthest from the light.

Stippling. An effective shadow technique is achieved by stippling. The surface to be shaded is covered with dots spaced at random. The darker areas have the dots closer together. Lighter areas have the dots farther apart. The dots are made freehand. The sharpness of the pencil point or size of the point on the inking device is important. A fine point is used on light areas and a larger point on darker areas. This method takes a lot of time but produces an attractive drawing, Fig. 17–9.

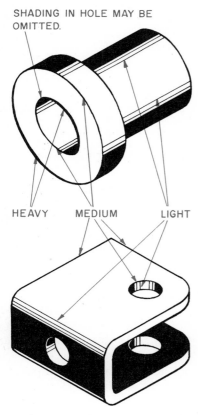

SHADING IN HOLE MAY BE OMITTED.

HEAVY   MEDIUM   LIGHT

SHADING IN HOLES MAY BE OMITTED.

17–7. *Solid line shading.*

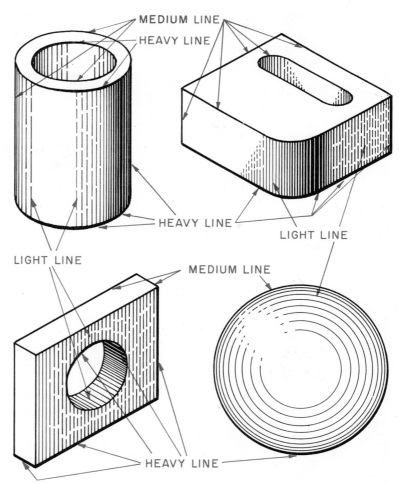

MEDIUM LINE

HEAVY LINE

HEAVY LINE

LIGHT LINE

LIGHT LINE

MEDIUM LINE

LIGHT LINE

HEAVY LINE

17–8. *Line shading is done by varying the line width and spacing between lines. The light is coming from the upper left.*

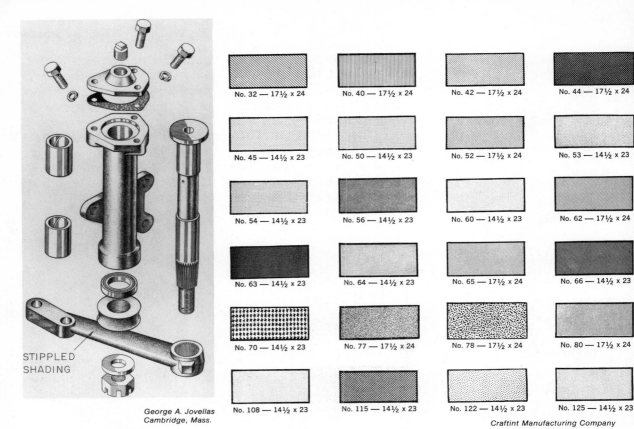

No. 32 — 17½ x 24	No. 40 — 17½ x 24	No. 42 — 17½ x 24	No. 44 — 17½ x 24
No. 45 — 14½ x 23	No. 50 — 14½ x 23	No. 52 — 17½ x 24	No. 53 — 14½ x 23
No. 54 — 14½ x 23	No. 56 — 14½ x 23	No. 60 — 14½ x 23	No. 62 — 17½ x 24
No. 63 — 14½ x 23	No. 64 — 14½ x 23	No. 65 — 17½ x 24	No. 66 — 14½ x 23
No. 70 — 14½ x 23	No. 77 — 17½ x 24	No. 78 — 17½ x 24	No. 80 — 17½ x 24
No. 108 — 14½ x 23	No. 115 — 14½ x 23	No. 122 — 14½ x 23	No. 125 — 14½ x 23

STIPPLED SHADING

George A. Jovellas
Cambridge, Mass.

17-9. A technical illustration utilizing strippled shading. Strippled shading is done by varying the size and spacing of the dots.

Craftint Manufacturing Company

17-11. Typical patterns of printed stick-on shading products.

17-10. A smudge shading.

Smudge Shading. If a grayed shading is desired, smudge shading is sometimes used. It can be used on one surface or a complete part to make it stand out. See Fig. 17-10.

An easy way to do smudge shading is to sand some carbon from the lead of a pencil. Put this on a cloth and rub it over the surface to be shaded. Artists' pastel sticks also can be used.

Colored leads can be used to call attention to parts of an object. An example would be to color the hydraulic lines on an illustration of an automobile chassis.

Stick-On Shading Products. Several companies sell shading patterns of dots, lines, stippling, and other patterns. They are available in black and colors. The patterns are printed on a transparent sheet having an adhesive back. See Fig. 17-11. One type must be cut to the shape of the surface. It is then pressed to stick on the drawing. See Fig. 17-12. Another type is placed over the area to be shaded and rubbed with a smooth object. The dots stick to the drawing and pull loose from the transparent sheet. See Fig. 17-13. Both are fast and effective.

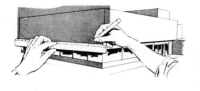
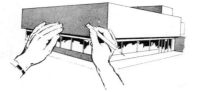

Craftint Manufacturing Company

17–12.　*How to use stick-on shading.*

Instantex by Letraset Limited

17–13.　*Place sheet over area to be shaded. Rub with a burnisher. Dots in the area burnished stick to the drawing below.*

How to Use Stick-On Shading

1. Make sure area to be shaded is free from pencil marks and eraser crumbs.

2. Remove a piece of stick-on shading material from the protective back.

3. Place shading material over area to be covered with adhesive side down.

4. Flatten shading material and press lightly so it sticks slightly.

5. Apply light pressure starting from the lower left corner and work upward and to the right. Make certain shading material is stretched flat.

6. Cut off excess shading material that runs beyond edges of area to be shaded. See Fig. 17–12A.

7. Peel off unwanted pieces of shading material.

8. Burnish entire area shaded so material is smooth and perfectly adhered. See Fig. 17–12B.

Airbrush Shading. An airbrush is illustrated in Fig. 17–14. It is used to shade illustrations for catalog copy and photo retouching. Compressed air is used to spray a coloring solution. The airbrush can be adjusted to produce different shading effects. See Fig. 17–15. The illustrations in Figs. 17–16 and 17–17 were prepared using the airbrush.

EXAMPLES OF TECHNICAL ILLUSTRATIONS

There are many ways an illustrator can show pictorially the technical information he must present. After a careful study of the object and the purpose of the illustration, the illustrator decides how to draw it. The most commonly used drawings are exploded, assembly, cutaway, or some combination of these.

Exploded Drawings. Exploded drawings show each part of an assembly drawn separately and in a position that shows how the parts fit together. Center lines sometimes are used to show the order in which parts are assembled. If the order is clear, center lines may be omitted. Fig. 17–18 shows an exploded view.

17–14.　*An airbrush is used to shade illustrations in color. The reservoir on the end holds the colored fluid. Compressed air is used to spray this fluid.*

Meisel Photochrome Corp.

17–15. *An illustrator at work. Notice how he holds the airbrush.*

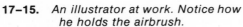

Combination Pump Valve Company

17–16. *A rendering for sales literature shaded with the airbrush.*

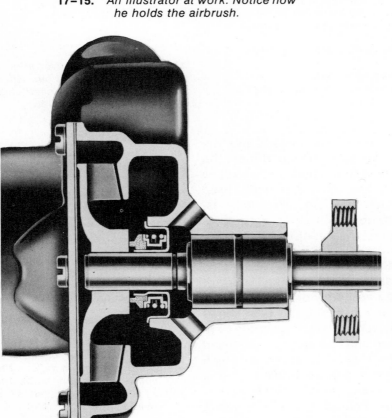

17–17. *An airbrush illustration. Notice the metal parts cut to make the section are not shaded. This makes them stand out against the darker, shaded metal parts.*

Chevrolet Motor Division
General Motors Corp.

450

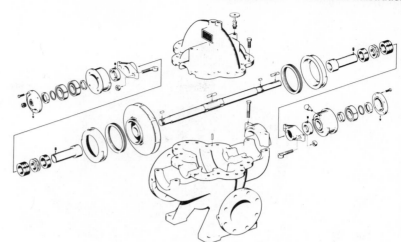

17–18. *An exploded view. Notice the line shading used to give a more realistic appearance. A center line is used to clarify the sequence of parts.*

A. Chandronhait
Raytheon Co.

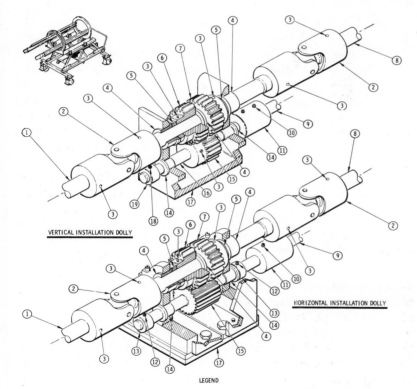

VERTICAL INSTALLATION DOLLY

HORIZONTAL INSTALLATION DOLLY

17–19. *A pictorial section through an assembled object. Notice the use of part numbers and a legend to identify each part.*

B. Machado
Lockheed Missiles and Space Corp.

LEGEND

1.	SHAFT 1914381 (HORIZONTAL DOLLY) 1914382 (VERTICAL DOLLY)	9. SHAFT 1914388	17. SUPPORT 1914377 (HORIZONTAL DOLLY) 1955902 (VERTICAL DOLLY)
2.	UNIVERSAL JOINT PS-1-2471-2	10. SETSCREW REF PS-1-2304-12	18. SETSCREW REF PS-2-1180-11
3.	PIN NAS561P6-26	11. COUPLING PS-1-2304-12	19. COLLAR PS-2-1180-11
4.	BEARING MS35772-84	12. SETSCREW REF PS-1-3267-11	20. BOLT NAS608-3-10
5.	SLEEVE 1914384	13. COLLAR PS-1-3267-11	21. PIN NAS561P6-20
6.	ELEVATION GEAR 1915767	14. BEARING MS35772-75	22. NUT MS21042L6
7.	PITCH GEAR 1915768	15. MAIN GEAR 1914380 (HORIZONTAL DOLLY) PS-2-3078-1 (VERTICAL DOLLY)	23. BOLT AN6-13
8.	SHAFT 1914383 (HORIZONTAL DOLLY) 1914385 (VERTICAL DOLLY)	16. SHAFT 1955903	24. WASHER NAS1099-6

451

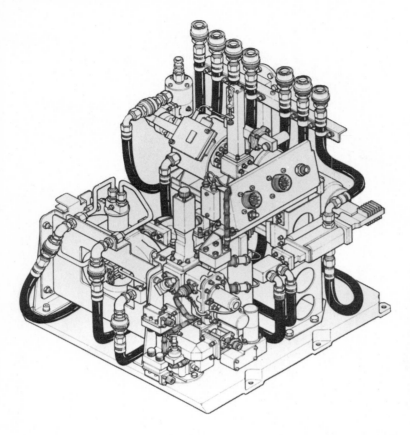

J. M. Graham
Raytheon Co.

17–20. *An exterior assembly drawing. This drawing would be useful to maintenance men or those assembling the device.*

An effort should be made to keep each part from overlapping with any other part. This is not always possible. Sometimes each part is identified with a number. A parts list gives details about the part. See Fig. 17–25.

Exploded views may be shaded if this is desirable. They can be drawn as isometric, dimetric, trimetric, or perspective views.

Cutaway Illustrations. If it is desired to show interior details of an object in its assembled position, a cutaway illustration can be used. This resembles a sectional view found on working drawings but is in pictorial form. The surfaces cut are usually shaded with slanting lines. See Fig. 17–19.

Exterior Assembly Drawings. Some illustrations picture the exterior of an object. To be most effective, they are usually shaded. Fig. 17–20 illustrates such a drawing. Notice how the flexible hoses are shaded to aid in reading the drawing. Examine it carefully and note how the use of different line widths adds to the appearance. The thicker lines give a shadow effect.

USES FOR TECHNICAL ILLUSTRATIONS

An illustration for an *installation manual* is shown in Fig. 17–21. It is to help workers understand how a device is assembled. It also shows where it fits on the aircraft. Notice that some parts are shown in an exploded position. Each part is identified by lettering the name on the drawing.

Fig. 17–22 shows an installation drawing for electrical wiring in an automobile. Notice how clearly it shows the worker where the wires are to run and how the connections are to be made. Frequently the wires are colored to match the actual color code on the wiring.

Fig. 17–23 shows another type of installation drawing. This drawing, when put into the hands of an automobile assembly plant worker, shows him how to fasten the unit together. The workers do not need to know how to read a working drawing. Notice the effective use of center lines.

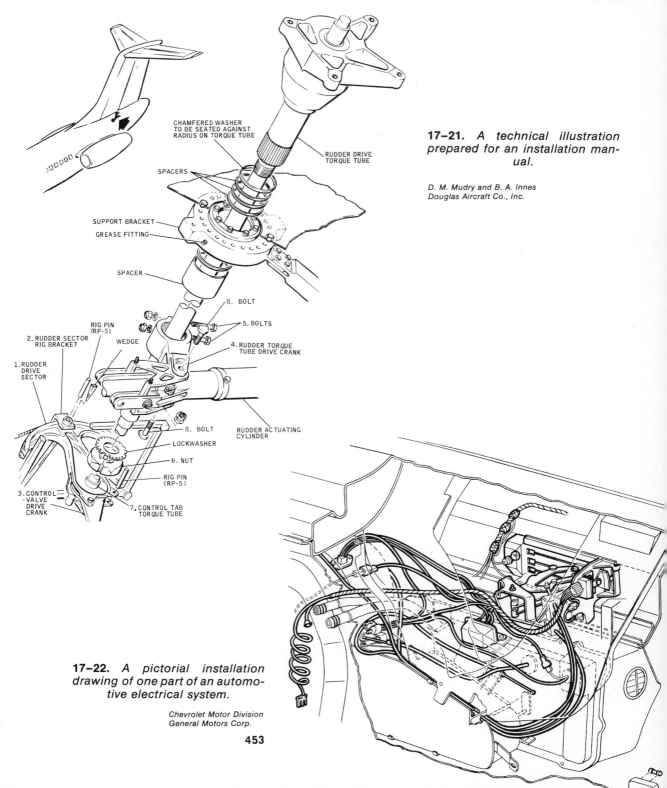

CHAMFERED WASHER
TO BE SEATED AGAINST
RADIUS ON TORQUE TUBE

RUDDER DRIVE
TORQUE TUBE

SPACERS

SUPPORT BRACKET

GREASE FITTING

SPACER

8. BOLT

5. BOLTS

4. RUDDER TORQUE
TUBE DRIVE CRANK

RIG PIN
(RP-5)

2. RUDDER SECTOR
RIG BRACKET

WEDGE

1. RUDDER
DRIVE
SECTOR

8. BOLT

RUDDER ACTUATING
CYLINDER

LOCKWASHER

6. NUT

RIG PIN
(RP-5)

3. CONTROL
-VALVE
DRIVE
CRANK

7. CONTROL TAB
TORQUE TUBE

17–21. *A technical illustration prepared for an installation manual.*

D. M. Mudry and B. A. Innes
Douglas Aircraft Co., Inc.

17–22. *A pictorial installation drawing of one part of an automotive electrical system.*

Chevrolet Motor Division
General Motors Corp.

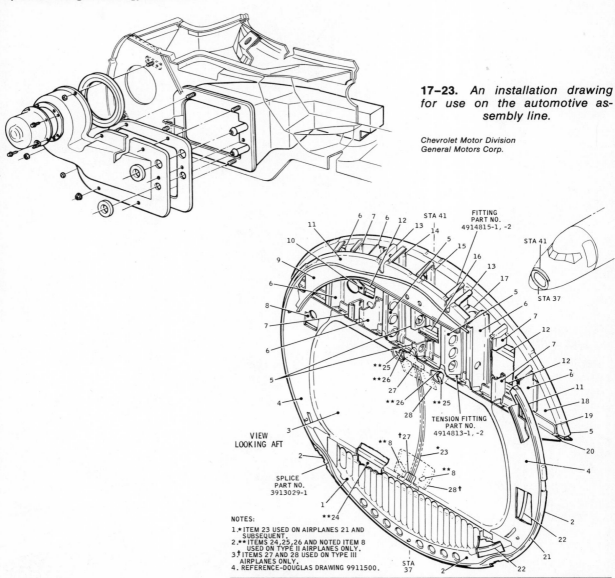

17–23. *An installation drawing for use on the automotive assembly line.*

*Chevrolet Motor Division
General Motors Corp.*

17–24. *An illustration for an aircraft repair handbook.*

*D. R. Peterson
Douglas Aircraft Co., Inc.*

STA 41
FITTING
PART NO.
4914815-1, -2

STA 41

STA 37

VIEW
LOOKING AFT

TENSION FITTING
PART NO.
4914813-1, -2

SPLICE
PART NO.
3913029-1

STA
37

NOTES:
1. ★ ITEM 23 USED ON AIRPLANES 21 AND
   SUBSEQUENT.
2. ★★ ITEMS 24,25,26 AND NOTED ITEM 8
   USED ON TYPE II AIRPLANES ONLY.
3. † ITEMS 27 AND 28 USED ON TYPE III
   AIRPLANES ONLY.
4. REFERENCE-DOUGLAS DRAWING 9911500.

ITEM	NOMENCLATURE	GAGE	MATERIAL	ITEM	NOMENCLATURE	GAGE	MATERIAL
1	WEB	.020	CLAD 2024-T42	15	TEE		1482282
2	FORMER	.050	CLAD 7075-T6	16	SUPPORT		1417096
3	ANTENNA DISH	.025	CLAD 2024-T42	17	INTERCOSTAL	.032	CLAD 7075-T6
4	WEB	.020	CLAD 2024-T3	18	PANEL	.040	CLAD 2014-T6
5	FRAME	.080	CLAD 7075-T6	19	TEE		2914150
6	BEAM		4594694	20	SUPPORT	.063	CLAD 7075-T6
7	FRAME	.071	CLAD 7075-T6	21	SEAL		4913109-1
8	DOUBLER	.040	CLAD 2024-T3	22	ANGLE	.040	CLAD 7075-T6
9	FRAME	.040	CLAD 7075-T6	23	★ STIFFENER		1093777
10	DOUBLER	.040	CLAD 2014-T6	24	★★ STIFFENER	.040	CLAD 7075-T6
11	ANGLE	.050	CLAD 7075-T6	25	★★ ANGLE	.040	CLAD 2024-T42
12	SUPPORT		1243092	26	★★ SUPPORT	.040	CLAD 2024-T42
13	INTERCOSTAL	.080	CLAD 7075-T6	27	† DOUBLER	.025	CLAD 2024-T3
14	ANGLE		1418201	28	† DOUBLER	.032	CLAD 2024-T3

17–25. *A pictorial assembly showing how the valve operates.*

G. Caldwell
Douglas Aircraft Co., Inc.

Pictorial drawings are used to help workers who repair devices. Fig. 17–24 shows a drawing for an aircraft *repair handbook.* Each part is identified by a number. A table is given listing the part number, name of the part, the thickness (gage) of the material, and type of material used in each part.

A *pictorial assembly* is shown in Fig. 17–25. It shows the assembly details of a complex valve and how it works. Notice the use of different line widths to give the major outline importance.

An *illustrated parts breakdown* is shown in Fig. 17–26. It is for use in a parts catalog. All parts of a transmission are shown. The auto mechanic can use this to secure the replacement parts needed. In a parts catalog, this drawing would have a list of each part by name and part order number. Fig. 17–27 shows another parts catalog illustration. Notice that each part is named and the part number given on the drawing. Notice the use of line shading.

17–26. *An illustrated parts breakdown for an auto parts catalog.*

J. Suvada
Ford Motor Company

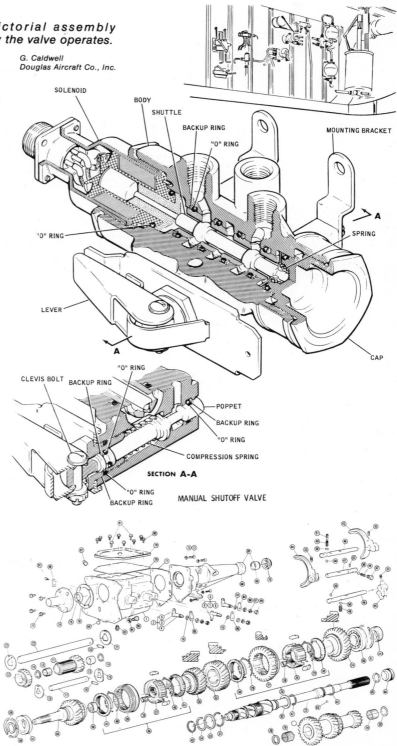

MANUAL SHUTOFF VALVE

SECTION A-A

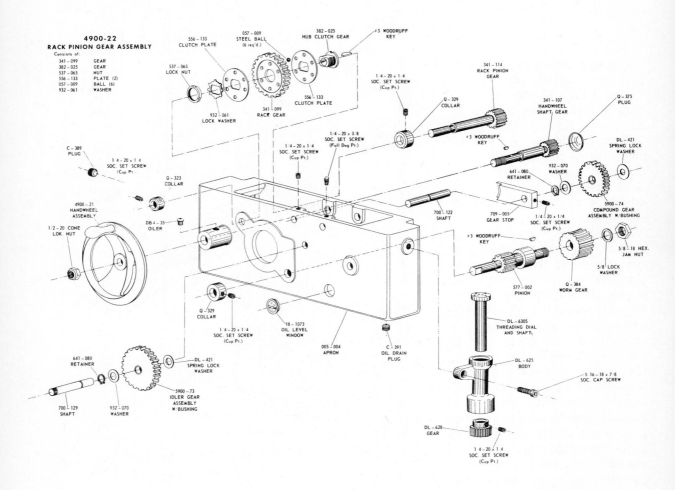

4900-22
RACK PINION GEAR ASSEMBLY
Consists of:
341 – 099   GEAR
382 – 025   GEAR
537 – 065   NUT
556 – 133   PLATE (2)
057 – 009   BALL (6)
932 – 061   WASHER

556 – 133
CLUTCH PLATE

537 – 065
LOCK NUT

057 – 009
STEEL BALL
(6 req'd.)

382 – 025
HUB CLUTCH GEAR

#3 WOODRUFF
KEY

341 – 114
RACK PINION
GEAR

932 – 061
LOCK WASHER

341 – 099
RACK GEAR

556 – 133
CLUTCH PLATE

1 4 – 20 x 1 4
SOC. SET SCREW
(Cup Pt.)

Q – 329
COLLAR

#3 WOODRUFF
KEY

341 – 107
HANDWHEEL
SHAFT GEAR

Q – 375
PLUG

C – 389
PLUG

1 4 – 20 x 1 4
SOC. SET SCREW
(Cup Pt.)

Q – 323
COLLAR

1/4 – 20 x 3/8
SOC. SET SCREW
(Full Dog Pt.)

1 4 – 20 x 1 4
SOC. SET SCREW
(Cup Pt.)

932 – 070
WASHER

DL – 421
SPRING LOCK
WASHER

641 – 080
RETAINER

5900 – 74
COMPOUND GEAR
ASSEMBLY W/BUSHING

4900 – 21
HANDWHEEL
ASSEMBLY

DB 4 – 35
OILER

700 – 122
SHAFT

709 – 005
GEAR STOP

1 4 – 20 x 1 4
SOC. SET SCREW
(Cup Pt.)

1/2 – 20 CONE
LOK. NUT

#3 WOODRUFF
KEY

5/8 – 18 HEX.
JAM NUT

5/8 LOCK
WASHER

Q – 329
COLLAR

1 4 – 20 x 1 4
SOC. SET SCREW
(Cup Pt.)

18 – 1073
OIL LEVEL
WINDOW

005 – 004
APRON

C – 391
OIL DRAIN
PLUG

577 – 002
PINION

Q – 384
WORM GEAR

DL – 630S
THREADING DIAL
AND SHAFT

641 – 080
RETAINER

DL – 421
SPRING LOCK
WASHER

5900 – 73
IDLER GEAR
ASSEMBLY
W/BUSHING

DL – 625
BODY

5 16 – 18 x 7 8
SOC. CAP SCREW

700 – 129
SHAFT

932 – 070
WASHER

DL – 628
GEAR

1 4 – 20 x 1 4
SOC. SET SCREW
(Cup Pt.)

Clausing, Kalamazoo, Mich.

17–27. *A parts catalog illustration of the rack pinion gear assembly of a metal lathe.*

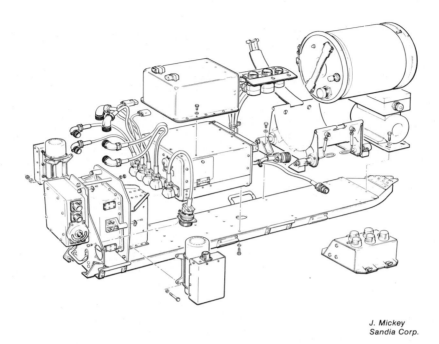

J. Mickey
Sandia Corp.

17–28. *A maintenance manual illustration.*

A pictorial for use in a *maintenance manual* is shown in Fig. 17–28. It shows each major part in an exploded position. This helps maintenance workers to understand how the device is assembled. It makes it easy for them to see how parts can be removed for servicing.

Another type of *maintenance manual* drawing is shown in Fig. 17–29. It gives help in adjusting a carburetor. Notice that it is not a pictorial view. The illustrator clearly shows what is needed with a simple one-view drawing. No-

tice that only the outline of the carburetor is drawn. Any details would add confusion to the drawing. The adjustment linkage is what is important. It is clearly shown.

Pictorial drawings are difficult to draw. They should be used only when they are the best way to present technical information.

Fig. 17–30 is from a machine tool maintenance manual. Notice it uses pictorial drawings to indicate clearly each lubrication point. It has notes telling how to lubricate and what lubricant to use.

FULL-COLOR ILLUSTRATIONS

Sometimes it is helpful to present an idea or concept by making a pictorial illustration in full color. Such drawings present the idea in a most realistic form. They are useful in feasibility studies and present the total idea before the actual device is much beyond the initial engineering stage. Usually they are rendered in watercolor, pastels, or oil colors. An airbrush is useful for this work. A knowledge of painting, color, and perspective is essential.

457

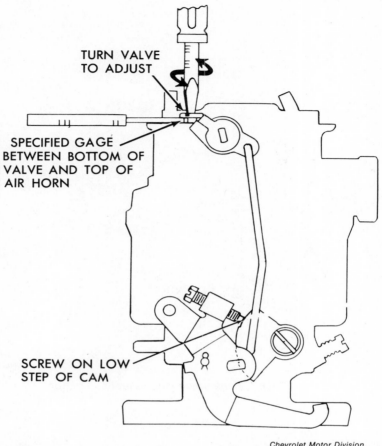

TURN VALVE
TO ADJUST

SPECIFIED GAGE
BETWEEN BOTTOM OF
VALVE AND TOP OF
AIR HORN

SCREW ON LOW
STEP OF CAM

Chevrolet Motor Division
General Motors Corp.

17-29. *A one-view drawing for a maintenance manual.*

Additional Reading

Batho, Robert. *A Practical Approach to Technical Illustration*. New York: Hart Publishing Co., 1968.

Gibby, Joseph C. *Technical Illustration*. Chicago: American Technical Society, 1962.

Thomas, T. A. *Technical Illustration*. New York: McGraw-Hill Book Co., 1968.

Build Your Vocabulary

Following are terms you should understand as a part of your working vocabulary. Write an explanation of what each means.

Technical illustration
Engineering illustrations
Publication illustrations
Shading
Airbrush

Solid line shading
Stippling
Surface line shading
Stick-on shading
Exploded drawings
Cutaway illustrations

LUBRICATION CHART - - - 5900 SERIES CLAUSING LATHES

CODE

D-DAILY oil with LUBRICANT X or equivalent.
WEEKLY
W1-Oil with LUBRICANT X or equivalent.
W2-Check oil level in window. Remove pipe plug and fill to mark with oil Y or equivalent.
W3-With motor running and variable dial turned to low speed, fill with oil Y or equivalent.
W4-Check oil level in window. Remove filler plug and fill to mark with oil Z or equivalent.
 * Remove plug.
 * * Remove plug and turn spindle until oiler shows.

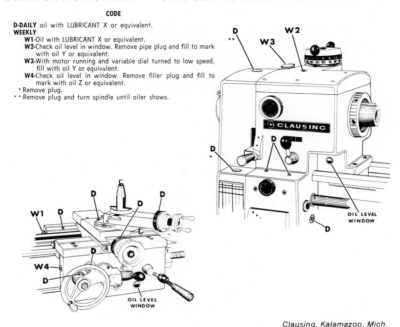

Clausing, Kalamazoo, Mich.

17-30. *Part of a maintenance manual illustration giving lubrication directions.*

Study Problems

1. Draw a simple exploded view of the conveyor hanging bolt, P17–1. Use the exterior line shading technique. Draw it full size.

2. Draw a pictorial illustration of the swivel pad screw, P17–2, that would be suitable for use in a sales catalog. Be certain to show clearly the details of the swivel tip. Draw twice actual size.

3. Draw an exploded view of the machine tool leveling pad, P17–3. Use any type of shading you desire. Draw twice actual size.

4. Draw an exploded view of the back gear shaft assembly, P17–4. Shade using the solid line shading technique. Select a scale that is large enough to show clearly all the parts.

5. Draw an exterior assembly of the control handle for a metal lathe, P17–5. Shade using the stippling technique. Draw on a two to one proportion so it can be reduced for printing to fit in a column three inches wide.

6. Draw a pictorial illustration of the steam pipe hanger, P17–6, to show a worker how

it is assembled. Select a scale large enough to show it clearly.

7. A. Draw a pictorial assembly of the electrical wiring pressure connector. Pick the dimensions off the drawing with dividers. Draw it at a two to one proportion so it will fit a column in a sales catalog four inches wide.

B. Draw a pictorial illustration of the conveyor carrier to show how it is to be assembled. Draw twice full size.

8. A. Draw an assembled pictorial view of the conveyor carrier, P17–8, with necessary

section to show hidden details. Shade using any method desired. Draw 1½ times full size.

B. Draw an exploded illustration of the electrical wiring pressure connector. Select any scale.

9. Make a parts catalog illustration of the adjustable clamp, P17–9. Identify each part by a number and give a parts list. Pick the dimensions off the drawing with dividers. Draw twice as large as the printed size.

10. A. Draw a pictorial section of the tumbler knob assembly, P17–10. Shade it so it could serve as an illustration for a sales catalog. Draw true size.

B. Draw a parts catalog illustration of the tumbler knob assembly. Shade it if it will

help clarify the drawing. Identify each part on the drawing. Draw true size.

11. A. Draw a pictorial section of the water connection for a rock drilling machine, P17–11. Shade it so it can be used in a sales catalog. Draw the section 50 per cent larger than it is printed.

B. Draw a pictorial assembly of the water connection for a rock drilling machine, showing exterior appearance. Shade using the stippling technique.

C. Draw pictorially the hose stem of the water connection for a rock drilling machine. Shade it using the solid line technique.

12. Draw a parts illustration of the auto radio antenna shown in P17–12. Select names for

each part. Identify each part using a parts list. Draw half true size.

13. A. Draw a pictorial illustration showing the exterior features of the penlight, P17–13. Draw twice actual size. Shade using the smudge shading technique.

B. Draw an exploded view of the penlight. Select a scale that will clearly show each part.

14. Bring to the classroom objects that you can take apart easily, such as a ball-point pen, a vise, a door lock, or water faucet. Prepare a technical illustration to serve a particular purpose—for a catalog or for the assembly worker. Secure dimensions by measuring the parts after they have been disassembled.

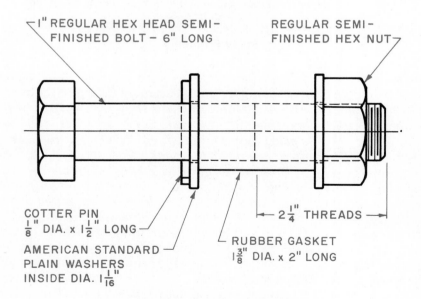

1" REGULAR HEX HEAD SEMI-FINISHED BOLT – 6" LONG

REGULAR SEMI-FINISHED HEX NUT

COTTER PIN
⅛" DIA. x 1½" LONG

AMERICAN STANDARD PLAIN WASHERS INSIDE DIA. 1 1/16"

RUBBER GASKET
1⅜" DIA. x 2" LONG

2¼" THREADS

P17–1. *Conveyor hanging bolt.*

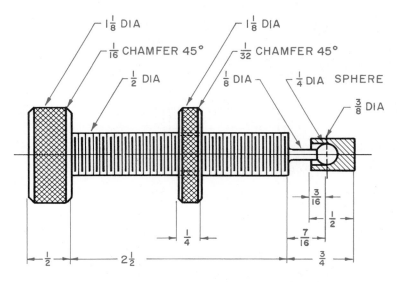

$1\frac{1}{8}$ DIA

$\frac{1}{16}$ CHAMFER 45°

$\frac{1}{2}$ DIA

$1\frac{1}{8}$ DIA

$\frac{1}{32}$ CHAMFER 45°

$\frac{1}{8}$ DIA

$\frac{1}{4}$ DIA   SPHERE

$\frac{3}{8}$ DIA

$\frac{3}{16}$

$\frac{1}{2}$

$\frac{7}{16}$

$\frac{1}{4}$

$\frac{1}{2}$

$2\frac{1}{2}$

$\frac{3}{4}$

Reid Tool Supply Co.

P17-2. *Swivel pad screw.*

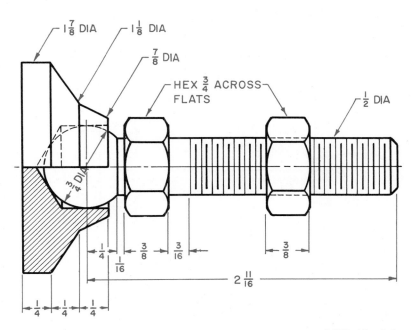

$1\frac{7}{8}$ DIA

$1\frac{1}{8}$ DIA

$\frac{7}{8}$ DIA

HEX $\frac{3}{4}$ ACROSS FLATS

$\frac{1}{2}$ DIA

$\frac{3}{4}$ DIA

$\frac{1}{4}$

$\frac{1}{16}$

$\frac{3}{8}$

$\frac{3}{16}$

$\frac{3}{8}$

$\frac{1}{4}$

$\frac{1}{4}$

$\frac{1}{4}$

$2\frac{11}{16}$

Reid Tool Supply Co.

P17-3. *Machine tool leveling pad.*

461

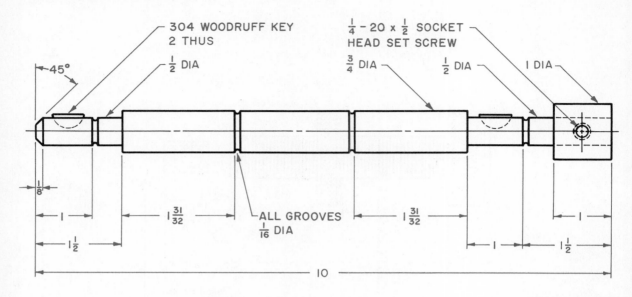

45°

304 WOODRUFF KEY
2 THUS

$\frac{1}{2}$ DIA

$\frac{1}{4}$ - 20 x $\frac{1}{2}$ SOCKET
HEAD SET SCREW

$\frac{3}{4}$ DIA

$\frac{1}{2}$ DIA

I DIA

$\frac{1}{8}$

I

$1\frac{31}{32}$

ALL GROOVES
$\frac{1}{16}$ DIA

$1\frac{31}{32}$

I

$1\frac{1}{2}$

I

$1\frac{1}{2}$

10

Clausing, Kalamazoo, Mich

P17–4. *Back gear shaft assembly for metal lathe.*

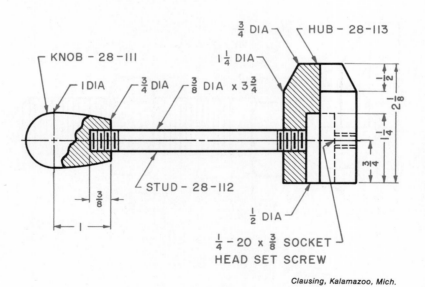

KNOB - 28-III

I DIA

$\frac{3}{4}$ DIA

$\frac{3}{8}$ DIA x $3\frac{3}{4}$

$\frac{3}{4}$ DIA

$1\frac{1}{4}$ DIA

HUB - 28-113

$\frac{1}{2}$

$2\frac{1}{8}$

$1\frac{1}{4}$

$\frac{3}{4}$

STUD - 28-II2

$\frac{3}{8}$

I

$\frac{1}{2}$ DIA

$\frac{1}{4}$ - 20 x $\frac{3}{8}$ SOCKET
HEAD SET SCREW

Clausing, Kalamazoo, Mich.

P17–5. *Control handle for metal lathe.*

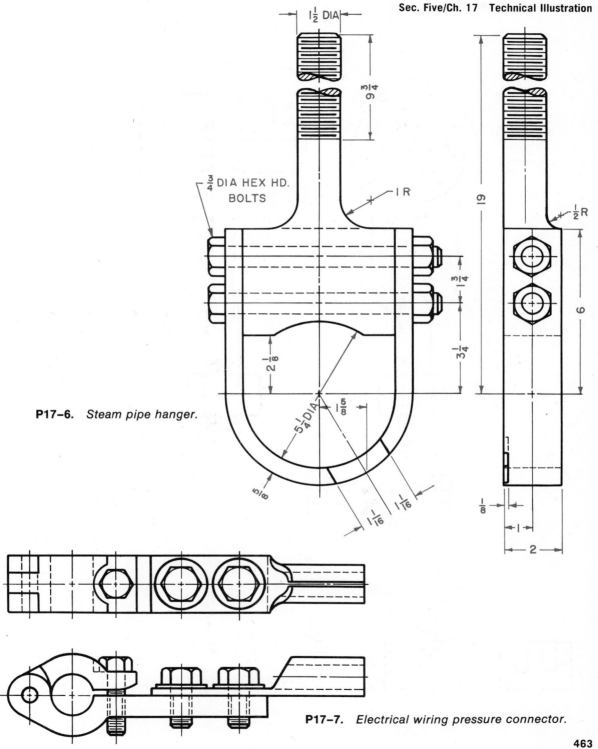

P17-6. *Steam pipe hanger.*

P17-7. *Electrical wiring pressure connector.*

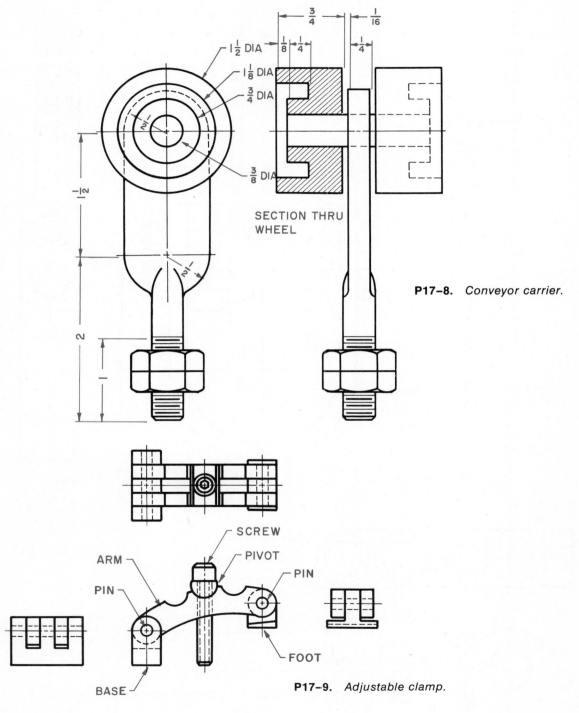

$1\frac{1}{2}$ DIA

$1\frac{1}{8}$ DIA

$\frac{3}{4}$ DIA

$\frac{3}{8}$ DIA

$\frac{3}{4}$

$\frac{1}{8}$ $\frac{1}{4}$

$\frac{1}{16}$

$\frac{1}{4}$

SECTION THRU WHEEL

P17–8. *Conveyor carrier.*

SCREW

ARM

PIVOT

PIN

PIN

FOOT

BASE

P17–9. *Adjustable clamp.*

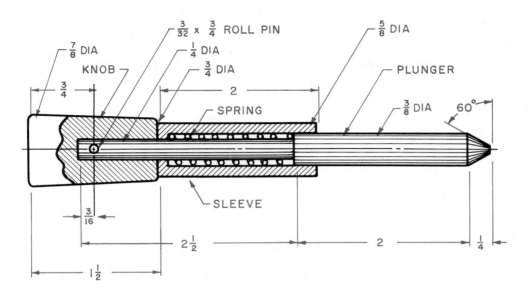

P17–10. *Tumbler knob assembly on metal lathe.*

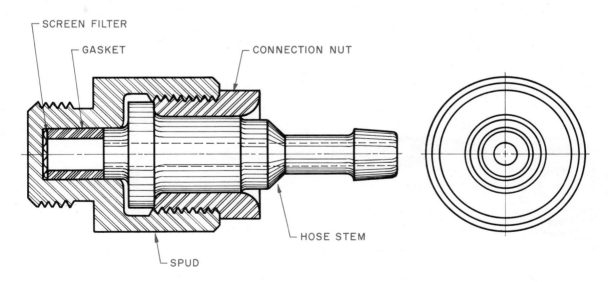

P17–11. *Water connection for rock drilling machine.*

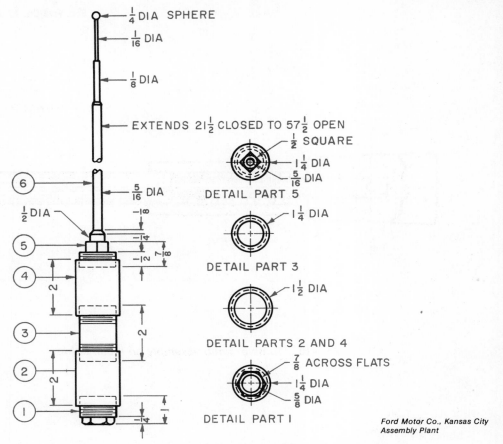

$\frac{1}{4}$ DIA SPHERE

$\frac{1}{16}$ DIA

$\frac{1}{8}$ DIA

EXTENDS $21\frac{1}{2}$ CLOSED TO $57\frac{1}{2}$ OPEN

$\frac{1}{2}$ SQUARE

$1\frac{1}{4}$ DIA

$\frac{5}{16}$ DIA

DETAIL PART 5

$\frac{5}{16}$ DIA

$\frac{1}{2}$ DIA

$-\frac{1}{8}$

$-\frac{1}{4}$

$-\frac{1}{2}$

$\frac{7}{8}$

$1\frac{1}{4}$ DIA

DETAIL PART 3

2

2

2

$\frac{1}{2}$ DIA

DETAIL PARTS 2 AND 4

$\frac{7}{8}$ ACROSS FLATS

$1\frac{1}{4}$ DIA

$\frac{5}{8}$ DIA

DETAIL PART I

$-\frac{1}{4}$

Ford Motor Co., Kansas City
Assembly Plant

P17–12. *Auto radio antenna.*

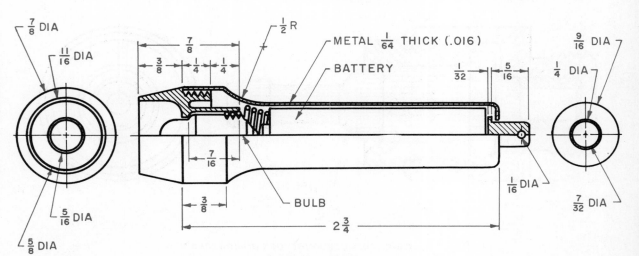

$\frac{7}{8}$ DIA

$\frac{11}{16}$ DIA

$\frac{7}{8}$

$\frac{3}{8}$ $\frac{1}{4}$ $\frac{1}{4}$

$\frac{1}{2}$ R

METAL $\frac{1}{64}$ THICK (.016)

BATTERY

$\frac{1}{32}$ $\frac{5}{16}$

$\frac{9}{16}$ DIA

$\frac{1}{4}$ DIA

$\frac{7}{16}$

$\frac{5}{16}$ DIA

$\frac{5}{8}$ DIA

$\frac{3}{8}$

BULB

$2\frac{3}{4}$

$\frac{1}{16}$ DIA

$\frac{7}{32}$ DIA

P17–13. *Penlight.*

Chapter 18

Piping Drafting

Piping drawings are the master plans for various types of piping projects. They show equipment locations, piping hookups, piping details, and necessary dimensions for field construction. These drawings include such items as pipe, vessels, valves, pumps, motors, and various pipe fittings required for any one project. These drawings should be complete and in sufficient detail to clearly indicate pipe clearances, connections, supports, and contain complete descriptive information for clear instructions to the fabricator.

THINGS DRAFTSMEN SHOULD KNOW ABOUT PIPING DESIGN

Designing piping installations is the work of mechanical engineers and design draftsmen. Engineers have the responsibility of planning and determining the basic design, based on what the installation is to produce. For a large plant or refinery, designing involves various problems to determine efficient layouts, specifications, flow sheets, process evaluation, equipment bids, pipe selections, pressure vessels, exchangers, storage tanks, instrumentation, mechanical equipment, heating, air conditioning, plumbing, machines, engines, compressors, and pumps. This work requires various types of drawings, specifications, and in some cases, building a scaled model of the plant. Fig. 18–1 shows a large refinery installation involving all phases of piping design and drafting.

Piping draftsmen convert the design plan into the required drawings, showing necessary detailed piping and supporting equipment in plans, elevations, and details.

Draftsmen should also be familiar with the operating principles of the process in each piping project. This is necessary, since many of the

18–1. *A close-up view of the complex piping in a refinery.*

Phillips Petroleum Co.

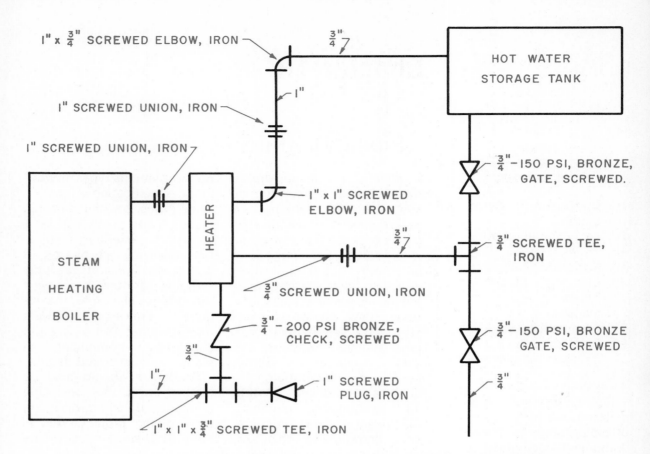

I" x ¾" SCREWED ELBOW, IRON

I" SCREWED UNION, IRON

I" SCREWED UNION, IRON

I"

HOT WATER
STORAGE TANK

¾"

STEAM

HEATING

BOILER

HEATER

I" x I" SCREWED
ELBOW, IRON

¾"

¾" SCREWED UNION, IRON

¾" - 200 PSI BRONZE,
CHECK, SCREWED

¾"

I"

I" SCREWED
PLUG, IRON

I" x I" x ¾" SCREWED TEE, IRON

¾" - 150 PSI, BRONZE,
GATE, SCREWED.

¾" SCREWED TEE,
IRON

¾" - 150 PSI, BRONZE
GATE, SCREWED

¾"

18-2. *Single-line piping diagram showing the heater installation in steam heating system.*

piping details and hookups are determined by draftsmen.

Working with piping problems requires a basic understanding of piping nomenclature and sizes of piping fittings for various working pressures. This requires reference checks with standard piping data books and manufacturer's catalogs.

PIPING DESIGN CONSIDERATIONS

1. Economy—Piping design is made on an economical basis with savings wherever possible

in sizing of lines and valves, and reduction of connections through proper location of fittings.

2. Safety—Industrial practice requires the insulation of hot lines for personnel protection. There must be overhead clearance of pipe in passage ways. Piping must clear all walkways, stairs, and ladders. Low piping should not block equipment areas that require regular controlling and servicing.

3. Supports—Provide adequate supports for all piping.

4. Assessibility—Passageways between equipment or equipment piping and adjacent equipment must be clear for servicing and repair. Pumps should be located for easy removal.

SINGLE-LINE DRAWINGS

Single-line piping drawings or sketches are used for small scale drawings. Fittings are shown by standard conventional symbols regardless of the pipe diameters. This type of drawing is used when dimensional accuracy is not important. An example is shown in Fig. 18–2.

Some of the common symbols for screwed, welded, and flanged fittings used in single-line drawings are shown in Fig. 18–3. A complete listing of conventional symbols is given in the American National Standards Institute Bulletin Z32.2.3–1949, *Graphical Symbols for Pipe Fittings, Valves, and Piping.*

DOUBLE-LINE DRAWINGS

Double-line drawings are used when dimensional accu-racy is important. The pipes and fittings are drawn to scale with location dimensions to center lines of pipes and fit-tings. This will show how much clearance there will be be-tween various items in the lay-out. Valves and equipment are located by measurements to their centers and the allow-ances for make-up is left to the pipe fitter on the job site. The sizes of pipe should be specified by notes or callouts telling the nominal diameters, never by dimension lines on the drawing. See Fig. 18–4.

Symbols used on double-line drawings should appear somewhat like the actual item. The common symbols for fit-tings are shown in Figs. 18–5 to 18–10. Common sizes for valves are shown in Figs. 18–11 to 18–17. These are used for drawing the symbols.

Double-line drawings should not be drawn to a scale less than $\frac{1}{4}'' = 1'0''$. They should show detailed information that is accurately dimensioned; therefore, they should be drawn to as large a scale as is practical.

18–3. *Single-line symbols for flanged, screwed, and welded pipe fittings and valves.*

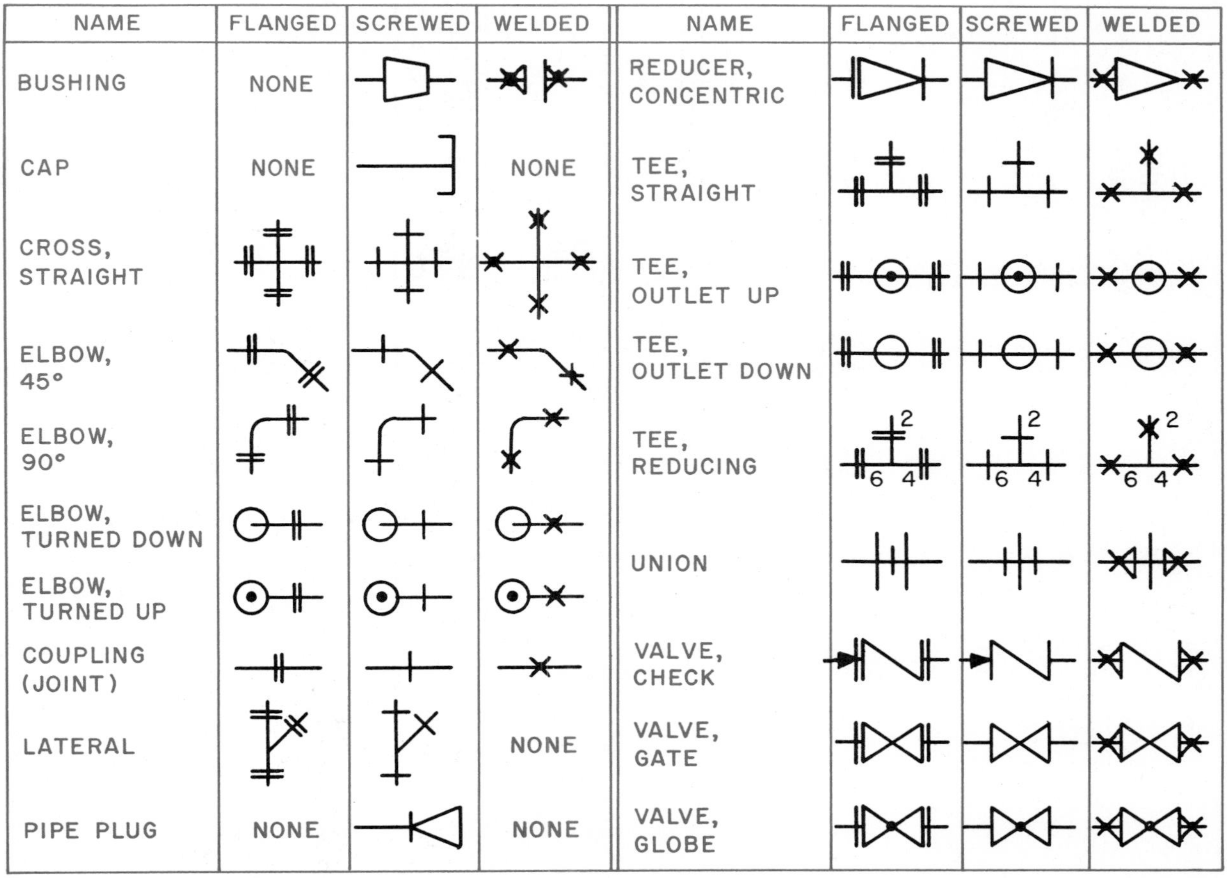

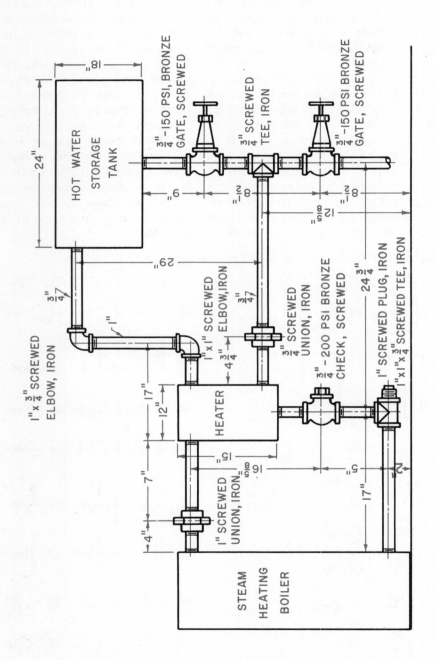

18-4. A double-line piping diagram of the heater installation, Fig. 18-2.

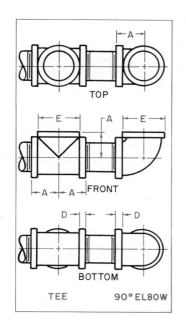

TOP

FRONT

BOTTOM

TEE   90° ELBOW

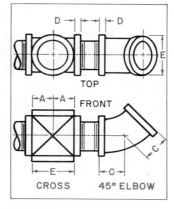

TOP

FRONT

CROSS   45° ELBOW

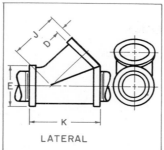

LATERAL

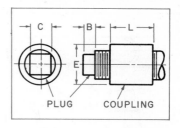

PLUG   COUPLING

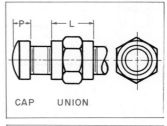

CAP   UNION

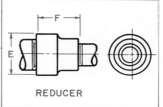

REDUCER

Approximate Malleable Iron Pipe Fitting Sizes (Screw Type)

Pipe Dia.	3/4	1	1½	2	2½
A	$1^5/_{16}$	$1^1/_2$	$1^{15}/_{16}$	$2^1/_2$	$2^{45}/_{64}$
B	$^1/_2$	$^{37}/_{64}$	$^{45}/_{64}$	$^3/_4$	$^{59}/_{64}$
C	1	$1^1/_8$	$1^7/_{16}$	$1^{11}/_{16}$	$1^{61}/_{64}$
D	$^9/_{32}$	$^5/_{16}$	$^3/_8$	$^{27}/_{64}$	$^{15}/_{32}$
E	$1^{29}/_{64}$	$1^{25}/_{32}$	$2^{27}/_{64}$	$2^{61}/_{64}$	$3^{19}/_{32}$
F	$1^7/_{16}$	$1^{11}/_{16}$	$2^5/_{16}$	$2^{13}/_{16}$	$3^1/_4$
G	1	$^{15}/_{32}$	$1^{21}/_{64}$	$1^{29}/_{64}$	$1^{45}/_{64}$
J	$2^3/_{64}$	$2^7/_{16}$	$3^9/_{32}$	$3^{15}/_{16}$	$4^{47}/_{64}$
K	$2^{25}/_{32}$	$3^9/_{32}$	$4^3/_8$	$5^{11}/_{64}$	$6^1/_4$
L	$1^1/_2$	$1^{43}/_{64}$	$2^5/_{32}$	$2^{17}/_{32}$	$2^7/_8$

Sizes selected from ANSI B16.3–1963. They are approximate and to be used for symbol drawing purposes only.

18–5. *Double-line symbols for malleable iron screw-type fittings.*

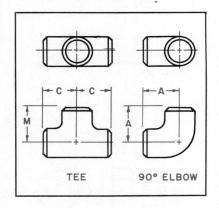

TEE   90° ELBOW

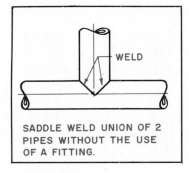

SADDLE WELD UNION OF 2
PIPES WITHOUT THE USE
OF A FITTING.

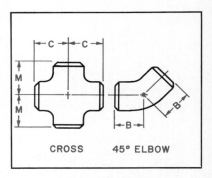

CROSS   45° ELBOW

CAP

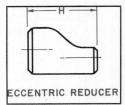

ECCENTRIC REDUCER

Steel Butt-Welded Reducers*

	Large End	Small End	H
³/₄ × ¹/₂	1.050	0.840	1¹/₂
1 × ³/₄	1.315	1.050	2
1¹/₂ × 1	1.900	1.315	2¹/₂
2 × 1¹/₂	2.375	1.900	3
2¹/₂ × 2	2.875	2.375	3¹/₂
3 × 2¹/₂	3.500	2.875	3¹/₂

*Dimensions in inches

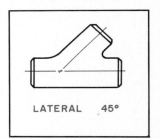

LATERAL   45°

Steel Butt-Welded Fittings

Nominal Pipe Size	Outside Dia. at Bevel	Elbows*		Tees and Crosses*		Caps*
		90° Elbow	45° Elbow	Center to Center		Length E
		A	B	Run C	Outlet M	
³/₄	1.050	1¹/₈	⁷/₁₆	1¹/₈	1¹/₈	1
1	1.315	1¹/₂	⁷/₈	1¹/₂	1¹/₂	1¹/₂
1¹/₂	1.900	2¹/₄	1¹/₈	2¹/₄	2¹/₄	1¹/₂
2	2.375	3	1³/₈	2¹/₂	2¹/₂	1¹/₂
2¹/₂	2.875	3³/₄	1³/₄	3	3	1¹/₂
3	3.500	4¹/₂	2	3³/₈	3³/₈	2

*Dimensions are in inches.

ANSI B16.9–1964

18–6. *Double-line symbols for steel butt-welded pipe fittings.*

18-7. *Double-line symbols for cast iron flanged fittings.*

ANSI A21.10–1964, American Water Works Assoc. and the American National Standards Assoc., by permission.

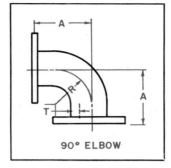

90° ELBOW

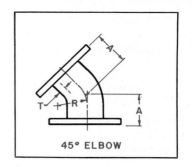

45° ELBOW

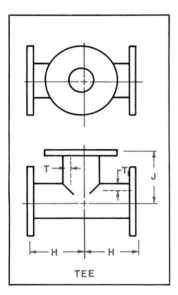

TEE

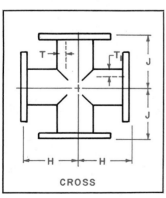

CROSS

Flanged Elbows — Cast Iron*

Nominal Pipe Size	T	90° Elbow		45° Elbow	
		A	R	A	R
2	0.35	4.5	3.0	2.5	2.44
3	0.48	5.5	4.0	3.0	3.62
4	0.52	6.5	4.5	4.0	4.81
6	0.55	8.0	6.0	5.0	7.25

*Dimensions in inches. For flange dimensions, see flange table.

Flanged Tees and Crosses — Cast Iron*

Nominal Pipe Size	T	T₁	H	J
2	0.35	0.35	4.5	4.5
3	0.48	0.35	5.5	5.5
4	0.52	0.35	6.5	6.5
6	0.55	0.35	8.0	8.0

*Dimensions in inches. For flange dimensions, see flange table.

Flange Dimensions for All Types of Flanged Fittings*

Nominal Pipe Size	Outside Diameter	Bolt Hole Circle	Flange Thickness	Bolt Hole Dia.	Bolt Dia. and Length	No. of Bolts
2	6.0	4.75	0.62	$3/4$	$5/8 \times 2 1/4$	4
3	7.50	6.00	0.75	$3/4$	$5/8 \times 2 1/4$	4
4	9.00	7.50	0.94	$3/4$	$5/8 \times 3$	8
6	11.00	9.50	1.00	$7/8$	$5/8 \times 3 1/4$	8

*Dimensions in inches.

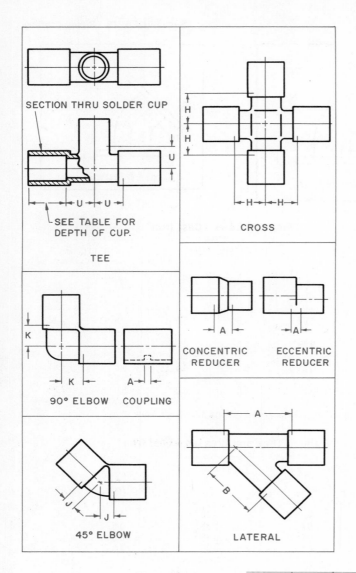

SECTION THRU SOLDER CUP

U

SEE TABLE FOR DEPTH OF CUP.

TEE

CROSS

H H

90° ELBOW   COUPLING

K

CONCENTRIC REDUCER   ECCENTRIC REDUCER

A

45° ELBOW

LATERAL

Depth of Solder Cup on Cast Brass Solder-Joint Fittings

Nominal Pipe Size	Solder Cup Depth in Inches
3/4	3/4
1	29/32
1 1/2	1 3/32
2	1 11/32
2 1/2	1 15/32
3	1 21/32

Cast Brass Solder-Joint Fittings

Nominal Pipe Size	Concentric Reducers	Eccentric Reducers
	A	A
3/4 to 1/2	5/16	7/32
1 to 3/4	3/8	1/4
1 1/2 to 1	3/8	3/16
2 to 1 1/2	1/2	9/32
3 to 2	9/16	7/16

Nominal Pipe Size	Tee	90° Elbow	45° Elbow	Coupling	Cross	Lateral	
	U	K	J	A	H	A	B
3/4	9/16	9/16	1/4	1/8	9/16	1 17/32	1 7/32
1	23/32	23/32	5/16	1/8	23/32	1 13/16	1 15/16
1 1/2	1	1	1/2	1/8	1	2 5/8	2 1/8
2	1 1/4	1 1/4	9/16	3/16	1 1/4	3 5/16	2 3/4
2 1/2	1 1/2	1 1/2	5/8	3/16	1 1/2	5	4 1/16
3	1 3/4	1 3/4	3/4	3/16	1 23/32	4 3/4	4 1/32

ANSI B16.18–1963 and Anaconda American Brass Co.

18–8.   *Double-line symbols for cast brass solder-joint fittings.*

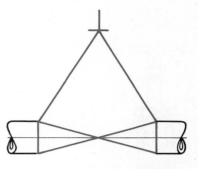

VALVE FOR INSTALLATIONS
WITH WELDED FITTINGS

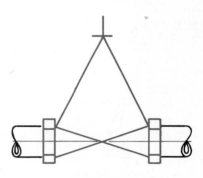

VALVE FOR INSTALLATIONS
WITH SCREWED FITTINGS

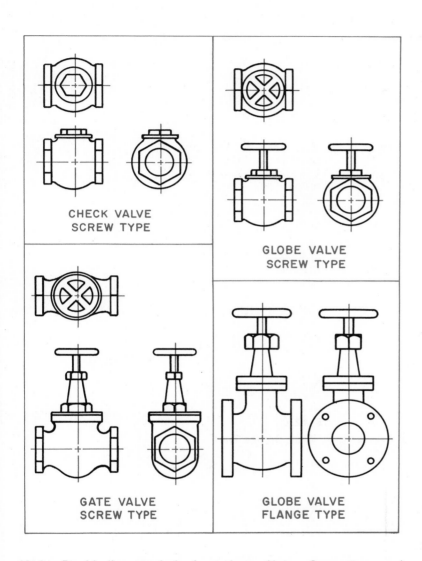

CHECK VALVE
SCREW TYPE

GLOBE VALVE
SCREW TYPE

GATE VALVE
SCREW TYPE

GLOBE VALVE
FLANGE TYPE

18-9. *Double-line symbols for valves. Note: Screw-type and flange-type valves are drawn exactly the same except for the hexagon ends or flange ends. See Figs. 18–11 through 18–17 for examples and dimensions.*

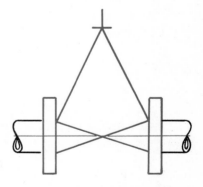

VALVE FOR INSTALLATIONS
WITH FLANGED FITTINGS

18-10. *Simplified double-line symbols for valves. The type of valve, such as gate or globe, is lettered beside the valve on the drawing.*

475

18–11.

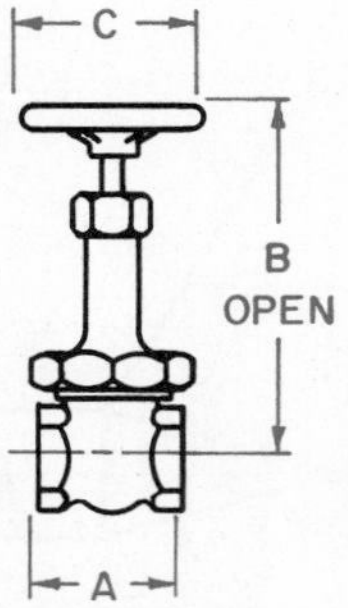

GATE VALVE
BRONZE

Gate Valves (Bronze)

DIMENSIONS (Inches)

Size	1/4	3/8	1/2	3/4	1	1 1/4	1 1/2	2	2 1/2	3
A.......	1 3/4	2	2 3/16	2 7/16	2 3/4	3	3 3/8	4	4 1/2	5
B.......	4 11/16	4 11/16	5 7/16	6 3/4	8	9 7/16	10 9/16	12 11/16	15 3/8	17 11/16
C.......	2 7/16	2 7/16	2 11/16	3	3 1/4	3 9/16	4 1/16	4 5/8	5 1/8	5 11/16
D.......	3 13/16	3 13/16	4 5/16	5 1/16	5 13/16	6 11/16	7 7/16	8 1/2	9 15/16	11 1/8

Pressure

125 PSI	Saturated Steam
200 PSI	Cold Water, Oil or Gas

Wm. Powell Co.

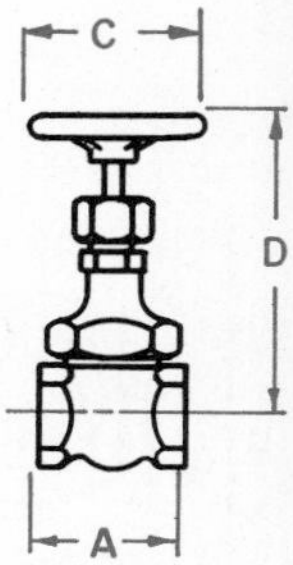

GATE VALVE
BRONZE

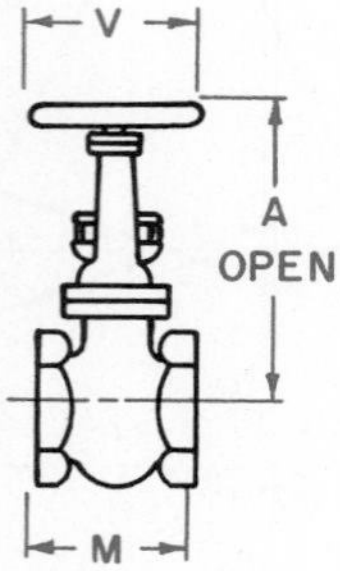

GATE VALVE
IRON

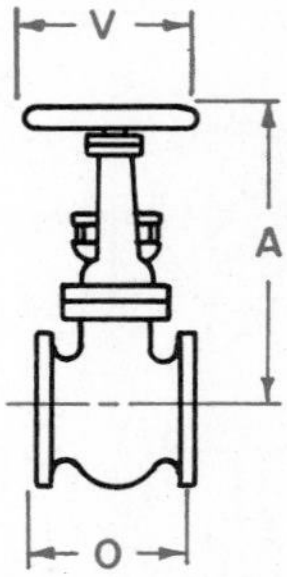

GATE VALVE
IRON

18–12.

Gate Valves (Iron Body, Bronze Trim)

DIMENSIONS (Inches)

Size	2	2 1/2	3	3 1/2	4	5	6	8	10	12	14	16	18	20	24	30
M......	5 7/16	5 7/8	6 1/8	6 1/2	6 7/8	7 3/8	7 3/4	8 3/4	—	—	—	—	—	—	—	—
O......	7	7 1/2	8	8 1/2	9	10	10 1/2	11 1/2	13	14	15	16	17	18	20	30
W......	6	7	7 1/2	8 1/2	9	10	11	13 1/2	16	19	21	23 1/2	25	27 1/2	32	38 3/4
E......	5/8	11/16	3/4	13/16	15/16	15/16	1	1 1/8	1 3/16	1 1/4	1 3/8	1 7/16	1 9/16	1 11/16	1 7/8	2 1/8
A......	14 1/4	15 3/4	18 1/4	21 1/4	23 1/2	28 1/2	32 1/2	41	50	57 1/4	65 1/2	75 1/4	84 1/4	92	107 3/4	134
V......	8	8	9	9	10	11	12	14	16	18	22	22	26	26	30	36

Pressure

	Screwed	Flanged	
2 to 12	125 PSI	125 PSI	Saturated Steam
	200 PSI	200 PSI	Cold Water, Gas or Oil

Wm. Powell Co.

18–13.
Globe and Angle Valves

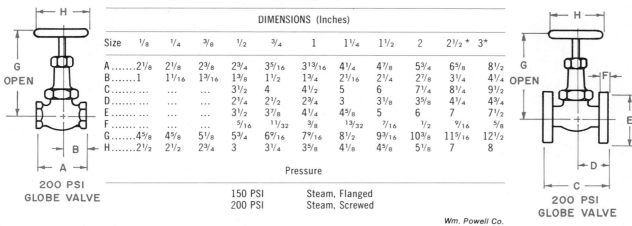

Size	$\frac{1}{8}$	$\frac{1}{4}$	$\frac{3}{8}$	$\frac{1}{2}$	$\frac{3}{4}$	1	$1\frac{1}{4}$	$1\frac{1}{2}$	2	$2\frac{1}{2}$*	3*
A	$2\frac{1}{8}$	$2\frac{1}{8}$	$2\frac{3}{8}$	$2\frac{3}{4}$	$3\frac{5}{16}$	$3\frac{13}{16}$	$4\frac{1}{4}$	$4\frac{7}{8}$	$5\frac{3}{4}$	$6\frac{5}{8}$	$8\frac{1}{2}$
B	1	$1\frac{1}{16}$	$1\frac{3}{16}$	$1\frac{3}{8}$	$1\frac{1}{2}$	$1\frac{3}{4}$	$2\frac{1}{16}$	$2\frac{1}{4}$	$2\frac{7}{8}$	$3\frac{1}{4}$	$4\frac{1}{4}$
C	...	...	...	$3\frac{1}{2}$	4	$4\frac{1}{2}$	5	6	$7\frac{1}{4}$	$8\frac{1}{4}$	$9\frac{1}{2}$
D	...	...	...	$2\frac{1}{4}$	$2\frac{1}{2}$	$2\frac{3}{4}$	3	$3\frac{1}{8}$	$3\frac{5}{8}$	$4\frac{1}{4}$	$4\frac{3}{4}$
E	...	...	...	$3\frac{1}{2}$	$3\frac{7}{8}$	$4\frac{1}{4}$	$4\frac{5}{8}$	5	6	7	$7\frac{1}{2}$
F	...	...	...	$\frac{5}{16}$	$\frac{11}{32}$	$\frac{3}{8}$	$\frac{13}{32}$	$\frac{7}{16}$	$\frac{1}{2}$	$\frac{9}{16}$	$\frac{5}{8}$
G	$4\frac{5}{8}$	$4\frac{5}{8}$	$5\frac{1}{8}$	$5\frac{3}{4}$	$6\frac{9}{16}$	$7\frac{9}{16}$	$8\frac{1}{2}$	$9\frac{3}{16}$	$10\frac{3}{8}$	$11\frac{5}{16}$	$12\frac{1}{2}$
H	$2\frac{1}{2}$	$2\frac{1}{2}$	$2\frac{3}{4}$	3	$3\frac{1}{4}$	$3\frac{5}{8}$	$4\frac{1}{8}$	$4\frac{5}{8}$	$5\frac{1}{8}$	7	8

DIMENSIONS (Inches)

Pressure

| 150 PSI | Steam, Flanged |
| 200 PSI | Steam, Screwed |

200 PSI
GLOBE VALVE

200 PSI
GLOBE VALVE

Wm. Powell Co.

18–14.
Globe and Angle Valves

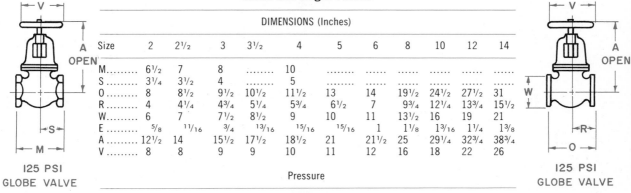

DIMENSIONS (Inches)

Size	2	$2\frac{1}{2}$	3	$3\frac{1}{2}$	4	5	6	8	10	12	14
M	$6\frac{1}{2}$	7	8	...	10	...	...	...	...	...	...
S	$3\frac{1}{4}$	$3\frac{1}{2}$	4	...	5	...	...	...	...	...	...
O	8	$8\frac{1}{2}$	$9\frac{1}{2}$	$10\frac{1}{2}$	$11\frac{1}{2}$	13	14	$19\frac{1}{2}$	$24\frac{1}{2}$	$27\frac{1}{2}$	31
R	4	$4\frac{1}{4}$	$4\frac{3}{4}$	$5\frac{1}{4}$	$5\frac{3}{4}$	$6\frac{1}{2}$	7	$9\frac{3}{4}$	$12\frac{1}{4}$	$13\frac{3}{4}$	$15\frac{1}{2}$
W	6	7	$7\frac{1}{2}$	$8\frac{1}{2}$	9	10	11	$13\frac{1}{2}$	16	19	21
E	$\frac{5}{8}$	$\frac{11}{16}$	$\frac{3}{4}$	$\frac{13}{16}$	$\frac{15}{16}$	$\frac{15}{16}$	1	$1\frac{1}{8}$	$1\frac{3}{16}$	$1\frac{1}{4}$	$1\frac{3}{8}$
A	$12\frac{1}{2}$	14	$15\frac{1}{2}$	$17\frac{1}{2}$	$18\frac{1}{2}$	21	$21\frac{1}{2}$	25	$29\frac{1}{4}$	$32\frac{3}{4}$	$38\frac{3}{4}$
V	8	8	9	9	10	11	12	16	18	22	26

Pressure

125 PSI Steam, Screwed and Flanged

125 PSI
GLOBE VALVE

125 PSI
GLOBE VALVE

Wm. Powell Co.

18–15.
Relief Valves

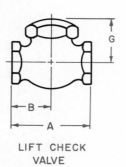

RELIEF
VALVE

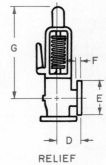

RELIEF
VALVE

DIMENSIONS (Inches)								
Size	1/2	3/4	1	1 1/4	1 1/2	2	2 1/2	3
B	1 11/16	1 9/16	2 1/16	2 5/16	2 11/16	3 1/8	3 1/2	3 15/16
D	2 1/2	2 1/2	2 5/8	3	3 1/4	3 7/8	4 1/2	5
E	3 1/2	3 7/8	4 1/4	4 5/8	5	6	7	7 1/2
F	11/32	11/32	7/16	1/2	9/16	5/8	11/16	3/4
G	7 13/16	7 7/8	9 3/8	10 5/8	12 1/4	14 5/8	17 1/2	18

Pressure

Adjustable from 10 to 300 PSI

Wm. Powell Co.

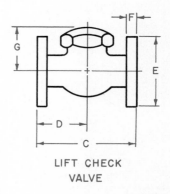

LIFT CHECK
VALVE

LIFT CHECK
VALVE

18–16.
Lift Check Valves

DIMENSIONS (Inches)											
Size	1/8	1/4	3/8	1/2	3/4	1	1 1/4	1 1/2	2	2 1/2	3
A	2 1/8	2 1/8	2 3/8	2 3/4	3 5/16	3 13/16	4 1/4	4 7/8	5 3/4	6 5/8	8 1/2
B	1 1/16	1 1/16	1 3/16	1 3/8	1 1/2	1 3/4	2 1/16	2 1/4	2 7/8	3 1/4	4 1/4
C		3	3	3 1/2	4	4 1/2	5	6	7 1/4	8 1/4	9 1/2
D		1 5/8	1 5/8	2 1/4	2 1/2	2 3/4	3	3 1/8	3 5/8	4 1/4	4 3/4
E		2 1/2	2 1/2	3 1/2	3 7/8	4 1/4	4 5/8	5	6	7	7 1/2
F		9/32	9/32	5/16	11/32	3/8	13/32	7/16	1/2	9/16	5/8
G	1 1/4	1 1/4	1 7/16	1 5/8	1 15/16	2 3/16	2 1/2	2 3/4	3 1/4	3 1/2	4 1/4

Pressure

Screwed | 200 PSI | Steam at 550°
| 400 PSI | Cold Water, Oil, or Gas

Flanged | 150 PSI | Steam at 500°
| 225 PSI | Cold Water, Oil, or Gas

Wm. Powell Co.

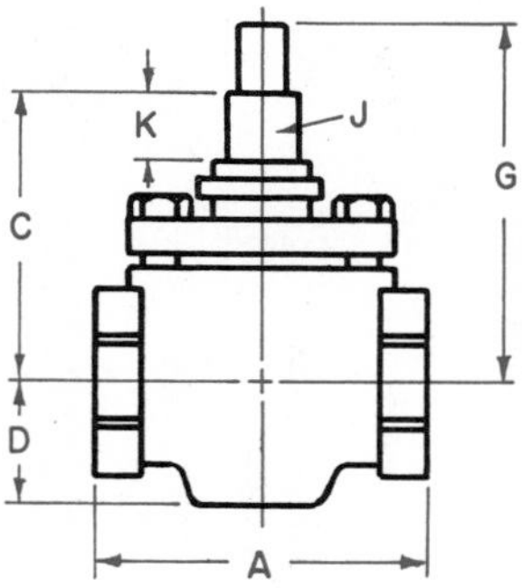

PLUG VALVE

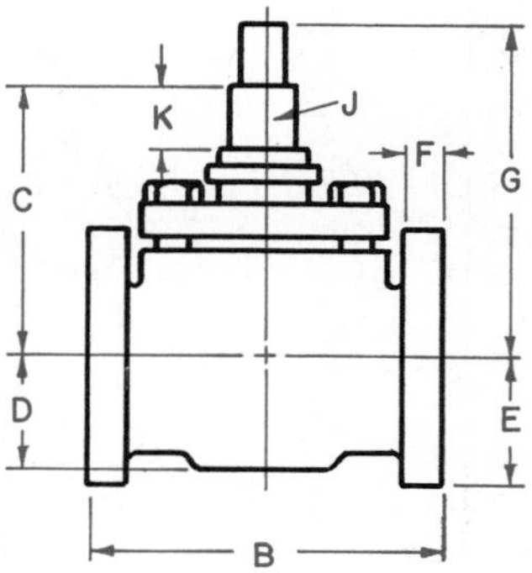

PLUG VALVE

18–17.
Plug Valves

DIMENSIONS (Inches)

Size	1	1 1/4	1 1/2	2	2 1/2	3	4
A End to End, screwed	4 1/2	5	5	5 7/8	7	7 5/8	9
B Face to Face, flanged	5 1/2	6 1/2	6 1/2	7	7 1/2	8	9
C Center line to top of shank (Approx.)	4 1/8	4 11/16	4 11/16	5 9/16	6 1/4	7	7 7/8
D Center line to bottom of body	1 9/16	2 1/16	2 1/16	2 5/16	2 9/16	3 3/16	3 3/4
E Diameter of flanges	4 1/4	4 5/8	5	6	7	7 1/2	9
F Thickness of flanges	7/16	9/16	9/16	5/8	11/16	3/4	15/16
G Clearance required to remove lubricant fitting	6 5/8	7 3/16	7 3/16	8 1/16	9 7/16	10 3/16	11 11/16
J Width of square of shank	13/16	29/32	29/32	1 1/16	1 1/4	1 7/16	1 19/32
K Height of square of shank	3/4	3/4	3/4	1 1/16	1 1/4	1 7/16	1 19/32
Number and size of flange bolts	4-1/2	4-1/2	4-1/2	4-5/8	4-5/8	4-5/8	8-5/8
Diameter of flange bolt circle	3 1/8	3 1/2	3 7/8	4 3/4	5 1/2	6	7 1/2
Size of straight pipe tap and lubricant fitting	1/4-18	1/4-18	1/4-18	1/4-18	3/8-18	3/8-18	1/2-14
Size of lubricant stick	B	B	B	B	C	C	D
Extreme width of body	2 3/8	3	3	3 9/16	4 3/16	4 7/8	6 3/8
Size of wrench	A-5	A-6	A-6	A-9	A-12	A-15	A-18
Length of standard wrench	5 1/2	6	6	9	12	15	18

Pressure

175 PSI Working Pressure

Wm. Powell Co.

Spence/Drafting Technology and Practice

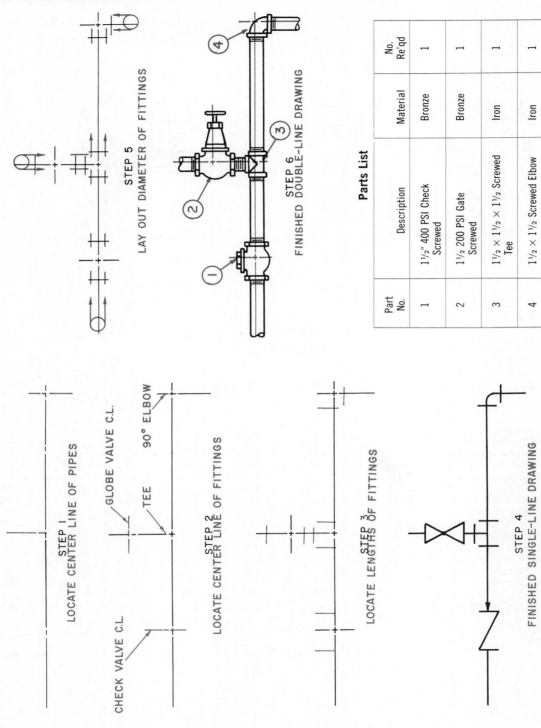

Parts List

Part No.	Description	Material	No. Re'qd
1	1½" 400 PSI Check Screwed	Bronze	1
2	1½ 200 PSI Gate Screwed	Bronze	1
3	1½ × 1½ × 1½ Screwed Tee	Iron	1
4	1½ × 1½ Screwed Elbow	Iron	1

STEP 5
LAY OUT DIAMETER OF FITTINGS

STEP 6
FINISHED DOUBLE-LINE DRAWING

GLOBE VALVE C.L.

90° ELBOW

TEE

CHECK VALVE C.L.

STEP 1
LOCATE CENTER LINE OF PIPES

STEP 2
LOCATE CENTER LINE OF FITTINGS

STEP 3
LOCATE LENGTHS OF FITTINGS

STEP 4
FINISHED SINGLE-LINE DRAWING

18–18. *How to make a piping drawing.*

480

MAKING A PIPING DRAWING

A proposed piping drawing is laid out in preliminary form by freehand sketching. This enables the draftsmen to plan, study, and determine the best layout for each problem. Once the entire drawing is sketched with required details and notes shown, a final decision or review is made to insure that an effective, complete layout has been developed. Then a final drawing can be made, using the following steps.

1. Lay out the center lines of all pipes and major equipment with fine lines. Both single and double-line drawings are started this way. See Fig. 18-18, Step 1.

2. Locate the center of each fitting on the pipe center line. See Fig. 18-18, Step 2.

3. Lay out the length of the fittings. See Fig. 18-18, Step 3.

4. If it is a single-line drawing, complete the single-line symbols. See Fig. 18-18, Step 4.

5. If it is a double-line drawing, lay out the length and diameter. See Fig. 18-18, Step 5. Data on size can be found in manufacturers' catalogs. Some sizes are given in Figs. 18-5 through 18-17.

6. Draw the double-line symbols for fittings and pipes. See Fig. 18-18, Step 6. Complete the parts list.

Other Drawing Considerations. Provisions for field assembly must be built into the piping layout. For example, threaded pipe must be longer than the distance it is drawn. This is because some of the pipe screws inside the fittings. Flanged fittings have gaskets to seal them and space must be allowed in the overall length for gaskets.

Flanged fittings are joined by bolts. This permits easy removal of valves and other parts for field repairs without removing sections of pipe as would be necessary with threaded fittings. Flanges are

18-20. *A pump installation with bolted fittings and butt-welded fittings. It has been designed with clearance to provide ease of maintenance.*

Buffalo Forge Co.

18-19. *A lift station with bolted, flanged fittings. They make it possible to remove easily sections of the installation.*

The Gorman-Rupp Co.

FLANGED FITTINGS

THIONYL STORAGE

WELDED FITTINGS

481

often used with welded piping around valves, pumps, and vessels. See Fig. 18–19.

Some piping is joined by welding. Usually large diameter pipe is welded. It can be fabricated in sections and then joined in place by welding. Welded fittings make a strong, permanent seal. Welding symbols used are those established by the American Welding Society. These symbols are explained in Chapter 26. Welded fittings are shown in Fig. 18–20.

When piping is joined by soldering, the fittings have a socket prepared to receive the end of the pipe. See Fig. 18–8. This connection system is commonly used for pipe sizes of 1/8 inch diameter to 12 inches in diameter. See Fig.

CAST BRASS SOLDER JOINT FITTINGS

BRASS TO CAST IRON

Anaconda American Brass Co.

18–21. *A water meter installation using cast brass solder-joint fittings. Notice the flanged cast iron fittings connected to the brass pipe.*

18–21 for a typical installation.

It is desirable for long runs of pipe to be grouped together. This is called a *pipe lane.* Piping in these lanes can be at various levels, depending on the equipment they serve. It is important to locate lanes where unnecessary sharp changes in directions can be avoided. This provides savings in materials and maintenance of the lines when in operation. Notice the pipe lanes show clearly in the aerial photo, Fig. 18–22.

Clearance — free space around pipes and equipment — is necessary so repairs can be made after the system is in operation. The clearance should be large enough to permit the use of wrenches to remove fittings, and to permit use of equipment for installations, repairs, and maintenance. Notice the clearances allowed around the pump installation, Fig. 18–23, allow for easy maintenance.

Plans should include supports for long runs of pipe or piping mounted on building frames or equipment. These supports have to be strong enough to support the pipe and withstand wind loads. Notice the supports shown in Figs. 18–1 and 18–22.

DIMENSIONING AND CALLOUTS

Piping drawings are dimensioned much the same as working drawings. Pipes are located by their center lines. Valves, traps, and other fittings are dimensioned by relating their center lines to the center line of another part. Pipe sizes are indicated by a note beside the pipe. A leader is sometimes used.

Sometimes the size and description of the fittings are not placed on the drawing. See Fig. 18–23. This means the worker installing the system must interpret the symbol. He must decide on the size of fittings and valves by the size of the pipe to be used. Much confusion can be avoided if each fitting is labeled.

Fittings should be identified by size, name, and material. The size is the nominal diameter of the pipe to join the fitting. See Figs. 18–2 and 18–4. For example, if the pipe is two inches in diameter, a two-inch elbow would be used. A typical fitting callout would read $2 \times 2 \times 2$ screwed tee, malleable iron. If the pressure rating is specified, it is added after the material, as $2 \times 2 \times 2$ screwed tee, malleable iron, 125 psi. Pressures for selected pipe can be found in Figs. 18–25 to 18–27.

Valves are indicated by giving the nominal size of the valve, the pressure rating, material, type of valve, and the type of connections used. A typical valve callout would read 2″, 150 psi., bronze, gate valve, screwed. Pressures for selected valves can be found in Figs. 18–11 to 18–17.

18–22. *Long runs of pipe, called pipe lanes, should be grouped together. Can you find the pipe lanes on the photo of the processing center?*

Some valves use two materials. A typical example is one with an iron body and bronze stems, wedges, and seats. The bronze parts are called trim. The material would be specified iron body, bronze trim.

Sometimes the fittings and valves are numbered on the drawing and the specifications are given on a parts list. See Fig. 18–18.

Usually the overall dimensions of fittings are not shown since they are standard. If clearance is a problem, they could be dimensioned. This is especially useful on detail drawings.

When dimensioning, it is best to locate all pipes and fittings from an established

location. In a building, this could be a wall. Figs. 18–4 and 18–23 show dimensioned drawings.

Some drawings used for illustrative purposes are not dimensioned but have each major part labeled by name. This makes it easier to read the drawing. See Fig. 18–24.

PIPE

Many different kinds of pipe are made. Things to consider when selecting pipe are heat, pressure, and chemical reactions between the pipe material and the gas or liquid to be run through it. Commonly used materials are copper, steel, cast iron, wrought iron, brass, lead, and plastics.

Copper pipe will withstand corrosion. It can carry oil, gas, steam, and sewage. It is good for carrying drinking water since minerals are less likely to stick to it.

Steel pipe will handle high temperatures and pressures. It is one of the most commonly used types. It is sometimes galvanized with zinc to prevent rust. When galvanized, it is used for carrying drinking water.

Cast iron pipe is used underground for water, gas, sewer, and low pressure steam lines. It resists corrosion. Strain and shock will break it.

Wrought iron pipe will withstand corrosion. It is used for the same applications as steel.

483

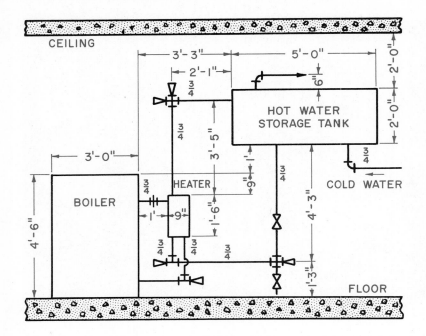

CEILING

3'-3"

2'-1"

3/4

5'-0"

2'-0"

6"

HOT WATER
STORAGE TANK

2'-0"

3/4

3'-0"

3/4

3'-5"

3/4

COLD WATER

3/4

BOILER

4'-6"

3/4

HEATER

6"

1'-6"

1'

9"

3/4

3/4

3/4

4'-3"

1'-3"

FLOOR

18-23. *General arrangement single-line piping diagram. It shows location of major items in relation to each other and the floor and ceiling.*

It is more expensive and is not as strong as steel pipe.

Brass pipe is used for handling corrosive liquids. It can be used underground. It will handle high pressures. Wrought iron and steel pipe are also used for this purpose. Brass pipe is more expensive than either of these.

Lead pipe is used to handle liquids containing acids. It is used in plumbing installations. It is flexible and will withstand vibrations and expansion. It is available in standard sizes.

Plastic pipe is used to carry drinking water and is resistant to acids and chemicals. It is flexible but will not withstand high pressure or heat. It is not used for oil. Plastic pipe can be used underground.

Pipe Sizing and Pressure. The most commonly used pipes for above-ground installations are wrought iron, steel, and copper. They will be the only types considered in this chapter.

The sizes of pipes under 14 inches are specified by the nominal inside diameter. For example, a three-inch pipe has a hole that is three inches in diameter. The outside diameter varies with the wall thickness.

Wrought iron and steel pipe are commonly made in three weights. These are standard, extra strong, and double extra strong. The stronger pipe has thicker walls. See Figs. 18-25 to 18-27. There are other weights that are used for special purposes.

Copper and red brass pipe are available in standard and extra strong weights. See Fig. 18-25.

The normal pressures for wrought iron, steel, copper, and red brass pipe are given in Figs. 18-25 through 18-27. These pressures are dependent upon a good job of installing the fittings. A poorly installed fitting can reduce the amount of pressure a system can carry.

Pipe Threads. Pipe threads are drawn using the same means of representation as other threads, Fig. 18-28. Two types used are taper and straight. A taper thread has a slope of 1/16 inch per foot. On industrial piping drawings, the threads are omitted on screwed fittings.

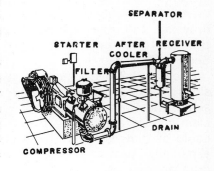

Compressed Air and Gas Institute

18-24. *A pictorial drawing used to show a typical installation.*

18–25.
Red Brass and Copper Pipe

Nominal Pipe Size	Outside Dia.	Standard Weight			Extra Strong
		Wall Thickness	Pressure-psi*		Wall Thickness
			Red Brass	Copper	
3/4	1.050	.114	1800	1350	.157
1	1.315	.126	1580	1190	.182
1 1/2	1.900	.150	1280	960	.203
2	2.375	.156	1050	790	.221
2 1/2	2.875	.187	1040	780	.280
3	3.500	.219	1000	750	.304
4	4.500	.250	880	660	.341
6	6.625	.250	600	450	.437

*Welded, brazed, or solder fittings.

Revere Copper and Brass Inc.

18–26.
Welded Steel Pipe

Nominal Pipe Size	Outside Dia.	Standard Weight		Wall Thickness	Extra Strong	
		Wall Thickness	Pressure-psi		Pressure-psi	
			Butt Weld		Butt Weld	Elec. Weld
3/4	1.050	.113	700	.154	850	
1	1.315	.133	700	.179	850	
1 1/2	1.900	.145	800	.200	1100	
2	2.375	.154	800	.218	1100	1500
2 1/2	2.875	.203	800	.276	1100	1500
3	3.500	.216	800	.300	1100	1500
4	4.500	.237	1200	.337	1700	1700
6	6.625	.280		.432		1700
8	8.625			.500		1700
10	10.750			.500		1600

Piping Handbook, 4th Edition
Republic Steel Corp.

18-27.
Wrought Iron Pipe

Nominal Pipe Size	Outside Dia.	Standard Weight			Extra Strong		
		Wall Thickness	Pressure-psi		Wall Thickness	Pressure-psi	
			Butt Weld	Lap Weld		Butt Weld	Lap Weld
¾	1.050	.115	700	—	.157	850	—
1	1.315	.136	700	—	.183	850	—
1½	1.900	.148	800	1000	.204	1100	1500
2	2.375	.158	800	1000	.223	1100	1500
2½	2.875	.208	—	1000	.282	—	1500
3	3.500	.221	—	1000	.306	—	1500
4	4.500	.242	—	1200	.344	—	1700
6	6.625	.286	—	1200	.441	—	1700
8	8.625	.329	—	1200	.510	—	1700
10	10.750	.336	—	1000	.510	—	1600

A. M. Byers Co.

Pipe 2½ inches in diameter or smaller, used under standard pressures and normal temperatures, usually have threaded fittings. Both internal and external threads are usually tapered. An internal thread is one cut inside a fitting. A thread inside an elbow is an example. External threads are cut on the outside of a member. A pipe threaded on the end is an example.

For certain uses, fittings are made having external threads tapered and internal threads straight. Common uses are for pressure-tight couplings, joints for fuel and oil fittings, mechanical fittings for fix-tures, and joints with lock nuts or hose connections.

Fittings. Pipe fittings are used to join pipe lengths, change direction, or connect pipes with different diameters. They are made from many different materials. The fittings are of the same material as the pipe used. Some permit the connection of pipe of different materials. For example, a fitting is available that permits connecting copper pipe to steel pipe.

A wide variety of pipe fittings are made. However, this chapter includes only a few selected types. Manufacturers' catalogs must be consulted for a complete listing. The following types of fittings

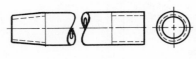

TAPERED   STRAIGHT
EXTERNAL THREADS

INTERNAL THREADS

18-28. *Conventional representation of pipe threads.*

are shown: selected malleable iron screwed fittings, Fig. 18–5; steel fittings that have been butt-welded, Fig. 18–6; cast iron flanged fittings, Fig. 18–7; and cast brass solder-joint fittings, Fig. 18–8.

CONTROLS

A piping system directs and controls the flow of liquids and gases. The material flowing through the pipes may be very cold or very hot, under low or great pressures. Valves are used for control purposes.

The basic functions of valves are to start, stop, and regulate flow, prevent back flow, regulate and relieve pressure.

There are several basic valve designs: gate, globe, relief, ball, butterfly, angle, plug, and check. Some of these are shown in Fig. 18–29.

A *gate valve* is a free-flow valve. When the gate valve is open, the liquid flows easily. It is not designed to regulate the rate of flow.

A *globe valve* permits closer regulation of flow and pressure by adjusting an opening through the seat to various sizes.

Angle valves serve the same purpose as a globe valve. They are used when the design requires the flow to leave the valve at a 90-degree angle.

A *ball valve* is a quick opening, on-and-off valve that permits a complete, unrestricted flow and is ideal for heavy consistency liquids.

A *cock valve* is used when it is necessary to control liquids through closed conduits under extreme conditions.

A *butterfly valve* derives its name from the flat disk flapper that closes against a resilient liner. It is an efficient and economical valve for most liquid and gas or air services.

Check valves permit flow in one direction. A liquid can flow through the valve but not be forced back through it.

A *relief valve* is a safety device. It remains closed until the pressure in a system rises to an unsafe level. It then is forced open by the pressure. Most are held closed by a spring.

PROCESS FLOW DIAGRAMS

A *process flow diagram* is the first drawing made when planning the piping of a factory or plant. It shows the basic items of major equipment and their relation to each other. The major flow lines are indicated as connecting these pieces of equipment. They help describe how the process operates.

Of great importance in process flow diagrams is simplicity in symbol forms. They preserve the general physical appearance of the equipment yet require a minimum of drafting effort.

Process flow diagrams are not drawn to scale. The symbols are drawn so they are in keeping with the overall size of the drawing.

The symbols are arranged on the drawing in a logical sequence of flow. They start where the material to be processed enters the system and follows the process through to the production of the main product. It is desirable to keep crossover of lines to a minimum. The diagram does not represent the actual location of the parts of the plant when it is built. The diagram represents the process. See Fig. 18–30.

Equipment outlines are drawn with a heavy line. Piping is drawn with a thin line. Usually instrumentation or electrical symbols are not placed on process flow diagrams. The direction of flow is shown by arrowheads.

Some commonly used symbols for process flow diagrams are shown in Fig. 18–31. For a complete listing, refer to American National Standards Institute Bulletin Y32.11–1961. Process quantities such as pressure, temperature, and flow are recorded inside symbols given a special shape. For example, a circle, the symbol for pressure, with 100 psi. lettered inside means a pressure of 100 pounds per square inch.

ORTHOGRAPHIC PROJECTION

An orthographic projection of a piping system includes a top, front, and side view and can be dimensioned. See Fig. 18–32. Usually at least two views must be drawn unless one view can show the necessary details.

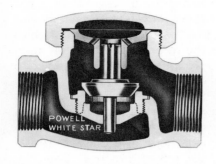

18-29A. A screwed-end horizontal lift check valve.

18-29B. A screwed-end plug.

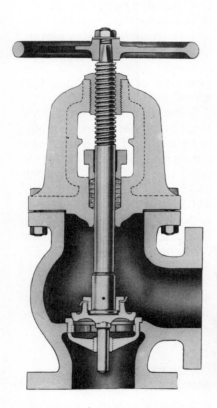

18-29C. A flanged-end angle valve.

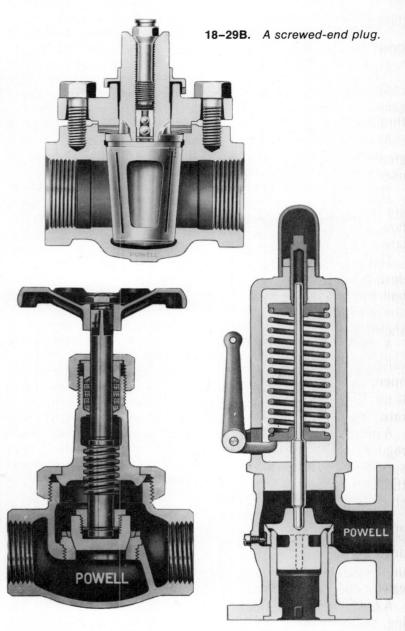

18-29D. A screwed-end globe valve.

18-29E. A flanged-end relief valve.

Wm. Powell Co.

18-29A-G. Interior details of some commonly used valves.

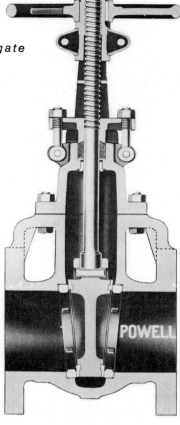

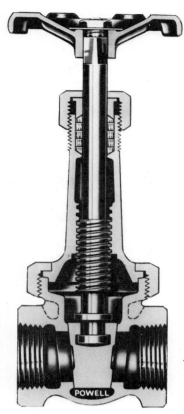

18–29G. *A flanged-end gate valve (iron).*

18–29F. *A screwed-end gate valve (bronze).*

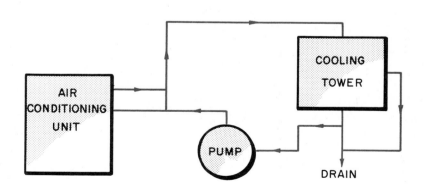

18–30. *Flow diagram for an air conditioning cooling tower installation. The arrows show the direction of flow.*

489

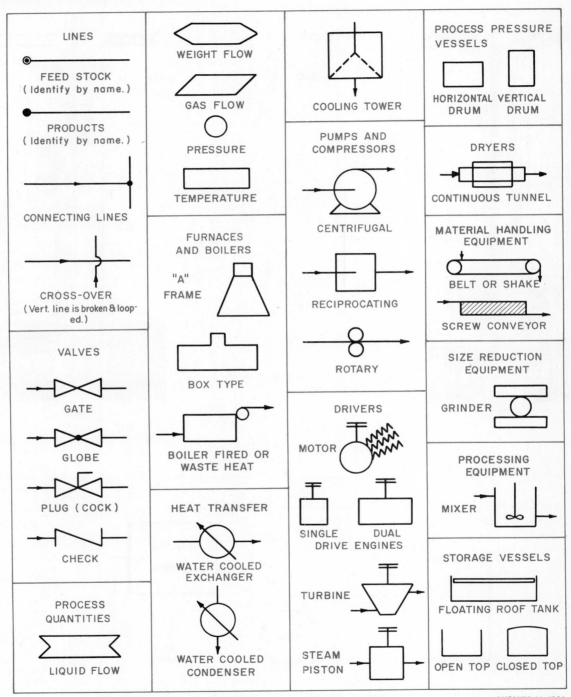

18–31. *Process flow diagram symbols.*

ANSI Y32.11–1961

A one-view drawing should show where each part is connected. The worker making the installation will take care of on-the-job details of the installation.

Some piping installations are very complex. The pipes and fittings overlap each other and make reading the drawing difficult. These installations can be clarified by drawing it pictorially or with a developed view.

Developed Views. Another plan used to clarify orthographic drawings that are confusing is to imagine the entire system flattened out on a plane. The vertical parts can be revolved to a horizontal plane, or horizontal parts can be revolved to a vertical plane. All fittings appear turned sideways, and all pipe is true length. Fig. 18–33 shows a developed drawing of the system in Fig. 18–32.

GENERAL ARRANGEMENT PIPING DIAGRAMS

A general arrangement drawing shows the location of machinery and the connecting pipes. It gives information needed to locate the pipes and machinery within the building. See Fig. 18–34.

General arrangement drawings can be single or double line. Single-line drawings are most commonly used. See Fig. 18–23. Detail drawings are then used to identify parts or clarify the assembly of one section of the larger diagram.

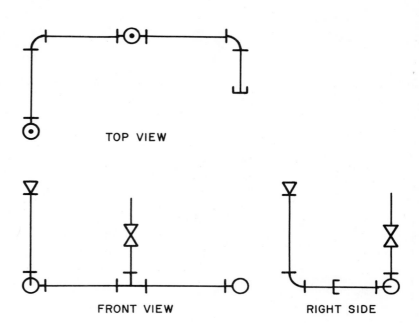

18–32. *Orthographic views of part of a piping system. This is a three-view detail drawing.*

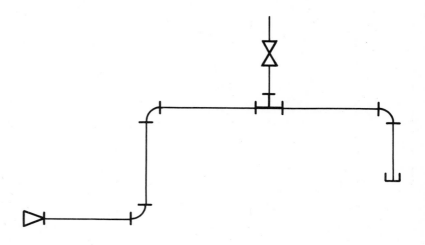

18–33. *A developed view on a horizontal plane of the piping system, Fig. 18–32.*

491

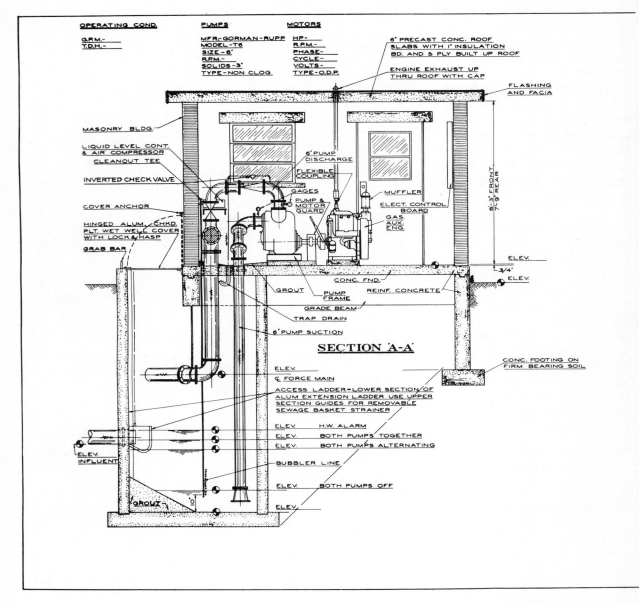

OPERATING COND.

G.P.M.-
T.D.H.-

PUMPS

MFR-GORMAN-RUPP
MODEL-T6
SIZE-6"
R.P.M.-
SOLIDS-3"
TYPE-NON CLOG

MOTORS

H.P.-
R.P.M.-
PHASE-
CYCLE-
VOLTS-
TYPE-O.D.P.

6" PRECAST CONC. ROOF
SLABS WITH 1" INSULATION
BD. AND 5 PLY BUILT UP ROOF

ENGINE EXHAUST UP
THRU ROOF WITH CAP

FLASHING
AND FACIA

MASONRY BLDG.

LIQUID LEVEL CONT.
& AIR COMPRESSOR
CLEANOUT TEE

INVERTED CHECK VALVE

COVER ANCHOR

HINGED ALUM. CHKD.
PLT. WET WELL COVER
WITH LOCK & HASP

GRAB BAR

6" PUMP
DISCHARGE

FLEXIBLE
COUPLING

GAGES

PUMP &
MOTOR
GUARD

MUFFLER

ELECT. CONTROL
BOARD

GAS.
AUX.
ENG.

6'-3" FRONT
7'-9" REAR

ELEV.
3/4"

ELEV.

CONC. FND.

GROUT
PUMP
FRAME

REINF. CONCRETE

GRADE BEAM

TRAP DRAIN

6" PUMP SUCTION

SECTION 'A-A'

CONC. FOOTING ON
FIRM BEARING SOIL

ELEV.
₵ FORCE MAIN

ACCESS LADDER-LOWER SECTION OF
ALUM EXTENSION LADDER USE UPPER
SECTION GUIDES FOR REMOVABLE
SEWAGE BASKET STRAINER

ELEV.   H.W. ALARM
ELEV.   BOTH PUMPS TOGETHER
ELEV.   BOTH PUMPS ALTERNATING

ELEV.
INFLUENT

BUBBLER LINE

ELEV.   BOTH PUMPS OFF

GROUT

ELEV.

Gorman-Rupp Co.

18–34. *A general arrangement drawing. It shows the top and front view, section AA, of a sewage pump station. The equipment is specified by name. The location of the equipment in the pump station is dimensioned.*

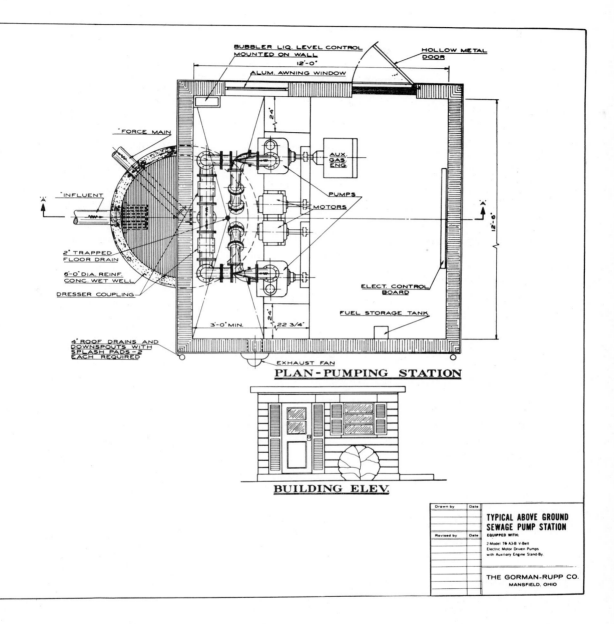

BUBBLER LIQ. LEVEL CONTROL
MOUNTED ON WALL
12'-0"
HOLLOW METAL DOOR
ALUM. AWNING WINDOW
"FORCE MAIN
24"
AUX. GAS. ENG.
"INFLUENT
PUMPS
& MOTORS
12'-6"
2" TRAPPED FLOOR DRAIN
6'-0" DIA. REINF. CONC. WET WELL
DRESSER COUPLING
ELECT. CONTROL BOARD
FUEL STORAGE TANK
3'-0" MIN.
24"
22 3/4"
4" ROOF DRAINS AND DOWNSPOUTS WITH SPLASH PADS - 2 EACH REQUIRED
EXHAUST FAN

PLAN - PUMPING STATION

BUILDING ELEV.

Drawn by	Date	**TYPICAL ABOVE GROUND SEWAGE PUMP STATION**
		EQUIPPED WITH:
Revised by	Date	2-Model T6 A3-B V-Belt Electric Motor Driven Pumps with Auxiliary Engine Stand-By.
		THE GORMAN-RUPP CO.
		MANSFIELD, OHIO

18–35.

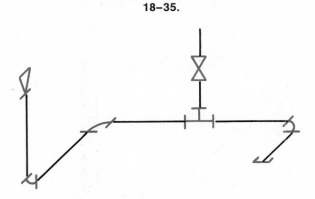

An oblique drawing of system shown in Fig. 18–32.

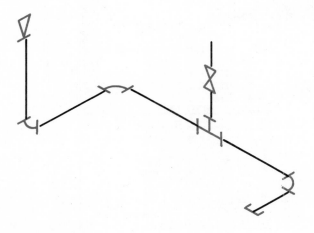

An isometric drawing of system shown in Fig. 18–32.

the general arrangement drawing or on separate sheets. A detail may be of one part or the assembly of several parts. Fig. 18–32 shows a detail drawing of one part of a larger system.

Detail drawings are usually drawn to the scale of $1'' = 1'0''$. A larger scale is better and should be used if possible.

PICTORIAL REPRESENTATION

Pictorial drawings are the easiest to read. Common methods used are oblique, isometric, dimetric, and trimetric. These methods, when used with orthographic views will clearly show how to fabricate and install piping.

When drawn in oblique, some parts of the system are parallel to the frontal reference plane. Other parts are drawn obliquely to these. The draftsman must lay out the system so it clearly shows the details. See Fig. 18–35.

Isometric piping drawings are made in the same way as any isometric drawing. All angles on the axis are equal. All vertical pipes are drawn vertical. Horizontal pipes recede to the right or left on a 30-degree angle. See Fig. 18–35.

Dimetric drawings have two angles equal and one different. Trimetric drawings have three angles different. They tend to give a more natural appearance to a piping drawing. Refer to Chapter 9, Pictorial Drawing, for directions on making pictorial drawings.

Pictorial drawings can be used to clarify parts of a general arrangement drawing. They can contain notes about pipe size, type of fittings, valves, or other information that would aid clarity.

General arrangement drawings have each part located

and related to a permanent part of the structure in which the system is to be installed by dimensions. See Fig. 18–34.
Detail Drawings. Detail drawings are made to clarify one portion of a complete general arrangement drawing. They may be on the same sheet as

Additional Reading

American National Standards Institute, 1430 Broadway, New York. Bulletins issued by the ANSI as follows:

Black and Hot Dipped Zinc Coated Welded and Seamless Steel Pipe for Ordinary Uses. B36.20–1964.

Cast Iron Pipe Flanges and Flanged Fittings. B16.1–1967.

Cast Iron Screwed Fittings, 125 and 250 Pounds. B16.4–1963.

Face to Face and End to End Dimensions of Ferrous Valves. B16.10–1957.

Graphical Symbols for Plumbing. Y32.4–1955.

Malleable Iron Screwed Fittings, 150 and 300 Pounds. B16.3–1963.

Symbols for Process Flow Diagrams. Y32.11–1961.

Steel Pipe Flanges and Flanged Fittings. B16.5–1968.

Welded and Seamless Steel Pipe. B36.1–1966.

Wrought Steel and Wrought Iron Pipe. B36.10–1959.

Babbitt, Harold E. *Plumbing.* New York: McGraw-Hill Book Co., 1960.

McGuinness, W. J., B. Stein, C. M. Gay, and C. DeVan Fawcett. *Mechanical and Electrical Equipment for Buildings.* New York: John Wiley and Sons, Inc., 1964.

Thompson, Charles H. *Fundamentals of Pipe Drafting.* New York: John Wiley and Sons, Inc., 1958.

Build Your Vocabulary

Following are terms that you should understand and use as a part of your working vocabulary. Write a brief explanation of what each means.

Single-line drawing
Double-line drawing
Symbols
Flanged fittings
Pipe lane
Fitting
Valve
Process flow diagram
Piping detail drawing

Study Problems

Flow Diagrams

1. Make a flow diagram for the piping system, P18–1.

Single-line Diagrams

2. Make a one-view, single-line drawing of the detail shown in P18–4. Select your own scale. The fittings are screwed type.

3. Examine the diagram of the cooling tower pump and condenser connections, P18–1. Draw this as a one-view, single-line diagram. Plan the drawing so the symbols are not crowded. Assume the fittings are the welded type. Select your own scale and pipe length.

4. Draw a one-view, single-line drawing of the industrial waste disposal pump and piping, P18–7. Select your own scale and pipe length.

Double-line Diagrams

5. Make a double-line drawing of the orthographic piping detail shown in P18–5. Select your own scale and pipe length.

6. Make a one-view, double-line drawing of the piping drawing, P18–3. Select the view that shows the system most clearly. Choose your own scale.

7. Draw a double-line drawing of the cooling tower pump and condenser, P18–1.

Assume the fittings to be the welded type. Select your own scale and length of pipe sections.

Orthographic Diagrams

8. Make a single-line orthographic drawing of the piping detail shown in P18–3. Draw twice as large as the printed diagram. Position the drawing so the fittings are clearly shown. Select your own scale and spacing between fittings. This system is used to circulate antifreeze solution through copper tubes cast in driveways and sidewalks. The solution is heated and melts the snow. Assume the fittings are the welded type.

9. Make an orthographic drawing of the detail of the cooling tower system, P18–2. Draw twice as large as the printed diagram.

10. Make a two-line orthographic drawing of the boiler installation, P18–6. Assume the fittings are the flanged type. Select a scale that will clearly show the details.

Developed Views

11. Make a developed pipe drawing of the system shown in P18–2. View from the direction that requires the least rotation. Select your own scale and pipe lengths.

Pictorial Drawing

12. Make a single-line pictorial drawing of the boiler installation, P18–6. Select a scale that is large enough to show clearly all the fittings.

13. Make a single-line pictorial drawing of the cooling tower pump and condenser installation, P18–1. Assume that the fittings are to be the welded type.

General Arrangement Drawing

14. P18–8 shows the piping details for a commercial hot

water installation. This must be installed in the room shown. The tank must be positioned so maintenance men can work on all sides. Pipes must be positioned so they can be removed and replaced. The fittings are to be the screwed type. Position the tank, locate the pipe, locate unions, and select the necessary elbows and other fittings. Make a general arrangement drawing showing the plan view, and a side view showing ceiling clearances. Dimension both views.

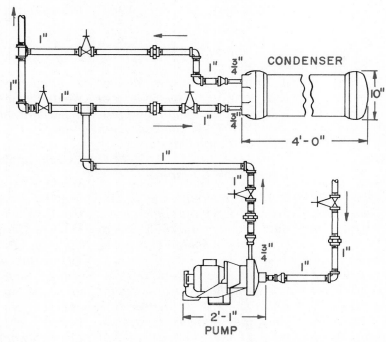

Bell and Gossett Co.

P18–1. *Cooling tower pump and condenser connections.*

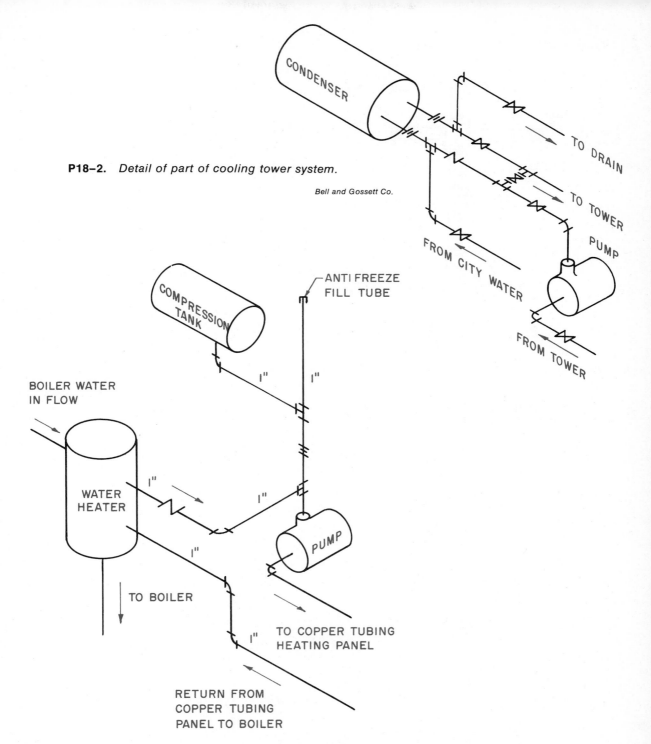

P18-2. *Detail of part of cooling tower system.*

Bell and Gossett Co.

ANTIFREEZE
FILL TUBE

COMPRESSION
TANK

CONDENSER

TO DRAIN

TO TOWER

PUMP

FROM CITY WATER

FROM TOWER

1"

1"

BOILER WATER
IN FLOW

WATER
HEATER

1"

1"

1"

PUMP

TO BOILER

TO COPPER TUBING
HEATING PANEL

1"

RETURN FROM
COPPER TUBING
PANEL TO BOILER

P18-3. *Detail of a system used to circulate antifreeze solution to a copper tubing panel buried in a concrete driveway.*

497

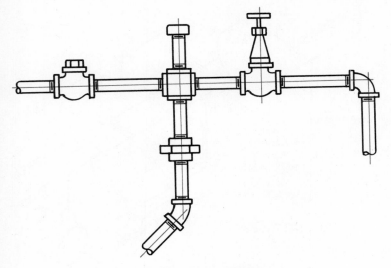

P18-4. *Piping detail.*

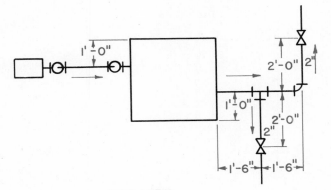

TOP VIEW

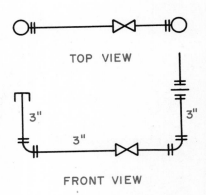

TOP VIEW

FRONT VIEW

P18-5. *Orthographic piping.*

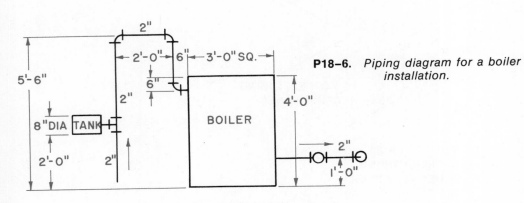

P18-6. *Piping diagram for a boiler installation.*

FRONT VIEW

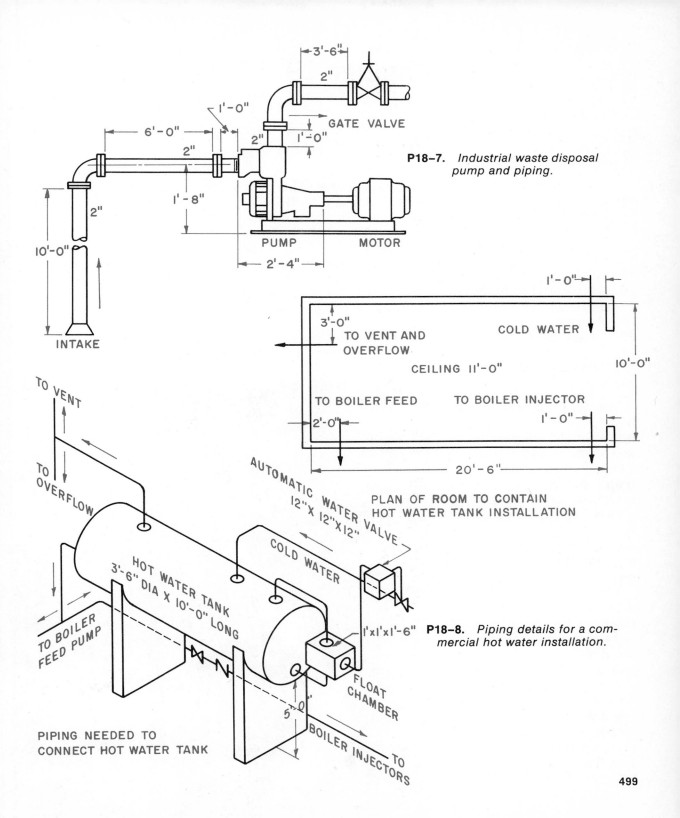

P18–7. *Industrial waste disposal pump and piping.*

3'-6"
2"
GATE VALVE
1'-0"
1'-0"
6'-0"
2"
2"
1'-8"
2"
10'-0"
INTAKE
PUMP   MOTOR
2'-4"

1'-0"
COLD WATER
3'-0"
TO VENT AND
OVERFLOW
CEILING 11'-0"
10'-0"
TO BOILER FEED   TO BOILER INJECTOR
2'-0"
1'-0"
20'-6"

PLAN OF ROOM TO CONTAIN
HOT WATER TANK INSTALLATION

TO VENT
TO OVERFLOW
AUTOMATIC WATER VALVE
12" X 12" X 12"
COLD WATER
HOT WATER TANK
3'-6" DIA X 10'-0" LONG
1'x1'x1'-6"
TO BOILER
FEED PUMP
FLOAT
CHAMBER
5'-0"
BOILER INJECTORS
TO

P18–8. *Piping details for a commercial hot water installation.*

PIPING NEEDED TO
CONNECT HOT WATER TANK

Chapter 19

Structural Drawings

Structural drawings show the design of the supporting framework of bridges, towers, and buildings. Fig. 19–1 shows a bridge with a structural steel frame and a concrete deck. Structural drawings give the details needed to make each member. (Member refers to any part of a structure.) They also show the fastening methods.

The engineering design is done by an engineer. He decides on the placement and sizes of the members. He also decides how they are to be joined.

The structural draftsman works from the design information given by the engineer. He prepares structural drawings of the entire framework. He also makes working drawings

for each member. The structural draftsman must understand basic design principles. A knowledge of fastening methods is needed. The structure is no stronger than the fastening devices used.

Structural drafting principles are much the same as those used in other industries. However, the method of presenting the information on structural drawings is different. Some of the methods are unique to this type of drawing.

Structural design and drafting is a very difficult and complex process. Presented

19–1. *The Golden Gate Bridge at San Francisco required hundreds of structural drawings.*

San Francisco Convention and Visitors Bureau

in this chapter are some of the easier and more common techniques used. Discussion is limited to selected wood, steel, and poured-in-place concrete drawings.

STRUCTURAL STEEL DRAWINGS

Fig. 19–2 shows the steel framework for a building. Before a draftsman can make the structural drawings, he must understand the terms used. The following sections give some of the more important things to know.

Terms Used in Structural Design. Commonly used structural terms are shown in Figs. 19–3 and 19–4.

Columns are vertical steel members used to support the roof and floor.

Girders are structural members running horizontally between columns.

Filler beams are structural members running horizontally between girders.

Pitch refers to the distance between the center lines of fasteners or holes.

Gage line is a continuous center line passing through holes or fasteners.

Gage distance is the distance the gage line is from the side of the structural member.

Edge distance is the distance the first hole or fastener is from the end of the member.

Slope is an indication of the angle a member has with the horizontal. It is shown by a triangle with the hypotenuse

19–2. *A building with a structural steel framework. An engineer decides the size and spacing of these members.*

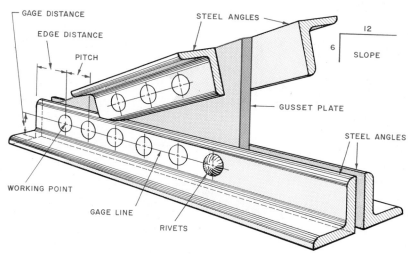

19–3. *Terms used in structural drafting.*

501

parallel with the structural member. The longest leg is always 12 inches. The slope symbol ₈⌐¹² means that for every 8 inches of rise, the member has 12 inches of run.

Gusset plates are used on trusses to join the structural members.

Steel angles are L-shaped metal structural members.

Working point is the point of intersection of the center lines of holes for rivets and bolts.

Structural Steel Shapes. Structural steel is made in a wide variety of standard sizes and shapes. The most commonly used shapes are shown in Fig. 19–5. These members are made to carry established loads. This information is

available from the companies manufacturing the members and from the *Manual of Steel Construction*, American Institute of Steel Construction.

Standard Abbreviations and Symbols. In order to simplify notes on drawings, a system of standard abbreviations has been developed by the American Institute of Steel Construction. This system does not use inch or pound marks. They are understood as the symbol is read. Selected standard abbreviations are shown in Fig. 19–6.

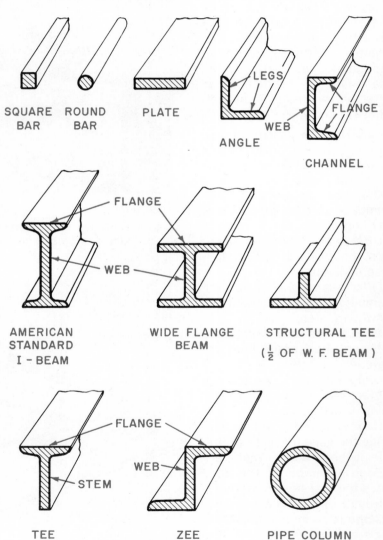

SQUARE BAR   ROUND BAR   PLATE   ANGLE   CHANNEL

AMERICAN STANDARD I – BEAM   WIDE FLANGE BEAM   STRUCTURAL TEE ($\frac{1}{2}$ OF W. F. BEAM)

TEE   ZEE   PIPE COLUMN

19–5. *Commonly used structural steel shapes.*

19–4. *Typical structural members.*

FILLER BEAM
GIRDER
COLUMN
COLUMN

Structural Shape	Symbol	Order of Presenting Data	Sample of Abbreviated Note
Square Bar	⊡	Size, Symbol, Length	$\frac{1}{2}$ ⊡ 5′–3
Round Bar	⌽	Size, Symbol, Length	$\frac{5}{8}$ ⌽ 7′–5
Plate	PL	Symbol, Width, Thickness, Length	PL 4 × $\frac{1}{4}$ × 11
Angle, Equal Legs	∠	Symbol, Leg 1, Leg 2, Thickness, Length	∠ 2 × 2 × $\frac{1}{4}$ × 9′–4
Angle, Unequal Legs	∠	Symbol, Long Leg, Short Leg, Thickness, Length	∠ 4 × 3 × $\frac{1}{4}$ × 6′–6
Channel	[	Depth, Symbol, Weight, Length	6 [10.5 × 12′–2
American Standard Beam	I	Depth, Symbol, Weight, Length	10 I 35.0 × 13′–4
Wide Flange Beam	WF	Depth, Symbol, Weight, Length	16 WF 64 × 20′–7
Structural Tee	ST	Symbol, Depth, Weight, Length	ST 18 WF 70
Tee	T	Symbol, Flange, Stem, Weight, Length	T 4 × 4$\frac{1}{2}$ × 11.2 × 10′–4
Zee	Z	Symbol, Web, Flange, Weight, Length	Z 4 × 3 × 12.5 × 7′–9
Pipe	Name of Pipe	Nominal Diameter, Name	3$\frac{1}{2}$″ extra strong pipe

American Institute of Steel Construction

19–6. *Symbols and abbreviations for structural steel members.*

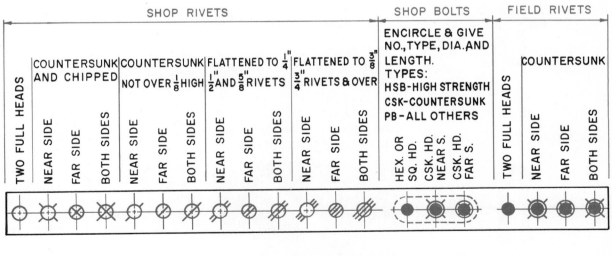

American Institute of Steel Construction

19–7. *Rivet and bolt symbols.*

Structural Rivets. One method for joining structural steel is by riveting. When parts are put together in a shop, the drawing calls for *shop rivets*. Rivets to be installed on the job in the field are called *field rivets*. Shop rivets are shown on the drawing as open circles. Field rivets are drawn solid black. See Fig. 19–7.

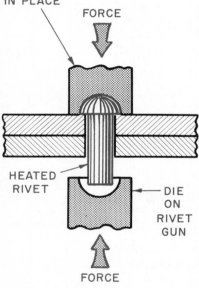

A DOLLY BAR HOLDS THE RIVET IN PLACE

FORCE

HEATED RIVET

DIE ON RIVET GUN

FORCE

HEATED RIVET BEING FORMED

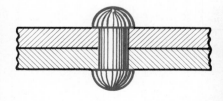

RIVET WITH HEAD FORMED

19–8. *Rivets are installed by heating the rivet, putting it in place, and forming a second head with a die on the rivet gun.*

Rivets are made in a large number of sizes. They have a variety of head shapes. The common head shapes are shown in Fig. 19–7. More information can be found in Chapter 26, Fastening Devices.

The holes to receive the rivets are made about 1/16-inch larger than the rivet. This permits the rivet to expand as it is driven into place.

Rivets are installed by heating them to a light cherry-red color, inserting them into the hole, and forming a head on the end. See Fig. 19–8.

High-Strength Steel Bolting. Another method for joining structural steel is by using high-tensile steel bolts. See Fig. 19–9. Bolting is rapidly replacing riveting for field and shop assembly. See Fig. 19–10.

Standards have been established for bolts suitable for this purpose. Details can be found in *Structural Joints Using ASTM A325 Bolts*, published by the American Institute of Steel Construction.

Russell, Burdsall and Ward Bolt and Nut Co.

19–9. *A high-tensile steel bolt used to assemble structural steel frameworks in buildings.*

It is important to get the proper size bolt to carry the load and stand the stress. The bolt must be tightened to the proper tension. The required tension is given in standardized form in the above publication. Hardened washers are placed under the bolt head and nut.

The symbols for drawing bolts are shown in Fig. 19–7.

Spacing Rivets and Bolts. The location of holes for rivets and bolts is controlled by the gage lines and the spacing of the holes on the gage lines. As stated before, the *gage line* is the center line of the hole that runs parallel to the long edge of the metal member. See Fig. 19–3.

The location of gage lines has been standardized for most connections. Fig. 19–11 shows the gage line distances for angle connections. These vary with the length of the leg. Notice that the gage line is located by measuring from the back of the angle.

The spacing between holes along a gage line of angle connectors is usually in units of 3 inches. This is for drilled and punched holes. See Fig. 19–12 for the spacing of holes on an angle connection. An *angle connection* is used to fasten a girder to a column.

Standard spacing of gage lines in I-beams is shown in Fig. 19–13. The gage lines are measured from the center of the flanges. See Fig. 19–14.

19-10. *High-tensile steel bolts can be rapidly fastened in place. The operation can be performed by workers with less skill than that required for riveting. The noise is reduced to a minimum.*

19-11.

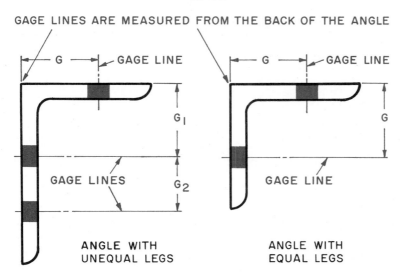

GAGE LINES ARE MEASURED FROM THE BACK OF THE ANGLE

ANGLE WITH UNEQUAL LEGS

ANGLE WITH EQUAL LEGS

Hole Gages for Angles in Inches

Leg	8	7	6	5	4	3½	2	2½	2	1¾	1½	1⅜	1¼	1
g	4½	4	3½	3	2½	2	1¾	1⅜	1⅛	1	⅞	⅞	¾	⅝
g₁	3	2½	2¼	2										
g₂	3	3	2½	1¾										

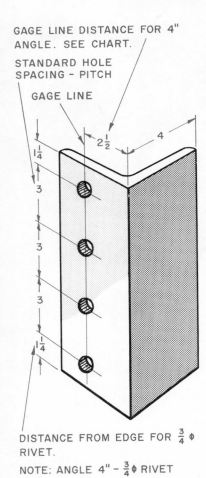

GAGE LINE DISTANCE FOR 4"
ANGLE. SEE CHART.

STANDARD HOLE
SPACING - PITCH

GAGE LINE

$2\frac{1}{2}$ 4

$1\frac{1}{4}$

3

3

3

$1\frac{1}{4}$

DISTANCE FROM EDGE FOR $\frac{3}{4}$ φ
RIVET.

NOTE: ANGLE 4" – $\frac{3}{4}$ φ RIVET

19–12. *Spacing of holes in
angle connections.*

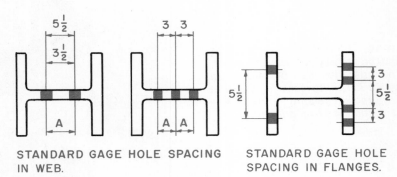

$5\frac{1}{2}$ 3 3

$3\frac{1}{2}$

$5\frac{1}{2}$

A   A A

3

$5\frac{1}{2}$

3

STANDARD GAGE HOLE SPACING
IN WEB.

STANDARD GAGE HOLE
SPACING IN FLANGES.

NOTE: MINIMUM DISTANCE FOR "A" IS 3".

STANDARD GAGES FOR DRILLED HOLES

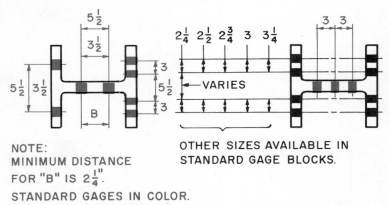

$5\frac{1}{2}$

$3\frac{1}{2}$

$5\frac{1}{2}$ $3\frac{1}{2}$

B

3

$5\frac{1}{2}$

3

$2\frac{1}{4}$ $2\frac{1}{2}$ $2\frac{3}{4}$ 3 $3\frac{1}{4}$

VARIES

3 3

OTHER SIZES AVAILABLE IN
STANDARD GAGE BLOCKS.

NOTE:
MINIMUM DISTANCE
FOR "B" IS $2\frac{1}{4}$".

STANDARD GAGES IN COLOR.

STANDARD GAGES FOR PUNCHED HOLES

American Institute of Steel Construction

19–13. *Standard spacing of gage lines in I-beams.*

Standard spacing is 3, 3½, and 5½ inches. The spacing of holes along these gage lines is in 3-inch units.

The minimum distance a hole can be from the edge is shown in Fig. 19–15. This distance is measured from the center line of the hole. It varies with the diameter of the rivet.

Structural Welding. Welding is another way of joining steel members. The fillet weld is most commonly used. Welding specifications are given in the *Manual of Steel Construction.* A study of welding symbols as used on drawings is found in Chapter 26, Fastening Devices. Some welded steel details are shown in Fig. 19–16.

Standard Beam and Column Connections. *Connections* are used to fasten beams to other structural members. These have been standardized by the American Institute of Steel Construction. The two common types of beam connections are framed and seated. They are designed to handle the loads and forces beams

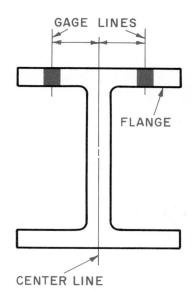

GAGE LINES

FLANGE

CENTER LINE

19–14. *Gage lines are located from the center of I-beam.*

19–15.

Rivet Diameter (inches)	Minimum Edge Distance (Inches) for Punched Holes		
	In Sheared Edges	In Rolled Edge of Plate	In Rolled Edge of Structural Shapes*
$\frac{1}{2}$	1	$\frac{7}{8}$	$\frac{3}{4}$
$\frac{5}{8}$	$1\frac{1}{8}$	1	$\frac{7}{8}$
$\frac{3}{4}$	$1\frac{1}{4}$	$1\frac{1}{8}$	1
$\frac{7}{8}$	$1\frac{1}{2}$	$1\frac{1}{4}$	$1\frac{1}{8}$
1	$1\frac{3}{4}$	$1\frac{1}{2}$	$1\frac{1}{4}$
$1\frac{1}{8}$	2	$1\frac{3}{4}$	$1\frac{1}{2}$
$1\frac{1}{4}$	$2\frac{1}{4}$	2	$1\frac{3}{4}$

* May be decreased $\frac{1}{8}$ inch when holes are near end of beam.

American Institute of Steel Construction

must carry. Some of these connections are shown in Fig. 19–16. For a complete description of connections see the *Manual of Steel Construction.*

The draftsman must be able to select the proper connections. This requires him to know the strength of the connections. He must know how to use the standards manual.

Scales For Structural Steel Drawings. The scale selected will vary with the size of the framing drawing or the erection drawing. They usually are from $\frac{1}{4}'' = 1'$-0'' to $1'' = 1'$-0''. See Figs. 19–17 and 19–19. Shop drawings are usually drawn $\frac{3}{4}'' = 1'$-0'' to $1'' = 1'$-0''. See Fig. 19–21.

Dimensioning. Structural steel drawings are dimensioned much the same as other drawings. Following are some recommended practices:

1. The dimension lines are unbroken. The numerals are placed above the line and near the center.

2. Smaller dimensions are placed close to the object outline. Longer dimensions are outside these.

3. Dimensions are given to the working points, as the gage line (center line) of a series of holes.

4. All dimensions over one foot are given in feet and inches except the width dimensions of plates and the depth of structural members. These are in inches.

5. The foot symbol (') is used on dimensions. Inch

marks (") are omitted unless needed for clarity.

6. Fasteners, such as rivets and holes, are dimensioned to their center lines.

7. Dimensions are given to the center line of the beams.

8. Dimensions are given to the back of angles and channels.

9. Shop drawings should be dimensioned so those making the members do not have to add or subtract dimensions.

10. Details, such as the size of fasteners, hole diameters, and painting instructions, are given as notes.

TYPES OF STRUCTURAL STEEL DRAWINGS

When plans are made for a steel structure, three types of drawings are commonly used. These are a steel framing drawing, an erection drawing, and shop drawing.

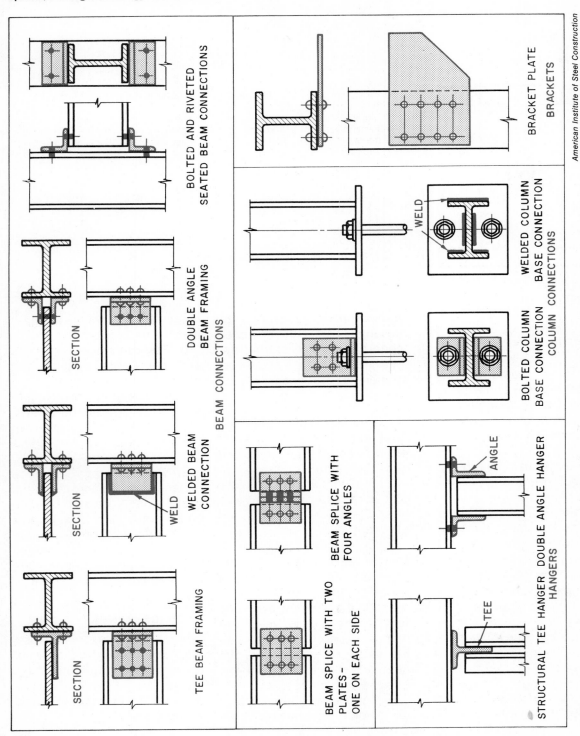

American Institute of Steel Construction

BOLTED AND RIVETED
SEATED BEAM CONNECTIONS

BRACKET PLATE
BRACKETS

SECTION

DOUBLE ANGLE
BEAM FRAMING

BEAM CONNECTIONS

WELD

WELDED COLUMN
BASE CONNECTION

BOLTED COLUMN
BASE CONNECTION

COLUMN CONNECTIONS

SECTION

WELD

WELDED BEAM
CONNECTION

SECTION

TEE BEAM FRAMING

BEAM SPLICE WITH
FOUR ANGLES

ANGLE

BEAM SPLICE WITH TWO
PLATES –
ONE ON EACH SIDE

TEE

STRUCTURAL TEE HANGER DOUBLE ANGLE HANGER

HANGERS

19-16. *Selected structural steel framing connections.*

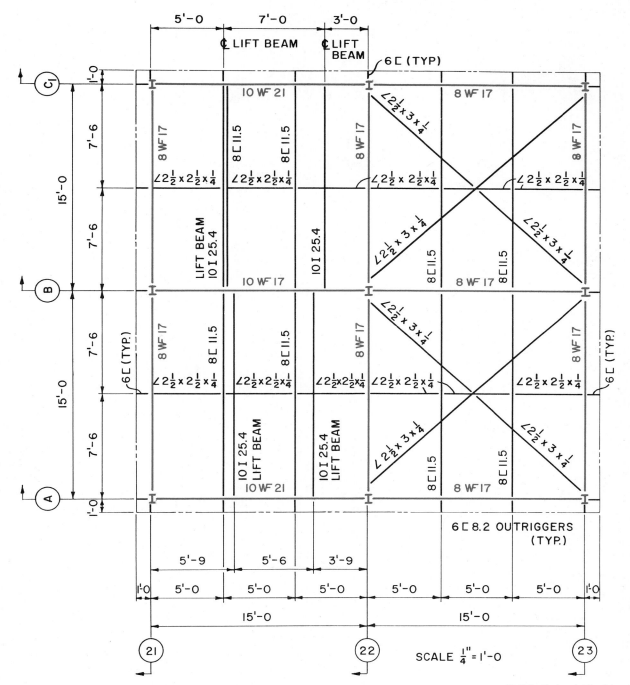

Roof Framing Plan Elevation 115'-0

McNally-Pittsburgh Mfg. Co.

19-17. *This is a roof framing plan for the car haul house, Fig. 19-18. The main structural system consists of I-beam columns and wide flange beams between these. These are printed in color to call attention to them. The rest of the members are angles and channels used to support the roof decking.*

19-18. *This is a partial view of a coal processing plant. The small building in the front is a car haul house. Some of the structural steel details for this building are shown in Figs. 19–17, 19–19, 19–20, 19–22, and 19–24.*

Steel Framing Drawings. The steel framing drawing is made by an engineer. It shows the location of beams, columns, and other steel members. The size and location of these members is shown. Examine the roof framing plan, Fig. 19–17. It shows the structural system used for the car haul house, Fig. 19–18.

Notice that the structural members are indicated by a solid line. The main framework is made from wide flange beams. These carry the roof load. In addition, they support two lift beams. Find these on the drawing. The lift beam is used with a hoist to pick up heavy loads. Notice that the wide flange beams in the area of the lift beams are heavier than those on the other side of the building.

The size of each beam is shown on the drawing. In addition, channels and angles are used to support the roof. These are indicated by a single line. Their sizes are also given. The actual length of the structural members is not shown. This information is found on the shop drawings.

The location of each member is carefully dimensioned. Dimensions are to the center line of the member. The I-beams used as columns are located with overall dimensions.

If a building is several stories high, a separate framing plan is made for each floor plus one for the roof. Framing drawings are made to scale.

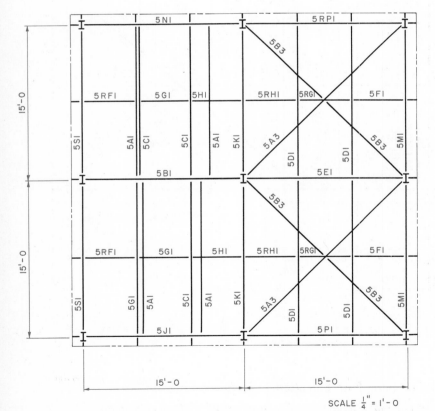

SCALE $\frac{1}{4}$" = 1'- 0

19-19. *Roof steel erection plan for car haul house roof.*

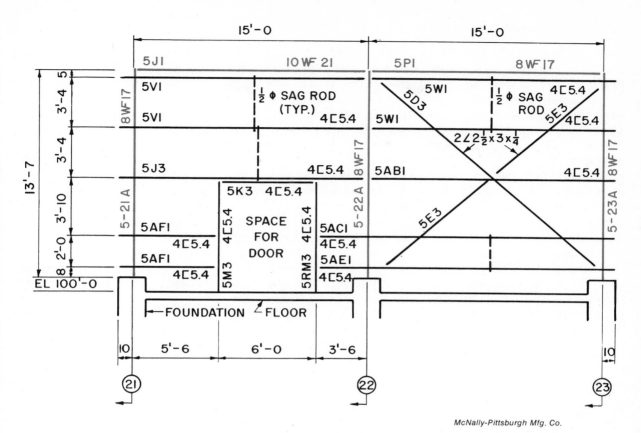

15'-0   15'-0

5J1   10 WF 21   5P1   8 WF 17

8WF17   5V1   ½ Φ SAG ROD (TYP.)   5D3 5W1   ½ Φ SAG ROD   4⊏5.4 5E3

5V1   4⊏5.4   5W1   4⊏5.4

2⊥2½×3×¼

5J3   4⊏5.4   8WF17   5AB1   5E3   4⊏5.4   8WF17

13'-7

3'-4   3'-4   3'-10   2'-0   8

5-21A   5K3 4⊏5.4   5-22A   5-23A

5AF1   4⊏5.4   SPACE FOR DOOR   5AC1 4⊏5.4   5E3

4⊏5.4   4⊏5.4   5AF1   5M3   5RM3   5AE1   4⊏5.4

EL 100'-0

FOUNDATION   FLOOR

10   5'-6   6'-0   3'-6   10

㉑   ㉒   ㉓

McNally-Pittsburgh Mfg. Co.

19-20. *An exterior wall steel framing and erection plan. This is a combination of the steel framing and erection plan of the car haul house. It was taken at column line A on Fig. 19-17. Each exterior wall requires a separate drawing. The main structural members are in color to call your attention to them.*

Erection Plan. The erection plan for the car haul house roof is shown in Fig. 19-19. It is an assembly drawing. It is drawn to scale.

The erection plan is used on the job by those putting the framework together. Each member has been made to the correct length and numbered. All the holes for riveting or bolting them together have been made. Those erecting the building only need to know where to put each piece shown on the erection plan.

Examine the plan. Notice it gives only the dimensions locating the columns. The only other information it gives is where each member is to be placed. Connections and other details are not needed. They appear on the shop drawings.

Some companies combine the framing plan and erection plan on one drawing, Fig. 19-20.

If a building is several stories high, a separate erection plan is made for each floor.

Exterior Walls. Framing and erection plans must be made for the exterior walls of some buildings. These are drawn in the same manner as explained for floor and roof plans. One wall of the car haul house is shown in Fig. 19-20. It has been drawn as a combination framing and erection drawing. Some companies prefer to combine these on one drawing.

The wall plan also shows the columns. Here their identification number is shown. This

511

does not appear on the erection plan because they are in end view on that drawing.

The other structural members on this elevation are steel channels. The building is covered with metal siding in large sheets. These sheets are fastened to these channels. The channels are fastened to the columns.

Shop Drawings. After the engineer designs the structure, the draftsman details each part. These are called shop drawings. The member is made from these drawings. See Figs. 19–21 through 19–24.

The draftsman uses the framing plan to secure the sizes. The actual length of each member is not shown on the framing plan. The draftsman must know how to figure this from the information given.

Shop drawings show the size of the member and the finished length. They show the connections used. The location of all holes is given. They must be dimensioned so the member can be made from the drawing. Usually they are dimensioned to the nearest $\frac{1}{16}$ inch. They include notes telling how the part is to be made. Instructions on how

to assemble the members are included.

Shop drawings should be drawn to a scale large enough so they are clear. They must not be crowded. The length of the members is not drawn to scale. The height is to scale. Since the drawings are completely dimensioned, an accurate scale drawing of the length is not necessary.

More information on making shop drawing is given in the paragraphs on drawing beams and columns.

Numbering Structural Members. Notice that the main structural members, 5JI and

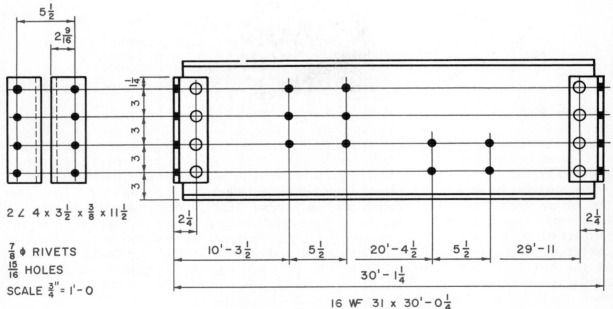

19–21. *A typical riveted steel beam shop drawing. All parts of this type of drawing are to scale except the length of the beam.*

5PI, shown in Fig. 19–20, appear on this drawing and the erection drawing.

Each part of a structural system is given a number. It is used wherever the piece appears on a drawing, whether it's the erection drawing or a shop drawing. The number is painted on the steel member after it is made. These numbers are helpful to the workmen erecting the building. If several members are the same size, they are given the same number.

Drawing Beams. Fig. 19–21 shows a shop drawing of a typical riveted beam. Notice the shop and field rivet symbols. Whenever possible, holes are lined up on common gage lines. It is necessary to locate the gage line for each angle connection. The edge distance must be shown as well as the pitch.

Connections usually extend beyond the beam on each end. This setback is the difference between the overall length of the assembled member and the length of the beam. In Fig. 19–21 the total unit length is 30'-1¼" and the beam is 30'-0¼". The setback is ½" on each end.

A left-end view is drawn to show the angle connection holes and their dimensions. If the right-end is the same, it need not be drawn. The beam is not drawn in the end view.

Fig. 19–22 shows a shop drawing for a welded beam. The details are the same as the riveted beam except for the angle connections. These have welding symbols to tell how they are to be joined. This beam is one from the car haul house, Fig. 19–17. Can you locate it on the plan?

Drawing Columns. Fig. 19–23 shows a shop drawing of a riveted steel column. Usually the column can be described

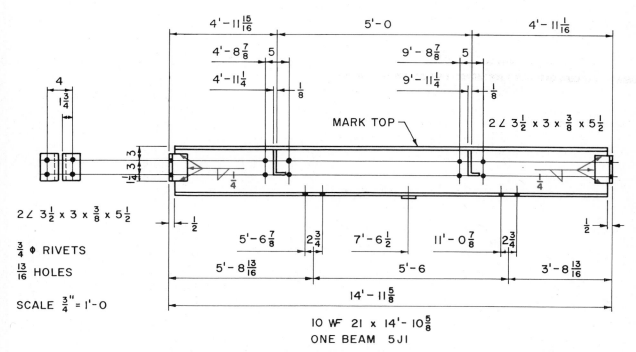

19–22. *A typical welded steel beam shop drawing. All parts of this type of drawing are to scale except the length of the beam. This is a beam from the car haul house, Fig. 19–17.*

with views of each face plus a view called a bottom section. The bottom section is actually a view from the top looking down on the column.

Notice on the column, Fig. 19–23, that the angles and plate are shop-riveted. The angles use buttonhead rivets. The plate uses a rivet with a countersunk head on the bottom and buttonhead on top. The length of the wide flange beam serving as a column is ½-inch shorter than the overall length with the angles in place.

The faces of columns are identified by a letter. The surface facing north on the structure is marked on the drawing.

When two faces are the same, only one need be drawn. The other can be shown with a center line and a letter. See Face C, Fig. 19–23.

Fig. 19–24 shows a shop drawing for a welded steel column. The details are the same as the riveted beam except for the welded plate and angles. Notice the use of welding symbols. This column is one from the car haul house, Fig. 19–17.

Drawing Steel Trusses. Trusses are structural units made from tees, wide flange beams, or angles with the long legs placed back to back. When a roof is built over a building with a span too great to use girders, trusses can be used economically. They give large areas of floor space free of columns.

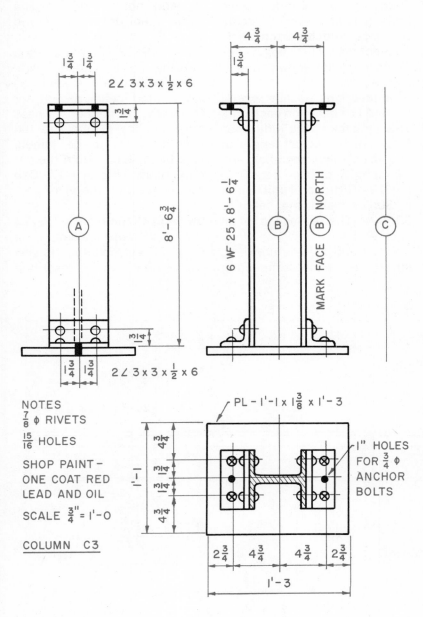

19–23. *A typical riveted steel column shop drawing.*

The designing of trusses is the work of an engineer. He must design them to absorb tension and compression stresses. *Tension* is a force that tries to stretch a member. *Compression* is a force that tries to shorten a member. See Fig. 19–25.

Fig. 19–26 shows the stresses on a Fink truss. Notice that the load on the roof, P, puts some members under tension. Other members are under compression.

Fig. 19–27 shows common types of steel trusses: the Warren, Pratt, Fink, bowstring, and scissors.

The Warren and Pratt trusses are used to support floor and roof loads. If a roof has to carry loads other than that of the roof, as an overhead crane, the Warren and Pratt are used. The Fink truss provides a roof with a good slope. It usually is not used to carry any other loads except the roof. The bowstring truss can span large distances. It provides a good slope to the roof. The scissors truss is used for high-pitched roofs. A church roof is a good example.

The draftsman must know the terms used in describing trusses. These are chord, web member, panel, and span. They are shown in Fig. 19–27. A *chord* is one of the principal structural members braced by web members. *Web members* are the internal braces running between chords. *Panels* are the distance between

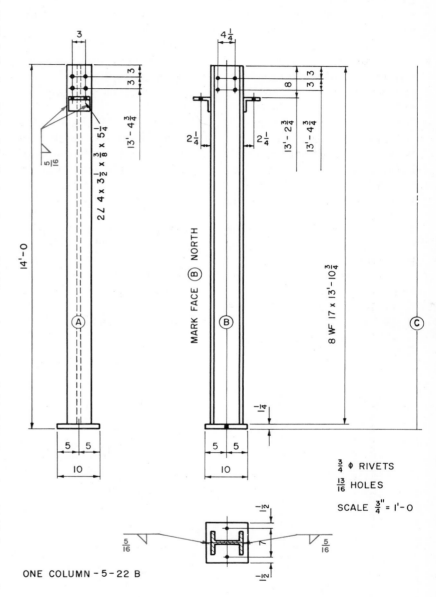

ONE COLUMN - 5 - 22 B

$\frac{3}{4}$ φ RIVETS

$\frac{13}{16}$ HOLES

SCALE $\frac{3''}{4}$ = 1'-0

McNally-Pittsburgh Mfg. Co.

19–24. *A typical welded steel column shop drawing. This is a column from the car haul house, Fig. 19–17.*

two web members measured along a chord. *Span* is the distance covered by the trusses with no support from columns or walls.

The engineer decides which type of truss is best for the job. He figures the stresses on the member. Then he makes a design drawing of the truss. The design drawing usually shows two things. It shows the tension and compression in pounds. Tension is shown by a − and compression by +. The size of the steel members to be used are shown. See Fig. 19–28. It is from this drawing that the draftsman prepares the shop drawing.

A shop drawing of a riveted truss is shown in Fig. 19–29. Since the individual parts are small and simple, each of

them is detailed in the assembled position. Only the left half of the truss is drawn when the other half is the same.

Notice the use of gusset plates to join the members. The draftsman locates the rivets according to standard design practice.

Study Fig. 19–29. Notice that the location of each rivet is dimensioned. The size of each structural member is shown. This includes the length. See Fig. 19–6 for the standard way to record this information. Each gusset plate size is given. The web members are located by dimensioning to the center line of the rivets. The distance of this center line from the top of the member is dimensioned.

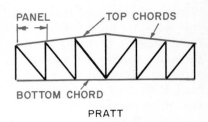

PRATT

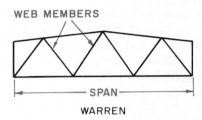

WARREN

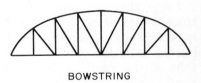

BOWSTRING

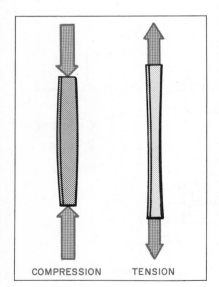

19–25. *Compression forces tend to shorten a member. Tension forces tend to lengthen a member.*

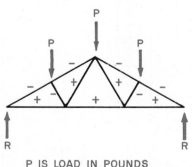

P IS LOAD IN POUNDS
R IS A POINT OF RESISTANCE
− INDICATES A TENSION
  STRESS
+ INDICATES A COMPRESSION
  STRESS

19–26. *Roof load places members of a truss under tension and compression stresses.*

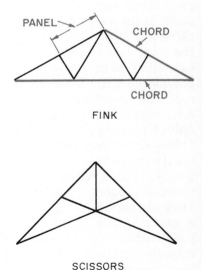

FINK

SCISSORS

19–27. *Typical types of steel trusses.*

Some truss drawings can become very crowded. See Fig. 19–30. The dimensioning of such a drawing must be carefully planned. The draftsman must work carefully so no details are left off.

Sections are used to clarify construction details and dimensions.

Notice that the dimensions parallel the edges of each part. The slope of members is shown with a triangle drawn on each sloped member. The longest leg is always 12 inches.

Trusses are usually shop assembled. If they are too large to ship, some parts are assembled in the shop. These subassemblies are then put together at the building site. Each individual member of a truss is not usually numbered.

Each truss is numbered as a single structural member, as T–1, T–2, and T–3.

All parts of the truss are drawn to the same scale. Detail drawings may be to a larger scale.

Fig. 19–31 shows a detail drawing for a welded truss. It is much the same as the riveted truss. One difference is that gusset plates are not needed.

Notice that the dimensions are measured along a line located just below the top of the angle. Fig. 19–31 shows this dimension as $1^{13}/_{16}$ below the top of the inclined chord. It is $1^{13}/_{16}$ inches from the edge of the horizontal chord. It is $^3/_4$ inch from the edge of the web members. This distance is found in engineering de-

sign manuals, which give this distance for the various size angles. The dimensions in a truss drawing refer to these lines. They do not refer to the outside edges of each member.

When designing welded trusses, the draftsman needs to know how many inches of weld are needed at each joint. He has to make certain the angles provide enough edge to give the specified length of weld.

Notice the use of welding symbols in Fig. 19–31. Study the section on welding in Chapter 26.

On the truss shown in Fig. 19–31 the web members are identified by letters. The size of each member is shown on a bill of material.

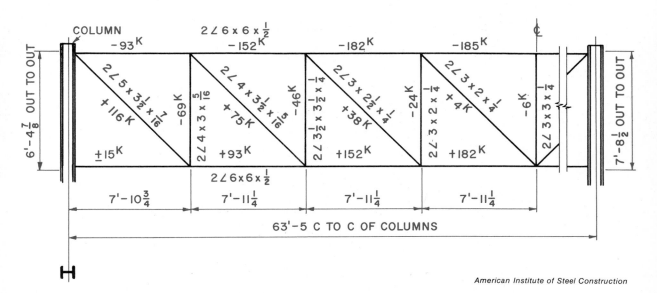

19–28. *A design drawing for a truss, as prepared by the engineer. The shop drawing for this truss is shown in Fig. 19–30. The term "out to out" on the height dimension refers to the outside surface of the top and bottom members.*

517

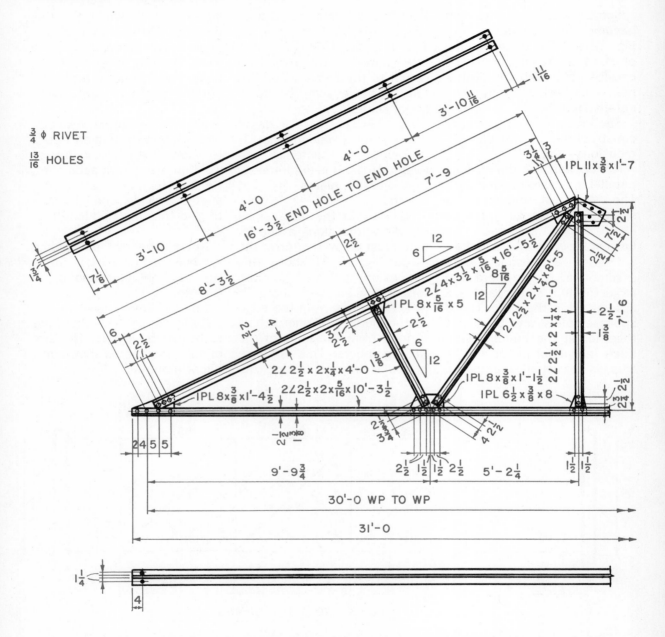

19–29A. *A simple riveted Fink steel truss that is made of steel angles. One of the gusset plate connections is shown larger to help show how they are drawn.*

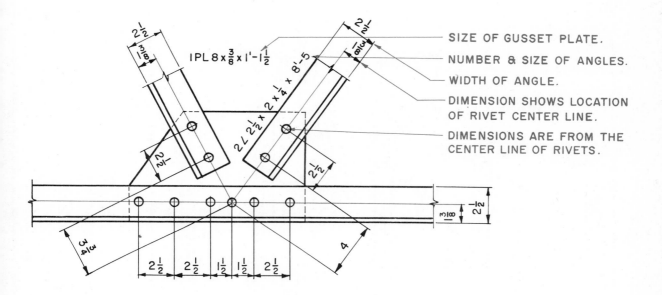

SIZE OF GUSSET PLATE.

NUMBER & SIZE OF ANGLES.

WIDTH OF ANGLE.

DIMENSION SHOWS LOCATION OF RIVET CENTER LINE.

DIMENSIONS ARE FROM THE CENTER LINE OF RIVETS.

19–29B.   *Enlarged detail of one gusset plate.*

Drawing Rigid Frames. Rigid frames are steel girder-like members. When they are assembled, they form the column and roof members. They can span large distances. This provides an area free of columns. See Fig. 19–32.

Rigid frames are made in the shop and shipped to the building site. Here they are bolted together. A large crane lifts them in place. The column is bolted to the concrete footing.

A drawing for a rigid frame is shown in Fig. 19–33. The frame is shown assembled. Detail drawings are used to show things that cannot be seen on this drawing. The details are drawn to a larger scale than the assembled drawing.

CONCRETE STRUCTURAL DRAWINGS

Concrete is made from portland cement, water, sand, and gravel. The strength varies with the quality of these items. The amount of each in the mix also influences the strength.

Concrete has a good compressive strength. *Compressive strength* is the ability of a material to withstand forces tending to shorten it. See Fig. 19–25. The tensile strength of concrete is low. *Tensile strength* is the ability of a material to withstand forces tending to lengthen it. See Fig. 19–25. To increase the tensile strength of concrete, steel reinforcing bars are cast in the concrete members. The steel resists tension forces. The concrete resists compression forces.

Concrete with no steel reinforcing is called *plain* concrete. If it has steel reinforcing bars, it is called *reinforced* concrete. Some structural members are cast with the steel reinforcing bars tightly stretched under tension before the concrete is poured. This is called *prestressed concrete.*

519

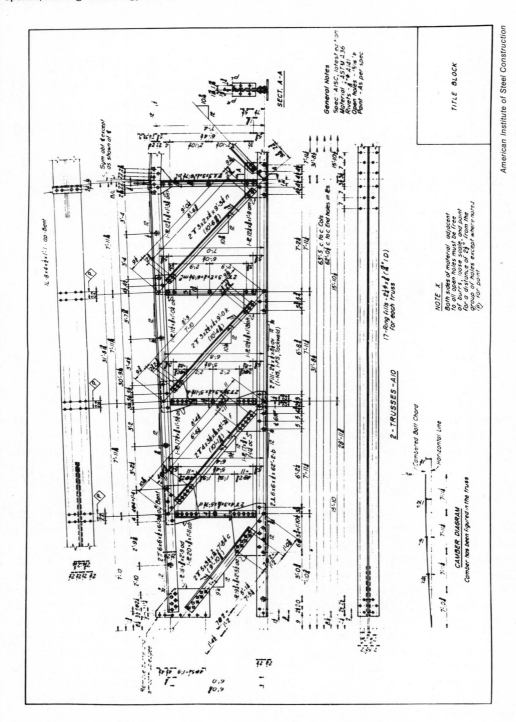

American Institute of Steel Construction

19-30. A detail drawing of a Pratt truss. Since such drawings are completely dimen-sioned assembly drawings, they become crowded.

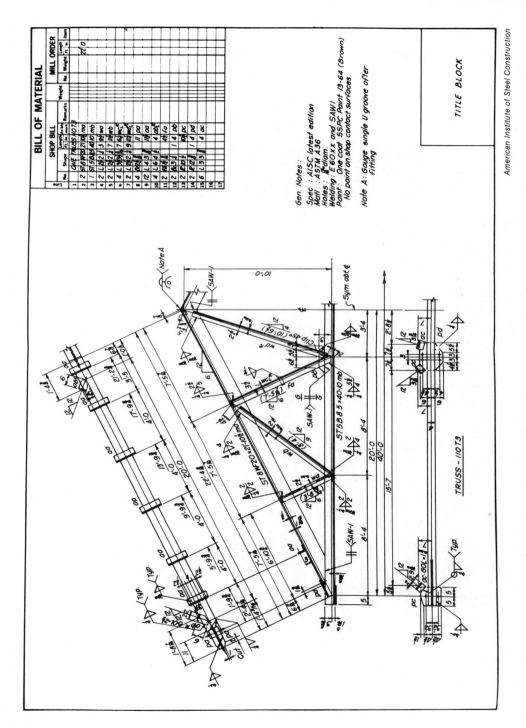

19-31. A welded truss drawing.

American Institute of Steel Construction

Butler Manufacturing Co.

19-32. Rigid frames provide a floor area free of columns.

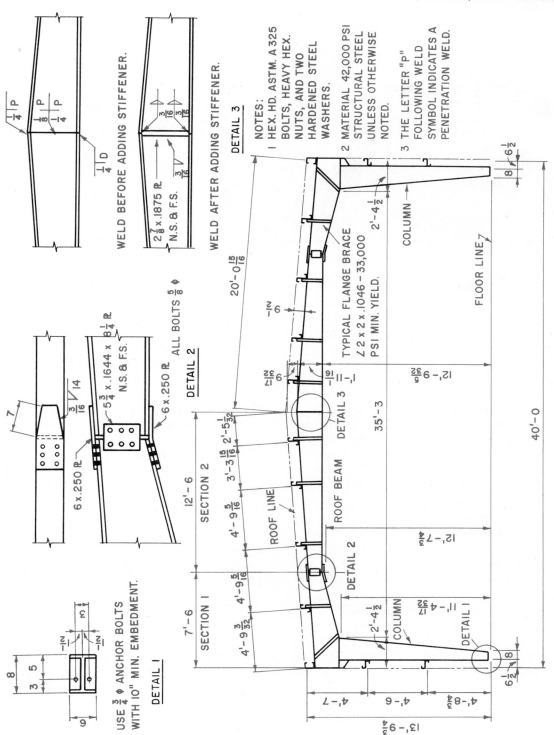

WELD BEFORE ADDING STIFFENER.

WELD AFTER ADDING STIFFENER.

DETAIL 3

DETAIL 2

ALL BOLTS $\frac{5}{8}$ Φ

DETAIL 1

USE $\frac{3}{4}$ Φ ANCHOR BOLTS
WITH 10" MIN. EMBEDMENT.

NOTES:
1 HEX. HD. ASTM. A 325
   BOLTS, HEAVY HEX.
   NUTS, AND TWO
   HARDENED STEEL
   WASHERS.

2 MATERIAL 42,000 PSI
   STRUCTURAL STEEL
   UNLESS OTHERWISE
   NOTED.

3 THE LETTER "P"
   FOLLOWING WELD
   SYMBOL INDICATES A
   PENETRATION WELD.

TYPICAL FLANGE BRACE
∠2 x 2 x .1046 – 33,000
PSI MIN. YIELD.

COLUMN

FLOOR LINE

ROOF BEAM

ROOF LINE

SECTION 2

SECTION 1

COLUMN

Butler Manufacturing Co.

19-33. *Drawing of a rigid frame. Notice the use of large scale detail drawings to show construction details.*

19-34. *Precast concrete floor and roof decking is placed on a bolted, steel frame.*

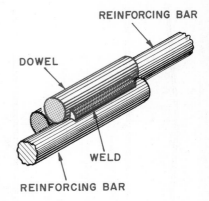

19-35. *Dowels are welded to bars to splice them together.*

Usually concrete structural members poured in place on the job are reinforced members. Those cast in a factory and shipped to the site as finished members are usually prestressed. Fig. 19-34 shows a steel-framed structure with a precast concrete floor and roof.

Definitions. *Bars* are the round steel reinforcing rods placed in a concrete member to strengthen it. Bars are specified by a number. The larger the number, the larger the bar diameter.

Dowels are round metal bars used to splice together two reinforcing bars in a column. They are usually the same diameter as the bars joined. They are welded in place. See Fig. 19-35.

Bar supports are round metal bars used to hold the reinforcing bars the proper distance above the bottom of the form. They prevent the bars from moving when the concrete is poured. There are several standard types in use. See Fig. 19-36.

Ties are made from a small diameter metal wire. They are used to hold bars in place in columns. They prevent the bars from moving when the concrete is poured. See Fig. 19-37.

Stirrups are round, metal bars usually bent in the shape of a U. They are used to hold bars in the proper position when the concrete is poured. See Fig. 19-38.

Marks is a term used for identifying the floor and type of member. See Fig. 19-39 for mark abbreviations. A typical mark would be 2B3. This indicates the second floor, beam number 3.

Beams are the main horizontal structural members. They run from column to column or column to foundation.

Joists are smaller horizontal structural members. They run between beams and provide support for the floor or roof slab.

Bent refers to bending reinforcing bars. These strengthen the structural members. See Fig. 19-40.

REINFORCED CONCRETE DRAWINGS

The design and drawing of reinforced concrete members is a complex task. It requires much study. The American Concrete Institute has prepared a book, *A Manual of Standard Practice for Detailing Reinforced Concrete Structures,* that gives standard drafting practices.

The following material covers concrete structural members that are formed and cast on the job.

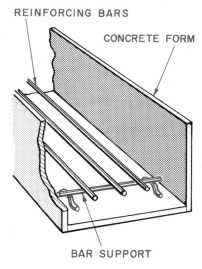

19–36. *Bar supports hold the reinforcing bars above the bottom of the form.*

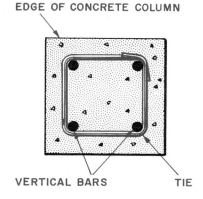

19–37. *Wire ties are used to hold reinforcing bars in place.*

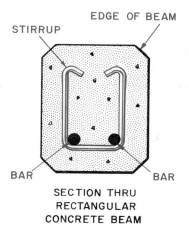

SECTION THRU
RECTANGULAR
CONCRETE BEAM

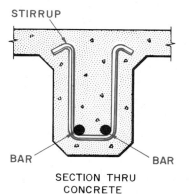

SECTION THRU
CONCRETE
T-BEAM

19–38. *Stirrups are used to locate and hold bars in place.*

Member	Mark
Beam	B
Column	C
Footing	F
Girder	G
Joist	J
Lintel	L
Slab	S
Wall	W

American Concrete Institute

19–39. *Marks used in building structural systems.*

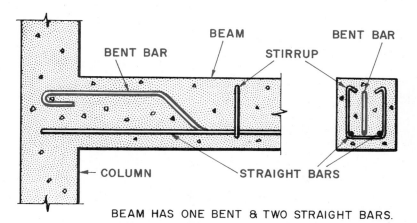

BEAM HAS ONE BENT & TWO STRAIGHT BARS.

19–40. *Bent bars strengthen the beam.*

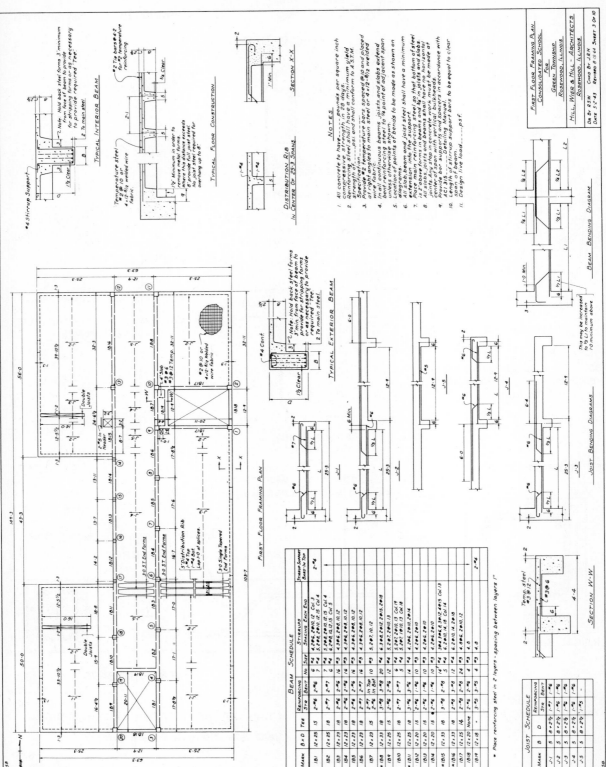

19-41. An engineering drawing for a concrete joist floor.

American Concrete Institute

526

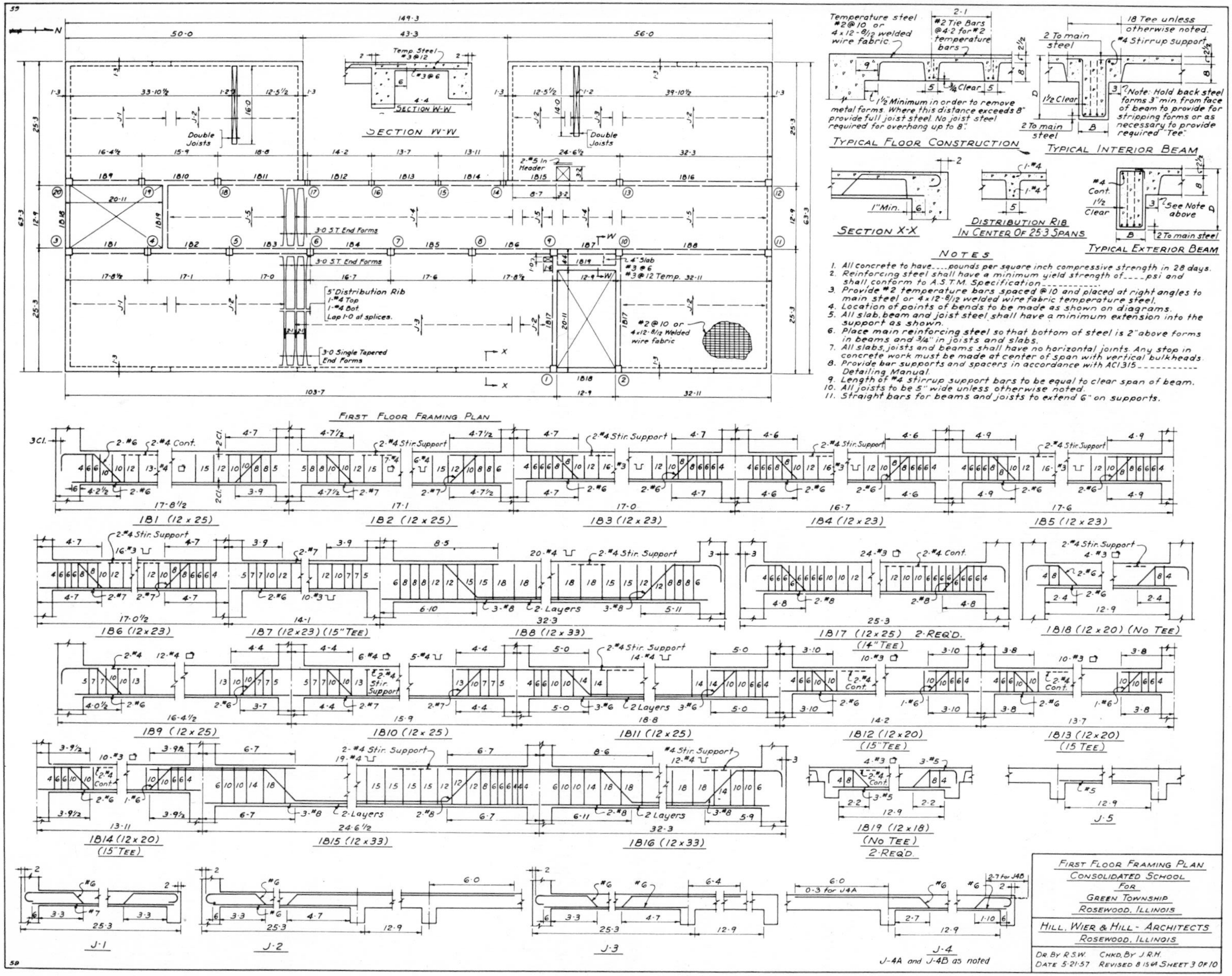

American Concrete Institute

19–42. *A placing drawing for the concrete joist floor shown in Fig. 19–41.*

Two types of drawings are commonly used. These are engineering drawings and placing drawings.

Engineering Drawings. Engineering drawings are prepared by the design engineer. They show the general arrangement of the structure, the size of the members, how they are to be reinforced, and notes needed for additional information. See Fig. 19–41.

These drawings should be complete in every detail. All design information must be included either by a drawing, a note, or a schedule. The engineer does not usually prepare complete drawings for each member. He does not usually schedule the bending of every bar. He does have to give the draftsman this information through notes and diagrams or refer him to standard manuals.

The engineering drawings include a typical concrete slab, joist, beam, and column detail. These must show how the reinforcing steel is arranged.

The engineer does not work out all the dimensions. He does show where the bars are to be bent and to what points they should be extended.

The engineer must show the quality of concrete to be used. The type and grade of reinforcing bars must be shown. He notes the live loads the structure will carry and other design data, such as the load-bearing capabilities of the soil upon which the structure is to be built.

Study the engineering drawing, Fig. 19–41, showing the plan of the first floor of a building. The dimensions of the building and column spacing are given. The direction of the joists is shown. The joists are identified by number.

The *beam schedule* shows all the information a draftsman needs for making the placing drawing. The following is an explanation of beam 1B1 in the schedule. The mark *1B1* means first floor, beam number 1. *B* × *D* gives the width and thickness of the beam. This beam is 12 inches by 25 inches.

The beam uses two No. 6 straight bars. It uses two No. 6 bars that are to be bent on the end. The beam contains six stirrups made from No. 4 bar. These six are for the end of the beam next to the No. 3 column. The first stirrup is 4 inches from the end of the beam. The next two are spaced 6 inches apart. The next two are spaced 10 inches apart. The last stirrup is 12 inches from the last 10-inch one.

From the end of the beam next to column 4 are seven stirrups made of No. 4 bar. The first stirrup is 5 inches from the end of the beam. The next two are spaced 8 inches apart. The next two are 10 inches apart. The next one is spaced 12 inches and the last is 15 inches.

The last entry for beam 1B1 means that two No. 4 bars are used to support the stirrups. These directions are detailed on the placing drawing shown in Fig. 19–42. See the detail for beam 1B1. Here the stirrups are drawn in their exact location and are dimensioned.

Notice the engineering drawing shows a typical section through a beam. It shows the two straight bars as round, black circles. The two bent bars are shown as round, black circles with the bent part shown by dashed lines.

Also shown are beam and joist bending diagrams, which indicate how the engineer wants the bars bent.

Sections are shown through the floor slab. This shows how the floor joins the exterior foundation (Section X–X). It also shows the slab in connection with the floor joists.

Examine the notes shown in Fig. 19–41. They give information that is not shown on the drawings.

Placing Drawings. *Placing drawings* are detail drawings. They show the floor plan and the details for making the structural members. They also show the placing of reinforcement bars. See Fig. 19–42. The only dimensions needed are those showing the location of the bars. They show the size, shape, and location of the bars.

The draftsman follows very detailed information on the

engineering drawing when making placing drawings.

The reinforcement of beams, joists, and girders is shown on a beam schedule. The placing drawing schedule gives the number, mark, and size of the member; number, size, and length of straight bars; number, size, total length, mark, and bending detail of bent bars and stirrups; spacing of stirrups; bar supports; and other placement information that may be needed.

Concrete slab reinforcement is drawn on the plan. A schedule is used to show the bars needed. The schedule is the same as the beam schedule. Sometimes the number of bars is not shown. This is found on the plan of the slab.

Column reinforcement is shown on a column schedule. This schedule has a section for showing ties and bent bars, and diagrams showing the arrangement and bending of the ties.

Bar supports are listed on placing drawings. They are usually a part of the beam schedule. The schedule shows the number, size, and length.

Ties are used in columns to prevent the bars from buckling. They hold them in place as the column is cast. Standards have been established giving the maximum spacing of ties. These are available in the American Concrete Institute standards manual.

Examine the placing drawing, Fig. 19–42. Of special im-portance are the detailed drawings showing bar and stirrup spacing. Notice that each beam and joist is completely dimensioned. The draftsman gets these dimensions from the floor framing plan. He also dimensions the bent bars. The information for figuring these dimensions comes from the *beam bending diagram* on the engineering drawing. The engineer does not calculate these dimensions.

The floor plan of a placing drawing can be traced from the plan on the engineering drawing.

Scale. All reinforced concrete drawings are drawn to scale. This includes both the engineering drawings and the placing drawings. Remember that structural steel shop drawings need not be to scale in their long direction.

The most commonly used scale is $\frac{1}{4}'' = 1'\text{-}0''$. A larger scale is usually used for sections and other detail drawings.

Dimensioning. Dimensions are placed in the same manner as structural steel drawings. They are in feet and inches. The foot mark is used. The inch mark is not used unless needed for clarity. Study Figs. 19–41 and 19–42 for standard dimensioning practices.

The Ceco Corporation

19–43. *Steel forms being set in place to cast as a single unit the beams, joists, and a floor slab.*

19-44. *The steel forms are removed from below after the concrete has hardened.*

Casting the Structural Members. Wood or steel forms are used to form the beams, joists, and floor slabs. One way this is done is to pour the columns first. Then forms are placed and beams, joists, and floor slab are poured together. In this way they form a single unit. This is called *monolithic casting.*

Fig. 19-43 shows steel forms being put into place. The reinforcing bars for the beams and joists are placed in the recesses left between the forms. The bars for the slab are placed on top of the forms. The concrete is poured over the entire thing.

After the concrete has set, the forms are removed from below, Fig. 19-44. Fig. 19-45 shows the underside of the floor with the joists and beams after the forms have been removed. Fig. 19-46 shows a section through the slab.

530

WOOD STRUCTURAL DRAWINGS

Research in wood technology has produced findings that have made wood more useful for structural members. Engineers can accurately figure tension and compression forces in wood members. They can design members to carry specified loads. Wood members can span wide distances. Two commonly used types are wood trusses and laminated wood members.

Wood Trusses. Two common types of wood trusses are shown in Fig. 19-47. These are the kingpost and the Fink.

The kingpost truss requires fewer pieces. On larger spans it usually requires 2" × 6" chords. The Fink truss has more pieces. It can span greater distances with 2" × 4" chords. These are used in house construction.

When designing wood trusses, the engineer must decide on the type of wood to use. Yellow pine and fir are often used. The grade of the wood must also be known. The various types and grades of wood have different load-carrying qualities. This information has been found by experimentation. It is available in table form. A good source is *Wood Structural Design Data,* published by National Lumber Manufacturers Association, Washington, D.C.

The method for fastening truss joints influences their strength. Two common methods are with plywood gussets and metal fasteners.

19-45. *The finished ceiling after the forms are removed and the surface is painted.*

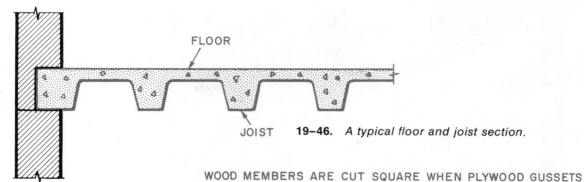

FLOOR

JOIST   **19–46.**   *A typical floor and joist section.*

The plywood gusset is glued and nailed or stapled. See Fig. 19–48. They are on both sides of each joint. The size of the gusset is carefully engineered. It must have the proper amount of gluing and nailing surface. The size of nails or staples must follow a standard. The spacing of the nails is specified. All of these must be correct if the truss is to carry the design load.

Notice that the web members are cut square on the end. This simplifies making the truss.

Metal gusset plates are shown in Fig. 19–47. They are 60-gage sheet steel. Special 8-penny nails, 1½ inches long, are used. Metal gusset plates are made in a wide variety of sizes.

Split-ring connectors are another method for strengthening trusses. See Fig. 19–49. They are inserted halfway into each member. When two web members join a chord, a split-ring is used for each one.

A drawing of a truss with split-ring connectors is shown in Fig. 19–50.

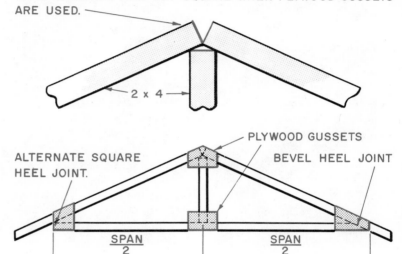

WOOD MEMBERS ARE CUT SQUARE WHEN PLYWOOD GUSSETS ARE USED.

2 x 4

ALTERNATE SQUARE HEEL JOINT.

PLYWOOD GUSSETS

BEVEL HEEL JOINT

$\dfrac{\text{SPAN}}{2}$ $\dfrac{\text{SPAN}}{2}$

SPAN

KING POST TRUSS WITH PLYWOOD GUSSETS

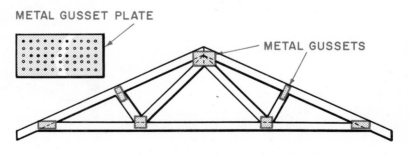

METAL GUSSET PLATE

METAL GUSSETS

FINK TRUSS WITH METAL GUSSET PLATES

19–47.   *Common wood trusses used in house construction.*

American Plywood Association

19–48. *Plywood gussets are glued and nailed on both sides of a truss.*

19–49. *Split-ring connectors are recessed into the two members to be joined.*

Timber Engineering Co.

DEAD LOAD + LIVE LOAD   35 LBS. SQ. FT.
DEAD CEILING FRAMING   10 LBS. SQ. FT.
TRUSSES SPACED   2'-0 O. C.

LUMBER SHALL BE OF A GROUP "B" CONNECTOR
LOAD SPECIES AND OF SUFFICIENT QUALITY TO
PERMIT ALLOWABLE UNIT STRESSES OF:
f = 900 P.S.I. EXTREME FIBER IN BENDING.
c = 900 P.S.I. COMPRESSION PARALLEL TO
      GRAIN.
E = 1,760,000 P.S.I. MODULUS OF ELASTICITY.

HARDWARE		
NO.	ITEM	SIZE
13	SPLIT RING	$2\frac{1}{2}$" DIA.
6	BOLTS	$\frac{1}{2}$" x 4"
2	BOLTS	$\frac{1}{2}$" x 6"
1	BOLT	$\frac{1}{2}$" x $7\frac{1}{2}$"
18	WASHERS	$\frac{1}{2}$"

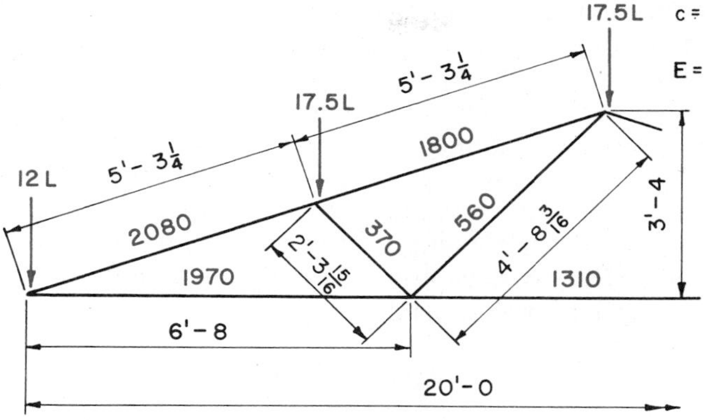

THIS IS A DESIGN DRAWING.

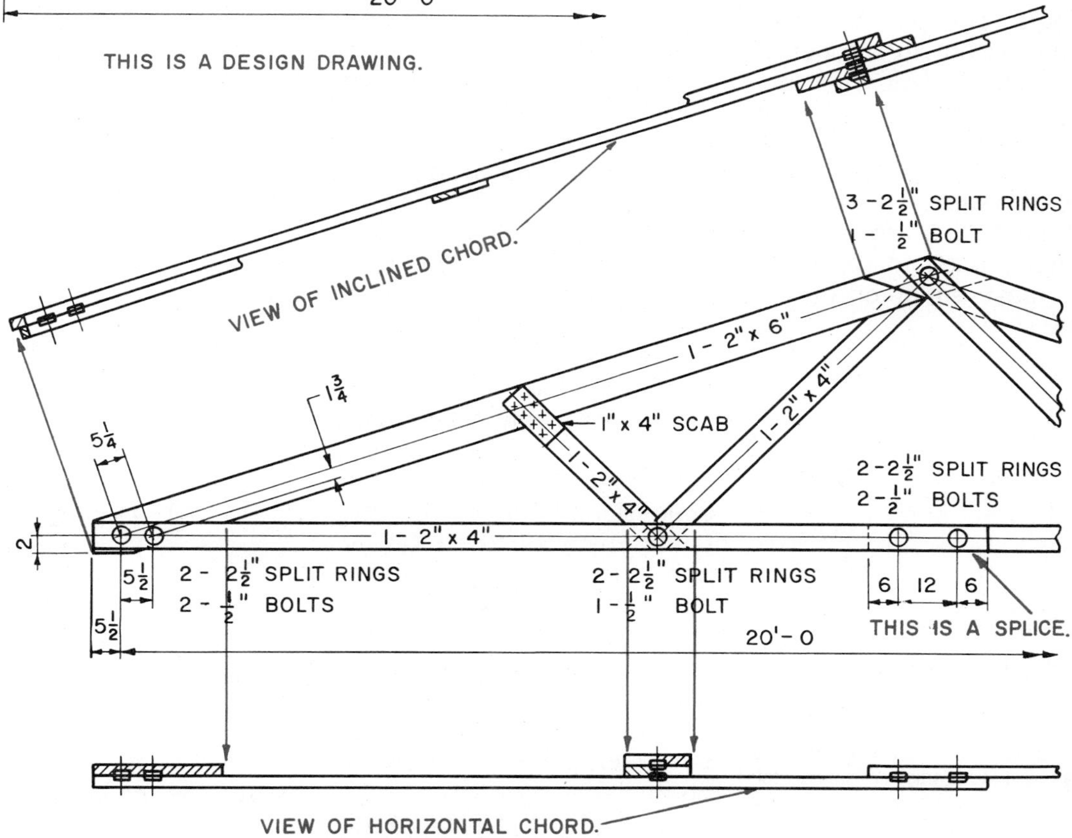

VIEW OF INCLINED CHORD.

$3 - 2\frac{1}{2}$" SPLIT RINGS
$1 - \frac{1}{2}$" BOLT

1 - 2" x 6"

1"x 4" SCAB

1- 2" x 4"

1 - 2" x 4"

$2 - 2\frac{1}{2}$" SPLIT RINGS
$2 - \frac{1}{2}$" BOLTS

1- 2" x 4"

$2 - 2\frac{1}{2}$" SPLIT RINGS
$2 - \frac{1}{2}$" BOLTS

$2 - 2\frac{1}{2}$" SPLIT RINGS
$1 - \frac{1}{2}$" BOLT

6   12   6

THIS IS A SPLICE.

20'- 0

VIEW OF HORIZONTAL CHORD.
THIS IS A PRODUCTION DRAWING.

Based on a Design by Timber Engineering Company

19–50.   *A wood truss with a slope of 4" in 12".*

Drawing Wood Trusses. A drawing for a wood truss is shown in Fig. 19–50. This truss is designed for house construction. The drawing is made of several parts—a design drawing, hardware list, wood specifications, and the truss production drawing.

Study the *design drawing.* The members are shown with a single line. This is the center line of each member. The load data computed by an engineer is given. Shown are the dimensions of the chords, web members, and panels. These dimensions are to the center line of the split-ring connectors. The connectors are on the center line of each member.

The allowable stresses the wood must carry are in a note. The wood used to make the truss must equal or exceed these.

Nominal Size*	Dressed Size**
2 × 2	1½ × 1½
2 × 3	1½ × 2⅝
2 × 4	1½ × 3⅝
2 × 6	1½ × 5½
2 × 8	1½ × 7½
2 × 10	1½ × 9½
2 × 12	1½ × 11½

*Nominal size is the rough size before the lumber is planed smooth. It is the size shown on the drawing.

**Dressed size is the actual size of the member after it is planed smooth. The members used in the truss are actually this size.

National Lumber Manufacturers Association

19–51. *Nominal and dressed lumber sizes.*

A hardware list is given. The length of the bolts depends upon the number of members the bolts must hold together.

The production drawing shows the members. It is drawn to scale. The thickness and width of each member before the wood is planed is given. Wood is rough-cut to these sizes in a mill. After it is dried, it is planed (smoothed). This makes the finished member smaller. The actual size of a 2″ × 4″ becomes 1½″ × 3⅝″ after planing. See Fig. 19–51 for standard sizes.

Split-ring connectors are used at critical connections. When two web members join a chord, each has a split ring. The split rings are shown on the drawing as solid circles. The number needed at each joint is given. The number of bolts is also noted. The bolt length is given in the hardware table.

Other connections are made with a 1″ × 4″ wood gusset, glued and nailed in place. The dimensions show the distances between the intersection of the center lines of the members. These points are also the center of the split ring. The dimensions are placed on the design drawing. The thickness and width of each member is placed on the production drawing.

Views of the horizontal and inclined chords are drawn. The horizontal chord is drawn with the truss sectioned just above the chord. Both views are drawn looking down on the chord from above. Colored projection lines show how the members are projected to these views.

When members are long, they must be made from two pieces that are joined by splicing. The standard spacing for connectors is shown on the drawing and must be carefully dimensioned.

Laminated Wood Structural Members. Laminated wood structural members are made by gluing together layers of wood. The most commonly used layers are ¾ inch and

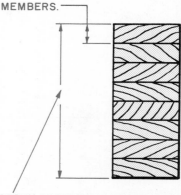

LAMINATIONS 1⅝″ IN LARGE MEMBERS, ¾″ IN SMALL MEMBERS.

DEPTH VARIES ACCORDING TO SPAN AND SHAPE.

19–52. *Typical section through laminated wood member.*

1⅝ inch. See Fig. 19–52. They are about one-third stronger than solid wood members of the same size. The layers can be seasoned better than solid members. Defects, as knots, can be reduced or eliminated.

An engineer designs laminated members. After he decides the loading and type of wood to be used, he computes the size of the member needed. The draftsman works with these data.

Fig. 19–53 shows a warehouse under construction. The structural system is made up entirely of laminated wood members. The structural drawings used to design this warehouse are shown in Figs. 19–54 through 19–57.

Fig. 19–54 shows a *roof framing plan* for the warehouse. It shows all the columns, girders, beams, and purlins that form the structure to support the roof decking. The dimensions have been left off this drawing because they are so complex. In Fig. 19–55 the complete *system of dimensioning* is shown. This is only a small corner of the building. The roof framing plan, when drawn to a large scale, would be completely dimensioned in this way.

Notice that each row of columns is indicated by either a letter or number. Each beam, girder, or purlin is identified by a letter.

The details of each structural member are shown by making sections through the building. One section is shown in Fig. 19–56. The section shown is taken only through the corner detailed, Fig. 19–55. On a large scale drawing, the section would continue through the entire building.

The size of the roof's structural members is shown on the section. This includes the thickness, height, and length. A complete listing of all beams is shown by a schedule. See Fig. 19–57.

Columns are numbered on the framing plan. Their sizes are given in a column schedule table. See Fig. 19–57.

Included on the drawing with the section are details. In Fig. 19–56 two details are shown: how the columns are connected to the beams and how the columns are connected at the base. Other details may be included in a drawing if the engineer decides they are needed.

Timber Structures, Inc.

19–53. *A warehouse being framed with laminated wood members. The columns, beams, and purlins are all wood. In this picture the beams have been painted white. The purlins are dark. The second row of columns and beams is shown in section view in Fig. 19–56. Compare the column base set in the corrugated metal drum shown in the photograph with the column base detail shown in Fig. 19–56.*

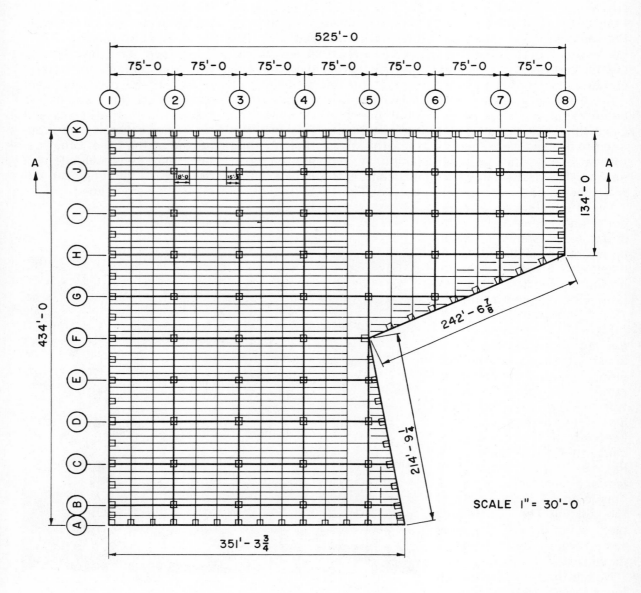

Timber Structures, Inc.

19–54. *This is the roof framing plan for the warehouse, Fig. 19–53. Dimensions and identification of parts have been left off because they cannot be shown on this reduced drawing. One section in color is shown in full detail in Fig. 19–55. This shows how the total drawing would be dimensioned.*

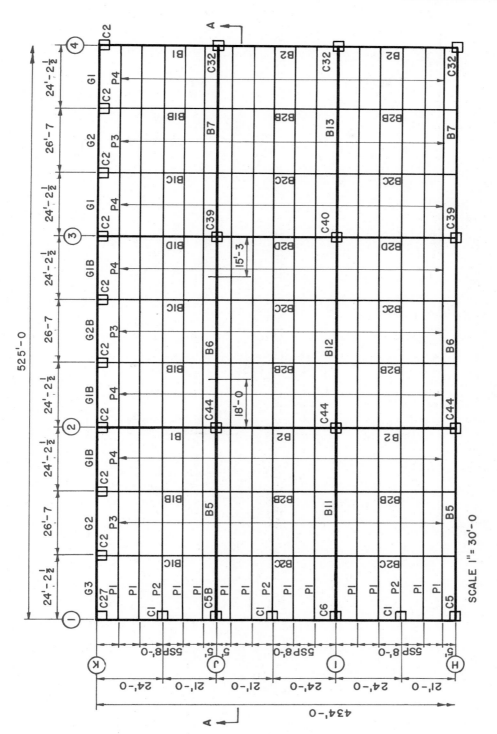

Timber Structures, Inc.

19–55. This is a roof framing plan for one corner of the warehouse, Figs. 19–53 and 19–54. Columns are marked with a "C", beams with a "B", purlins with a "P", and girders with a "G".

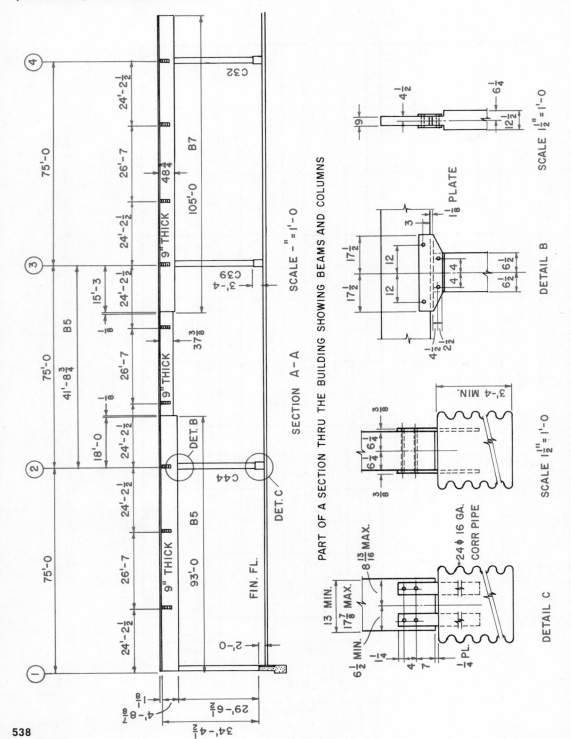

SECTION A-A

SCALE — " = 1'-0

PART OF A SECTION THRU THE BUILDING SHOWING BEAMS AND COLUMNS

DETAIL B

SCALE $1\frac{1}{2}$" = 1'-0

DETAIL C

SCALE $1\frac{1}{2}$" = 1'-0

19-56. *This drawing of the warehouse, Fig. 19-53, shows a part of Section A-A. It contains column and beam connections.*

19–57. *This is how the sizes of beams, girders, purlins, and columns are shown on the drawing.*

Column Schedule

Col. No.	Size	Length	No. Reqd.	Top Fab	Bottom Fab
C1	$9'' \times 17\tfrac{7}{8}''$	$33'\text{–}9\tfrac{3}{16}''$	5	Detail of C1	and P2
C2	$9'' \times 16\tfrac{1}{4}''$	$32'\text{–}8\tfrac{11}{16}''$	15	Detail of C2	and B1
C5	$9'' \times 17\tfrac{7}{8}''$	$29'\text{–}6\tfrac{1}{2}''$	3	Typ. of C1	Typ. of C1
C6	$9'' \times 17\tfrac{7}{8}''$	$30'\text{–}8\tfrac{9}{16}''$	6	" "	" "
C27	$9'' \times 9\tfrac{3}{4}''$	$35'\text{–}0\tfrac{5}{16}''$	1	Det. T	Corner Detail
C32	$12\tfrac{1}{2}'' \times 14\tfrac{5}{8}''$	$28'\text{–}5\tfrac{13}{16}''$	8	Typ. Int. Col.	Cap and Anchor
C39	$12\tfrac{1}{2}'' \times 14\tfrac{5}{8}''$	$27'\text{–}9\tfrac{3}{4}''$	2	"	"
C44	$12\tfrac{1}{2}'' \times 16\tfrac{1}{4}''$	$28'\text{–}9\tfrac{11}{16}''$	4	"	"

A partial schedule for the columns shown in Fig. 19–55.

Beam Schedule

Beam No.	Size	Length	No. Required
B1	$9'' \times 30''$	$45'\text{–}0''$	3
B2	$9'' \times 30''$	$45'\text{–}0''$	3
B5	$9'' \times 56\tfrac{7}{8}''$	$75'\text{–}0''$	3
B6	$9'' \times 37\tfrac{3}{8}''$	$41'\text{–}9''$	3
B7	$9'' \times 48\tfrac{3}{4}''$	$75'\text{–}0''$	2
B11	$9'' \times 56\tfrac{7}{8}''$	$75'\text{–}0''$	5
B13	$9'' \times 48\tfrac{3}{4}''$	$75'\text{–}0''$	3

A partial schedule for the beams shown in Fig. 19–55. Purlin and girder sizes are shown the same way.

19–58. *Open-haunch Tudor laminated wood arches used to support a wide, clear-span roof. Notice the laminated purlins running perpendicular to the arches.*

Timber Structures, Inc.

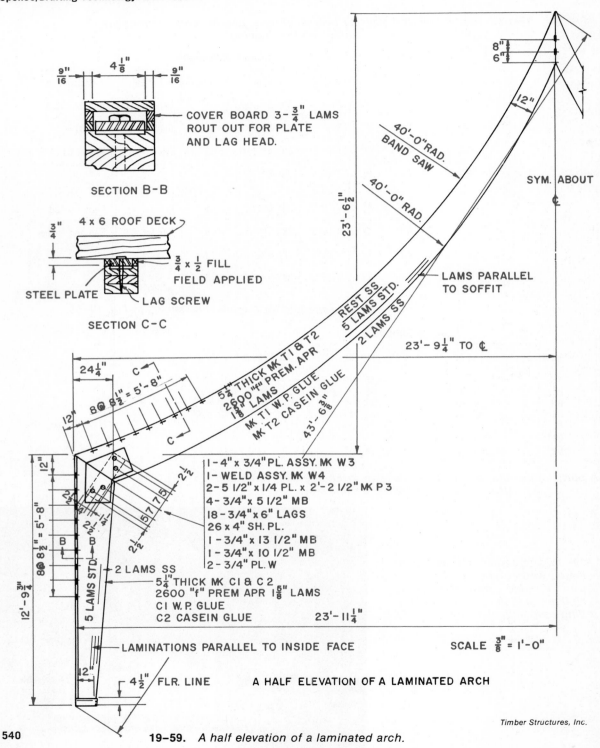

SECTION B-B

COVER BOARD 3-$\frac{3}{4}$" LAMS
ROUT OUT FOR PLATE
AND LAG HEAD.

$\frac{9}{16}$" 4 $\frac{1}{8}$" $\frac{9}{16}$"

4 x 6 ROOF DECK

$\frac{3}{4}$"

$\frac{3}{4}$ x $\frac{1}{2}$ FILL
FIELD APPLIED

STEEL PLATE

LAG SCREW

SECTION C-C

8"
6"

12"

40'-0" RAD.
BAND SAW

40'-0" RAD.

SYM. ABOUT
$\mathcal{C}$

23'-6$\frac{1}{2}$"

LAMS PARALLEL
TO SOFFIT

REST SS
5 LAMS STD.

2 LAMS SS

23'-9$\frac{1}{4}$" TO $\mathcal{C}$

5$\frac{1}{4}$" THICK MK T1 & T2
2600 "f" PREM. APR
1$\frac{5}{8}$" LAMS
MK T1 W.P. GLUE
MK T2 CASEIN GLUE

43'-6$\frac{3}{8}$"

24$\frac{1}{4}$"

C

12"

8 @ 8$\frac{1}{2}$" = 5'-8"

C

1-4" x 3/4" PL. ASSY. MK W 3
1-WELD ASSY. MK W 4
2-5 1/2" x 1/4 PL. x 2'-2 1/2" MK P 3
4-3/4"x 5 1/2" MB
18-3/4"x 6" LAGS
26 x 4" SH. PL.
1-3/4"x 13 1/2" MB
1-3/4"x 10 1/2" MB
2-3/4" PL. W

12"

2$\frac{2}{3}$"
2$\frac{2}{3}$"
2$\frac{1}{4}$"
2$\frac{2}{3}$"
2$\frac{1}{2}$"
2$\frac{1}{2}$"
5
7
5

B B

8 @ 8$\frac{1}{2}$" = 5'-8"

5 LAMS STD.

2 LAMS SS

5$\frac{1}{4}$" THICK MK C1 & C2
2600 "f" PREM APR 1$\frac{5}{8}$" LAMS
C1 W. P. GLUE
C2 CASEIN GLUE

23'-11$\frac{1}{4}$"

12'-9$\frac{3}{4}$"

LAMINATIONS PARALLEL TO INSIDE FACE

SCALE $\frac{3}{8}$" = 1'-0"

12"

4$\frac{1}{2}$" FLR. LINE

A HALF ELEVATION OF A LAMINATED ARCH

Timber Structures, Inc.

19-59. *A half elevation of a laminated arch.*

Another type of laminated wood structural member is the arch. One type is shown in Fig. 19–58. Arches are made in a wide variety of shapes. They can span large distances.

Fig. 19–59 is the detailed drawing for a three-hinged Tudor arch. How the arch is used in a church is shown in Fig. 19–60.

Each part of the arch must be dimensioned so it can be laid out and cut to shape. The width at each end is given. The curved member has the radius of the curve given. The thickness is shown with a note on the member.

The connection of the two members is of great importance. The location of each fastener is carefully dimensioned. Notice the sections through the members in Fig. 19–59. This reveals that a steel plate is set in a groove. On the vertical member, the plate is covered with wood. On the roof member, the decking covers the steel plate.

Since the arch is symmetrical, only one half need be drawn. The center line is used for dimensions locating the vertical member.

Timber Structures, Inc.

19–60. *Three-hinged Tudor wood-laminated arches used to form the roof and walls of a church. The roof decking is several inches thick and spans the distance between the arches without support of purlins.*

Additional Reading

CRSI Design Handbook. Concrete Reinforcing Steel Institute, 38 S. Dearborn St., Chicago, Ill.

CRSI Recommended Practice for Placing Reinforcing Bars. Concrete Reinforcing Steel Institute.

Ketchum, Milo S. *Handbook of Standard Structural Details for Buildings.* Prentice-Hall, Inc., Englewood Cliffs, N.J., 1956.

Manual of Standard Practice for Detailing Reinforced Concrete Structures. American Concrete Institute, Box 4754, Redford Station, Detroit, Mich., 1965.

Manual of Steel Construction. American Institute of Steel Construction, Inc., 101 Park Ave., New York, N.Y.

National Design Specification for Stress-Grade Lumber and Its Fastenings. National Lumber Manufacturers Association, 1719 Massachusetts Ave., N.W., Washington, D.C.

Saxe Manual for Structural Welding Practice. Saxe Welded Connections, 1701 St. Paul St., Baltimore, Md.

Structural Shop Drafting, Vol. 1, 2, and 3. American Institute of Steel Construction.

Structural Steel Detailing. American Institute of Steel Construction.

Time-Saver Standards. F. W. Dodge Corp., New York, N.Y.

Wood Handbook No. 72. U.S. Government Printing Office, Superintendent of Documents, Washington, D.C.

Wood Structural Design Data. National Lumber Manufacturers Association.

Build Your Vocabulary

Following are some terms you should understand and use as part of your working vocabulary. Write a brief explanation of what each means.

Column
Girder
Pitch
Gage line
Shop rivets
Field rivets
Angle connection
I-beam
Steel framing drawing
Steel erection drawing

Steel shop drawing
Truss
Chord
Rigid frame
Compressive strength
Tensile strength
Reinforced concrete
Stirrups
Bar supports
Beam schedule
Concrete placing drawings
Monolithic casting
Gusset
Wood truss web members
Laminated wood structural member

Study Problems

1. P19–1 shows an angle connector used to join steel beams. Notice it has unequal legs. Make a two-view drawing of this connector. Locate the rivet holes and dimension them. The holes have a 3/4-inch diameter.

2. P19–2 shows the connection of a 12-inch and 14-inch I-beam. The 14-inch I-beam has two rivet holes in the bottom flange. These are to have standard spacing. The beams are joined by a $3\frac{1}{2} \times 3\frac{1}{2} \times 8$-inch angle connector.

Make a full-size drawing of this detail. Locate all the holes in the 14-inch beam and the angle connector. Use standard spacing. Dimension the hole locations.

The rivets have a 3/4-inch diameter. The members have rolled edges.

3. Dimension the WF beam shown in P19–3. All rivets have standard spacing.

4. Draw a combined roof framing plan and erection drawing of the building shown in P19–4.

All beams are 8 WF 24 ($8'' \times 6\frac{1}{2}''$). They are connected to the columns with angle connectors prepared to receive two 3/4-inch diameter bolts. The angle connectors are welded to the beam. The top of the beam is flush with the top of the column.

All columns are 8 WF 17 ($8'' \times 5\frac{1}{2}''$) 8'-0" long. The WF member is 12'-3½" long. The base is a 10" × 10" plate, 1¼-inches thick. It has two ⅞-inch diameter holes. The base is welded to the column.

The channels are 8⌷ 18–25 (8" × 2½"). They are connected to the beams with angle connectors. The connectors have two ¾-inch diameter holes for bolting to the beams. They are welded to the channel.

5. Make shop drawings for all the beams in the structure shown in P19–4.

6. Make shop drawings for the columns in the structure shown in P19–4.

7. Draw a completely dimensioned detail drawing of the welded truss shown in Fig. 19–30. Draw to the scale 1" = 1'-0".

8. Study the riveted truss shown in Fig. 19–29. Prepare a parts list giving the size of each member. This includes gusset plates, structural members, and rivets. Record the number of each part needed to make one complete truss.

9. Study the riveted truss shown in Fig. 19–29. Find and record the answers to the following questions:

A. What is the rivet spacing on the web members?
B. What is. the distance between the center lines of the columns to hold this truss?
C. What is the overall height of the truss at the center line? This is the outside-to-outside dimension.

D. What is the overall height of the truss at the end that connects to the column? This is the outside-to-outside dimension.

10. Draw a completely dimensioned detail drawing of the riveted Fink truss shown in Fig. 19–29. Draw to the scale 1" = 1'-0".

11. Make a dimensioned drawing of the rigid frame shown in P19–5.

12. Study the concrete engineering shown in P19–6. Make a placing drawing. Draw to the scale ¼" = 1'-0". Include a beam and joist schedule. Dimension the location of each stirrup and bend. Identify each bar.

13. Make a production drawing of the wood truss shown in the design drawing, in P19–7. Tables are provided giving data for various spans. Select one of the spans and make the dimensioned production drawings for it. This design uses split-ring connectors.

14. Make a detail drawing of the laminated wood arch shown in Fig. 19–58. Draw to the scale ⅜" = 1'-0" or ½" = 1'-0".

15. Make a roof framing plan for the building shown in P19–8. Draw the section shown. Prepare a column, beam, girder, and purlin schedule.

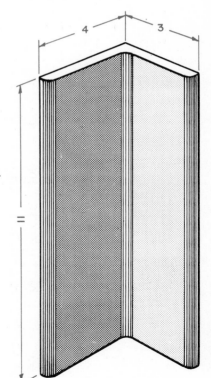

P19–1. *An angle connector.*

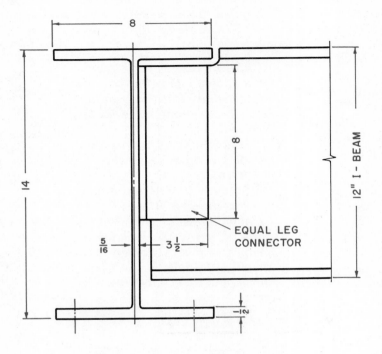

P19-2. *A connection between two I-beams.*

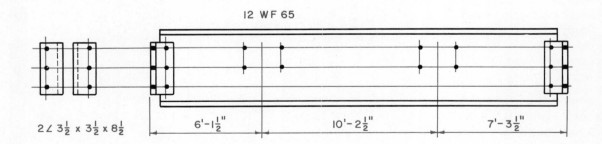

P19-3. *A steel wired flange beam.*

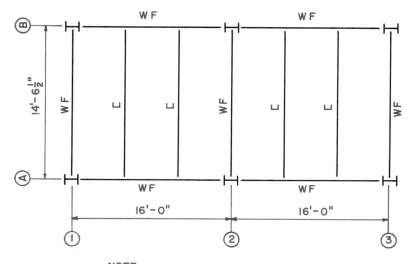

NOTE:
CHANNELS EQUALLY SPACED.

P19–4. *Roof framing plan for a steel framed building.*

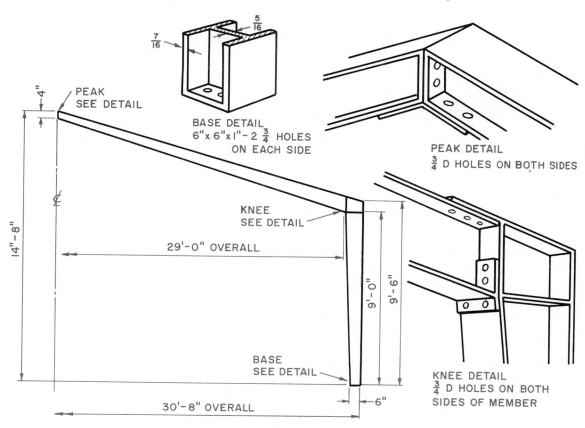

P19–5. *Steel structural members.*

545

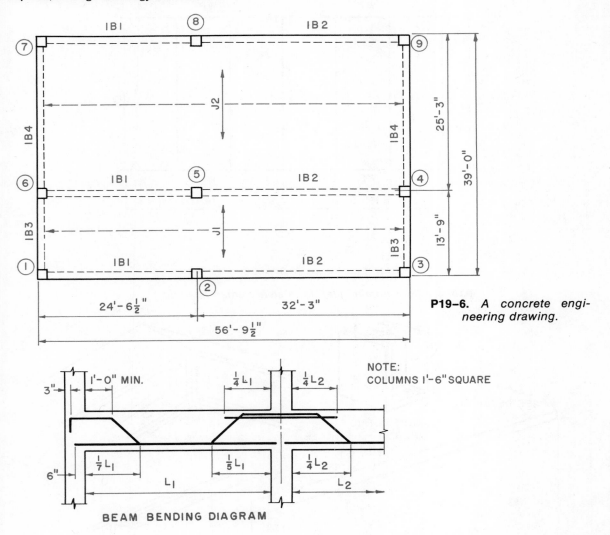

P19-6. *A concrete engineering drawing.*

NOTE:
COLUMNS 1'-6" SQUARE

BEAM BENDING DIAGRAM

Beam Schedule

| Mark | B × D | Reinforcing | | Stirrups | | | Stirrup Support Bars in Top |
		Str.	Bent	No.	Size	Spacing Each End	
1B1	12 × 33	3–#8	2–#8	14	#4	3@ 4, 3@ 6, 3@ 8, 3@ 12	2–#4
1B2	12 × 33	2–#8	3–#8	12	#4	2@ 10, 2@ 14, 2@ 18	2–#4
1B3	12 × 20	2–#6	2–#6	4	#3	1@ 4, 1@ 8	2–#4
1B4	12 × 33	3–#8	2–#8	14	#4	3@ 4, 3@ 6, 2@ 8, 2@ 12	2–#4

Joist Schedule

Mark	B	D	Reinforcing	
			Str.	Bent
J1	5	8 + 2½	1–#6	1–#6
J2	5	8 + 2½	1–#6	1–#6

P19–6. *continued*

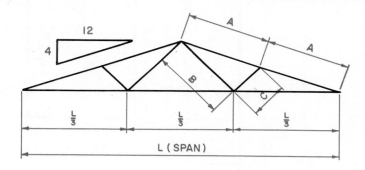

Design by Timber Engineering Co.

P19–7. *A wood truss design.*

Span	Lumber		
	Inclined Chords	Horizontal Chord	Web
20'–0"	2 × 6	2 × 4	2 × 4
26'–0"	2 × 6	2 × 4	2 × 4
32'–0"	2 × 8	2 × 6	2 × 4

Span	Dimensions		
	A	B	C
20'–0"	5'–3½"	4'–8³⁄₁₆"	2'–3¹⁵⁄₁₆"
26'–0"	6'–10³⁄₁₆"	6'–1³⁄₁₆"	3'–0⁷⁄₁₆"
32'–0"	8'–5³⁄₁₆"	7'–6³⁄₁₆"	3'–8⁷⁄₈"

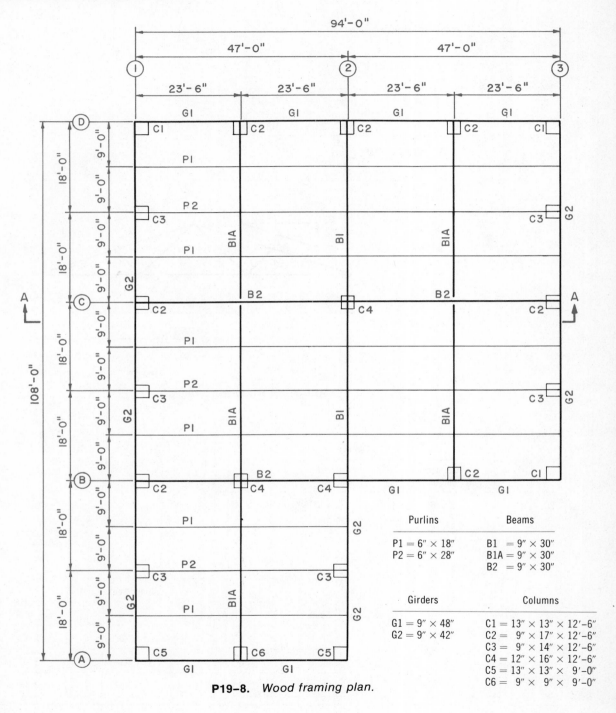

P19-8. *Wood framing plan.*

Purlins

P1 = 6″ × 18″
P2 = 6″ × 28″

Beams

B1 = 9″ × 30″
B1A = 9″ × 30″
B2 = 9″ × 30″

Girders

G1 = 9″ × 48″
G2 = 9″ × 42″

Columns

C1 = 13″ × 13″ × 12′–6″
C2 = 9″ × 17″ × 12′–6″
C3 = 9″ × 14″ × 12′–6″
C4 = 12″ × 16″ × 12′–6″
C5 = 13″ × 13″ × 9′–0″
C6 = 9″ × 9″ × 9′–0″

Chapter 20

Electrical–Electronics Diagrams

An important part of the plans for the manufacture of machines, communications equipment, and controlling devices is drawings showing the electrical or the electronic requirements. Electrical and electronics drawings show the wiring diagrams or circuits. These drawings are called *schematic diagrams.* They show the components used and how they are to be connected. A *component* is a single unit, such as a tube.

The design of electrical and electronics devices is done by electrical and electronics engineers. Their work is recorded as freehand sketches. The draftsman makes finished drawings from these sketches and notes.

Fig. 20–2 shows a sketch and the finished drawing (schematic) for a RF impedance transformer.

The draftsman is more effective in his work if he understands the fundamentals of electricity and electronics. He should have a basic idea of circuity. He should understand the purpose served by the components in the system. He must also know the symbols used to show the components. It is the purpose of this chapter to show the commonly used symbols and show how to draw schematics.

WHAT IS ELECTRICAL ENERGY?

The source of electrical energy is the *atom.* The atom is the smallest part of an *element*. There are over one hundred elements known to man. Examples are copper, silver, and iron. When an atom is broken down, it is found to have a core called a *nucleus.* The nucleus contains positive particles of electricity called *protons* and neutral particles of electricity called *neutrons.* Revolving around the nucleus are negatively charged particles of electricity called *electrons.* See Fig. 20–3.

An *electric current* is a stream of electrons moving along a conductor, as a copper wire. See Fig. 20–4. The electrical energy is transferred through a conductor by the free movement of electrons from one atom to the next

20–1. *The Gordon Evans Generating Station, northwest of Wichita, Kansas. This installation required thousands of drawings of all types. Architectural, structural, electrical, piping, and mapping drawings were needed.*

549

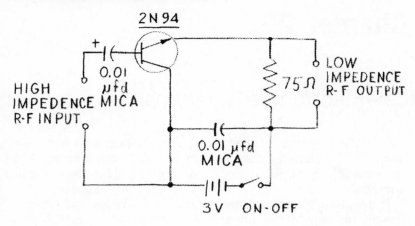

GTE-Sylvania Electric Products, Inc.

20–2A. *Rough sketch of proposed circuit of a R-F impedence transformer.*

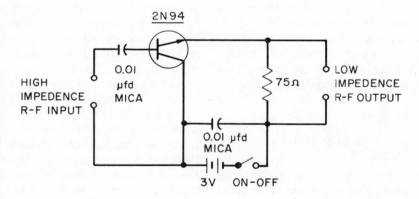

20–2B. *Finished schematic of a R-F impedence transformer.*

atom. As an electron enters one end of a conductor, another leaves the opposite end of the conductor. The movement of electrons through a conductor is called *electrical current.*

Electrical Measurements.
Electrical current and power can be measured. The units used to do this are volts, ohms, amperes, and watts. A *volt* is a measure of electrical *pressure.* It is the force that causes electricity to flow along a conductor. This pressure is sometimes called electromotive force. The symbol for voltage is E.

An *ohm* is a measure of *resistance* to the flow of electrical current. The symbol for an ohm is R.

An *ampere* is a measure of the amount of electrical *current* flowing. The symbol for the ampere is I.

One ohm of resistance will allow one ampere of current to pass when one volt of pressure is applied. This can be shown by the formula, $R = \dfrac{E}{I}$. If two of these are known, the third can be found by entering them into the formula and solving for the unknown.

A *watt* is a unit of electrical *power.* Electrical power is the rate at which electrical energy is delivered to a device. The power (wattage) is equal to the current (amperes) multiplied by the force (voltage). This is represented by the formula, $P = I \times E$.

Horsepower is a measure of how much work is done. One horsepower is equal to 746 watts of electrical power.

Electrical Conductors.
A *conductor* is a material that permits electrical current to pass through it. Some materials permit the flow of electrical power to pass easier than others. In other words, they offer little resistance to the current flow. The best conductors are silver, copper, and aluminum. Materials, such as tungsten, iron, and carbon, offer considerable resistance to the flow of electricity, and are therefore poor conductors.

Other materials will not conduct electrical energy. They are called insulators. Glass and plastics are examples.

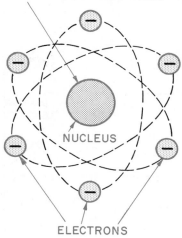

PROTONS AND NEUTRONS

NUCLEUS

ELECTRONS

20–3. *An atom has electrons revolving about a nucleus.*

A CONDUCTOR AS A COPPER WIRE.

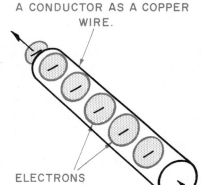

ELECTRONS

20–4. *Electric current is the movement of energy through a conductor by electron movement.*

BLOCK DIAGRAMS

A *block diagram* is one using rectangular blocks to represent the major parts of a circuit. The part is identified by labeling inside the block. It differs from a schematic diagram. In a schematic diagram, graphic symbols are used to show the individual components making up the device.

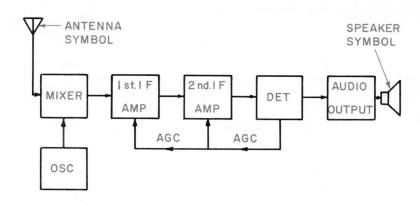

ANTENNA SYMBOL

SPEAKER SYMBOL

MIXER | 1st. I F AMP | 2nd. I F AMP | DET | AUDIO OUTPUT

AGC   AGC

OSC

20–5. *A block diagram of a transistor radio receiver circuit.*

The block diagram in Fig. 20–5 shows the overall circuit plan of a transistor radio receiver. It presents in a simple way the relationships between each major part of the circuit. It does not give details about the design of each major part. The diagram explains that the signal is received through the antenna (represented by a symbol), moves through the mixer, first and second intermediate-frequency stages (IF), and on to the speaker. It shows a feedback circuit (AGC) and an oscillator (OSC) connected into the main circuit.

A block could represent a complete unit in an installation, as a television camera or video tape recorder in a television studio.

Drawing a Block Diagram. The symbol generally used to represent the major units on a block diagram is a rectangular block. Sometimes squares and triangles are used. Some components are indentified by a graphic symbol rather than a

block. These are units that are located on the end of the drawing, such as a speaker or antenna. See Fig. 20–5.

The draftsman must decide on the relationship between the blocks. He must decide the best way to present these on a drawing. Usually a number of sketches on graph paper are made before a good solution is found.

The size of the blocks will vary with the lettering that must be placed on them and the space available for the drawing. The blocks are usually made the same size throughout the diagram.

As the plan is developed, the flow of current or the signal should be made to run from left to right. The input is usually at the upper left corner and the output at the lower right corner. The flow of the current or signal from one block to another is represented by a single line connecting the blocks. The direction of the flow is indicated by an arrow. The line weight

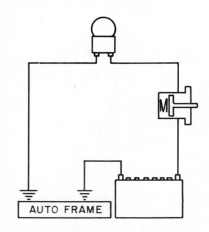

20-6. *A pictorial drawing of the circuit for the dome light in an automobile.*

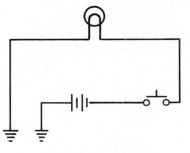

20-7. *A schematic diagram of the circuit for the dome light in an automobile.*

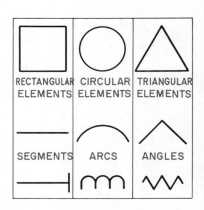

20-8. *Symbols are based on rectangular, circular, and triangular elements.*

for flow should be as heavy as that used for the block. This is usually a rather thick line similar to a visible line on working drawings.

SCHEMATIC DIAGRAMS

Electrical circuits may be illustrated by a pictorial wiring diagram or by a schematic drawing. The *schematic drawing* uses symbols to represent the electrical parts of the circuit. Fig. 20-6 shows a pictorial drawing of a simple circuit for a dome light in an automobile.

The battery, door switch, and light are the components. The lines connecting them are leads (wires). While this is a simple circuit, it shows how more complex circuits are drawn pictorially.

The auto battery supplies the electrical current. When the auto door is opened, the push button closes. The current runs through the push button, through the dome light, and through the frame of the auto back to the battery. This makes a complete circuit.

This same circuit is shown in Fig. 20-7 as a schematic diagram. Graphic symbols are used to represent the parts. This greatly simplifies the drawing. The person using the schematic diagram is not interested in what each part looks like but only in the design of the circuit.

Graphic Symbols. An electrical or electronics schematic diagram involves the use of symbols to represent the parts of the circuit. They are a type of technical shorthand enabling a complex circuit to be represented as clearly and briefly as possible. They usually do not represent a picture of how the part really looks. Industry and government agencies have developed standard symbols to be used on drawings. Three commonly used

sets of symbols are found in: ANSI Y32.2, *Graphic Symbols for Electrical and Electronics Diagrams.*

Mil. Std. 15-1, *Graphic Symbols for Electrical and Electronic Diagrams* (Military Standards).

IEC *Graphical Symbols* (International Electrotechnical Commission).

Graphic symbols are made of several basic elements. The size and location of each element is such that the purpose served by the symbol is clear. Fig. 20-8 shows the basic elements used to form graphic symbols.

Most symbols can be placed on the drawing in any position desired. Studying schematic diagrams is a good way to learn how symbols are placed.

Line Width. Generally the weight of lines used to draw

552

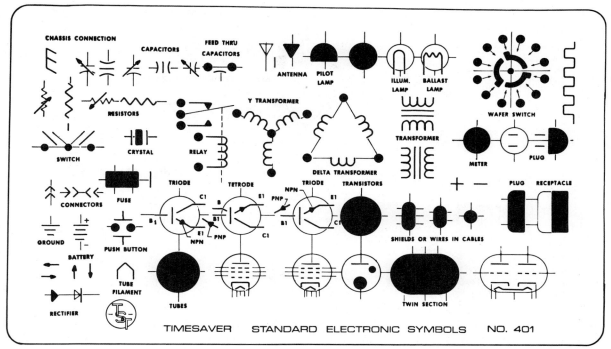

CHASSIS CONNECTION · CAPACITORS · FEED THRU CAPACITORS · ANTENNA · PILOT LAMP · ILLUM. LAMP · BALLAST LAMP · WAFER SWITCH · RESISTORS · Y TRANSFORMER · TRANSFORMER · METER · PLUG · SWITCH · CRYSTAL · RELAY · DELTA TRANSFORMER · PLUG · RECEPTACLE · CONNECTORS · FUSE · TRIODE · TETRODE · TRIODE · TRANSISTORS · NPN · PNP · NPN · E1 · E1 · B1 · C1 · B · PNP · NPN · C1 · SHIELDS OR WIRES IN CABLES · GROUND · PUSH BUTTON · BATTERY · TUBE FILAMENT · TUBES · TWIN SECTION · RECTIFIER · TIMESAVER   STANDARD ELECTRONIC SYMBOLS   NO. 401

20–9.   *An electronic symbol template.*

symbols is the same as for the leads. A *lead* is the line used to show the connections between symbols. If it is desired to make the symbols stand out, they can be drawn with a heavier line than the leads. The width of the line should be such that it is in proportion to the size of the drawing and the lettering to be used.

Symbol Size. Standard practice is to use templates to draw symbols. See Fig. 20–9. This sets a standard size for all symbols on the drawing. The symbol should be the same size wherever it appears on the drawing. A good size for symbols is shown in Fig.

20–10. These can be drawn a little smaller or larger depending upon the size of the drawing. If templates are not available, copy symbols on vellum, slide under your drawing, and trace them.

It is sometimes necessary to letter some type of identification with each symbol. This might show the type of part, size, electrical value, or other necessary data. A standard symbol system and an approved abbreviation should be used.

The symbols shown in Fig. 20–10 are the most commonly used for electrical and electronics diagrams. A standards manual will give a more com-

plete listing. Symbols for house wiring can be found in Chapter 22, Architectural Drawing.

Drawing a Schematic. The location of the components when a device is manufactured is different from that given on a schematic. The schematic shows the components and how they are connected. When the device is made, the location is influenced by such things as the need for it to be very small or to be a desired shape. For example, a radio could be made that is long and low or tall and thin. The same schematic could be used for both.

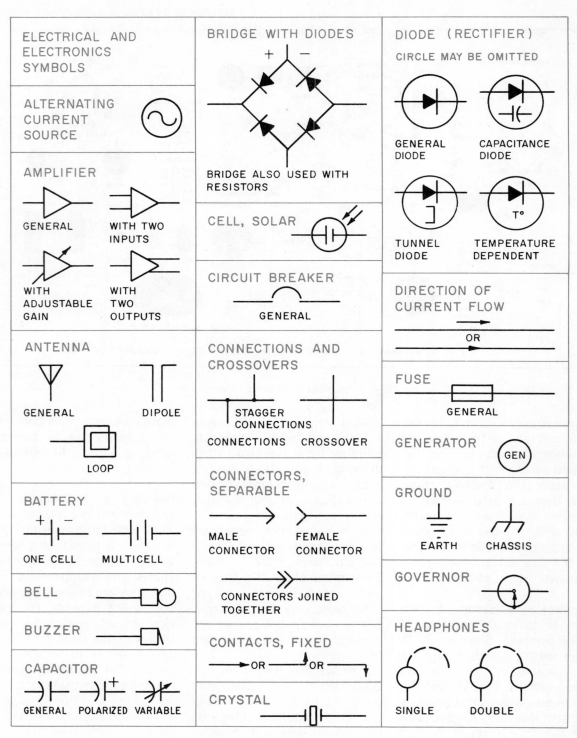

20-10A. *Symbols most commonly used for electrical and electronics diagrams.*

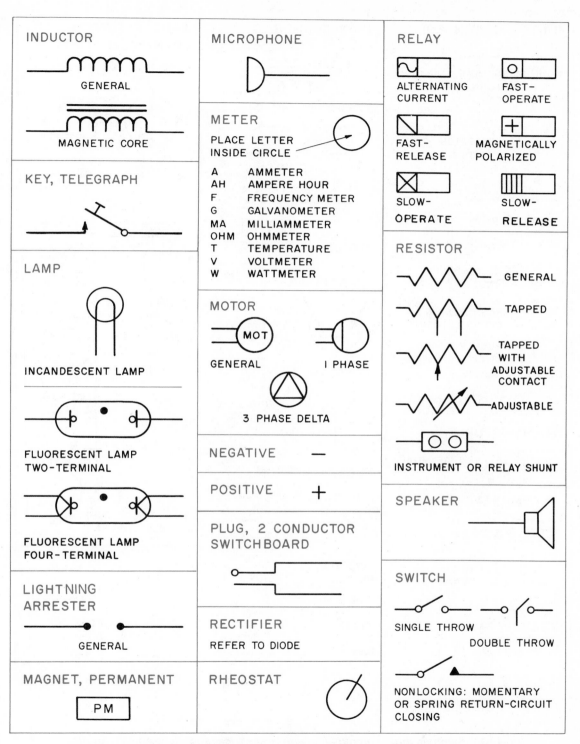

20-10B. *Symbols most commonly used for electrical and electronics diagrams.*

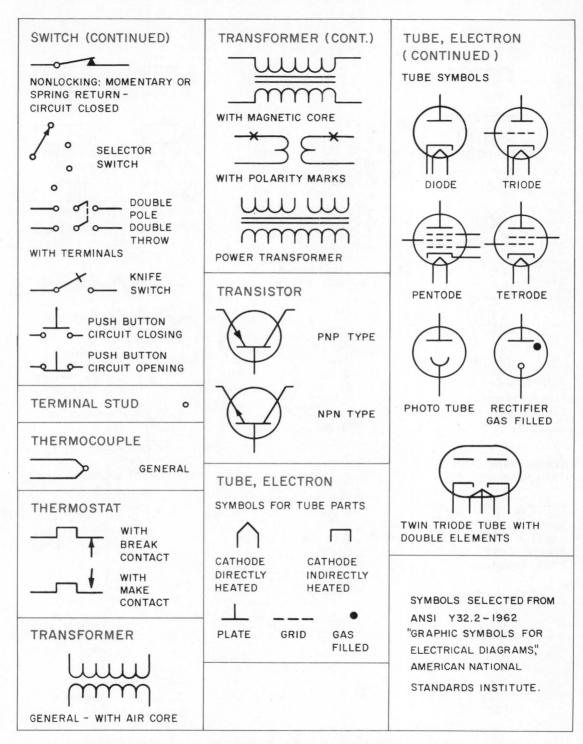

20-10C. *Symbols most commonly used for electrical and electronics diagrams.*

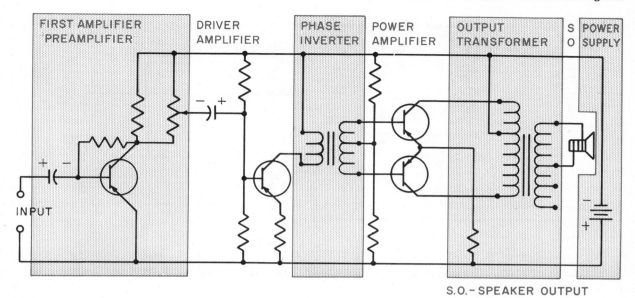

S.O. – SPEAKER OUTPUT

GTE-Sylvania Electric Products, Inc.

20–11. *Schematic for an audio amplifier showing symbol grouping in subunits.*

The things desired in a schematic diagram are: (1) correct grouping of the symbols, (2) uniform alignment of the symbols, leads, and lettering, (3) leads that are as short as possible to make, (4) as few leads crossing as possible, (5) uncrowded lettering, and (6) a balanced drawing.

Remember that there is more than one satisfactory way to lay out any schematic diagram.

It is difficult to notice errors on schematic diagrams; therefore, care and accuracy in drawing are important.

Correct Grouping of Symbols. Electronic devices are made of subunits. When these are put together, they make a complete device. The symbols

for each unit are grouped together on the drawing. Before starting to make the drawing, the symbols that are to be grouped must be marked. Then it must be decided how they are to be placed on the drawing. This is done by studying the first sketches.

Shown in Fig. 20–11 is the schematic of an audio amplifier. On this figure are seven subunits. The amplifier is made up of a first amplifier, driver amplifier, phase inverter, power amplifier, output transformer, speaker output, and power supply.

The symbols for each subunit are kept together on one section of the diagram. The blocks are then related to each other so the connections between them can be shown.

This requires an understanding of electricity and electronics. Since most students do not have this understanding, the schematics in this book will be already properly grouped. As engineers design circuits, they tend to properly group these subunits. However, a draftsman should try to improve the diagram by rearranging it so the final drawing is easier to read.

Alignment of Symbols, Leads, and Lettering. Symbols should be located according to a plan and placed so there is enough space for lettering the data needed. Two plans may be used. The symbols may be grouped in alignment or staggered. Fig. 20–12 shows symbols that are in *alignment*.

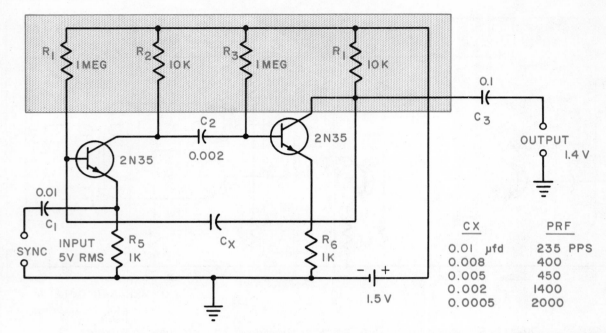

GTE-Sylvania Electric Products, Inc.

20-12. *A multivibrator. Notice the alignment of the resistors.*

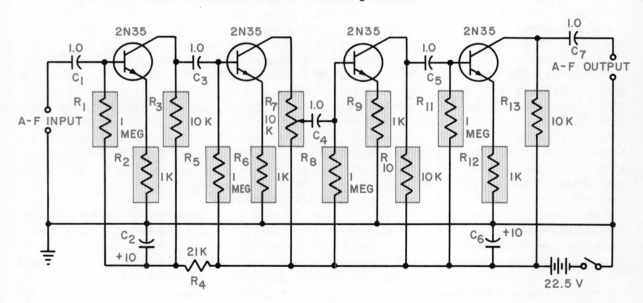

GTE-Sylvania Electric Products, Inc.

20-13. *Cascaded R-C coupled audio amplifier. Notice how the staggered resistors aid in recording values.*

558

Notice the resistors are lined up horizontally. They are *staggered* so data can be lettered without crowding, Fig. 20–13.

Schematic diagrams should be kept as simple as possible. One means of doing this is to record any data other than that specifically related to a symbol on a specification chart. This is on the drawing but not on the diagram itself. If this data is placed on the diagram, it becomes more difficult to read. See Fig. 20–12. Here some data is grouped to the side of the diagram.

The lettering to be on the diagram should be placed in the largest open space that is close to the symbol.

Another good practice is to use a *common lead* instead of drawing a number of parallel leads. Fig. 20–14 shows this simplification. Notice that the four parallel leads have been replaced by one lead.

Sometimes several leads can be drawn to a single point. This point serves as a common terminal for related circuits. This is shown in Fig. 20–15. This is not always possible because of design requirements but should be used whenever possible.

Ground leads are usually drawn short. This simplifies the drawing and saves space.

Notice that leads change direction by making a 90-degree corner. This makes the diagram easier to read.

Balance the Schematic Diagram. The spacing between symbols must be planned so that one side of a drawing is about the same as the other. A typical error is to begin the drawing on one side of a sheet and then crowd the symbols on the other side because enough space was not left. This error is shown in Fig. 20–16.

In order to avoid crowding, examine the sketch of the circuit to be drawn. Look for a pattern or a series of blocks in the design. Make a trial layout on graph paper before starting to make the final drawing. It is best to make several trial plans. The circuit of a simple preamplifier is shown in Fig. 20–16. It can be built small enough to be placed, with a battery, in the handle of a dynamic microphone. Notice that the layout of the schematic is in three major blocks. The space allowed for this drawing should be divided in three approximately equal areas. This allows room for the needed data, and it will be balanced in appearance.

20–14.

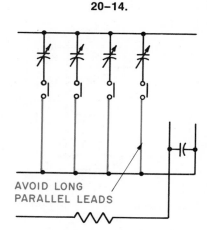

AVOID LONG PARALLEL LEADS

Long parallels add confusion to a schematic.

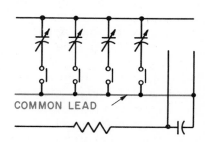

COMMON LEAD

Common lead replaces long parallel leads and simplifies schematic.

20–15. *Use a common ground when design permits.*

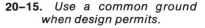
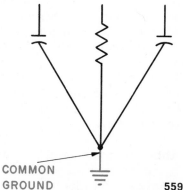

COMMON GROUND

559

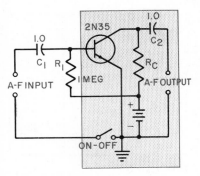

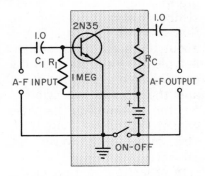

This schematic is crowded on the right side.

The revised schematic. Notice the crowding is reduced.

Another problem occurs when a schematic is stretched over too long an area. While it is not as bad to stretch a drawing as to crowd it, the appearance suffers. The proper spacing to make a pleasing drawing comes from experience and through studying properly drawn diagrams.

The symbols are used to show the relationship between the parts in a circuit. They are drawn the same size every place they occur on the diagram, even though the actual parts vary in physical size. For example, the symbol for a tube is the same size everywhere it appears on a diagram, even though some tubes are actually larger than others. Fig. 20–17 shows a complex schematic in good balance. Notice the tube symbols are drawn the same size.

Abbreviations and Letter Symbols. If a schematic diagram calls for identification of

various parts of the circuit, these must be labeled. These identifying terms are shown by abbreviations and letter symbols. See Fig. 20–18 for the most commonly used terms:

Labeling Electrical Units on Schematic Diagrams. Electrical values are indicated on drawings by using letters to represent that value. The schematics in this chapter show the use of letter symbols. Fig. 20–19 gives the symbols for electrical units.

If it is desired to give the actual electrical values of the units on a drawing, the numerical values can be used instead of the letter symbol. This is of special value to those servicing electrical and electronic units.

Color Codes. The use of a color code is another way of presenting information on schematics.

The wires are color-coded to assist in identifying the circuits in an automobile, Fig. 20–40. This is especially useful information to those doing the actual wiring. Also see Fig. 20–22. One standard color code system is used by the Electronic Industries Association. See Fig. 20–20. This system enables letter abbreviations or numbers to be used to identify wire colors. This simplifies the drawing.

Electronic components also use a color code to indicate their characteristics. For example, the values of capacitors and resistors are shown by the colored bands or dots on the outside. See Fig. 20–21. These systems vary somewhat among manufacturers. Detailed information is available in electronic standards publications. See Fig. 20–24 for actual size.

PICTORIAL PRODUCTION DRAWINGS

Drawings have to be made for use by production workers who build electronic devices. The schematic diagram does not tell how they are assembled. Several different kinds of drawings are used. The most common is point-to-point connection drawings.

A *point-to-point connection drawing* shows each component in its place and the desired wiring connections. In Fig. 20–22 is shown a pictorial production connection drawing. This is used by the worker as he assembles the device.

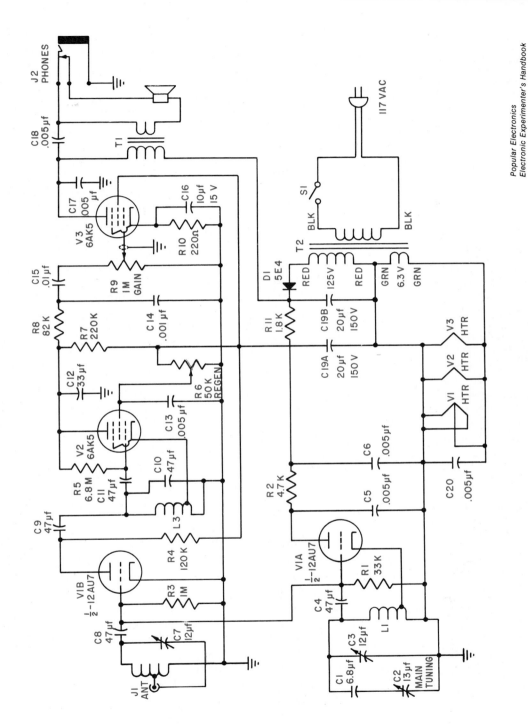

Popular Electronics
Electronic Experimenter's Handbook

20-17.   Two-meter amateur band receiver.

20–18.

Abbreviations of Electricity and Electronics Terms

Alternating Current	A.C.	Low Frequency	L.F.
Audio Frequency	A.F.	Radio Frequency	R.F.
Direct Current	D.C.	Ultra High Frequency	U.H.F.
High Frequency	H.F.	Very High Frequency	V.H.F.
Intermediate Frequency	I.F.		

20–19.

Electrical Values

Unit	Letter Symbol
Ampere (current)	I
Ohm (resistance)	R
Volt (electromotive force)	E
Farad (capacitance)	C
Henry (self-inductance)	L
Henry (mutual inductance)	M
Ohm (inductive reactance)	X_L
Ohm (capacitance reactance)	X_C
Ohm (reactance)	X
Ohm (impedance)	Z
Watt (power)	P
Cycles (frequency)	F
Kilocycles (frequency)	F
Megacycles (frequency)	F

20–20.

Electronic Industries Association Standard Color Code

Color	Abbreviation	Number
Black	Blk	0
Brown	Brn	1
Red	Red	2
Orange	Orn	3
Yellow	Yel	4
Green	Grn	5
Blue	Blu	6
Violet	Vio	7
Gray	Gra	8
White	Wht	9

Sometimes the regular symbols are used to indicate some of the components. Color coding is important on this kind of drawing.

PRINTED CIRCUIT DRAWINGS

The invention of the transistor and the development of the printed circuit have enabled the electronics industry to advance rapidly into miniaturization. Without this development our small radios, hearing aids, and other miniature electronic devices could not be built. Our rockets and space vehicles would be much heavier.

A *printed circuit* is one that is reproduced upon a base made of an insulating material with a foil surface. See Figs. 20–23 and 20–24. It replaces hand wiring. Circuit characteristics can be closely maintained. They can be mass produced. This reduces the cost of making the circuit.

Conductors in Printed Circuits. Many things must be considered when drawing the layout for a printed circuit. The size of the board upon which it will be made must be decided. This is influenced by the purpose of the device and by the size of the components to be fastened to the board. The draftsman must know the basic conductor shapes. These are shown on Fig. 20–25. A *conductor* is a thin flat metal strip. It serves in place of a wire. It carries the electric current.

The actual circuit is difficult to design. The conductors should be kept as *short as possible.* However, they should not cross over each other. Crossovers are used only as a last resort. This is done by installing a short jumper over the etched conductor. Usually these must be insulated.

The *width and thickness* of the conductor must be decided. This allows the conductor to carry the required current. For suggested sizes, see Fig. 20–26.

The *spacing* between conductors is also important. The

greater the voltage potential difference, the farther apart they should be. For example, conductors with a potential difference of 0 to 150V should be at least 0.031 inches apart. This is generally a safe minimum for circuit wiring. For other suggested spacing, see Fig. 20–27.

The longer conductors should be placed near the edge of the circuit board.

The diameter of the pad should be large enough to give at least 0.025 inches of conductor material around the mounting hole. The size of the mounting hole varies with the size of the wire to be put through it. A common size hole is 0.020. The most frequently used pad sizes are 0.100 and 0.125. The pad must be large enough to provide a good soldering base.

If an eyelet is to be placed through the pad, the pad diameter should be 0.025 larger than the flange of the eyelet.

Conductors should meet pads or join another conductor smoothly by using a radial fillet or straight fillet. See Fig. 20–25. Sharp corners should always be avoided. Tees and elbows should have rounded corners.

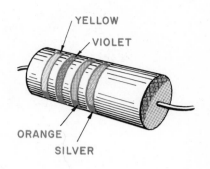

20–21. *Color code on a capacitor.*

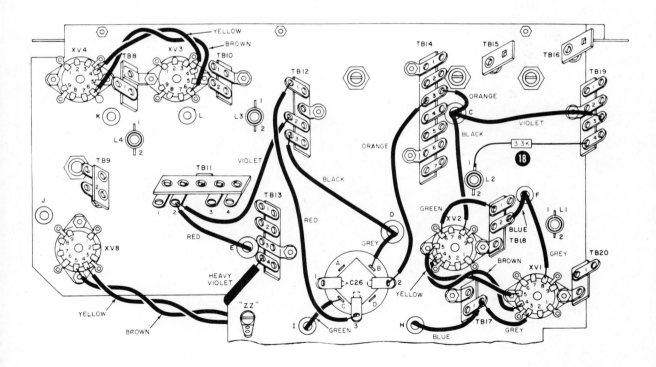

20–22. *A point-to-point pictorial production drawing. Each part is pictured as it appears to the worker.*

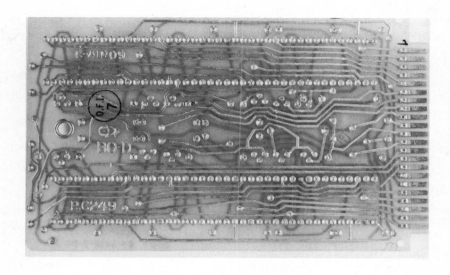

20-23. *The bottom of a printed circuit board showing the etched board, the foil conductors, and the soldered connections. This is the actual size.*

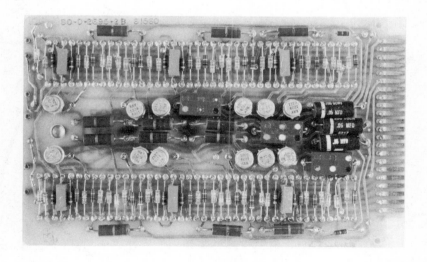

20-24. *This is the top side of the printed circuit board shown in Fig. 20-23. It shows the components in place. This is the actual size.*

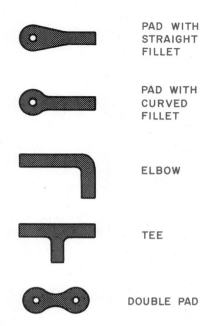

PAD WITH
STRAIGHT
FILLET

PAD WITH
CURVED
FILLET

ELBOW

TEE

DOUBLE PAD

20–25. *Typical conductor
shapes.*

20–26.

Width and Thickness
of Conductors

Current (amperes)	0.00135 Inch Wide Conductor	0.0027 Inch Wide Conductor
1.5	0.015 wide	—
2.5	0.031 wide	0.015 wide
3.5	0.062 wide	0.031 wide

20–27.

Conductor Spacing Standards

Minimum Spacing Between Conductors	Voltage Potential Difference
0.031 inches	0–150V
0.062 inches	151–300V
0.125 inches	301–500V

From "Electrical and Electronics Drawing" by C. J. Baer,
Copyright 1966, McGraw-Hill Book Co., used by permission.

Sometimes brass eyelets are placed in the holes in the pads. The eyelet can also be used to connect conductors on both sides of the circuit board. See Fig. 20–28.

How Are Printed Circuit Drawings Made? A regular schematic drawing of a circuit gives information about the components needed and the proper connections. See Fig. 20–29. From this information, a drawing is made which revises the circuit plan so it can be used for producing a printed circuit assembly drawing. This is first sketched freehand. Then an instrument drawing is made. It serves as an *assembly layout.* See Fig.

20–30. After all design problems are solved on the assembly layout, a *master layout* is drawn. See Fig. 20–31. It is drawn on a dimensionally stable medium such as polyester film. This drawing must be *extremely accurate.* It shows all the pads, elbows, tees, and the patterns of the conductors. It is usually drawn much larger than the final size of the circuit. Usual sizes are 2:1, 4:1, and 8:1, with 4:1 most commonly used. A 4:1 drawing is made 4 times larger than the finished drawing. It is reduced to the proper size by photographic reduction. All the items and conductor paths are drawn solid with ink or a stick-down tape.

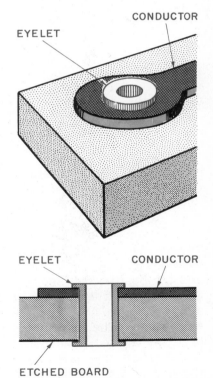

20–28. *Eyelet in printed circuit board.*

565

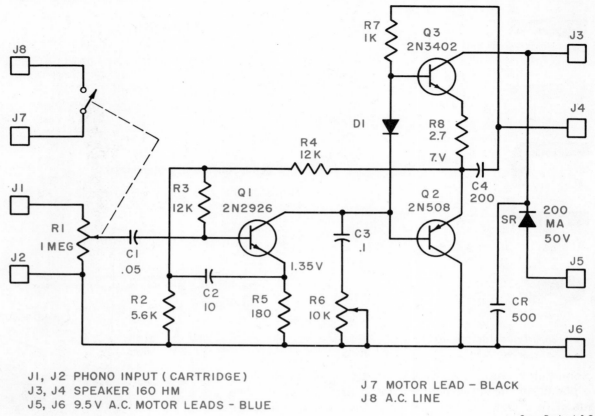

J1, J2 PHONO INPUT (CARTRIDGE)
J3, J4 SPEAKER 16 OHM
J5, J6 9.5 V A.C. MOTOR LEADS - BLUE

J7 MOTOR LEAD – BLACK
J8 A.C. LINE

Sears Roebuck & Co.

20–29. *Schematic of a phonograph.*

From the master layout, other needed drawings can be made. An example is a *drilling drawing.*

Sometimes a drawing of the board is made, giving the exact locations of holes, hole sizes, and the board sizes. See Fig. 20–32. Usually the dimensions are given to two or three decimal places. The drawing is used by those making the board. Since they only make the board, they do not need to know about the circuits that are planned.

Holes are numbered starting in the upper left corner and numbered in vertical columns. The locations of the holes are given using X and Y coordinates. See Fig. 20–32. The X dimensions are measured from the left edge of the board. The Y are given from the bottom edge.

Another drawing made from the master layout is the *component drawing.* It is used to show how to assemble the components on the board. The drawing shows the top side of

the board with the components in place. The conductor paths are also shown, even though they are on the bottom of the board. The conductors are lightly shaded, since they are actually hidden. See Fig. 20–33.

Making a Printed Circuit Layout

1. Study the schematic diagram. Notice the groups of components having common connections. Also notice places having high potential

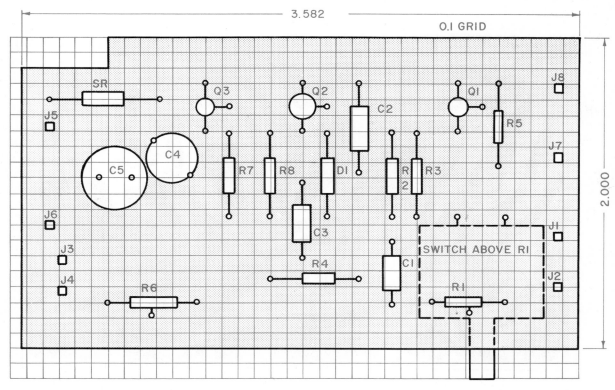

20–30. *An assembly layout of components of the phonograph on the circuit board. This shows the locations of the actual components. This is the top side of the board.*

points and those having heavy current requirements. Mark these on the schematic. Colored pencil is helpful.

2. Draw a *freehand sketch* of the wiring diagram, rearranging the components. The required connections must be made, crossovers must be removed. The components can be represented by rectangular blocks, circles can represent transistors, and small circles can represent tube connections. It is helpful to remember that the components,

when installed, can span conductors. Generally several sketches will have to be made before a good solution is found.

3. Draw an *assembly layout.* The components will be drawn to scale on standard grid paper. The actual sizes of the components must be known. See Fig. 20–34. The common spacing for the grid is 0.10 inches. If the drawing is made to a 4:1 scale, the grid and components must be enlarged this same amount. Compo-

nents are usually placed horizontally or vertically. This drawing locates parts such as fasteners, jumpers, and terminal pads. Now the circuit paths can be drawn. This is done on another sheet laid over the layout of the components. See Fig. 20–30.

4. Check the assembly layout to see if all special clearances have been included. These could be areas of high potential or heavy current requirements. Usually changes are necessary at this time.

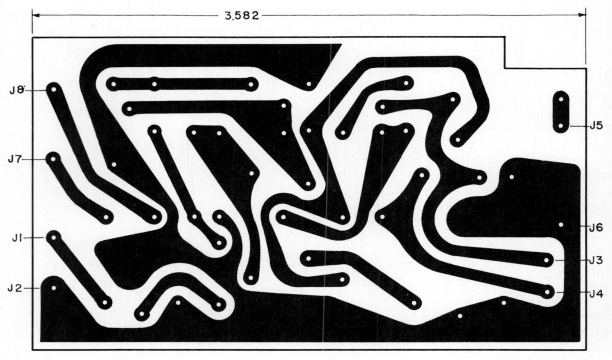

Sears Roebuck & Co.

20–31. *The master layout of the printed circuit for the phonograph. This is the bottom side of the board.*

5. Check the assembly layout with the schematic to see that it is correct.

6. Draw the *printed circuit master.* See Fig. 20–31. The master layout is made by drawing the circuit paths between the components. This has to be in *reverse* of the component drawing since the foil conductors are on the bottom of the board and the components are on the top. An easy way to do this is to draw the circuits on the back side of the assembly drawing. The master layout is drawn using the assembly drawing with the conductors located on the back as a guide.

The master layout is a black drawing on a white background. This is necessary since the pattern is copied by photographic processes. The master layout is drawn larger than the finished size and reduced photographically. The scale should be given on the drawing. One critical dimension is also given. This is usually the longest dimension.

The terminals located on the assembly drawing can be transferred to plastic drafting film by tracing, placing it under the film so it shows through and copying it, or by redrawing directly on the film. Remember to turn the assembly drawing with the face side down before using it to make the master layout. The assembly drawing is the top of the board. The master layout is the bottom. The master layout will be in reverse of the assembly drawing. See the *notch* in the

568

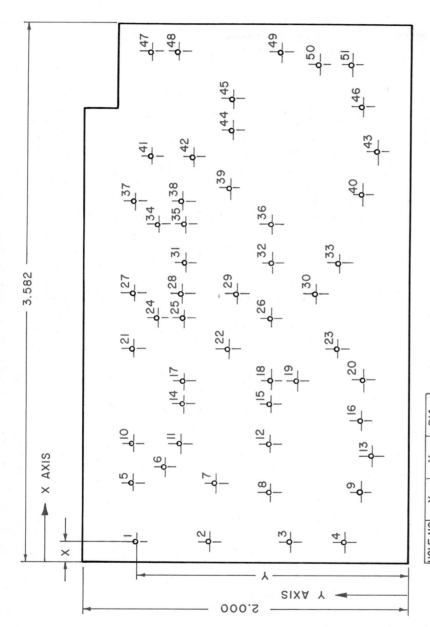

HOLE NO	X	Y	DIA.
1	.250	1.624	.031
2	.250	1.166	.031
3	.250	.708	.031
4	.250	.375	.031
5	.500	1.624	.144
6	.708	1.500	.113

PARTIAL TABLE OF HOLE
COORDINATES & SIZES

20-32. *A drilling drawing that shows board size, hole size, and hole location for the phonograph.*

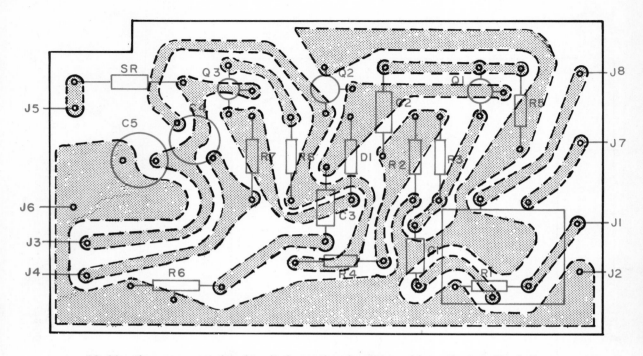

20–33. *A component drawing. It shows the circuit board from the top. The foil conductors are on the bottom. They are hidden.*

upper left corner of the assembly drawing, Fig. 20–30. Then notice the location of the notch shown in Fig. 20–31. The conductor pattern, pads, tees, and elbows are drawn in black India ink or with black adhesive drafting tape.

Terminal pads are drawn first. These have small holes in the center to indicate the location of eyelets. Elbows and tees are drawn next. The conductors are drawn last. The outline of the circuit board is shown by thin lines.

If a circuit board has conductors on both sides, separate drawings of each side are made.

Manufacturing the Printed Circuit Board. The actual printed circuit is made on specially prepared boards. These are commonly laminated paper, phenolic, or lucite. These boards are coated on one side or both sides with a thin layer of conducting material called *foil.* See Fig. 20–35A. The boards are available in standard thicknesses from 1/32 to 1/4 inches. The foil is usually copper, aluminum, copper-clad aluminum, or brass. The standard thicknesses of the foil are one ounce—0.00135 inches; two ounce—0.0027 inches; and three ounces—0.0040 inches.

From the master drawing the pads, elbows, tees, conductors, and other parts are reproduced directly on the foil on the circuit board. This is usually done by offset printing, photoengraving, or silk screen. See Fig. 20–35B. The circuit is reproduced on the board with an acid-resisting coating. The board is placed in a solution which etches away the foil not covered with the acid-resisting coating. See Fig. 20–35C. All areas covered remain, thus forming a conductor. See Fig. 20–35D.

After the board is etched, the necessary holes are drilled to receive the components.

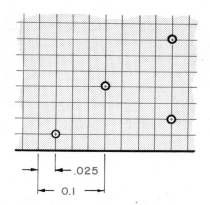

20–34. *Typical grid pattern for developing a printed circuit layout.*

The components are set in place and soldered. The entire board can then be coated with a protective layer to keep it free from dust and moisture.

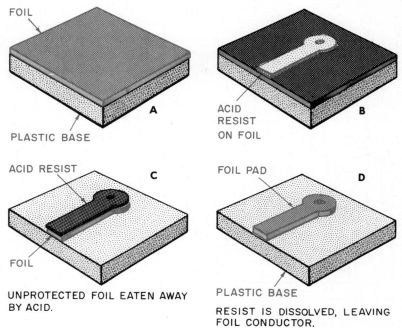

20–35. *Steps in making a printed circuit board.*

INDUSTRIAL CONTROLS

A study of industrial controls is very complex and lengthy. Presented here are a few examples to show some of the types of schematics used. Many types of industrial controls are not electronic. In other words, they do not use tubes or transistors. The important uses of industrial controls are starting, stopping, running, speed control, and overload protection.

Controlling Devices. Controlling devices govern the electric power available to the equipment to which they are connected. Some typical controlling devices are circuit breakers, contactors, motor-circuit switches, and relays.

Circuit breakers interrupt the flow of power in a circuit. Their primary function is to protect the line against a fault. Most frequently this is an overload on the line. The circuit breaker opens, thus breaking the circuit and protecting the motor or other device to which it is connected.

Contactors are used to repeatedly open and close electric power circuits. They are usually operated magnetically. Sometimes they are operated remotely by pilot devices or relays. Fig. 20–36 shows a diagram using a contact maker unit.

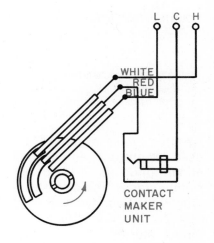

The Foxboro Co.

20–36. *Schematic internal control circuit diagram for intermittent-contact, nonrecording controllers.*

571

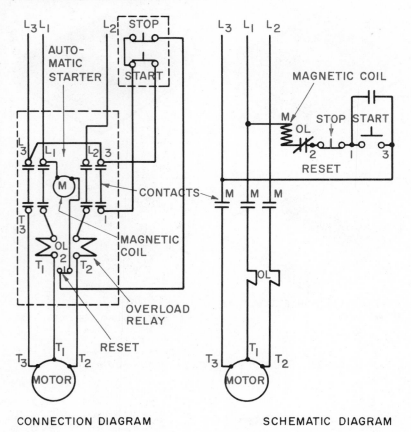

CONNECTION DIAGRAM   SCHEMATIC DIAGRAM

How to Read Electrical Blueprints, Heine, Dunlop, and Jones,
American Technical Society. With permission.

20–37. *An across-the-line automatic starter connection diagram and schematic diagram. The connection diagram shows the circuit connections. The schematic diagram shows the plan and operation of the circuits of the automatic starter.*

Motor-circuit switches are used to interrupt an overload of current or remove the power from a starter and motor running under an overload. Fig. 20–37 shows an automatic motor starter with a reset that opens the circuit, should an overload occur. The reset button remains closed until an overload occurs. It then opens to break the circuit. The

contacts, labeled L₁, L₂, and L₃ on the connection diagram, are drawn open. They are closed by the magnetic coil.

Examine the schematic drawing. It shows how the automatic starter operates. Push the start button. Current flows from L₁ through the magnetic coil. The coil closes the contacts marked M. The motor starts running. The

start button can be released. The motor continues to run because the circuit is completed through contacts 3 to 1.

When the stop button is pressed, the circuit is broken. The magnetic coil looses its magnetism. The contacts marked M open.

Relays are electrically operated switches used to control electric current from a distance. The relay operates on a small amount of current. It can use smaller wires than the main circuit it controls.

Speed control serves to give a machine operator a rapid and accurate means of regulating the speed at which his machine runs. This could be the revolutions per minute a lathe turns. It can be done in several ways. A drawing with one solution is shown in Fig. 20–38.

Fig. 20–39 shows a schematic for an electric heating control panel with an automatic temperature controller. This diagram uses both symbols and pictorial means for representing components in a control schematic.

Symbols for Industrial Control Diagrams. Industry has not accepted one final set of symbols for industrial controls. While many of the standard symbols listed in Fig. 20–10 are used, some differences between companies do exist. The symbols that are not standard are somewhat similar to the standard symbols.

20-38. *A speed control system for large metal lathes. It was designed to give a wide range of speed ratios with high accuracy regulation. A section of the manufacturer's literature is reproduced with the drawing. This will help you to understand the complexity of the problem which was solved with the help of this drawing.*

Honeywell Inc.

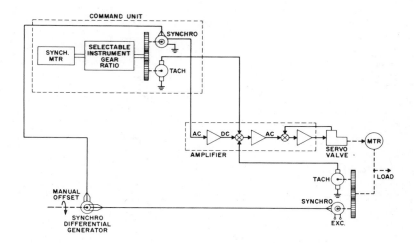

THE PROBLEM:

A large lathe, used for machining rollers in a steel mill, needed a wide range of speed ratios, with high-accuracy regulation. A conventional system, using motor control in a velocity loop, would give a ratio range of from 20:1 to 100:1, with regulation of from 1% to ½% of maximum speed. To increase this ratio range, heavy power-level shifting and velocity-transmission equipment would be needed, resulting in a cumbersome and expensive arrangement, with poor regulation. The problem was to widen the ratio range without sacrificing regulation.

THE SOLUTION:

The solution was an electronic gear train concept, with all speed ratio shifting done at instrument level. The shifting device could be either a variable ratio transmission, a stepped mechanism with selectable fixed speed ratios, or, for very wide ratio ranges, a combination of the two.

In this particular system, a variable ratio mechanism was used, giving a speed ratio range from 20:1 to 1000:1. A wider range was not needed here; however, experimental systems have been operated at from 10,000:1 to 100,000:1 ratios, using electronic gear train principles.

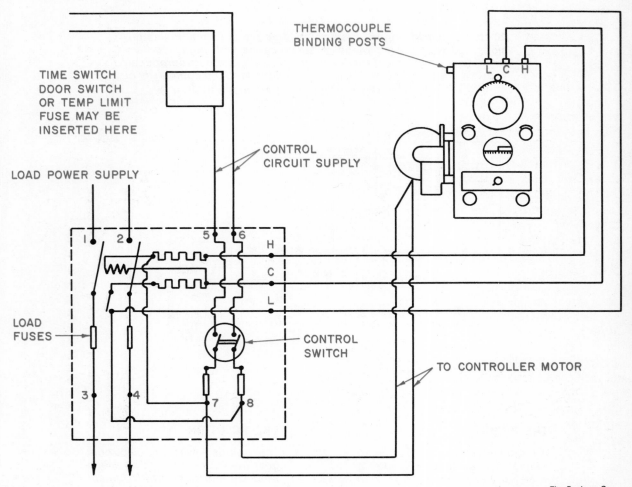

TIME SWITCH
DOOR SWITCH
OR TEMP LIMIT
FUSE MAY BE
INSERTED HERE

THERMOCOUPLE
BINDING POSTS

CONTROL
CIRCUIT SUPPLY

LOAD POWER SUPPLY

LOAD
FUSES

CONTROL
SWITCH

TO CONTROLLER MOTOR

The Foxboro Co.

20–39. *Electric heating control panel and automatic temperature controller.*

ELECTRICAL DRAWINGS

Some of the common types of electrical drawings are: (1) point-to-point, (2) highway, (3) baseline, and (4) single-line drawings.

A *point-to-point* diagram shows the components pictorially as they appear to someone looking at them. They are connected with lines to represent the conductors. See Figs. 20–22 and 20–40. This drawing is used to give assembly information to workers. Each conductor runs from a terminal on one component to the connecting terminal on another component. Each conductor is drawn entirely separate from the others.

The *highway diagram* is much like the point-to-point diagram. The components are drawn pictorially rather than by symbol. The main difference is that on a highway diagram all the conductors leading from one component to another are joined into a single line, called a highway. They are separated when they reach the component to which each conductor is connected. See Fig. 20–41. In this figure, find the conductors 12 Wht–D1, 12 Blk–D2, and 12 Red–D3. Notice how they join to form a highway and are separated when they reach the motor. This lets each conductor be connected to the motor at the proper terminal. It is not necessary to draw three long conductors running parallel to the same component. Highways simplify a drawing. Notice that the conductors are drawn curved whenever they connect to a highway.

It is necessary to carefully label each conductor on highway drawings so they can be easily followed. Read the next section on baseline drawings for directions on how to read the lead codes.

Baseline drawings are much like highway drawings. They are built around a horizontal or vertical line, called the baseline. This is an imaginary line not an actual cable. It is used to simplify the drawing. Lead lines from the components run to the baseline. They meet and leave it at right angles. Examine the baseline drawing, Fig. 20–42. Find component B. Notice one lead is 12 Wht–D1. This lead leaves component B and connects with the baseline. It leaves the baseline and connects to terminal 1 on component D.

Identification of each component, terminal, and lead is vital to all types of electrical

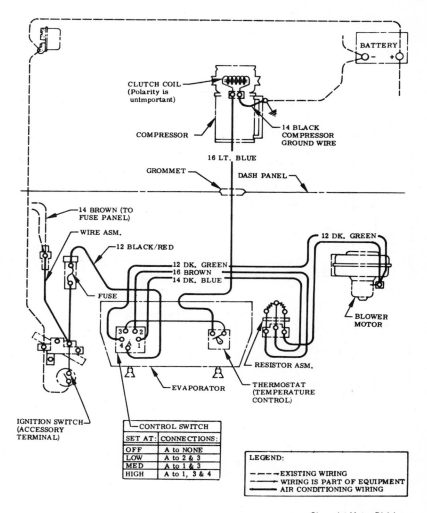

Chevrolet Motor Division
General Motors Corp.

20–40. *Automotive schematic with leads color coded. This is a form of a point-to-point pictorial production drawing.*

drawings. In Fig. 20–40 the wire size and color are given for each lead. Since the leads all run directly to the terminal on each component, these are not numbered.

Highway and baseline drawings must have complete identification since each lead does not go directly from one component to another.

One system in use labels each component with a large letter. Each terminal is numbered. The leads are identified by color and size. For example, in Fig. 20–41 the motor shown has three leads. One of these is 12 Red–B1. This means the lead is wire size 12, red in color, goes to component B, and connects with

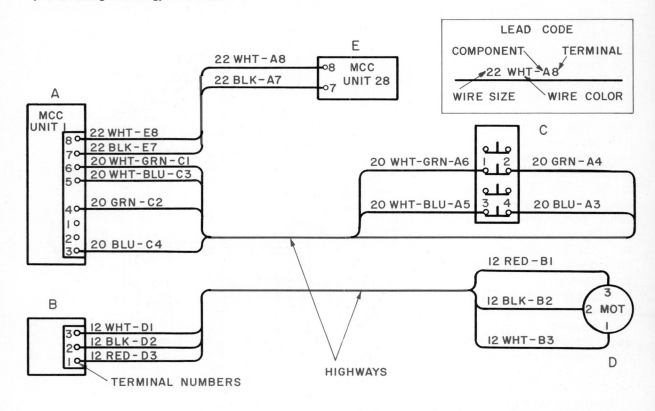

20–41. *A highway line-type connection diagram. Notice how several conductors are merged into one highway and are divided when they reach the component.*

National Aeronautics and Space Administration

terminal 1. Examine the lead identification at component B. It reads 12 Wht–D3. This means it is number 12 wire, white in color, and connects with component D at terminal 3.

A *single-line diagram* shows in simple form the major pieces of equipment for a complete project, such as an electrical power substation. It includes information about each major component, such as voltages of potential transformers, ampere rating of

current transformers, interrupting capacity and trip ratings of circuit breakers, and motor horsepower ratings. They can include wire and cable information and descriptions of the major components.

Symbols are used to represent the components. The single line shows the order of connection. It can represent several conductors in a cable.

A typical single-line diagram is shown in Fig. 20–43. When drawing line diagrams, the

main circuits are usually shown in the most direct path. They are placed in the order in which they occur in the circuit. The single line connecting the symbols runs either in a vertical or horizontal direction. Spacing is the same as that on schematics. Enough background is needed to avoid crowding the symbols. Special attention must be given to allow room for notes. Notes are an important part of a single-line diagram. They contain information about types, ratings, and other data related to the parts of the circuit.

DRAWINGS FOR ELECTRICAL POWER FIELD

Many different types of drawings are used in the electrical power field. Fig. 20-1 shows an electrical power generating station. It took thousands of drawings to design this station. Some of the types of drawings needed were architectural, structural, electrical, piping, and mapping.

Typical small substations are shown in Figs. 20-44 and 20-45. When a small substation is in the planning stage, a one-line diagram is one of the first drawings made. Fig. 20-43 shows such a drawing. It gives the overall plan for the substation. Notice it is located on the building site by a reference to the north.

The details of the substation are made up in the form of very detailed, schematic diagrams. Fig. 20-46 shows a portion of such a drawing. This is a control circuit for opening a high voltage circuit breaker in the substation. The symbols used are in most cases like those in the standards.

Many other detail drawings are necessary. Fig. 20-47 shows a plan view (top view) of the towers and connections. Fig. 20-48 shows an elevation of the towers and the power unit in place.

Many other detail drawings are necessary. One example

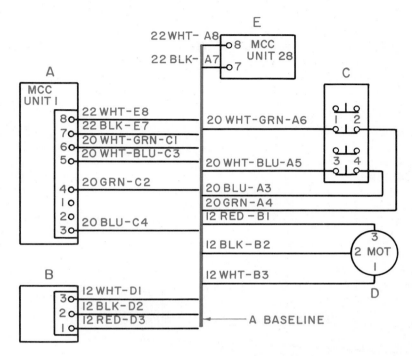

National Aeronautics and Space Administration

20-42.   *This is a baseline connection diagram. It is the same circuit illustrated in Fig. 20-35.*

is shown in Fig. 20-49. This gives the details of part of the equipment in a substation.

Additional Reading

Automotive Electrical Circuits —Principles and Problem Diagnosis. Dearborn, Mich.: Ford Motor Co.

Baer, C. J. *Electrical and Electronics Drawing.* New York: McGraw-Hill Book Co.

Kyle, James. *Transistor Etched-Circuit Projects.* New York: Howard W. Sams and Co.

Mark, D. *How to Read Schematic Diagrams.* New York: John F. Rider Inc., Publisher.

Raskhodoff, H. M. *Electronic Drafting Handbook.* New York: The Macmillan Co.

Shiers, George. *Electronic Drafting.* Englewood Cliffs, N.J.: Prentice-Hall.

Transistor Circuit Handbook. New York: GTE-Sylvania Electric Products.

Twenty-Eight Uses for Junction Transistors. New York: GTE-Sylvania Electric Products.

Build Your Vocabulary

Following are terms that you should understand and use as a part of your working vocabulary. Write a brief explanation of what each means.

Schematic diagram
Component
Electric current
Volt
Ohm
Ampere
Watt
Conductor
Block diagram
Lead
Circuit
Color code
Point-to-point diagram
Printed circuit
Circuit breaker
Relay
Highway diagram
Baseline drawing

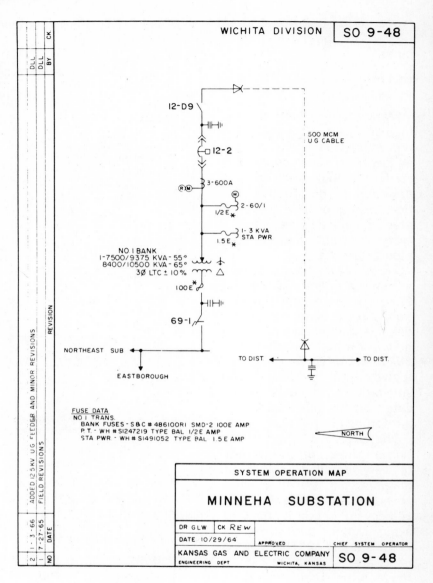

Kansas Gas and Electric Co.

20–43. *A single-line drawing of a substation. Reading from the bottom, the units are a switch, arc resisting (69–1); lightning arrester and ground; switch, fused disconnect; power transformers; non-disconnecting fuse and current transformer; another non-disconnecting fuse and current transformer; three current transformers at 600 amperes each; air circuit breaker electrically operated (12–2); lightning arrester and ground; disconnect switches (12–D9); and a cable termination at which an underground cable is connected.*

Kansas Gas and Electric Co.

20-44. *A low profile substation. This design is used in residential areas where appearance is important. The photo was made with a gate open to show the inside. This type of installation is more expensive than the one shown in Fig. 20-45.*

Kansas Gas and Electric Co.

20-45. *A unit substation in which the transformer and control circuits are contained in a single cabinet. This type of installation is used in rural and industrial areas.*

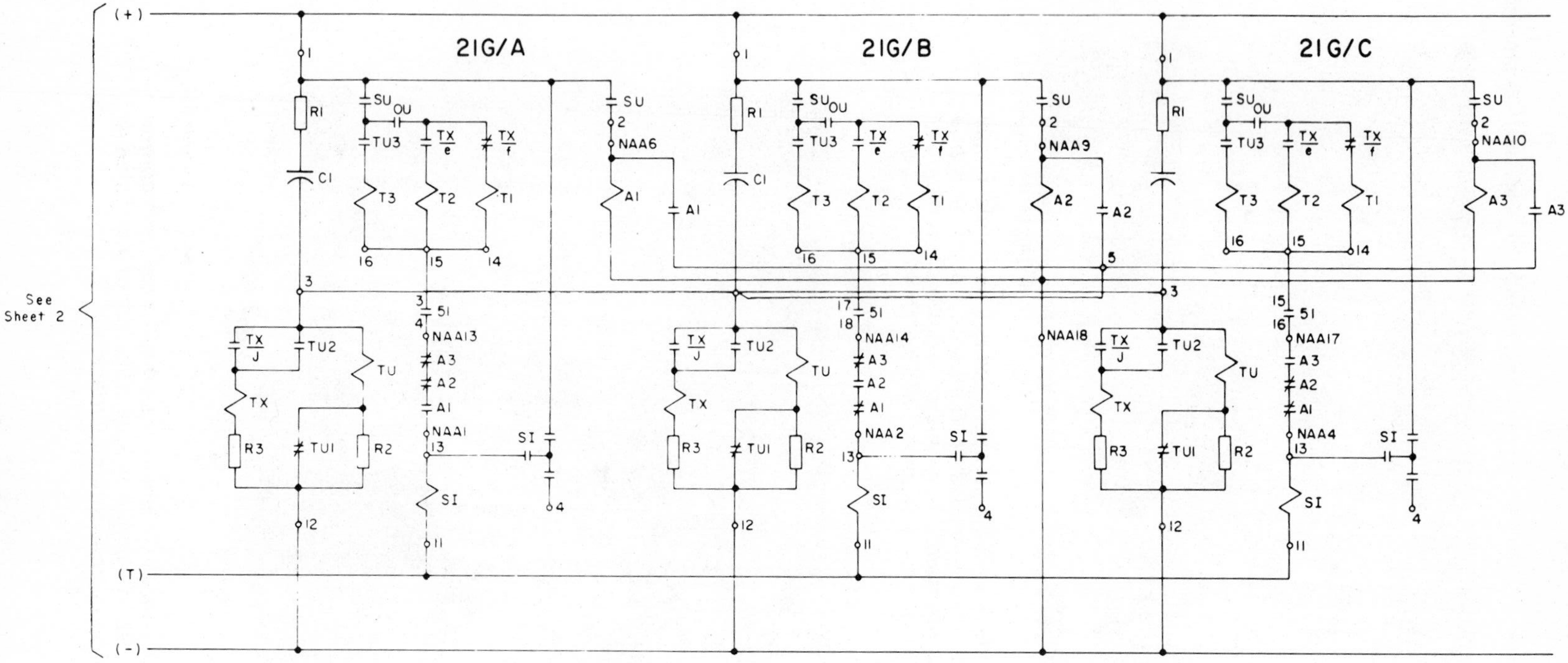

69-6 TRIP CIRCUIT
GCXI5 GROUND RELAYS

20-46. *A schematic of part of a substation. This is a control circuit for a high voltage circuit breaker.*

Kansas Gas and Electric Co.

580

See Sheet 2

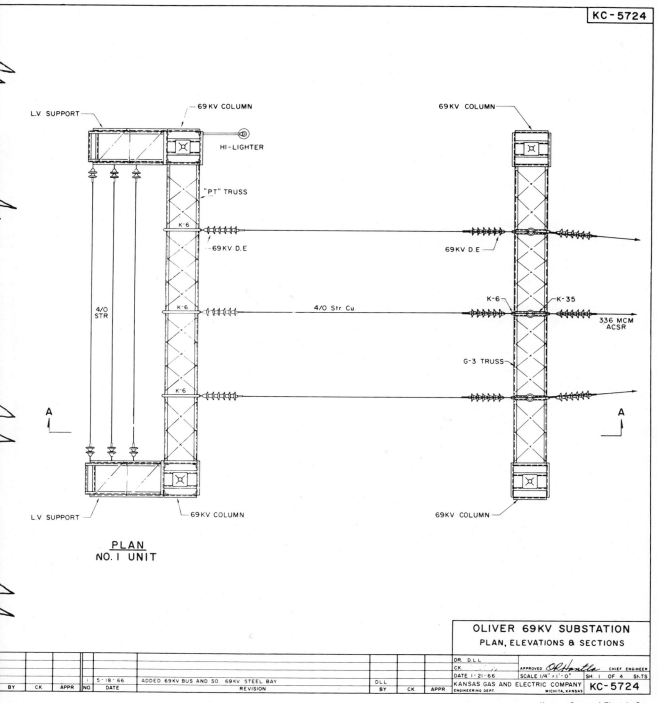

20–47.　*A plan view of a substation.*

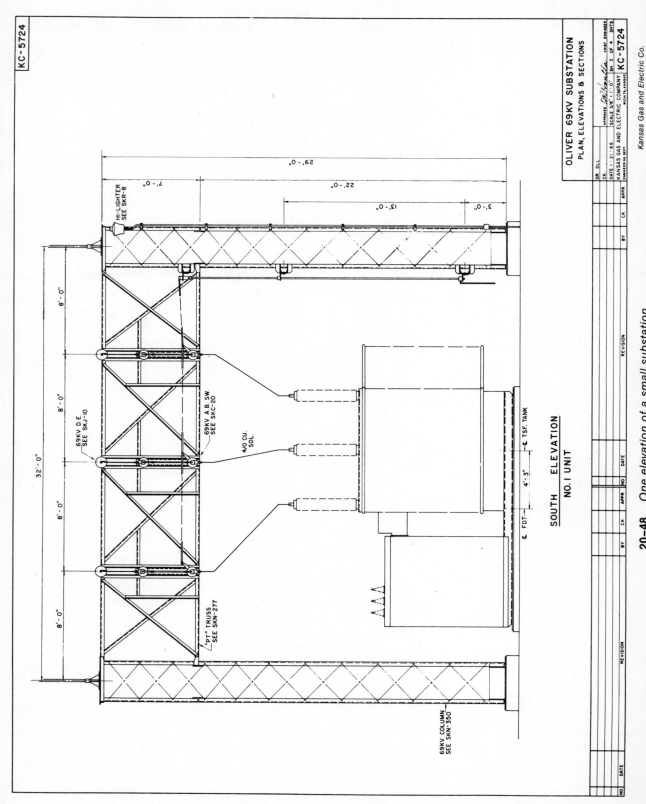

20-48. One elevation of a small substation.

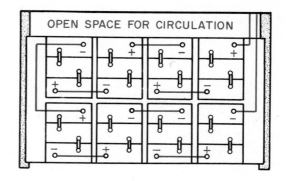

SECTION A-A

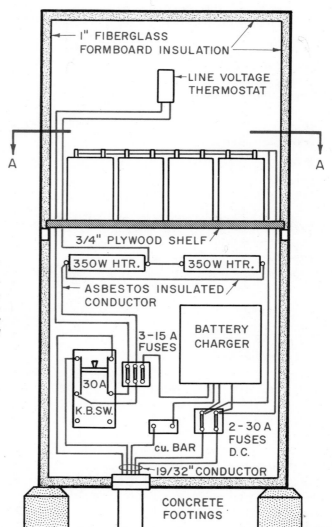

1" FIBERGLASS FORMBOARD INSULATION

LINE VOLTAGE THERMOSTAT

A

A

3/4" PLYWOOD SHELF

350W HTR.   350W HTR.

ASBESTOS INSULATED CONDUCTOR

3-15 A FUSES

BATTERY CHARGER

30A

K.B.SW.

cu. BAR

2-30 A FUSES D.C.

19/32" CONDUCTOR

CONCRETE FOOTINGS

Kansas Gas and Electric Co.

20–49. *One type of detail drawing needed when drawing a substation plan. This shows the battery installation with the battery charger, heating elements, and electrical connections.*

EQUIPMENT ARRANGED ON A HALF PANEL

Study Problems

Block Diagrams

1. Study the connections shown on the pictorial drawing of a part of an auto electrical system, P20–1. Draw a block diagram showing the system.
2. Make a block diagram of the pictorial drawing of the turn table, preamplifier, amplifier, and speaker, P20–2.

Reading Electronic Schematics.
Following are some schematics used to illustrate solutions to typical design problems. Study these and answer the questions about each drawing.

3. The receiver, P20–3, using a single audio transistor and a diode detector, enables the receiver to be built in a unit smaller than a cigarette package. Study the diagram and answer the following questions:

1. What is the power source?
2. What is used to enable a person to hear the signal?
3. What type of switch is specified?
4. How many resistors are used?
5. Does it use tubes or transistors?
6. What is the tube or transistor specification number?
7. How many capacitors are used?
8. What is the diode specification number?

9. What type of transformer is used?
10. What is the rating of the variable capacitor?

4. The transmitter shown in P20–4 is an 80-meter crystal oscillator-type C-W transmitter. Study the diagram and answer the following questions:

1. How many transistors are needed to build this transmitter?
2. What type of transistors are used?
3. How many resistors are in the circuit?
4. Where is it grounded?
5. What type of tuner is used?
6. What voltage is required for operation?
7. What device is used to send the signal?
8. How many capacitors are used?
9. How many chokes are used?
10. What type of crystal is used?

Symbol Drawing.
Draw the following schematics. Enlarge them so normal size symbols can be used.

5. Make a freehand sketch of the automatic safety flasher, P20–5. Complete it by inserting the proper symbols.
6. Make an instrument drawing of the schematic of the battery charger and complete it by inserting the proper symbols. See P20–6. By each symbol give the value.
7. Make an instrument drawing of the schematic of the S meter, P20–7, and complete it by inserting the proper symbols.

Planning Schematic Diagrams.
The schematics in the following problems are correctly designed. They have been purposely drawn in a poor arrangement. There is no need to redesign the plan of circuitry. The assignments do require that the indicated connections and components be drawn to show more clearly the schematic and present a pleasing, balanced drawing.

8. Revise the schematic of the constant current generator, P20–8. The placing of components on the diagram is poor and should be improved.
9. Revise the schematic of the TV control box, P20–9. The placing of the components on the diagram is poor and should be improved.
10. Revise the schematic for the light amplifier, P20–10. This device increases the light intensity. Use is made of a sun battery to convert light energy to electrical energy. The electrical energy is amplified by the device and used to furnish power for a light bulb.

Draw the proper component symbols and rearrange the diagram so it is pleasing in appearance. Place the components so they are in a better relationship.

11. Revise the schematic of the class A audio power amplifier, P20–11. Rearrange the diagram so it is pleasing in appearance and the components are in a better relationship.

Printed Circuit Drawings

12. Study the schematic for the volter, P20–12. It consists of two resistors, a penlight battery cell, and a battery holder. An assembly drawing gives the size of the circuit board and the location of the components on the board. Design a master layout for a printed circuit for this unit. Draw it at a four to one proportion.

13. Make a component drawing of the volter, P20–12. Draw it at a four to one proportion.

14. Make a drilling drawing of the volter, P20–12. Draw it at a four to one proportion. Include all data to locate each hole on the X and Y coordinates.

15. Study the schematic for the code practice oscillator, P20–13. Notice the required connections. Given is a component assembly drawing. It is drawn on a grid of 0.10. Make a master layout printed circuit, drawing the actual size indicated on the drawing.

Be certain each pad is located exactly as shown on the drawing. The proper connections are shown with solid lines. Draw it at a four to one proportion.

16. Make an assembly drawing of the code practice oscillator, P20–13. Properly identify all components. Draw it at a four to one proportion.

17. Make a master layout printed circuit board drawing from the schematic diagram of the microphone preamplifier, P20–14. The outside dimensions of the finished board are 1¼ by 2 inches.

Industrial Electricity Drawing

18. A small portion of a schematic for a business office machine is shown in P20–15. Simplify by redrawing it as a baseline-type diagram. Identify each unit and lead, using a standard system.

19. Simplify the partial schematic of an industrial control, P20–16. Make it a highway-type diagram. Identify each component, terminal, and lead, using a standard system.

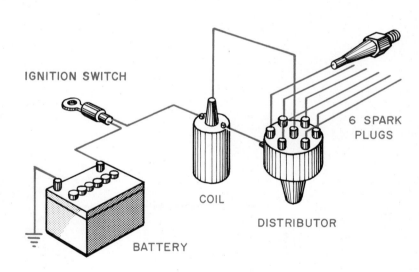

P20–1.   *Part of an auto electrical system.*

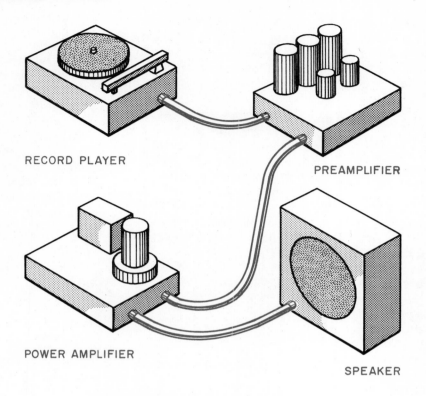

P20–2. *Record amplification system.*

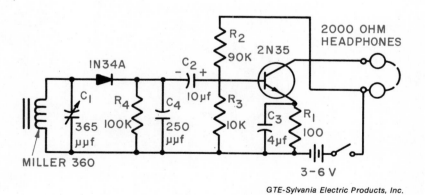

GTE-Sylvania Electric Products, Inc.

P20–3. *One-transistor pocket receiver.*

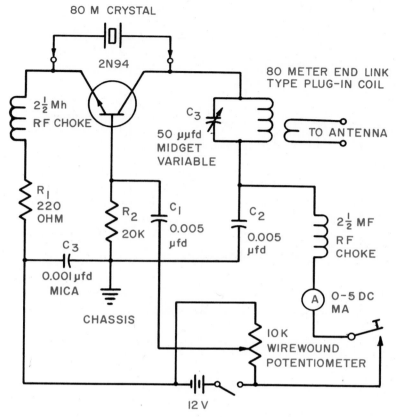

80 M CRYSTAL

2N94

80 METER END LINK
TYPE PLUG-IN COIL

$2\frac{1}{2}$ Mh
RF CHOKE

C_3
50 μμfd
MIDGET
VARIABLE

TO ANTENNA

R_1
220
OHM

R_2
20K

C_1
0.005
μfd

C_2
0.005
μfd

$2\frac{1}{2}$ MF
RF
CHOKE

C_3

0.001 μfd
MICA

CHASSIS

A   0-5 DC
MA

10 K
WIREWOUND
POTENTIOMETER

12 V

GTE-Sylvania Electric Products, Inc.

P20-4. *Novice C-W transmitter.*

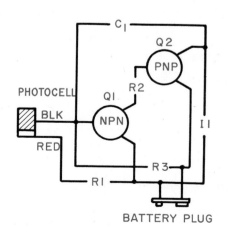

C_1

Q 2

PNP

PHOTOCELL

Q1   R2

BLK

NPN

RED

R 1

R 3

I 1

BATTERY PLUG

Parts List

C1 — 30 μf, 10 volt capacitor
I1 — Pilot light
Q1 — 2N229 NPN transistor
Q2 — 2N187 PNP transistor
R1 — 1,200 ohm, ½ watt resistor
R2 — 470 ohm, ½ watt resistor
R3 — 47,000 ohm, ½ watt resistor

P20-5. *Automatic safety flasher.*

*Popular Electronics
Electronic Experimenter's Handbook*

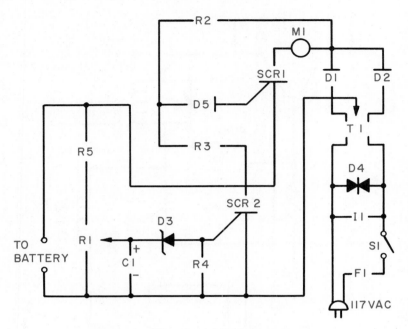

Popular Electronics
Electronic Experimenter's Handbook

Parts List

C1 — 100 μf, 25 volt capacitor
D1, D2 — 15 amp, 50 volt silicon rectifier
D3 — 8.2 volt, 1 watt zener diode
D4 — Transient voltage suppressor
D5 — 100 volt, 600 MA silicon rectifier
F1 — 2 amp fuse
I1 — 120 volt neon indicator light
M1 — 0–10 amp meter
R1 — 500 ohm, 2 watt linear scale potentiometer
R2, R3 — 27 ohm, 3 watt resistor
R4 — 1,000 ohm, ½ watt resistor
R5 — 47 ohm, 2 watt resistor
S1 — S.P.S.T. toggle switch
SCR1 — Silicon controlled rectifier
SCR2 — Silicon controlled rectifier
T1 — Power transformer, primary 117V, secondary 24V

P20–6. *Auto battery charger.*

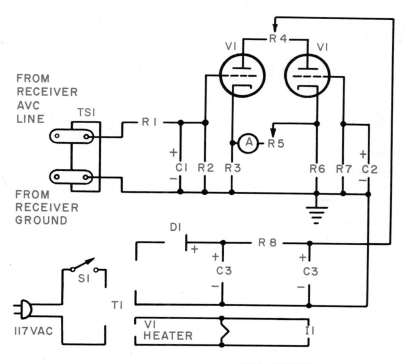

Popular Electronics
Electronic Experimenter's Handbook

Parts List

C1, C2 — 0.0.1 μf., 150 volt capacitor
C3 — Dual 20 μf, 150 w.v.d.c. capacitor
D1 — 1N 2069 diode
I1 — 6.3 volt pilot lamp
M1 — 2⅜" square illuminated S meter, 1 MA movement
R1 — 2.2 megohms
R2, R7 — 10 megohms, ½ watt resistor
R3, R6 — 3,300 ohms, ½ watt resistor
R4, R5 — 20,000 ohm potentiometer
R8 — 1,000 ohm, 5 watt resistor
S1 — S.P.S.T. switch
T1 — Power transformer: primary 117 volts, A.C., secondaries, 125 volts
TS1 — 2 lug terminal strip
V1 — 12 AU 7 tube

P20–7. *S meter to measure signal strength of communications'*
receivers.

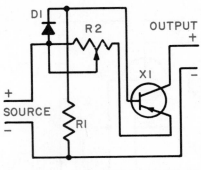

Howard W. Sams and Co., Inc.

Parts List

D1 — 1N456 silicon diode
R1 — 1,000 ohm, ½ watt resistor
R2 — 1,000 ohm potentiometer
X1 — PNP transistor 2N404

P20–8. *Constant current generator circuit purposely arranged in a poor layout.*

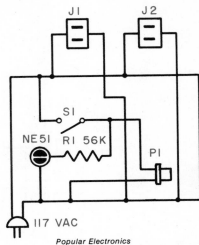

Popular Electronics
Electronic Experimenter's Handbook

Parts List

J1, J2 — Accessory sockets, A.C.
S1 — Toggle switch
NE51 — Indicator lamp
R1 — 56,000 ohm resistor
P1 — Plug to fit connection on rear of
television set

P20–9. *Schematic for a television control box used when servicing the set with the back removed. It provides power to operate the set and two outlets for accessories, such as a soldering iron. The circuit is purposely arranged in a poor layout.*

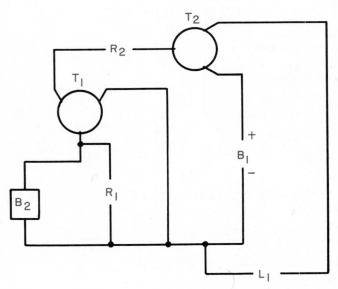

GTE-Sylvania Electric Products, Inc.

Parts List

B1 — 6 volt battery
B2 — Sun battery
L1 — 6 volt pilot lamp
R1 — Resistor, 4.7 K, ½ watt
R2 — Resistor, 100 Ω, ½ watt
T1 — Transistor, 2N229, NPA type
T2 — Transistor, 2N255, PNP type

P20–10. *A light amplifier circuit purposely arranged in a poor layout.*

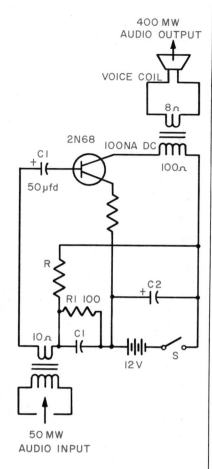

400 MW
AUDIO OUTPUT

VOICE COIL

8Ω

2N68   IOONA DC

100Ω

C1
+
50µfd

R

R1 100

10Ω   C1

12V   S

C2
+

50 MW
AUDIO INPUT

GTE-Sylvania Electric Products, Inc.

P20–11. *Class audio power amplifier purposely arranged in a poor layout.*

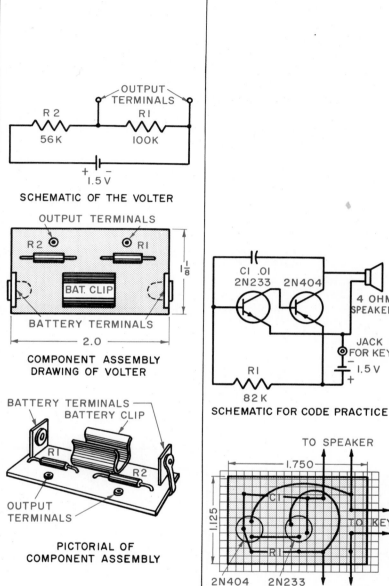

R2
56K

OUTPUT
TERMINALS

R1
100K

+   −
1.5 V

SCHEMATIC OF THE VOLTER

OUTPUT TERMINALS

R2   R1

BAT. CLIP

BATTERY TERMINALS

2.0

1 1/8

**COMPONENT ASSEMBLY
DRAWING OF VOLTER**

BATTERY TERMINALS
BATTERY CLIP

R1

R2

OUTPUT
TERMINALS

**PICTORIAL OF
COMPONENT ASSEMBLY**

Howard W. Sams and Co., Inc.

P20–12.   *Printed circuit problem.*

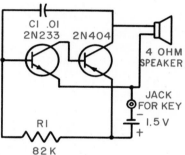

C1 .01
2N233   2N404

4 OHM
SPEAKER

JACK
FOR KEY

1.5 V
+

R1

82 K

SCHEMATIC FOR CODE PRACTICE

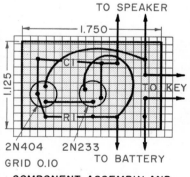

TO SPEAKER

1.750

C1

TO KEY

1.125

R1

2N404   2N233

GRID 0.10   TO BATTERY

**COMPONENT ASSEMBLY AND
CONDUCTOR CONNECTIONS**

Howard W. Sams and Co., Inc.

P20–13.   *A code practice oscillator.*

591

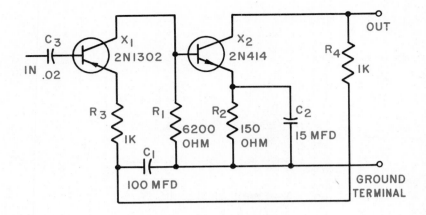

Howard W. Sams and Co., Inc.

P20–14. *Schematic of microphone preamplifier.*

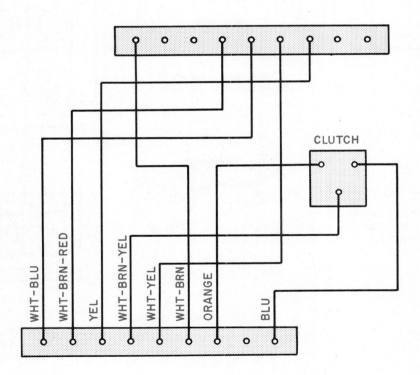

P20–15. *Partial schematic for office machine.*

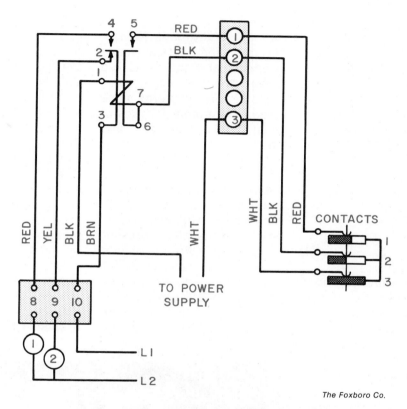

The Foxboro Co.

P20–16. *Partial schematic of an industrial control (highway).*

Chapter 21

Mapping

A map is a drawing showing a portion of the surface of the earth. Maps are also made of other bodies, such as the moon, and the vastness of space is mapped. There are many different kinds of maps, each serving a special purpose. The features on maps vary a great deal, depending upon the purpose.

A person who designs and draws maps is called a cartographer. Throughout this chapter the cartographer's role in mapping will be discussed. (See Chapter Two for additional description of the cartographer's work).

Geographers and astronomers make extensive use of maps. Surveyors and builders use land survey maps. Navigation and aeronautical maps have special features to aid navigators and pilots.

This chapter presents some of the more common types of maps, their features, and uses.

SCALE

Maps are drawn to *scale*. Scale refers to the distance one inch on the map represents. It is usually expressed in fractional form in inches. Consider the scale 1:63,360. This means that one inch on the map equals one mile on the ground. The figure, 63,360, is the number of inches in a mile. A mile is 5,280 feet. Multiply this by 12 inches to the foot to complete the fraction. This fraction is called the natural scale.

Most maps also show graphic scale. See Fig. 21–1. This is a line drawn to the same scale as the map. It is divided into lengths that equal distances on the map. Straight line distances can be measured this way. Curves, such as presented by a highway, can be measured by setting dividers to a small distance, as one-half mile. It can then be stepped along the curve. Each step is equal to one-half mile. A special tool, called a carto-

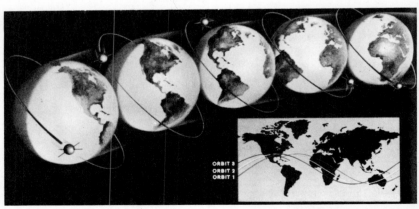

ORBIT 3
ORBIT 2
ORBIT 1

Great World Atlas
Reader's Digest Association

Flight path of an orbiting satellite.

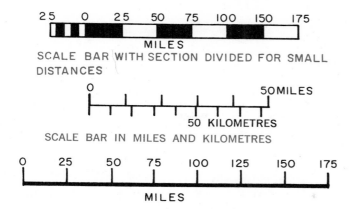

SCALE BAR WITH SECTION DIVIDED FOR SMALL DISTANCES

SCALE BAR IN MILES AND KILOMETRES

MILES

21-1. *Typical examples of scale bars. Always mark the scale on a strip of paper and use it to measure distances.*

meter, is used for this purpose also. See Fig. 21-2.

The scale is sometimes shown by indicating the number of inches that represent a mile or several miles. This is stated *one inch to the mile*. For example, one inch to 39 miles means one inch on the map *represents* 39 miles. It does not mean $1'' = 39$ miles.

Large scale maps are those having scales larger than four miles to the inch (1:250,000). Three miles to the inch is a larger scale than four miles to the inch.

Medium scale maps are those with scales from 4 to 16 miles to the inch. These are 1:250,000 to 1:1,000,000. The larger the scale the more detail that can be shown.

Small scale maps are those with more than 16 miles to the inch. Such maps can only show major features, such as rivers and mountains.

FEATURES OF MAPS

The features found on a map vary with the purpose of the map and its scale. Features that are more important than others are emphasized to call attention to them. For example, see Fig. 21-31. Emphasis is given to airports and air-

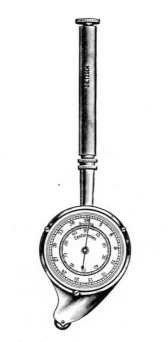

Eugene Dietzgen Co.

21-2. *A cartometer is used to measure mileage on maps.*

markers in this aeronautical map. These features are the ones most important to the pilot using the map.

Features that do not contribute to the usefulness of the map should be omitted. For example, a contour map will not show highways.

The cartographer must decide which features to show and emphasize. He must locate every feature on the map accurately in relation to the other features. Features that are included should not be crowded to the extent that the map is difficult to read.

Features are given emphasis by using heavy lines, drawing large symbols, or using bright colors. Sometimes special lettering is used. Arrows can be drawn pointing to these features.

The details on a map depend upon the scale. Small scale maps show general features. Large scale maps can show more detail. For example, houses that show on a large scale map are not shown on small scale maps.

USING SYMBOLS

Maps use symbols to represent various features. The smaller the scale, the more symbols are needed. For example, on a large scale map a city could be shown by drawing its actual outline. See Fig. 21-11. Some major features may be shown. On a small scale map, a city may be represented by a circle.

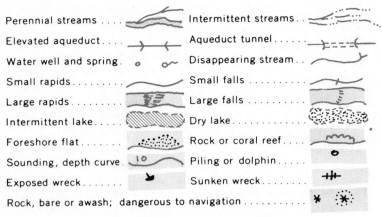

21–3. *Water feature symbols.*

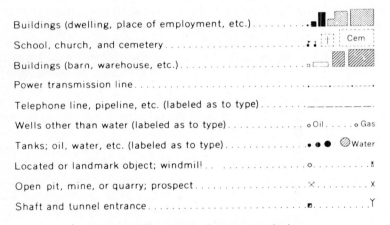

21–4. *Man-made feature symbols.*

4. Those representing vegetation. These are usually green.

5. Special symbols for unusual features, such as a shrine or ruins. These often have a legend to call attention to them. These are usually a bright color such as orange.

Water Feature Symbols. Water feature symbols are shown in Fig. 21–3. Usually a cartographer will draw rivers, lakes, or shore line first. From these, the other features are located. Water feature symbols are usually blue. Black can be used.

The thickness of the lines must vary. For example, the thickness of the lines representing rivers varies. Branches of the river are thinner than the main stream. The main stream gets wider as more branches connect to it.

Some rivers are dry part of the year. The intermittent river symbol shows this. Some rivers have a series of dams that control the flow of river traffic. These are called canalized rivers. The location of the dam should be drawn on the river symbol.

It is not usually possible to show every turn in a river. The general features and major changes in direction should be shown.

Man-Made Feature Symbols. There are a large number of standard symbols in use to show man-made features. See Fig. 21–4. These are very sim-

Small scale maps use simpler symbols than those on large scale maps. For example, on a large scale map, a railroad is shown by parallel lines with short crosslines. On the small scale map, it is a single line with crosslines.

Symbols should be easy to draw. They should be of simple design, yet their meaning should be easily recognized. Most of these are standardized.

Symbols may be divided into several types:

1. Those representing water features. These are usually in blue.

2. Those representing man-made features, such as buildings. These are usually black or red.

3. Those representing relief features, such as mountains. These are usually brown. Relief features show the varying land heights.

ROCK CLIFF

SHORE LINE

MOUNTAIN

PASS

SAND DUNES

ERODED
HILLSIDE

21–5. *Some relief symbols.*

ple in design. Whenever possible, they are made to look like a typical feature. For example, a tank is shown by a picture of a cylinder. The actual tank may be somewhat different in shape, but the meaning of this symbol is clear. These symbols can be drawn as viewed from directly above or in pictorial form.

Relief Feature Symbols. Relief feature symbols show contour lines and land forms. Relief maps cannot show every hill and valley. They show the general type of typography with symbols. See Fig. 21–5. The symbols are drawn viewing the land from about 45 degrees above the surface. Land form symbols are used only on small scale maps. Larger scale maps use the block diagram method of showing relief features. Other examples are shown in Fig. 21–25.

Vegetation and Surface Feature Symbols. There are a number of standardized vegetation feature symbols. These try, in a very simple way, to represent the actual vegetation. They are drawn in color. If they are drawn in black, they should be used sparingly. If too many symbols are placed on an area, the map becomes crowded and hard to read. The symbol patterns show the vegetation from the side view. Examples are shown in Fig. 21–6.

Road, Railroad, Canal and Dam Symbols. Road symbols show the number of lanes and the type of surface. They are drawn with black lines. Hard surfaced roads use red on the symbol. Road and railroad bridge symbols are black. Railroad symbols are drawn in black. Dams are shown with a solid black line. Canals are drawn in blue. Examples of these symbols are shown in Fig. 21–7.

Index contour . . . . Intermediate contour . .

Supplementary contour . . Depression contours . .

Fill . . . . Cut . . . .

Levee . . . . Levee with road . . . . .

Mine dump . . . . Wash . . . .

Tailings . . . . Tailings pond . . . . . .

Strip mine . . . . Distorted surface . . . .

Sand area . . . . Gravel beach . . . .

Submerged marsh . . . . Marsh (swamp) . . . . . .

Mangrove . . . . Wooded marsh . . . . . .

Orchard . . . . Woods or brushwood . .

Scrub . . . . Vineyard . . . .

Urban area . . . . Inundation area . . . . .

21–6. *Vegetation and surface features.*

Hard surface, heavy duty road, four or more lanes . . . . . .

Hard surface, heavy duty road, two or three lanes . . . . . .

Hard surface, medium duty road, four or more lanes . . . .

Hard surface, medium duty road, two or three lanes . . . . .

Improved light duty road . . . .

Unimproved dirt road and trail . . . .

Dual highway, dividing strip 25 feet or less . . . .

Dual highway, dividing strip exceeding 25 feet . . . .

Road under construction . . . .

Railroad, single track and multiple track . . . .

Railroads in juxtaposition . . . .

Narrow gage, single track and multiple track . . . .

Railroad in street and carline . . . .

Bridge, road and railroad . . . .

Drawbridge, road and railroad . . . .

Footbridge . . . .

Tunnel, road and railroad . . . .

Overpass and underpass . . . .

Important small masonry or earth dam . . . .

Dam with lock . . . .

Dam with road . . . .

Canal with lock . . . .

21–7. *Road, railroad, canal, and dam symbols.*

Boundary Symbols. Boundary symbols are combinations of long and short dashes or dots. The thickness of these lines varies with the symbol. The type of boundary is indicated by various symbols. For example, a state boundary uses a different symbol than a county boundary. The line thickness also varies. Some symbols are black, while others are red. See Fig. 21–8.

Special Feature Symbols. In addition to the previously mentioned symbols, some special symbols are used to show features not occurring commonly. These are generally on maps used for a special purpose. For example, on aeronautical maps, air markers are noted to aid the pilot. Often the cartographer has to develop a symbol to show these special features. They are usually in a bright color. Their meaning is often lettered beside the symbol.

Legends. Many symbols are clear with no additional explanation. Others require a legend to explain the meaning. For example, the size of cities is often shown by a circular symbol. The size of the symbol is used to show the size of the city. This is explained on the title portion of the map. See Fig. 21–9. The legend is used to explain the use of colored areas. Often the heights of land surfaces are given different shading or different colors. A legend explains what each represents. See Fig. 21–10. Legends are developed by the cartographer.

HOW MUCH DETAIL?

Even on large scale general maps, it is not possible to show every house or garage separately. Generally, large scale maps can only show the area of a farm or park. Details, such as park buildings, would be so small they might be just a dot. On medium scale maps, an entire city might be shown as a shaded outline area. Only main roads and railroads can be shown.

Small scale maps do not even show roads. They can only show the location of cities by a square or circular symbol. It is not possible to show the shape of the area of the city. See Fig. 21–11, pages 600 and 601.

Boundary, national . . . .
State . . . .
County, parish, municipio . . . .
Civil township, precinct, town, barrio . . . .
Incorporated city, village, town, hamlet . . . .
Reservation, national or state . . . .
Small park, cemetery, airport, etc. . . . .
Land grant . . . .
Township or range line, United States land survey . . . . . .
Township or range line, approximate location . . . .
Section line, United States land survey . . . .
Section line, approximate location . . . .
Township line, not United States land survey . . . .
Section line, not United States land survey . . . .
Section corner, found and indicated . . . .
Boundary monument: land grant and other . . . .
United States mineral or location monument . . . .

21–8. *Boundary symbols.*

● UNDER 5000

○ 5000 TO 25,000

◉ 25,000 TO 50,000

◉ 50,000 TO 100,000

◎ OVER 100,000

☼ CAPITOL

21–9. *A typical legend showing the size of cities.*

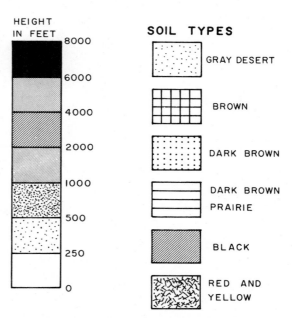

21–10. *Legends are used to explain the meaning of shading patterns.*

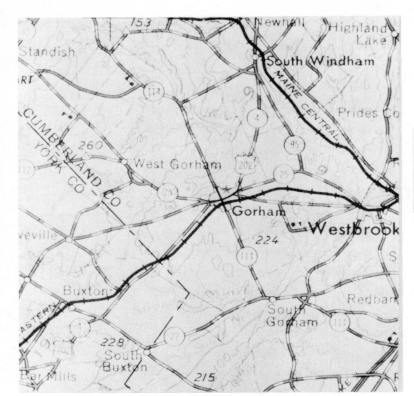

21–11A

1:250,000 scale,
1 inch = nearly 4 miles.
Area shown,
107 square miles.

21–11A, B, C. *The details shown
vary with the scale of the map.
(C is on the next page.)*

21–11B

1:62,500 scale,
1 inch = nearly 1 mile.
Area shown,
6¾ square miles.

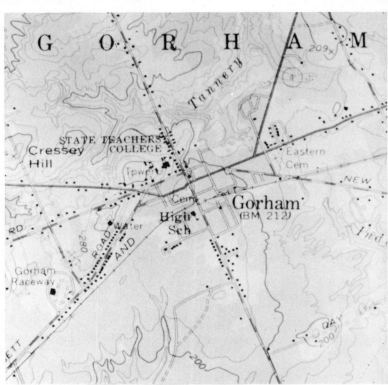

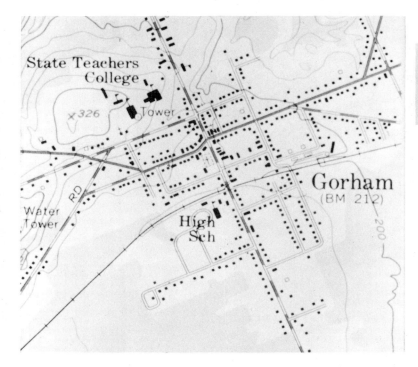

21-11C

1:24,000 scale,
1 inch = 2000 feet.
Area shown,
1 square mile.

The cartographer must use judgment in deciding on details. He decides how to combine features so the map serves its intended purpose. Some details he omits entirely.

About the only time all buildings and all streets can be shown is when a very small area or several city blocks are drawn to a large scale. This is a common practice when recording the actual land sizes as found by a careful survey of an area. These maps are used for legal descriptions of property. City engineers also use these maps.

COLOR ON MAPS

Most maps are in color. Proper use of colors greatly helps in the reading of maps. For example, color can be used to emphasize the important features of a map. As stated in the discussion of symbols, water is shown in blue, forests and vegetation in green. Ice is shown in white, deserts in reddish brown.

The cartographer must understand the principles of color and use them to advantage. He must try to avoid too much contrast between large areas of color, so that these areas do not compete. For example, strong, bright greens and browns are not generally used. They are toned down so too much contrast is not developed. The use of pastel colors is more effective.

Cities and roads are usually shown in red or black. They stand out clearly against the usual green and brown land areas. They are small and must be strong so they can be easily seen. Lines used to indicate boundaries, as between two states, are usually red or black.

The cartographer will vary the hue of the color in areas that are similar. For example, he may use a green for corn and a lighter green for hay. A wheat field might be indicated by light brown or yellow-brown.

Color can be used effectively to separate features, such as the countries of Europe. The color, in this case, has no reference to the actual ground. It is used to show clearly the shape and size of each country. Such maps do not generally use a different color for each country. The colors might be limited to two or three. The intensity is varied for contrast. See Fig. 21–12.

Another effective approach is to color a narrow strip along each border. The map does not appear like a gaily colored pin wheel. The colors chosen should blend harmoniously and not clash. Notice the gray areas in Fig. 21–13.

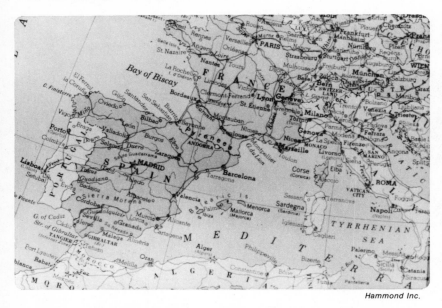

Hammond Inc.

21-12. *In color form this map uses one color over the entire area of each country to emphasize the shape and size.*

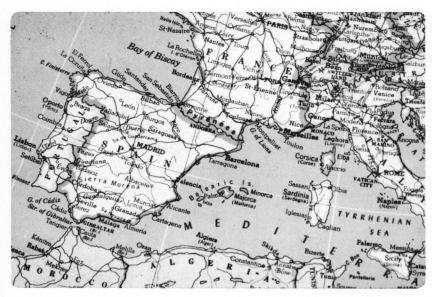

Hammond Inc.

21-13. *This map uses color (shown as gray here) along the borders to emphasize the shape and size.*

One effective medium used to color maps is transparent water colors. The beginner should practice on scrap paper before trying to color a map. To be effective, much experience is needed. Generally water colors are applied starting at the top and sides of the area. Paint toward the center and down to the bottom of the area. Keep the brush well filled with paint. A dry brush does not produce effective results. Be careful not to overlap areas. It is best to let an area dry before painting the one next to it.

Colored pencils can be used also. Pastel sticks are good for shading.

TOOLS

The tools used by the cartographer are the same as those used by draftsmen. Usually soft drafting pencils are preferred. The H and HB are popular. Various types of inking pens are used. The T-square or parallel rule with drafting triangles form the basic layout equipment. Water colors and water color brushes are necessary.

A pantograph, Fig. 21-14, is helpful for enlarging or reducing a map. It has scales on the arms. These are set to enlarge or reduce the drawing in the ratio wanted. For example, it can be set to enlarge a map to a ratio of ten to one. The enlarged map will be ten times the size of the original. One arm has a tracer point.

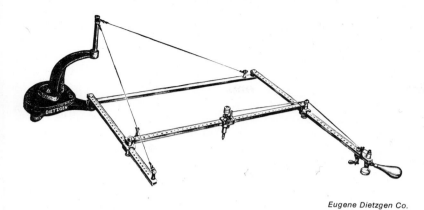

Eugene Dietzgen Co.

21–14. *A pantograph is used to enlarge or reduce maps.*

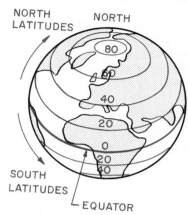

21–15. *Latitude is measured north and south of the equator.*

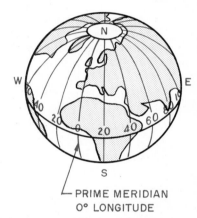

21–16. *Longitude is measured east and west of the prime meridian.*

It is moved along the lines of the original map. The other arm has a lead holder. It draws the enlarged or reduced map.

Papers used should be of best quality. They should have a high rag content. The surface should not be too slick or glossy. Transparent plastic sheets are used for some maps. They require special inks.

Of considerable use in map making are the various shading materials available. They are printed on transparent sheets. They have an adhesive back. Pieces are cut to the shape of the area to be shaded and are pressed into place. See Chapter 17, Technical Illustration, for more information.

PARALLELS AND MERIDIANS

The earth's surface has been divided into a system of coordinates. These are parallels of latitude and meridians of longitude. *Latitude* is an arc distance measured in degrees from the equator. Those parallels north of the equator are called north latitude. Those south are called south latitude. The equator is zero degrees latitude. See Fig. 21–15.

Longitude is the arc distance measured in degrees from the prime meridian. The prime meridian is a specially selected longitude running through Greenwich, England. It is designated as zero degrees longitude. Longitude is measured east and west of the prime meridian. See Fig. 21–16. The United States is in west longitude. Russia is in east longitude.

Latitude and longitude are measured in degrees, minutes, and seconds. For example, New York City is 74°0′ west longitude and 40°45′ north latitude. See Fig. 21–17.

21–17. *Latitude and longitude are used to locate points on the surface of the earth.*

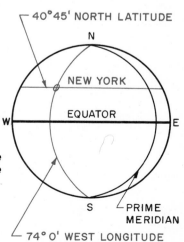

Latitude and longitude are essential in mapping. They provide an accurate means for locating features on a map.

DESIGNING MAPS

There are many questions to be raised before drawing a map. The cartographer must do considerable study and research. He must make many decisions. Following are some of the types of things he must know.

1. Purpose of the map.
2. Scale.
3. Best layout to use for the drawing space available.
4. Features to be shown.
5. Amount of detail needed.
6. The kind of map to use.
7. Method to be used to reproduce the map.
8. Land area to be included.
9. Features that need special emphasis.
10. Is a legend necessary?
11. What lettering will be needed? Is there room for it?
12. What is the best way to show features—standard symbols, relief, land form symbols?
13. Should it be in color? How many colors? What colors?
14. How much contrast in shaded areas?
15. What shading medium should be used?
16. Where to find material on the area to be drawn.

Making the Layout. Always make a preliminary layout on

tracing paper or graph paper. This should be in pencil. It should be to the same scale as the finished drawing. Plan space for the title area. Freehand letter the words to be in the title. The more important words are larger than the others. The title usually includes the name of the area, the type of map, the cartographer's name, date, scale, legend, and any other remarks considered necessary. The title is placed in one of the corners. It must be in an area where no important features are located.

Plan for a border. Most maps are enclosed with some type of border. These can be very simple. Two parallel lines are commonly used. Sometimes part of the map will cut through the border. The border should be broken to permit this. The inside line of the border is the actual edge of the map. It is called the *neat line.*

If a map has an inset plan, where is it to go? A typical example is an enlarged map showing the main streets of a city on a highway map.

Maps have a symbol indicating north. The top of the map is always north. Plan an area for this symbol. If the parallels and meridians are shown, this is sometimes omitted.

After the rough layout is made and all details are planned, the finished map can be drawn. This is drawn carefully in pencil. Use a sharp, soft pencil. It is not inked un-

til the map is completely drawn in pencil. Following is a suggested general order for laying out the parts of the map:

1. Draw the neat line. This sets the area of the map.
2. Draw the parallels and meridians.
3. Draw the water features such as rivers, lakes, and shore lines.
4. Draw roads and railroads.
5. Draw hills and mountains.
6. Locate the symbols for cities and other special features.
7. Letter any words needed. Place in clear, uncluttered areas as near the symbol or location as possible. These can be lettered lightly in freehand. They will be inked more accurately with lettering templates.
8. Draw guidelines for the lettering in the title area.
9. Now ink the drawing. It is recommended by some cartographers that inking be done in the reverse order of the pencil drawing. Ink the title, legend, and words first, then the symbols, and on through the list in reverse order. This reverse inking procedure can help prevent mistakes. For example, a parallel or meridian can be broken if it crosses a word or other important feature. A river can be broken if it is necessary to letter a name across it. If a map is to be printed in color, it is not always necessary to break these lines. Those items that might

conflict can be printed in a light color so the darker words or symbols will print clearly over them.

10. After the inking is finished, all pencil lines should be removed. This must be done carefully so the inked lines are not damaged.

11. If areas are to be shaded, this can now be applied. Water color areas can be painted.

12. It is wise to mount the map on heavy cardboard to prevent damage.

TOPOGRAPHIC MAPS

Topographic maps are general maps of a large area. They show all the important land features. Since they show a large area, they are drawn to a large scale. Some are developed at a medium scale.

Topographic maps are in relief. They involve the elevation above sea level, height of an area above adjacent areas, and the steepness of slopes. There are several ways to show relief — with contour maps, land form maps, and photographs of scale models.

Planographic maps show the same features as a topographic map. The difference is that they are not in relief. Color is used to show relief.

CONTOUR MAPS

A contour map is difficult to read unless you understand what each line represents. Fig. 21–18 shows how to read contours. Examine the hill, Fig. 21–18. It is cut in slices

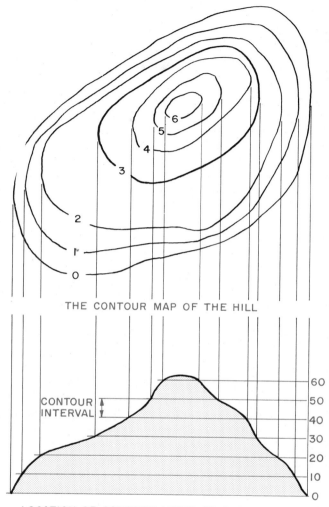

THE CONTOUR MAP OF THE HILL

LOCATION OF CONTOUR LINES ON THE HILL

21–18. *A contour drawing.*

at regularly spaced distances. The shape of the hill at each cut is the contour at the height above sea level. The cuts are horizontal. This contour map shows the hill from above with each contour line drawn. The height above sea level is lettered on the contour line. The contour line passes through all the points on the hill that are that particular height.

Contour maps are made to show the depth of water as well as height of hills and mountains.

The distance between contour levels varies with the scale of the map. Intervals of

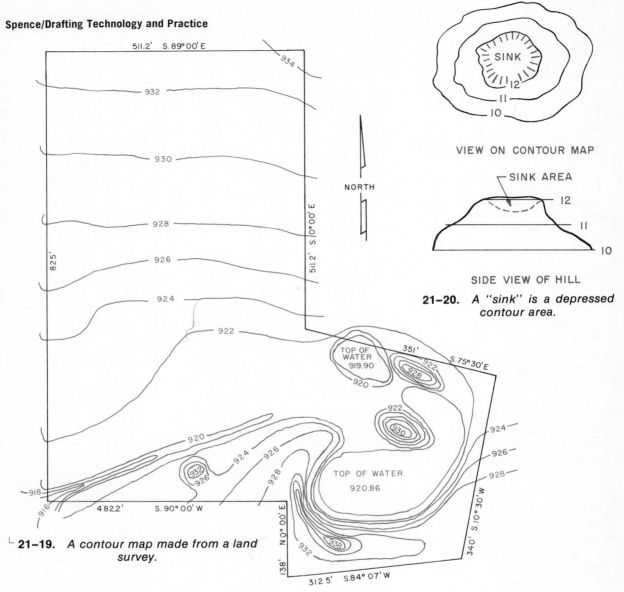

21–20. *A "sink" is a depressed contour area.*

VIEW ON CONTOUR MAP

SIDE VIEW OF HILL

21–19. *A contour map made from a land survey.*

10 to 20 feet are commonly used. Larger intervals are used on high, steep mountains. Smaller intervals are used on rolling and flat land.

Contour data are sometimes difficult to obtain. A surveying crew may measure elevations of the most important points and estimate the contours in between. This re-

quires actual field drawings made as they observe the area. Accurate contour maps can be made from aerial photographs using stereovisual equipment. This special equipment enables the photograph to be seen in three dimensions.

The drawing of contours should be done in pencil.

Every fourth or fifth contour is drawn with a heavy line. The level of the contour is lettered on the line. The line is broken at the number. Contour lines do not cross each other. Fig. 21–19 shows a contour map made from a land survey. A hollow or depressed area is shown with a special contour line. It has short lines drawn perpendicular to it. Often it is labeled "sink" to call attention to it. See Fig. 21–20.

PROFILES

Profiles are vertical cuts through a land area. They are much like a section through a machine part. Profiles are often made from contour maps. If a contour map is used, the location of the cut is marked. This is much like making a section on a machine drawing. A profile that has the vertical scale equal to the horizontal scale produces the actual profile. With these scales even high hills and mountains appear quite small. See Fig. 21–21. To emphasize the height, cartographers use a larger vertical scale. In mountainous areas, less exaggeration is needed than in flatter areas. See Fig. 21–22.

Fig. 21–23 is an engineer's drawing showing a profile of a sewer line. It was plotted on graph paper from an engineering survey. The surveyor's field notes were used to plot the profile.

RELIEF MODELS

Cartographers build scale relief models. They clearly show the hills, mountains, valleys, and rivers. To get good photographs of these, they are lighted from one side. This creates light and dark areas. The photographs show the relief effect clearly. See Fig. 21–24. Models are often

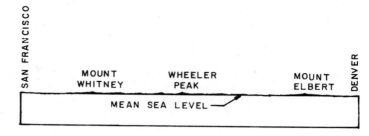

21–21.   *This profile has the same vertical and horizontal scale. It is of little value in showing the land profile.*

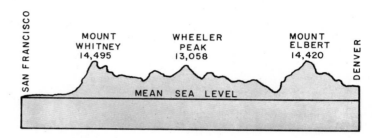

21–22.   *A profile through part of the Rocky Mountains. The vertical scale is 40 times greater than the horizontal scale.*

colored. The colors are used to emphasize the different heights. Low areas can be green, plains and hill areas tan to yellow, and mountains a red-brown.

LAND FORMS

A land form map shows basically the same thing as a contour map. It is in pictorial form and easier for most people to understand. No attempt is made to show every hill and valley. It shows the general typography of the area. See Fig. 21–25. Land form maps are used for small scale drawings. Larger scale drawings are drawn as block diagrams.

The size of the symbols drawn will vary with size of the mountain. Mountains are not drawn in proportion to each other. Very high mountains are usually drawn half of their actual height in relationship to surrounding smaller hills and mountains. If true proportion were used, the smaller hills and mountains would be very small. Some land forms would not be seen at all. The height above sea level is sometimes given at different levels. They are usually lettered in black ink.

It takes a great deal of work to develop the information needed for a land form map. The cartographer secures information from many sources.

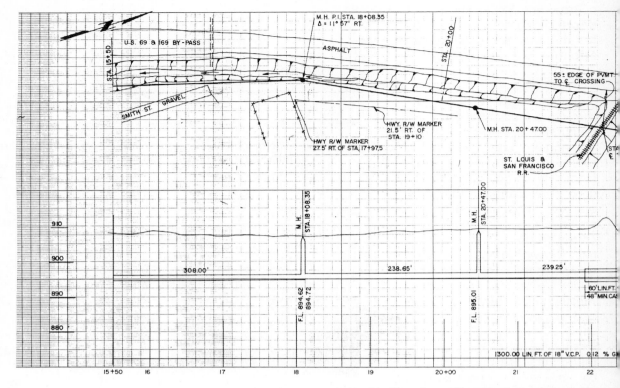

21-23. *A profile drawing of a sewer installation. In the top left side of the drawing a pictorial view is given of the line. In the top right is a plan view. The drawing at the bottom is the profile. It was drawn from the field notes of the*

21-24. *A raised relief map showing the earth's surface as it would appear to an astronaut. The relief on this map was magnified 20 times greater than its actual height.*

A. J. Nystrom and Co., Chicago

608

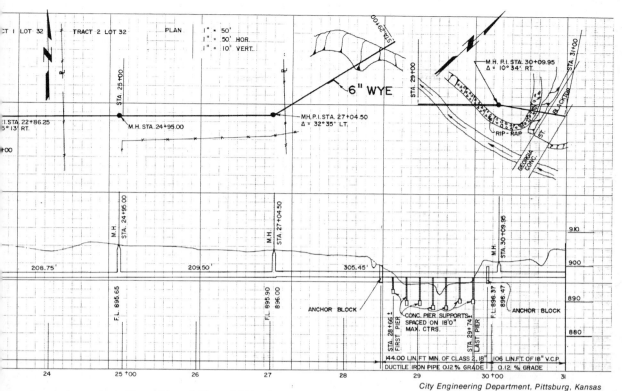

City Engineering Department, Pittsburg, Kansas

surveying crew. The abbreviation, M.H.,
refers to manholes. Notice the construction of supports to carry the sewer pipe
across a creek. Notice the difference in
the horizontal and vertical scales.

21–25. *Part of a phasiographic diagram using
land form symbols.*

Reproduced by permission from Mapping *by David Greenhoon,
The University of Chicago Press*

Portion of a physiographic diagram (Cities designated by initials)*
* Drawn by A. K. Lobeck. Courtesy of The Geographic Press.

A LAND FORM DRAWING

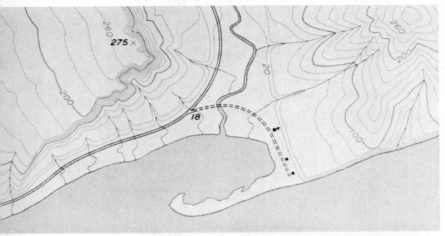

THE CONTOUR DRAWING

United States Department of the Interior, Geological Survey

21-26. *A land form drawing using shading to give emphasis to the contour. Can you relate the land form drawing to the contour drawing below?*

He may fly over the area to be drawn. Photographs can be examined. Topographic maps and other special maps of the area can be examined.

Common land form symbols, Fig. 21-5, are formed by drawing lines in the direction of the slope. This is the direction rain would drain when it falls on the land. These lines are called *hachure* lines. Notice how the hachure lines are spaced. Each one is carefully located.

Flat plains can be left blank. If plains are rolling, short horizontal or curved lines can be drawn. Hachure lines are used to show the earth sloping to a river. The rivers are drawn first. Then the hachure lines are adjusted to them.

Study the symbol for flat plateaus. Notice the bluff surface common to these formations. The slopes usually have valleys carved by erosion. Some plateaus tend to be rounded and worn.

Some mountains are old and worn. While they have high peaks, they tend to be rounded. Mountains formed by glaciers tend to have sharper peaks and ridges. The surfaces are concave. They have deep troughs running to glacial lakes or rivers.

Deserts have a wide variety of land forms. They can have rock plateaus and rounded mountainous areas. For the most part, they have large basins and areas of sand dunes.

A shaded land form drawing, Fig. 21-26, uses shading instead of hachure lines. This drawing was made from the contour drawing shown below it.

BLOCK DIAGRAMS

Block diagrams are another way to show the earth's surface in relief. They show a section of the earth removed. The drawing is viewed from above and at an angle. The more commonly used block diagrams are drawn in isometric, one-point, and two-point perspectives. See Chapter 9, Pictorial Drawing, for details on how to draw these.

The isometric block diagram is the easiest to draw. It is reasonably accurate. The information for a block diagram is taken from a contour map. Draw a light grid of squares over it. Fig. 21-27, Step 1. Next draw the contour map in isometric, Fig. 21-27, Step 2. This places all of the contour lines in isometric. Now draw an isometric box. Locate each contour height, Step 3.

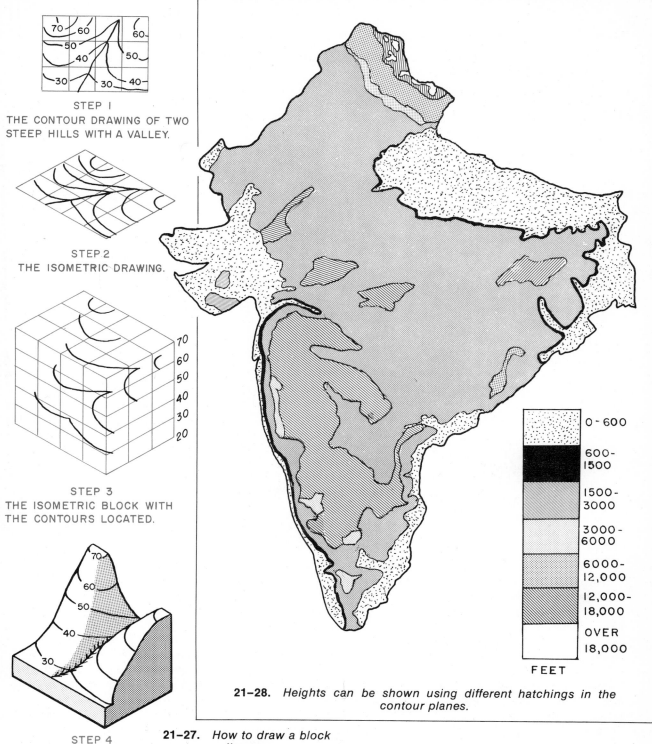

STEP I
THE CONTOUR DRAWING OF TWO
STEEP HILLS WITH A VALLEY.

STEP 2
THE ISOMETRIC DRAWING.

STEP 3
THE ISOMETRIC BLOCK WITH
THE CONTOURS LOCATED.

STEP 4
THE BLOCK DIAGRAM

21-27. *How to draw a block
diagram.*

	FEET
	0 - 600
	600-1500
	1500-3000
	3000-6000
	6000-12,000
	12,000-18,000
	OVER 18,000

21-28. *Heights can be shown using different hatchings in the
contour planes.*

Start the finished drawing by laying out the highest contour first. If the drawing is on transparent paper, this can be traced from the isometric drawing made in Step 2. Slide the isometric contour drawing down and copy the contour at each level. Connect the ends of the contour lines. Label each with the height above sea level, Fig. 21-27, Step 4.

PHYSICAL MAPS

Physical maps show altitudes by tinting various areas different colors. The colors selected should be representative of the type of surface. For example, areas that are suitable for farming can be shown in green. If a map is not in color, the various heights can be shown by using different hatching on each one. (Hatching means to shade with fine lines or dots). See Fig. 21-28. A legend is placed on the map to show the meaning of each color or hatched symbol. See map, page 611.

NAUTICAL AND AERONAUTICAL CHARTS

Nautical maps are used by navigators to guide their ships and avoid dangerous conditions. They show such things as accurate shore line contours, features of the coast line, depth indications, navigation lights, tides, currents, shoals and other obstructions, and harbor details. See Fig. 21-29. The scales used vary considerably. A very detailed harbor map will be on a large

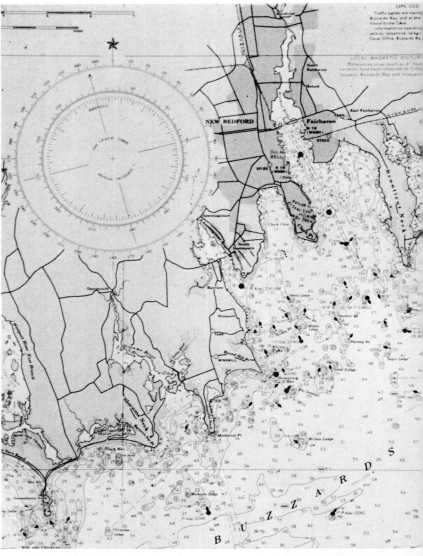

21-29. *A small part of a nautical map. Notice the symbols used.*

scale. One of an entire ocean would be on a small scale. Some of the symbols used are shown in Fig. 21-30.

Aeronautical charts are used by aircraft navigators. They show such things as the altitude of mountains, naviga-

tion beacons, airports, visual landmarks such as lakes, location and extent of navigation beacons, high obstructions such as a tower, and the general shape of cities. Generally the scale is small. See Fig. 21-31.

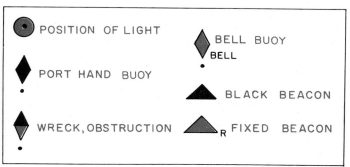

21–30. *A few or the many symbols used on nautical maps.*

POSITION OF LIGHT

PORT HAND BUOY

WRECK, OBSTRUCTION

BELL BUOY
BELL

BLACK BEACON

FIXED BEACON R

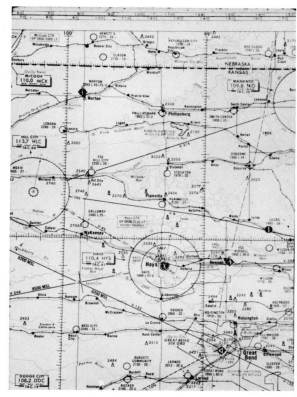

21–31. *An aeronautical map has special navigational symbols.*

CULTURE

CITIES ....

HIGHWAYS ....

RAILROADS .... Single / Double / Abandoned

OIL WELLS

MINES and QUARRIES

RACE TRACK

LANDMARKS

TERRAIN ELEVATION .... .3492

Wildlife Areas: ....
"The unauthorized operation of aircraft at low altitudes over, or the unauthorized landing of aircraft on a wildlife refuge area is prohibited, except in the event of emergency." Consult Fish and Wildlife regulations.

AIRPORTS

	PAVED	UNPAVED
PUBLIC USE .... Attended during daylight hours. Aviation gasoline available.		
PUBLIC USE .... No facilities.		
WARNING: UNCERTAIN OR UNKNOWN .... This airport may have some unknown or hazardous condition for which current information should be obtained prior to use.	W	W
MILITARY		
HELIPORT		H
ABANDONED		

GROUND ELEVATION IN FEET

Over 3000	2000 to 3000	1000 to 2000	Sea Level to 1000

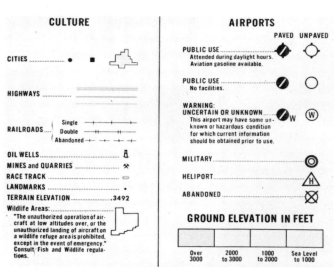

RADIO FACILITIES

VHF OMNI RANGE (VOR)
 Radials are magnetic FROM the station.

FAN MARKER (75 MHz.)

TACAN

AIRPORT CONTROL ZONE (CTR)

AIRPORT INFORMATION BOX
 FSS - FLIGHT SERVICE STATION
 NAME
 ELEVATION
 LIGHTS (* Part Time and/or On Request)
 LONGEST RUNWAY in hundreds of feet
  (figure in parens. hard surface if less than longest rwy)
 UNICOM - 122.8
  (123.0 with control tower or FSS)
 TOWER FREQUENCY (* Part Time)

ROTATING BEACON AT AIRPORT

INSTRUMENT LANDING SYSTEM (ILS)
 Magnetic bearing TO the station.
 Feathered side is blue.

ILS INFORMATION BOX

RESTRICTED AREA ....

EAGLE

FSS
GARDEN CITY
617* L 28(26)U
119.9

ILS
013° 109.9 (LNK)

COLUMBIA
113.3
368 ELM
TAC 80

VOR FREQUENCY

NDB FREQUENCY

TACAN CHANNEL

NON-DIRECTIONAL RADIO BEACON (NDB)

ENROUTE AIRWAY
 Airways have a floor of 1200' above ground level (AGL) unless otherwise designated.

TRANSITION AREAS
 Areas shown in BLUE have a floor of 1200' above ground level (AGL) in conjunction with airways, unless otherwise designated
 Areas shown in MAGENTA have a floor of 700' above ground level (AGL) around an airport with an approved Instrument Approach, unless otherwise designated.

COMPULSORY REPORTING POINT

ALTERNATE AIRWAY

ON - REQUEST REPORTING POINT

MAN MADE OBSTRUCTIONS
 Elevation above mean sea level

SPECIMEN STATION MODEL

Cloud type. (Middle altocumulus.)

Barometric pressure at sea level. Initial 9 or 10 omitted. (1014.7 millibars.)

Amount of barometric change in past 3 hours. (In tenths of millibars.)

Barometric tendency in past 3 hours. (Rising.)

Sign showing whether pressure is higher or lower than 3 hours ago.

Time precipitation began or ended. (Began 3 to 4 hours ago.)

Weather in past 6 hours. (Rain.)

Amount of precipitation in last 6 hours.

Total amount of clouds. (Sky completely covered.)

Wind speed. (18–22 knots.)

Direction of wind. (From the northwest.)

Temperature in degrees Fahrenheit.

Present weather. (Continuous slight snow in flakes.)

Visibility. (¾ miles.)

Dewpoint in degrees Fahrenheit.

Cloud type. (Low fractostratus and/or fractocumulus.)

Height of cloud base. (300 to 599 feet.)

Part of sky covered by lowest cloud. (Seven or eight tenths.)

Abridged from International Code

SURFACE WEATHER MAP AND STATION WEATHER AT 7:00 A.M., E.S.T.

POLAR STEREOGRAPHIC PROJECTION TRUE AT LATITUDE 60°

SCALE OF NAUTICAL MILES AT VARIOUS LATITUDES

21-32. A daily surface weather map. It shows conditions at selected stations across the country.

OTHER MAPS

There are many other types of maps in use. There are too many to cover in this brief chapter. Following is a sampling of some of these. Maps are designed to serve hundreds of purposes. As stated before, the cartographer must decide which form to use and what information is needed.

Regional maps show large areas such as a country, continent, or the entire world. They are drawn to a medium or small scale. Detail is very limited.

Weather maps show the temperature, precipitation, wind, pressure fronts, snow, and other such data. They are issued daily. Separate maps are developed for surface weather, temperatures and precipitation. A system of standard symbols is used. See Fig. 21–32.

Single purpose maps are designed to show the special features for which they are to be used. The special features are emphasized. All unnecessary details are eliminated. For example, a map showing the streets of a city does not need to show the contours of the land. Some examples of special purpose maps are geological, vegetation, statistical, land use, city, transportation, political, science, and land ownership.

There are many types of *statistical maps.* They are used to report data such as rainfall, population, and farm production. One example is shown in

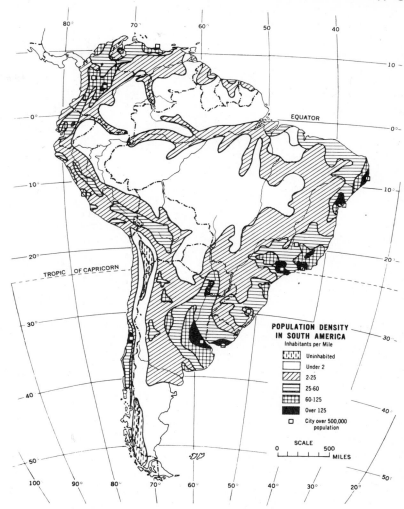

Reproduced with permission from World Geography, *J. W. Morris and O. W. Freeman. McGraw-Hill Book Company.*

21–33. *A statistical map. It shows the population density in various regions in South America.*

Fig. 21–33. Color is used extensively on statistical maps.

Maps are in use that show the bottom of the ocean, soil use, erosion, and vegetation. They are used to show the wealth and education of an area. The distribution of races can be mapped. A common map that everyone uses is the auto road map. The number and variety of maps is almost limitless.

Fig. 21–34 shows a map used by the National Aeronautics and Space Administration to show coverage of the earth by orbiting spacecraft.

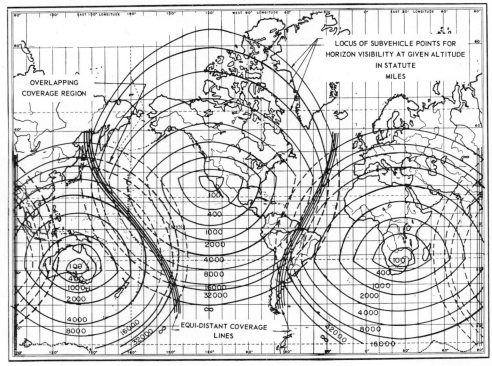

National Aeronautics and Space Administration

21-34. *This is a map developed to show the coverage of a space-craft by three stations as the spacecraft increases its altitude. The three stations are Goldstone, California; Woomera, Australia; and Johannesburg, South Africa.*

LAND SURVEY

One common type of map that almost everyone uses is the land survey. When a house is sold, the land included in the sale is mapped. Entire cities are mapped. Farms and industrial properties are mapped.

The survey process requires too much explanation to go into much detail in this text. Basically, a land survey provides two types of data needed to draw the map. The compass direction of each side of the land is needed. The length of each side must be measured.

The compass direction is taken using the north-south axis as the base. The angle formed is called the *bearing angle.* Study the bearing angles, Fig. 21–35. They are located in the four quadrants formed by the N-S and E-W axes. For example, the bearing angle is 30 degrees east of north, Fig. 21–35A. It is read N 30° E. If the 30-degree bearing angle was in each of the other quadrants, it would read as shown in Fig. 21–35 B, C, and D. The bearing angle cannot be over 90 degrees.

Another way the direction of a line can be shown is with azimuth angles. The *azimuth* is an angle measured from the north or south axis in a clockwise direction. The axis is zero. All angles are measured from it in degrees in a clockwise direction. The azimuth angle is shown measured from the north axis in Fig. 21–36. The north axis is generally used for azimuth angles, but

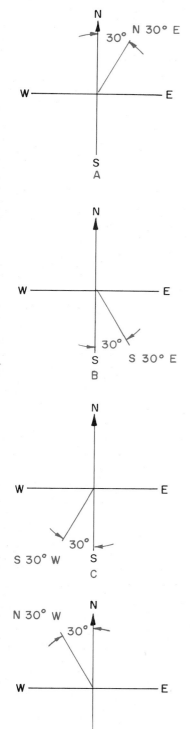

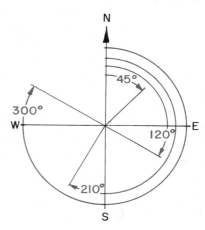

21-36. *Azimuth angles are measured clockwise in degrees from the north axis.*

the south axis can also be used. Then the south axis is zero and the angles are measured clockwise from it, as shown in Fig. 21-37.

The map using azimuth angles should always tell if the north or south axis is used.

In beginning a survey, the surveyor must accurately locate one corner. This is usually found by starting from a point that is known and working to a corner of the land to be surveyed. From this point, he sights with a transit the bearings for each side. He then measures the distance of the side. A stake is driven here. He then moves the transit to this stake and measures the angle and distance to the next side. These angles and distances are usually obtained from the legal description of the land. This information is part of the deed to the land. Many landowners drive metal stakes at each corner to preserve these locations. A typical land survey map is shown in Fig. 21-38. It is sometimes called a plat. It should contain the following:

Direction of each side.
Length of each side.
Acreage in the property.
Location and description of each permanent marker.
Location of roads, streams, and right of ways.
Division lines within the tract, if any exist.

21-35. *The bearing angle is measured from the north-south axis.*

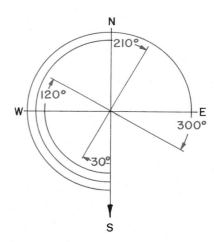

21-37. *The south axis can be used to measure azimuth angles.*

Names of adjoining property owners.

A dedication. This is a word description of the bearings and length of sides of the property.

A title, scale, date of survey, name of surveyor, official seal of surveyor, descriptive information, such as a lot number,

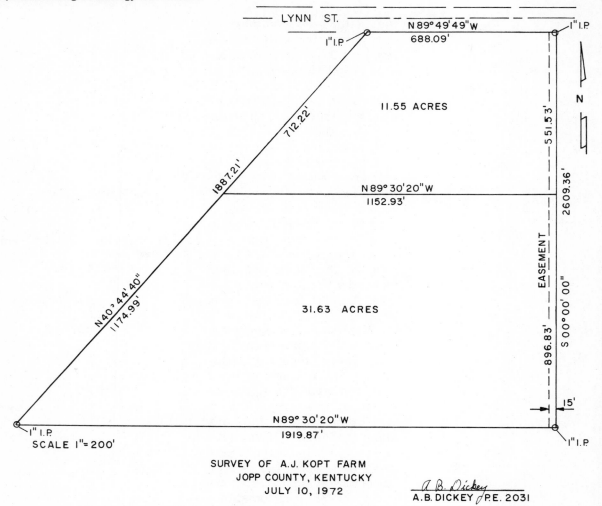

LYNN ST.

N 89°49'49"W
688.09'

1" I.P.

1" I.P.

11.55 ACRES

712.22'

551.53'

1887.21'

N 89°30'20"W
1152.93'

2609.36'

EASEMENT

S 00°00'00"

N40°44'40"
1174.99'

31.63 ACRES

896.83'

15'

1" I.P.
SCALE 1"= 200'

N 89°30'20"W
1919.87'

1" I.P.

SURVEY OF A.J. KOPT FARM
JOPP COUNTY, KENTUCKY
JULY 10, 1972

A.B. Dickey
A.B. DICKEY P.E. 2031

21–38. *A land survey of acreage. Bearings are given using a bearing angle. Each corner is located with an iron pipe. This is shown on the drawing as I. P.*

location of property, such as county or city and state, and name of the owner.

The bearing angles and length of sides of a tract of land are recorded in the deed. The deed is filed with the recorder of deeds in a city or county. This makes a permanent record of this informa-

tion. When such a tract is surveyed and the corners located, the surveyor often does not repeat the bearing angles. He sometimes gives the angles between the sides in degrees. See Fig. 21–39A.

After a land survey is made, the engineer writes a legal description. See Fig. 21–39B. It

gives the general location of the land area. Then it cites the point of beginning. This is the point from which all measurements were taken. The description gives the direction and lengths of the sides of the tract of land. Try to match the legal description with the land survey, Fig. 21–39A.

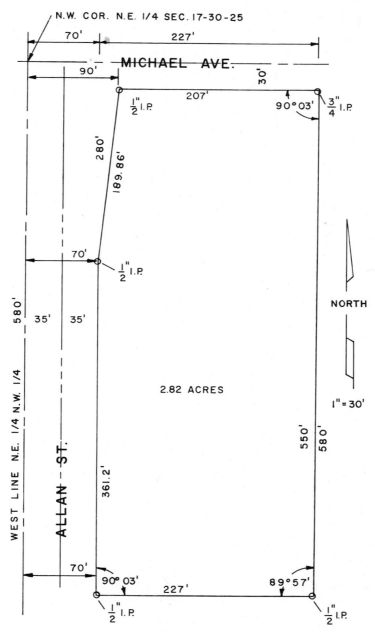

21-39A. *A land survey showing the direction of the sides using corner angles.*

DESCRIPTION

A PART OF THE N.E. 1/4 OF THE N.W. 1/4 OF SECTION 17, TOWNSHIP 30 SOUTH, RANGE 25 EAST, PITTSBURG, CRAWFORD COUNTY, KANSAS, MORE PARTICULARLY DESCRIBED AS FOLLOWS:

BEGINNING AT A POINT 580 FEET SOUTH AND 70 FEET EAST OF THE N.W. CORNER OF SAID N.E. 1/4 OF THE N.W. 1/4, THENCE EAST AND PARALLEL TO THE NORTH LINE OF SAID N.E. 1/4 OF THE N.W. 1/4 227.0 FEET, THENCE NORTH AND PARALLEL TO THE WEST LINE OF SAID N.E. 1/4 OF THE N.W. 1/4 550.0 FEET TO A POINT 30 FEET SOUTH AND 297 FEET EAST OF THE N.W. CORNER OF SAID N.E. 1/4 OF THE N.W. 1/4, THENCE WEST AND PARALLEL TO THE NORTH LINE OF SAID N.E. 1/4 OF THE N.W. 1/4 207.0 FEET, THENCE SOUTHWESTERLY 189.86 FEET TO A POINT 218.8 FEET SOUTH AND 70 FEET EAST OF THE N.W. CORNER OF SAID N.E. 1/4 OF THE N.W. 1/4, THENCE SOUTH AND PARALLEL TO THE WEST LINE OF SAID N.E. 1/4 OF THE N.W. 1/4 361.2 FEET TO THE POINT OF BEGINNING. CONTAINING 2.82 ACRES MORE OR LESS.

CERTIFICATE

THIS IS TO CERTIFY THAT I HAVE SURVEYED AND MARKED THE CORNERS OF THE ABOVE TRACT, AND THE ABOVE IS A TRUE RECORD THEREOF.

Gene E. Russell
GENE E. RUSSELL P.E. 2322

21-39B. *Legal description of a land survey.*

Another type of land survey is a plat of land divided into lots. This is done whenever acreage is divided to start a subdivision. This plat is filed with city or county recorder. It shows the size and location of each lot. All permanent stakes are noted. These stakes are called monuments. Each lot is numbered. They are referred to by this number. For example, it would be recorded that you purchased Lot 10, Baker's Subdivision. See Fig. 21-40A.

Cities make plats of all lots and acreage. They show streets and all major buildings. These are also used to show the location of gas, sewer, and water lines. They are used to plan improvements, such as new or rebuilt streets. Sometimes contours are shown. Two types of city plats are shown in Figs. 21-41 and 21-42.

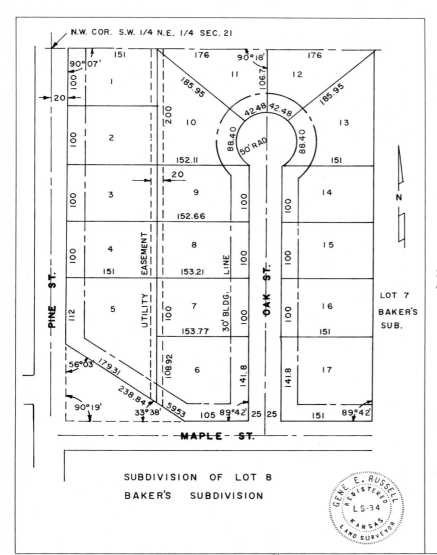

21-40A. *A land survey dividing acreage into lots. The size of each lot is given. They are numbered.*

21-40B. *Legal description of a lot in the subdivision.*

DESCRIPTION

A TRACT OF LAND IN LOT 8 BAKER'S SUBDIVISION, SECTION 21, T.W.P. 30 SOUTH, RANGE 25 EAST, LAWTON COUNTY, MISSOURI, MORE PARTICULARLY DESCRIBED AS FOLLOWS: BEGINNING AT A POINT 20' EAST OF THE NORTHWEST CORNER SOUTHWEST 1/4, NORTHEAST 1/4 OF SAID SECTION 21, THENCE EAST 503' TO THE NORTHEAST CORNER OF SAID LOT 8, THENCE SOUTH 641.8' TO THE SOUTHEAST CORNER OF SAID LOT 8, THENCE WEST 306' ALONG SOUTH LINE OF LOT 8, THENCE NORTHWESTERLY 238.84' TO A POINT ON THE WEST LINE OF SAID LOT 8 130' NORTH OF THE SOUTHWEST CORNER, THENCE NORTHERLY 512' MORE OR LESS TO THE POINT OF BEGINNING 7.14 ACRES MORE OR LESS.

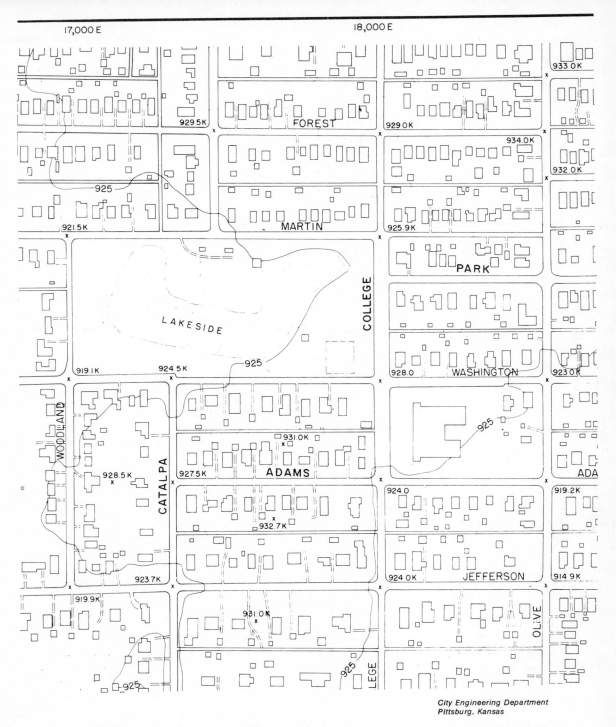

21-41. *Part of a city map that shows the location of every building. Contour lines are given. This basic drawing is used to show utilities, such as sewer and water lines. Copies of this basic drawing have these recorded on them.*

City Engineering Department
Pittsburg, Kansas

21-42. *Another form of a city map shows each lot in each block. The lot sizes are given. Each lot is identified by a number. This is a lot plan of the same area shown in Fig. 21-41.*

FIRST PHOTO · SECOND PHOTO · THIRD PHOTO

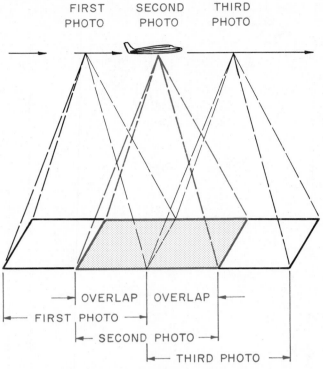

OVERLAP — OVERLAP

FIRST PHOTO

SECOND PHOTO

THIRD PHOTO

21–43. *How aerial photographs are overlapped.*

21–44. *The stereoscope (field type) enables the viewer to get a three-dimensional effect when it is placed over stereo pairs of aerial photographs.*

AERIAL MAPPING

Aerial photographs are a major factor in the work of a cartographer. The photographs cover a small area of land surface. A large area is mapped by photographing a series of land areas. The photographs are taken so they overlap. See Fig. 21–43. Photographs taken perpendicular to the earth's surface gives a rather true record of directions and distances. These are taken through a hole in the floor of the aircraft. The aircraft must fly as straight a course as possible. When it doubles back for a parallel run, it must be in a position so the photographs overlap. Usually the end overlaps are from 50 to 60 per cent. Side overlaps are from 15 to 50 per cent.

After photographs are taken and developed, the cartographer begins to work. He interprets what they show. He selects and arranges them into one complete photo of the land area. This is called a *photomosaic.* It is rephotographed to produce the final product.

Interpretation of the photos requires practice. The cartographer must learn how things appear in a photograph. A lake may show up as a white or black area. If it is reflecting light, it will appear white. Otherwise, it will appear dark. A round feature may be an empty pond or a crater.

Interpretation also involves identification of the three

624

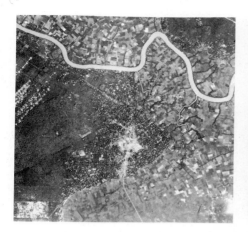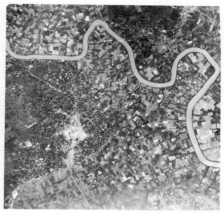

21–45. *Stereo pair of aerial photographs of an area in Vietnam. The photo covers the same areas as the line map and photomosaic.*

dimensions of the land's surface. To assist in viewing this third dimension, stereoscopic devices are used. With these instruments, the points and lines on the photograph can be plotted in relief. Fig. 21–44 shows a simple stereoscope. These are used in the field to view stereo pairs of photographs as shown in Fig. 21–45. Viewing the stereo pairs of photographs through the stereoscope produces a three-dimensional effect. Stereoscopes used in the laboratory by cartographers are much more complex. The stereoplanograph is a device used for plotting the relief of land and the heights of objects by inspecting pairs of aerial photographs.

Fig. 21–46 shows a photo of an area in Vietnam. It is the same area shown in the stereo pair of photographs. From this was developed the line map, Fig. 21–47. Study the photo-

graph and see if you can identify the important features. Then study the map for the interpretation. Find the rivers and streams. Notice the villages and roads. One area is wooded. A large area is farm and grass land.

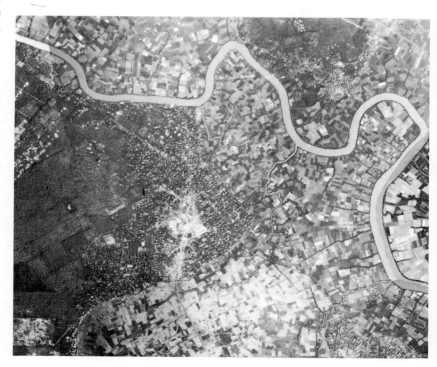

21–46. *Photomosaic of an area in Vietnam.*

625

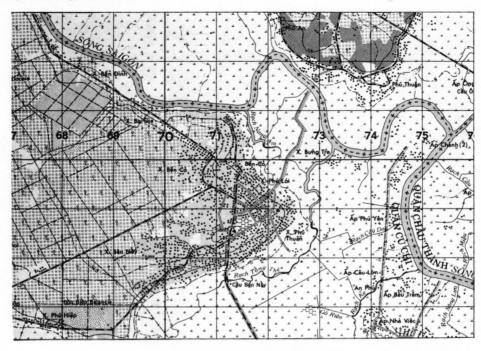

21-47. *Line map of an area in Vietnam. This photo covers same area as stereo pair and photomosaic.*

Additional Reading

Birch, T. W. *Maps, Topographical and Statistical.* London, England: Oxford, Clarendon Press, 1964.

Greenhood, David. *Mapping.* Chicago: The University of Chicago Press, 1964.

Raisz, Erwin. *Principles of Cartography.* New York: McGraw-Hill Book Co., 1962.

Trewartha, Glenn, A. H. Robinson, and E. H. Hammond. *Elements of Geography.* New York: McGraw-Hill Book Co., 1967.

Build Your Vocabulary

Following are some terms you should understand and use as a part of your working vocabulary. Write a brief explanation of what each means.

Cartographer
Scale
Symbols
Legend
Pantograph
Topographic maps
Parallels
Meridians

Longitude
Latitude
Contour
Profile
Land forms
Hachure lines
Block diagram
Physical maps
Charts
Land survey
Bearing angle
Azimuth
Stereoscope

Study Problems

Accompanying this list of problems are six drawings. Line drawings of the states and countries are given. These are generalized. They can be drawn by using the grid method. More detailed information can be found in a world atlas.

1. Get a highway map from your local service station.

A. List all the information it shows.
B. Find your county on the map. Draw it so the finished map will fit on an 8 × 10 inch paper. Locate all cities, roads, rivers, airports, and special features. Record the scale used. Use color to improve the clarity of the map.

2. Find the parallel of latitude and longitude of your town or city.
3. Find the parallel of latitude of the northern and southern borders of your state. What countries in other parts of the world are located within these parallels?
4. Make a map of your state. Locate each county. Find the population for each county. Using color or shading, indicate the population densities. Develop your own legend.
5. Draw a map of South America, P21–1. Locate each country. Develop a legend to show the national average food consumption in calories per day. Following is the data to be charted.

Brazil 2,400 to 2,900
Bolivia under 2,400
Peru under 2,400
Equador under 2,400
Colombia under 2,400
Venezuela under 2,400
British Guiana under 2,400
Surinam under 2,400
Paraguay under 2,400
Chile 2,400 to 2,900
French Guiana under 2,400
Argentina 2,900 and over
Uruguay 2,900 and over

6. Secure a highway map of your state. Make a map showing the river system. Also show the major highways. Locate any special features shown such as parks, camping grounds, and historical sites. Locate all cities over 25,000 population. Locate the state capital. Develop a legend to identify the special features and size of the cities.
7. Cut the weather map from your daily newspaper. Make a large drawing of this map. Use color to emphasize the important parts.
8. A map of the United States is shown in P21–2. Make a map showing any statistical or economic data you desire. This could be the population of each state. It could be the dollar volume of manufacturing. Data can be secured from an atlas or encyclopedia. Develop a legend. Use color.
9. A map of Arkansas is shown in P21–3. The heights of various areas are shown. Draw a relief map using land form

symbols. This map is greatly simplified.
10. Find a map of your state in an encyclopedia or geography book. Study it to find the heights of various parts of the state. Draw a relief map using land form symbols. Use color to give emphasis to the various heights. Draw a color legend in a clear corner.
11. Copy the map of Libya, P21–4. Make a planographic map to show elevations. Carefully select the colors used to represent the heights.
12. Data for a contour map is given in P21–5. Enlarge this map to use most of an 8 × 10 inch drawing area. The contour intervals are to be 10 feet. Make a contour map connecting points of elevation so the 10-foot contour intervals are shown.
13. Make a profile drawing using the data for the contour map in Problem 12. The location of the profile cut is shown on the drawing. Select a scale that clearly shows the profile.
14. Make a line map like the one shown in Fig. 21–43. The data for the map is given in P21–6. Use standard symbols to indicate the features. To find distances, use the scale with the line drawing layout.
15. Following is data from a surveyor's field notes. Make a plot of this acreage. Start the drawing by locating a point one inch from the left edge of your paper and one inch from the bottom. The bearings

and distances are from this point. It is Point One. The angles given are bearing anglos. Use the scale 1" = 100'.

16. The following data is to be used to draw a plot for a subdivision. Start the drawing in the lower left corner of the paper. The boundary of the area has the following specifications:

After the boundary of the area is drawn, draw a 50-foot wide street right of way from the center of the east boundary. Draw it perpendicular to the boundary. Next divide the two areas on each side of this street into two equal parts parallel to the street. This gives four areas to be divided into lots. Lay out these areas into

nine lots each. Draw their sides perpendicular to the 50-foot right of way. Make all the lots 100 feet wide except those on the west end. Any extra footage add to these lots. Using your scale, find the lengths of all the sides of the lots. Letter these on the drawing.

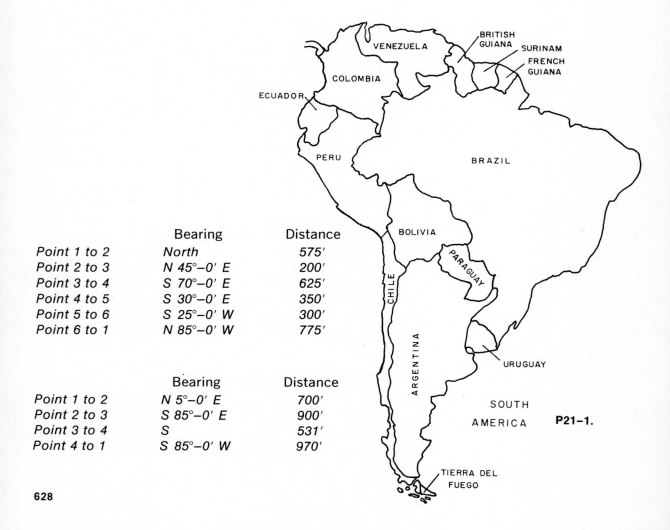

	Bearing	Distance
Point 1 to 2	North	575'
Point 2 to 3	N 45°–0' E	200'
Point 3 to 4	S 70°–0' E	625'
Point 4 to 5	S 30°–0' E	350'
Point 5 to 6	S 25°–0' W	300'
Point 6 to 1	N 85°–0' W	775'

	Bearing	Distance
Point 1 to 2	N 5°–0' E	700'
Point 2 to 3	S 85°–0' E	900'
Point 3 to 4	S	531'
Point 4 to 1	S 85°–0' W	970'

P21–1.

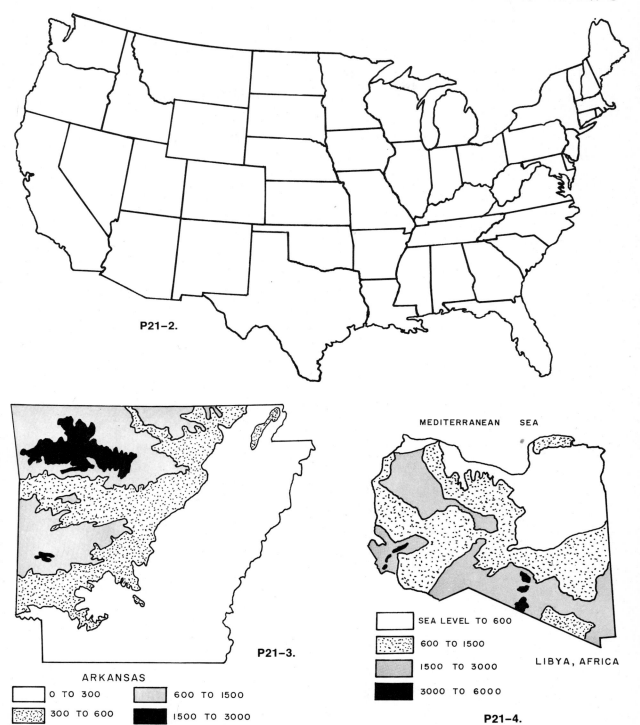

P21-2.

P21-3.

ARKANSAS

	0 TO 300		600 TO 1500
	300 TO 600		1500 TO 3000

MEDITERRANEAN   SEA

LIBYA, AFRICA

	SEA LEVEL TO 600
	600 TO 1500
	1500 TO 3000
	3000 TO 6000

P21-4.

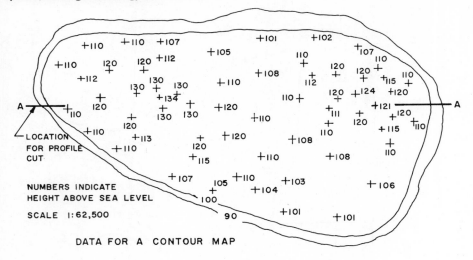

DATA FOR A CONTOUR MAP

P21-5.

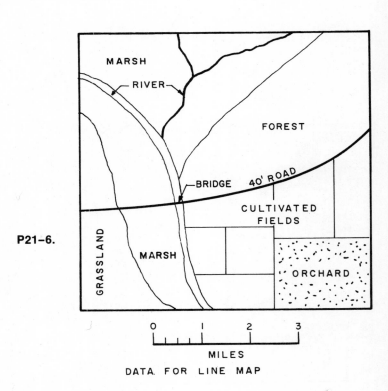

P21-6.

DATA FOR LINE MAP

Chapter 22

Architectural Drawing

The design of houses and commercial buildings is the work of the architect. The architectural draftsman makes the actual drawings. He must know how buildings are constructed. He must be able to use the graphic symbols of architecture.

PLANNING A HOUSE

The architect is responsible for designing a house to meet the needs of his client. In order to do this, he must find out the following things about the family:

a. How much the family can afford to spend.

b. The style of house they like.

c. The size of the family.

d. The ages of the members of the family.

e. Recreational and social activities they enjoy.

f. The shape and slope of the lot upon which the house is to be built.

g. Building codes and deed restrictions.

To plan the house, many other decisions must be made. Some of these are:

a. The type of house — one-story, two-story, or split-level.

b. Basement, concrete slab floor, or crawl space.

c. The number and types of rooms.

d. Concrete block or poured concrete foundation.

e. Frame or masonry wall construction.

f. Type of roof and roofing material.

g. Type of windows.

h. Type of heat and cooling.

i. Size of garage.

The house is designed around three areas: eating, living and sleeping. See Fig. 22-2. When the architect designs a floor plan, these sections of a house must be planned so they are separate areas. The eating area involves food storage, preparation, cleanup, and serving. All these activities involve the movement of food and dishes and a certain amount of odor and mess. Therefore, these activities should not conflict with the living and sleeping activities. The living area also includes noisy activities. The living room, family room, game room, or hobby area are typically found in this area. This noise, while a necessary part of living, would tend to disrupt those desiring to rest in the sleeping area unless careful planning is followed. Bathrooms must be located so they can be easily reached from all areas. It should not be necessary to disrupt activity in any area to reach a bathroom.

The traffic patterns in a house should be planned. Those entering the front door should be able to go to any part of the house without going through one of the areas. For example, it should not be necessary to walk through the living room to get to the bedroom area. Groceries should be carried directly into the kitchen. The stairs to the

22–1. An open floor plan. Notice the use of low room dividers to separate areas. The house appears much larger because of the lack of walls blocking off each area.

basement should be centrally located, and reached easily from any area in the house. Notice the traffic patterns on the plan shown in Fig. 22–2.

Small houses can be made to appear larger by reducing the number of walls. This is called "open planning." For example, the kitchen and dining area can be one large room. The family room and laundry can be together. The living room and dining area can be combined. See Fig. 22–1.

Planning the Kitchen. Kitchens are the center of many activities—food preparation, dining, laundry, and sewing. Kitchens are usually arranged in "I," "L," "U," and corridor shapes. See Fig. 22–3. They are planned around three appliances, the refrigerator, stove, and sink. Cabinets are built between these units.

The sink is usually placed near the food storage area. It is used to prepare food and clean up afterwards. It should not be too far from the dining area.

The stove is placed near the dining area. The hot food is served from the stove to this area.

The refrigerator is the center of the food storage area. It should be near the sink. It is best if it is near the door through which groceries enter the kitchen.

The total walking distance between the three major appliances should not be more than 22 feet. If it is more than this, the kitchen plan should be revised.

Ideally, the kitchen should have at least 15 feet of free counter space. This is in addition to the surfaces of the sink and stove. The sink and stove should have two to three feet of counter space on each side. It is helpful to have some counter space beside the refrigerator. Study the kitchen in the plans in this chapter.

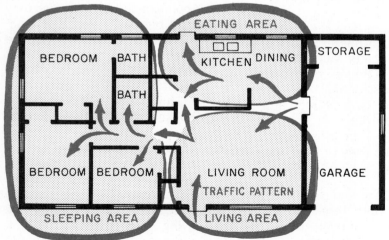

22–2. A house is designed around three areas: sleeping, eating, and living.

If laundry facilities are in a kitchen, they should not interfere with the areas for food preparation.

A small eating area is often part of the kitchen. It is used for light snacks and breakfasts. See Fig. 22–4.

The size of kitchen cabinets is standardized. Cabinet details and sizes are given in Chapter 23, Drafting in the Cabinetmaking Industry.

Planning the Dining Area. The dining area can be a separate room near the kitchen, as shown in Fig. 22–5, or part of the kitchen. Sometimes it is combined with a family room or living room.

A major factor in dining area design is to make it large enough to hold the furniture planned. Room must be left behind each chair so the person can be seated easily. See Fig. 22–6.

The dining area should have a pleasant atmosphere. Large windows are helpful.

Plans should include storage space for linens, china, and silverware in or near the dining area.

Planning the Living Area. The living room is one of the most used rooms in a house. It can serve many purposes. For some families, it is a sitting room, music room, study, and library. It can even serve as an extra bedroom for short periods of time.

Usually this room is placed facing the best outside view.

633

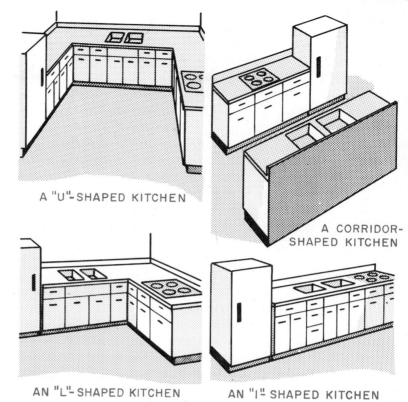

A "U"-SHAPED KITCHEN

A CORRIDOR-SHAPED KITCHEN

AN "L"-SHAPED KITCHEN

AN "I" SHAPED KITCHEN

22–3. *Common kitchen shapes.*

22–4. *A small eating area in a kitchen. The atmosphere is bright and cheerful.*

Armstrong Cork Company

22–5. *An attractive dining room. Notice that the table and chairs are not crowded. The atmosphere is pleasant.*

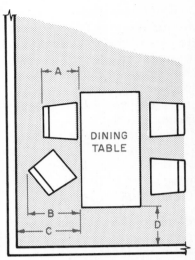

A- PERSON SEATED NEEDS 1'-6" TO
   1'-10" FROM CHAIR BACK TO TABLE
B- AVERAGE PERSON NEEDS 2'-6"
   TO 3'-1" TO RISE FROM TABLE.
C- TABLE SHOULD BE 3'-6" FROM
   WALL IF AN AISLE IS PLANNED
   BEHIND THE CHAIRS.
D- AISLE SHOULD BE 2'-0" IF NO
   ONE IS TO BE SEATED AT END
   OF TABLE.

22–6. *Adequate dining space.*

Large areas of windows are popular. They do reduce the amount of wall space available for furniture. The living room should have a central focal point. This is often a fireplace or a picture window. Furniture is grouped so conversation can be held easily. Persons seated over six feet apart have difficulty carrying on a conversation. See Fig. 22–7. It is best if the front entrance does not open directly into the living room. It makes part of the room into a hall and reduces the space available for use.

Do not crowd the furniture. Leave aisles of at least 3'-6". This helps people move about the room.

Planning the Family Room. The family room is used for hobbies and relaxation. It is an informal room. It can be used for television, laundry, and games. It can include a dining area. It can open onto a patio or screened porch. Fireplaces are often placed in the family room. See Fig. 22–8.

22–7. *A living room designed for comfort and conversation. The seating is grouped around a circular table. Windows provide a pleasant outside view.*

Armstrong Cork Company

634

Planning the Bedrooms. Bedrooms are tending to become smaller. They must be large enough to hold the furniture planned. Aisles for traffic must be provided. Aisles should be at least 24 inches wide. Space for dressing is needed. Closets are essential. The area in front of closets should be free of furniture. See Fig. 22–9 for minimum spacing of furniture.

Windows are needed so proper ventilation is obtained. Some prefer windows placed four to five feet above the floor. This insures privacy and makes the wall available for placing furniture below the window.

Bedrooms should be placed so they are entered from a hall. It should never be necessary to go through one bedroom to get to another.

It is convenient to have a full or half bath off the master bedroom.

Planning Closets. Closets are essential in every house. They are used in every bedroom. Hall closets are used for linens and other storage. A closet is needed near the front entrance. This gives a place for visitors to hang their coats.

Bedroom closets should be at least two feet deep. See Fig. 22–10. It is best if three to four feet of closet space is provided for each person using the bedroom. Sliding and folding doors are generally used.

Each bedroom closet should have a coat rod. At least one

Armstrong Cork Company

22–8.   *A family room provides for a variety of activities. It should have durable furnishings and building materials that will withstand hard use.*

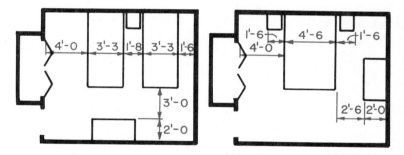

22–9.   *Minimum spacing for twin and double beds.*

shelf should be above the rod. It is desirable to have a light in each closet.

Linen closets can be as shallow as one foot. The length can vary. The average house needs a linen closet at least two feet long.

The closet doors should open the entire width. See Fig. 22–10. If they do not, space is wasted.

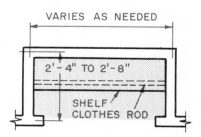

22–10.   *Typical closet size.*

Planning a Bath. Most new houses have a bath and a half. Two full baths are common. A four-bedroom house should have two full baths. Houses built on several levels should have a bath or water closet on each level.

Fig. 22–11 gives the most commonly used fixture sizes. When planning a house, the fixtures are selected. The actual sizes of the units are then known.

Fig. 22–12 gives the minimum spacing for bath fixtures. Each fixture should be easily accessible.

It is economical to put two baths side by side. See Fig. 22–13. A bath can be backed up to a kitchen. This saves in plumbing costs.

A second bath can use a shower instead of a tub. This saves floor space.

A bath can be planned so two people can use it at the same time. The water closet is placed in a separate room. See Fig. 22–14.

The door to the bath should be placed so the water closet is not seen when the door is open. Windows are not usually placed over a tub. This prevents drafts. Such windows are difficult to reach and open.

If a house has one bath, it should be placed near the bedrooms. It should be easily reached from other parts of the house without going through another room.

Planning Garages and Carports. Carports are used in

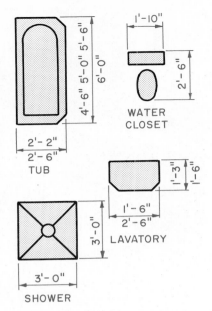

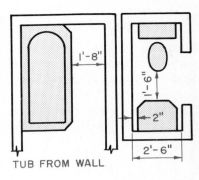

22–11. *Standard bath fixture sizes.*

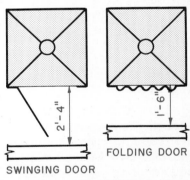

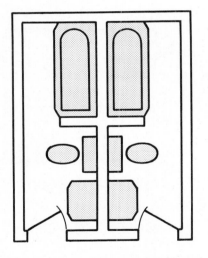

22–13. *Two baths can be placed side by side to lower plumbing costs.*

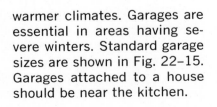

SHOWER WITH SWINGING AND FOLDING DOORS

22–12. *Minimum spacing for bath fixtures.*

warmer climates. Garages are essential in areas having severe winters. Standard garage sizes are shown in Fig. 22–15. Garages attached to a house should be near the kitchen.

Space should be provided for storage of garden equipment. This space is in addition to the sizes shown in Fig. 22–15.

The garage should be lo-

American Olean Tile Co.

22–14. *A bath planned for use by several persons. Two lavatories are available. The water closet is in a room by itself, just beyond the tub.*

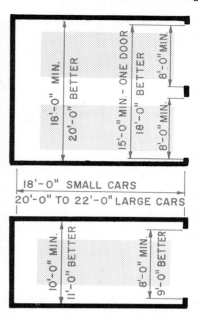

22–15. *Typical garage sizes.*

cated on the plan so it is easy to enter. Sharp turns make it difficult to park the cars.

Two-car garages can have two single doors or one large double door. Usually a door that slides overhead is used. It is easy to open and is out of the way.

Planning Stairs. The three most commonly used plans for stairs are straight, "L," and "U." See Fig. 22–16.

The straight stair is the easiest to make. It costs less than the others. It can have a landing.

The "L" and "U" stairs require a landing. The landing is used to change the direction of the stair. When planning a building, floor area must be saved for stairs. The amount of space needed depends on the size and type of stair.

The main stairway in a house should be at least 2'-8" wide, clear of the handrail. It should have 6'-8" clear headroom. Basement stairs should be at least 2'-6" wide, with 6'-4" clear headroom.

The parts of a stairway are shown in Fig. 22–17. The size

of treads varies with the height of the riser. Commonly used tread and riser sizes are shown in Fig. 22–18. For example, an eight-inch tread requires a nine-inch riser.

To find the number of risers, the total rise of the stair is divided by the height of the riser selected. The total rise is the distance from one floor to another.

Study the following example. The total rise is eight feet (96 inches). The stair riser wanted is nine inches. Divide the total rise, 96 inches, by the height of one riser, nine inches. This gives $10\frac{2}{3}$ risers. This shows that 11 risers are needed. However, the riser will be slightly shorter than nine inches. The sizes given in Fig. 22–18 are for guidance.

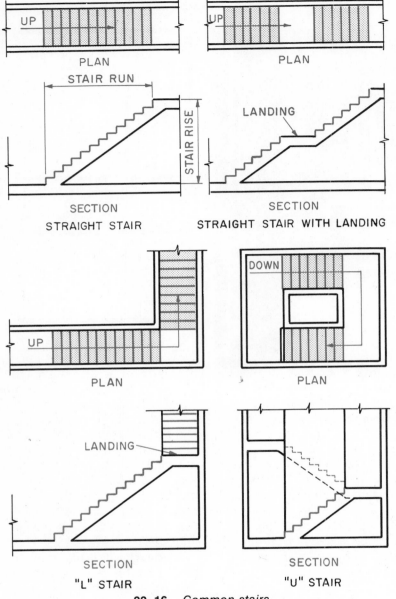

22-16. *Common stairs.*

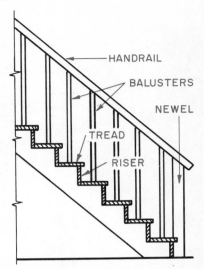

22-17. *Parts of a stair.*

Tread Width	Riser Height
8"	9"
9"	8"
10"	7½"
11"	7"

22-18. *Commonly used tread and riser sizes.*

Any solution that is near the sizes given is usable.

The actual size of the riser will be found by dividing the total rise, 96 inches, by the number of risers, 11. In this case the actual riser will be 8.73 inches.

Orientating the House. The comfort and enjoyment of a house can be improved by proper orientation. *Orientation* refers to the placement of the house on the lot and the location of rooms as related to the sun, winds, and view.

In the northern hemisphere, the south wall of the house receives the hot summer sun most of the day. The north wall receives very little sunlight. The west wall receives much sun late in the day.

The living room can face east or south. If it faces south, it will become rather warm in the evening. Generally, the garage, kitchen, or utility room is placed on the north side. The kitchen and dining areas can be placed on the east side. Any rooms facing west will be very hot in the summer. A porch or garage

638

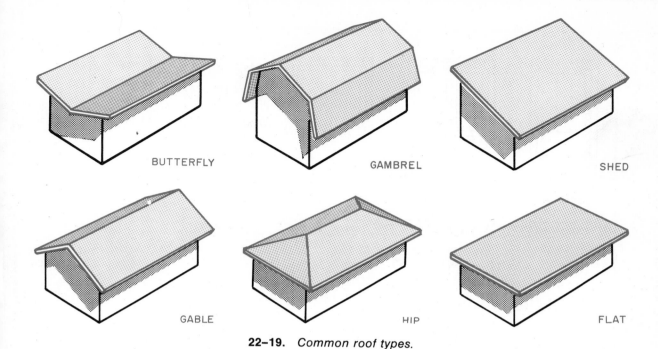

22–19. *Common roof types.*

BUTTERFLY   GAMBREL   SHED

GABLE   HIP   FLAT

can be placed here to help break the suns rays.

It is important to consider some means of shielding the house on the hot sides. Sometimes a large roof overhang is used. Trees are also good.

The direction of the prevailing wind must be considered. The summer breeze can be used to cool the bedroom area, if it is on that side of the house. Winter winds can be broken by placing the garage on that side.

PRELIMINARY SKETCHES

The house is first planned by making freehand sketches of the floor plan. These are on graph paper. Paper with ¼-inch squares could be used. Each square represents one foot. Sketch the rooms planned for the house. Relate them to each other. Check to see if the principles of good planning have been followed. Change the plan until it is satisfactory.

Make certain the rooms are large enough to hold the furniture. It is helpful to make paper templates to scale of the furniture and appliances. Place these on the drawing and see if the rooms are large enough.

As the floor plan is developed, the exterior of the house must be considered. Whenever one room is moved, this changes the outside appearance. Make sketches of the exterior of the house. It may be desirable to change the plan slightly to improve the appearance. The location of windows is important to appearance.

The type of roof used greatly affects the appearance of a house. The common roof types are shown in Fig. 22–19.

When a workable floor plan has been developed and a pleasing exterior exists, the working drawings can be made. Before these can be drawn, it is necessary to know the symbols used in architectural drawing. It is also necessary to understand the construction details of a house.

Symbols in Architecture. The working drawings for a building are complex. Items such as doors, windows, and electrical fixtures cannot be drawn on the plan in detail. They are represented by symbols so the drawing is easier to make.

Selected symbols are shown in Figs. 22–20 to 22–22. More detailed symbol lists can be found in architectural standards manuals.

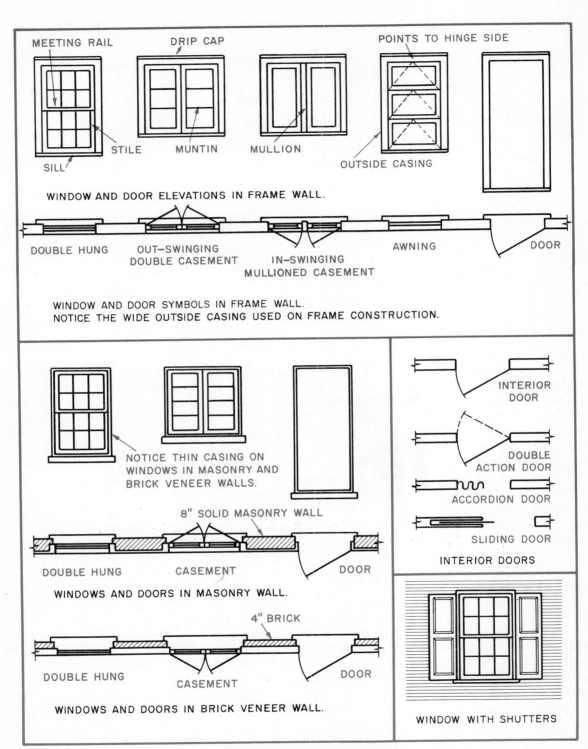

MEETING RAIL

DRIP CAP

POINTS TO HINGE SIDE

STILE

SILL

MUNTIN

MULLION

OUTSIDE CASING

WINDOW AND DOOR ELEVATIONS IN FRAME WALL.

DOUBLE HUNG

OUT–SWINGING
DOUBLE CASEMENT

IN–SWINGING
MULLIONED CASEMENT

AWNING

DOOR

WINDOW AND DOOR SYMBOLS IN FRAME WALL.
NOTICE THE WIDE OUTSIDE CASING USED ON FRAME CONSTRUCTION.

NOTICE THIN CASING ON
WINDOWS IN MASONRY AND
BRICK VENEER WALLS.

INTERIOR
DOOR

DOUBLE
ACTION DOOR

ACCORDION DOOR

SLIDING DOOR

INTERIOR DOORS

8" SOLID MASONRY WALL

DOUBLE HUNG

CASEMENT

DOOR

WINDOWS AND DOORS IN MASONRY WALL.

4" BRICK

DOUBLE HUNG

CASEMENT

DOOR

WINDOWS AND DOORS IN BRICK VENEER WALL.

WINDOW WITH SHUTTERS

22–20. *Window and door symbols.*

ELECTRICAL SYMBOLS		PLUMBING & PIPING SYMBOLS	
BELL		AIR LINE	
BUZZER		CLEAN-OUT	
		COLD WATER, TANK PRESSURE	
CIRCUIT BREAKER		COLD WATER, STREET PRESSURE	
CONVENIENCE OUTLET DUPLEX		DRAIN	
CONVENIENCE OUTLET OTHER THAN DUPLEX 1- SINGLE 3- TRIPLEX		GAS LINE	
		GAS OUTLET	
		HOSE BIB	
CONVENIENCE OUTLET WEATHER PROOF		HOT WATER, FLOW	
DROP CORD		HOT WATER, RETURN	
LIGHTING PANEL		REFRIGERANT, SUPPLY	
POWER PANEL		REFRIGERANT, RETURN	
		SEWER VENT	
OUTLET, FLOOR		SOIL AND WASTE	
		SOIL AND WASTE, UNDERGROUND	
OUTLET, LIGHT		WATER HEATER	
OUTLET, PULL CHAIN		BATH TUB	
OUTLET, RANGE		LAVATORY	
PUSH BUTTON		SHOWER	
OUTLET, SPECIAL PURPOSE, DESCRIBE IN SPECIFICATIONS		DRINKING FOUNTAIN	
SWITCH, SINGLE POLE		WATER CLOSET	
SWITCH, DOUBLE POLE		HEATING SYMBOLS	
SWITCH, THREE WAY		DUCT, HEAT SUPPLY (SECTION)	
		DUCT, HEAT SUPPLY (PLAN)	
SWITCH, PULL		DUCT, EXHAUST (SECTION)	
SWITCH AND CONVENIENCE OUTLET		DUCT, EXHAUST (PLAN)	
		DUCT, SUPPLY OUTLET	
TELEPHONE		DUCT, EXHAUST INLET	
TRANSFORMER		RADIATOR	

22–21. *Electrical, plumbing and piping, and heating symbols.*

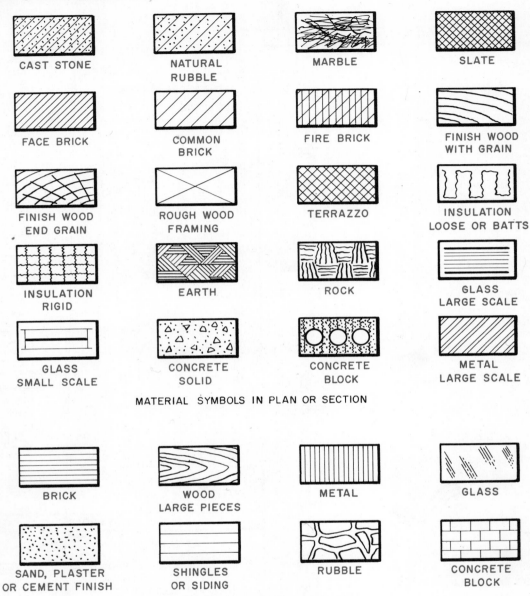

CAST STONE

NATURAL RUBBLE

MARBLE

SLATE

FACE BRICK

COMMON BRICK

FIRE BRICK

FINISH WOOD WITH GRAIN

FINISH WOOD END GRAIN

ROUGH WOOD FRAMING

TERRAZZO

INSULATION LOOSE OR BATTS

INSULATION RIGID

EARTH

ROCK

GLASS LARGE SCALE

GLASS SMALL SCALE

CONCRETE SOLID

CONCRETE BLOCK

METAL LARGE SCALE

MATERIAL SYMBOLS IN PLAN OR SECTION

BRICK

WOOD LARGE PIECES

METAL

GLASS

SAND, PLASTER OR CEMENT FINISH

SHINGLES OR SIDING

RUBBLE

CONCRETE BLOCK

MATERIAL SYMBOLS IN ELEVATION

22-22. *Material symbols.*

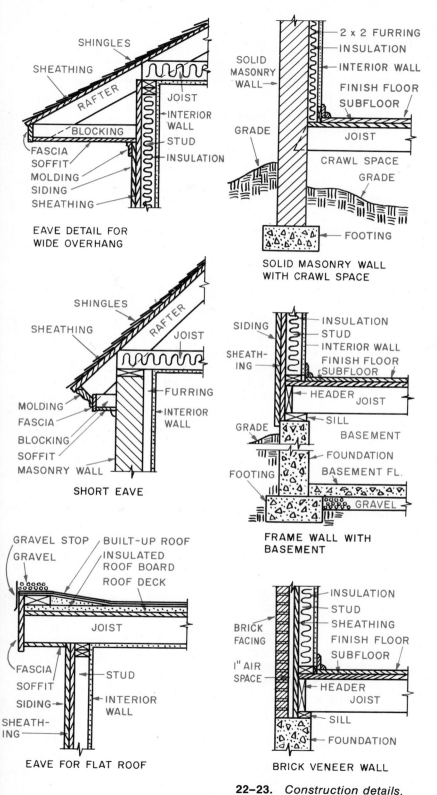

EAVE DETAIL FOR WIDE OVERHANG

SHINGLES
SHEATHING
RAFTER
JOIST
INTERIOR WALL
BLOCKING
STUD
FASCIA
SOFFIT
MOLDING
SIDING
SHEATHING
INSULATION

SOLID MASONRY WALL WITH CRAWL SPACE

2 x 2 FURRING
INSULATION
INTERIOR WALL
FINISH FLOOR
SUBFLOOR
SOLID MASONRY WALL
GRADE
JOIST
CRAWL SPACE
GRADE
FOOTING

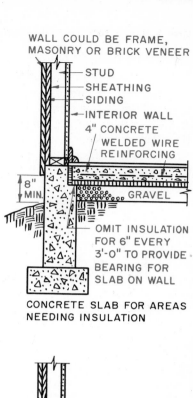

CONCRETE SLAB FOR AREAS NEEDING INSULATION

WALL COULD BE FRAME, MASONRY OR BRICK VENEER
STUD
SHEATHING
SIDING
INTERIOR WALL
4" CONCRETE
WELDED WIRE REINFORCING
8" MIN.
GRAVEL
OMIT INSULATION FOR 6" EVERY 3'-0" TO PROVIDE BEARING FOR SLAB ON WALL

SHORT EAVE

SHINGLES
SHEATHING
RAFTER
JOIST
MOLDING
FASCIA
BLOCKING
SOFFIT
MASONRY WALL
FURRING
INTERIOR WALL

FRAME WALL WITH BASEMENT

SIDING
SHEATHING
INSULATION
STUD
INTERIOR WALL
FINISH FLOOR
SUBFLOOR
HEADER
JOIST
SILL
GRADE
BASEMENT
FOOTING
FOUNDATION
BASEMENT FL.
GRAVEL

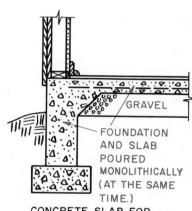

CONCRETE SLAB FOR AREAS NOT REQUIRING PERIMETER INSULATION

GRAVEL
FOUNDATION AND SLAB POURED MONOLITHICALLY (AT THE SAME TIME.)

EAVE FOR FLAT ROOF

GRAVEL STOP
GRAVEL
BUILT-UP ROOF
INSULATED ROOF BOARD
ROOF DECK
JOIST
FASCIA
SOFFIT
SIDING
SHEATHING
STUD
INTERIOR WALL

BRICK VENEER WALL

BRICK FACING
1" AIR SPACE
INSULATION
STUD
SHEATHING
FINISH FLOOR
SUBFLOOR
HEADER
JOIST
SILL
FOUNDATION

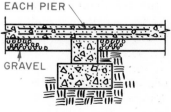

PIER DETAIL FOR SLABS REQUIRING INTERIOR SUPPORT

4'-0" SQUARE WIRE OVER EACH PIER
GRAVEL

22–23. *Construction details.*

CONSTRUCTION DETAILS

As a building is being planned, it is necessary to consider construction details. A plan may require construction that is very expensive. Knowledge of construction details enables the architect to develop the desired building at a minimum cost.

Selected construction details are shown in Fig. 22–23. Three common ways to construct walls are shown: frame, solid masonry, and brick veneer. A frame wall is usually five inches thick. A solid masonry wall is usually eight inches thick. A brick veneer wall is usually ten inches thick.

Construction details for concrete slab construction are shown in Fig. 22–23. If the slab is built in a cold climate, the edge of the floor must be insulated. Slabs can be supported by the ground or piers. If the ground has had much fill, piers should be used. Piers eight inches square should be spaced six feet on center throughout the floor.

Windows. The common types of windows are shown in Fig. 22–20. The type of window selected depends somewhat upon the style of house. Casement windows would look strange on a colonial style house.

The placement of windows is important. They are the major source of air to ventilate the house. The location of the windows influences where furniture can be placed in a room. The exterior appearance is also affected by window location. Windows should be balanced to give a pleasing exterior.

The Electrical System. The electrical system is shown on the floor plan of small buildings. Larger buildings require a separate electrical drawing.

Standard symbols are used to represent lights, outlets, switches, and other electrical units, Fig. 22–21. The floor plan, Fig. 22–27A, shows how these are placed on a drawing.

Convenience outlets in the wall should be spaced so no point along the wall is over six feet from an outlet. Light switches should be placed on the latch side of doors.

Lettering. Most architectural draftsmen use capital Gothic letters. These can be vertical or slanted. Lettering is usually 1/8 to 3/16-inch high. See the lettering section in another chapter.

Dimensioning. The working drawings for a building are made to a small scale. The only accurate way size is shown is with dimensions.

Study the dimensioning on the drawings in this chapter. Notice how they are placed. Distances over 12 inches are shown in feet and inches, such as 10'-6". Dimension lines are not broken. The dimensions are placed above the line.

Dimensions are placed on all sides of the foundation and floor plan. The overall length is given on each side. See Fig. 22–24 for the methods used to locate the corners in the different types of construction.

Windows, doors, interior partitions, and beams are located by their center lines. Each room is identified by name. The number of stair treads and rises is shown. The size and direction of ceiling joists is on the floor plan. The floor joists are shown on the foundation plan.

Scale. Foundation, floor plans, and elevations for small buildings are drawn to the scale 1/4" = 1'-0". Larger buildings are drawn 1/8" = 1'-0". Construction details are drawn to the scale 3/4" = 1'-0". Cabinet details are drawn 3/8" = 1'-0".

Title Blocks. The design of title blocks will vary from company to company. The one shown in Fig. 22–25 is typical of those in use.

MAKING THE WORKING DRAWINGS

After the house has been planned, final working drawings can be made. These include the following:

a. Plot plan.
b. Foundation plan.
c. Floor plan.
d. Elevations.
e. Details.

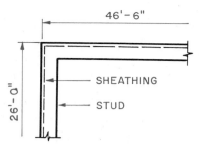

FRAME BUILDINGS ARE DIMEN-
SIONED TO THE CORNER OF
THE FOUNDATION. THIS INCLUD-
ES STUD AND SHEATHING. SINCE
THE SIDING OVERLAPS THE
FOUNDATION, IT IS NOT INCLUDED
IN THIS DIMENSION.

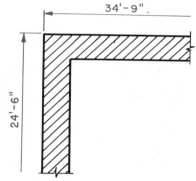

SOLID MASONRY BUILDINGS ARE
DIMENSIONED TO THE CORNER
OF THE FOUNDATION.

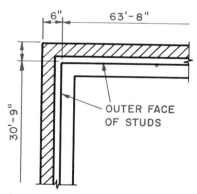

MASONRY VENEER BUILDINGS
ARE DIMENSIONED TO THE
OUTER FACE OF THE STUDS
AND THE FOUNDATION.

22–24. *Methods used to locate corners.*

645

JOHN C. KING AND ASSOCIATES

ENGINEERS AND ARCHITECTS

KANSAS CITY   MISSOURI

MEMORIAL STATE HOSPITAL
STATE BUREAU OF HEALTH
JEFFERSON CITY ~ MISSOURI

REVISIONS NONE		DATE	DRAWN BY B. S. H.	SHEET
		JOB NUMBER 50 - 13245		1

22–25. *A typical title block.*

A set of drawings for a small house are shown in Figs. 22–26 through 22–29.

Floor Plan. The floor plan is a horizontal view of the house as seen from above. The house is sectioned through the windows. The floor plan shows:

a. Room size and arrangement

b. Window and door size and location

c. Exterior and interior wall thickness

d. Stairs

e. Electrical system

f. Kitchen cabinets and fixtures

g. Bath fixtures

h. Fireplace

i. Any special built-in items, such as shelving and counters.

Usually the floor plan is the first drawing made. It provides the information needed for the foundation plan.

Study the floor plan in Fig. 22–27A. How many duplex outlets are in the house? What type of windows are used? What is the size of the living room?

The size of doors and windows is shown with a door and window schedule. This is drawn on the same sheet as the floor plan. Another system is shown on Figs. 22–27 through 22–29. Here the door sizes are shown on the floor plan. The window sizes are on the elevations. When drawing the floor plan:

1. Lay out the overall length and width of the house.

2. Draw the wall thickness.

22–26. *An architectural rendering of the house shown in Figs. 22–27 through 22–29.*

Home Planners, Inc.

3. Locate and draw the interior walls.

4. Locate and draw windows.

5. Draw the stairs, fireplace, and kitchen cabinets.

6. Draw porches and steps.

7. Indicate materials on walls.

8. Locate electrical outlets and fixtures.

9. Draw doors.

10. Dimension outside of house.

11. Dimension inside of house.

12. Place code for windows and doors.

13. Complete door and window schedule.

14. Label items needing identification as stove, refrigerator.

15. Letter any other notes needed.

Foundation Plan. A foundation plan is a top view of the foundation of the house. It shows the foundation before the house is built on it. This plan gives:

a. Size of footings.

b. Size of foundation wall.

c. Location and size of beams and columns.

d. Wall material (concrete or concrete block).

e. Wall openings (vents, windows, or doors).

f. Electrical outlets and lights if it is a basement.

g. Stairs, furnace, hot water heater, or any other such items.

h. Size and direction of floor joists.

i. Dimensions for all these items.

Study the foundation plan, Fig. 22-27B. Try to identify each part. Notice that the dimensions are to the corners of the foundation wall. The front of the building is drawn facing the bottom of the drawing.

Elevations. Elevations are views of the outside of the house. Normally a view is made of each side. See Fig. 22-28A and 22-28B. They are usually labeled according to the direction they face.

Elevations show the roof line and pitch. The windows and doors are drawn. The type of wall material is shown with the proper symbol. Sometimes it is also written on the drawing. The entire surface need not be covered with the symbol. It is drawn in a few places on the large surfaces.

Very few dimensions are given. Usually the finished floor and ceiling lines are dimensioned. Overhangs are sometimes dimensioned.

The finished grade is shown. This is the level to which the soil is to be graded.

When drawing elevations:

1. The foundation and floor plan should be completed.

2. Make a sectional drawing through the wall of the building up through the eave.

3. Draw the end elevations first. This locates the lines of the roof.

4. Draw the front and rear elevations.

5. Lay out the finished floor line.

6. Locate the finished ceiling height.

7. Draw in the grade line.

8. Draw the extreme right and left sides of the house.

9. Locate any breaks in the exterior wall, as in an "L"-shaped house.

10. Block in the roof. The pitch must be found on the side elevation. This distance can be transferred to the front elevation.

11. Locate the doors and windows and draw them.

12. Locate and draw chimneys, porches, and other items.

13. Draw the foundation and footing.

14. Indicate materials using the proper symbols.

Details. It is necessary to show details whenever something on the plans is not clear. Usually the following details are needed:

a. A section through the wall of the house. This runs from the footing through the eave. It shows the size of the footing, foundation thickness, wall construction, and how the eave is to be made. It is dimensioned. All parts are labeled. See Fig. 22-29A.

b. Elevations of the kitchen cabinets. This shows how the cabinet doors and drawers are to be arranged. An elevation is made of each wall with cabinets. See Fig. 22-29B. These are usually on the same sheet as the floor plan.

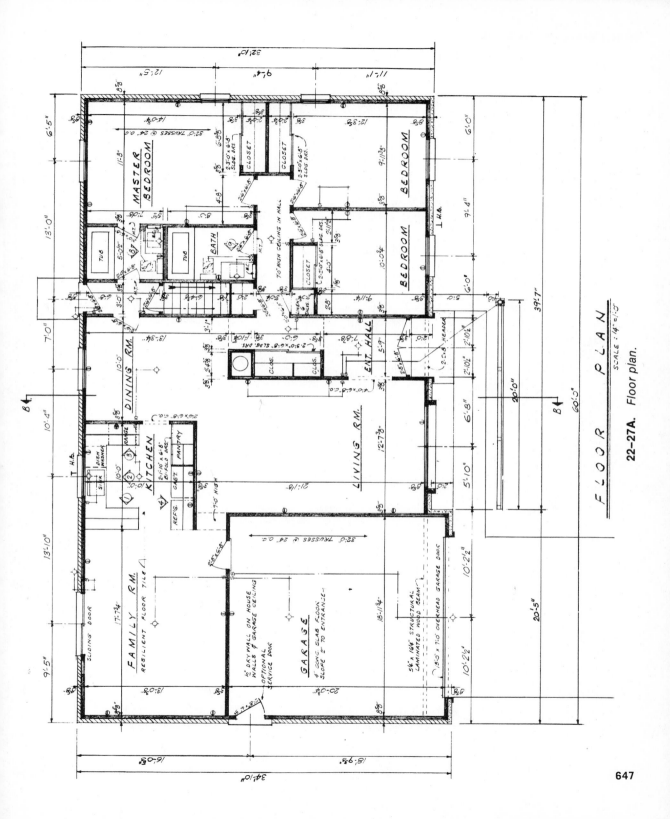

FLOOR PLAN
SCALE: 1/4" = 1'-0"

22-27A. Floor plan.

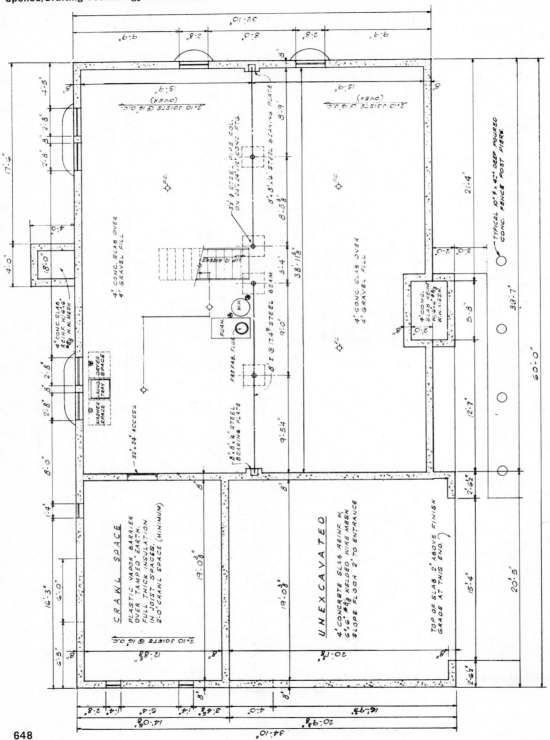

22-27B. Basement foundation plan.

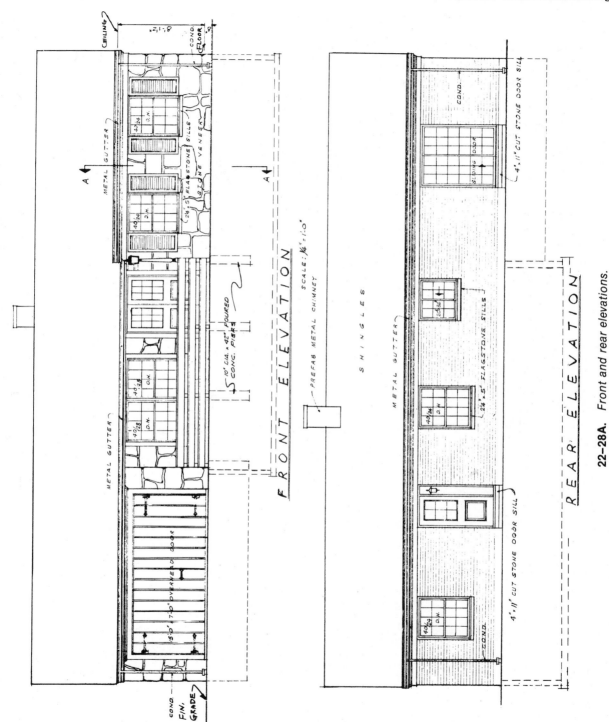

FRONT ELEVATION

SCALE: ¼" = 1'-0"

REAR ELEVATION

22-28A. *Front and rear elevations.*

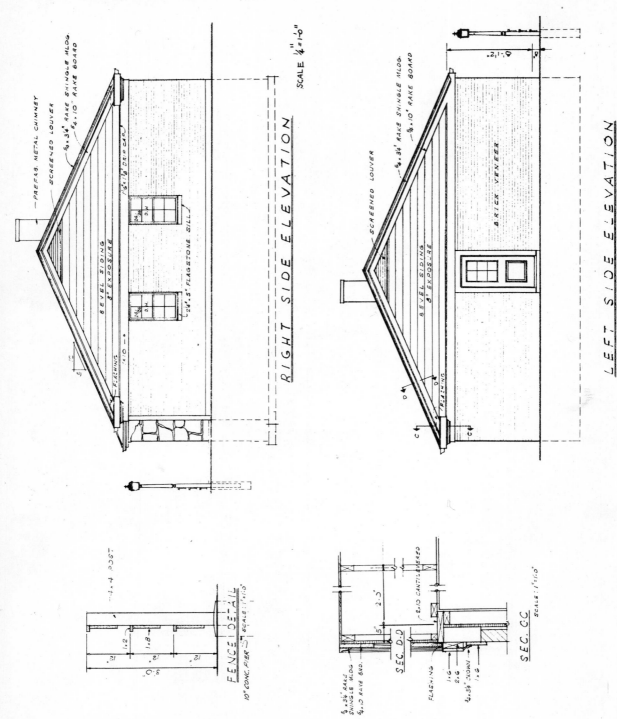

RIGHT SIDE ELEVATION

SCALE ¼" = 1'-0"

- PREFAB. METAL CHIMNEY
- SCREENED LOUVER
- ¾ x 3¾" RAKE SHINGLE MLDG.
- ⅝ x 10" RAKE BOARD
- 1 x 10 DRIP CAP
- BEVEL SIDING 8" EXPOSURE
- 2 x 4 .5" FLAGSTONE SILL
- DH
- DH
- FLASHING

LEFT SIDE ELEVATION

- SCREENED LOUVER
- ¾ x 3¾" RAKE SHINGLE MLDG.
- ⅝ x 10" RAKE BOARD
- BRICK VENEER
- BEVEL SIDING 8" EXPOSURE
- FLASHING

22-28B. Left and right side elevations.

FENCE DETAIL
SCALE: 1"=1'-0"
- 4 x 4 POST
- 10" CONC. PIER

SEC. D.D
SCALE: 1"=1'-0"
- ¾ x 3¾" RAKE SHINGLE MLDG.
- ⅝ x 10" RAKE BRD.
- 2 x 10 CANTILEVERED
- FLASHING
- 2 x 4 .34 JOIST

SEC. CC
SCALE: 1"=1'-0"

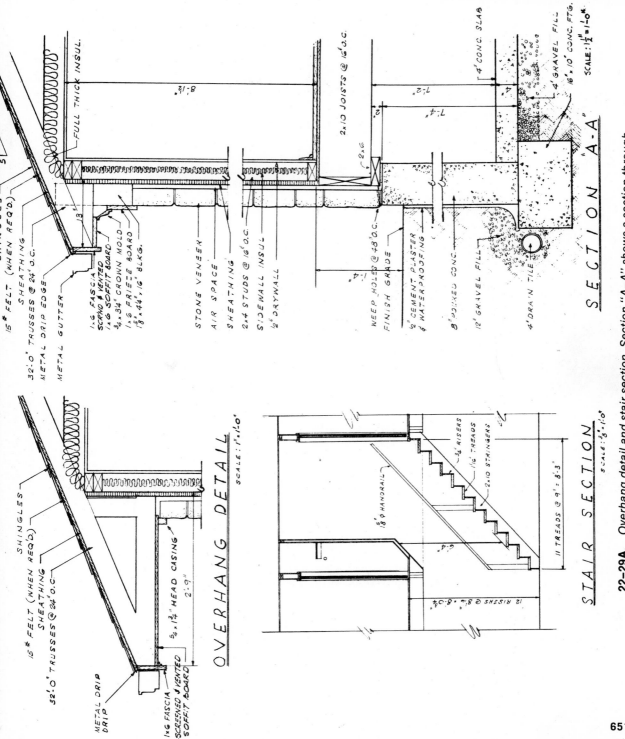

SECTION "A-A"

SCALE: $1\frac{1}{2}$"=1'-0"

Labels in Section A-A:
- SHINGLES (WHEN REQ'D.)
- 15# FELT (WHEN REQ'D.)
- 32'-0 TRUSSES @ 24" O.C.
- METAL DRIP EDGE
- METAL GUTTER
- FULL THICK INSUL.
- 8'-1½"
- 1×6 FASCIA SCRND & VENTED
- 1×6 SOFFIT BOARD
- ¾×3¾" CROWN MOLD
- 1×6 FRIEZE BOARD
- 1⅛"×4¼", 16' BLKG.
- STONE VENEER
- AIR SPACE
- SHEATHING
- 2×4 STUDS @ 16" O.C.
- SIDEWALL INSUL.
- ½" DRYWALL
- 2×6
- 2×10 JOISTS @ 16" O.C.
- 7'-2"
- 7'-4"
- 1'-4"
- WEEP HOLES @ 48" O.C.
- FINISH GRADE
- 10" CEMENT PLASTER & WATERPROOFING
- 8" POURED CONC.
- 12" GRAVEL FILL
- 4" DRAIN TILE
- 4" CONC. SLAB
- 4' GRAVEL FILL
- 16"×10" CONC. FTG.

OVERHANG DETAIL

SCALE: 1"=1'-0"

Labels in Overhang Detail:
- SHINGLES
- 15# FELT (WHEN REQ'D.)
- SHEATHING
- 32'-0 TRUSSES @ 24" O.C.
- 5½×1¼" HEAD CASING
- 2'-9"
- METAL DRIP
- 1×6 FASCIA SCRENED & VENTED
- SOFFIT BOARD

STAIR SECTION

SCALE: $\frac{3}{8}$"=1'-0"

Labels in Stair Section:
- 5" HANDRAIL
- 1⅛" Ø HANDRAIL
- 3½" RISERS
- 1×6 TREADS
- 2×10 STRINGERS
- 11 TREADS @ 9" = 8'-3"
- 6'-4"
- 12 RISERS @ 8¼" = 8'-0¼"

22-29A. Overhang detail and stair section. Section "A-A" shows a section through the wall of the house.

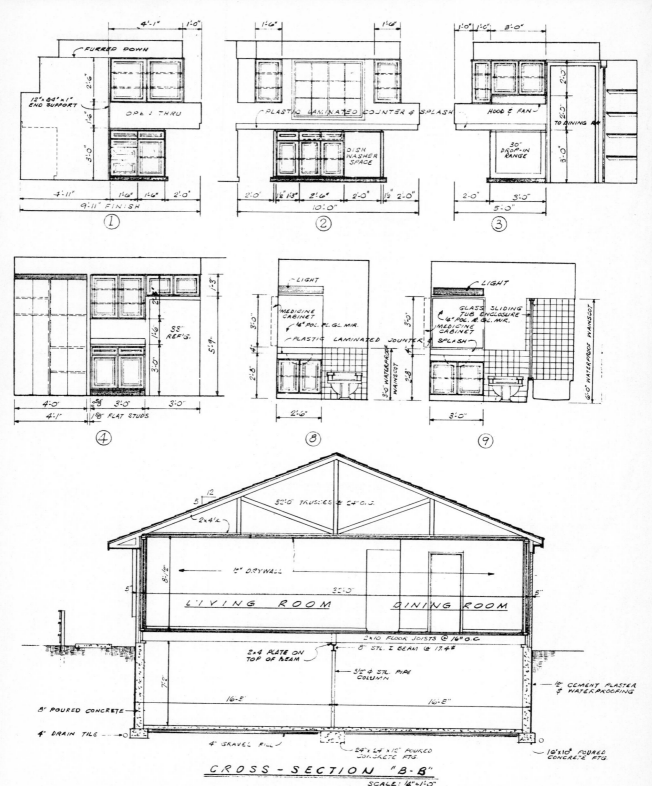

CROSS — SECTION "B-B"

SCALE: 1/4"=1'-0"

22-29B. *Elevations of kitchen cabinets. Cross section "B-B" helps clarify construction.*

A	B	C	D	E	F	G	H	I	J	K	L
24	11	16	12	8	14	18	$8\frac{3}{4}$	24	19	32	10
30	17	16	12	8	14	18	$8\frac{3}{4}$	29	24	38	13
36	23	16	12	12	14	23	$8\frac{3}{4}$	29	27	44	16

DESIGN SIZES FOR FIREPLACES
DONLEY BROTHERS CO.

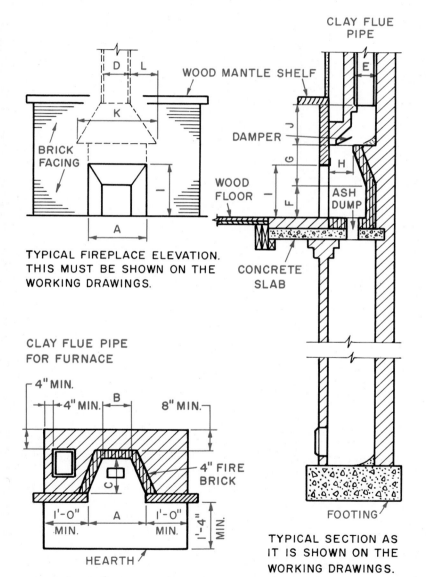

TYPICAL FIREPLACE ELEVATION. THIS MUST BE SHOWN ON THE WORKING DRAWINGS.

CLAY FLUE PIPE FOR FURNACE

THIS IS HOW A FIREPLACE IS DRAWN ON THE FLOOR PLAN.

TYPICAL SECTION AS IT IS SHOWN ON THE WORKING DRAWINGS.

22–30. *Details needed to show a fireplace on the working drawings.*

c. Elevations and sections through the fireplace. These are dimensioned. See Fig. 22–30.

d. Stair details. See Fig. 22–29A.

e. Elevations or sections of any other cabinets or unusual construction details.

f. Sections to clarify construction. A section through the house is shown in Fig. 22–29B.

THE PLOT PLAN

A plot plan is a top view of the building site. See Fig. 22–31. It shows the location of the house on the site. Also shown are sidewalks, driveways, porches, and carports. The size of the site is dimensioned. The building line is noted. This is the minimum distance a house can be built from the street.

The plot plan shows the elevations of the contours of the land. The elevation is the number of feet above the local datum level. The datum level is an assumed basic level above sea level. It is used in measuring heights.

Study the plot plan, Fig. 22–31. Notice the existing contours are shown with a solid line. This site has 110 feet as its highest point. It slopes to 107 feet in the front and 104 feet in the rear.

The new contours are shown with a dashed line. These are the desired levels after the site is graded. The dimensions and contours for the plot plan are obtained by a surveyor.

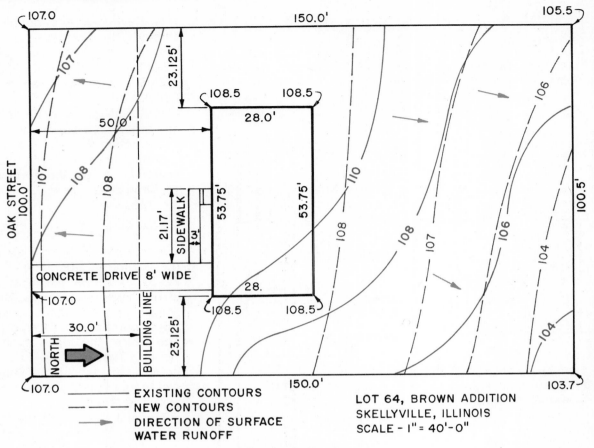

22-31. *A plot plan.*

Legend for plot plan:
- ——— EXISTING CONTOURS
- — — — NEW CONTOURS
- ——→ DIRECTION OF SURFACE WATER RUNOFF

LOT 64, BROWN ADDITION
SKELLYVILLE, ILLINOIS
SCALE - 1" = 40'-0"

DISPLAY DRAWINGS

Display drawings are made to show the proposed building in its finished form. They are used to sell a client on the plan developed by the architect. Usually they are only made for large, expensive, commercial buildings.

These drawings have a perspective of the proposed building. They show the floor plan but with little detail. The plans are made attractive by drawing trees and shrubs around the house. See Fig. 22-40.

654

Water colors and ink are used. This gives the building a realistic effect.

TYPES OF HOUSES

The following figures show elevations and floor plans for typical types of houses. Study these. Notice how the architect related the various areas. Compare the front elevation with the floor plan.

One-Story House. Fig. 22-32 shows a one-story house. This style is sometimes called a

ranch house because of a resemblance to original ranch houses. However, they tended to have a more rambling, irregular floor plan. Small city lots have forced the original ranch house into a tight floor plan.

The house, Fig. 22-32, has a brick veneer exterior. The posts on the front are like those used on early Spanish-style homes found in the southwestern United States. This house can be built with or without a basement.

National Homes Corp.

22–32A. *A one-story ranch house.*

22–32B. *Floor plan.*

National Homes Corp.

PATIO

BEDROOM
15 X 11

FAMILY ROOM
16 X 12

KITCHEN

DINING ROOM
13 X 10

STORAGE

DOWN

LINEN

LINEN

PANTRY

STORAGE

BATH

LIVING ROOM
17 X 15

GARAGE
28 X 20

BEDROOM
12 X 12

BEDROOM
12 X 10

FOYER

FLOOR PLAN FOR A ONE-STORY HOUSE

Another one-story house, Fig. 22–33, has very simple lines. Some call this contemporary styling, meaning it is designed for modern living. The planning does not follow that of older houses.

One-and-One-Half Story House.

This house is like the one-story house except the roof is higher. This gives space to build several rooms on the second floor. See Fig. 22–34. Only part of the floor area on the second floor can be used for living space. See Fig. 22–35. Usually the side walls are made five feet high.

Light and air are brought to these second floor rooms by dormer windows. These are windows cut through the roof. They have sides and a small roof built around the window.

This type of house offers a lot of floor space at a low cost. The cost is lower because the roof also serves as side walls. It takes less material than a two-story house.

The house shown in Fig. 22–34 is Cape Cod style.

Two-Story House. The two-story house, Fig. 22–36, is of frame construction. The first floor front elevation is brick veneer. It is designed after the style of colonial American houses.

Notice on the floor plan how the stairs to the basement and second floor are located. This style house has a balanced front elevation. This fact must be remembered by the architect as the room locations are planned. This type house can be built with or without a basement.

A contemporary two-story house is shown in Fig. 22–37. Notice the differences between it and the more traditional style shown in Fig. 22–36.

House with Split-Foyer Entrance. The house, Fig. 22–38, is the same as a one-story house with a basement. The basement level is raised several feet out of the ground.

This makes it possible to use regular windows in this area. It is pleasant and useful.

The entrance is about on the level of the ground. It is between the upper and lower levels. To reach the upper level, a half flight of stairs is needed. A half flight of stairs is needed to reach the lower level. See Fig. 22–39.

Split-Level House. Fig. 22–40 shows one type of split-level house. This house has three levels. The first level is the garage. It is out of the ground because the ground slopes. The second level is the living area. The third level is the sleeping area, located above the garage.

In Fig. 22–41 the common levels used on split-level houses are shown. The basement can be omitted. This was done in the house shown in Fig. 22–40.

The front entrance can be on the garage level, Fig. 22–40. It is often located on the living level.

Home Planners, Inc.

22–33A. *A one-story contemporary house.*

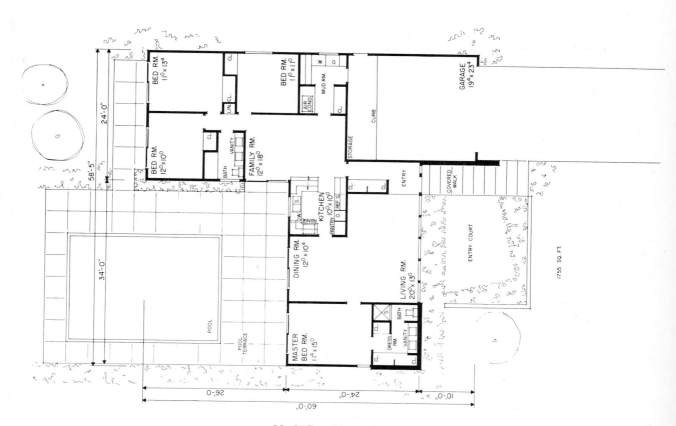

22–33B. *Floor plan.*

Home Planners, Inc.

22-34A. *A one-and-half story house. Notice the dormer windows in the roof. They provide air and light to the second floor living area.*

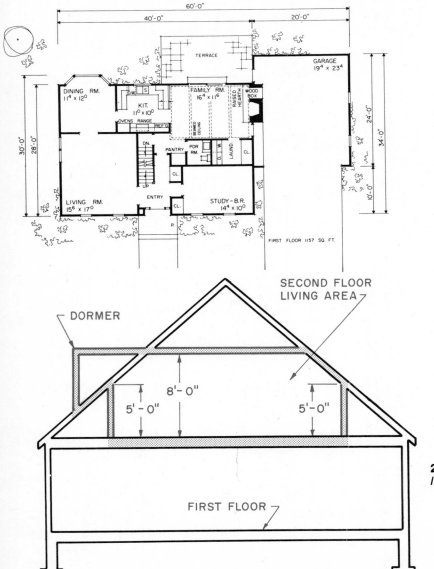

FIRST FLOOR 1157 SQ. FT.

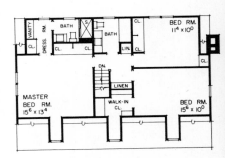

SECOND FLOOR 875 SQ. FT.

Home Planners, Inc.

22-34B. *Floor plan.*

SECOND FLOOR LIVING AREA

DORMER

8'-0"

5'-0"   5'-0"

FIRST FLOOR

22-35. *Attic space is used for extra living area in a one-and-one-half story house.*

658

22-36A. *A traditional two-story house.*

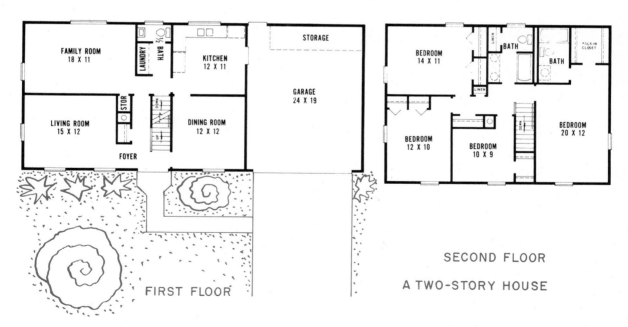

FAMILY ROOM
18 X 11

LAUNDRY

1/2 BATH

KITCHEN
12 X 11

STORAGE

STOR

LIVING ROOM
15 X 12

DINING ROOM
12 X 12

GARAGE
24 X 19

FOYER

UP

DOWN

FIRST FLOOR

BEDROOM
14 X 11

LINEN

BATH

WALK-IN CLOSET

BATH

LINEN

BEDROOM
12 X 10

BEDROOM
10 X 9

DOWN

BEDROOM
20 X 12

SECOND FLOOR

A TWO-STORY HOUSE

22-36B. *Floor plan.*

22–37A. *A contemporary two-story house.*

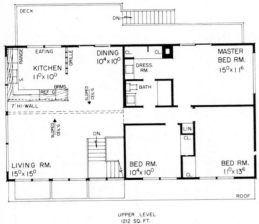

UPPER LEVEL
1212 SQ. FT.

22–37B. *Upper level floor plan.*

22–37C. *Lower level floor plan.*

LOWER LEVEL
733 SQ. FT.

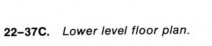

RICHARD B. POLLMAN DESIGNER · IRVING E. PALMQUIST, A.I.A.

home planners, inc.

16310 GRAND RIVER AVENUE   DETROIT 27, MICHIGAN

22-38A. *A house with a split-level entrance.*

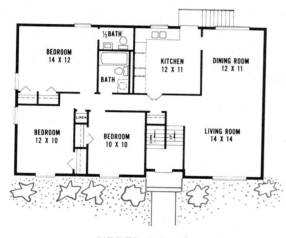

UPPER LEVEL

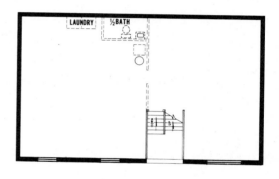

LOWER LEVEL

A HOUSE WITH A SPLIT-LEVEL ENTRANCE

22-38B. *Floor plan.*

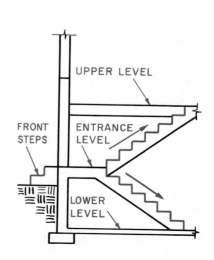

22-39. *Stairs in a split-level entrance house.*

Home Planners, Inc.

22–40A. *Front elevation of split-level house with three levels.*

22–40B. *Rear elevation.*

22–40C. *Floor plan of main and lower levels.*

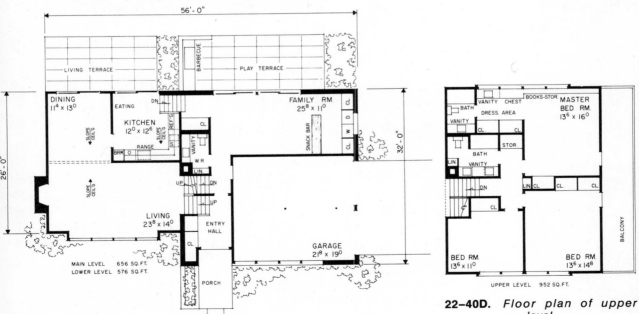

56'-0"

LIVING TERRACE

BARBECUE

PLAY TERRACE

DINING
11⁴ x 13⁰

EATING

DN.

FAMILY RM.
25⁸ x 11⁰

CL

KITCHEN
12⁰ x 12⁶

SLOPE CEIL'G

REF'G

CL

VANITY

SNACK BAR

CL

W D

RANGE

BRM O

W R

LIN

26'-0"

SLOPE CEIL'G

UP

DN.

UP

32'-0"

LIVING
23⁸ x 14⁰

ENTRY HALL

GARAGE
21⁸ x 19⁰

CL

MAIN LEVEL   656 SQ.FT.
LOWER LEVEL   576 SQ.FT.

PORCH

Floor plan of upper level. — **22–40D.**

VANITY   CHEST   BOOKS-STOR.   MASTER BED RM.
13⁶ x 16⁰

BATH   DRESS. AREA

VANITY

CL   CL

STOR.

BATH

LIN   VANITY

DN.

LIN CL   CL   CL

CL

BALCONY

BED RM.
13⁶ x 11⁰

BED RM.
13⁶ x 14⁸

UPPER LEVEL   952 SQ.FT.

RICHARD B. POLLMAN, DESIGNER • IRVING E. PALMQUIST, A.I.A.

home planners, inc.

16310 GRAND RIVER AVENUE   DETROIT 27, MICHIGAN

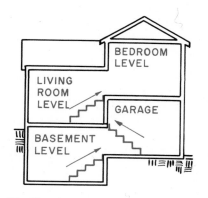

22–41. *Levels in a split-level house. The basement level can be omitted, giving a three-level house.*

Build Your Vocabulary

Following are terms that you should understand and use as a part of your working vocabulary. Write a brief explanation of what each means.

Architect
Traffic patterns
Orientation
Preliminary sketches
Floor plan
Foundation plan
Plot plan
Architectural symbols
Brick veneer
Slab construction
Elevations
Details
Display drawing
Open planning
Stair riser

Additional Reading

Ramsay, Charles G., and Harold R. Sleeper. *Architectural Graphic Standards.* New York: John Wiley and Sons, Inc., 1970.

Spence, William P. *Architecture—Design, Engineering, Drawing.* Bloomington, Ill.: McKnight and McKnight Publishing Co., 1972.

Stegman, George K., and Harry J. Stegman. *Architectural Drafting.* Chicago: American Technical Society.

Wyatt, William E. *General Architectural Drawing.* Peoria, Ill.: Chas. A. Bennett Co., Inc., 1969.

Study Problems

1. *Plan a bathroom that can be used by two people at the same time. Keep the floor area required as small as possible. Draw the floor plan for this solution. Dimension it, giving the exact location of each fixture.*

2. *Study the floor plan for the house, Fig. 22–32. On graph paper, sketch two different front elevations for this plan. You may change window size, type, and location if it will not hurt the use of the rooms.*

3. *Draw the details for the fireplace in the house, Fig. 22–34. The opening is to be four feet. The hearth is raised eight inches. Design the front elevation and mantel to suit yourself.*

4. *Study the kitchen in the house shown in Fig. 22–36. Try to redesign the kitchen so the dining area is in front of the windows. Draw the plan and elevations of the cabinets.*

5. *Design and prepare a full set of drawings for a small lakeside cottage that has the following features:*

A. Low cost materials that require little upkeep.
B. Sleeping, eating, and living area for four people.
C. Floor area as small as possible yet large enough to serve four people.
D. A minimum bathroom.
E. An inexpensive source of heat.
F. Storage for items commonly found at the lake, such as fishing gear, life preservers, and lawn furniture.
G. Electricity and water service are available.
H. Lot slopes toward the water on a ten-degree angle.

6. *Design a carport that can be built next to the cottage in Problem 5. Include a storage wall as part of the structure.*

7. *Draw a full set of working drawings for the one-story house, Fig. 22–34. The front walls are stone veneer. It is built with a concrete slab floor. The roof is tar and gravel.*

Chapter 23

Drafting in the Cabinetmaking Industry

The cabinetmaking industry is divided into several branches. *Millwork* plants include those making kitchen cabinets, window sash, and doors. Some also produce lumber for building construction purposes.

Fixture plants manufacture store and office fixtures. The term fixture is used to describe units used in stores, offices, and businesses. Examples are display shelving, tellers' windows in banks, and merchandise display cases. Some plants specialize in only one type of fixture, such as those for restaurants, clothing stores, or retail food stores.

Furniture plants manufacture household and office furniture. They make items such as living room, dining room, and bedroom suites. Some specialize in single items, as chairs.

Drafting practices in the furniture and cabinet industry differ very little from those in other industries. A common practice is to follow the drafting standards published by the Society of Automotive Engineers. Any draftsman familiar with these standards could work in the cabinetmaking industry.

There are special things about this industry a draftsman should know. He should become familiar with the methods of joining material and the finishes used. Production methods must be understood. A knowledge of fabrics, upholstery, and hardware is helpful. A wide variety of different materials are used – wood and wood products, such as particle board; plastics; and metal. The draftsman needs to know their uses and limitations. See Figs. 23–1 and 23–2.

The draftsman must become familiar with the drafting standards of the company. These can vary with the design and manufacturing techniques used.

23–1. *An executive office desk and chair. The legs are metal. The desk is a hardwood veneer with an oil finish. The top can be wood veneer or plastic laminate. All drawers use metal slides. The chair has a formed plastic shell that is upholstered.*

Herman Miller, Inc.

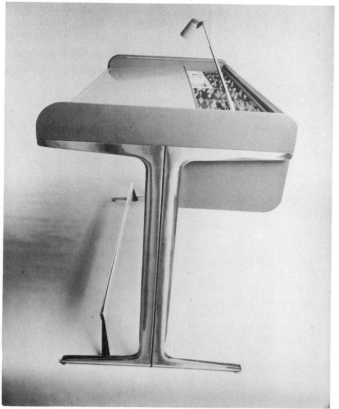

23–2. *An office desk designed for efficient use. The sides and writing top are covered with plastic laminate glued to a particle board core. The legs are polished aluminum. The file racks are hung on metal brackets. The bin into which the files hang is plywood. It has a wood top that closes over the writing surface. This has metal tracks. The edges of the top and sides are covered with an extruded plastic molding. The desk has three drawers made of molded plastic.*

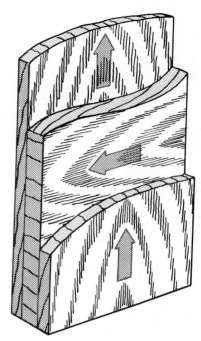

23–3. *Plywood is made by gluing thin layers of wood with the grain at 90-degree angles.*

MATERIALS

The variety of materials used is extensive. Following are some of the more common types:

Plywood is a sheet stock made by gluing thin layers of wood together. Each layer has the grain at 90 degrees to the next. Common thicknesses are $1/8$, $3/16$, $1/4$, $5/16$, $3/8$, $1/2$, $5/8$, $3/4$, 1, $1\frac{1}{8}$, and $1\frac{3}{16}$ inches. Plywood is available with a variety of hardwoods on the surface. See Fig. 23–3.

Particle board is a sheet stock made by gluing wood chips into a solid sheet. It is strong and resists warping. It is commonly available in thicknesses of $1/4$, $3/8$, $1/2$, $5/8$, $11/16$, $3/4$, 1 and $1\frac{1}{8}$ inches.

Plastic laminates are thin sheets of plastic. They are glued over plywood or particle board. They are made in a variety of colors and patterns. They are also made with simulated wood grain patterns. See Fig. 23–6.

Veneers are very thin wood sheets. They are generally $1/28$ inch thick. They are glued over plywood or particle board. Sheets made with veneer over plywood or particle board tend to warp less than solid stock. See Fig. 23–4.

Extrusions are metal or plastic parts made in long lengths.

665

23-4. *This executive desk has wood veneer sides and top. The desk and chair frames are solid wood. Notice the soft effect given by the upholstery fabric.*

23-5. *The chair frames are cast polished aluminum. They are upholstered with a plastic material or fabric. The pads are a vinyl foam. The table has an aluminum base. The top is a plastic laminate with extruded plastic molding on the edge.*

They are cut to fit the furniture. An example is the metal or plastic edging on plastic-covered cabinet tops.

Castings are parts cast from molten metal. Detailed information on castings can be found in the chapter on production drawing. The table and chair shown in Fig. 23-5 have many parts that are cast in aluminum.

Plastic panels are a new development in the furniture industry. The panel has a foamed-in-place polyurethane core. The outer surfaces are plastic laminates. The foam is injected between the laminates and expands to fill the void.

The panels are lightweight and have high strength and rigidity. The foam serves as a sound deadner.

Panels are also made with steel skins. Desk tops are made with steel skins on the top and bottom. A plastic laminate covers the edges. The foam is injected into this hollow panel. When it solidifies, the panel is rigid. Fig. 23-6 shows filing cabinets made of this material.

Molded plastic parts are used a great deal. These are made from rigid polyurethane foams, high impact polystyrenes, and polypropylenes. These are used for decorative parts and also for structural parts such as drawers. The chair shells shown in Fig. 23-7 are molded plastic. Fig. 23-8 shows molded plastic stacking chairs.

23-6. *Drawers are being installed in new filing cabinets, using panels made with urethane foam core for high strength. The outer layer is a plastic laminate.*

23-7. *The table and chair bases are aluminum with a fused plastic finish. The seat shell is plastic. The seat is upholstered.*

23-8. *Molded plastic stacking chairs with metal legs.*

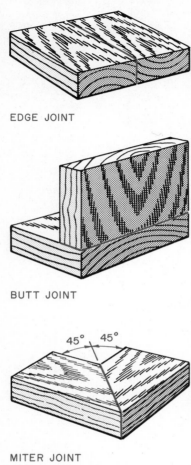

23–9. *Edge, butt, and miter joints.*

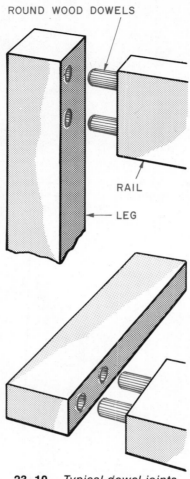

23–10. *Typical dowel joints.*

Miter joints are made by cutting each member on a 45-degree angle. No end grain is shown when it is assembled. See Fig. 23–9.

Dowel joints are rather strong. They are not as strong as a mortise and tenon. The diameter of the wood dowel pin is usually one half the thickness of the wood. They usually enter the wood to a depth that is three times their diameter. See Fig. 23–10.

The *mortise-and-tenon* joint, Fig. 23–11, is a strong joint. A common use is to join legs and rails of tables, chairs, and dressers. The *blind mortise and tenon* gives a joint that is completely hidden. The *open mortise-and-tenon* joint is easier to make but the end grain does show.

Another type of tenon is a *round tenon*. It is formed on the end of the rail. It serves the same purpose as a large dowel. However, the round tenon has a diameter almost as large as the thickness of the wood. It is not possible to do this with dowels.

A variety of *lap joints* are used. The recess is cut halfway through the member. When assembled, the surfaces are flush. They are usually glued, but they can be nailed or fastened with screws. See Fig. 23–12.

A *groove* is a rectangular slot cut with the grain of a board. The size depends upon the thickness of the member to be placed in the groove. See Fig. 23–13.

Molded plastic parts make it possible to produce period furniture having sculptured designs. They can be made to look like almost any type of wood. Production is fast and low in cost.

JOINING WOOD MEMBERS

The common joints used in wood construction are the edge, butt, miter, dowel, mortise and tenon, lap, groove, tongue, rabbet, dado, dovetail,

and box. These are used in many different ways. They are combined for certain uses. For example, a rabbet can be inserted in a dado.

The *edge joint* consists of placing two square edges together. It can be glued. Dowels can be used on the joint. A tongue and groove can be cut on the edges. See Fig. 23–9.

A *butt joint* is simple to make. It can be nailed or fastened with screws, Fig. 23–9.

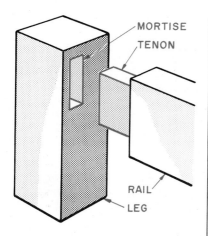

BLIND MORTISE-AND-TENON

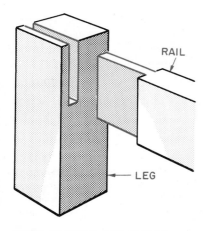

OPEN MORTISE-AND-TENON

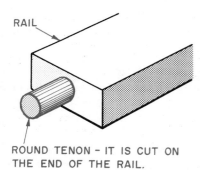

ROUND TENON – IT IS CUT ON
THE END OF THE RAIL.

23–11.   *Mortise-and-tenon joints.*

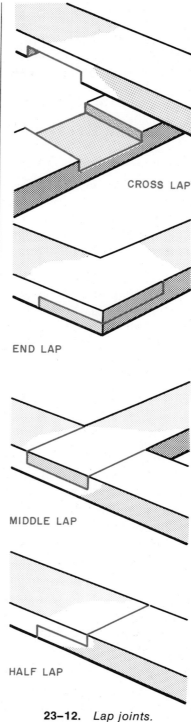

CROSS LAP

END LAP

MIDDLE LAP

HALF LAP

23–12.   *Lap joints.*

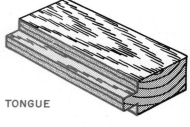

GROOVE

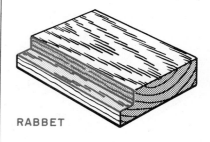

TONGUE

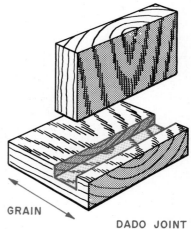

RABBET

DADO JOINT

23–13.   *Groove, tongue, rabbet, and dado joints.*

669

A *dado* is a rectangular slot cut across the grain of a board. Dadoes are usually glued. They can be strengthened with nails or screws. See Fig. 23-13.

A *tongue* is rectangular in shape. It is machined to fit in a groove. See Fig. 23-13.

A *rabbet* is made by cutting a rectangular section from the end of a board. See Fig. 23-13. It is easy to make. Rabbets are used in making boxes, cases, and drawers. The backs of cabinets are installed in rabbets.

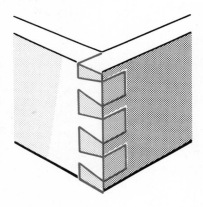

OPEN DOVETAIL

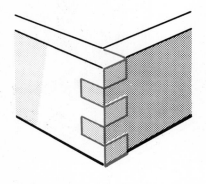

BOX JOINT

23-14. *Dovetail and box joints.*

670

Dovetail joints are used in drawer construction. They are expensive to make and very strong. High quality furniture uses this type of joint. See Fig. 23-14.

Box joints are used to join stock at the corners. They are strong and easy to make. They are more likely to slip than dovetail joints. See Fig. 23-14.

Edge Trim. Plywood and particle board with veneer or laminated plastic overlays are used in furniture construction. The edges of these are usually covered. This improves the appearance. The common ways of trimming the edges are shown in Fig. 23-15. The thin veneer shows very little from the top. Wider edging frequently is used as a part of the design of the top.

TYPICAL CONSTRUCTION DETAILS

Each manufacturing company will have standard construction practices. These are generally used by the draftsman as he details a unit. The following construction details are in general use in many companies.

Framed panels are used to form large surfaces, such as the end of a dresser. They consist of a solid stock frame built around a thin panel. The horizontal solid members are called *rails*. The vertical members are *stiles*. This panel can be hardboard, plywood, or some other thin sheet stock. The rails and stiles are

WOOD VENEER OR PLASTIC LAMINATE.

GLUE VENEER OR PLASTIC LAMINATE.

WOOD EDGING WITH TONGUE AND GROOVE JOINT.

WOOD BLOCK EDGE GLUED. CAN BE SQ. OR A MOLDING.

MOLDING WITH TONGUE AND GROOVE JOINT.

GROOVED MOLDING.

MITERED BLOCK WITH VENEER OR PLASTIC LAMINATE.

A WIDE VARIETY OF METAL AND PLASTIC MOLDINGS ARE AVAILABLE.

23-15. *Some methods of trimming the edges of plywood and particle board.*

grooved to hold the panel. See Fig. 23-16 for typical details.

Back panels are made from thin sheet stock. They are set in a rabbet. A common use is on the back of a dresser. They keep dust from entering the drawers. See Fig. 23-17.

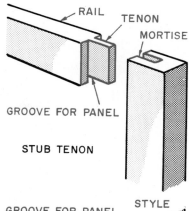

STUB TENON

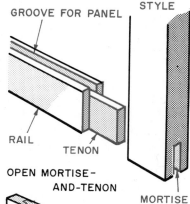

OPEN MORTISE-
AND-TENON

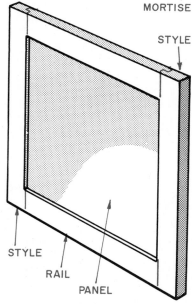

FRAMED PANEL

23-16. *Framed panel construction.*

Rails are solid stock members used to join end panels. This makes the basic shell of a unit. Drawers or doors are placed between the rails. See Fig. 23–18 for typical details. Table legs are also joined with rails.

Solid end panels are usually made from plywood or particle board sheets. They can be made from solid stock. See Fig. 23–19.

Web frames are used to connect framed or solid end panels. These form the basic unit in a manner similar to rails. They can be joined using a groove or mortise. On high quality furniture the web frame has a panel. This protects each shelf or drawer from dust. See Fig. 23–19.

A *face frame* is made from solid stock. It is placed over the end grain of solid end panels. Door hinges can be fastened to this frame. See Fig. 23–19.

Drawers can be made many ways. Some of the most common details are shown in Fig. 23–20. The two common types of drawers are lip and flush.

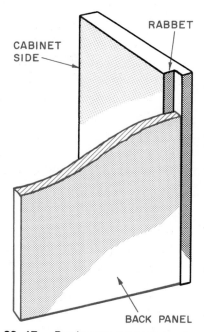

23-17. *Backs are set into a rabbet.*

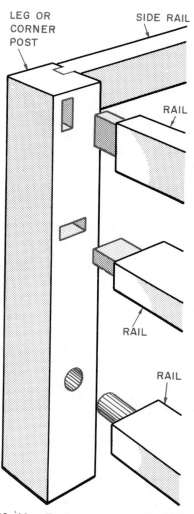

23-18. *Typical ways to join rails to legs or corner posts.*

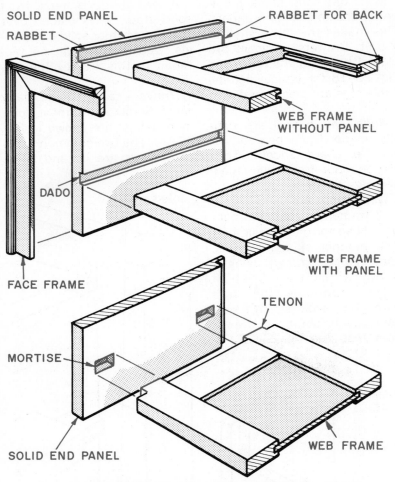

SOLID END PANEL

RABBET

RABBET FOR BACK

WEB FRAME
WITHOUT PANEL

DADO

FACE FRAME

WEB FRAME
WITH PANEL

TENON

MORTISE

SOLID END PANEL

WEB FRAME

ALTERNATE METHOD FOR JOINING WEB FRAME
TO SOLID END PANEL.

23–19. *Web frames can join solid end panels by cutting a dado or mortise in the end panel. Web frame can be used with or without an insert.*

Drawers usually slide on the web frame. Metal-roller-bearing drawer runners are available for carrying heavy loads. See Fig. 23–21. Usually drawers without metal slides have guides in the center. This helps them slide without binding. See Fig. 23–22.

There are three common

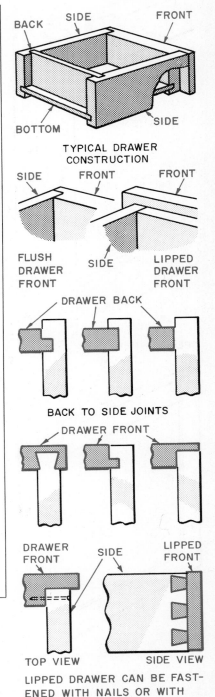

BACK

SIDE

FRONT

SIDE

BOTTOM

SIDE

TYPICAL DRAWER
CONSTRUCTION

SIDE

FRONT

FRONT

FLUSH
DRAWER
FRONT

SIDE

LIPPED
DRAWER
FRONT

DRAWER BACK

BACK TO SIDE JOINTS

DRAWER FRONT

DRAWER
FRONT

SIDE

LIPPED
FRONT

TOP VIEW

SIDE VIEW

LIPPED DRAWER CAN BE FAST-
ENED WITH NAILS OR WITH
DOVETAIL JOINTS.

FRONT TO SIDE JOINTS

23–20. *Typical methods of drawer construction.*

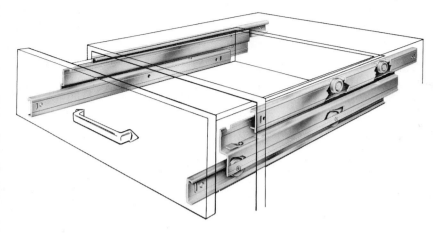

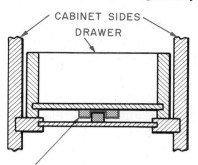

WOOD DRAWER GUIDE RUNS
THE LENGTH OF THE DRAWER.
IT IS GLUED TO THE DRAWER
BOTTOM.

CENTER DRAWER GUIDE

Knape and Vogt Manufacturing Co.

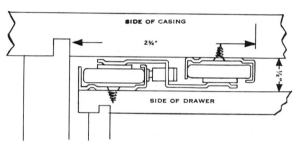

23–21. *One type of metal drawer slide. These units permit the drawer to slide easily and quietly. There are many other types available.*

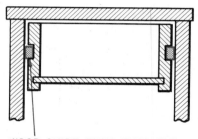

WOOD GUIDE RUNS THE LENGTH
OF THE DRAWER. DRAWER SIDE
IS GROOVED.

SIDE DRAWER GUIDE

23–22. *Typical wood drawer guides.*

types of doors. These are flush, overlay, and lip. See Fig. 23–23. The lip is cut to fit the type of hinge to be used.

Legs and rails are joined in many ways. Some of the common ways are with mortise-and-tenon joints and with dowel joints. See Fig. 23–24. Corner brackets with a hanger bolt are also used. See Fig. 23–25.

A number of types of metal fasteners are used to join

legs to a cabinet frame. Fig. 23–26 shows one of these.

Tops can be attached in many ways. Some of the common methods are shown in Fig. 23–27.

Sliding doors are used a great deal in furniture and cabinet construction. The doors can be lipped and slid in grooves cut in the wood. A variety of metal and plastic slides are also available. See Fig. 23–28.

Laminated and Molded Wood Parts. Curved wood parts can be formed by cutting them from solid stock. This usually wastes material. The part produced is often weak. See Fig. 23–29.

Many curved parts are formed by softening solid stock and bending it to shape. The stock is clamped in a form until it has dried.

Another method is to laminate the part and form it.

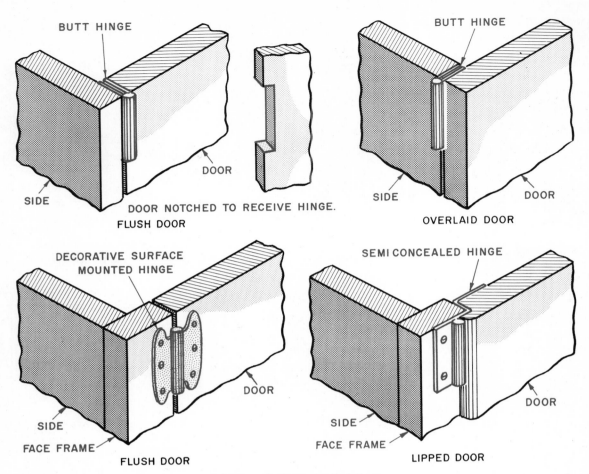

BUTT HINGE

SIDE

DOOR

DOOR NOTCHED TO RECEIVE HINGE.

FLUSH DOOR

BUTT HINGE

SIDE

DOOR

OVERLAID DOOR

DECORATIVE SURFACE
MOUNTED HINGE

SIDE

FACE FRAME

DOOR

FLUSH DOOR

SEMI CONCEALED HINGE

SIDE

FACE FRAME

DOOR

LIPPED DOOR

23-23. *Typical types of doors.*

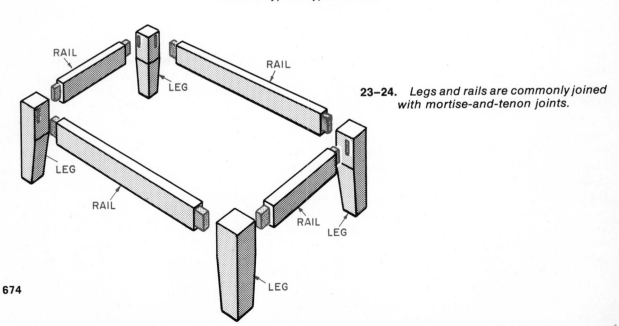

RAIL

LEG

RAIL

LEG

LEG

RAIL

RAIL

LEG

LEG

23-24. *Legs and rails are commonly joined with mortise-and-tenon joints.*

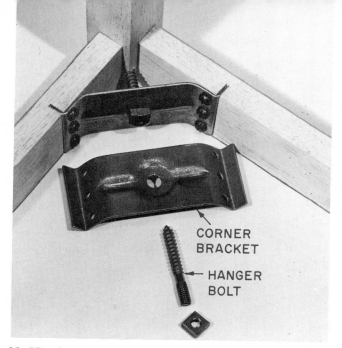

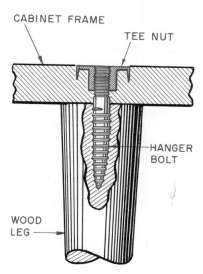

23-26A. *Wood legs can be fastened with a hanger bolt and tee nut.*

23-25. *Legs and rails can be joined with a corner bracket and hanger bolt.*

23-26B. *A hanger bolt and tee nut.*

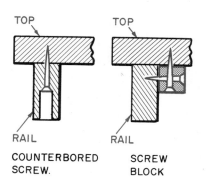

COUNTERBORED SCREW.

SCREW BLOCK

RABBETED BLOCK WITH GROOVE IN RAIL.

METAL BRACKET WITH GROOVE RAIL.

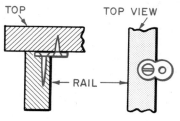

METAL BRACKET IN RECESS SET IN RAIL.

23-27. *Typical methods of joining a top to rails.*

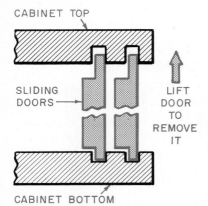

CABINET TOP

SLIDING DOORS

LIFT DOOR TO REMOVE IT

CABINET BOTTOM

SECTION THRU SLIDING CABINET DOORS.

THE GROOVES IN THE TOP ARE CUT DEEPER SO THE DOORS CAN BE REMOVED BY LIFTING THEM OUT OF THE BOTTOM GROOVE.

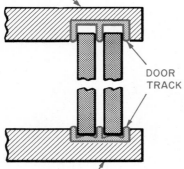

CABINET TOP

DOOR TRACK

CABINET BOTTOM

DOOR TRACKS ARE MADE OF PLASTIC OR METAL

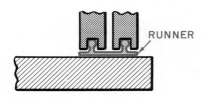

RUNNER

RUNNER FOR BOTTOM OF SLIDING DOORS.
TOP OF DOOR CAN SLIDE IN A GROOVE OR TRACK.

23-28. *Methods for installing sliding doors.*

676

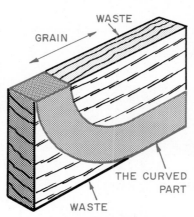

WASTE

GRAIN

THE CURVED PART

WASTE

LAYOUT OF CURVED PART FROM SOLID STOCK.

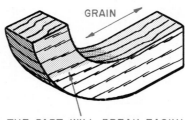

GRAIN

THE PART WILL BREAK EASILY WITH THE GRAIN.

THE FINISHED PART

23-29. *Curved parts can be cut from solid stock.*

Laminating is the process of forming parts by gluing several thin layers of wood together. The grain of the wood in each layer runs in the same direction. See Fig. 23-30.

The thin layers of wood are coated with glue and clamped in a form before they dry. The form plus the pressure of the clamps holds the part in shape. After the glue has dried, the part keeps this shape. The formed part can

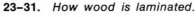

23-31. *How wood is laminated.*

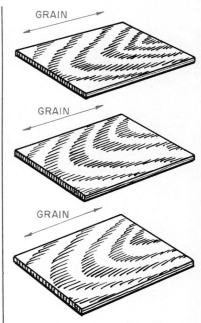

GRAIN

GRAIN

GRAIN

23-30. *Wood laminate layers have the grain in the same direction.*

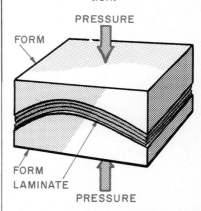

PRESSURE

FORM

FORM

LAMINATE

PRESSURE

THE LAYERS TO BE FORMED ARE GLUED AND PLACED IN FORMS UNDER PRESSURE TO DRY.

THE FINISHED FORMED WOOD LAMINATE.

Knoll Associates, Inc.

23-32. *Chair with a laminated, formed wood frame. Notice how the back leg joins the frame and bends at the top of the chair. The seat and back are upholstered over a molded plastic shell.*

GRAIN

GRAIN

GRAIN

23-33. *Molded layers have the grain in 90-degree angles to the previous layer.*

23-34. *The seat and back on this chair are molded plywood.*

Herman Miller, Inc.

then be sawed and smoothed to final shape. See Fig. 23–31. The chair in Fig. 23–32 has a laminated wood frame.

Some parts are formed by a molding process. Thin layers of wood are glued together. Each layer has the grain direction turned at right angles to the next. This is the same as plywood. See Fig. 23–33. The entire part is clamped in a mold under great pressure. This process is used to form large pieces, as chair seats. See Fig. 23–34. The unit is formed with curves in any direction needed. Laminated parts only bend in simple curves.

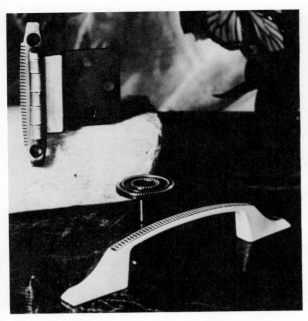

23–35A. *Classically styled cabinetware for use on a wide range of furniture and cabinets.*

23–35B. *Traditional-design cabinetware.*

Amerock Corporation

23–35C. *Cabinetware reflecting the simple, graceful Mediterranean motif.*

23–35D. *These units utilize the old Spanish designs for today's furniture and cabinets.*

23–35. *Furniture and cabinet hardware is available in a wide variety of styles.*

Hardware. A wide variety of mass-produced hardware is available. If a furniture manufacturer has enough need to justify the expense, he can design and make special hardware. Hardware for furniture includes handles, knobs, drawer pulls, hinges, door catches, drawer slides, and shelf brackets.

A few examples of handles, knobs, and hinges are shown in Fig. 23–35.

One type of metal drawer slide is shown in Fig. 23–21. This is especially useful if a drawer is large or is meant to hold heavy loads.

Shelf brackets are made so shelves can be easily adjusted to a variety of heights. Two types are shown in Figs. 23–36 and 23–37.

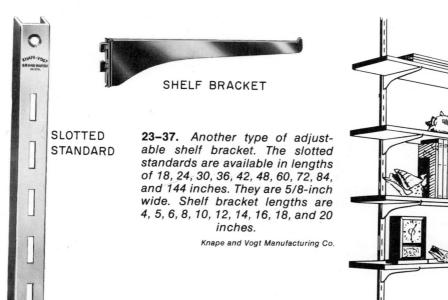

23–36. *Metal adjustable shelf standards and clips can be installed on the surface or recessed. The metal clips that hold the shelf are easily moved. The standards are available in lengths of 24, 30, 36, 42, 48, 60, 72, 84, and 144 inches. They are 13/16-inch wide. The clips are one inch long and 13/16-inch wide.*

Knape and Vogt Manufacturing Co.

SURFACE MOUNTED STANDARD

SHELF CLIP

RECESSED STANDARD

SHELF BRACKET

SLOTTED STANDARD

23–37. *Another type of adjustable shelf bracket. The slotted standards are available in lengths of 18, 24, 30, 36, 42, 48, 60, 72, 84, and 144 inches. They are 5/8-inch wide. Shelf bracket lengths are 4, 5, 6, 8, 10, 12, 14, 16, 18, and 20 inches.*

Knape and Vogt Manufacturing Co.

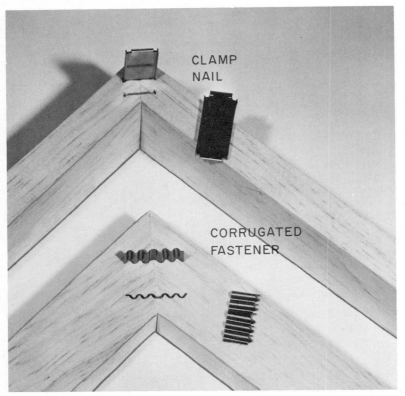

CLAMP NAIL

CORRUGATED FASTENER

23-38. *Corrugated fasteners and clamp nails are used to join miter and butt joints.*

23-39. *The back support strip on a sofa frame is stapled in position. The pneumatic staple nailer drives a two inch, coated, 16-gage staple.*

Fastener Corp.

Metal Fasteners. The most common metal fasteners are *wood screws* and *nails.* These are discussed in the chapter on fastening devices.

Corrugated fasteners are used to join miter joints and butt joints. They are quickly applied. Since they show, they are used on rough work or on a joint that is hidden. See Fig. 23-38.

Steel clamp nails are used to hold butt and miter joints. A saw cut is made to prevent splitting the wood. See Fig. 23-38.

Hanger bolts have wood screw threads on one end and machine threads on the other. See Fig. 23-26. A common use is to fasten legs in place. The wood screw threads are screwed into the leg. The machine thread end is placed through a hole in the cabinet bottom. A nut is used to hold the leg to the cabinet. Special nuts, called tee nuts, can be used to fasten legs in place. These are set on a counterbored hole. They are then flush with the surface. See Fig. 23-26.

Corner brackets are used to join rails to legs. See Fig. 23-25. The corner bracket is set into the rail in a saw cut. It is joined to the rail with wood screws. A hanger bolt is used to join it to the leg.

Power-driven hand-operated staplers are used a great deal. They drive staples as long as two inches. The staples can be driven rapidly and are strong. See Fig. 23-39.

Nailing machines are used in mass production of furniture. They hold the wood members together while they drive the nail. See Fig. 23–40.

Finishes. The designer must be familiar with the many finishes available. This includes wood and metal finishing. The finish is a very important part of the total design. It has a great influence on the final appearance.

Fabrics. The choice of fabrics available to the furniture designer is almost unlimited.

Various types of plastic upholstery material are used. These have a wide variety of colors and patterns. See Fig. 23–41.

Cloth fabrics in common use include wool, cotton, and nylon. Many fabrics are a combination of these. The colors and patterns used are carefully considered by the designer.

A designer can choose the fabric from hundreds of materials already produced; or he can follow the trend of designing special fabrics for the unit being planned. Some furniture manufacturing companies have their own special line of fabric designs.

FURNITURE SIZES

Size is an important consideration in designing furniture or cabinets. Units must be large enough to serve their purpose, but no larger than needed. For example, a table

Auto-Nailer Company

23–40. *This nailing machine face-nails narrow wood and plastic molding and thin overlays. It makes and drives small nails from coils of threaded wire. It is operated by air cylinders.*

23–41. *These chairs are upholstered with a plastic material. The material is strong and easily cleaned.*

Jens Risom Design, Inc.

designed for typing must be the proper height for typewriter operation. The top must be large enough to hold the typewriter and written material used by the typist.

A chair must be designed to fit the human body. The height of the seat from the floor is important. The size of the seat must be decided. The angle of the back must be comfortable. A chair designed for dining might need a different back angle than one for lounging.

Base cabinets for use in a kitchen must be designed for the average person. Usually work performed at these cabinets is done standing. This influences the height of the top.

It can be seen from these examples that the size of the human body must be considered by the designer. Fig. 23–42 shows the sizes of men and women in sitting and standing positions.

These sizes help the designer decide such questions as chair and table height, cabinet top height, and depth of a cabinet. Industrial units, such as a telephone switchboard, use these same design sizes. The switchboard operator must have the switchboard within easy reach.

When seated, the torso of the average man is 16 to 19 inches from the floor. See Fig. 23–42. Thus, the recommended seat height is 16 to 18 inches. Notice in Fig. 23–42 that a seat height of 17 inches is best for 80 per cent of the male population. If the seat can be adjusted one inch up and down, it will suit 95 per cent of the male population. The designer must use these basic sizes in reaching a decision on the height he wants to use. The items being designed must fit not only the average man but be usable by the large and small man and woman.

After a unit has been designed, the best test is to build a full-scale model and have people of various sizes try it out. In this way, the two-dimensional solution presented by a drawing can be checked.

Fig. 23–43 gives some examples of acceptable design sizes. They show typical overall sizes of selected units. These were designed by experienced designers. They do not represent standards, but can be used as a guide when designing furniture.

FURNITURE DESIGN PROCEDURE

The procedures used to design furniture will vary within the industry. The following is a typical plan:

The management makes a decision to produce a new unit. This could be a single item such as a chair. It could be an entire series of furniture of the same style.

The designer prepares *sketches* of possible units. These are examined by the merchandising and sales staff. After a design has been approved, specifications are written. Study Fig. 23–44, showing a designer's sketch that includes specifications. Notice the items in the specifications.

Next, the designer prepares a *design drawing,* Fig. 23–45. This is a full-size drawing on which the designer interprets his design sketch. He decides on the size of the unit, the seat height, and depth. The angle of the back is drawn. The curve of the arm is finalized. Some construction details are shown. These are not complete. The draftsman preparing the production drawing has to make some of these decisions. This design drawing is used as a guide for the engineering staff when they prepare construction details.

Some companies send this design drawing to their experimental shop. Here a full-size model is built. During this process, the designer may make some changes. These may be styling changes, or changes that will make the chair easier to build.

After the model is complete and approved, the *production drawings* are made. The draftsman works from the model and design drawings. Many dimensions are measured directly from the model. The design drawings are used as a second reference. Production drawings for the chair in Fig. 23–44 are shown in Fig. 23–46.

Fig. 23–47 shows a group of completed furniture units as they are placed on the market. This group is shown on pages 692–694.

PRODUCTION DRAFTING PROCEDURES

The production drawings made will vary with the standards of the company. Some make their production drawings *to scale.* Others make them *full size.* Both methods are explained on the following pages. See pages 695–702.

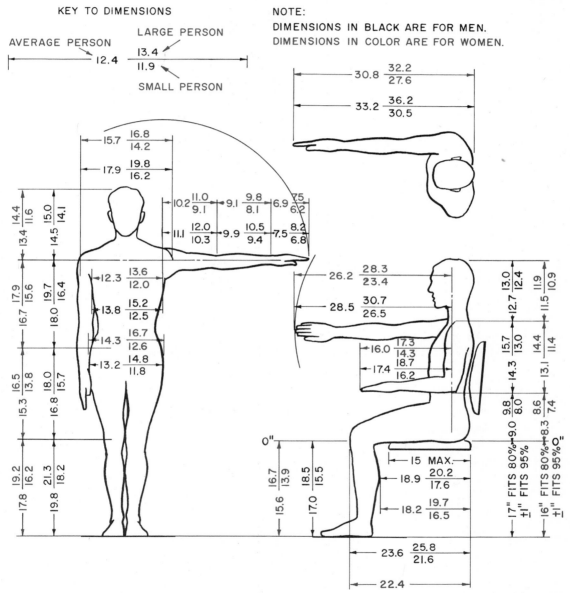

Data *from* The Measure of Man *by Henry Dreyfuss, Whitney Library of Design, New York*

23–42. *Measurements for men and women when standing and seated.*

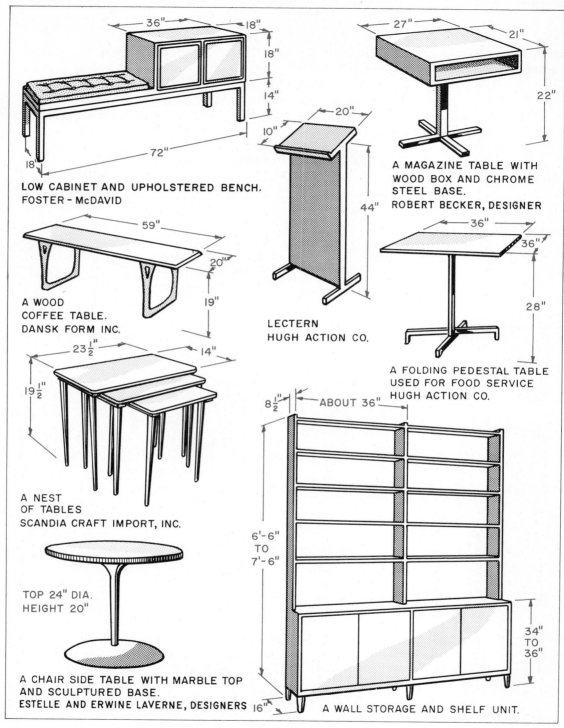

36"
18"
18"
14"
72"
18"
10"

LOW CABINET AND UPHOLSTERED BENCH.
FOSTER – McDAVID

27"
21"
22"

A MAGAZINE TABLE WITH
WOOD BOX AND CHROME
STEEL BASE.
ROBERT BECKER, DESIGNER

59"
20"
19"

A WOOD
COFFEE TABLE.
DANSK FORM INC.

20"
44"

LECTERN
HUGH ACTION CO.

36"
36"
28"

A FOLDING PEDESTAL TABLE
USED FOR FOOD SERVICE
HUGH ACTION CO.

23½"
14"
19½"

A NEST
OF TABLES
SCANDIA CRAFT IMPORT, INC.

8½"
ABOUT 36"
6'-6"
TO
7'-6"
34"
TO
36"

TOP 24" DIA.
HEIGHT 20"

A CHAIR SIDE TABLE WITH MARBLE TOP
AND SCULPTURED BASE.
ESTELLE AND ERWINE LAVERNE, DESIGNERS
16"

A WALL STORAGE AND SHELF UNIT.

The Furniture Forum

23–43A. *Size is an important factor in designing furniture units.*

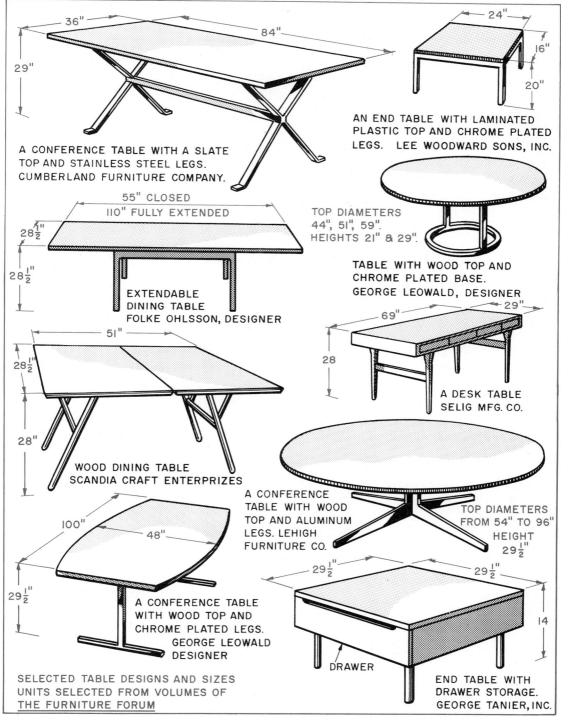

36"

84"

29"

A CONFERENCE TABLE WITH A SLATE
TOP AND STAINLESS STEEL LEGS.
CUMBERLAND FURNITURE COMPANY.

24"

16"

20"

AN END TABLE WITH LAMINATED
PLASTIC TOP AND CHROME PLATED
LEGS. LEE WOODWARD SONS, INC.

55" CLOSED

110" FULLY EXTENDED

$28\frac{1}{2}$"

$28\frac{1}{2}$"

EXTENDABLE
DINING TABLE
FOLKE OHLSSON, DESIGNER

TOP DIAMETERS
44", 51", 59".
HEIGHTS 21" & 29".

TABLE WITH WOOD TOP AND
CHROME PLATED BASE.
GEORGE LEOWALD, DESIGNER

51"

$28\frac{1}{2}$"

28"

WOOD DINING TABLE
SCANDIA CRAFT ENTERPRIZES

69"   29"

28

A DESK TABLE
SELIG MFG. CO.

100"

48"

$29\frac{1}{2}$"

A CONFERENCE TABLE
WITH WOOD TOP AND
CHROME PLATED LEGS.
GEORGE LEOWALD
DESIGNER

A CONFERENCE
TABLE WITH WOOD
TOP AND ALUMINUM
LEGS. LEHIGH
FURNITURE CO.

TOP DIAMETERS
FROM 54" TO 96"
HEIGHT
$29\frac{1}{2}$"

$29\frac{1}{2}$" $29\frac{1}{2}$"

14

DRAWER

END TABLE WITH
DRAWER STORAGE.
GEORGE TANIER, INC.

SELECTED TABLE DESIGNS AND SIZES
UNITS SELECTED FROM VOLUMES OF
THE FURNITURE FORUM

23–43B. *Selected table designs and sizes.*

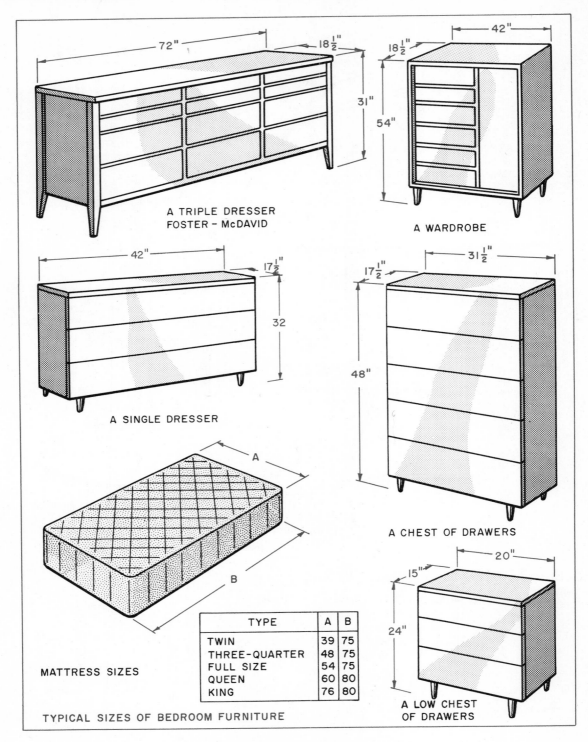

A TRIPLE DRESSER
FOSTER – McDAVID

72"

18½"

31"

A WARDROBE

42"

18½"

54"

A SINGLE DRESSER

42"

17½"

32"

A CHEST OF DRAWERS

31½"

17½"

48"

MATTRESS SIZES

A

B

TYPE	A	B
TWIN	39	75
THREE-QUARTER	48	75
FULL SIZE	54	75
QUEEN	60	80
KING	76	80

A LOW CHEST
OF DRAWERS

20"

15"

24"

TYPICAL SIZES OF BEDROOM FURNITURE

23–43C. *Typical sizes of bedroom furniture.*

WOOD LEGS, UPHOLSTERED
SEAT AND BACK.
BRENDAN REILLY ASSOCIATES

SEAT 17½" HIGH
BACK 39" HIGH

WOOD FRAME ROCKING CHAIR.
HANS OLSEN, DESIGNER

WOOD FRAME SIDE CHAIR.
SCANDIA CRAFT IMPORT, INC.

AN "EGG" CHAIR UPHOLSTERED
WITH ALUMINUM BASE.
FRITZ HANSEN, INC.

HIGH BACK LOUNGE CHAIR WITH
WOOD FRAME AND UPHOLSTERED
SEAT, BACK AND ARMS.
BORGE MOGENSEN, DESIGNER

STEEL FRAME
STACKING CHAIR
SEAT UPHOLSTERED.
G. STEINER, DESIGNER

SEAT 19½" HIGH

UPHOLSTERED CHAIR WITH
WOOD LEGS.
SYLVE STENQUIST, DESIGNER

WOOD FRAME SIDE CHAIR.
TORBJORN AFDAL, DESIGNER.

WOOD FRAME LOUNGE CHAIR.
TORBJORN AFDAL, DESIGNER.

A VARIETY OF CHAIR
STYLES AND SIZES

The Furniture Forum

23–43D. *Chairs in various styles and sizes.*

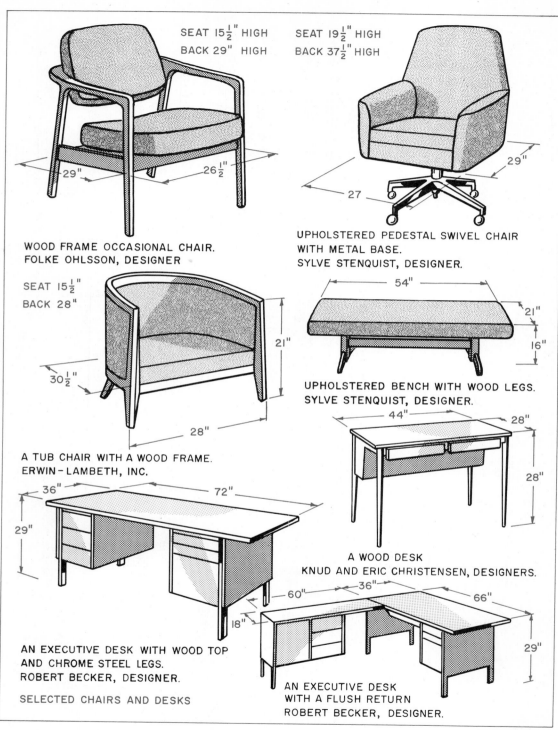

SEAT 15½" HIGH
BACK 29" HIGH

SEAT 19½" HIGH
BACK 37½" HIGH

29"

26½"

29"

27

WOOD FRAME OCCASIONAL CHAIR.
FOLKE OHLSSON, DESIGNER

UPHOLSTERED PEDESTAL SWIVEL CHAIR
WITH METAL BASE.
SYLVE STENQUIST, DESIGNER.

SEAT 15½"
BACK 28"

54"

21"

16"

21"

30½"

28"

UPHOLSTERED BENCH WITH WOOD LEGS.
SYLVE STENQUIST, DESIGNER.

A TUB CHAIR WITH A WOOD FRAME.
ERWIN – LAMBETH, INC.

44"

28"

28"

36"

72"

29"

A WOOD DESK
KNUD AND ERIC CHRISTENSEN, DESIGNERS.

60"

36"

66"

18"

29"

AN EXECUTIVE DESK WITH WOOD TOP
AND CHROME STEEL LEGS.
ROBERT BECKER, DESIGNER.

SELECTED CHAIRS AND DESKS

AN EXECUTIVE DESK
WITH A FLUSH RETURN
ROBERT BECKER, DESIGNER.

The Furniture Forum

23–43E. *Selected chairs and desks.*

FAMILY ROOM NO. 764
OCCASIONAL BRUNCH CHAIR
ATTACHED PILLOW BACK - GROUND POLY
FAB - H-29616-8 NPZ - 54 YDS.

TEMP. NO. *764*   PERM NO. *B 176*

APPROVED SALES DIV.	DATE	APPROVED DESIGN DIV. *E.S.C.*	DATE *8-24-67*
RAIL LENGTH *23"*	FLOUNCE —	STRIPPED DECK —	
BACK CONST. *NO-SAG*	BUTTON SIZE *NO. 22*	SELF DECK *OPTIONAL (NO)*	
BACK FILLING *GROUND POLY*	FABRIC *H-29616-8NP2*	PADDED O.S. BACK *YES*	
SEAT CONST. *NO-SAG*	CUSHION *K FORM*	PADDED O.S. ARM *NO*	
ARM CONST. *COTTON*	LEG HEIGHTS *SAME AS A-40-18*	FINISH *GUM NO. 20*	

DIVISION *SIGNATURE CHAIRS*

23–44. *This designer's sketch includes specifications. They are a guide for the preparation of the full-size design drawing. See Fig. 23–45.*

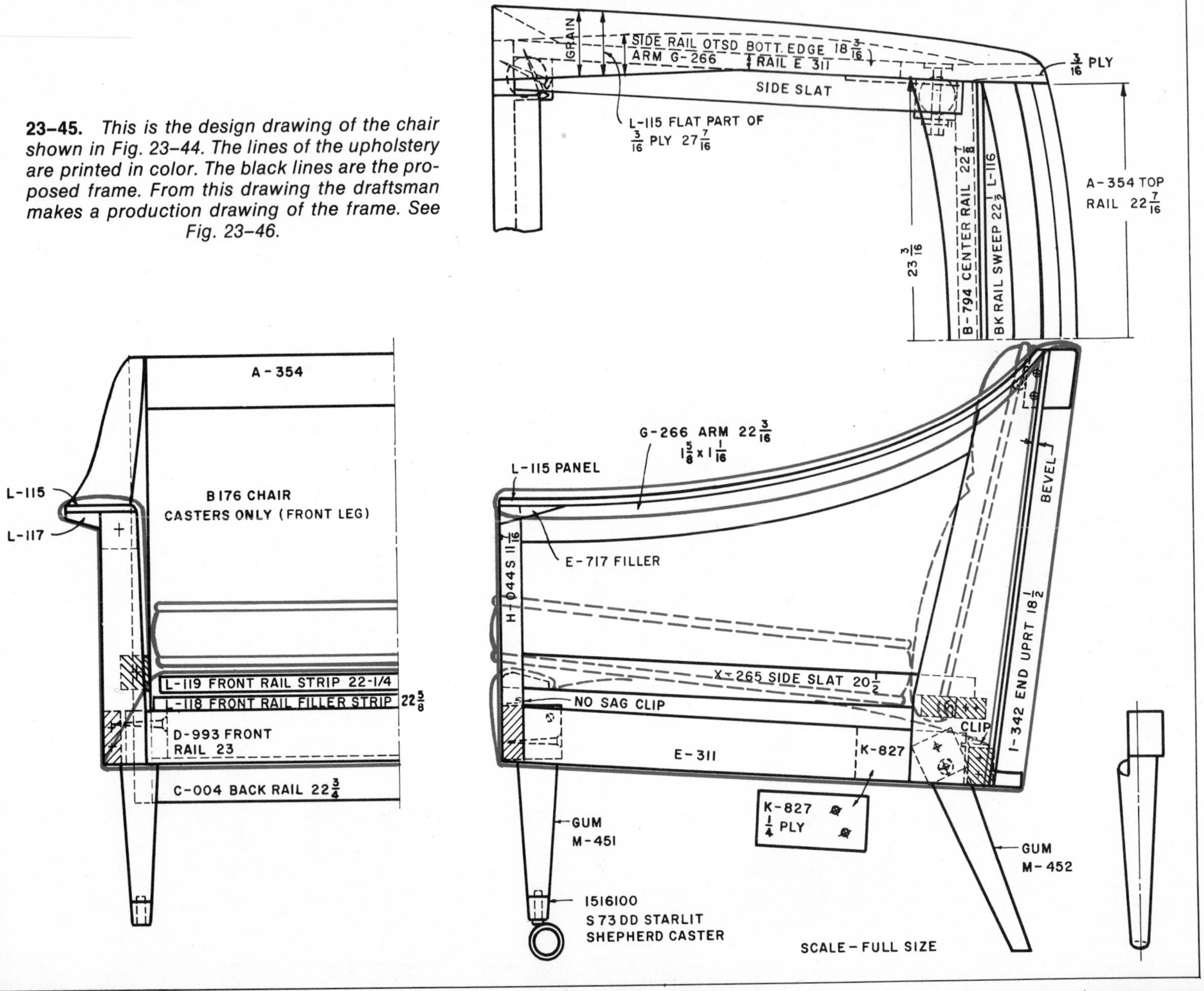

23-45. This is the design drawing of the chair shown in Fig. 23-44. The lines of the upholstery are printed in color. The black lines are the proposed frame. From this drawing the draftsman makes a production drawing of the frame. See Fig. 23-46.

Kroehler Manufacturing Co.

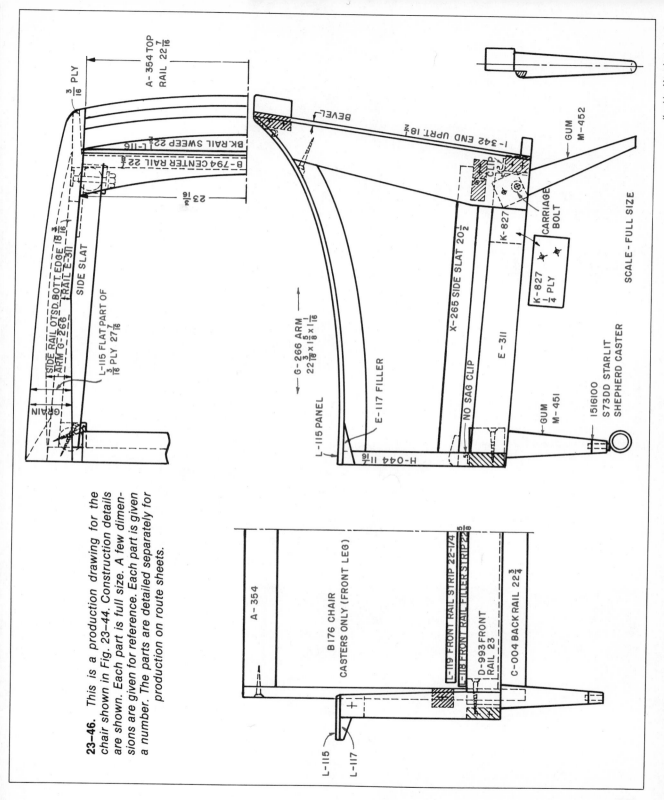

23-46. This is a production drawing for the chair shown in Fig. 23-44. Construction details are shown. Each part is full size. A few dimensions are given for reference. Each part is given a number. The parts are detailed separately for production on route sheets.

A-354 TOP RAIL 22 7/16

$\frac{3}{16}$ PLY

BK.RAIL SWEEP 22 $\frac{1}{16}$ L-116

B-794 CENTER RAIL 22 $\frac{1}{16}$

SIDE SLAT

RAIL E-311

23 $\frac{1}{16}$

SIDE RAIL OTSD.BOTT.EDGE 18 $\frac{3}{16}$

L-115 FLAT PART OF $\frac{3}{16}$ PLY 27 $\frac{7}{16}$

ARM G-266

SIDE RAIL ARM G-266

GRAIN

BEVEL

1-342 END UPRT 18 $\frac{3}{4}$

GUM M-452

CLIP

K-827

CARRIAGE BOLT

X-265 SIDE SLAT 20 $\frac{1}{2}$

K-827 $\frac{1}{4}$ PLY

SCALE - FULL SIZE

G-266 ARM 22 $\frac{3}{18}$ × 1 $\frac{5}{8}$ × 1 $\frac{1}{16}$

E-117 FILLER

L-115 PANEL

NO SAG CLIP

E-311

GUM M-451

H-044 11 $\frac{1}{16}$

L-115

1516100 S73DD STARLIT SHEPHERD CASTER

A-354

B176 CHAIR CASTERS ONLY (FRONT LEG)

L-119 FRONT RAIL STRIP 22-174

L-118 FRONT RAIL FILLER STRIP 22 $\frac{5}{8}$

D-993 FRONT RAIL 23

C-004 BACK RAIL 22 $\frac{3}{4}$

L-115

L-117

Knoll Associates, Inc.

23-47A. An office storage unit designed to serve in an executive office. It contains a file drawer, shallow drawers, and shelving. The end panels are solid. The square legs are steel. It has sliding doors.

23-47B. A durable, distinctive table and chair designed for use on a patio. The table has a wood top and cast aluminum base. The chair frame is of round welded bar stock. The seat and back are of formed, welded, round steel wire. The seat cushion is removable. It is covered with a vinyl fabric.

Knoll Associates, Inc.

23-47C. A conference table with a wood veneer top. The base is cast, polished aluminum. The bases of the chairs in the background are designed to match the table. They are also cast aluminum.

Herman Miller, Inc.

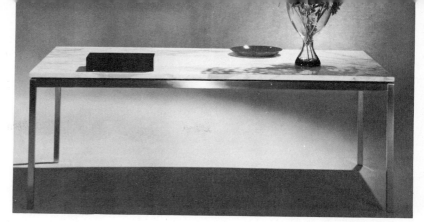

23–47D. *The legs and frame of this table are square steel bars. The top is marble.*

23–47E. *A six-place library table with study carrels. The table is framed of wood. The top and the carrel shelves are covered with a plastic laminate. The chairs have wood frames and nylon uphol-stered seats.*

23–47F. *A pleasing union of wood and metal makes a strong yet delicate table. The top is redwood. The legs are steel pipe.*

693

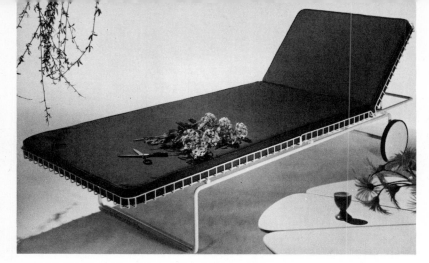

23–47G. *Outdoor furniture must be able to withstand hard use and weather. This unit has a pipe frame. The bed is of welded metal wire. The cushions are removable.*

Knoll Associates, Inc.

23–47H. *The bases of these tables are made of steel welded wire. It is strong and graceful. The glass top on the small table permits the wire pattern to be visible.*

Knoll Associates, Inc.

Jens Risom Design, Inc.

23–47I. *Secretarial typing desk. The supplies are stored in a swing-out compartment. This unit matches the executive cabinet.*

Jens Risom Design, Inc.

23–47J. *An executive cabinet. It provides storage and filing space in an executive office. The cabinet is made of walnut. It has an oil finish.*

PRODUCTION DRAFTING PROCEDURES

The following production drafting procedures are based on those used by Herman Miller, Inc., Zeeland, Michigan.

HERMAN MILLER INC.
TECHNICAL CENTER

STANDING INSTRUCTIONS FOR _____ PRODUCTION

SUBJECT: _____ DRAFTING PROCEDURE

1. Technical Center shall be responsible for production drawings and specifications for all parts, both purchased and manufactured.
2. Prints of drawings shall be furnished to the various departments according to the Technical Center distribution lists and to the Scheduling Department as requested. Prints shall be issued as parts are released for production and as revisions and changes are made to the drawings.
3. The drawing of a part shall be as it appears after it has gone through the last machining operation.
4. *Manufactured Parts*
   4.1   The process routing, unit equivalent, and rough sizes will be furnished by the Production Engineering Department on a separate routing sheet which will accompany the production drawing as the part is manufactured or processed.
   4.2   Drawings will not be made of veneers, crossbands, cores, edge bands, and core strips. These parts will be shown on the composite drawing.
   4.3   The "B" size paper (11 × 17) will be used whenever possible as long as crowding can be avoided.
   4.4   All drawings shall include a part number as specified in company standards. In the case of a laid-up panel, the individual parts of the panel shall have the part number plus one of the following suffixes:
   A. First ply, veneer, and first and last ply when identical
   B. Last ply, veneer (when different than 1st ply)
   C. Center ply, veneer (less than 1/4" thick)
   D. 2nd and 4th plies
   E. 2nd and 6th plies
   F. 3rd and 5th plies
   J. First ply, high pressure laminate
   K. Last ply, high pressure laminate
   P. Reverse Hand
   R. Core, Wood (1/4" or more thick), or commercial plywood
   S. Core, Particle Board
   U. Core Strip, Length
   V. Core Strip, Width
   W. 2nd and 8th plies (when plies are identical)
   X. 3rd and 7th plies (when plies are identical)
   Y. 4th and 6th plies (when plies are identical)
   Z. Composite Core (5 ply before applying high pressure laminate)

Where a part number is used in both a right hand unit and a left hand unit, or is made in pairs, the number of the part shown will be the basic part number and this same part number with the suffix "P" will also be shown on the drawing as the reverse hand.

4.4.1 Veneered panels. Drawings of laid-up panels will show the number of plies, the materials and the thickness of the core and crossbands. Grain direction will be shown as ↔. Drawing of the veneer for matched panels will be shown on a separate drawing.

Due to differences in the thickness of face veneer of various species, the thickness of face veneers will not be shown on the drawings. The basic panel thickness is governed by the .750 ± .005 thickness of particle board as received in the plant and the .740 thickness as planed. By adding two sheets of $\frac{1}{36}$ (.028″) veneer faces, the panel thickness before sanding becomes .795. Assuming one-half of each face veneer is sanded off, the finished panel thickness becomes .767. This will, therefore, be the nominal panel thickness after sanding. The thickness of the five panel core will be adjusted to arrive at the .767 thickness also. In this case, the core thickness will be .640.

Where $\frac{1}{28}$ inch thick face veneers are specified and it is important to maintain the panel thickness, the laid-up panel will be sanded to the .767 dimension.

4.4.2 Materials. Drawings will show the wood species or materials from which the part will be made.

4.4.3 Finishing Specifications. Drawings will show the surface and edge finishes of the part.

    4.4.3.1 Any line will be a machined surface such as shaped, drilled, routed, planed, miter sawn, etc., unless otherwise indicated.

    4.4.3.2 Finish symbols:

        $\overset{S}{\diagdown}$ = Sanded surface
        $\overset{P}{\diagdown}$ = Polished surface
        $\overset{R}{\diagdown}$ = Rough or sawn surface

4.4.4 Special Instructions. Will be noted on drawing as required and will include such as:

    Glue-up allowed
    No glue-up allowed
    Sawn surface (or edge) permissible
    Glue-paint edges black
    Putty

4.4.5 Dimensions. Will be shown after the part has gone through the last operation with the exception of the final sanding and/or polishing. Where significant, the drawing will also list the dimensions after sanding and/or polishing. On parts with edge veneer, the edge veneer will be shown as a double line and dimensions both with and without edge veneer will be shown.

4.4.6 Pieces per unit. "Unit" is defined as the complete product and the drawing will not show the number of pieces required but this will be found in the stock list.

4.4.7 Tolerances: When tolerances are stated, they shall be "bilateral", that is, in two directions with equal or unequal plus and minus values, as for example:

(1) $2.620 \, {}^{+\,.030}_{-\,.062}$ (2) $2.620 \pm .010$ (3) $2.620 \, {}^{+\,.000}_{-\,.010}$

The following discussion will explain the preceding drafting procedures:

1. The Technical Center receives the design drawings and the model, if one was made. It prepares the production drawings and specifications for all parts.
2. The Technical Center provides copies of the finished production drawings as needed. They make needed revisions on these drawings. A set of scaled production drawings is shown in Figs. 23–48 to 23–51.
3. Each drawing shows the size of the finished part after it has been machined to size.
    4.1 The Technical Center prepares the drawings; the Production Engineering Department receives these drawings and decides how the part is to be made. It plans the order of steps to be performed to make the part. This is placed on a routing sheet that goes to the production area.
    4.2 Detail drawings are not made of the items listed. They are shown on the drawing of the assembled unit.
    4.3 Herman Miller, Inc. uses 11×17 polyester film (B size) for all original drawings. A, C, and D size sheets are used if necessary. The filing system is designed for these sizes.
    4.4 Each drawing is given a part number. This helps identify the part. It also helps in filing and retrieving the drawing. The system in use is complex. It is designed for use with a computer. See Figs. 23–48 to 23–50.
    The layers of a laid-up panel are identified on each drawing by a special code used on a drawing of the panel. A laid-up panel is one made by gluing together thin layers of wood to form a thicker panel. Plywood is an example. See Fig. 23–49.
        4.4.1 This gives the directions for indicating laid-up panels and grain direction. It states the manufacturing tolerances. Notice the panels are held to ±.005 inches. See the typical title block, Fig. 23–49.
        4.4.2 This tells the draftsman to indicate the species of wood on the drawing. See Fig. 23–48.
        4.4.3 Finishing specifications refer to the type of wood surface needed, as a sanded surface. It does *not* refer to painting or lacquering the surface.
            4.4.3.1 Each line drawn on the drawing means that some type of surface exists. This surface must have a specified finish.
            4.4.3.2 Symbols for indicating finish are given. These are placed on the edge view of the surface to which they refer. See Figs. 23–48 and 23–49.
        4.4.4 This section tells how to note special instructions needed by the production staff.
        4.4.5 Drawings are dimensioned in the same manner as machine drawings. See Chapter 7, Dimensioning, for detailed instructions. Fig. 23–52 shows how to dimension a part when edge veneer is shown.
        4.4.6 A unit is one completed piece of furniture.
        4.4.7 The system for indicating tolerances is the same as for machine drawing. Tolerances are always shown in a plus and minus direction.

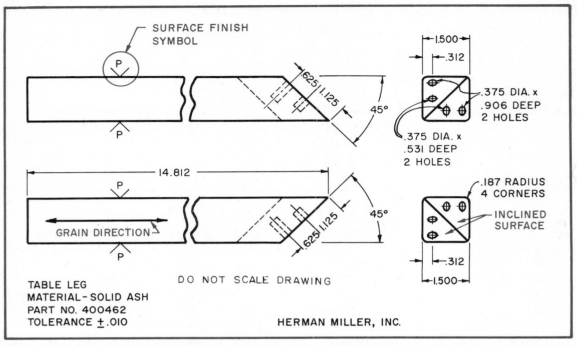

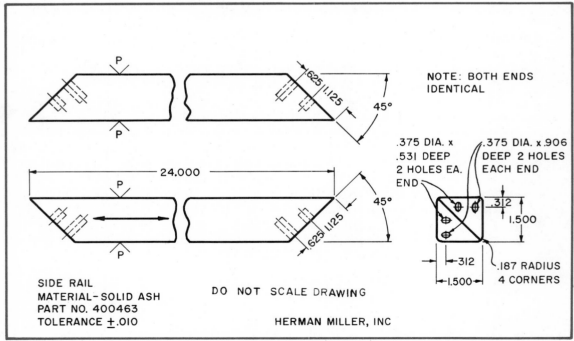

23–48. *Coffee table leg and side-rail detail drawings.*

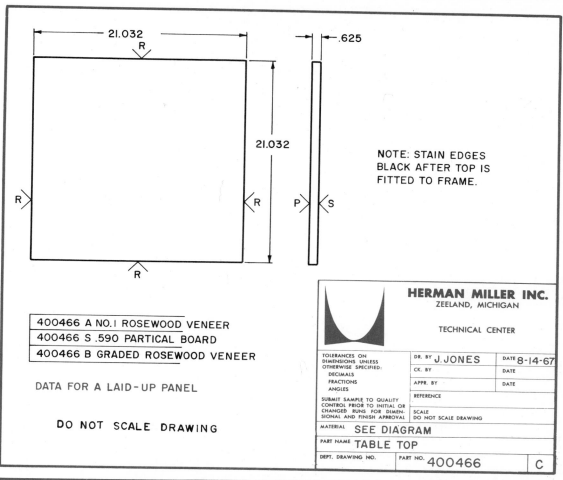

21.032

R

21.032

R   R

R

.625

P   S

NOTE: STAIN EDGES
BLACK AFTER TOP IS
FITTED TO FRAME.

400466 A NO.I ROSEWOOD VENEER
400466 S .590 PARTICAL BOARD
400466 B GRADED ROSEWOOD VENEER

DATA FOR A LAID-UP PANEL

DO NOT SCALE DRAWING

HERMAN MILLER INC.
ZEELAND, MICHIGAN

TECHNICAL CENTER

TOLERANCES ON DIMENSIONS UNLESS OTHERWISE SPECIFIED: DECIMALS FRACTIONS ANGLES	DR. BY J. JONES	DATE 8-14-67
	CK. BY	DATE
	APPR. BY	DATE
SUBMIT SAMPLE TO QUALITY CONTROL PRIOR TO INITIAL OR CHANGED RUNS FOR DIMENSIONAL AND FINISH APPROVAL	REFERENCE	
	SCALE DO NOT SCALE DRAWING	
MATERIAL SEE DIAGRAM		
PART NAME TABLE TOP		
DEPT. DRAWING NO.	PART NO. 400466	C

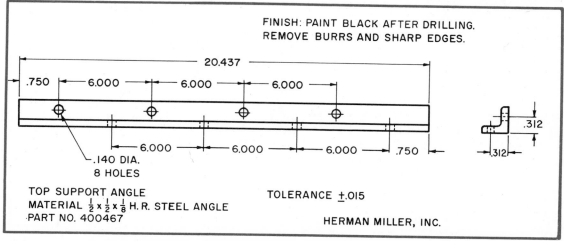

FINISH: PAINT BLACK AFTER DRILLING.
REMOVE BURRS AND SHARP EDGES.

20.437

.750   6.000   6.000   6.000

.140 DIA.
8 HOLES

6.000   6.000   6.000   .750

.312

.312

TOP SUPPORT ANGLE
MATERIAL $\frac{1}{2}$ x $\frac{1}{2}$ x $\frac{1}{8}$ H.R. STEEL ANGLE
PART NO. 400467

TOLERANCE ±.015

HERMAN MILLER, INC.

23-49. *Coffee table top and top-support-angle detail drawings.*

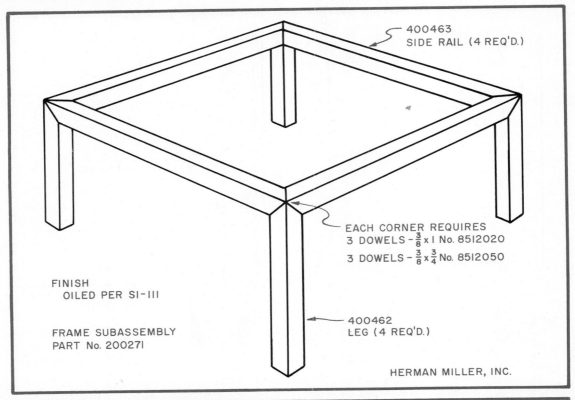

400463
SIDE RAIL (4 REQ'D.)

EACH CORNER REQUIRES
3 DOWELS – $\frac{3}{8}$ x I No. 8512020
3 DOWELS – $\frac{3}{8}$ x $\frac{3}{4}$ No. 8512050

FINISH
OILED PER SI-III

FRAME SUBASSEMBLY
PART No. 200271

400462
LEG (4 REQ'D.)

HERMAN MILLER, INC.

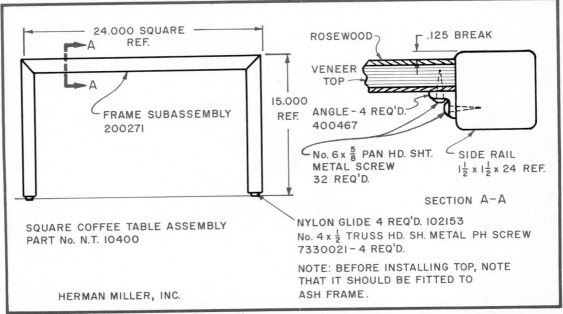

24.000 SQUARE
REF.

A

A

FRAME SUBASSEMBLY
200271

15.000
REF.

ROSEWOOD

.125 BREAK

VENEER
TOP

ANGLE – 4 REQ'D.
400467

No. 6 x $\frac{5}{8}$ PAN HD. SHT.
METAL SCREW
32 REQ'D.

SIDE RAIL
$1\frac{1}{2}$ x $1\frac{1}{2}$ x 24 REF.

SECTION A-A

SQUARE COFFEE TABLE ASSEMBLY
PART No. N.T. 10400

NYLON GLIDE 4 REQ'D. 102153
No. 4 x $\frac{1}{2}$ TRUSS HD. SH. METAL PH SCREW
7330021 – 4 REQ'D.

NOTE: BEFORE INSTALLING TOP, NOTE
THAT IT SHOULD BE FITTED TO
ASH FRAME.

HERMAN MILLER, INC.

23-50. *Two drawings used to give assembly information.*

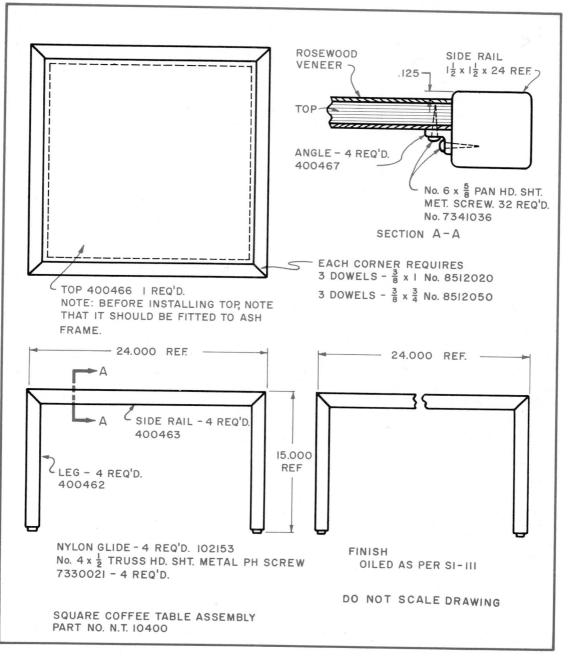

ROSEWOOD VENEER

TOP

SIDE RAIL
$1\frac{1}{2}$ x $1\frac{1}{2}$ x 24 REF.

.125

ANGLE – 4 REQ'D.
400467

No. 6 x $\frac{5}{8}$ PAN HD. SHT.
MET. SCREW. 32 REQ'D.
No. 7341036

SECTION A–A

TOP 400466 I REQ'D.
NOTE: BEFORE INSTALLING TOP, NOTE
THAT IT SHOULD BE FITTED TO ASH
FRAME.

EACH CORNER REQUIRES
3 DOWELS – $\frac{3}{8}$ x I No. 8512020
3 DOWELS – $\frac{3}{8}$ x $\frac{3}{4}$ No. 8512050

24.000 REF.

A

A

SIDE RAIL – 4 REQ'D.
400463

LEG – 4 REQ'D.
400462

15.000
REF

24.000 REF.

NYLON GLIDE – 4 REQ'D. 102153
No. 4 x $\frac{1}{2}$ TRUSS HD. SHT. METAL PH SCREW
7330021 – 4 REQ'D.

FINISH
OILED AS PER SI-III

DO NOT SCALE DRAWING

SQUARE COFFEE TABLE ASSEMBLY
PART NO. N.T. 10400

Herman Miller, Inc.

23–51.   *An alternate way to make a furniture assembly drawing. This follows the
principles of orthographic projection.*

23–52. *Veneers are shown with a double line. Dimensions are given with and without the veneer.*

USE OF SCALED DRAWINGS

If a part can be drawn full size on 11 × 17 paper, this is done. If the part is complex, it is drawn full size even if a larger sheet is needed. Most drawings are reduced to fit on an 11 × 17 sheet. The most commonly used scales are ½″ = 1″, and ¼″ = 1″.

The actual scale used to reduce a drawing is frequently not indicated. Since drawings require revision, the practice is to change the dimension but not the drawing. Many drawings get out of scale. Since the part is made by using the dimensions, an accurate scale drawing is not needed. The production staff must never measure the actual drawing to get a dimension. Herman Miller, Inc. clearly marks each drawing, "DO NOT SCALE DRAWING."

Another factor influencing scale is the use of microfilm. When drawings placed on microfilm are reproduced, they are slightly smaller than the original. This makes them out of scale. The sizes cannot be scaled from the printout made from the microfilm.

It should be clearly understood that this scale system is not used by every company.

Some companies make their assembly drawings full size.

The system used for furniture drawings that are not full size is shown in Figs. 23–48 through 23–51. Notice that each part is detailed separately. The assembly information is shown in Figs. 23–50 and 23–51.

Fig. 23–48 shows two detail drawings. These are for the leg and side rails. These drawings give the dimensions needed to make the part. They give the overall sizes, angles of cuts, and hole locations. The holes are drawn as ellipses because they are on inclined surfaces. Notice the indication of surface finish. The grain direction is shown with a symbol.

Fig. 23–49 shows two more detail drawings. One shows the top support angle. Even though this is metal, it is detailed in the same way as wood parts.

The other drawing shows the table top. This is made from particle board. Notice the data given for a laid-up panel. On each layer the part number is given. Then the letter indicating the ply is given. The top layer is marked "A." This is the first ply. The drawing shows this to be No. 1 rose-

wood veneer. The second layer is also identified. The "S" means it is a particle board core. The thickness of the core is given. The bottom layer is identified by the letter "B." This means it is the last ply and the veneer required is different from that on the top. In this case, it is a lower grade rosewood veneer than that on the "A" ply.

Notice the tolerances on this and the other drawings, page 698.

A typical title block is included on this drawing. These blocks were left off the other drawings to save space.

Fig. 23–50 gives the assembly information. One drawing is a perspective view of the assembled frame. This clearly shows how it is to be assembled. The parts are identified by the part number. Dimensions are not needed.

The other drawing shows a front view of the assembled frame. It has a section telling how the top support angle is installed. Only the overall dimensions are shown. Pictorials are used only if they make the drawing easier to understand. A more complex piece of furniture would not use a pictorial. Instead, the assembly drawing would be made as shown in Fig. 23–51, page 701.

FULL-SIZE ASSEMBLY DRAWING

Typical assembly furniture drawings using full scale are shown in Fig. 23–54A and B.

Such drawings must be studied carefully before an inexperienced person can learn to read them. Study the photo of the table, Fig. 23–53. The photo will help you understand the drawing. The original full-size drawing was 36 inches wide and 48 inches long.

Since the table is symmetrical, only half of a front elevation is needed. This shows the shape of the leg. It shows the drawers and the over-laid panels. The drawer pulls are located. The major dimensions are shown.

At the right of the front elevation is a section through the back. It shows the top rim shape and the top and bottom molding. This view projects from the front elevation. This is the same as any orthographic drawing.

To the left of the front elevation are two more sections. One of these is a section through the front. It shows the details of the drawer fronts. It also shows how the carved fill block is placed on the leg. A revolved section shows that the legs are square.

The third section is through the side of the table. It shows the drawer runners and the drawer end. The solid side rail is shown. The top and bottom molding as it is used on the side is located.

These sections are not identified. Those using the drawings understand what they represent.

The top view is also full size. This drawing does not project

Hekman Furniture Company

23–53. *A lamp table with two drawers. It is made of cherry. The top is 24 × 26 inches. The height is 21 inches. It is available with a cherry or marble top. Full-scale drawings are shown in Figs. 23–54A and B.*

from the front view. If the front view was a full view, the top view would project from it.

The top view is actually several views on one drawing. Notice it is built around a center line. The lower portion is drawn with the top removed. It shows the leg and side rails. A top view of the drawer is included. The drawer runner is also shown. Such views are difficult to read because it is a common practice to show parts as visible even though they are hidden. Examine the drawer runner. A part of it is really hidden below the drawer. However, it is drawn as though it is visible. When this practice is understood, it is easier to read the drawing.

The upper right corner of the top view shows the top of the table. This shows the two-inch wide top rim and the top

surface. This change is not indicated on the drawing. Again, it is understood by those using the drawing that this part is not a section.

The top view of this table shows that it can be made with two different tops. Notice that the size of a wood top is given. It is 22 inches by 20 inches without the tongue. The table is also made with a marble top. The marble is 21$^{15}/_{16}$ inches by 19$^{15}/_{16}$ inches. A $^{1}/_{32}$-inch space is left on all sides. If the table has a marble top, a different top rim is needed. Notice the alternate detail for the top rim for a marble top.

It is important to notice that each part is not fully dimensioned. Only the major dimensions are given. These give enough information so a draftsman can detail each part from the drawing. He would

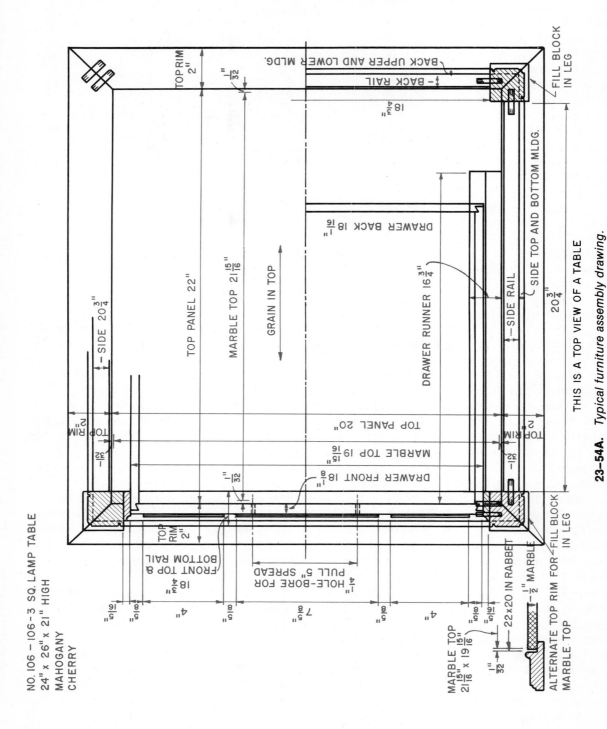

NO. 106 – 106-3 SQ. LAMP TABLE
24" x 26" x 21" HIGH
MAHOGANY
CHERRY

TOP RIM 2"

SIDE 20¾"

TOP PANEL 22"

MARBLE TOP 21¹⁵⁄₁₆

GRAIN IN TOP

BACK UPPER AND LOWER MLDG.

BACK RAIL

18 ⁴³⁄₃₂"

¹⁄₃₂"

DRAWER BACK 18¹⁵⁄₁₆"

DRAWER RUNNER 16¾"

SIDE RAIL

SIDE TOP AND BOTTOM MLDG.

FILL BLOCK IN LEG

20¾"

THIS IS A TOP VIEW OF A TABLE

TOP RIM 2"

TOP RIM 2"

¹⁄₃₂"

¹⁄₃₂"

TOP PANEL 20"

MARBLE TOP 19¹⁵⁄₁₆

DRAWER FRONT 18¹⁄₈"

¹⁄₃₂"

TOP RIM 2"

FRONT TOP & BOTTOM RAIL

18 ⁴³⁄₃₂"

HOLE-BORE FOR 4¾" PULL 5" SPREAD

FILL BLOCK IN LEG

22x20 IN RABBET

ALTERNATE TOP RIM FOR MARBLE TOP

½ MARBLE

MARBLE TOP 21¹⁵⁄₁₆ x 19¹⁵⁄₁₆

¹⁄₃₂"

⁵⁄₁₆"

⁵⁄₈"

4"

⁵⁄₈"

7⁵⁄₈"

⁵⁄₈"

4"

⁵⁄₈"

⁵⁄₁₆"

23-54A. *Typical furniture assembly drawing.*

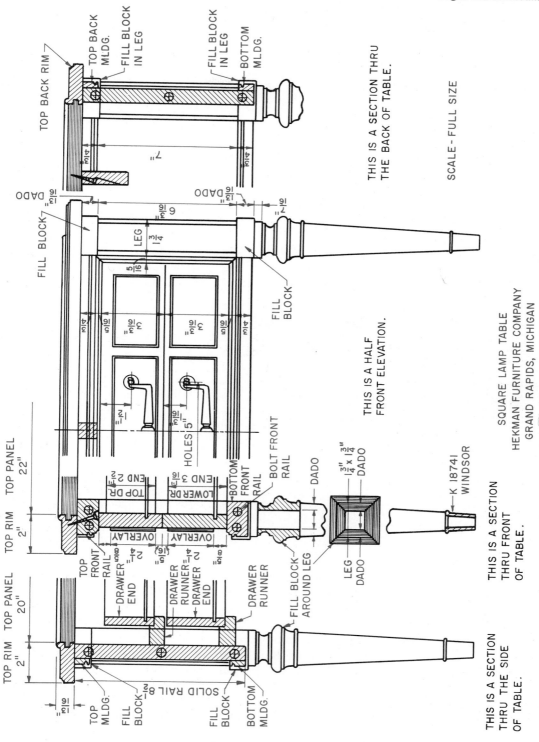

THIS IS A SECTION THRU
THE BACK OF TABLE.

SCALE - FULL SIZE

THIS IS A HALF
FRONT ELEVATION.

SQUARE LAMP TABLE
HEKMAN FURNITURE COMPANY
GRAND RAPIDS, MICHIGAN

THIS IS A SECTION
THRU FRONT
OF TABLE.

THIS IS A SECTION
THRU THE SIDE
OF TABLE.

23-54B. *Typical furniture assembly drawing.*

decide the depth of joints to be used. The tongue-and-groove joint between the top and the top rim is an example. These dimensions would be given on a detail drawing, called a routing drawing.

It should be made clear that even though the drawing shown in Fig. 23–54 appears crowded, this is not the case when it is drawn full size. The crowding is due to the reduction of the drawing from its original 36 × 48 inch size to fit the page of this book.

Routing Drawings. A routing drawing is used by those making furniture parts. The drawing is planned by the engineering department, and gives the information needed to make the part. Included are instructions that give the order in which the operations are to be performed.

The routing drawing accompanies the part through the plant. The workmen use the routing drawing rather than the full set of production drawings, since they only need information on the part they are making. A typical routing drawing is shown in Fig. 23–55.

DESIGNING KITCHEN CABINETS

When a home is designed, the kitchen cabinets can be secured in several ways. One way is to select units from a catalog. Many companies mass produce a wide variety of sizes and styles. The home owner chooses the cabinets he wants from those available.

Another way is to have the cabinets custom built. This means they are built on order for a particular job. This is also necessary if the job requires a special type or style of cabinet. See Fig. 23–56.

The architect prepares elevations of the cabinets wanted. A draftsman then makes the production drawings.

Standard Kitchen Cabinet Sizes. The typical base cabinet for a kitchen is 34½-inches high without the top. Since top thickness can vary,

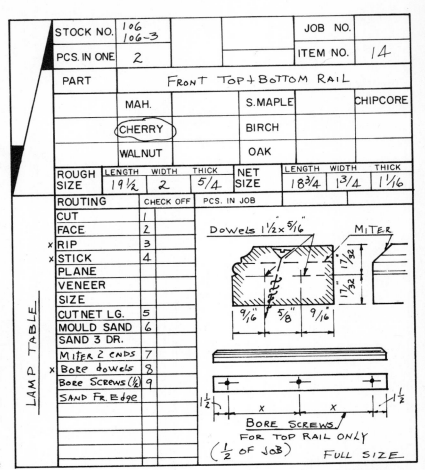

Hekman Furniture Company

23–55. *A routing drawing for the front and bottom rails for the table, Fig. 23–54. The stock number identifies the table on which the rails are used. The item number identifies the specific part. The rough and net (finished) size of the part are recorded. A dimensioned drawing is included. The routing section lists the order of the operations which must be performed to make the part.*

the usual total height is 35¼ to 36 inches. The depth is 24 inches. This will allow the installation of most dishwashers under the cabinet. See Fig. 23–57. The toe space on a base cabinet is usually 3½-inches high and 1½-inches deep. The arrangement of drawers and doors in base cabinets depends upon the desires of the owner. Some commonly used arrangements are shown in Fig. 23–58.

Wall cabinets vary a great deal in height. Usually the bottom of the wall cabinet is 18 inches above the top of the base cabinet. The actual height can extend to the ceiling of the room if desired. Fig. 23–57 gives some common wall cabinet heights.

The higher the wall cabinet, the more shelves it will have. See the typical examples, Fig. 23–57. Wall cabinets are usually 12 to 13 inches deep.

Cabinet Construction Details.
There are many ways to make a cabinet. Figs. 23–59 and 23–60 show details for a simple method, called *box construction*. The sides are solid pieces of plywood. They are usually ¾-inch thick. On the base cabinet the bottom and shelf are set in a dado. The facing on the front is made of ¾-inch thick stock. It is usually 1⅝ to 1⅞-inches wide. The stiles and rails are joined with dowel joints. The face is joined to the end panels with a tongue-and-groove joint.

Coppes, Inc.

23–56. *A circular cabinet base containing the cooking surface unit. The hood carries away the fumes from cooking.*

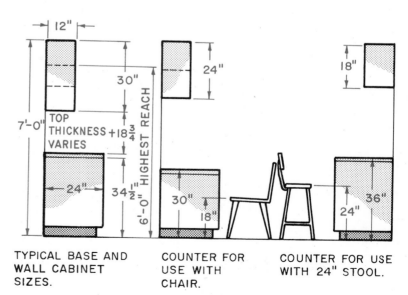

TYPICAL BASE AND WALL CABINET SIZES.

COUNTER FOR USE WITH CHAIR.

COUNTER FOR USE WITH 24" STOOL.

23–57. *Typical kitchen cabinet and counter sizes.*

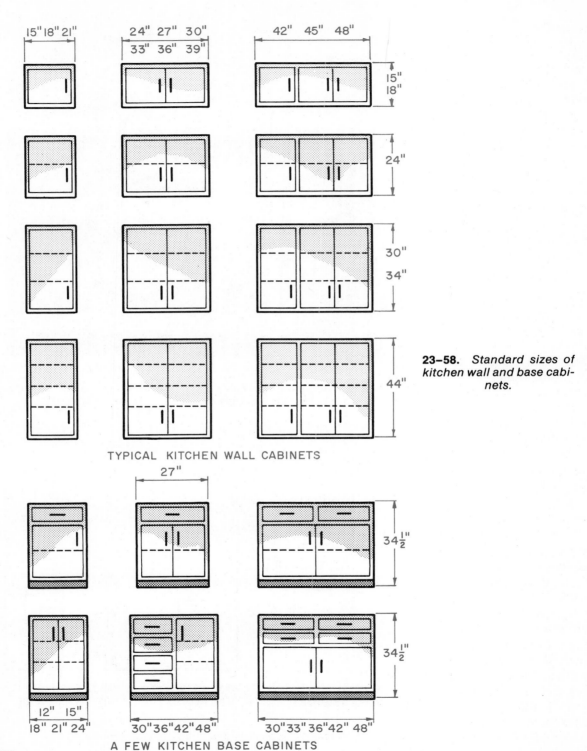

15" 18" 21" 24" 27" 30" 42" 45" 48"
33" 36" 39"

15"
18"

24"

30"
34"

44"

TYPICAL KITCHEN WALL CABINETS

27"

$34\frac{1}{2}$"

$34\frac{1}{2}$"

12" 15" 30"36"42"48" 30"33"36"42" 48"
18" 21" 24"

A FEW KITCHEN BASE CABINETS

23–58. *Standard sizes of kitchen wall and base cabinets.*

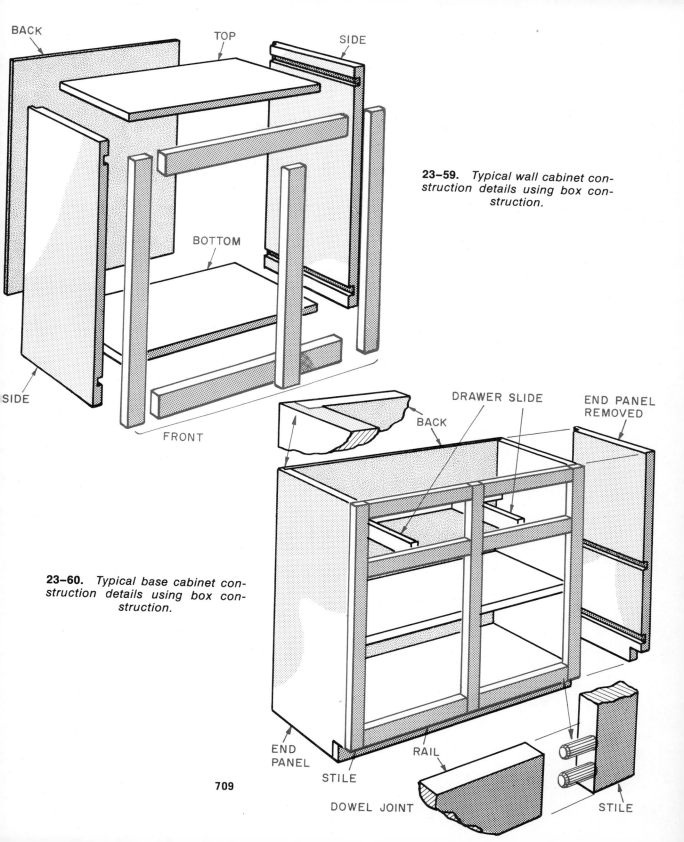

23–59. *Typical wall cabinet construction details using box construction.*

BACK

TOP

SIDE

SIDE

BOTTOM

FRONT

23–60. *Typical base cabinet construction details using box construction.*

BACK

DRAWER SLIDE

END PANEL REMOVED

END PANEL

STILE

RAIL

DOWEL JOINT

STILE

709

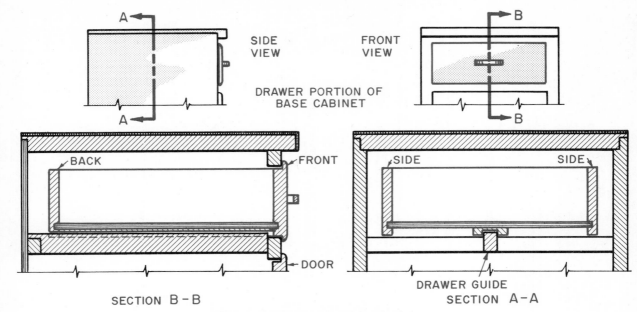

SIDE VIEW

FRONT VIEW

DRAWER PORTION OF BASE CABINET

BACK — FRONT

SIDE — SIDE

DOOR

DRAWER GUIDE

SECTION B-B

SECTION A-A

23–61. *Section through a typical drawer.*

23–62. *A molding can be applied to cabinet doors to give a decorative appearance.*
Coppes, Inc.

Sometimes a butt joint is used, with nails through the stiles and rails.

The shelves in the wall cabinets can be adjustable or fixed. The construction details are the same as for the base cabinet.

Fig. 23–61 shows details for making a simple drawer. The drawer sides are commonly $7/16$ to $1/2$-inch thick. The drawer bottom can be $3/16$ to $1/4$-inch plywood or hardboard.

Cabinet doors can be decorated by cutting a design into the wood. This makes it appear like a paneled door. Another easy way to decorate is to apply moldings to the door as shown in Fig. 23–62.

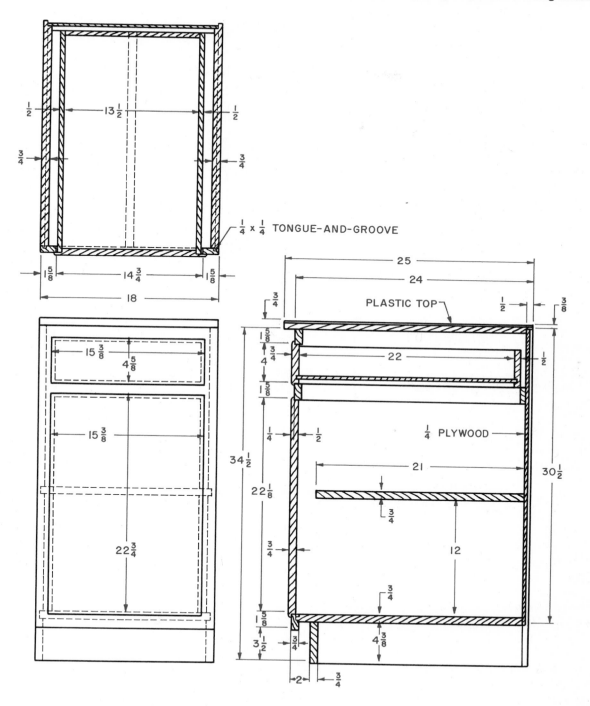

23–63. *Typical production drawing for a base cabinet unit for a kitchen.*

23-64. *Metal cabinets designed for use in a classroom. Extensive home and commercial use is made of metal for cabinet construction.*

Cabinet Production Drawings. The exact details for making cabinet production drawings varies somewhat from one company to another. The following discussion presents a typical system.

The drawings are positioned much the same as regular working drawings. Most cabinets can be detailed with a regular front view and two sections. The right side view is a typical section. The top view is also a typical section. Cutting-plane lines are not needed. It is understood that these are typical sections. See Fig. 23–63.

The drawings can be any scale desired. Many companies prefer to make them full size. Special details, such as the shapes of moldings or lipped doors, do not require special details since they are full size. Construction details that need clarification are drawn as partial details. These are full size.

Dimensions are placed in the same manner as other working drawings.

Some companies use a parts list, as shown in Fig. 23–73 and explained in the section on fixture drawings.

METAL CABINETS

Metal is used a great deal for cabinet construction in homes and commercial buildings. Fig. 23–64 shows metal units designed for use in a classroom. They are strong and easily cleaned.

Metal Cabinet Drawings. One system for detailing metal cabinets uses an isometric view of the assembled unit. It has the major overall dimensions. A typical section is then shown. See Fig. 23–65.

Each piece of the cabinet is detailed separately and drawn as a development. A *development* is the pattern of the part before it is formed to the final shape. This pattern is called a *blank.* The drawing is fully dimensioned. In addition to the blank drawing, each part is shown after it is formed. Usually a side view and typical section of the formed part are shown. See Fig. 23–66. Dimensions are given to show the size after forming. This is placed next to the blank drawing on the sheet.

Some units require other details. An example is the assembly of the formed blanks for a drawer slide. Each metal part of the drawer slide would be detailed separately, then drawn assembled.

Notice that the thickness of the metal is drawn when the metal sheet appears as an edge. It is not section lined on sectional views unless it aids in reading the drawing. The scale used varies with the size of the unit. Common scales for the pictorial and section drawings are $1\frac{1}{2}'' = 1'\text{-}0''$ and $3'' = 1'\text{-}0''$. The blank drawings are made full size whenever possible. If the part is large, it can be drawn $\frac{3}{4}'' = 1''$ or $\frac{1}{2}'' = 1''$.

TYPES OF FIXTURES

The fixtures used in retail stores tend to fall into several general groups. The design varies a great deal. The common types are wall cases, display cases, gondolas, counters, and island fixtures.

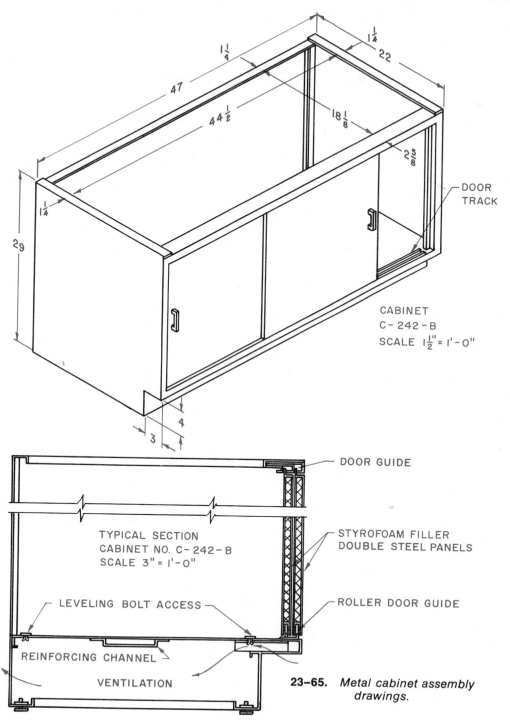

CABINET
C-242-B
SCALE $1\frac{1}{2}$" = 1'-0"

DOOR
TRACK

TYPICAL SECTION
CABINET NO. C-242-B
SCALE 3" = 1'-0"

DOOR GUIDE

STYROFOAM FILLER
DOUBLE STEEL PANELS

LEVELING BOLT ACCESS

ROLLER DOOR GUIDE

REINFORCING CHANNEL

VENTILATION

23-65.   *Metal cabinet assembly drawings.*

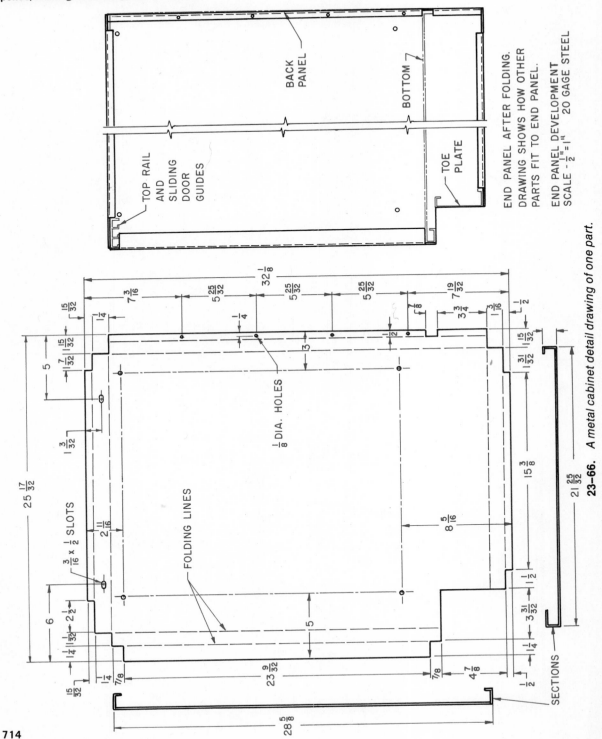

BACK PANEL

BOTTOM

TOP RAIL AND SLIDING DOOR GUIDES

TOE PLATE

END PANEL AFTER FOLDING. DRAWING SHOWS HOW OTHER PARTS FIT TO END PANEL.

END PANEL DEVELOPMENT
SCALE - $\frac{1}{2}$" = 1" 20 GAGE STEEL

$\frac{1}{8}$ DIA. HOLES

FOLDING LINES

$\frac{3}{16} \times \frac{1}{2}$ SLOTS

SECTIONS

23-66. A metal cabinet detail drawing of one part.

A *wall case* is built to back up to a wall. It is open on one side. It usually has a wood or metal base. The base is used for storage. Shelving extends above the base. This can be open so the customer can handle the items. It can be closed with glass doors. See Fig. 23–67.

Display cases may have two parts. The lower is used for storage. The upper displays the items for sale. The upper section is usually enclosed with glass. This protects the merchandise. The clerk works behind the case. See Fig. 23–68A. Another type of display case, Fig. 23–68B, uses all space for merchandise.

23–67. *A wall case can have display shelving and storage below. It can have display shelving to the floor.*

Reflector Hardware Corp.

715

23–68A. *A display case with storage below.*

Reflector Hardware Corp.

23–68B. *A display case with all the space for merchandise.*

Reflector Hardware Corp.

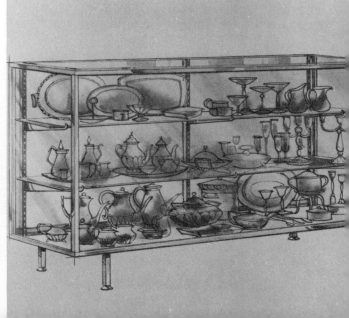

Reflector Hardware Corp.

23–69. *A gondola unit. These units are placed back to back to provide shelving on two aisles.*

23–70A. *An island fixture with display shelving and storage.*

Reflector Hardware Corp.

Gondolas are long sections of shelving. They are open on both sides. The customer can handle the merchandise. Shelving in a food store is an example. See Fig. 23–69.

Island fixtures are much like a display case. However, they are usually placed so the customer can handle the merchandise. The customer can approach the fixture from all sides. An island fixture can have storage in the bottom or use the entire space for display. See Fig. 23–70A and B.

Counters are used to provide service of some kind. They are not used for display purposes. Examples are wrapping counters or food service counters. See Fig. 23–71. Typical sizes for merchandising fixtures are shown in Fig. 23–72.

DESIGNING FIXTURES

Fixtures are designed in much the same way as explained for furniture. The purpose of the unit must be understood. The designer must understand the best ways to display the articles to be held by the fixture. For example, very small pieces of jewelry are usually stored in a fixture that clearly displays the articles. Yet the articles must be protected from shoplifters. Usually, this fixture contains a glass top and front. Lighting is built-in to increase the effectiveness of the display.

The size and type of items to be displayed are important.

A fixture to display men's suits would be different from one for bakery goods.

The size of the human form must be considered. A check-writing counter for a bank permits the person to write in a standing position. The teller's counter also is designed for a standing customer. The cashier is seated. This difference in positions might be handled by a tall chair for the teller. The fixture could have a raised platform on the teller's side. This would permit the use of a regular chair.

The average reach of a person is considered in the design. Shelving to display canned goods cannot be too deep. Height is sometimes limited if it is desirable for the customer to see over the entire store. These examples show the type of information a fixture designer must consider. There are many other factors to be considered.

The styling of fixture units is often decided by the owner and architect. The draftsman works from their drawings and sketches. If a company offers a standard line of fixtures, the styling is decided in the same manner as for furniture.

FIXTURE PRODUCTION DRAWINGS

Fixture production drawings vary within the industry. They are about the same as cabinet drawings. The drafting standards that follow represent a typical system.

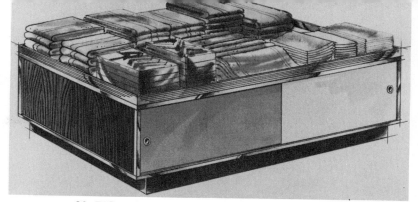

23–70B.   *An island fixture with storage below.*

Scale. Front and side views of the unit are drawn. These are usually very small. A common scale is $\frac{1}{8}'' = 1''$. For very large units a smaller scale can be used. See Fig. 23–73.

All details and sections showing construction details are drawn full size.

Size of Drawings. Drawings are made on large sheets of vellum. Usually rolls 36 inches wide and 50 yards long are used. The drawing is long enough to show the full size details. These drawings can be six to ten feet long or longer.

The draftsman places a roll of vellum in a rack at the right end of his bench. He unrolls a piece long enough to cover his bench top. On this paper he makes his drawing. If he needs additional space, he rolls up the section with the drawing. It is placed in a rack at the left of the bench. As he rolls up the finished drawing, fresh paper is unrolled. He then continues his drawing. In this way, the drawing length is limited only by the amount of paper on the roll. The drawing is cut to length after it is finished.

23–71.   *A checkout counter for a supermarket.*

Views Needed. Most fixtures can be detailed by showing a small scale front and side elevation. In addition, sections through the length and width of the unit are made. Partial sectional views are needed for showing construction details that otherwise would not be clear. If a construction detail is not clear on the large section views, a detail must be made. See Fig. 23–73 and 23–74 for examples of views used.

Drawing the Views. The small scale elevations or pictorial views are placed on the right side of the paper near the parts list. See Figs. 23–73 and 23–74. The full-size section views through the length and width of the unit are placed side by side. In this way they project to each other the same as any orthographic view.

The partial sections showing construction details are placed as near the area sectioned as possible. The section lining is sometimes left off the section view. This reduces the number of lines on the drawing. The production workers understand that the views are sections. Parts serving as blocking are sectioned with an "X". *Blocking* is a wood part used to fill voids or strengthen a case. It is not visible on the finished units.

23–72. *Typical fixture sizes.*

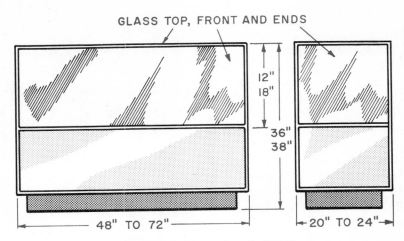

GLASS TOP, FRONT AND ENDS

TYPICAL DISPLAY CASE SIZES

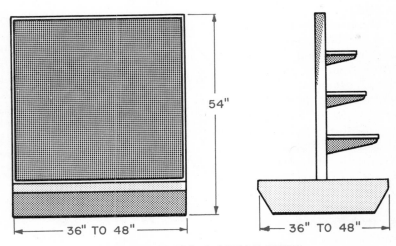

TYPICAL GONDOLA DESIGN SIZES

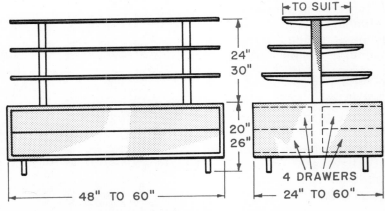

TO SUIT

4 DRAWERS

TYPICAL ISLAND FIXTURE SIZES
THE BASE CAN BE USED WITHOUT THE SHELVES TO DISPLAY ITEMS AS SHIRTS OR TOWELS.

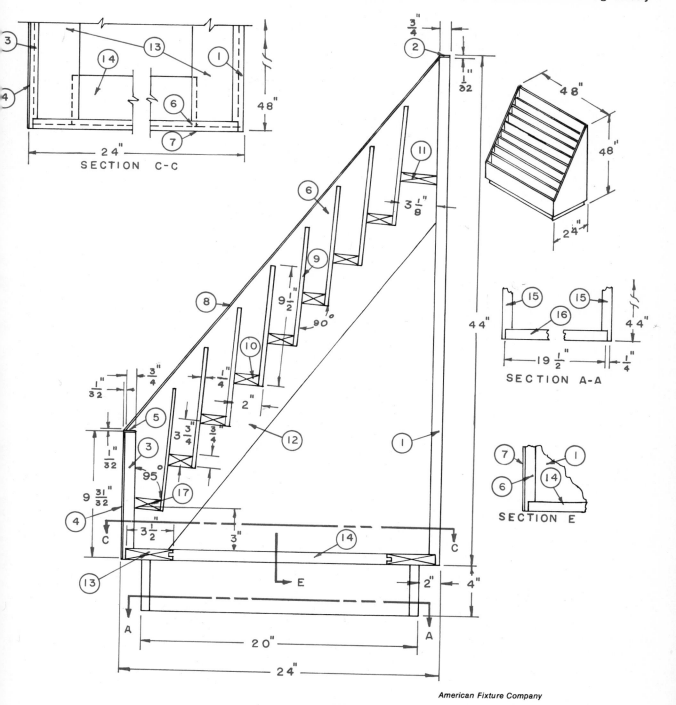

③

⑬ ①

⑭

⑥

⑦

48"

24"

SECTION C-C

$\frac{3}{4}$"

②

$\frac{1}{32}$"

48"

48"

24"

⑪

$3\frac{1}{8}$"

⑥

⑨

$9\frac{1}{2}$"

90°

⑧

⑮ ⑮

⑯

44"

$19\frac{1}{2}$" $\frac{1}{4}$"

SECTION A-A

⑩

$\frac{3}{4}$"

$\frac{1}{4}$"

$\frac{1}{32}$"

2"

⑤

③

$3\frac{3}{4}$" $\frac{3}{4}$"

95°

⑫

44"

①

⑦ ①

⑥ ⑭

SECTION E

$\frac{1}{32}$"

$\frac{1}{32}$"

$9\frac{31}{32}$"

⑰

④

C

$3\frac{1}{2}$"

3"

⑭

C

⑬

E

2" 4"

A

20"

A

24"

American Fixture Company

23–73. *A magazine display case.*

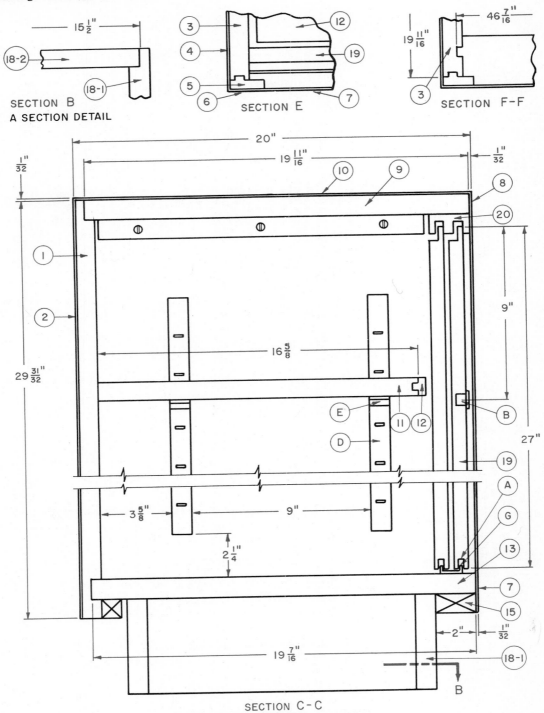

SECTION B
A SECTION DETAIL

SECTION E

SECTION F-F

SECTION C-C

23–74. *A fixture production drawing.*

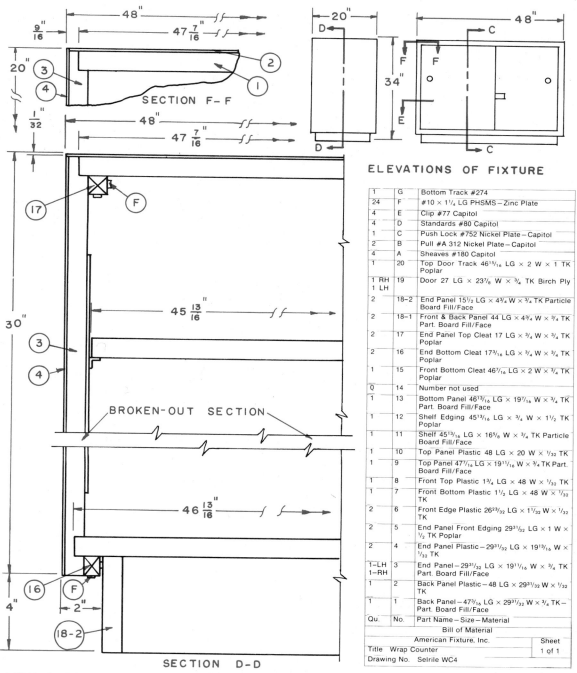

ELEVATIONS OF FIXTURE

SECTION F—F

SECTION D—D

BROKEN-OUT SECTION

ONLY ONE HALF OF A SYMMETRICAL OBJECT
NEED BE DRAWN

Qu.	No.	Part Name – Size – Material	
1	G	Bottom Track #274	
24	F	#10 × 1¼ LG PHSMS – Zinc Plate	
4	E	Clip #77 Capitol	
4	D	Standards #80 Capitol	
1	C	Push Lock #752 Nickel Plate – Capitol	
2	B	Pull #A 312 Nickel Plate – Capitol	
4	A	Sheaves #180 Capitol	
1	20	Top Door Track 46¹⁵⁄₁₆ LG × 2 W × 1 TK Poplar	
1 RH 1 LH	19	Door 27 LG × 23⅞ W × ¾ TK Birch Ply	
2	18-2	End Panel 15½ LG × 4¾ W × ¾ TK Particle Board Fill/Face	
2	18-1	Front & Back Panel 44 LG × 4¾ W × ¾ TK Part. Board Fill/Face	
2	17	End Panel Top Cleat 17 LG × ¾ W × ¾ TK Poplar	
2	16	End Bottom Cleat 17³⁄₁₆ LG × ¾ W × ¾ TK Poplar	
1	15	Front Bottom Cleat 46⁷⁄₁₆ LG × 2 W × ¾ TK Poplar	
0	14	Number not used	
1	13	Bottom Panel 46¹³⁄₁₆ LG × 19⁷⁄₁₆ W × ¾ TK Part. Board Fill/Face	
1	12	Shelf Edging 45¹³⁄₁₆ LG × ¾ W × 1½ TK Poplar	
1	11	Shelf 45¹³⁄₁₆ LG × 16⅝ W × ¾ TK Particle Board Fill/Face	
1	10	Top Panel Plastic 48 LG × 20 W × ¹⁄₃₂ TK	
1	9	Top Panel 47⁷⁄₁₆ LG × 19¹¹⁄₁₆ W × ¾ TK Part. Board Fill/Face	
1	8	Front Top Plastic 1¾ LG × 48 W × ¹⁄₃₂ TK	
1	7	Front Bottom Plastic 1½ LG × 48 W × ¹⁄₃₂ TK	
2	6	Front Edge Plastic 26²³⁄₃₂ LG × 1¹⁄₃₂ W × ¹⁄₃₂ TK	
2	5	End Panel Front Edging 29³¹⁄₃₂ LG × 1 W × ½ TK Poplar	
2	4	End Panel Plastic – 29³¹⁄₃₂ LG × 19¹³⁄₁₆ W × ¹⁄₃₂ TK	
1-LH 1-RH	3	End Panel – 29³¹⁄₃₂ LG × 19¹¹⁄₁₆ W × ¾ TK Part. Board Fill/Face	
1	2	Back Panel Plastic – 48 LG × 29³¹⁄₃₂ W × ¹⁄₃₂ TK	
1	1	Back Panel – 47³⁄₁₆ LG × 29³¹⁄₃₂ W × ¾ TK – Part. Board Fill/Face	
Qu.	No.	Part Name – Size – Material	
		Bill of Material	
	American Fixture, Inc.		Sheet
Title	Wrap Counter		1 of 1
Drawing No.	Selrile WC4		

23–74. *A fixture production drawing.*

Since the full-size section views are large, they can be drawn with sections broken out. For example, if a cabinet is four feet high and the center section has no special details, this section could be removed to make the drawing smaller. See Fig. 23–74.

If a unit is symmetrical, only one half need be drawn. See Fig. 23–74.

Dimensioning. The small scale elevations have only the overall dimensions given. The section views have overall and location dimensions. Location dimensions show the position of a part. Dimensions are shown to the edges of the wood parts. If parts have a wood veneer or plastic laminate, these are dimensioned as a separate piece.

The actual size of each piece is given on the parts list. The order of dimensions is length, width, and thickness. Sometimes the following abbreviations are used after the dimension: Length, LG; Width, W; and Thickness, TK. Each wood piece is given a number. Each metal or plastic piece is identified by a letter. The letters are used to key the part to the parts list. This method is the same as part identification on a machine assembly drawing.

The parts list shows the number of each part needed for one unit, the part identification number, the size of the part, the name of the part, and the material. See Fig. 23–74.

Some companies have a standardized system of joint sizes. These companies do not dimension the joints on their drawings. Fig. 23–74 shows an example of this practice.

All dimensions are given in inches. The inch symbol is generally used.

The manufacturing tolerance is usually shown in the title block. A common tolerance of $\pm 1/32$ inch is used on wood parts. Metal parts dimensioned to three decimal places have a tolerance of $\pm .010$. Those to two decimal places are $\pm .030$.

Build Your Vocabulary

Following are some terms you should understand and use as part of your working vocabulary. Write a brief explanation of what each means.

Fixture
Plywood
Particle board
Plastic laminate
Framed panels
Back panels
Rails
Web frames
Face frame
Laminating
Design drawing
Routing drawing

Additional Reading

Dreyfuss, Henry. *Designing for People.* New York: Grossman Publishers, Inc., 1967.

Dreyfuss, Henry. *The Measure of Man.* New York: Whitney Library of Design, 1967.

Feirer, John L. *Cabinetmaking and Millwork.* Peoria, Ill.: Chas. A. Bennett Co., Inc., 1970.

Furniture Forum. Sarasota, Fla.: Furniture Forum, Inc., Publishers.

Green, Lois W. *Interiors' Book of Offices.* New York: Whitney Publications.

Interiors Magazine. New York: Whitney Publications. Kirk, Paul D., and E. D. Sternburg. *Doctors' Offices and Clinics.* New York: Reinhold Publishing Co.

Ketchum, Morris Jr. *Shops and Stores.* New York: Reinhold Publishing Co., 1957.

Manasseh, Leonard, and Roger Cunliffe. *Office Buildings.* New York: Reinhold Publishing Co.

Ramsey, C. G., and H. R. Sleeper. *Architectural Graphic Standards.* New York: John Wiley and Sons, Inc., 1970.

Spence, William P. *Architecture—Design, Engineering, and Drawing.* Bloomington, Ill.: McKnight and McKnight Publishing Co., 1972.

Study Problems

Following are a variety of furniture drafting problems based on the work of experienced designers. Some of the things you need to show on the drawing are missing. An example is the type of joint to use. You must make these decisions.

Before starting the final production drawing, all decisions concerning size and methods of construction must be made. It is helpful to make a series of freehand sketches to try out and record the solutions. This will save much time when you start the final drawings.

These problems will give you the chance to study good design and learn about furniture construction and drafting.

1. Study the drawing of the lectern, Fig. 23–43A, designed by the Hugh Action Company. The inclined top is solid walnut. It has a metal strip on the lower edge to keep the papers from sliding off. The upright part is 1½-inches thick. It is made from ¼-inch walnut veneer plywood, fastened to a walnut frame. It is hollow in the center. The feet are metal. Prepare a complete set of scaled production drawings of the lectern, including a parts list.

2. Examine the slatted-top table, Fig. 23–47F. Prepare a complete set of drawings. Use the scaled furniture drawing system. The top is red-wood. The legs are steel pipe. They are welded and painted black. You will have to decide how to fasten the legs to the top and how to hold the top boards together. Make the table any size you want. Prepare a complete parts list. Identify each part by a name and number.

3. Using the full-size method of furniture drawing, prepare an assembly drawing of the end table with drawer storage designed by George Tanier, Inc., shown in Fig. 23–43B. Prepare routing sheets for the top, sides, and back of the case. The back has the same surface as the sides. The table is made of particle board with a plastic laminate over it. Design the drawer to use the metal drawer slides shown in Fig. 23–21.

4. Draw a set of production drawings for a bookcase. It has two adjustable shelves. Use metal shelf supports. Set them flush with the surface. The back is ¼-inch plywood. The sliding doors are ¾-inch thick. The sides and shelves are solid mahogany. The doors are particle board with a mahogany veneer on the outside face only. The bookcase should hold 7½ lineal feet of books. You must decide how deep it should be made. Design a base or legs to hold it off the floor. Make the drawings full size. Include a parts list.

5. The low cabinet and upholstered bench designed by Foster-McDavid, Fig. 23–43A, has square metal legs and frame. The cabinet case is made of plastic laminate over particle board. The thickness of these parts is ¾ inch. The doors are of the flush type. They are made with wood frame and a ¼-inch plywood panel. Prepare scaled production drawings of this furniture unit.

6. A low chest of drawers is shown in Fig. 23–43C. The case is made from particle board with birch veneer. The sides of the drawers are lipped to cover the edge grain of the side of the case. The drawers slide on a web frame with a plywood panel. Prepare full-size production drawings of this chest.

7. Chairs are very difficult to design. Study the wood frame lounge chair, Fig. 23–43D, designed by Torbjorn Afdal. Decide on a comfortable seat height. How much should the seat slope to the rear? What is a comfortable angle for the back? What is a good height for the arms? Prepare a full-size cardboard or plywood model of the seat and back, and experiment with different angles. After these decisions are made, prepare full-size production drawings of the chair.

8. P23–1 gives the overall sizes for typical wall and base

cabinets for a kitchen. Develop a set of working drawings from which these could be made. Use lipped drawers and doors.

Fixture Design Problems. Following are several fixture design problems. They are rather difficult and will require extra study. The size of the items to be displayed must be found. Suitable construction materials must be selected. A visit to a store having these units in use will give some ideas concerning design, size, materials, and construction methods to use. The examples given in Figs. 23–67 to 23–72 are general suggestions. The actual size and design of these units will vary from store to store.

When solving the following problems, several things are required. First, a sketch of the proposed solution must be made. This sketch can be on graph paper. The proposed dimensions should be placed on the sketch. Construction details must be decided. This may require a number of freehand sections or details. After the solution is clear, prepare full-size working drawings for the fixture.

9. Design a gondola for displaying pasteries in a self-service food store. The items to be displayed on one side include such things as pies, cakes, cookies, and donuts. The other side of the gondola will be used for bread.

Measure loaves of bread, cakes, and pies to find typical sizes. This will help you decide on the width of the fixture and shelf width.

10. Design an island-type fixture for displaying china, such as dinner plates, cups, saucers, dessert dishes, and serving bowls. The display should be visible from all sides. The customer should be able to examine the china. Consider making the shelves adjustable. Shelf brackets, such as those in Fig. 23–37, may be used.

11. A jeweler wants to display a collection of expensive rings. It is necessary that they be protected from theft. Design a display case for these rings. The lower part of the case will be used for storage. It should have a lock. The case should be loaded from the rear only. Consider building an artificial light into the case.

12. Visit a self-service food store. Examine the checkout counter. Make a list of the functions the counter must perform. For example, it must hold the cash register and bags for the groceries. Make a freehand drawing of a counter of your own design. Record the sizes needed. Select durable materials. Then prepare a set of production drawings.

Original Design Problems. The following design problems will give you the chance to try to develop original solutions.

Keep the solutions as simple as possible. Examine the pictures presented in this chapter. Visit furniture stores. Read magazines. Design ideas surround you.

13. Design a durable, light weight chair or bench for use on a patio or by a swimming pool. It must be able to withstand the weather and rough use.

14. Design a stand to hold a portable television. Select a particular television set and measure the cabinet. Examine the base to see how the stand must be built to hold the set. The stand should be light weight. Its appearance must add to the attractiveness of the set. The stand should be designed so the television and stand can be moved easily about a room.

15. Design a typing table. It must be strong enough to hold the weight of the machine yet be light enough to move easily. The top must hold a typewriter and have space for the copy to be typed.

16. Design a small bedside table. Select the type of styling you like. The table must hold a table lamp and a clock. It should contain at least one drawer for storage purposes.

17. Design a picnic table for use in your back yard. It should seat four people. A major defect in most picnic tables is that they tip over easily. Make certain your table permits two people to sit on one side when no one is

on the other without tipping. Keep the weight down so it can be moved by two people.
18. Design an end table or coffee table with a glass or marble top. The frame and legs should be metal. Select a size that will fit into your living room at home.
19. Design a conference table to seat 12 persons. Make the top and legs from any material you desire.
20. Design a check-writing

table for use in a bank. It can be on legs to stand in the center of the lobby or be a wall-hung unit without legs. Include storage for blank checks and deposit slips. Since these units get hard use, the materials must be durable.

Visit several banks and study the type of tables they use for this purpose.

21. Design a unit for holding coats, hats, and umbrellas

that can be used in a business office. It may be a wall-hung unit or have a floor stand. Attractive appearance is important. It should hold ten coats, four to six hats, and three or four umbrellas.
22. Restaurants are using small tables that seat two people. These can be placed side by side to seat four, six, or more persons. Design a table of this type that can be moved easily yet not tip over.

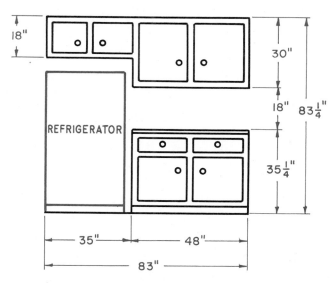

P23–1. Wall and base cabinets for a kitchen.

Chapter 24

Charts

Charts are used to describe facts, statistics, and relationships in a manner that is easy to understand. They use lines, bars, and other descriptive drawings.

The terms charts, graphs, and diagrams, actually have the same meaning. They tend to be used interchangeably in industry.

Charts have become increasingly important in all areas of business and industry. The stock broker keeps a record of stock prices with charts. A medical doctor records the characteristics of a heartbeat with a chart. The engineer uses charts to record test results. See Fig. 24–1.

BAR CHARTS

Bar charts are used to show comparisons of amounts. The numerical value is shown by the length of the bar. They are used to present data for the nontechnical reader. See Fig. 24–2.

There are several commonly used types of bar charts. These are shown in Figs. 24–3 through 24–10. A *single-bar chart* is the simplist form. Each bar shows a value. The bars can be vertical or horizontal. See Figs. 24–3 and 24–4. In Fig. 24–3 a grid is used to show the value of each bar. Another way to show values, Fig. 24–4, is to letter the values at the end of the bar. No grid is used. The values can be lettered on the bar, if desired.

A *multiple-bar chart* shows comparisons of several factors. Bars for each factor are drawn next to each other. Fig. 24–5 shows the comparison of iron ore and coking coal consumption over a two-year period.

A *divided-bar chart* is shown in Fig. 24–6. The total length of the bar shows the total tons used per year. For example, in 1963 120 million net tons were used. The bar is divided to show how much of the total was secured from three parts of the world. The bars in this example were placed touching. Space could have been left between them if the draftsman desired.

24–1. *Charts are widely used in business and industry.*

Chart-Pak

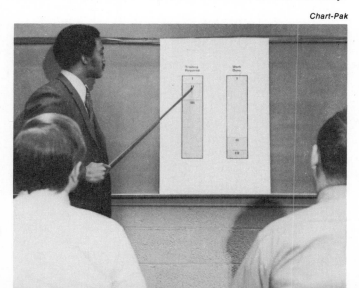

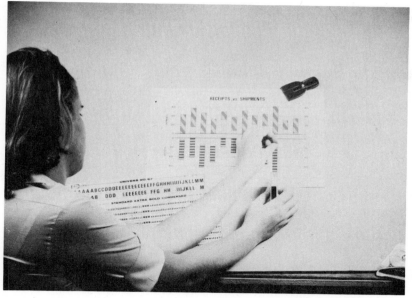

Chart-Pak

24-2.   A bar chart presents data for the nontechnical reader. This chart is being made with pressure-sensitive tape.

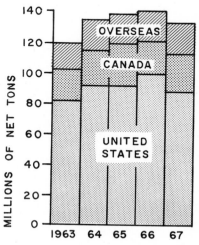

FINISHED STEEL SHIPMENTS

American Iron and Steel Institute

24-3.   A single-bar chart with vertical bars.

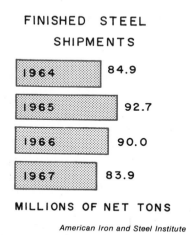

FINISHED STEEL SHIPMENTS

1964	84.9
1965	92.7
1966	90.0
1967	83.9

MILLIONS OF NET TONS

American Iron and Steel Institute

24-4.   A single-bar chart with horizontal bars.

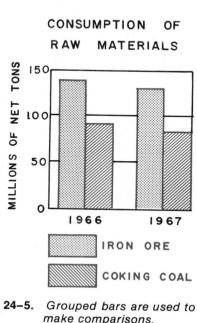

CONSUMPTION OF RAW MATERIALS

IRON ORE

COKING COAL

24-5.   Grouped bars are used to make comparisons.

SOURCES OF IRON ORE FOR U.S. MILLS

OVERSEAS

CANADA

UNITED STATES

American Iron and Steel Institute

24-6.   A divided-bar chart.

727

INCOME AND TAXES

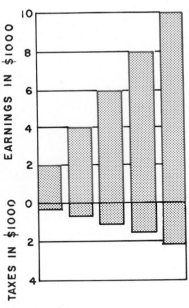

24-7. *An over-and-under bar chart.*

REVENUE DISTRIBUTION

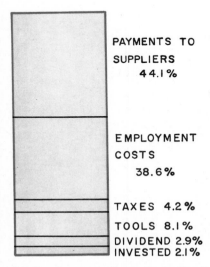

PAYMENTS TO
SUPPLIERS
44.1%

EMPLOYMENT
COSTS
38.6%

TAXES 4.2%

TOOLS 8.1%
DIVIDEND 2.9%
INVESTED 2.1%

24-9. *A one-column bar chart used to represent 100 per cent.*

Fig. 24–7, an over-and-under bar chart, shows the relationship between two series of numbers. This chart shows salary as related to the income tax paid.

A *range-bar chart,* Fig. 24–8, shows the high and low averages over a period of time. The difference between a high and a low is called the range. Examine the column for 1965. The year's low is 840 and the high is 975. The range is 135. This figure also shows another way to place bars on a chart.

Another type of bar chart is shown in Fig. 24–9. The single bar equals 100 per cent. It is divided into parts to represent the percentages of each item shown. The bar can be divided in many ways. Fig. 24–9 shows the percentages divided by lines. Each area could be shown by using different shading patterns or different colors.

There are many other variations that can be made when drawing bar charts. Some things to consider are:

1. The bars should be equal width.

2. A scale can be used along the side of the chart to show values.

3. Values can be lettered on or next to the bar.

4. Bars showing different things should have different surface indications. Color is very useful for this purpose.

5. The scale selected should show the differences in values. It can be placed on the horizontal or vertical axis.

SELECTED STOCK AVERAGES

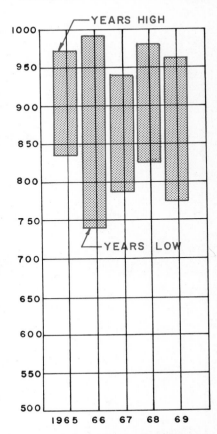

24-8. *A bar chart showing the range of values. This chart shows the high and low averages of a selected group of stocks for a five-year period.*

6. A title should be lettered on each chart. It is usually above the chart and off the grid. The title should be larger than the lettering on the chart.

7. The meaning of the chart should be clear even though it is separated from the text material it is illustrating.

8. Sometimes a key is needed to tell what each bar represents. See Fig. 24–5.

INDUSTRIAL USE OF STEEL

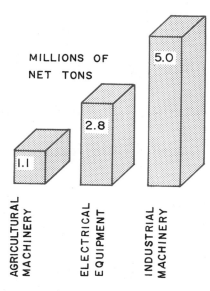

MILLIONS OF NET TONS

5.0

2.8

1.1

AGRICULTURAL MACHINERY — ELECTRICAL EQUIPMENT — INDUSTRIAL MACHINERY

American Iron and Steel Institute

24–10. *A pictorial bar chart.*

9. The bars can be drawn pictorially. See Fig. 24–10.

LINE CHARTS

A line chart is an easy way to show change. It is built around two axes. The vertical axis is called the Y axis. The horizontal axis is the X axis. See Fig. 24–11, Step 1. A scale is laid out on each axis. The scale represents a quantity such as pounds or time. See Fig. 24–11, Step 2. This produces a grid pattern. Next locate the value for each quantity. In this example, it is the millions of net tons produced each year. See Fig. 24–11, Step 3. After the points are located, connect them with

a line. For simple charts this is a straight line. See Fig. 24–11, Step 4.

The Variables. Generally the independent variable is placed on the X axis. The *independent variable* is that which is controlled. In a study of stock market prices, the months are controlled. The prices vary. The prices are the *dependent variable.* They vary from month to month. Dependent variables are placed on the Y axis. See Fig. 24–11, Step 1.

Scale. The selection of the scale is very important. The shape of the grid will influence the shape of the curve. Fig. 24–12 shows the values on each axis are equal. With this condition, the line should be on a 45° angle. Notice what happens to the angle on the line as the scales are changed. Angles greater than 40 degrees give the impression of an important, rapid use. Those 10 to 15 degrees indicate an unimportant rise. Study the significance of the change to be reported. Choose a scale that shows the importance of the change.

Grids for charts involving time are usually rectangular. Proportions of two to three or three to four are commonly used. A grid that is wider than it is high is best for time charts

24–11. *How to lay out a line chart.*

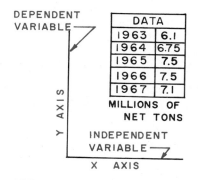

DATA	
1963	6.1
1964	6.75
1965	7.5
1966	7.5
1967	7.1

STEP I. LAY OUT THE AXES.

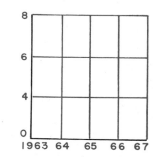

STEP 2. DEVELOP THE SCALE.

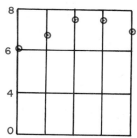

STEP 3. LOCATE THE VALUES.

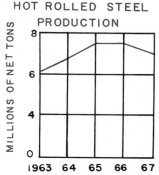

HOT ROLLED STEEL PRODUCTION

STEP 4. CONNECT THE VALUES. LETTER THE TITLES.

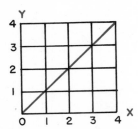

THIS IS THE WAY THE
GRAPH SHOULD APPEAR

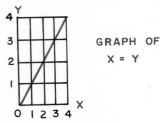

GRAPH OF
X = Y

X AXIS SCALE IS HALF
THE Y AXIS SCALE

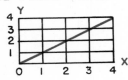

Y AXIS SCALE IS HALF

THE X AXIS SCALE

24-12. *The selection of the scale for the grid influences the impression given by the plotted data.*

VALUE OF A SHARE OF STOCK

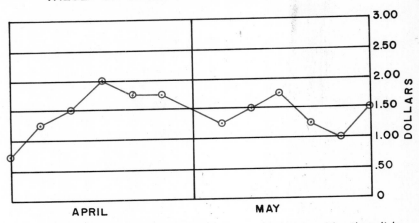

24-13. *Charts with many points use a grid that is wider than it is high.*

TRENDS IN STEEL
IMPORTS

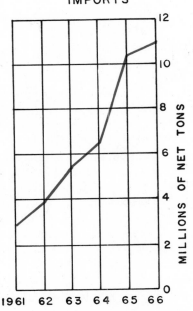

American Iron and Steel Institute

24-14. *Charts with a rapid rise use a grid that is higher than it is wide.*

INTEREST RATE

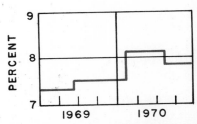

24-15. *This is a step-line chart.*

having many points to be plotted. It is also used when it covers a long period of time. See Fig. 24-13. A grid that is higher than it is wide is used for time charts covering short periods or having a rapid change. See Fig. 24-14.

Plotted Points. Line charts which are used for general information purposes do not have each point noted. The points are located where the ends of the plotted lines meet.

See Fig. 24-14. The line used to show the curve is usually drawn thick.

If the value of each point is important, a dot within a small circle is used. The dot accurately locates the value. The line forming the curve touches the circle and stops. See Fig. 24-13. Usually this curve is drawn with a thin line.

Stepped-Line Chart. A *stepped-line chart,* Fig. 24-15,

LASER TEMPERATURE RISE

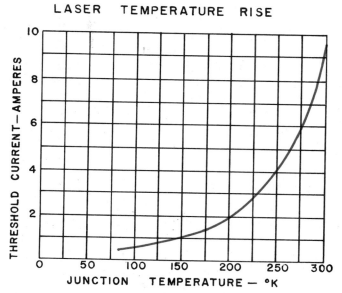

24–16. *This is a smooth curve graph. Changes in one variable cause changes in the other variable.*

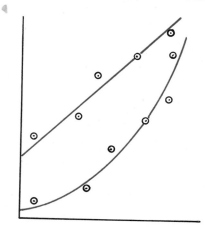

24–17. *Drawing a smooth curve for continuous data when all points do not fall on the curve.*

is used when data shown retains a fixed value over a period of time. Fig. 24–15 shows the interest rate was 7.3 per cent at the beginning of 1969. It stayed at this rate for almost half the year. It then increased to 7.5 per cent. It stayed at this rate into 1970. During 1970 it raised above 8 per cent and then dropped.

The Smooth-Curve Graph. Graphs showing continuous data use a smooth curve to connect the points. In continuous data we find a mathematical relationship. As one variable changes, the other also changes. For example, Fig. 24–16 shows the change in temperature as the amperes are increased. The more amps

731

of electric current, the higher the temperature generated.

In some cases, the points will not fall exactly on a smooth curve. However, the curve is still drawn smoothly even though some points will be missed. Generally the curve is drawn so some points fall on each side. See Fig. 24–17.

Multiple-Line Charts. A line chart can be used to show more than one set of related data, but the lines used to show the data should be different. They can be solid, dotted, or dashed. They can be different colors. See Fig. 24–18.

24–18. *Multiple-line charts show several sets of related data. This chart was made with pressure-sensitive tape.*

Chart-Pak

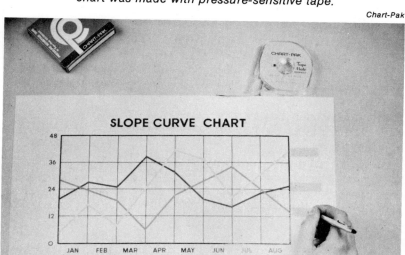

LOAD CARRYING CAPACITIES
OF SOLID WOOD GIRDERS

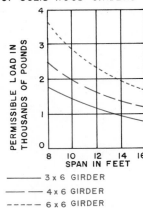

3 x 6 GIRDER
4 x 6 GIRDER
6 x 6 GIRDER

24–19. *This is a multiple-line chart. Several sets of related data can be shown on one chart.*

STEEL PRODUCTION IN CANADA

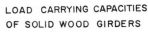

American Iron and Steel Institute

24–20. *Line chart curves can be identified by lettering along the curve.*

It is necessary to explain what each curve represents. This can be done with a key. See Fig. 24–19. Another way is to letter a caption by each curve. See Fig. 24–20.

The data on a multiple-line chart should be related in some way. In Fig. 24–19 the lines represent the load-carrying capabilities of solid wood girders. The scales are common to all three girders. The strength of the girders is compared with each other.

Surface Charts. A *surface chart* shows values by the height of a shaded surface. They can be used to give emphasis to a single line chart. See Fig. 24–21.

Surface charts can show the difference between several sets of data. See Fig. 24–22. This figure shows the raw steel

AGE OF STEELWORKERS

American Iron and Steel Institute

24–21. *A surface chart gives added emphasis to data.*

RAW STEEL PRODUCTION

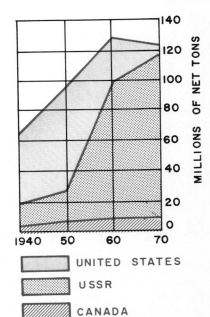

UNITED STATES

USSR

CANADA

INDUSTRIAL USE OF STEEL

American Iron and Steel Institute

24–22. *A surface chart can show the difference between several sets of data.*

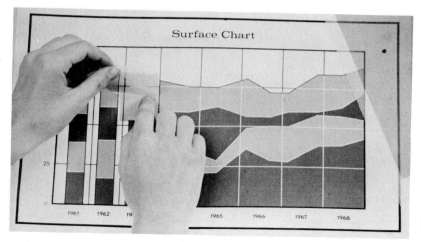

Surface Chart

Chart-Pak

24–23. *Pressure-sensitive sheets are used to make surface charts.*

COST AND SALES PRICE

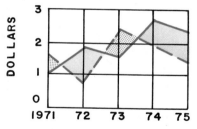

—————— COST

— — — — SALE PRICE

▓▓▓▓ PROFIT

▢▢▢▢ LOSS

24–24. *Surface charts can show the net difference between data.*

production for three countries. Each set of data is related to the zero point. The shaded area is the difference between these countries. Fig. 24–23 shows a surface chart made with pressure-sensitive sheets.

Surface charts also show the net difference between two series. For example, it can show the gain and loss. It can compare performance with a standard or average. See Fig. 24–24. In this figure, the difference between the cost and selling price is shown. The difference, shown by the shaded area, is the profit and loss.

Surface charts can be used to show the cumulation of data. It is the addition of data graphically. See Fig. 24–25. In this figure, the lower shaded area is the actual value of a stock. The stock earns dividends. This is shown by the top shaded area. The total value of the stock is the actual value plus the dividends.

The layout of surface charts is much the same as for line charts. The shading helps to emphasize the points the chart is intended to show.

VALUE OF 100 SHARES OF STOCK

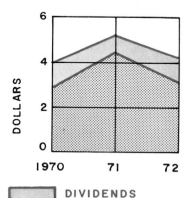

▢▢▢▢ DIVIDENDS

▓▓▓▓ VALUE OF STOCK

24–25. *Surface charts can be used to show the addition of data. Each area is a separate amount. The top line shows their total value.*

Quadrants. Most line charts are drawn in the first quadrant. See Fig. 24–26. Some data, especially mathematical data, require the use of several quadrants. Each quadrant represents a positive and negative

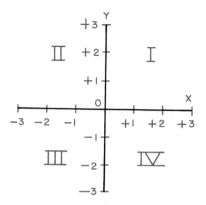

24–26. *Line charts can be drawn in any of the four quadrants. The quadrants used depend on the data to be plotted.*

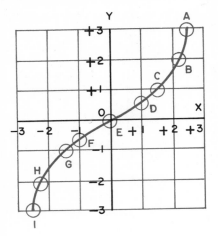

DATA		
	X	Y
A	2.5	3.0
B	2.25	2.0
C	1.5	1.0
D	1.0	0.6
E	0	0
F	−1.0	−0.6
G	−1.5	−1.0
H	−2.25	−2.0
I	−2.5	−3.0

24–27. *This line chart shows data having positive and negative values. The first and third quadrants were used.*

value of X and Y. The point of intersection of axes X and Y is zero. The values to the right of the Y axis are positive. Those to the left are negative.

The Y values above the X axis are positive. Those below are negative. Each point has an X and Y value. Study the data shown in Fig. 24–27. Notice how each point was plotted using its X and Y value. It is essential to pay attention to the positive or negative sign on each value.

Lettering on the Chart. Titles and captions on the scales are placed outside the axes. The scale captions are read in the aligned system. In other words, they are read from the right and bottom of the chart. The vertical axis caption is read from the right. The vertical axis scale is read from the bottom. Both the scale and caption on the horizontal axis are read from the bottom.

The caption and scale for the vertical axis can be placed on the left or right side of the chart.

The title for the line chart is usually placed above the chart.

In many drafting rooms, all lettering is typed on the chart. If the chart is on vellum, a piece of carbon paper is placed beneath the sheet. This produces a darker image because the carbon leaves an impression on the back of the sheet. Pressure-sensitive letters can be used on charts.

PIE CHARTS

Pie charts are sometimes referred to as percentage charts. These are similar to a bar chart because there is a visual comparison between areas. A pie chart shows the amount or percentage by a sector of a circle. These charts show how a given amount is divided into specific parts or percentages. Fig. 24–28 shows a pie chart. The circle represents 100 per cent or any specified amount.

Pie charts are most effective when there are no more

RAW STEEL PRODUCTION BY TYPES OF FURNACES IN THE U.S.

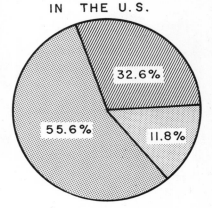

24–28. *A pie chart.*

than eight divisions. If more than eight parts are used in the chart, it becomes difficult to understand. The circle should be large enough to permit lettering without crowding. It is standard practice to place all lettering horizontally on the sector of the circle. If several of the sectors are too small, place the lettering on the outside of the circle. The lettering is then connected to the sector with a leader.

It is a simple matter to convert percentage to degrees of a sector. Since there are 360 degrees in a circle, simply multiply 360 degrees by the percentage. This will give the number of degrees to be laid off. For example, to find the

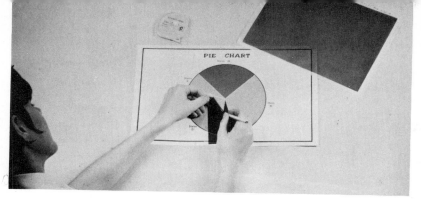

24–29. *Pie charts are easily made using sheets of pressure-sensitive shading materials.*

Chart-Pak

WORLD STEEL PRODUCTION

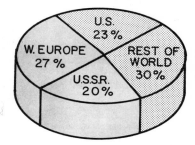

A PICTORIAL PIE CHART

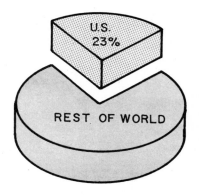

SECTORS CAN BE REMOVED FOR ADDED EMPHASIS

American Iron and Steel Institute

24–30. *Pictorial pie charts can be drawn in many ways.*

MOTOR VEHICLE OUTPUT

EACH SYMBOL EQUALS ONE MILLION VEHICLES

24–31. *A pictorial chart. This is in bar chart form.*

number of degrees in 35 per cent:

First, convert the per cent to a decimal, 35% = .35.

Second, multiply 360° by the per cent, $.35 \times 360° = 126°$.

Areas of the pie chart are usually shaded. Each sector should be shaded in a different manner. The use of color is especially effective. See Fig. 24–29.

In some cases, a pie chart may be made in pictorial form. See Fig. 24–30. When a pictorial pie chart is drawn, the sectors of the circle may be separated. The pictorial form is not effective for charts having over five or six segments.

PICTORIAL CHARTS

Pictorial charts are sometimes referred to as pictographs. These are classed as nonmathematical charts. A pictorial chart is a visual form of presenting information to the nontechnical reader. These charts give comparisons by the means of pictures. Each picture represents a specific unit or quantity. See Fig. 24–31. The pictorial chart is a form of a bar chart. The row of representative symbols makes up the bar.

The pictorial chart has a limited degree of accuracy. Usually the amount represented by the symbols is given next to the symbols. In Fig. 24–31 each symbol shown equals one million vehicles. A

735

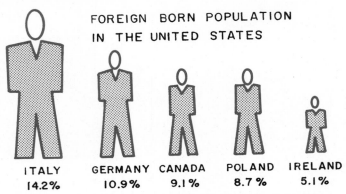

24-32. *A picture chart in which the size of the figure is used to make a comparison of the data.*

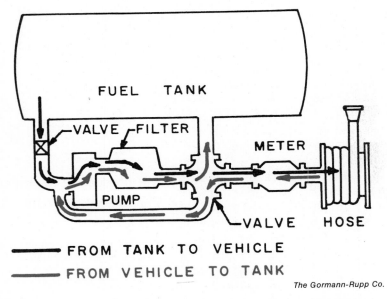

The Gormann-Rupp Co.

24-33. *A flow chart shows the movement of flow involved in a process. This diagram shows the flow of fuel from a tank truck to an aircraft and the reverse flow to remove fuel from it.*

key should always be given on the chart. The key shows the symbol and its value.

Some forms of pictorial charts use pictorial symbols to represent a quantity. The comparison is made by the difference in size of the sym-

bols. See Fig. 24-32. These charts can be misleading. The reader frequently does not know whether the draftsman intended the symbol to be concerned with area or linear size. As an example, a symbol measuring 2″ × 2″ would have an

area of 4 square inches. If the draftsman wanted to show twice the amount, he might draw the symbol 4″ × 4″. But then the area is 16 square inches or four times the original size, rather than twice the amount. This would be misleading to the reader. Therefore, it is necessary to give the value each symbol represents.

FLOW AND ORGANIZATION CHARTS

Both the flow and organization charts are similar in purpose. These are nonmathematical charts. The advantage of a flow or organizational chart is that it replaces lengthy verbal or written description. See Fig. 24-33. A *flow* chart shows a schematic representation of a process. It may be used to represent the flow or distribution of materials as they pass through a manufacturing plant. For example, the processing, production, packaging, and marketing of a product can be easily described. Frequently, pictures or schematic symbols are used to describe the process. A simple form of a flow chart uses only a series of rectangles. The name of the process is lettered in the rectangle. See Fig. 24-34. More complex forms use pictures that resemble the units involved. See Fig. 24-33.

Organization charts show the line of authority. They show the responsibility of the executive to a division and to a divisional head. Interrelated and interlocking activities of an organization can be easily diagrammed. An organization chart may also be used to show the separate functions of a company. The names of divisions or functions are enclosed in rectangles. For variety, use both rectangles and circles. See Fig. 24–35.

COORDINATE PAPER

The easiest way to lay out a chart is on coordinate paper. Some of the types available have rectangular rulings 4, 5, 6, 8, 10, 12, 16 and 20 spaces to the inch. They are available on opaque paper and vellum.

Special types of graph paper are available for pie charts which represent percentages. Polar paper is used for plotting curves by polar coordinates. Triangular coordinate paper is used for plotting a curve based on three variables. Logarithmic graph paper is divided on the basis of logarithms. This is used when a percentage of change is more important than the quantity of change.

DRAFTING AIDS

In recent years drawing of charts has been greatly simplified. Pressure-sensitive tapes, shading films, and other components have been developed that reduce the time it takes to make a graph. Adhesive-backed tapes can be used on charts. Tapes range in size from 1/16 to 2 inches in width. They are available in assorted colors, densities and surfaces.

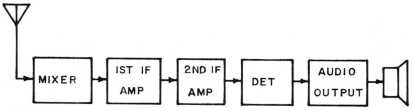

24-34. *This flow chart shows the major components of a transistor radio receiver.*

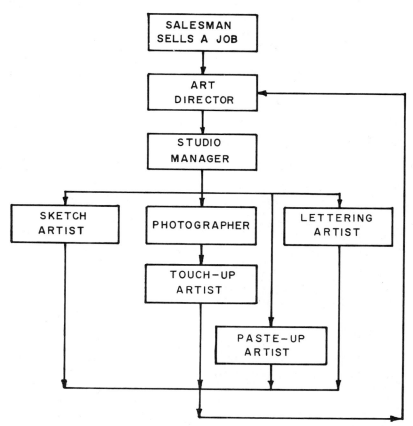

24-35. *This is an organization or flow chart. It shows the relationship between parts of an organization or process. This chart shows how a job flows through an advertising agency.*

See Fig. 24–2. If a chart is to be used on an overhead projector, colored transparent tape can be used for grids, curves, and bars. Glossy tapes are used primarily for direct presentation.

Curves can be "drawn" by using thin tape. The tape is in rolls. It is held in a special con-

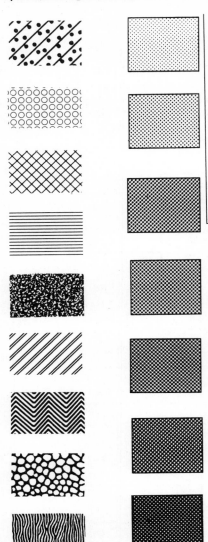

24–36. *These are a few of the shading patterns available on pressure-sensitive sheets.*

transfer lettering

a s *G* H h

C F *k* S

G C SAM p

M HOW ROTEX

Chart-Pak

24–37. *A few of the sizes and styles of pressure-sensitive letters.*

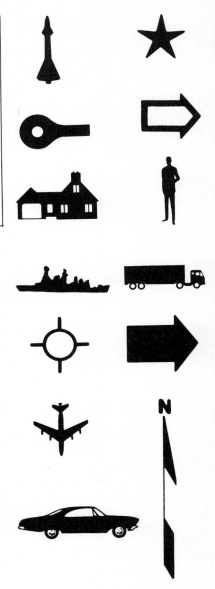

Chart-Pak

24–38. *A few of the symbols available on pressure-sensitive tape.*

tainer that makes it easy to lay the tape on the curves. See Fig. 24–18.

Pressure-sensitive screening, Fig. 24–36, is also used to make charts. Various dot and line pattern screens are used to identify different parts of a pie or multiple-bar chart. Screens may also be used for bar charts. Most of the charts in this chapter were shaded with this material.

Pressure-sensitive lettering, Fig. 24–37, is sometimes used for identification on all types of charts. This material is applied directly to the chart. It is available in a wide variety of sizes and faces.

Symbols are available for use in pictorial charts. These symbols are used when figures must be repeated. Examples of these symbols are shown in Fig. 24–38.

When any type of pressure-sensitive material is used, errors may be corrected very easily. The tape or screen can be lifted off with an "X-Acto" knife or razor blade.

Build Your Vocabulary

Following are some terms you should understand and use as part of your working vocabulary. Write a brief explanation of what each means.

Bar chart
Line chart
Variables
Grid
Smooth curve graph
Multiple-line chart
Surface chart
Quadrants
Pie chart
Pictorial chart
Flow chart
Organization chart
Pressure-sensitive tape
Shading film

Study Problems

1. Make a bar chart showing the telephones in service, P24–1.

2. Make a line chart showing the trend of U.S. investments abroad, P24–2.

3. Make a multiple-line chart comparing the tourist arrivals, P24–3.

4. Make a pie chart or percentage bar showing the world production of nickel, P24–4.

5. Make a chart comparing the ten largest nations by area, P24–5.

6. Make a double-bar chart showing the monetary reserves of the government of West Germany, P24–6.

7. Make a high-low bar chart showing the frequency ranges of several radio bands, P24–7.

8. Plot the curve using the data shown in P24–8.

9. Develop an over-and-under bar chart to compare the accounts paid and the delinquent accounts, P24–9.

10. Make an organization chart showing the line of authority in your school system. Your teacher or principal can help trace the line of authority, beginning with the voters who elect the school board.

11. Make a flow chart showing the generation and distribution of electricity to residential areas. The electricity is generated in a generating plant. It then goes to a step-up transformer. This increases the voltage and sends it over high voltage wires to a step-down substation. This lowers the voltage for distribution in a local area, such as a city. The current then travels over wires to a small distribution station. This station sends the current over wires to local residential areas. In these areas are small step-down transformers. They lower the voltage and send it to a residence.

P24–10 through P24–18 contain a variety of data to be charted. Select the type of chart you think will best show the data. Draw the chart accurately.

P24–1.

Telephones in Service			
Australia	2,978,336	West Germany	9,532,417
Brazil	1,431,653	Japan	16,011,745
Canada	7,880,471	U.S.S.R.	8,400,000
Finland	8,923,000	United States	98,789,000
France	6,554,441	United Kingdom	11,376,000

P24–2.

U.S. Investments Abroad*	
1967	59,491
1968	64,983
1969	71,016
1970	78,090

*Millions of Dollars.

P24–3.

International Tourist Arrivals				
Country	1966	1967	1968	1969
Africa	1,833,000	1,926,000	2,000,500	2,105,000
Europe	95,975,000	100,322,000	100,501,000	101,397,000
Central & N. America	23,753,000	27,379,000	29,479,000	29,970,000
Middle East	3,138,000	2,510,000	2,102,000	2,000,000

P24–4.

World Production of Nickel*	
Canada	247
U.S.S.R.	105
Rest of World	132
World Total	484

*Thousand Tons.

P24–5.

Ten Largest Nations by Area*	
U.S.S.R.	8,649,489
Canada	3,851,809
China	3,691,502
United States	3,615,211
Brazil	3,286,470
Australia	2,967,877
India	1,232,560
Argentina	1,072,156
Sudan	967,491
Algeria	919,590

*Square Miles.

P24–6.

Monetary Reserves — West Germany*		
	Gold	Foreign Exchange
Dec. 1958	2.5	3.1
Dec. 1967	4.2	2.8
March 1968	3.95	3.3
Sept. 1968	4.4	3.1

*Billions of Dollars.

P24–7.

Frequency Ranges of Radio Bands	
Radio Frequency Bands	Frequency Ranges*
Very Low Frequency — VLF	3 to 30
Low Frequency — LF	30 to 300
Medium Frequency — MF	300 to 3000

*Cycles Per Second.

P24–8.

Points on a Curve When X = Y		
Point	X	Y
1	-5	$+5$
2	-4.0	$+3.0$
3	-3.0	$+1.8$
4	-1.5	$+.75$
5	0	0
6	$+1.5$	$-.75$
7	$+3.0$	-1.8
8	$+4.0$	-3.0
9	$+5.0$	-5.0

P24–9.

Yearly Financial Report*				
	Jan.–March	April–June	July–Sept.	Oct.–Dec.
Accounts Paid	10.2	7.5	9.5	13.8
Accounts Delinquent	1.5	5.7	4.2	0.9

*Thousands of Dollars.

P24–10.

Voltages Produced by Different Sources	
Source	Voltage Range
Batteries	6–12
Auto Generator	6–12
Train Generators	32–60
Residential Transformers	120–440
City Distribution Transformers	2,300–4,200
Power Transmission Lines	13,200–287,500

P24–11.

Ten Largest Nations by Population	
China	850,000,000
India	524,080,000
U.S.S.R.	234,396,000
United States	201,750,000
Indonesia	111,000,000
Pakistan	109,519,831
Japan	100,501,000
Brazil	89,376,000
West Germany	60,165,000
Nigeria	61,450,000

P24–12.

U.S. Corporate Bond Prices*	1968	1967	
January	77.2	85.9	
February	77.5	86.4	
March	76.9	85.6	
April	76.2	85.4	
May	75.3	83.4	
June	76.6	81.7	
July	76.1	81.1	
August	78.1	80.3	
September	78.4	80.0	
October	77.0	78.5	
November	75.7	76.8	
December	75.0	75.9	

*Average Price in Dollars Per $100 Bond.

P24–13.

World Production of Tin*	
Malaysia	80,766
Bolivia	30,117
Thailand	25,118
China	22,000
U.S.S.R.	28,000
Indonesia	15,229
Rest of World	40,802

*Tons.

Plastics Production*		
	1967 Monthly Average	1968 Monthly Average
United States	475	510
Japan	340	390
West Germany	230	250
U.S.S.R.	90	110
United Kingdom	90	95

P24–14.

*Thousand Metric Tons.

World Petroleum Production*	
North America	11.74
South America	4.54
Western Europe	0.40
Africa	3.13
Middle East	10.05
Asia	0.78
Eastern Europe	6.33
World Total:	36.97

P24–15.

*Millions of Barrels Daily.

P24–16.

Strength of Various Steels	
Steel Number	Tensile Strength PSI
C1018	69,000
B1112	82,500
B1113	83,500
Ledloy 375	79,000
C1045	103,000
C1095	145,000

P24–17.

Composition of German Silver	
Copper	50%
Nickel	20%
Zinc	30%

P24–18.

Melting Points of Metals	
Metal	Temperature Degrees Fahrenheit
Pewter	420
Tin	449
Lead	621
Zinc	787
Aluminum	1218
Bronze	1675
Gold	1945
Cast Iron	2200
Steel	2500

Section Six
Related Technical Information

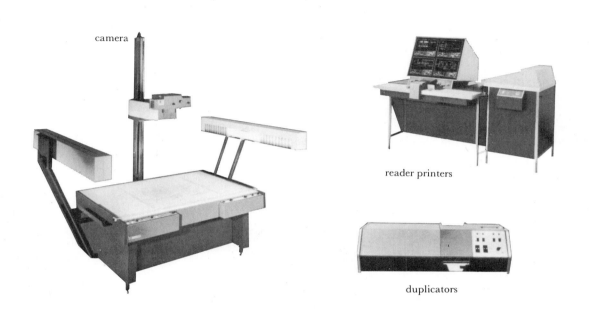

camera

reader printers

duplicators

Links to Learning

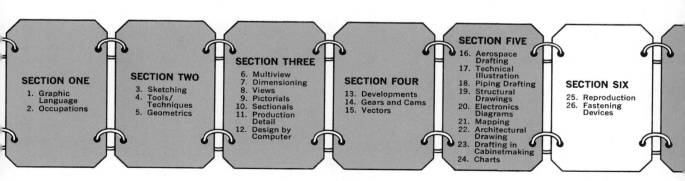

Chapter 25

Reproduction of Drawings

Most drawings are made on vellum, tracing cloth, or polyester drafting film. Since these are transparent, it makes it possible to produce exact copies. These copies are called *prints.* After a print is made, the original drawing can be stored. This protects it from damage. There are several ways copies can be made. The most common are blueprints and whiteprints.

BLUEPRINTS

A blueprint has a blue background and white lines. See Fig. 25–1. It is made on a very strong paper that will withstand hard use. The blue color fades very little.

Blueprint paper has a chemical coating on one surface. This coating is made from a solution of potassium dichromate and ferrocyanide of potassium. It dries a light yellow. When ultraviolet light hits this chemical, it turns blue.

A blueprint is made by placing the original drawing next

to a piece of blueprint paper. See Fig. 25–2. Together they are exposed to an ultraviolet light. This causes a chemical reaction to take place wherever the light hits the blue paper. The lines on the drawing protect parts of the blueprint paper. Here the light does not touch the chemical coating. After the paper has been ex-

posed to the light the proper length of time, the blueprint paper is separated from the drawing.

Next the blueprint paper is developed by washing it in water. The areas protected by the lines wash away. The white paper shows through. The areas not protected are blue. The blueprint is then placed

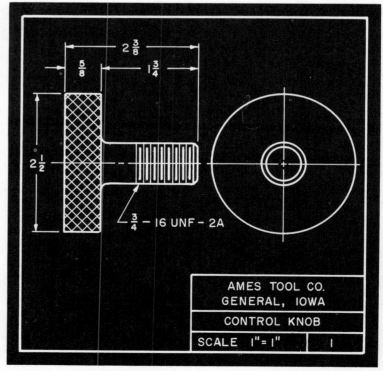

25–1. *A blueprint has white lines and a blue background.*

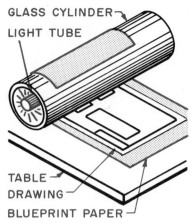

GLASS CYLINDER
LIGHT TUBE

TABLE
DRAWING
BLUEPRINT PAPER

STEP I
THE BLUEPRINT PAPER IS
EXPOSED WITH THE DRAW-
ING TO A LIGHT.

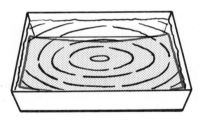

STEP 2
THE BLUEPRINT IS WASHED
IN WATER.

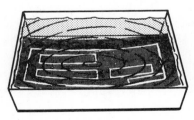

STEP 3
THE BLUEPRINT IS BATHED
IN A POTASSIUM BICHRO-
MATE SOLUTION.

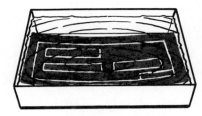

STEP 4
THE BLUEPRINT IS RINSED
IN WATER.

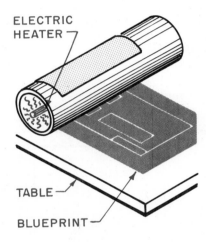

ELECTRIC
HEATER

TABLE

BLUEPRINT

STEP 5
THE BLUEPRINT IS DRIED.

25–2. *How a blueprint is made.*

in a solution of potassium bichromate. This fixes the print so it is rather permanent. It is then washed again in water and dried.

Industry makes blueprints on machines that do all of these operations automatically. Fig. 25–3 shows a machine with a large roll of blueprint paper. It is protected from light. The drawing is fed into the machine. The blue-

print paper unwinds as the drawing enters. Then it goes through all the steps to make a final blueprint.

WHITEPRINTS

Whiteprints have colored lines on a white background. The lines are usually blue, black, or maroon. See Fig. 25–4.

Whiteprint paper has an azo dye coating. The process is commonly called the diazo process.

This process is much like the blueprint process. See Fig. 25–5. The whiteprint paper has the drawing placed on top. It is fed into a diazo printing machine, where it is exposed to ultraviolet light. The light destroys the azo dye, leaving the white paper. The lines on the drawing keep the light from destroying the dye under them.

The exposed print passes through an ammonia vapor. This vapor develops the azo dye left on the surface. The lines appear in color on the white background.

The process described above is a dry process. The finished print is dry. The ammonia is heated to a vapor for developing.

The same process is used for wet diazo prints. These are the same as dry prints except a liquid developer is used. The exposed print runs through rollers in a liquid developer. The wet print then runs over electric heater coils for drying.

Revolute Blueprint Machine
Charles Bruning Co.

25-3. *This is an industrial blueprint machine. It exposes and develops the blueprint.*

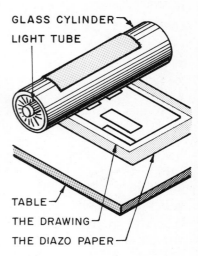

GLASS CYLINDER
LIGHT TUBE

TABLE
THE DRAWING
THE DIAZO PAPER

STEP I
THE DIAZO PAPER IS EXPOSED WITH THE DRAWING TO THE LIGHT.

THE AMMONIA VAPORS ARE EVAPORATED BY THE HEATER, GO THROUGH THE HOLES AND DEVELOP THE PRINT.

THE DIAZO PRINT PASSES OVER HOLES.

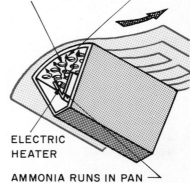

ELECTRIC HEATER

AMMONIA RUNS IN PAN

STEP 2
THE PRINT IS DEVELOPED WITH AMMONIA VAPORS.

25-5. *Steps for printing a dry diazo whiteprint.*

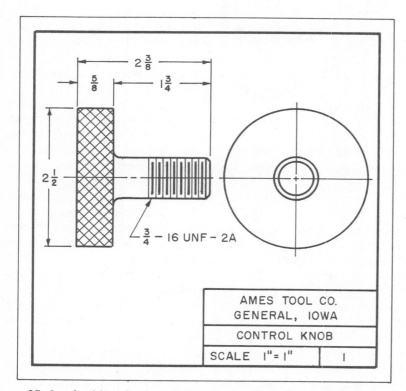

$2\frac{3}{8}$

$\frac{5}{8}$

$1\frac{3}{4}$

$2\frac{1}{2}$

$\frac{3}{4}$ - 16 UNF - 2A

AMES TOOL CO.
GENERAL, IOWA

CONTROL KNOB

SCALE 1" = 1" | 1

25-4. *A whiteprint has colored lines on a white background.*

A whiteprinter is shown in Fig. 25–6.

VANDYKE PRINTS

A vandyke print is a negative copy of the original drawing. The lines on the drawing appear white against a dark brown background. It is made in the same manner as a blueprint. The negative is then used to make blueprints instead of the original drawing. It produces a print that is the reverse of a blueprint. The lines are blue and the background is white. These are called blue-line prints.

INTERMEDIATES

An intermediate is a copy of the original drawing developed by the diazo process on a special translucent paper. The intermediate copy is used to produce additional whiteprint copies instead of the original drawing. This saves wear and tear on the original. Often the intermediate will produce better copies than the original drawing.

Changes can be made on the intermediate without changing the original. Lines to be removed are painted out with a special correction fluid. New lines can be drawn on the intermediate copy. It is possible to cut away a section of the intermediate and glue in a revised section drawn on vellum.

MICROFILM

A microfilm can record a drawing on a small strip of film

25–6. *A whiteprint machine.*

Charles A. Bruning Co.

which, when magnified, gives a full-size conventional drawing. The film images can be on rolls of film or placed individually in aperture cards. See Fig. 25–7. The aperture card securely holds the film. It has punched into it information about the drawing. This information will vary from one company to another.

Microfilm systems in use vary with their purpose. Most systems are designed especially to fit the needs of a particular company. Following is such a system that shows the basic features. See Fig. 25–7.

The original drawing is photographed. A specially designed microfilm camera is used, Fig. 25–7. This reduces the drawing to a small nega-

tive. The film is processed in a film processor, Fig. 25–7. See page 750.

After the negative is developed, the original negative is mounted in an aperture card, Fig. 25–8. If additional negatives are needed, they are produced on a card duplicator. See Figs. 25–7 and 25–9.

The aperture card can be placed in a reader-printer. This machine projects an enlarged image of the drawing on a screen. It also produces enlarged prints. See Figs. 25–7 and 25–10.

Microfilm systems are found in three sizes, 16mm, 35mm, and 105mm.

There are many advantages to having a microfilm system for engineering drawings. One big advantage is storage.

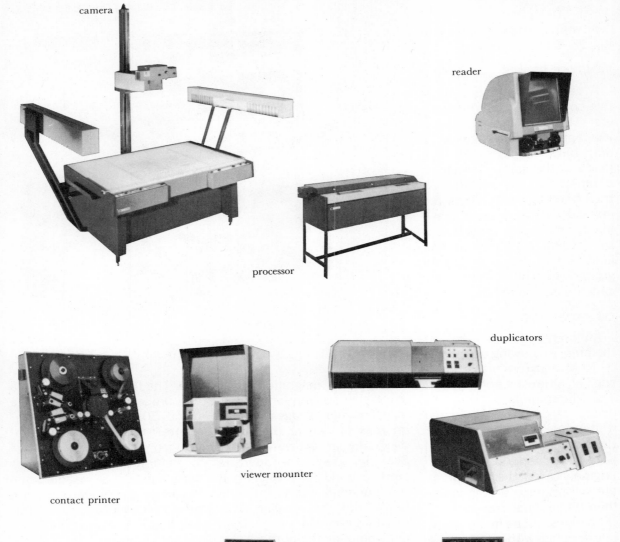

camera

reader

processor

duplicators

contact printer

viewer mounter

cards and copycards

reader printers

25-7. *Machines used in a microfilm system.*

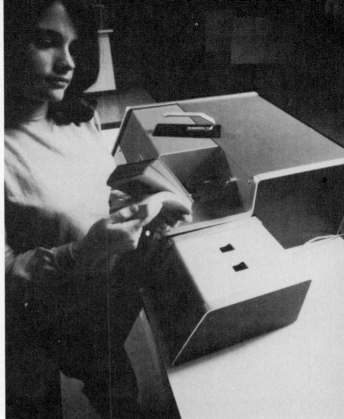

25-8. *This viewer mounter is used to mount both pressure-sensitive and thermal-bonded film cards. It has a large viewing screen.*

25-9. *This film card duplicator feeds and stacks up to 150 diazo copycards. It uses a sealed anhydrous ammonia system.*

25-10. *This reader-printer uses both 16mm and 35mm film. Images are magnified 14.5 times on the 18-inch viewing screen.*

Full-size drawings are difficult to store. They are stored flat or rolled. Since the drawing is reduced on microfilm to a small negative, hundreds of microfilmed drawings can be stored in a small cabinet.

Microfilm can help preserve valuable original drawings. After the microfilm is made, the original drawing can be stored. The drawing can be easily sent to other companies by mailing a copy of the microfilm. This is much easier than mailing a large bundle of drawings.

Drawings stored in a microfilm system are easy to locate. The aperture card has specific information punched into it which helps find the drawing wanted. Microfilm in rolls can be rapidly examined by running the roll through a machine that projects the image on a screen. This is much faster than sorting through the full-size original drawings.

Reproduction of prints from microfilm is rapid. Machines are available that will produce copies automatically. Some of these machines can control automatically the number of copies produced. They also can reproduce the drawing in a variety of sizes.

Preparing Drawings for Microfilm. A drawing to be microfilmed must be carefully made. Two critical parts are line uniformity and density, and lettering.

All lines of a particular type must be of the same width. For example, all visible lines should be the same width. Lines should be as dense as possible. Density refers to a line's ability to keep light rays from passing through it. The degree of density should be uniform for all lines.

Ink drawings produce the best microfilm. The inking tools permit control of line width. The ink dries with a uniform density. Ink lines are usually blacker, sharper, and have a higher, more uniform density than pencil lines. The surface of ink lines dries flat. There is less reflection. The reduction of reflection produces an image on microfilm that has a higher line-to-background density ratio. Microfilms made from ink drawings are easier to read on a reader machine and make better prints.

Pencil drawings can be used for microfilm. Before this is done, the draftsman should experiment with a variety of vellums and pencils. Some produce better microfilm than others. Once a satisfactory selection is made, do not vary from it. Do not use several different kinds of pencils on the same drawing, or it will be difficult to get uniform line width. The density of the line will vary considerably as well.

Lettering must be large enough to be reduced photographically and then enlarged without loss of clarity. Most companies establish standards for lettering heights that will fit their microfilm systems. Fig. 25–11 shows sizes recommended by the American National Standards Institute. The size varies with the sheet size. The larger the sheet size, the larger the lettering required.

		35mm Film Reduction	Blowback (% of Original Size)	Recommended Lettering Height	
ANSI Drafting Size				Fractions	Decimals
A	8½ × 11	16	.919	1/8	.125
B	11 × 17	20	.735	1/8	.125
C	17 × 22	24	.613	5/32	.150
D	22 × 34	28	.525	3/16	.170
E	34 × 44	28 or 30	.490	3/16	.170

American National Standards Institute

25–11. *Microfilm reductions, blowback sizes, and recommended lettering heights.*

These recommendations are for the 35mm microfilm system.

Notice that the A size sheet, Fig. 25–11, has a reduction of 16 times while the E size has a reduction of 28 or 30 times. The A size sheet is blown back by the microfilm printer to a drawing .919 per cent of the original size. The E size sheet is blown back to .490 per cent of the original size.

All letters must be carefully formed. The open spaces in each letter must be sharp and clear. Letters should not be crowded together. If they are too close, they may blend together when reduced. The space between lines of lettering is important. It should be at least equal to one half the height of the letters. Extra spacing is usually left between paragraphs. A space equal to one line of lettering is needed.

The most easily read lettering is capital Gothic letters. They are clean cut letters and free of unnecessary lines.

Often lettering is typed on drawings to be microfilmed. Care must be used when typing to ensure a solid, uniform, dense image. Do not use a ribbon with too much ink for this produces a smeared image. Worn ribbons, worn type, or a worn platen on the typewriter produce poor images. Carbon ribbons generally produce the best lettering. They are used only once and then discarded.

Care must be exercised when erasing. No smears can be permitted. All smears and dirt are recorded on the microfilm. The background of the drawing must be of uniform density. (Background refers to the surface of the drawing paper.) Erasures tend to produce an area that is cleaner than the surrounding surface.

If a portion of the drawing is cut away and a new surface inserted, the new surface usually is cleaner and brighter. These differences will appear on the microfilm. One practice used to overcome this is to dirty the bright area so it matches the original background. If a correction is inserted by cutting away part of the original drawing, an old piece of vellum can be used.

Any of the normal ways to keep a drawing clean can be used. Keep the drawing covered when it is not being worked on. When working on it, cover the area not needed with vellum. Use a dry cleaning pad and draftsman's brush. Be especially careful to keep perspiration from the hands and arms off the drawing.

If drawings are to be microfilmed, store flat. Never roll them. If a drawing has been rolled, microfilming is difficult. It is recommended that only flat sheet stock be used. Stock sold in rolls is difficult to flatten.

Never use the original drawing for reference. Always keep it stored. If a reference to the drawing is needed, make a print from the microfilm. It is best if original drawings are not used to make blueprints or whiteprints.

PHOTOGRAPHIC REPRODUCTION OF DRAWINGS

There are several ways to reproduce a microfilm or original drawing on film photographically. The photographic process produces a full-size copy of the drawing on a tough polyester film. Portions of the drawings can be removed and the revisions drawn on the film. Three of these systems are:

Method 1. This procedure uses Eastman Kodak Estar base film. See Fig. 25–12, Method 1.

1. Make a copy on Kodagraph Autopositive Projection Film. This can be made from a microfilm or negative of the original drawing. The Autopositive copy is a negative.

2. Opaque off the Autopositive negative the portions of the drawing to be revised.

3. Make a positive print from this negative on Kodagraph Wash-off Contact Film or Kodak Contact Film. On this print, draw the revisions. Either pencil or ink can be used to make the revisions. This copy is now ready to use as a revised original to make copies by any process desired.

Method 2. This procedure is shown in Fig. 25–12, Method 2.

FROM MICROFILM TO DRAFTING BOARD

There are many ways to reproduce a drawing on film photographically so that further drafting can be done. But sometimes original drawings or good-quality prints are not available, so that it is necessary to make a photo reproduction from a microfilm or reduced-size film record. Three methods for doing this are shown here. The best method to use depends on the requirements that must be met.

METHOD 1

This procedure uses durable, dimensionally stable Estar Base film in each step.

1. Make a copy on KODAGRAPH AUTO-POSITIVE Projection Film. (The copy will have a negative image, the same as the microfilm or reduced-size negative from which it is made.)

2. Opaque to remove the unwanted portions of the drawing.

3. Using the enlarged negative film copy, make a print on KODAGRAPH Wash-Off Contact Film or KODAGRAPH Contact Film. This print is ready for drafting.

4. Using either pencil or ink, make the necessary additions or revisions.

Enlarge

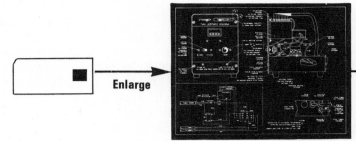

KODAGRAPH AUTOPOSITIVE
Projection Film, EAP4

METHOD 2

This procedure consists simply of enlarging the microfilm or reduced-size negative directly onto KODAGRAPH Projection Film. Remove any unwanted portions of the photographic image with KODAGRAPH Eradicator. Use pencil or ink to make additions or revisions on either side of the film.

Enlarge

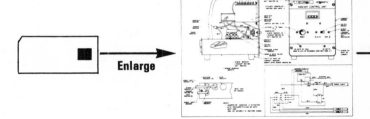

KODAGRAPH Projection Film, EP4/7
KODAGRAPH Ortho Matte Film, EOM4

METHOD 3

This procedure combines the economy of paper and the scissors drafting technique.

1. Make a reverse-reading copy by enlarging on KODAGRAPH Projection Paper, KODAGRAPH Ortho Paper, or KODAGRAPH Projection Q Paper.

2. With scissors, cut out the unwanted portions of the drawing.

3. From the paper print with the unwanted portion removed, make a print on KODAGRAPH AUTOPOSITIVE Film. This print is ready for drafting.

4. Using either pencil or ink, make the necessary additions or revisions.

Enlarge

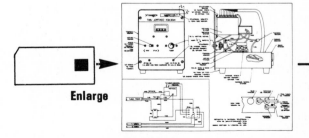

Scissors Draft To Remove Unwanted Portions

KODAGRAPH Projection Paper, P1
KODAGRAPH Projection Q Paper, PQ1
KODAGRAPH Ortho Paper, OP3

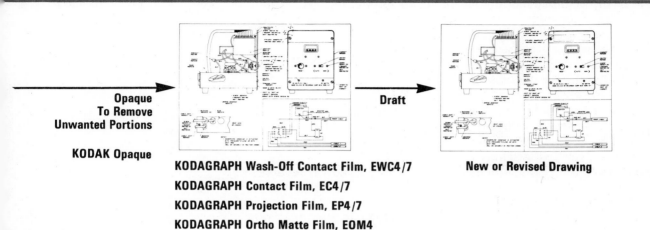

**Opaque
To Remove
Unwanted Portions**

KODAK Opaque

Draft

KODAGRAPH Wash-Off Contact Film, EWC4/7

KODAGRAPH Contact Film, EC4/7

KODAGRAPH Projection Film, EP4/7

KODAGRAPH Ortho Matte Film, EOM4

New or Revised Drawing

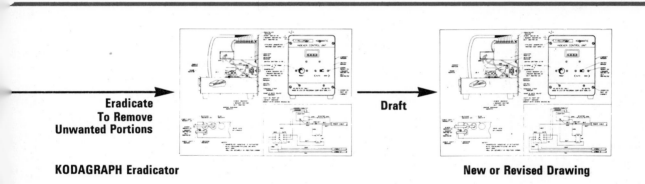

**Eradicate
To Remove
Unwanted Portions**

KODAGRAPH Eradicator

Draft

New or Revised Drawing

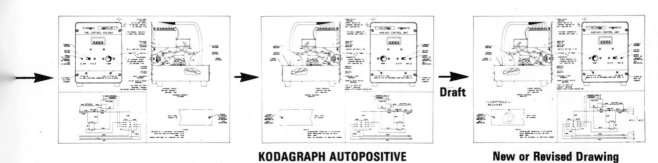

**KODAGRAPH AUTOPOSITIVE
Film, EA4/7**

**KODAGRAPH Wash-Off
AUTOPOSITIVE Film, EWA4**

Draft

New or Revised Drawing

Eastman Kodak Company

1. Make a true-size positive copy on Kodagraph Projection Film from the microfilm negative or the negative made from the original drawing.

2. On this positive copy, remove any portions of the drawing needing revision. Kodagraph Eradicator is used to remove necessary lines.

3. Using pencil or ink, draw the revisions on this positive copy. This copy is now ready to use as a revised original to make copies by any process desired.

Method 3. This procedure, Fig. 25–12, Method 3, uses the scissors drafting technique.

1. Make a reverse-reading copy of the negative of the original drawing or a microfilm on Kodagraph Projection Paper, Otho Paper, or Projection Q Paper.

2. With a scissors, cut out the portions of the drawing to be revised.

3. From this reverse print, make a print on Kodagraph Autopositive Film.

This print is not in reverse. Draft changes on this copy. Use pencil or ink. This copy is now ready to use as a revised original to make copies by any process desired.

Build Your Vocabulary

Following are terms that you should understand and use as a part of your working vocabulary. Write a brief explanation of what each means.

Blueprints
Whiteprints
Diazo process
Vandyke prints
Intermediates
Microfilm
Density
Polyester film

Chapter 26

Fastening Devices

As a product is designed, one important decision to be made is the method of fastening or joining the parts. Some parts must be fastened so they can be easily separated. Others can be permanently joined. The type of materials to be joined influences the type of fastening method to be used. Sometimes the stress upon a joint requires a special method of fastening.

The draftsman must know how to draw the various types of fasteners. He must know how to give the specification of the fastener on the drawing.

Many fastening devices have been standardized for use in all industries in the United States. They have been approved by the American National Standards Institute. They are referred to by the abbreviation ANSI.

HEAD TYPES

Threaded fastening devices are manufactured with a wide variety of head types, Fig. 26–1.

The flat head is used when the head must be flush with the surface. The oval head, used when the head is exposed to view, improves the appearance. The pan and truss heads are used to cover large clearance holes.

Hexagon heads are strong and easy to drive, but they are not pleasing in appearance. If appearance is impor-

tant, the cross recess head is used. The cross recess head is easier to drive than the slotted head. The clutch recess head is also easy to drive.

TYPES OF THREADS

A thread is in the form of a helix (spiral). It is a ridge formed on the surface of a cylinder or a cone-shaped surface. A thread on a cylinder is called a *straight thread*. A thread on a cone-shaped surface is called a *taper thread*.

Thread Terminology. Since threads are of great importance in the design of products, the draftsman should understand the parts, Fig. 26–2.

An *external thread* is on the external surface of a cone or cylinder.

An *internal thread* is on the internal surface of a hollow cylinder or cone.

A *right-hand thread* is one, when viewed axially, winds in a clockwise and receding direction. See Fig. 26–3.

A *left-hand thread* is one, when viewed axially, winds in a counterclockwise and receding direction. See Fig. 26–3.

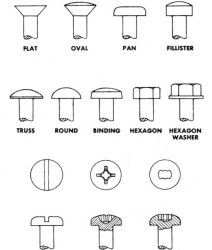

26–1. *Head types and common driver recess types.*

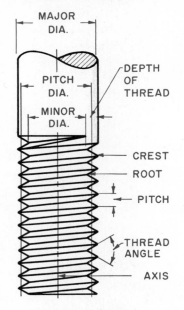

EXTERNAL THREAD

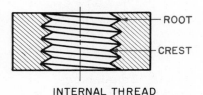

INTERNAL THREAD

26-2. *Thread terminology.*

The *root of the thread* is the surface that joins the sides of the thread form.

Thread *pitch* is the distance, measured parallel to the axis of the thread, between a point on one crest to a similar point on the next crest.

The *lead* is the distance a point on a threaded shaft moves parallel to the axis in one complete revolution.

The *threads per inch* is the number of threads in one linear inch along a threaded shaft.

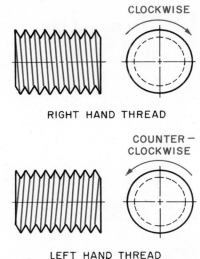

26-3. *Right-hand and left-hand threads.*

The *crest* is the surface joining the sides of the thread. It is the farthest part of the thread from the cylinder or cone from which the thread projects.

A *single thread* has a lead equal to the pitch.

A *multiple thread* has a lead equal to more than one pitch. For example, a double thread has a lead equal to two pitch distances. See Fig. 26-4.

A multiple thread permits a more rapid advance of a part without using a coarser thread form. A double thread will advance twice as far in one revolution as will the same form in a single thread.

The *thread form* is the profile of the thread in cross section. See Fig. 26-5.

The *depth of thread* is the distance from the crest to the

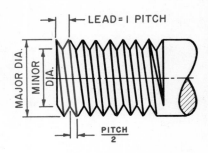

SINGLE THREAD

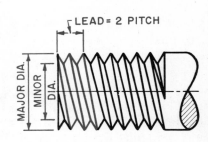

DOUBLE THREAD

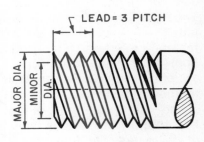

TRIPLE THREAD

26-4. *Single and multiple threads.*

root measured perpendicular to the axis.

The *major diameter* is the largest diameter of the thread.

The *minor diameter* is the smallest diameter of a thread.

The *pitch diameter* is an imaginary cylinder cutting the thread form where the width of the thread and groove are equal.

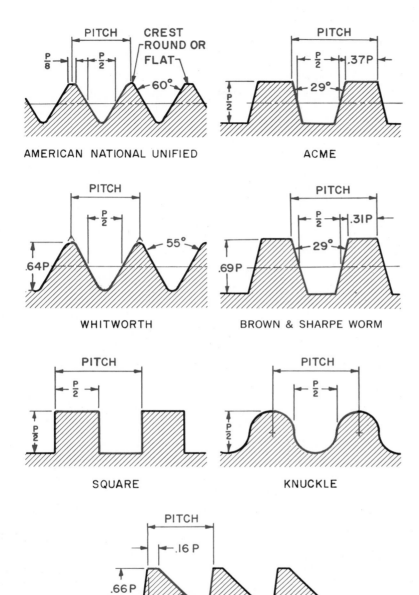

AMERICAN NATIONAL UNIFIED

ACME

WHITWORTH

BROWN & SHARPE WORM

SQUARE

KNUCKLE

BUTTRESS

26–5. *Common thread forms.*

The *thread angle* is the angle formed between the sides of a thread.

THREAD FORMS

Threads are used for fastening, adjusting, and transmitting power. Some threads perform these functions better than others. Fig. 26–5 shows the more commonly used thread forms.

The American National and the Unified thread series are used for fasteners and adjustments. They are in common use in the United States. The Whitworth thread serves the same purpose as the American National thread. It still has limited use in Great Britain, where the system was developed.

The square, Acme, and Brown and Sharpe worm threads are used primarily for transmitting power. The buttress thread transmits power in only one direction.

The knuckle thread is made by casting or is rolled into thin metal. A common use is on a light bulb.

American National Thread Series. As stated above, the majority of fastening devices in the United States use the American National and Unified thread series. Data for these are shown in Figs. 26–6 and 26–7. Notice the relationship between the thread diameter and the number of threads per inch. These are classified into six basic series.

26–6.

Standard Unified and American National Threads

Nominal Diameter	Coarse[1] NC UNC		Fine[1] NF UNF		Extra Fine[2] NEF UNEF		Nominal Diameter	Coarse[1] NC UNC		Fine[1] NF UNF		Extra Fine[2] NEF UNEF	
	Thds. per in.	Tap Drill	Thds. per in.	Tap Drill	Thds. per in.	Tap Drill		Thds. per in.	Tap Drill	Thds. per in.	Tap Drill	Thds. per in.	Tap Drill
0 (.060)			80	3/64	......	......	1	8	7/8	12	59/64	20	61/64
1 (.073)	64	No. 53	72	No. 53	......	......	1 1/16	......	......	......	......	18	1
2 (.086)	56	No. 50	64	No. 50	......	......	1 1/8	7	63/64	12	1 3/64	18	1 5/64
3 (.099)	48	No. 47	56	No. 45	......	......	1 3/16	......	......	......	......	18	1 9/64
4 (.112)	40	No. 43	48	No. 42	......	......	1 1/4	7	1 7/64	12	1 11/64	18	1 3/16
5 (.125)	40	No. 38	44	No. 37	......	......	1 5/16	......	......	......	......	18	1 17/64
6 (.138)	32	No. 36	40	No. 33	......	......	1 3/8	6	1 7/32	12	1 19/64	18	1 5/16
8 (.164)	32	No. 29	36	No. 29	......	......	1 7/16	......	......	......	......	18	1 3/8
10 (.190)	24	No. 25	32	No. 21	......	......	1 1/2	6	1 11/32	12	1 27/64	18	1 7/16
12 (.216)	24	No. 16	28	No. 14	32	No. 13	1 9/16	......	......	......	......	18	1 1/2
1/4	20	No. 7	28	No. 3	32	7/32	1 5/8	......	......	......	......	18	1 9/16
5/16	18	F	24	I	32	9/32	1 11/16	......	......	......	......	18	1 5/8
3/8	16	5/16	24	Q	32	11/32	1 3/4	5	1 9/16	......	......	......	......
7/16	14	U	20	25/64	28	13/32	2	4 1/2	1 25/32	......	......	......	......
1/2	13	27/64	20	29/64	28	15/32	2 1/4	4 1/2	2 1/32	......	......	......	......
9/16	12	31/64	18	33/64	24	33/64	2 1/2	4	2 1/4	......	......	......	......
5/8	11	17/32	18	37/64	24	37/64	2 3/4	4	2 1/2	......	......	......	......
11/16	......	......	......	......	24	41/64	3	4	2 3/4	......	......	......	......
3/4	10	21/32	16	11/16	20	45/64	3 1/4	4		......	......	......	......
13/16	......	......	......	......	20	49/64	3 1/2	4		......	......	......	......
7/8	9	49/64	14	13/16	20	53/64	3 3/4	4		......	......	......	......
15/16	......	......	......	......	20	57/64	4	4		......	......	......	......

[1]Classes 1A, 2A, 3A, 1B, 2B, 3B, 2, and 3.
[2]Classes 2A, 2B, 2, and 3.

Extracted from *Standard Unified Screw Threads* (ANSI B1.1–1960), with permission of the publisher, The American Society of Mechanical Engineers.

American National Coarse. This is abbreviated on drawings, NC. It is a coarse thread used for all general fasteners. It permits rapid assembly and disassembly.

American National Fine. This is abbreviated on drawings, NF. It is a fine thread. It will not vibrate loose as easily as the NC Thread and is stronger.

8-Pitch American National Thread. This is abbreviated on drawings, 8N. It has eight threads per inch regardless of the diameter of the part. It serves much the same purpose as the NC thread for parts with diameters above one inch.

12-Pitch American National Thread. This is abbreviated on drawings, 12N. It has 12 threads per inch regardless of the diameter of the part. While it is used for parts with a diameter as small as 1/2 inch, it is

ISO METRIC THREAD FORM

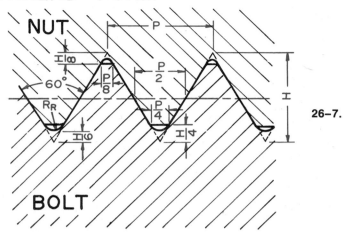

P = PITCH in mm

$H = 0.86603P$

$\dfrac{H}{4} = 0.21651P$

$\dfrac{H}{6} = 0.14434P$

$\dfrac{H}{8} = 0.10825P$

$R_R = 0.14434P$

DEPTH OF THREAD IN SCREW =
$\dfrac{17}{24}H = 0.61343P$

DEPTH OF THREAD IN NUT =
$\dfrac{5}{8}H = 0.54127P$

26–7.

ISO Metric Thread—Coarse Series

Diameter	2	2.5	3	4	5	6	8	10
Pitch	0.4	0.45	0.5	0.7	0.8	1.0	1.25	1.5
Basic effective diameter	1.740	2.208	2.675	3.545	4.480	5.350	7.188	9.026
Depth of thread in screw	0.25	0.28	0.31	0.43	0.49	0.61	0.77	0.92
Area of Root dia. (mm^2)	1.79	2.98	4.47	7.75	12.7	17.9	32.8	52.3
Diameter of tapping drill	1.6	2.05	2.5	3.3	4.2	5.0	6.8	8.5

Diameter	12	16	20	24	30	36	42	48
Pitch	1.75	2.0	2.5	3.0	3.5	4.0	4.5	5.0
Basic effective diameter	10.863	14.701	18.376	22.051	27.727	33.402	39.077	44.752
Depth of thread in screw	1.07	1.23	1.53	1.84	2.15	2.45	2.76	3.07
Area of Root dia. (mm^2)	76.2	144	225	324	519	759	1050	1380
Diameter of tapping drill	10.2	14.0	17.5	21.0	26.5	32.0	37.5	43.0

OMFS Thread Data

Diameter	1.6	2.0	2.5	3.0	3.5	4.0	5.0	6.3	8.0	10.0	12.0	14.0	16.0
Pitch	0.35	0.4	0.45	0.5	0.6	0.7	0.8	1.0	1.25	1.75	1.75	2.0	2.0

Diameter	20	24	30	36	42	48	56	64	72	80	90	100
Pitch	2.5	3.5	3.5	4	4.5	5.0	5.5	6.0	6.0	6.0	6.0	6.0

ISO (International Organization for Standardization) Metric Thread form and data and The Optimum Metric Fastener System (OMFS) data.

generally used for diameters above 1½ inches.

16-Pitch American National Thread. This is abbreviated on drawings, 16N. It has 16 threads per inch regardless of the diameter of the part. It is a fine thread and is used on large diameter parts.

National Extra Fine Thread. This is abbreviated on drawings, NEF. It is finer than the NF thread. It uses the same thread form as the NF. It is used where severe vibrations are expected or on parts under great stress.

Special Thread Series. If a thread series is not one of the standard series, it is a special thread shown on a drawing by the abbreviation NS.

Unified Screw Thread. The Unified screw thread series is much the same as the American National series. It is a standard series agreed upon by the United States, Canada, and Great Britain. It is replacing the American National Standard series. The roots and crests of the external threads are more rounded than the American National series. The internal threads have flatter roots and crests. Fig. 26-6. These are used for the same purpose as the American National series. On a drawing they are abbreviated as follows: Unified National Coarse, UNC; Unified National Fine, UNF; Unified National Extra-Fine, UNEF; and the Unified 8, 12 and 16 thread series, 8 UN, 12 UN and 16 UN.

Metric Threads. The ISO metric thread and Optimum Metric Fastener Series (OMFS) standards are in Fig. 26-7.

Acme and Square Threads. American Standard Acme thread data are given in Fig. 26-8. Square threads have not been standardized.

THREAD CLASSES

The thread in a nut must be slightly larger than those on the screw to be used with the nut. For some purposes this difference can be great. For other uses, they must fit rather closely, in other words, have a close tolerance. The *class of a thread* is the standard tolerance or the amount of tolerance and allowance at the pitch diameter.

The Unified and American screw thread standards have three classes. External threads have classes 1A, 2A, and 3A. Internal threads have classes 1B, 2B, and 3B.

Classes 1A and 1B have a large allowance and permit a loose fit. They are used where a part must be easily and quickly assembled or disassembled. Classes 2A and 2B are used for general purpose fastening. Most bolts, screws, and nuts are of this class. Classes 3A and 3B have closer tolerances than 2A and 2B. They are used where closer tolerances are needed. Still in use are classes 2 and 3. These are from the old class system that has been replaced by the Unified and American screw thread classes.

DRAWING THREADS

The draftsman has three ways of representing threads on a drawing. These are the detailed, schematic, and simplified methods.

The draftsman must decide which best suits his purpose. This depends upon the purpose and use of the drawings. All three methods can be used on the same drawing.

Detailed Representation. Detailed representation presents the threads about as they actually appear. In Fig. 26-9 the American National and Unified threads are shown. The pitch does not have to be to exact scale. If a part has nine threads to the inch, it can be drawn as eight threads because it is easier to layout. The curved crest and root lines are drawn as straight lines. The crest and root are drawn to a sharp "V" shape, even though on the actual thread they are flat or slightly rounded. The steps for draw-

26-8.

Standard General Purpose Acme Threads

Size	Threads per Inch	Size	Threads per Inch	Size	Threads per Inch	Size	Threads per Inch
1/4	16	3/4	6	1 1/2	4	3	2
5/16	14	7/8	6	1 3/4	4	3 1/2	2
3/8	12	1	5	2	4	4	2
7/16	12	1 1/8	5	2 1/4	3	4 1/2	2
1/2	10	1 1/4	5	2 1/2	3	5	2
5/8	8	1 3/8	4	2 3/4	3	—	—

Extracted from *Standard Acme Screw Threads* (ANSI B1.5–1952), with the permission of the publisher, The American Society of Mechanical Engineers.

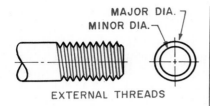
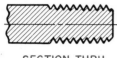

EXTERNAL THREADS

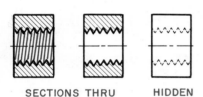

SECTION THRU
EXTERNAL THREADS

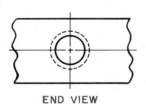

SECTIONS THRU
INTERNAL THREADS

HIDDEN
VIEW

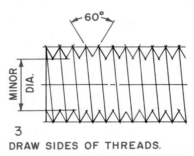

END VIEW

26–9. *Detailed representation of American National and Unified threads.*

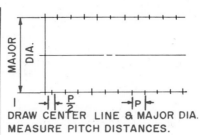

1
DRAW CENTER LINE & MAJOR DIA.
MEASURE PITCH DISTANCES.

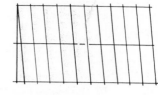

2
DRAW CREST LINES.

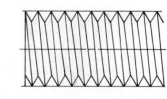

3
DRAW SIDES OF THREADS.

4
DRAW ROOT LINES.

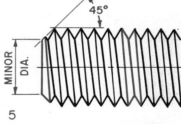

5
END THREAD CHAMFERED AT 45°.
REMOVE CONSTRUCTION LINES.

26–10. *How to draw a detailed representation of American National and Unified threads.*

ing detailed National thread representation are shown in Fig. 26–10.

How to draw a detailed representation of Acme threads is shown in Fig. 26–11. How to draw a detailed representation of Square threads is shown in Fig. 26–12.

Schematic Representation. Schematic representation is easier to draw than detailed representation. It still gives an effective appearance. See Fig. 26–13. The crest and root lines are spaced by eye. Practice will enable a draftsman to space them so they will give a pleasing appearance. They

are usually spaced wider on larger diameter shafts.

It is common practice to draw the root line heavier than the crest line. They are usually drawn perpendicular to the axis of the part threaded. The steps for drawing schematic thread representation are shown in Fig. 26–14.

763

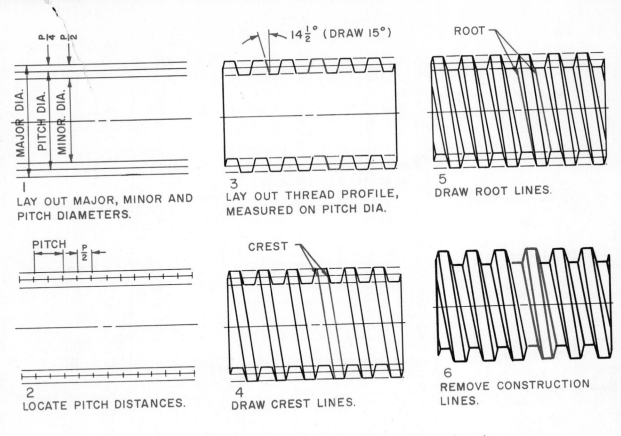

P/4 ‖ P/2

1
LAY OUT MAJOR, MINOR AND
PITCH DIAMETERS.

PITCH ‖ P/2

2
LOCATE PITCH DISTANCES.

$14\frac{1}{2}°$ (DRAW 15°)

3
LAY OUT THREAD PROFILE,
MEASURED ON PITCH DIA.

CREST

4
DRAW CREST LINES.

ROOT

5
DRAW ROOT LINES.

6
REMOVE CONSTRUCTION
LINES.

26-11. *How to draw detailed representation of Acme threads.*

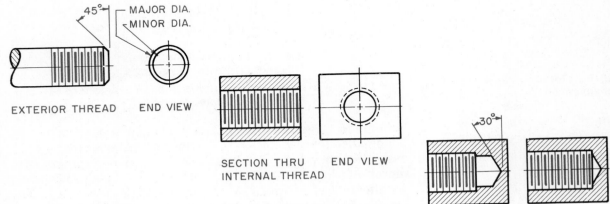

45°   ┌ MAJOR DIA.
      └ MINOR DIA.

EXTERIOR THREAD   END VIEW

SECTION THRU   END VIEW
INTERNAL THREAD

26-13. *Schematic representation of screw threads.*

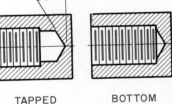

30°

TAPPED   BOTTOM
HOLE   TAPPED HOLE

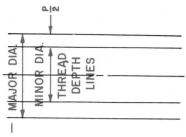

1

LAY OUT MAJOR DIAMETER
AND MINOR DIAMETER, OR
THREAD DEPTH LINES.

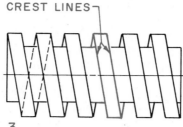

3

DRAW CREST LINES.
DOTTED LINES SHOW HOW
TO LOCATE REAR SIDE OF
THREAD.

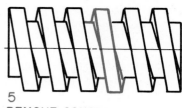

5

REMOVE CONSTRUCTION
LINES.

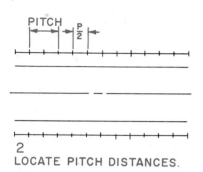

2

LOCATE PITCH DISTANCES.

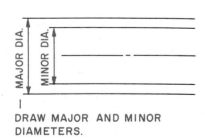

4

DRAW ROOT LINES.
HIDDEN LINES SHOW HOW
TO LOCATE NEEDED POINTS.

26–12. *How to draw detailed representation of square threads.*

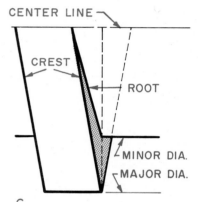

6

ENLARGEMENT SHOWING
HOW TO LOCATE ROOT
LINES.

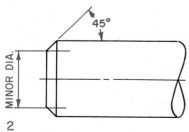

1

DRAW MAJOR AND MINOR
DIAMETERS.

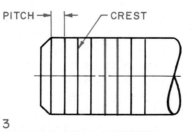

3

MEASURE PITCH DISTANCES
AND DRAW CREST LINES.

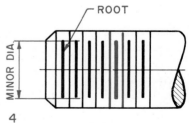

4

DRAW ROOT LINES.

26–14. *How to draw schematic representation of screw threads.*

2

CHAMFER END AT 45° FROM
MINOR DIAMETER.

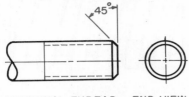

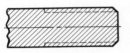

EXTERNAL THREAD   END VIEW

SECTION THRU
EXTERNAL
THREAD   END VIEW

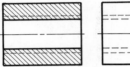

SECTION THRU   HIDDEN VIEW
INTERNAL   OF INTERNAL
THREADS   THREADS

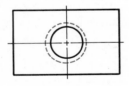

END VIEW

SECTION THRU   EXTERIOR VIEW
TAPPED HOLE   OF TAPPED
HOLE

26–15. *Simplified representation of screw threads.*

Simplified Representation. Simplified representation is the easiest and quickest method of indicating threads. It should not be used if the dashed lines might be confused with other hidden lines on the drawing. See Fig. 26–15. Drawing simplified thread representation is shown in Fig. 26–16.

Thread Notes. The information about a screw thread is given on a drawing with a note. The note must tell the screw diameter or screw number, number of threads per inch, thread series, and class of the screw thread. See Fig. 26–17. All threads are assumed to be right-hand single threads. Left-hand threads are noted by adding LH after the class. If a thread is a double or triple thread, this is noted by lettering the work, DOUBLE or TRIPLE, after the class of fit.

Thread notes for internal threads are indicated on the circular view of the threaded hole. Thread notes for external threads are given in side view of the part threaded.

The depth of a drilled hole and threaded hole is given at the end of the note.

Templates are available so that threads can be drawn rapidly. They should be used whenever possible to speed up the drawing of the threads.

BOLTS

A bolt is a fastening device with a head on one end

1
DRAW MAJOR AND MINOR DIAMETERS.

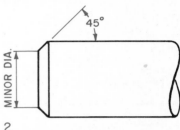

2
CHAMFER END AT 45° FROM MINOR DIAMETER.

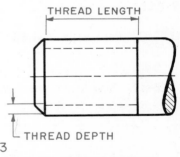

3
MEASURE THREAD LENGTH. DRAW HIDDEN LINES FOR THREAD DEPTH.

26–16. *How to draw simplified representation of screw threads.*

and threads on the other. It joins by passing through holes in a part and being secured with a nut. See Fig. 26–18 for the common types of bolts.

The proper type of bolt to use is determined by the job. The need for strength, rapid assembly, and cost are considered. Bolts have two types

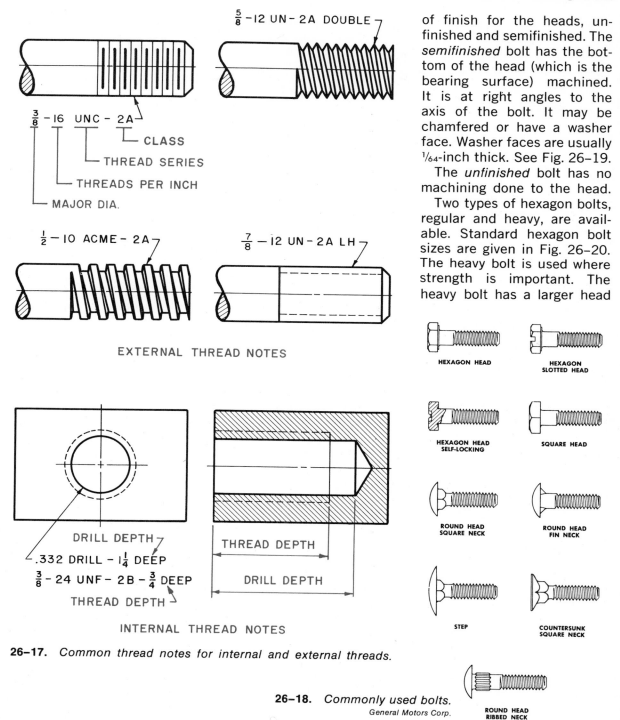

$\frac{5}{8}$ – 12 UN – 2A DOUBLE

$\frac{3}{8}$ – 16 UNC – 2A
- CLASS
- THREAD SERIES
- THREADS PER INCH
- MAJOR DIA.

$\frac{1}{2}$ – 10 ACME – 2A

$\frac{7}{8}$ – 12 UN – 2A LH

EXTERNAL THREAD NOTES

DRILL DEPTH
.332 DRILL – 1$\frac{1}{4}$ DEEP
$\frac{3}{8}$ – 24 UNF – 2B – $\frac{3}{4}$ DEEP
THREAD DEPTH

THREAD DEPTH
DRILL DEPTH

INTERNAL THREAD NOTES

26–17. *Common thread notes for internal and external threads.*

26–18. *Commonly used bolts.*
General Motors Corp.

HEXAGON HEAD

HEXAGON
SLOTTED HEAD

HEXAGON HEAD
SELF-LOCKING

SQUARE HEAD

ROUND HEAD
SQUARE NECK

ROUND HEAD
FIN NECK

STEP

COUNTERSUNK
SQUARE NECK

ROUND HEAD
RIBBED NECK

of finish for the heads, unfinished and semifinished. The *semifinished* bolt has the bottom of the head (which is the bearing surface) machined. It is at right angles to the axis of the bolt. It may be chamfered or have a washer face. Washer faces are usually 1/64-inch thick. See Fig. 26–19.

The *unfinished* bolt has no machining done to the head.

Two types of hexagon bolts, regular and heavy, are available. Standard hexagon bolt sizes are given in Fig. 26–20. The heavy bolt is used where strength is important. The heavy bolt has a larger head

767

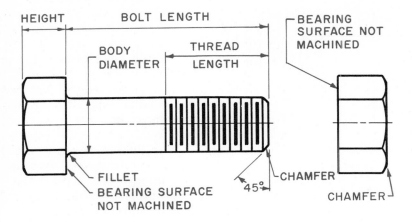

HEIGHT | BOLT LENGTH

BODY DIAMETER | THREAD LENGTH

BEARING SURFACE NOT MACHINED

FILLET
BEARING SURFACE NOT MACHINED

45° CHAMFER

CHAMFER

UNFINISHED BOLT AND NUT

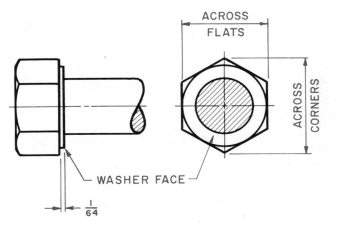

ACROSS FLATS

ACROSS CORNERS

WASHER FACE

$\frac{1}{64}$

SEMIFINISHED BOLT HEAD

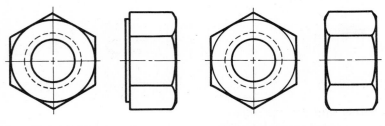

SEMIFINISHED HEXAGON NUT WITH WASHER FACE

SEMIFINISHED HEXAGON NUT WITH CHAMFER FACE

26–19. *Semifinished and unfinished hexagon bolts and nuts.*

than the regular and offers a larger bearing surface.

Square head bolts and nuts are available in the regular series only. They are available only in the unfinished type. They are used for less accurate work. Standard square bolt sizes are given in Fig. 26–20.

Round head bolts are used when a smooth head is needed. They have a square, finned, or ribbed neck. This locks the bolt in place when it is tightened.

NUTS

Nuts are most commonly made in hexagon and square shapes. They are available in unfinished and semifinished types. They are made in the regular and heavy series. ANSI sizes are given in Figs. 26–21 through 26–23.

A jam nut is used for locking another nut in place. They have the same dimensions as the regular nut except they are thinner. They are made in regular, heavy, unfinished, and semifinished types. They are available only in the hexagon shape. ANSI sizes are given in Figs. 26–21 and 26-22.

Drawing Bolts and Nuts. When drawing bolts and nuts where accuracy is not important, the sizes of each can be taken from tables in Figs. 26–20 through 26–23. If greater accuracy is required, the decimal sizes are available in the American National Standards Institute publications.

26-20.

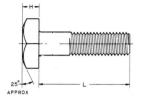

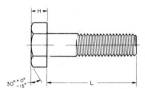

Standard Regular Hexagon, Heavy Hexagon, and Regular Square Bolts

Nominal Size or Major Thread Dia.	Regular Hexagon[2]			Heavy Hexagon[2]			Regular Square[3]	
	F Across Flats	H Unfin.	H Semifin.	F Across Flats	H Unfin.	H Semifin.	F Across Flats	H Unfin.
$1/4$	$7/16$	$11/64$	$5/32$				$3/8$	$11/64$
$5/16$	$1/2$	$7/32$	$13/16$				$1/2$	$13/64$
$3/8$	$9/16$	$1/4$	$15/64$				$9/16$	$1/4$
$7/16$	$5/8$	$19/64$	$9/32$				$5/8$	$19/64$
$1/2$	$3/4$	$11/32$	$5/16$	$7/8$	$11/32$	$5/16$	$3/4$	$21/64$
$5/8$	$15/16$	$27/64$	$25/64$	$1 1/16$	$27/64$	$25/64$	$15/16$	$27/64$
$3/4$	$1 1/8$	$1/2$	$15/32$	$1 1/4$	$1/2$	$15/32$	$1 1/8$	$1/2$
$7/8$	$1 5/16$	$37/64$	$35/64$	$1 7/16$	$37/64$	$35/64$	$1 5/16$	$19/32$
1	$1 1/2$	$43/64$	$39/64$	$1 5/8$	$43/64$	$39/64$	$1 1/2$	$21/32$
$1 1/8$	$1 11/16$	$3/4$	$11/16$	$1 13/16$	$3/4$	$11/16$	$1 11/16$	$3/4$
$1 1/4$	$1 7/8$	$27/64$	$25/32$	2	$27/32$	$25/32$	$1 7/8$	$27/32$
$1 3/8$	$2 1/16$	$29/32$	$27/32$	$2 3/16$	$29/32$	$27/32$	$2 1/16$	$29/32$
$1 1/2$	$2 1/4$	1	$15/16$	$2 3/8$	1	$15/16$	$2 1/4$	1
$1 3/4$	$2 5/8$	$1 5/32$	$1 3/32$	$2 3/4$	$1 5/32$	—		
2	3	$1 11/32$	$1 7/32$	$3 1/8$	$1 11/32$	—		
$2 1/4$	$3 3/8$	$1 1/2$	$1 3/8$	$3 1/2$	$1 1/2$	—		
$2 1/2$	$3 3/4$	$1 21/32$	$1 17/32$	$3 7/8$	$1 21/32$	—		
$2 3/4$	$4 1/8$	$1 13/16$	$1 11/16$	$4 1/4$	$1 13/16$	—		
3	$4 1/2$	2	$1 7/8$	$4 5/8$	2	—		

[1] Finished hexagon bolts and hexagon cap screws are considered the same. See Fig. 26–34 for sizes.

[2] Regular and heavy hexagon sizes through 2 inches nominal size are unified dimensionally with British and Canadian standards.

[3] Square sizes through $1 1/2$ inches nominal size are unified dimensionally with British and Canadian standards.

Extracted from *Standard Square and Hex Bolts and Screws* (ANSI B18.2.1–1965), with permission of the publisher, The American Society of Mechanical Engineers.

26–21.
Standard Square and Hexagon Nuts

Diameter	Unfinished Regular F	Unfinished Nut H	Unfinished Jam Nut H	Semifinished Regular F	Semifinished Nut H	Semifinished Jam Nut H
$1/4$	$1/2$	$1/4$	—	$1/2$	$15/64$	$11/64$
$5/16$	$9/16$	$5/16$	—	$9/16$	$19/64$	$13/64$
$3/8$	$11/16$	$3/8$	—	$11/16$	$23/64$	$15/64$
$7/16$	$3/4$	$7/16$	—	$3/4$	$27/64$	$17/64$
$1/2$	$7/8$	$1/2$	—	$7/8$	$31/64$	$19/64$
$9/16$	—	—	—	$15/16$	$35/64$	$21/64$
$5/8$	$1\,1/16$	$5/8$	—	$1\,1/16$	$39/64$	$23/64$
$3/4$	$1\,1/4$	$3/4$	—	$1\,1/4$	$47/64$	$27/64$
$7/8$	$1\,7/16$	$7/8$	—	$1\,7/16$	$55/64$	$31/64$
1	$1\,5/8$	1	—	$1\,5/8$	$63/64$	$35/64$
$1\,1/8$	$1\,13/16$	$1\,1/8$	$5/8$	$1\,13/16$	$1\,7/64$	$39/64$
$1\,1/4$	2	$1\,1/4$	$3/4$	2	$1\,7/32$	$23/32$
$1\,3/8$	$2\,3/16$	$1\,3/8$	$3/16$	$2\,3/16$	$1\,11/32$	$25/32$
$1\,1/2$	$2\,3/8$	$1\,1/2$	$7/8$	$2\,3/8$	$1\,15/32$	$27/32$
$1\,5/8$	—	—	—	$2\,9/16$	$1\,19/32$	$29/32$
$1\,3/4$	$2\,3/4$	$1\,3/4$	1	$2\,3/4$	$1\,23/32$	$31/32$
$1\,7/8$	—	—	—	$2\,15/16$	$1\,27/32$	$1\,1/32$
2	$3\,1/8$	2	$1\,1/8$	$3\,1/8$	$1\,31/32$	$1\,3/32$
$2\,1/4$	$3\,1/2$	$2\,1/4$	$1\,1/4$	$3\,1/2$	$2\,13/64$	$1\,13/64$
$2\,1/2$	$3\,7/8$	$2\,1/2$	$1\,1/2$	$3\,7/8$	$2\,29/64$	$1\,29/64$
$2\,3/4$	$4\,1/4$	$2\,3/4$	$1\,5/8$	$4\,1/4$	$2\,45/64$	$1\,37/64$
3	$4\,5/8$	3	$1\,3/4$	$4\,5/8$	$2\,61/64$	$1\,45/64$

Sizes overprinted in color are those unified dimensionally with British and Canadian standards. $9/16$, $1\,5/8$, $1\,7/8$ not available in unfinished series.

Extracted from *Standard Square and Hex Nuts* (ANSI B18.2.2–1965), with permission of the publisher, The American Society of Mechanical Engineers.

SEMIFINISHED HEXAGON NUT

UNFINISHED
HEXAGON NUT

SEMIFINISHED
HEXAGON JAM
NUT

UNFINISHED
JAM NUT

26–22.

Standard Regular Hexagon Nuts[1]

Diameter	Unfinished[2]			Semifinished[2]		
	Regular	Nut	Jam Nut[3]	Regular	Nut	Jam Nut[3]
	F	H	H	F	H	H
1/4	7/16	7/32	5/32	7/16	7/32	5/32
5/16	9/16	17/64	3/16	1/2	17/64	3/16
3/8	5/8	21/64	7/32	9/16	21/64	7/32
7/16	3/4	3/8	1/4	11/16	3/8	1/4
1/2	13/16	7/16	5/16	3/4	7/16	5/16
9/16	7/8	1/2	5/16	7/8	31/64	5/16
5/8	1	35/64	3/8	15/16	35/64	3/8
3/4	1 1/8	21/32	27/64	1 1/8	41/64	27/64
7/8	1 5/16	49/64	31/64	1 5/16	3/4	31/64
1	1 1/2	7/8	35/64	1 1/2	55/64	35/64
1 1/8	1 11/16	1	5/8	1 11/16	31/32	39/64
1 1/4	1 7/8	1 3/32	3/4	1 7/8	1 1/16	23/32
1 3/8	2 1/16	1 13/64	13/16	2 1/16	1 11/64	25/32
1 1/2	2 1/4	1 5/16	7/8	2 1/4	1 9/32	27/32

[1] Refer to Fig. 26–21 to find location of sizes marked F and H.
[2] All sizes are unified with British and Canadian standards.
[3] Jam nut "across the flats" dimension is the same as the regular nut.

Extracted from *Standard Square and Hex Bolts and Screws* (ANSI B18.2.1.–1965), with the permission of the publisher, The American Society of Mechanical Engineers.

26–23.

Standard Regular and Heavy Square Nuts

Diameter		Regular		Heavy	
		F	H	F	H
1/4	0.2500	7/16	7/32	1/2	1/4
5/16	0.3125	9/16	17/64	9/16	5/16
3/8	0.3750	5/8	21/64	11/16	3/8
7/16	0.4375	3/4	3/8	3/4	7/16
1/2	0.5000	13/16	7/16	7/8	1/2
5/8	0.6250	1	35/64	1 1/16	5/8
3/4	0.7500	1 1/8	21/32	1 1/4	3/4
7/8	0.8750	1 5/16	49/64	1 7/16	7/8
1	1.0000	1 1/2	7/8	1 5/8	1
1 1/8	1.1250	1 11/16	1	1 13/16	1 1/8
1 1/4	1.2500	1 7/8	1 3/32	2	1 1/4
1 3/8	1.3750	2 1/16	1 13/64	2 3/16	1 3/8
1 1/2	1.5000	2 1/4	1 5/16	2 3/8	1 1/2

Extracted from *Standard Square and Hex Bolts and Screws* (ANSI B18.2.1–1965), with the permission of the publisher, The American Society of Mechanical Engineers.

The steps for drawing hexagon and square bolts and nuts are shown in Figs. 26–24 and 26–25. It shows how to draw these without templates. Nuts and bolt templates should be used if available.

The washer face on semi-finished bolts and nuts is drawn 1/64- to 1/32-inch thick.

To draw hexagon head bolts, see Fig. 26–24.

1. Draw the top view of the bolt head, Fig. 26–24A. Draw a circle equal to the "across the flats" dimension, F.

2. Draw a hexagon outside the circle. The sides must have a tangent to the circle. The sides are drawn on a 60 degree angle.

3. Locate the bearing surface, Fig. 26–24B.

4. Measure the head height, H. If it is a semifinished bolt, the washer face is included in the thickness.

5. Project the width of the head and visible corners from the top view.

6. Locate the radius, R, of the curve on the narrow surfaces, Fig. 26–24C.

7. Draw the arcs on the small surfaces.

8. Locate the radius, R, by trial and error. It should meet the small arcs and have a tangent with the top of the bolt.

9. Draw the chamfer on each corner. The chamfer must have a tangent to the arc. It is drawn on an angle of 30 degrees.

10. Remove construction

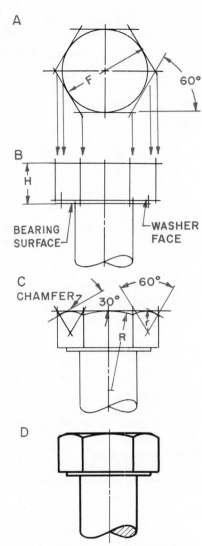

26–24. *How to draw a hexagon bolt "across corners."*

lines and darken object lines, Fig. 26–24D.

To draw a square bolt, see Fig. 26–25.

1. Draw the top view of the bolt head, Fig. 26–25A. Draw a circle equal to the "across the flats" dimension.

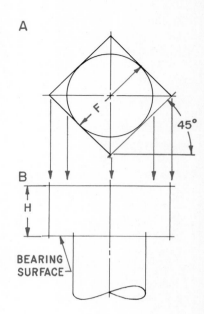

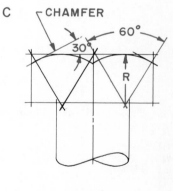

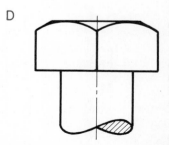

26–25. *How to draw a square bolt "across corners."*

2. Draw a square outside the circle. The sides must have a tangent to the circle and on a 45-degree angle.

3. Locate the bearing surface, Fig. 26–25B.

4. Measure the head height, H.

5. Project the width of the head from the top view.

6. Locate the radius, R, of the curve, Fig. 26–25C.

7. Draw the arcs on the bolt head.

8. Draw the chamfer on each corner. The chamfer must be tangent to arc. It is drawn on an angle of 30 degrees.

9. Remove construction lines and darken the object lines, Fig. 26–25D.

Bolt and Nut Specifications. Standard fasteners are described on a drawing or a parts list by a note. The information is presented in the following order:

1. Diameter of the body of the bolt.

2. Number of threads per inch.

3. Type of thread.

4. Class of thread fit.

5. Length of the bolt.

6. Finish of the bolt.

7. Type of head.

8. Name of the item, such as bolt, screw, or nut.

When a note does not show the bolt or nut to be regular or heavy, it is assumed to be *regular.* If it is meant to be heavy, the word *heavy* is lettered before the name of the fastener.

If an *unfinished* bolt is to be specified, nothing is stated in the note. If the fastener is to be *semifinished,* this is included in the note before the name of the fastener.

If the fastener is made from steel, nothing is indicated in the note. If it is made from another metal, the type of metal is lettered before the name of the fastener.

Following are examples of notes for bolts and nuts: 7/8 diameter, 9 threads per inch, National Coarse thread, Class 2A fit, 3 inches long, brass material, heavy type, semifinished, hexagon bolt. This is abbreviated on a drawing in the following manner: 7/8–9NC–2A × 3 brass heavy Semifin. Hex. bolt.

A note for a nut follows: Nut for 7/8 diameter bolt, 9 threads per inch, Unified National Coarse thread, Class 2B fit, steel material, regular type, unfinished, square nut. This is abbreviated on a drawing in the following manner: 7/8–9UNC–2B Sq. Nut.

STUDS

A stud is a rod that is threaded on each end. It usually passes through a clearance hole in one piece and screws into a tapped hole in a second piece. See Fig. 26–26. A nut is screwed on the other end. This holds the two pieces together. It usually has coarse threads on one end and fine threads on the other. Studs are not standardized. A stud is dimensioned by giving

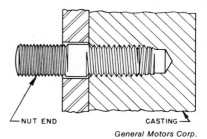

General Motors Corp.

26–26. *A stud.*

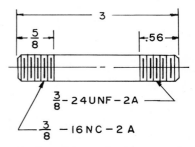

26–27. *Dimensioning a stud.*

the length of the stud, length of thread, and thread notes. See Fig. 26–27. One industrial standard is shown in Fig. 26–28.

TAPPING SCREWS

Tapping screws are designed to cut their own threads as they are screwed into place. They are used to fasten sheet metal and plastic parts. Some are designed to fasten thin materials to soft metal castings.

The most common types of tapping screws are shown in Fig. 26–29. Notice that type AB has a pointed end and a coarse thread. Type B has a blunt end. The threads are finer. It is used for joining thicker sheet metal. Type C is used for joining thicker metal sheet stock than AB or B.

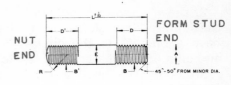

FORM STUD END

NUT END

26–28.

Selected Stud Sizes

	Nominal Size—A	1/4	5/16	3/8	7/16	1/2	9/16	5/8	3/4
B	Threads per Inch	20	18	16	14	13	12	11	10
D	Full Thread Length, Maximum	.3750	.5000	.5625	.6875	.7500	.8750	.9375	1.1250
B¹	Threads per Inch	28	24	24	20	20	18	18	16
D¹	Minimum Thread Length	7/16	9/16	5/8	3/4	13/16	7/8	15/16	1 1/16
R	Radius of Crown	1/4	5/16	3/8	7/16	1/2	9/16	5/8	3/4

Threads
Nut end (B¹) Unified National Fine, Class 2A.
Form stud end (B) United National Coarse.
Extracted from General Motors Engineering Standards.

TYPE AB

TYPE B

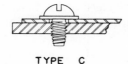

TYPE C

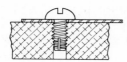

TYPE D, F, G, T

26–29. *Common tapping screws.*

Types D, F, G, and T, have blunt ends. They have a machine screw thread. They are used to join thin material to thick material. They cut threads in thick material. One advantage is greater holding power. The sizes of the thread forms for these screws are shown in Fig. 26–30.

Tapping screws are available in a wide variety of heads. The most common are flat, oval, round, pan, fillister, truss, and hexagon heads. They are made both slotted and recessed. Fig. 26–31 shows some head types and sizes.

Tapping screws are dimensioned by giving the length, wire diameter, type, right-hand or left-hand threads, type of fastener, and finish. A typical note would read 1/2–No. 4 Type B–RH Tapping Screw–Nickel Finish.

CAP SCREWS

A cap screw passes through a clearance hole in one part and screws into a tapped hole in the other. It does not use a nut. See Fig. 26–32.

The common head types are round, flat, hexagon, fillister, and socket. On shorter lengths they are threaded to the head. Longer lengths are threaded only part way to the head.

ANSI cap screw data is given in Figs. 26–33 through 26–35. Cap screws are specified by giving the diameter, number of threads, thread type, class of fit, length, head type, and the type of screw. A typical note would read 3/8–24UNF–2A × 3/4 Flat HD Cap Scr. See Fig. 26–32.

Notice that the slot is drawn on a 45-degree angle. This is common practice for all slotted fasteners.

26–30.

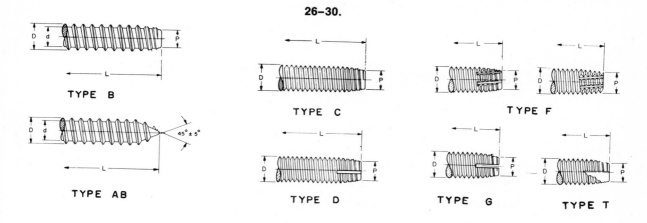

Thread Forms for Standard Tapping Screws, Types AB, B, C, D, F, G, T

Screw Size	Threads Per Inch	Type B			Type AB		Screw Size	Threads Per Inch	Type C		Type D, F, G, T	
		D	d	P	D	d			D	P	D	P
0	48	0.060	0.036	0.031	0.060	0.036	2	56	0.0860	0.067	0.0860	0.067
1	42	0.073	0.049	0.044	0.075	0.049	2	64	0.0860	0.070	0.0860	0.070
2	32	0.086	0.064	0.058	**0.088**	**0.064**	3	48	0.0990	0.077	0.0990	0.077
3	28	0.099	0.075	0.068	0.101	0.075	3	56	0.0990	0.080	0.0990	0.080
4	24	0.112	0.086	0.079	**0.114**	**0.086**	4	40	0.1120	0.086	0.1120	0.086
5	20	0.125	0.094	0.087	0.130	0.094	4	48	0.1120	0.090	0.1120	0.090
6	20	0.138	0.104	0.095	**0.139**	**0.104**	5	40	0.1250	0.099	0.1250	0.099
7	19	0.151	0.115	0.105	0.154	0.115	5	44	0.1250	0.101	0.1250	0.101
							6	32	0.1380	0.106	0.1380	0.106
							6	40	0.1380	0.112	0.1380	0.112
8	18	0.164	0.122	0.112	**0.166**	**0.122**	8	32	0.1640	0.132	0.1640	0.132
10	16	0.190	0.141	0.130	**0.189**	**0.141**	8	36	0.1640	0.135	0.1640	0.135
12	14	0.216	0.164	0.152	0.215	0.164	10	24	0.1900	0.147	0.1900	0.147
							10	32	0.1900	0.158	0.1900	0.158
							12	24	0.2160	0.173	0.2160	0.173
							12	28	0.2160	0.179	0.2160	0.179
1/4	14	0.250	0.192	0.179	**0.246**	**0.192**	1/4	20	0.2500	0.198	0.2500	0.198
5/16	12	0.3125	0.244	0.230	0.315	0.244	1/4	28	0.2500	0.213	0.2500	0.213
3/8	12	0.375	0.309	0.293			5/16	18	0.3125	0.255	0.3125	0.255
7/16	10	0.4375	0.359	0.343			5/16	24	0.3125	0.269	0.3125	0.269
1/2	10	0.500	0.423	0.407			3/8	16	0.3750	0.310	0.3750	0.310
							3/8	24	0.3750	0.332	0.3750	0.332

Extracted from *Standard Slotted and Recessed Head Tapping Screws and Metallic Drive Screws* (ANSI B18.6.4–1966), with the permission of the publisher, The American Society of Mechanical Engineers.

26–31.

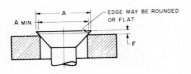

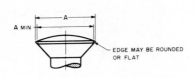

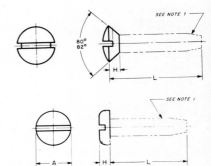

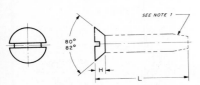

Standard Slotted Head Tapping Screws[2]

Nominal Size[2] or Basic Screw Diameter	Applicable to Screw Types[1]	Flat Head		Oval Head		Pan Head	
	Code Symbols	A	H	A	H	A	H
0 0.0600	● ▲	0.119	0.035	0.119	0.035	0.116	0.039
1 0.0730	● ▲	0.146	0.043	0.146	0.043	0.142	0.046
2 0.0860	● ▲ ■	0.172	0.051	0.172	0.051	0.167	0.053
3 0.0990	● ▲ ■	0.199	0.059	0.199	0.059	0.193	0.060
4 0.1120	● ▲ ■	0.225	0.067	0.225	0.067	0.219	0.068
5 0.1250	● ▲ ■	0.252	0.075	0.252	0.075	0.245	0.075
6 0.1380	● ▲ ■	0.279	0.083	0.279	0.083	0.270	0.082
7 0.1510	● ▲	0.305	0.091	0.305	0.091	0.296	0.089
8 0.1640	● ▲ ■	0.332	0.100	0.332	0.100	0.322	0.096
10 0.1900	● ▲ ■	0.385	0.116	0.385	0.116	0.373	0.110
12 0.2160	● ▲ ■	0.438	0.132	0.438	0.132	0.425	0.125
1/4 0.2500	● ▲ ■	0.507	0.153	0.507	0.153	0.492	0.144
5/16 0.3125	● ▲ ■	0.635	0.191	0.635	0.191	0.615	0.178
3/8 0.3750	▲ ■	0.762	0.230	0.762	0.230	0.740	0.212
7/16 0.4375	▲	0.812	0.223	0.812	0.223	—	—
1/2 0.5000	▲	0.875	0.223	0.875	0.223	—	—

[1] Indicates type of thread for screw size. See Fig. 26–30 for thread dimensions.

● Type AB thread forming
▲ Type B thread forming
■ Type C thread forming and types D, F, G, and T thread cutting

[2] See the standard for other slotted types and sizes and for all recessed slot types.

Extracted from *Standard Slotted and Recessed Head Tapping Screws and Metallic Drive Screws* (ANSI B18.–6.4–1966), with the permission of the publisher, The American Society of Mechanical Engineers.

26–32. *A cap screw with clearance hole.*

$\frac{3}{8}$ – 24 UNF – 2A x $\frac{3}{4}$ FLAT HEAD CAP SCR.

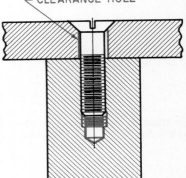

CLEARANCE HOLE

26–33.

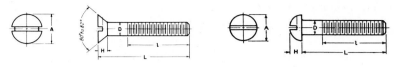

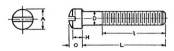

Standard Slotted Head Cap Screws

Nominal Size	D — Diameter of Screw	Flat Head		Round Head		Fillister Head		
		A	H	A	H	A	H	O
1/4	0.250	0.500	0.140	0.437	0.191	0.375	0.172	0.216
5/16	0.3125	0.625	0.177	0.562	0.245	0.437	0.203	0.253
3/8	0.375	0.750	0.210	0.625	0.273	0.562	0.250	0.314
7/16	0.4375	0.8125	0.210	0.750	0.328	0.625	0.297	0.368
1/2	0.500	0.875	0.210	0.812	0.354	0.750	0.328	0.413
9/16	0.5625	1.000	0.244	0.937	0.409	0.812	0.375	0.467
5/8	0.625	1.125	0.281	1.000	0.437	0.875	0.422	0.521
3/4	0.750	1.375	0.352	1.250	0.546	1.000	0.500	0.612
7/8	0.875	1.625	0.423			1.125	0.594	0.720
1	1.000	1.875	0.494			1.312	0.656	0.803
1 1/8	1.125	2.062	0.529					
1 1/4	1.250	2.312	0.600					
1 3/8	1.375	2.562	0.665					
1 1/2	1.500	2.812	0.742					

Extracted from *Standard Hexagon Head Cap Screws, Slotted Head Cap Screws, Square Head Set Screws, and Slotted Headless Set Screws* (ANSI B18.6.2–1956), with the permission of the publisher, The American Society of Mechanical Engineers.

26–34.

Standard Hexagon Head Cap Screws

Nominal Size or Basic Major Diameter of Thread		F — Width Across Flats	H — Height
1/4	0.2500	7/16	5/32
5/16	0.3125	1/2	13/64
3/8	0.3750	9/16	15/64
7/16	0.4375	5/8	9/32
1/2	0.5000	3/4	5/16
9/16	0.5625	13/16	23/64
5/8	0.6250	15/16	25/64
3/4	0.7500	1 1/8	15/32
7/8	0.8750	1 5/16	35/64
1	1.0000	1 1/2	39/64
1 1/8	1.1250	1 11/16	11/16
1 1/4	1.2500	1 7/8	25/32
1 3/8	1.3750	2 1/16	27/32
1 1/2	1.5000	2 1/4	15/16

Sizes above dark line are unified dimensionally with British and Canadian standards. Bearing surface always chamfered or has washer face.

Thread Length
Screw lengths up to 6 inches have thread length equal to 2D + 1/4 inch.
Threads may be National coarse, National fine, or 8-Thread series, class 2A fit.

Extracted from *Standard Square and Hex Bolts and Screws* (ANSI B18.2.1–1965), with the permission of the publisher, The American Society of Mechanical Engineers.

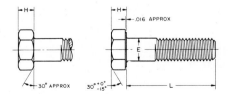

26–35.

Standard Hexagon and Spline Socket Head Cap Screws

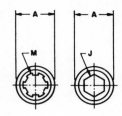

Nominal Size	D Body Diameter	A Head Diameter	H Head Height	M Spline Socket Size	J Hexagon Socket Size	T Key Engagement
	Max	Max	Max	Nom	Nom	Min
0	0.0600	0.096	0.060	0.062	0.050	0.025
1	0.0730	0.118	0.073	0.074	1/16	0.031
2	0.0860	0.140	0.086	0.098	5/64	0.038
3	0.0990	0.161	0.099	0.098	5/64	0.044
4	0.1120	0.183	0.112	0.115	3/32	0.051
5	0.1250	0.205	0.125	0.115	3/32	0.057
6	0.1380	0.226	0.138	0.137	7/64	0.064
8	0.1640	0.270	0.164	0.173	9/64	0.077
10	0.1900	5/16	0.190	0.188	5/32	0.090
1/4	0.2500	3/8	1/4	0.221	3/16	0.120
5/16	0.3125	15/32	5/16	0.298	1/4	0.151
3/8	0.3750	9/16	3/8	0.380	5/16	0.182
7/16	0.4375	21/32	7/16	0.463	3/8	0.213
1/2	0.5000	3/4	1/2	0.463	3/8	0.245
5/8	0.6250	15/16	5/8	0.604	1/2	0.307
3/4	0.7500	1 1/8	3/4	0.631	5/8	0.370
7/8	0.8750	1 5/16	7/8	0.709	3/4	0.432
1	1.0000	1 1/2	1	0.801	3/4	0.495
1 1/8	1.1250	1 11/16	1 1/8	......	7/8	0.557
1 1/4	1.2500	1 7/8	1 1/4	......	7/8	0.620
1 3/8	1.3750	2 1/16	1 3/8	......	1	0.682
1 1/2	1.5000	2 1/4	1 1/2	......	1	0.745

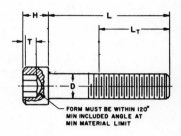

Screw head shall be flat and chamfered. Chamfer E shall be at an angle of 30°± with surface of the flat.

Screw lengths:
Screw lengths 1/8 to 1 inch available in 1/8 inch increments.
Screw lengths 1 to 3 1/2 inches available in 1/4 increments.
Screw lengths 3 1/2 to 6 inches available in 1/2 increments.

Thread lengths:
National Coarse threads — thread length, L_T, equal 2D + 1/2 inch.
National Fine Threads — thread length, L_T, equal 1 1/2 D + 1/8 inch.
Thread class of fit is 3A.

Extracted from *Standard Socket Cap, Shoulder, and Set Screws* (ANSI B18.3–1961), with the permission of the publisher, The American Society of Mechanical Engineers.

MACHINE SCREWS

Machine screws use a nut to fasten parts together. They are much like a cap screw. They can be used in a tapped hole like a cap screw. The common head types are flat, oval, round, fillister, pan, and hexagon. See Fig. 26–36.

Machine screws are usually used to join thin materials. They have sizes smaller than cap screws. Sizes below 1/4 inch in diameter are specified by numbers. ANSI machine screw data is given in Fig. 26–36 and 26–37.

Machine screws are noted on drawings by giving the diameter, number of threads, type of thread, class of fit, length, head type, and the type of fastener. A typical note would read 1/4–28UNF–2A × 1/2 Flat Head Mach. Scr.

26–36.

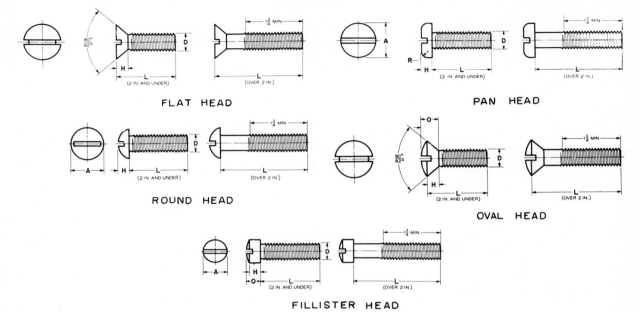

FLAT HEAD   PAN HEAD

ROUND HEAD

OVAL HEAD

FILLISTER HEAD

Standard Machine Screws

Nominal Size	D Diameter of Screw	Flat Head		Round Head		Pan Head		Oval Head			Fillister Head		
		A	H	A	H	A	H	A	H	O	A	H	O
0	0.0600	0.119	0.035	0.113	0.053	0.116	0.039	0.119	0.035	0.056	0.096	0.045	0.059
1	0.0730	0.146	0.043	0.138	0.061	0.142	0.046	0.146	0.043	0.068	0.118	0.053	0.071
2	0.0860	0.172	0.051	0.162	0.069	0.167	0.053	0.172	0.051	0.080	0.140	0.062	0.083
3	0.0990	0.199	0.059	0.187	0.078	0.193	0.060	0.199	0.059	0.092	0.161	0.070	0.095
4	0.1120	0.225	0.067	0.211	0.086	0.219	0.068	0.225	0.067	0.104	0.183	0.079	0.107
5	0.1250	0.252	0.075	0.236	0.095	0.245	0.075	0.252	0.075	0.116	0.205	0.088	0.120
6	0.1380	0.279	0.083	0.260	0.103	0.270	0.082	0.279	0.083	0.128	0.226	0.096	0.132
8	0.1640	0.332	0.100	0.309	0.120	0.322	0.096	0.332	0.100	0.152	0.270	0.113	0.156
10	0.1900	0.385	0.116	0.359	0.137	0.373	0.110	0.385	0.116	0.176	0.313	0.130	0.180
12	0.2160	0.438	0.132	0.408	0.153	0.425	0.125	0.438	0.132	0.200	0.357	0.148	0.205
1/4	0.2500	0.507	0.153	0.472	0.175	0.492	0.144	0.507	0.153	0.232	0.414	0.170	0.237
5/16	0.3125	0.635	0.191	0.590	0.216	0.615	0.178	0.635	0.191	0.290	0.518	0.211	0.295
3/8	0.3750	0.762	0.230	0.708	0.256	0.740	0.212	0.762	0.230	0.347	0.622	0.253	0.355
7/16	0.4375	0.812	0.223	0.750	0.328			0.812	0.223	0.345	0.625	0.265	0.368
1/2	0.5000	0.875	0.223	0.813	0.355			0.875	0.223	0.354	0.750	0.297	0.412
9/16	0.5625	1.000	0.260	0.938	0.410			1.000	0.260	0.410	0.812	0.336	0.466
5/8	0.6250	1.125	0.298	1.000	0.438			1.125	0.298	0.467	0.875	0.375	0.521
3/4	0.7500	1.375	0.372	1.250	0.547			1.375	0.372	0.578	1.000	0.441	0.612

Extracted from *Standard Slotted and Recessed Head Machine Screws and Machine Screw Nuts* (ANSI B18.6.3–1962), with the permission of the publisher, The American Society of Mechanical Engineers.

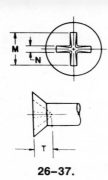

26–37.

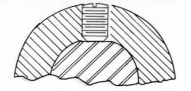

HEADLESS SET SCREW

26–38. *Common set screws.*

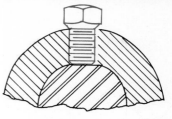

SQUARE HEAD
SET SCREW

Cross Recess Dimensions of Flat Head Machine Screws

	M		T		N
Nom-inal Size	Diameter of Recess		Depth of Recess		Width of Recess
	Max	Min	Max	Min	Min
0	0.083	0.073	0.042	0.031	0.021
1	0.102	0.091	0.054	0.042	0.024
2	0.120	0.109	0.066	0.054	0.027
3	0.139	0.127	0.079	0.066	0.030
4	0.157	0.145	0.088	0.075	0.032
5	0.176	0.162	0.101	0.087	0.035
6	0.195	0.180	0.113	0.098	0.038
8	0.232	0.216	0.132	0.117	0.043
10	0.269	0.251	0.156	0.140	0.048
12	0.307	0.287	0.181	0.163	0.054
1/4	0.355	0.334	0.204	0.186	0.061
5/16	0.444	0.420	0.262	0.242	0.074
3/8	0.523	0.495	0.314	0.291	0.086
7/16	0.568	0.540	0.343	0.320	0.092
1/2	0.608	0.578	0.370	0.345	0.098
9/16	0.656	0.623	0.400	0.374	0.104
5/8	0.656	0.623	0.400	0.374	0.104
3/4	0.656	0.623	0.400	0.374	0.104

This type of recess consists of two intersecting slots with parallel sides converging to a slightly truncated apex at bottom of recess.

Extracted from *Standard Slotted and Recessed Head Machine Screws and Machine Screw Nuts* (ANSI B18.6.3–1962), with the permission of the publisher, The American Society of Mechanical Engineers.

SET SCREWS

Set screws are used to keep one part in position on another. One common use is to prevent a part from slipping on a shaft. An example would be to keep a pulley from slipping on a shaft.

It screws into tapped threads in one part. The point touches the shaft, keeping the parts from slipping. See Fig. 26–38. The shaft usually has a flat spot or cone-shaped hole to receive the point of the set screw.

Some set screws are headless and set flush with the surface. They are safer than those with heads. The head when

26–39. **Standard Square Head Set Screws**

		F	H		G
Nominal Size		Width Across Flats	Nom.	Max	Width Across Corners
#10	0.190	0.1875	9/64	0.148	0.247
#12	0.216	0.216	5/32	0.163	0.292
1/4	0.250	0.250	3/16	0.196	0.331
5/16	0.3125	0.3125	15/64	0.245	0.415
3/8	0.3750	0.375	9/32	0.293	0.497
7/16	0.4375	0.4375	21/64	0.341	0.581
1/2	0.500	0.500	3/8	0.389	0.665
9/16	0.5625	0.5625	27/64	0.437	0.748
5/8	0.6250	0.625	15/32	0.485	0.833
3/4	0.750	0.750	9/16	0.582	1.001
7/8	0.875	0.875	21/32	0.678	1.170
1	1.000	1.000	3/4	0.774	1.337
1 1/8	1.125	1.125	27/32	0.870	1.505
1 1/4	1.250	1.250	15/16	0.966	1.674
1 3/8	1.375	1.375	1 1/32	1.063	1.843
1 1/2	1.500	1.500	1 1/8	1.159	2.010

Square head set screw points are the same types and sizes as shown for slotted headless set screws.

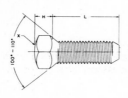

Extracted from *Standard Hexagon Head Cap Screws, Slotted Head Cap Screws, Square Head Set Screws, and Slotted Headless Set Screws* (ANSI B18.6.2–1956), with the permission of the publisher, The American Society of Mechanical Engineers.

26-40.

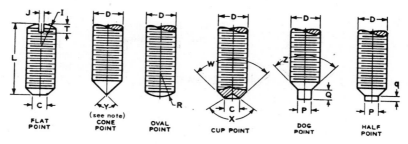

FLAT POINT CONE POINT (see note) OVAL POINT CUP POINT DOG POINT HALF POINT

Standard Slotted Headless Set Screws

D		I	J	T	R	C	P	Q	q
Nominal Size		Radius of Headless Crown	Width of Slot	Depth of Slot	Oval Point Radius	Diameter of Cup and Flat Points	Diameter of Dog Point	Length of Dog Point (see note)	
						Max	Max	Full	Half
5	0.125	0.125	0.023	0.031	0.094	0.067	0.083	0.060	0.030
6	0.138	0.138	0.025	0.035	0.109	0.074	0.092	0.070	0.035
8	0.164	0.164	0.029	0.041	0.125	0.087	0.109	0.080	0.040
10	0.194	0.190	0.032	0.048	0.141	0.102	0.127	0.090	0.045
12	0.216	0.216	0.036	0.054	0.156	0.115	0.144	0.110	0.055
1/4		0.250	0.045	0.063	0.188	0.132	0.156	0.125	0.063
5/16		0.313	0.051	0.078	0.234	0.172	0.203	0.156	0.078
3/8		0.375	0.064	0.094	0.281	0.212	0.250	0.188	0.094
7/16		0.438	0.072	0.109	0.328	0.252	0.297	0.219	0.109
1/2		0.500	0.081	0.125	0.375	0.291	0.344	0.250	0.125
9/16		0.563	0.091	0.141	0.422	0.332	0.391	0.281	0.140
5/8		0.625	0.102	0.156	0.469	0.371	0.469	0.313	0.156
3/4		0.750	0.129	0.188	0.563	0.450	0.563	0.375	0.188

All dimensions given in inches.
Where usable length of thread is less than the nominal diameter, half-dog point shall be used.
When L (length of screw) equals nominal diameter or less, = 118 deg ± 2 deg; when L exceeds nominal diameter, Y = 90 deg ± 2 deg.
Point Angles. W = 80 deg to 90 deg; X = 118 deg ± 5 deg; Z = 100 deg to 110 deg.

Extracted from *Standard Hexagon Head Cap Screws, Slotted Head Cap Screws, Square Head Set Screws, and Slotted Headless Set Screws* (ANSI B18.6.2–1956), with the permission of the publisher, The American Society of Mechanical Engineers.

rotating is dangerous. ANSI square head set screw sizes are shown in Fig. 26–39.

The length of a set screw is the distance under the head to the point.

The headless set screw is made with slotted, recess and fluted socket type heads. They are available with a variety of points—cup, flat, oval, cone, full dog, and half dog. See Figs. 26–40 and 26–41. If parts are to be under great stress, keys are used instead of set screws.

Set screws are specified by giving the diameter, number of threads, thread type, class of fit, length, type of head, type of point, and type of screw. A typical note would read 3/8–16UNC–2A × 5/8 Slotted Flat PT Set Scr.

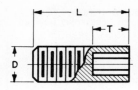

26-41.

HEXAGON
SOCKET

SPLINE
SOCKET

26-42. *Drawing keys and key-seats.*

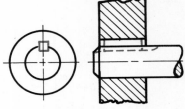

DRAWING PARALLEL AND
GIB HEAD KEYS AND
KEYSEATS.

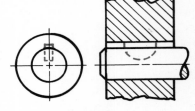

DRAWING WOODRUFF KEYS
AND KEYSEATS.

Standard Hexagon and Spline Socket Set Screws

	D	J	M	T
	Nominal Size	Hex Socket Width Across Flats	Spline Socket Diameter	Key Engagement
5	0.1250	$1/16$	0.074	0.050
6	0.1380	$1/16$	0.074	0.050
8	0.1640	$5/64$	0.098	0.062
10	0.1900	$3/32$	0.115	0.075
$1/4$	0.2500	$1/8$	0.149	0.100
$5/16$	0.3125	$5/32$	0.188	0.125
$3/8$	0.3750	$3/16$	0.221	0.150
$7/16$	0.4375	$7/32$	0.256	0.175
$1/2$	0.5000	$1/4$	0.298	0.200
$5/8$	0.6250	$5/16$	0.380	0.250
$3/4$	0.750	$3/8$	0.463	0.300

Points used are the same as those given for slotted headless set screws.

Lengths: $1/4$ to $5/8$ by $1/16$ inch increments.
$5/8$ to 1 inch by $1/8$ increments.
1 to 4 inches by $1/4$ inch increments.

Extracted from *Standard Hexagon Head Cap Screws, Slotted Head Cap Screws, Square Head Set Screws, and Slotted Headless Set Screws* (ANSI B18.6.2–1956), with permission of the publisher, The American Society of Mechanical Engineers.

KEYS AND KEYSEATS

A *key* is a metal part which is placed in a keyseat to prevent a hub from slipping on a shaft. A *keyseat* is a rectangular groove machined into a shaft or a hub. See Fig. 26–42.

The common types of keys are square, rectangular, plain taper, Gib head, and Woodruff. See Fig. 26–43.

Square keys are preferred for use on shafts $6\frac{1}{2}$ inches in diameter or smaller. See Fig. 26–44 for key dimensions and tolerances.

26-43. *Common types of keys.*

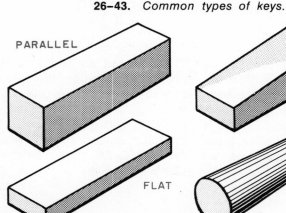

PARALLEL

FLAT

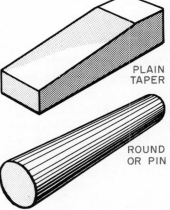

PLAIN
TAPER

ROUND
OR PIN

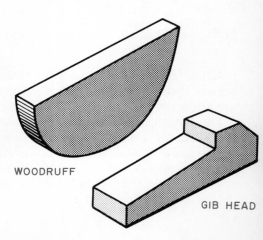

WOODRUFF

GIB HEAD

26–44.
Parallel, Plain Taper, and Gib Head Key Dimensions and Tolerances[2]

Key			Nominal Key Size Width, W		Tolerance	
			Over	To (Incl)	Width, W	Height, H
Parallel	Square	Bar Stock[1]	—	3/4	+0.000 −0.002	+0.000 −0.002
			3/4	1 1/2	+0.000 −0.003	+0.000 −0.003
			1 1/2	2 1/2	+0.000 −0.004	+0.000 −0.004
			2 1/2	3 1/2	+0.000 −0.006	+0.000 −0.006
		Keystock[1]	—	1 1/4	+0.001 −0.000	+0.001 −0.000
			1 1/4	3	+0.002 −0.000	+0.002 −0.000
			3	3 1/2	+0.003 −0.000	+0.003 −0.000
	Rectangular	Bar Stock[1]	—	3/4	+0.000 −0.003	+0.000 −0.003
			3/4	1 1/2	+0.000 −0.004	+0.000 −0.004
			1 1/2	3	+0.000 −0.005	+0.000 −0.005
			3	4	+0.000 −0.006	+0.000 −0.006
			4	6	+0.000 −0.008	+0.000 −0.008
			6	7	+0.000 −0.013	+0.000 −0.013
		Keystock[1]	—	1 1/4	+0.001 −0.000	+0.005 −0.005
			1 1/4	3	+0.002 −0.000	+0.005 −0.005
			3	7	+0.003 −0.000	+0.005 −0.005
Taper	Plain or Gib Head Square or Rectangular		—	1 1/4	+0.001 −0.000	+0.005 −0.000
			1 1/4	3	+0.002 −0.000	+0.005 −0.000
			3	7	+0.003 −0.000	+0.005 −0.000

[1] Two types of stock are used for parallel keys. One is a bar stock with a negative tolerance. Another is a key stock with a close plus tolerance.
[2] All dimensions are in inches.

Extracted from *Standard Keys and Keyseats* (ANSI B17.1–1967), with the permission of the publisher, The American Society of Mechanical Engineers.

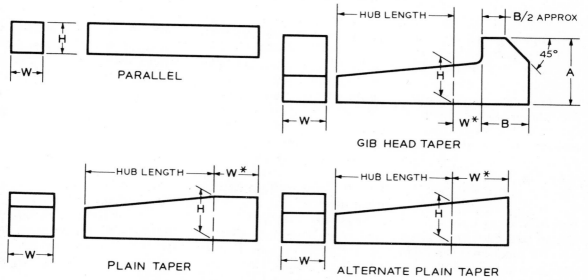

Plain and Gib Head Taper Keys Have a 1/8" Taper in 12"

Key Size for Shaft Diameters for Parallel, Plain Taper, and Gib Head Keys

Nominal Shaft Diameter		Nominal Key Size			Nominal Keyseat Depth	
			Height, H		$H/2$	
Over	To (Incl)	Width, W	Square	Rectangular	Square	Rectangular
5/16	7/16	3/32	3/32		3/64	
7/16	9/16	1/8	1/8	3/32	1/16	3/64
9/16	7/8	3/16	3/16	1/8	3/32	1/16
7/8	1 1/4	1/4	1/4	3/16	1/8	3/32
1 1/4	1 3/8	5/16	5/16	1/4	5/32	1/8
1 3/8	1 3/4	3/8	3/8	1/4	3/16	1/8
1 3/4	2 1/4	1/2	1/2	3/8	1/4	3/16
2 1/4	2 3/4	5/8	5/8	7/16	5/16	7/32
2 3/4	3 1/4	3/4	3/4	1/2	3/8	1/4
3 1/4	3 3/4	7/8	7/8	5/8	7/16	5/16
3 3/4	4 1/2	1	1	3/4	1/2	3/8
4 1/2	5 1/2	1 1/4	1 1/4	7/8	5/8	7/16
5 1/2	6 1/2	1 1/2	1 1/2	1	3/4	1/2
6 1/2	7 1/2	1 3/4	1 3/4	1 1/2	7/8	3/4
7 1/2	9	2	2	1 1/2	1	3/4
9	11	2 1/2	2 1/2	1 3/4	1 1/4	7/8
11	13	3	3	2	1 1/2	1
13	15	3 1/2	3 1/2	2 1/2	1 3/4	1 1/4
15	18	4		3		1 1/2
18	22	5		3 1/2		1 3/4
22	26	6		4		2
26	30	7		5		2 1/2

Sizes and dimension in the unshaded area are preferred.

Extracted from *Standard Keys and Keyseats* (ANSI B17.1–1967), with the permission of the publisher, The American Society of Mechanical Engineers.

Rectangular keys are used on shafts with a diameter larger than 6½ inches. Fig. 26–44 gives key dimensions and tolerances.

Plain taper keys are made in two forms. One tapers the distance the key is in contact with the part. This is the hub length, Fig. 26–44. The other tapers the entire length of the key.

Gib head keys have a large head on one end. They are a form of a taper key. The head helps remove the key. Key dimensions and tolerances are shown in Fig. 26–44.

The size of the key used varies with the shaft diameter. Standard sizes are shown in Fig. 26–45. If the shaft is stepped, use the diameter of the shaft at the point where the key is to be placed. The sizes in the unshaded areas, Fig. 26–45, are those that are most used.

Keys are specified by size and type. Some examples are ⅛ × ¾ *square key*. The key is ⅛-inch square and ¾-inch long. A rectangular key is specified width × height × length. An example is ¼ × ³⁄₁₆ × 1½ *rectangular key*. Taper and Gib head are specified in the same manner as square and rectangular. The taper is a standard ⅛ inch in 12 inches and need not be specified. Woodruff keys are specified by number. An example is *No. 204 Woodruff key*.

Keyseat depths for square, rectangular, plain taper, and Gib head keys are given in fractional sizes, Fig. 26–45.

26–46.

Standard Woodruff Keys and Keyseats

Key No.[1]	Nominal Sizes			Maximum Sizes				Key No.[1]	Nominal Sizes			Maximum Sizes			
	$W \times B$	E	F	D	A	C	H		$W \times B$	E	F	D	A	C	H
204	$1/16 \times 1/2$	$3/64$	$1/32$	.194	.1718	.203	.0422	808	$1/4 \times 1$	$1/16$	$1/8$	.428	.3130	.438	.1360
304	$3/32 \times 1/2$	$3/64$	$3/64$	.194	.1561	.203	.0579	809	$1/4 \times 1 1/8$	$5/64$	$1/8$	.475	.3590	.484	.1360
305	$3/32 \times 5/8$	$1/16$	$3/64$	.240	.2031	.250	.0579	810	$1/4 \times 1 1/4$	$5/64$	$1/8$	.537	.4220	.547	.1360
404	$1/8 \times 1/2$	$3/64$	$1/16$	.194	.1405	.203	.0735	811	$1/4 \times 1 3/8$	$3/32$	$1/8$	.584	.4690	.594	.1360
405	$1/8 \times 5/8$	$1/16$	$1/16$	.240	.1875	.250	.0735	812	$1/4 \times 1 1/2$	$7/64$	$1/8$	.631	.5160	.641	.1360
406	$1/8 \times 3/4$	$1/16$	$1/16$	.303	.2505	.313	.0735	1008	$5/16 \times 1$	$1/16$	$5/32$	.428	.2818	.438	.1672
505	$5/32 \times 5/8$	$1/16$	$5/64$	.240	.1719	.250	.0891	1009	$5/16 \times 1 1/8$	$5/64$	$5/32$	.475	.3278	.484	.1672
506	$5/32 \times 3/4$	$1/16$	$5/64$	.303	.2349	.313	.0891	1010	$5/16 \times 1 1/4$	$5/64$	$5/32$	.537	.3908	.547	.1672
507	$5/32 \times 7/8$	$1/16$	$5/64$	.365	.2969	.375	.0891	1011	$5/16 \times 1 3/8$	$3/32$	$5/32$	.584	.4378	.594	.1672
606	$3/16 \times 3/4$	$1/16$	$3/32$	.303	.2193	.313	.1047	1012	$5/16 \times 1 1/2$	$7/64$	$5/32$	.631	.4848	.641	.1672
607	$3/16 \times 7/8$	$1/16$	$3/32$	.365	.2813	.375	.1047	1210	$3/8 \times 1 1/4$	$5/64$	$3/16$	.537	.3595	.547	.1985
608	$3/16 \times 1$	$1/16$	$3/32$	.428	.3443	.438	.1047	1211	$3/8 \times 1 3/8$	$3/32$	$3/16$	.584	.4065	.594	.1985
609	$3/16 \times 1 1/8$	$5/64$	$3/32$	.475	.3903	.484	.1047	1212	$3/8 \times 1 1/2$	$7/64$	$3/16$	.631	.4535	.641	.1985
807	$1/4 \times 7/8$	$1/16$	$1/8$	.365	.2500	.375	.1360	......	....	...	...	.....	......	.....	....

[1] The last two numbers of the key number indicate the diameter (B) in eighths of an inch. The other one or two numbers in front of these indicate the width of the key (A) in thirty-seconds of an inch. For example, key number 608 means the diameter is $8/8$ or 1 inch and the thickness is $6/32$ or $3/16$ inch.

Extracted from *Standard Woodruff Keys and Keyseats* (ANSI B17.2–1967), with the permission of the publisher, The American Society of Mechanical Engineers.

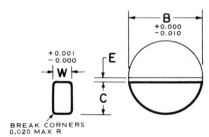

FULL RADIUS TYPE

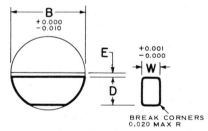

FLAT BOTTOM TYPE

KEYSEAT-SHAFT

KEY ABOVE SHAFT

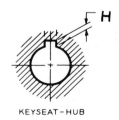

KEYSEAT–HUB

26–47.

Woodruff Key Sizes for Shaft Diameters

Shaft Diameter	$\frac{5}{16}$ to $\frac{3}{8}$	$\frac{7}{16}$ to $\frac{1}{2}$	$\frac{9}{16}$ to $\frac{5}{8}$	$\frac{11}{16}$ to $\frac{3}{4}$	$\frac{13}{16}$
Key Numbers	204	304 305	404 405	404 405 406	505 506
Shaft Diameter	$\frac{7}{8}$ to $\frac{15}{16}$	1	$1\frac{1}{16}$ to $1\frac{1}{8}$	$1\frac{3}{16}$	$1\frac{1}{4}$ to $1\frac{5}{16}$
Key Numbers	505 506 507	606 607 608	606 607 608 609	607 608 609	607 608 609 810
Shaft Diameter	$1\frac{3}{8}$ to $1\frac{7}{16}$	$1\frac{1}{2}$ to $1\frac{5}{8}$	$1\frac{11}{16}$ to $1\frac{3}{4}$	$1\frac{13}{16}$ to $2\frac{1}{8}$	$2\frac{3}{16}$ to $2\frac{1}{2}$
Key Numbers	608 609 810	808 809 810 812	809 810 812	1011 1012	1211 1212

Selected sizes from General Motors Engineering Standards.

These are accurate enough for drafting purposes but not for most manufacturing. Notice that half the key is in the shaft and half in the hub. Formulas for calculating depth control of keyseats can be found in ANSI B17.1–1967, *Keys and Keyseats.* Keyseats are drawn as shown in Fig. 26–42.

Woodruff keys are almost half round. They fit into a semicircular slot machined into the shaft. The top of the key fits into a keyseat machined into the part to be held to the shaft. Standard key sizes are shown in Fig. 26–46. The size Woodruff key to use for various shaft diameters has not been standardized. A suggested guide is given in Fig. 26–47.

PINS

Pins are used to hold parts, as a collar, to a shaft. They are only good for light work. They pass through the collar and shaft. See Fig. 26–48.

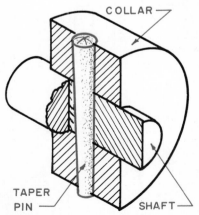

26–48. *Taper pins pass through the shaft.*

Taper pins fit into a tapered hole. They are held in place by friction. They are easily removed. See Fig. 26–49 for standard sizes.

The size of pin to use for various shaft diameters is not standardized. The engineer selects the size to use. Some suggested sizes are shown in Fig. 26–50.

Straight pins are not tapered. They are available with chamfered or straight ends. They can be removed easily. Straight pins are cheaper to use since a tapered hole is not necessary. Fig. 26–51 gives standard sizes of straight pins.

Dowel pins are basically the same diameter their entire length. The chamfered end is slightly smaller than the round end. They are used to hold parts in position or alignment. Fig. 26–52 gives sizes of dowel pins.

A *clevis pin* is shown in Fig. 26–53. A cotter pin is used to hold clevis pins in place. Fig. 26–53 gives standard sizes.

Cotter pins are used to keep parts from separating. They are placed through a hole drilled in the part. The ends are bent out to secure it in place. See Fig. 26–54 for standard sizes.

WASHERS

The common types of washers are plain, lock, and tooth lock.

The *plain washer* is flat. It is used under the head of a bolt or a nut. It spreads the load over a larger area.

26–49.   *Taper pins.*

Standard Taper Pins

Number	7/0	6/0	5/0	4/0	3/0	2/0	0	1	2	3	4	5	6	7	8	9	10
Size (Large End)	0.0625	0.0780	0.0940	0.1090	0.1250	0.1410	0.1560	0.1720	0.1930	0.2190	0.2500	0.2890	0.3410	0.4090	0.4920	0.5910	0.7060
Length, L																	
0.375	X	X															
0.500	X	X	X	X	X	X	X										
0.625	X	X	X	X	X	X	X										
0.750		X	X	X	X	X	X	X	X	X							
0.875				X	X	X	X	X	X	X							
1.000			X	X	X	X	X	X	X	X	X	X					
1.250						X	X	X	X	X	X	X	X				
1.500							X	X	X	X	X	X	X	X			
1.750								X	X	X	X	X	X				
2.000								X	X	X	X	X	X	X	X		
2.250									X	X	X	X	X	X	X		
2.500									X	X	X	X	X	X	X		
2.750										X	X	X	X	X	X	X	
3.000										X	X	X	X	X	X	X	
3.250													X	X	X	X	
3.500													X	X	X	X	X
3.750													X	X	X	X	X
4.000													X	X	X	X	X
4.250															X	X	X
4.500															X	X	X
4.750															X	X	X
5.000															X	X	X
5.250																X	X
5.500																X	X
5.750																X	X
6.000																X	X

Extracted from *Standard Machine Pins* (ANSI B5.20–1958), with permission of the publisher, The American Society of Mechanical Engineers.

26–50.

Suggested Shaft Diameters to Use with Taper Pins

Pin No.	7/0	6/0	5/0	4/0	3/0	2/0	0	1	2	3	4	5	6	7	8
Suggested Shaft Dia.		7/32	1/4	5/16	3/8	7/16	1/2	9/16	5/8	3/4	13/16	7/8	1	1 1/4	1 1/2

26–51.

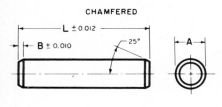

CHAMFERED

SQUARE END

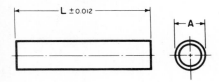

26–52.

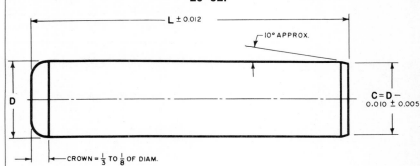

Hardened and Ground Dowel Pins

Length, L	Nominal Diameter D									
	1/8	3/16	1/4	5/16	3/8	7/16	1/2	5/8	3/4	7/8
	Diameter of Standard Pins ±0.0001									
	0.1252	0.1877	0.2502	0.3127	0.3752	0.4377	0.5002	0.6252	0.7502	0.8752
	Diameter Oversize Pins ±0.0001									
	0.1260	0.1885	0.2510	0.3135	0.3760	0.4385	0.5010	0.6260	0.7510	0.8760
1/2	X	X	X	X						
5/8	X	X	X	X						
3/4	X	X	X	X	X					
7/8	X	X	X	X	X	X				
1	X	X	X	X	X	X				
1 1/4		X	X	X	X	X	X	X		
1 1/2		X	X	X	X	X	X	X	X	
1 3/4		X	X	X	X	X	X	X	X	
2		X	X	X	X	X	X	X	X	X
2 1/4				X	X	X	X			
2 1/2			X		X	X	X	X	X	X
3							X	X	X	X
3 1/2							X	X		
4							X	X	X	X
4 1/2								X	X	X
5									X	X
5 1/2									X	X

All dimensions are given in inches.

These pins are extensively used in the tool and machine industry and a machine reamer of nominal size may be used to produce the holes into which these pins tap or press fit. They must be straight and free from any defects that will affect their serviceability.

Extracted from *Standard Machine Pins* (ANSI B5.20–1958), with the permission of the publisher, The American Society of Mechanical Engineers.

Standard Straight Pins

Nominal Diameter	Diameter A		Chamfer B
	Max	Min	
0.062	0.0625	0.0605	0.015
0.094	0.0937	0.0917	0.015
0.109	0.1094	0.1074	0.015
0.125	0.1250	0.1230	0.015
0.156	0.1562	0.1542	0.015
0.188	0.1875	0.1855	0.015
0.219	0.2187	0.2167	0.015
0.250	0.2500	0.2480	0.015
0.312	0.3125	0.3095	0.030
0.375	0.3750	0.3720	0.030
0.438	0.4375	0.4345	0.030
0.500	0.500	0.4970	0.030

All dimensions are given in inches.

These pins must be straight and free from burrs or any other defects that will affect their serviceability.

Extracted from *Standard Machine Pins* (ANSI B5.20–1958), with permission of the publisher, The American Society of Mechanical Engineers.

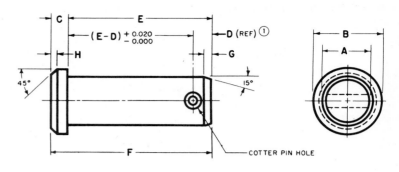

26-53.

Standard Clevis Pins

Diameter of Pin A			Diameter of Head B	Height of Head C	Distance From Center of Hole to End of Pin D	Length of Pin Under Head ② E	Total Length of Pin F	Chamfer of Pin G	Chamfer of Head H	Hole Sizes +.010 -.005	Drill Sizes
Nom	Max	Min									
0.188	0.186	0.181	5/16	1/16	7/64	19/32	21/32	3/64	1/64	0.0781	5/64
0.250	0.248	0.243	3/8	3/32	1/8	31/64	57/64	1/16	1/32	0.0781	5/64
0.312	0.311	0.306	7/16	3/32	5/32	31/32	1 1/16	5/64	1/32	0.1094	7/64
0.375	0.373	0.368	1/2	1/8	5/32	1 3/32	1 7/32	5/64	1/32	0.1094	7/64
0.438	0.436	0.431	9/16	5/32	11/64	1 5/32	1 25/64	3/32	3/64	0.1094	7/64
0.500	0.496	0.491	5/8	5/32	7/32	1 27/64	1 37/64	7/64	3/64	0.1406	9/64
0.625	0.621	0.616	13/16	13/64	1/4	1 23/64	1 59/64	9/64	1/16	0.1406	9/64
0.750	0.746	0.741	15/16	1/4	19/64	2 3/64	2 19/64	5/32	5/64	0.1719	11/64
0.875	0.871	0.866	1 1/16	5/16	21/64	2 11/32	2 21/32	3/16	3/32	0.1719	11/64
1.000	0.996	0.991	1 3/16	11/32	23/64	2 5/8	2 31/32	7/32	7/64	0.1719	11/64

All dimensions are given in inches.

26-54.

Standard Cotter Pins

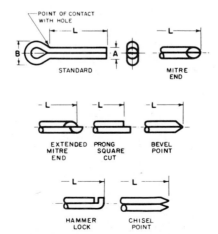

Diameter Nominal A	Outside Eye Diameter B Min.	Hole Sizes Recommended
0.031	1/16	3/64
0.047	3/32	1/16
0.062	1/8	5/64
0.078	5/32	3/32
0.094	3/16	7/64
0.109	7/32	1/8
0.125	1/4	9/64
0.141	9/32	5/32
0.156	5/16	11/64

Diameter Nominal A	Outside Eye Diameter B Min.	Hole Sizes Recommended
0.188	3/8	13/64
0.219	7/16	15/64
0.250	1/2	17/64
0.312	5/8	5/16
0.375	3/4	3/8
0.438	7/8	7/16
0.500	1	1/2
0.625	1 1/4	5/8
0.750	1 1/2	3/4

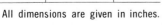
All dimensions are given in inches.

Extracted from *Standard Machine Pins* (ANSI B5.20–1958), with the permission of the publisher, The American Society of Mechanical Engineers.

26–55.

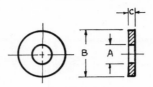

Standard Plain Washers[1]

Nominal Washer Size			Inside Dia.	Outside Dia.	Nominal Thickness
			A	B	C
......	......		0.078	0.188	0.020
......	......		0.094	0.250	0.020
......	......		0.125	0.312	0.032
No. 6	0.138		0.156	0.375	0.049
No. 8	0.164		0.188	0.438	0.049
No. 10	0.190		0.219	0.500	0.049
3/16	0.188		0.250	0.562	0.049
No. 12	0.216		0.250	0.562	0.065
1/4	0.250	N	0.281	0.625	0.065
1/4	0.250	W	0.312	0.734	0.065
5/16	0.312	N	0.344	0.688	0.065
5/16	0.312	W	0.375	0.875	0.083
3/8	0.375	N	0.406	0.812	0.065
3/8	0.375	W	0.438	1.000	0.083
7/16	0.438	N	0.469	0.922	0.065
7/16	0.438	W	0.500	1.250	0.083
1/2	0.500	N	0.531	1.062	0.095
1/2	0.500	W	0.562	1.375	0.109
9/16	0.562	N	0.594	1.156	0.095
9/16	0.562	W	0.625	1.469	0.109
5/8	0.625	N	0.656	1.312	0.095
5/8	0.625	W	0.688	1.750	0.134
3/4	0.750	N	0.812	1.469	0.134
3/4	0.750	W	0.812	2.000	0.148
7/8	0.875	N	0.938	1.750	0.134
7/8	0.875	W	0.938	2.250	0.165
1	1.000	N	1.062	2.000	0.134
1	1.000	W	1.062	2.500	0.165
1 1/8	1.125	N	1.250	2.250	0.134
1 1/8	1.125	W	1.250	2.750	0.165
1 1/4	1.250	N	1.375	2.500	0.165
1 1/4	1.250	W	1.375	3.000	0.165
1 3/8	1.375	N	1.500	2.750	0.165
1 3/8	1.375	W	1.500	3.250	0.180
1 1/2	1.500	N	1.625	3.000	0.165

[1] Additional sizes in standards.

Extracted from *Standard Plain Washers* (ANSI B27.2–1965), with the permission of the publisher, The American Society of Mechanical Engineers.

They are specified on a drawing by giving the inside diameter, outside diameter, and thickness. A typical note would read .406 × .812 × .065 type A plain washer. Fig. 26–55 gives some standard sizes.

Lock washers are split and bent in a helical shape. They provide a force due to the spring action that helps hold a part, as a nut, in place. They also make it easier to unscrew bolted parts. They are available in regular, extra duty, and high-collar series. They are specified by giving the nominal size (hole diameter) and the series. A typical note would read 3/8" regular helical spring lock washer. See Fig. 26–56 for standard sizes.

Tooth lock washers are made in three types. These are external, internal, and internal-external. They are hardened. The hardened teeth are twisted so they can grip the bolt head or nut and keep it from vibrating loose. They are specified by giving nominal size, description, and type. A typical note would read 1/4 inch, internal tooth, type A.

See Fig. 26–57 for standard sizes of tooth lock washers.

RIVETS

Rivets are permanent fasteners. They are used to fasten sheet metal parts and steel plate. They have a head on one end. They are placed through holes in the parts to be fastened. The straight end is then formed into another head. See Fig. 26–58.

26–56.

Standard Helical Spring Lock Washers

Nominal Washer Size		Inside Diameter A		Outside Diameter B	Washer Section	
					Width W	Thickness $\frac{T+t}{2}$
		Min	Max	Max**	Min	Min
No. 2	0.086	0.088	0.094	0.172	0.035	0.020
No. 3	0.099	0.101	0.107	0.195	0.040	0.025
No. 4	0.112	0.115	0.121	0.209	0.040	0.025
No. 5	0.125	0.128	0.134	0.236	0.047	0.031
No. 6	0.138	0.141	0.148	0.250	0.047	0.031
No. 8	0.164	0.168	0.175	0.293	0.055	0.040
No. 10	0.190	0.194	0.202	0.334	0.062	0.047
No. 12	0.216	0.221	0.229	0.377	0.070	0.056
1/4	0.250	0.255	0.263	0.489	0.109	0.062
5/16	0.312	0.318	0.328	0.586	0.125	0.078
3/8	0.375	0.382	0.393	0.683	0.141	0.094
7/16	0.438	0.446	0.459	0.779	0.156	0.109
1/2	0.500	0.509	0.523	0.873	0.171	0.125
9/16	0.562	0.572	0.587	0.971	0.188	0.141
5/8	0.625	0.636	0.653	1.079	0.203	0.156
11/16	0.688	0.700	0.718	1.176	0.219	0.172
3/4	0.750	0.763	0.783	1.271	0.234	0.188
13/16	0.812	0.826	0.847	1.367	0.250	0.203
7/8	0.875	0.890	0.912	1.464	0.266	0.219
15/16	0.938	0.954	0.978	1.560	0.281	0.234
1	1.000	1.017	1.042	1.661	0.297	0.250
1 1/16	1.062	1.080	1.107	1.756	0.312	0.266
1 1/8	1.125	1.144	1.172	1.853	0.328	0.281
1 3/16	1.188	1.208	1.237	1.950	0.344	0.297
1 1/4	1.250	1.271	1.302	2.045	0.359	0.312
1 5/16	1.312	1.334	1.366	2.141	0.375	0.328
1 3/8	1.375	1.398	1.432	2.239	0.391	0.344
1 7/16	1.438	1.462	1.497	2.334	0.406	0.359
1 1/2	1.500	1.525	1.561	2.430	0.422	0.375

Extracted from *Standard Lock Washers* (ANSI B27.1–1965), with the permission of the publisher, The American Society of Mechanical Engineers.

26–57.

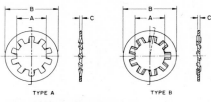

TYPE A TYPE B

Internal Tooth Lock Washers

Dimensions of Internal Tooth Lock Washers

Nominal Washer Size		A Inside Diameter	B Outside Diameter	C Thickness
		Max	Max	Max
No. 2	0.086	0.095	0.200	0.015
No. 3	0.099	0.109	0.232	0.019
No. 4	0.112	0.123	0.270	0.019
No. 5	0.125	0.136	0.280	0.021
No. 6	0.138	0.150	0.295	0.021
No. 8	0.164	0.176	0.340	0.023
No. 10	0.190	0.204	0.381	0.025
No. 12	0.216	0.231	0.410	0.025
1/4	0.250	0.267	0.478	0.028
5/16	0.312	0.332	0.610	0.034
3/8	0.375	0.398	0.692	0.040
7/16	0.438	0.464	0.789	0.040
1/2	0.500	0.530	0.900	0.045
9/16	0.562	0.596	0.985	0.045
5/8	0.625	0.663	1.071	0.050
11/16	0.688	0.728	1.166	0.050
3/4	0.750	0.795	1.245	0.055
13/16	0.812	0.861	1.315	0.055
7/8	0.875	0.927	1.410	0.060
1	1.000	1.060	1.637	0.067
1 1/8	1.125	1.192	1.830	0.067
1 1/4	1.250	1.325	1.975	0.067

Dimensions of Heavy Internal Tooth Lock Washers

Nominal Washer Size		A Inside Diameter	B Outside Diameter	C Thickness
		Max	Max	Max
1/4	0.250	0.267	0.536	0.045
5/16	0.312	0.332	0.607	0.050
3/8	0.375	0.398	0.748	0.050
7/16	0.438	0.464	0.858	0.067
1/2	0.500	0.530	0.924	0.067
9/16	0.562	0.596	1.034	0.067
5/8	0.625	0.663	1.135	0.067
3/4	0.750	0.795	1.265	0.084
7/8	0.875	0.927	1.447	0.084

Extracted from *Standard Lock Washers* (ANSI B27.1–1965), with the permission of the publisher, The American Society of Mechanical Engineers.

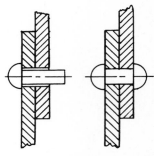

RIVET IN PLACE BEFORE FORMING

RIVET AFTER HEAD IS FORMED

26–58. *Fastening parts with a rivet.*

Small rivets are made with pan, button, countersunk, and flat heads. Standard sizes are given in Fig. 26–59.

Large rivets are made with button, high button, pan, flat top countersunk, and cone heads. Standard sizes are shown in Fig. 26–60.

Rivets on a drawing are shown as in Fig. 26–61.

Rivets are specified on a drawing by giving body diameter, length of the rivet, and head type. A typical note would read 1/4 × 1/2 pan head.

WOOD SCREWS

Wood screws are used to fasten two pieces of wood together or secure some other material to wood. One part has a hole large enough so the screw passes freely through it. The other part has a hole approximately the diameter of the core of the screw. The threads cut into the sides of the hole, forming internal threads.

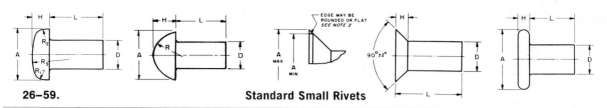

26–59.   **Standard Small Rivets**

Nominal Size or Basic Shank Diameter		Diameter of Shank D	Pan Head					Button Head			Countersunk Head		Flat Head	
			A	H	R_1	R_2	R_3	A	H	R	A	H	A	H
1/16	0.062	0.064	0.118	0.040	0.019	0.052	0.217	0.122	0.052	0.055	0.118	0.027	0.140	0.027
3/32	0.094	0.096	0.173	0.060	0.030	0.080	0.326	0.182	0.077	0.084	0.176	0.040	0.200	0.038
1/8	0.125	0.127	0.225	0.078	0.039	0.106	0.429	0.235	0.100	0.111	0.235	0.053	0.260	0.048
5/32	0.156	0.158	0.279	0.096	0.049	0.133	0.535	0.290	0.124	0.138	0.293	0.066	0.323	0.059
3/16	0.188	0.191	0.334	0.114	0.059	0.159	0.641	0.348	0.147	0.166	0.351	0.079	0.387	0.069
7/32	0.219	0.222	0.391	0.133	0.069	0.186	0.754	0.405	0.172	0.195	0.413	0.094	0.453	0.080
1/4	0.250	0.253	0.444	0.151	0.079	0.213	0.858	0.460	0.196	0.221	0.469	0.106	0.515	0.091
9/32	0.281	0.285	0.499	0.170	0.088	0.239	0.963	0.518	0.220	0.249	0.528	0.119	0.579	0.103
5/16	0.313	0.317	0.552	0.187	0.098	0.266	1.070	0.572	0.243	0.276	0.588	0.133	0.641	0.113
11/32	0.344	0.348	0.608	0.206	0.108	0.292	1.176	0.630	0.267	0.304	0.646	0.146	0.705	0.124
3/8	0.375	0.380	0.663	0.225	0.118	0.319	1.286	0.684	0.291	0.332	0.704	0.159	0.769	0.135
13/32	0.406	0.411	0.719	0.243	0.127	0.345	1.392	0.743	0.316	0.358	0.763	0.172	0.834	0.146
7/16	0.438	0.443	0.772	0.261	0.137	0.372	1.500	0.798	0.339	0.387	0.823	0.186	0.896	0.157

Small rivets are available in length increments of 1/32.
Extracted from *Standard Small Solid Rivets* (ANSI B18.1–1965)

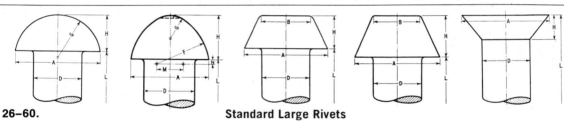

26–60.   **Standard Large Rivets**

Body Diameter of Rivet		Button Head			High Button Head				Pan Head			Cone Head			Flat Top	
D		A	H	G	A	H	F	G	A	B	H	A	B	H	A	H
1/2	0.500	0.922	0.344	0.484	0.875	0.375	0.562	0.375	0.844	0.500	0.328	0.922	0.469	0.406	0.905	0.281
5/8	0.625	1.141	0.438	0.594	1.062	0.453	0.672	0.453	1.047	0.625	0.406	1.141	0.594	0.516	1.131	0.343
3/4	0.750	1.375	0.516	0.719	1.250	0.531	0.797	0.531	1.266	0.750	0.484	1.375	0.703	0.625	1.358	0.406
7/8	0.875	1.594	0.609	0.844	1.438	0.609	0.922	0.609	1.469	0.875	0.578	1.594	0.828	0.719	1.584	0.469
1	1.000	1.828	0.688	0.953	1.625	0.688	1.031	0.688	1.687	1.000	0.656	1.828	0.938	0.828	1.810	0.531
1 1/8	1.125	2.062	0.781	1.078	1.812	0.766	1.156	0.766	1.891	1.125	0.734	2.063	1.063	0.938	2.036	0.609
1 1/4	1.250	2.281	0.859	1.188	2.000	0.844	1.266	0.844	2.094	1.250	0.812	2.281	1.172	1.031	2.262	0.672
1 3/8	1.375	2.516	0.953	0.312	2.188	0.938	1.406	0.938	2.312	1.375	0.906	2.516	1.297	1.141	2.489	0.751
1 1/2	1.500	2.734	1.031	1.438	2.375	1.000	1.500	1.000	2.516	1.500	0.984	2.734	1.406	1.250	2.715	0.813
1 5/8	1.625	2.969	1.125	1.547	2.562	1.094	1.641	1.094	2.734	1.625	1.062	2.969	1.531	1.344	2.941	0.875
1 3/4	1.750	3.203	1.203	1.672	2.750	1.172	1.750	1.172	2.938	1.750	1.141	3.203	1.641	1.453	3.168	0.938

Head dimensions are for manufactured head after driving. Large rivets are available in length increments of 1/8 inch.
Extracted from *Standard Large Rivets* (ANSI B18.4–1960), with the permission of the publisher, The American Society of Mechanical Engineers.

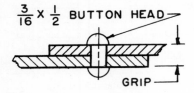

$\frac{3}{16} \times \frac{1}{2}$ BUTTON HEAD

GRIP

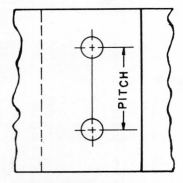

PITCH

**SINGLE RIVETED
LAP JOINT**

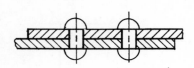

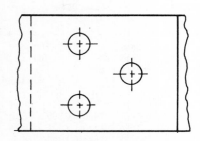

**STAGGERED RIVETED
LAP JOINT**

26-61. *Drawing riveted joints.*

Screws are easy to install and remove. They can be removed and replaced several times before the threads in the wood are damaged. See Fig. 26-62.

Wood screws are made from steel, brass, and aluminum alloy. They have flat, round, and oval heads. The heads may be slotted or recessed. ANSI sizes are given in Figs. 26-63 and 26-64.

Screw sizes are noted by giving length, diameter of the body (wire diameter), type of head, type of screw, and the material. A typical note reads 1½-No. 8 FH wood screw-steel.

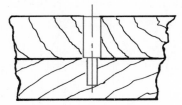

**HOLES PREPARED FOR
SCREW**

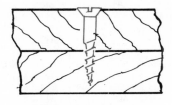

SCREW IN PLACE

26-62. *Fastening with wood screws.*

SPRINGS

Springs are devices used to apply energy. They can either push or pull. In common use are two types, coil and flat springs.

Coil springs are formed in the shape of a helix. The common types are compression, extension and torsion.

Compression springs apply energy (push) when they are squeezed (compressed). In their natural state, the coils are not touching. See Fig. 26-65.

They are made with plain open ends, plain closed ends, ground open ends, and ground closed ends. See Fig. 26-66.

Extension coil springs apply energy when they are stretched (pulled). In their natural state, the coils are usually touching. See Fig. 26-65.

They are commonly made with some type of hook or loop on each end. A screen door spring is an extension spring.

Torsion coil springs apply energy when one end is moved in circular direction. See Fig. 26-65.

Flat springs are made from spring steel. They are commonly used to hold something in place. They permit easy removal or movement of the part. A common use is in door catches on cabinets. Such a spring is shown in Fig. 26-67.

They are designed to serve a special purpose. They are not standardized.

26–63.

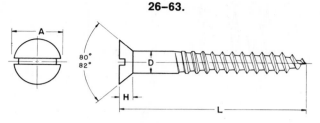

FLAT HEAD

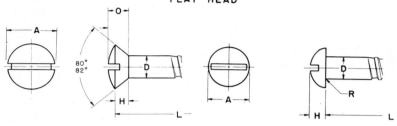

OVAL HEAD   ROUND HEAD

Standard Slotted Head Wood Screws

Nom-inal Size	Threads per Inch	Screw Dia. D	Flat and Oval Head			Round Head	
			A	H	O	A	H
0	32	0.060	0.119	0.035	0.056	0.113	0.053
1	28	0.073	0.146	0.043	0.068	0.138	0.061
2	26	0.086	0.172	0.051	0.080	0.162	0.069
3	24	0.099	0.199	0.059	0.092	0.187	0.078
4	22	0.112	0.225	0.067	0.104	0.211	0.086
5	20	0.125	0.252	0.075	0.116	0.236	0.095
6	18	0.138	0.279	0.083	0.128	0.260	0.103
7	16	0.151	0.305	0.091	0.140	0.285	0.111
8	15	0.164	0.332	0.100	0.152	0.309	0.120
9	14	0.177	0.358	0.108	0.164	0.334	0.128
10	13	0.190	0.385	0.116	0.176	0.359	0.137
12	11	0.216	0.438	0.132	0.200	0.408	0.153
14	10	0.242	0.491	0.148	0.224	0.457	0.170
16	9	0.268	0.544	0.164	0.248	0.506	0.187
18	8	0.294	0.597	0.180	0.272	0.555	0.204
20	8	0.320	0.650	0.196	0.296	0.604	0.220
24	7	0.372	0.756	0.228	0.344	0.702	0.254

Screw lengths: 1/4 to 1 inch by 1/8 inch increments.
1 to 3 inches by 1/4 inch increments.
3 to 5 inches by 1/2 inch increments.
Thread lengths equal to 2/3 screw length.

Extracted from *Standard Slotted and Recessed Head Wood Screws* (ANSI B18.6.1–1961), with the permission of the publisher, The American Society of Mechanical Engineers.

26–64.

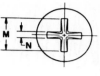

Standard Type Two Recessed Flat Head Wood Screws

Nom-inal Size	M	T	N
0	0.083	0.042	0.021
1	0.102	0.054	0.024
2	0.120	0.066	0.027
3	0.139	0.079	0.030
4	0.157	0.088	0.032
5	0.176	0.101	0.035
6	0.195	0.113	0.038
7	0.213	0.125	0.040
8	0.232	0.132	0.043
9	0.251	0.145	0.046
10	0.269	0.156	0.048
12	0.307	0.181	0.054
14	0.344	0.197	0.059
16	0.381	0.221	0.064
18	0.418	0.245	0.070
20	0.455	0.269	0.075
24	0.529	0.318	0.086

Head dimensions are the same as those of slotted head wood screws.

Extracted from *Standard Slotted and Recessed Head Wood Screws* (ANSI B18.6.1–1965), with the permission of the publisher, The American Society of Mechanical Engineers.

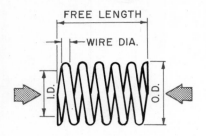

WIRE DIA.

FREE LENGTH

I.D.   O.D.

COMPRESSION COIL SPRING

FREE LENGTH

EXTENSION COIL SPRING

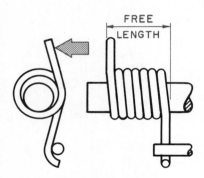

FREE
LENGTH

TORSION COIL SPRING

26-65. *Common types of coil springs.*

Terminology of Springs. It is important to know the terms used to indicate springs. *Free length* is the length of the spring when it is in its natural, unloaded condition. The free length of a compression spring includes the entire spring.

796

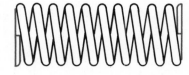

OPEN PLAIN END

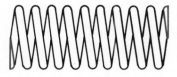

CLOSED GROUND END

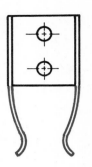

OPEN GROUND END

26-66. *Common ends for coil springs.*

Extension spring free length is measured inside the hooks. See Fig. 26-65. The loaded length is the length under a known load.

Solid length is the length of a compression spring when all the coils are compressed so they touch. *Outside diameter* is the overall diameter of the outside of the coil. The *inside diameter* is measured inside the coil. *Wire size* is the diameter of the wire used to make the spring. A *coil* is one turn of the wire through 360 degrees.

Coil springs are available wound right hand or left hand.

FLAT SPRING FOR CABINET DOOR CATCH

26-67. *A flat spring.*

This is the same as a right-hand or left-hand thread.

How to Draw Coil Springs. Coil springs are shown on drawings in detailed or schematic representation. The springs, Fig. 26-65, are drawn in detailed form.

To draw detailed representation of a coil spring, see Fig. 26-68. Even though the spring wire forms a helix, the sides of the coils are drawn with straight lines. This is easier and faster.

1. Locate the center line of the coil, Fig. 26-68A.
2. Draw the outside diameter minus the wire diameter (OD-WD).
3. Draw the coil length.
4. Lay out the pitch distance of the coils.
5. Draw light circles for wire diameters.
6. Connect wire diameter circles with straight lines, Fig. 26-68B. These are the coils on the front of the spring.

7. Connect the sides of the coils on the rear of the spring, Fig. 26–68C.

8. Remove construction lines and darken object lines.

Schematic coil representation is shown in Fig. 26–69. The coils are indicated by straight lines. The distance from one point to the next point represents the pitch or one coil.

Notice that the ten-coil compression spring is drawn with eight points. Each end

of the coil is closed or squared off. They do not serve as active coils. Two coils are lost this way. The eight active coils are indicated on the drawing.

In the extension spring, all coils are active. The number

of points indicates the number of coils.

KNURLING

A knurled surface is one with a series of grooves machined into it. One purpose of knurling is to provide a better handgrip of a part. Another use is for fastening two parts together. A knurled shaft held in a smooth hole by a pressed fit has considerable holding power.

Two common types of knurling are straight and diamond, Fig. 26–70.

Information needed for knurling for handgrip purposes is the pitch (distance between grooves), types of knurl, and

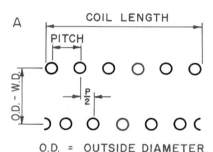

A

O.D. = OUTSIDE DIAMETER
W.D. = WIRE DIAMETER

B

C

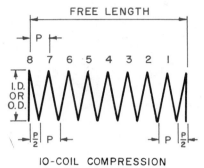

IO-COIL COMPRESSION SPRING

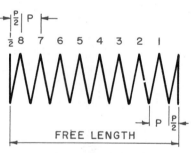

IO½-COIL COMPRESSION SPRING

4-COIL EXTENSION SPRING

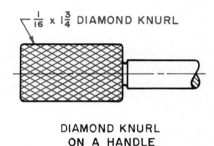

DIAMOND KNURL ON A HANDLE

STRAIGHT KNURL ON A SHAFT FOR PRESS FIT

26–68. *How to draw detailed representation of coil springs.*

26–69. *Schematic representation of coil springs.*

26–70. *Common types of knurling.*

length of the area to be knurled.

Parts knurled for fastening purposes need the tolerance diameter of the shaft before knurling, the pitch, type of knurl, and minimum diameter after knurling.

Standards for knurling are found in ANSI B5.30–1958.

WELDING

Welding is one way that metal parts can be fastened together. Steel, aluminum, and magnesium can be welded.

The two basic welding processes are fusion welding and resistance welding. *Fusion welding* involves the melting of a metal rod, called welding rod, and combining it with the metal parts to be joined. When the melted rod cools, the parts are permanently joined together. The rod can be melted by electricity or gas. See Fig. 26–71.

Resistance welding is done by passing an electric current through the spot to be welded. This is done under pressure. The current heats the metal parts. The pressure plus heat welds them together. This is commonly done on sheet metal parts. See Fig. 26–72.

Fusion Welding. Some common types of joints used in fusion welding are shown in Fig. 26–73. Information on the drawing about the type of weld to be used is given with standardized symbols. Some of these symbols are given in

798

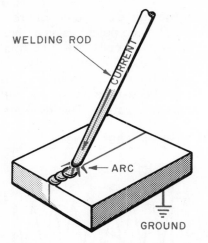

26–71. *Fusion welding joins by combining a melted rod with melted metal in the parts joined.*

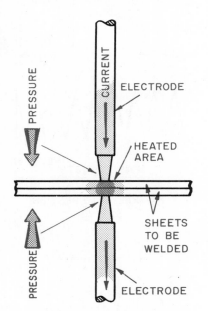

26–72. *Resistance welding joins by heat and pressure.*

26–73. *Common types of welded joints in fusion welding.*

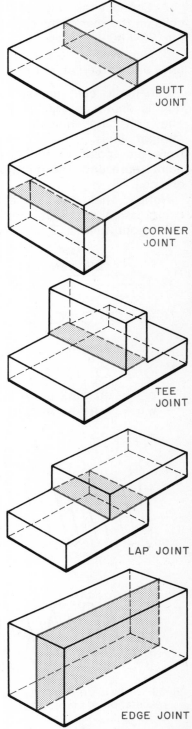

BUTT JOINT

CORNER JOINT

TEE JOINT

LAP JOINT

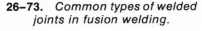

EDGE JOINT

TYPE OF WELD								SUPPLEMENTARY SYMBOLS			
BACK OR BACKING	FILLET	PLUG OR SLOT	GROOVE					WELD ALL AROUND	FIELD WELD	CONTOUR	
			SQUARE	V	BEVEL	U	J			FLUSH	CONVEX
⌒	◺	�054	‖	∨	⋁	∪	⋃	◯	●	—	⌢

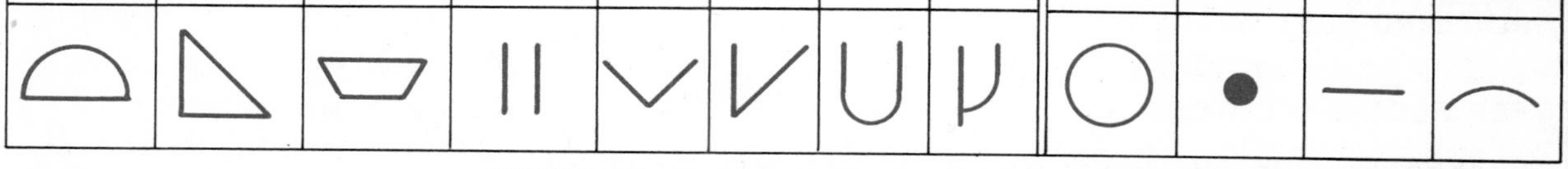

ANSI Y32.3–1959, American Welding Society and the American National Standards Association, by permission.

26–74. *Fusion welding symbols.*

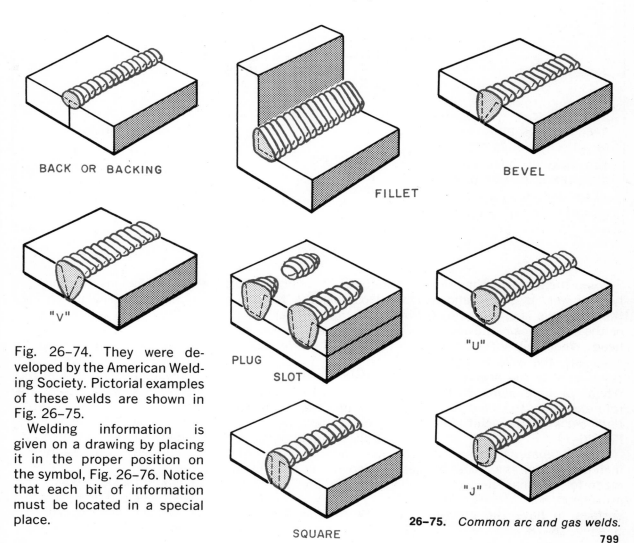

BACK OR BACKING

FILLET

BEVEL

"V"

PLUG

SLOT

"U"

Fig. 26–74. They were developed by the American Welding Society. Pictorial examples of these welds are shown in Fig. 26–75.

Welding information is given on a drawing by placing it in the proper position on the symbol, Fig. 26–76. Notice that each bit of information must be located in a special place.

"J"

SQUARE

26–75. *Common arc and gas welds.*

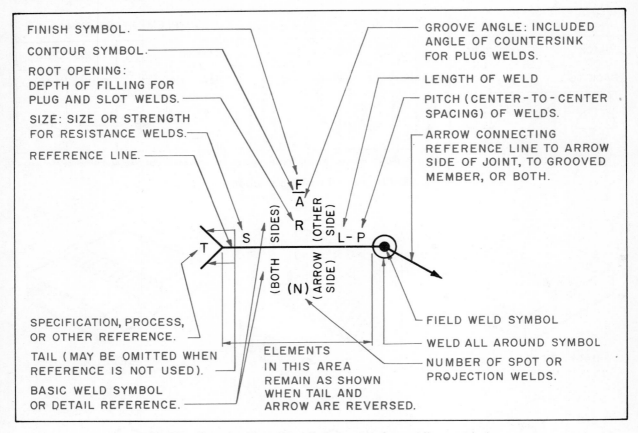

FINISH SYMBOL.

CONTOUR SYMBOL.

ROOT OPENING:
DEPTH OF FILLING FOR
PLUG AND SLOT WELDS.

SIZE: SIZE OR STRENGTH
FOR RESISTANCE WELDS.

REFERENCE LINE.

GROOVE ANGLE: INCLUDED
ANGLE OF COUNTERSINK
FOR PLUG WELDS.

LENGTH OF WELD

PITCH (CENTER-TO-CENTER
SPACING) OF WELDS.

ARROW CONNECTING
REFERENCE LINE TO ARROW
SIDE OF JOINT, TO GROOVED
MEMBER, OR BOTH.

F
A
R
(OTHER SIDE)
S
(BOTH SIDES)
(ARROW SIDE)
(N)
T
L-P

SPECIFICATION, PROCESS,
OR OTHER REFERENCE.

TAIL (MAY BE OMITTED WHEN
REFERENCE IS NOT USED).

BASIC WELD SYMBOL
OR DETAIL REFERENCE.

ELEMENTS
IN THIS AREA
REMAIN AS SHOWN
WHEN TAIL AND
ARROW ARE REVERSED.

FIELD WELD SYMBOL

WELD ALL AROUND SYMBOL

NUMBER OF SPOT OR
PROJECTION WELDS.

26–76. *Standard location of elements of a welding symbol.*

When the symbol giving the type of weld is placed *above* the reference line, the weld is on the side *opposite* the arrowhead. When *below* the reference line, the weld is on the *same side* as the arrowhead. When on *both sides* of the reference line, the weld is on *both sides* of the joint. See Fig. 26–77.

Usually material over $1/8$-inch thick requires a groove be made for welding. The symbol arrowhead points *toward* the surface to have the groove. See Fig. 26–78.

The type of weld symbols are placed on the reference line with the vertical leg always on the *left* as you view the symbol. See Fig. 26–77.

Field Weld Symbol. A solid round circle at the end of the reference line means the weld is to be a field weld. This means the welding occurs on the job where the device is assembled rather than in the shop. See Fig. 26–79.

Weld All Around Symbol. A circle at the end of the refer-

ence line means the joint is welded on all sides. See Fig. 26–79.

Symbol Tail. If a weld has a special specification, the abbreviation for this is in the tail. The symbols used are available in ANSI Y32.3–1958. If no specifications are needed, the tail is not used. See Fig. 26–76 for examples.

Indicating Weld Sizes. The size of welds is indicated on the welding symbol. The information given is the size,

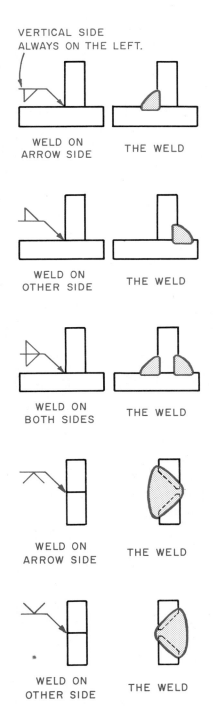

VERTICAL SIDE
ALWAYS ON THE LEFT.

WELD ON
ARROW SIDE

THE WELD

WELD ON
OTHER SIDE

THE WELD

WELD ON
BOTH SIDES

THE WELD

WELD ON
ARROW SIDE

THE WELD

WELD ON
OTHER SIDE

THE WELD

26-77. *Use of terms arrow side and other side on the welding symbol.*

strength, groove angle, length, pitch, and number of spot and projection welds. The location of these is shown in Fig. 26–76.

Fillet Welds. The location of the size of fillet welds with both legs the same length is shown in Fig. 26–80A.

Fillet welds that are the same size on both sides can be shown as in Fig. 26–80B. If the welds are different sizes, they are shown as in Fig. 26–80C. If the length of the weld is the same as the length of the part being welded, no length dimension is given.

If the legs of a fillet weld are not the same size, they are shown as in Fig. 26–80D.

Groove Welds. The proper way to indicate the size of groove welds is shown in Fig. 26–81. The depth of the groove is given to the left of the weld symbol. If the groove is from both sides and is the same size, only one side need be dimensioned. If the welds differ, both must be dimensioned.

Groove angles are given outside the opening of the weld symbol.

The root opening is given inside the weld symbol. A root opening is the space between the two members being welded. See Fig. 26–81.

If all grooves have the same angle, they need not be dimensioned. A note, *All V-Groove Welds 60° groove angle unless otherwise specified,*

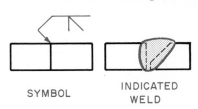

SYMBOL

INDICATED WELD

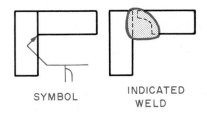

SYMBOL

INDICATED WELD

26-78. *The arrowhead points toward the surface that will have a groove.*

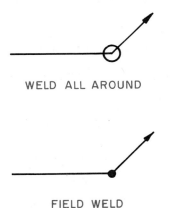

WELD ALL AROUND

FIELD WELD

26-79. *Location of the field weld and weld all around symbol.*

can be used to give this information.

Resistance Welding. Two types of resistance welding are spot and seam. The ANSI resistance symbols are shown in Fig. 26–82.

A *spot-weld* is the fusion of parts in which the area of joining is a small circular shape. The spot-weld symbol

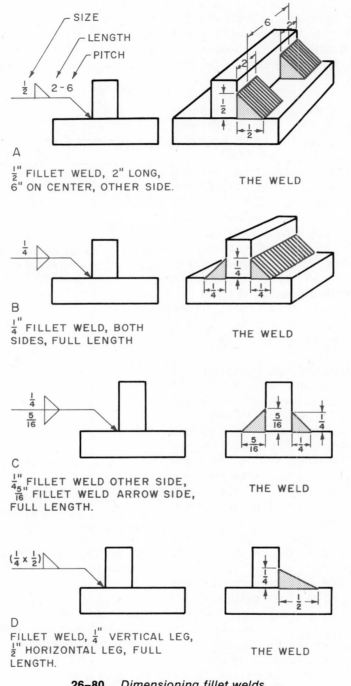

A

SIZE
LENGTH
PITCH

$\frac{1}{2}$" FILLET WELD, 2" LONG,
6" ON CENTER, OTHER SIDE.

THE WELD

B

$\frac{1}{4}$" FILLET WELD, BOTH
SIDES, FULL LENGTH

THE WELD

C

$\frac{1}{4}$" FILLET WELD OTHER SIDE,
$\frac{5}{16}$" FILLET WELD ARROW SIDE,
FULL LENGTH.

THE WELD

D

FILLET WELD, $\frac{1}{4}$" VERTICAL LEG,
$\frac{1}{2}$" HORIZONTAL LEG, FULL
LENGTH.

THE WELD

26-80. *Dimensioning fillet welds.*

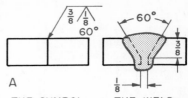
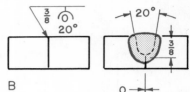
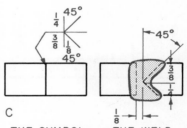
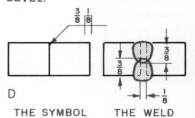

A

THE SYMBOL   THE WELD

V GROOVE, $\frac{3}{8}$" DEEP, 60° ANGLE,
$\frac{1}{8}$" ROOT OPENING, ARROW SIDE.

B

THE SYMBOL   THE WELD

U GROOVE, ARROW SIDE
$\frac{3}{8}$" DEEP, 20° ANGLE,
0" ROOT OPENING.

C

THE SYMBOL   THE WELD

BEVEL GROOVE, ARROW SIDE,
$\frac{3}{8}$" DEEP, 45° ANGLE; OTHER
SIDE, $\frac{1}{4}$" DEEP, 45° ANGLE,
$\frac{1}{8}$" ROOT OPENING, ARROW
POINTS TO PIECE TO HAVE THE
BEVEL.

D

THE SYMBOL   THE WELD

SQUARE GROOVE,
$\frac{3}{8}$" PENETRATION FROM EACH
SIDE, $\frac{1}{8}$" ROOT OPENING.

26-81. *Dimensioning groove
welds.*

	TYPE OF WELD			SUPPLEMENTARY SYMBOLS			
RESISTANCE-SPOT	PROJECTION	RESISTANCE-SEAM	FLASH OR UPSET	WELD ALL AROUND	FIELD WELD	CONTOUR	
						FLUSH	CONVEX
✳	╱	⋙	|	◯	●	—	⌒

26–82. *Basic resistance weld symbols.*

is shown in Fig. 26–83. The size of the spot-weld is the diameter. It can be in hundredths of an inch or fractions. It is always to the left of the weld symbol. If it is necessary to show the shear strength in pounds per square inch instead of the weld diameter, this is shown at the left of the weld symbol. The pitch is given in inches at the right of the weld symbols. The number of welds (spots) on a joint is indicated above or below the weld symbol. The number is enclosed in parentheses.

When the exposed surface of a resistance spot-weld is to be flush, the surface is indicated by adding the flush-contour weld symbol. If the bar is placed below the symbol, the flush surface is on the arrow side. If above the symbol, it is on the other side.

A *seam weld* is one in which the resistance weld is continuous the entire length. It does not leave unwelded spaces between welds as spot welding does.

The placement of data on the seam welding symbol is shown in Fig. 26–84. The symbols are much like those used

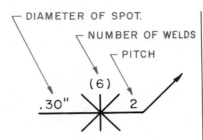

SPOT-WELD, .30" SPOT DIA., 6 SPOTS SPACED 2" APART, CENTER-TO-CENTER.

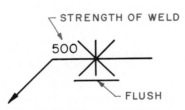

STRENGTH, 500 POUNDS/SQUARE INCH. SURFACE OF WELD IS FLUSH.

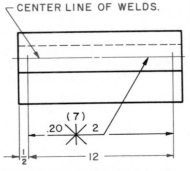

TYPICAL PLACEMENT OF SPOT-WELD SYMBOL.

26–83. *Dimensioning spot welds.*

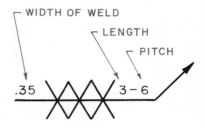

.35" WIDE SEAM WELD, 3" LONG, 6" ON CENTER.

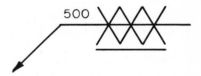

SEAM WELD, 500 POUNDS/SQUARE INCH, FLUSH SURFACE.

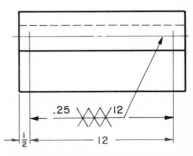

TYPICAL PLACEMENT OF SEAM WELD SYMBOL.

26–84. *Dimensioning seam welds.*

803

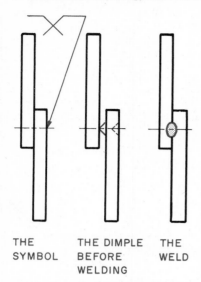

THE
SYMBOL

THE DIMPLE
BEFORE
WELDING

THE
WELD

26–85. *A projection weld.*

for spot welding. The weld symbol is centered on the reference line because there is no arrow side or other side.

Projection Welds. In projection welding, one member has a dimple stamped into it, forming a pointed projection. This is placed in contact with the member to be joined to it. Electric current is passed through the projection. This plus pressure welds the two parts together. See Fig. 26–85.

The projection welding symbol is shown in Fig. 26–86. Notice the location of the data on the symbol. These are much like those for spot welding.

If the projection weld symbol is below the reference line, the dimple is made in the arrow side part. If it is above the reference line, it is on the other side.

Flash and Upset Butt Welds. Flash butt welds are made by passing an electric current through two members that are spaced so they do not quite touch. An electric arc is formed between the members, melting the edge. They are then forced together under pressure. Upon cooling, they are welded. See Fig. 26–87.

Upset butt welds are made by passing an electric current through two members that are in firm contact with each other. Heat for the weld comes from passing the current through the parts. This heat plus pressure welds the pieces together. See Fig. 26–87.

There is no arrow side or other side in flash and upset butt welds. The symbols used are shown in Fig. 26–88. The weld symbol is placed in the center of the reference line. The tail contains information about the process. FW is the flash weld symbol. UW the upset weld symbol. No dimensions are necessary on the welding symbol.

Flash and upset welds can be made flush to the surface of the part. The material forming a bulge at the weld can be removed by machining, rolling, grinding, chipping, or hammering. The first letter of each of these processes is used on the symbol to indicate how the bulge created is to be removed. For example, M means machine the bulge away.

If the surface is to be flush, the flush symbol with the

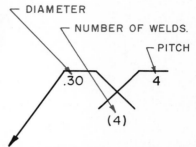

PROJECTION WELD, .30" DIA., 4 WELDS, 4" ON CENTER, DIMPLE ON ARROW SIDE OF PART.

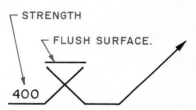

PROJECTION WELD, STRENGTH 400 POUNDS/SQUARE INCH, FLUSH, DIMPLE ON OTHER SIDE.

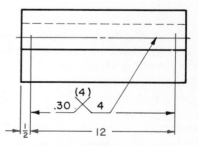

TYPICAL PLACEMENT OF THE PROJECTION WELD SYMBOL.

26–86. *Dimensioning projection welds.*

process to be used is given. If the surface is to be convex, the convex symbol and process is given.

804

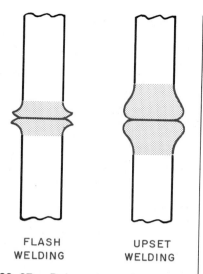

FLASH
WELDING

UPSET
WELDING

26-87. *Bulges caused by weld-ing process.*

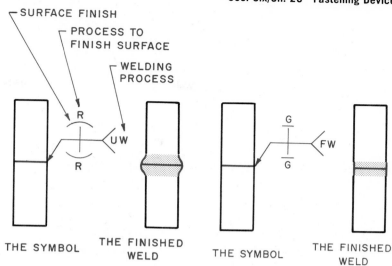

SURFACE FINISH

PROCESS TO
FINISH SURFACE

WELDING
PROCESS

THE SYMBOL

THE FINISHED
WELD

UPSET WELD, CONVEX CONTOUR
FORMED BY ROLLING BULGE.

THE SYMBOL

THE FINISHED
WELD

FLASH WELD, FLUSH SURFACE
FORMED BY GRINDING.

26-88. *Typical flash and upset welds.*

Build Your Vocabulary

Following are some terms you should understand and use as part of your working vocabulary. Write a brief explanation of what each means.

Internal thread
External thread
Major thread diameter
NC thread

NF thread
Unified screw threads
Bolt
Nut
Stud
Tapping screw
Cap screw
Machine screw
Set screw

Key
Keyseat
Pin
Washer
Rivet
Wood screw
Spring
Knurling
Welding

Index

A

Acme threads, 759, 762, 763
"Across the corners" distance, 83, 85
"Across the flats" distance, 83, 85
Acute angle, 77
Addendum (of a gear), 365, 366
Aerial mapping, 624, 625
Aerodynamics, 394–396
Aeronautical charts, 612, 613
Aerospace drawings
  bend allowance in, 411–413
  drafting techniques of, 404, 405
  for sheet metal parts, 398, 400
  layout of, 407–409
  numbering of, 407
  study problems, 431–440
  types of, 396
Airbrush, 449
Aircraft design
  fastening methods, 400, 404
  general discussion, 396, 397
  plaster models of, 409–411
Airfoil, 424–429
Aligned dimensioning, 138, 139, 236
Alphanumeric characters, 330
American Concrete Institute, 524
American Institute of Steel Construction, 502, 504
American National Standards Institute (ANSI)
  organization of, 20
  sizes, recommended for
    lettering for microfilm, 752
    locational fits, 170
    machine screws, 778
    nuts and bolts, 768
  standards for
    dimensioning, 131
    fasteners, 757
    gears, 372
  symbols
    electronic, 552
    piping, 469, 487
American National thread series,
  759–761
American Society of Mechanical
  Engineers, 70
American Standards Association
  (ASA), 20
American Welding Society, 482, 799
Ampere, definition of, 550
Amplifier, audio, 557
Angle
  azimuth, 616, 617
  bearing, 617, 618

Angle *(cont'd)*
  definition of, 77
  dihedral, 207, 208
  isometric, 230, 231
  steel, 502
  true size of, 203, 204
Angle valve, 487
Angular perspective, 244, 245
ANSI. See American National Standards Institute.
Aperture card, for microfilm, 749,
  752
Arch, 539–541
Architect, 25, 26, 631
Architect's scale, 54
Architectural drawing
  definition of, 16
  display type, 654
  elevations, 646, 649, 650
  lettering, dimensioning, and title
    block, 644
  of houses, 654–662
  plans
    floor, 645–647
    foundation, 646, 648
    plot, 653
  preliminary sketches, 639
  study problems, 663
  symbols, 639–643
Arcs
  definition of, 79
  dimensioning of, 143
  drawing of, 86–89
  in perspective, 245, 247
  isometric, 231, 232
Arrowheads (of lines), 132, 133
Artist, commercial, 27
ASA. See American Standards
  Association.
Assembly drawings
  check, 300
  definition of, 16, 296
  design, 283
  furniture, 702–706
  general, 300–306
  orthographic, 296–299
Assembly sections, 266
Atom, 549
Auxiliary
  plane, 191
  sections, 200, 262
Auxiliary views
  general discussion, 191–205
  study problems, 216–225
Axis
  in numerical control systems, 326,
    327

Axis *(cont'd)*
  isometric, 227–229
  of a chart, 729, 734
  of revolution, 212–214
Axonometric drawings, 226, 227
Azimuth angles (in land survey), 616,
  617
Azo dye, 747

B

Back panels, in cabinetmaking, 670
Ball valve, 487
Bar chart, 726–729
Bars, to reinforce concrete, 524, 528
Base circle of a gear, 366
Baseline drawings, electrical, 575
Bathroom, planning for, 636
Beam compass, 57
Beams
  drawing of, 513
  filler, 501
  steel, 524, 525
Bearing angles (in land surveys), 616,
  618
Bedroom, planning for, 635
Bend allowance of metal
  definition of, 294
  in aircraft, 411–413
  pattern development for, 359
Bend, open- and closed-bevel, 413
Bent, steel, 524
Bevel gear, 367–369
Bilateral system of dimensioning,
  200
Bisecting lines and angles, 80, 81
Blank (pattern), in cabinetmaking,
  712
Blanking of metal, 293
Blind hole, 150
Block diagram
  for electronic circuitry, 551, 552
  used in maps, 610–612
Blocking, in fixtures, 718
Blueprint, 444, 746, 747
Bolts
  for joining steel, 504, 506
  general discussion, 767–773
  hanger, 680
Bow's notation, 387, 388
Bowstring truss, 515
Box joints, 670
Brackets, corner, 680
Break lines, 69, 110–112
Brick veneer, 644
Broken-out section, 235, 236, 260
Brown and Sharpe thread, 759

Metric Equivalents of Two-Place Decimals

				Millimetre Equivalent					Millimetre Equivalent
		.02		.508			.52		13.208
			.03	.762				.53	13.462
		.04		1.016			.54		13.716
			.05	1.270				.55	13.970
	.06			1.524		.56			14.224
		.08		2.032			.58		14.732
			.09	2.286				.59	14.986
.10				2.540	.60				15.240
	.12			3.048		.62			15.748
		.14		3.556			.64		16.256
			.15	3.810				.65	16.510
		.16		4.064			.66		16.764
	.18			4.572		.68			17.272
			.19	4.826				.69	17.526
.20				5.080	.70				17.780
		.22		5.588			.72		18.288
	.24			6.096		.74			18.796
			.25	6.350				.75	19.050
		.26		6.604			.76		19.304
		.28		7.112			.78		19.812
.30				7.620	.80				20.320
			.31	7.874				.81	20.574
	.32			8.128		.82			20.828
		.34		8.636			.84		21.336
			.35	8.890				.85	21.590
		.36		9.144			.86		21.844
			.37	9.398				.87	22.098
	.38			9.652		.88			22.352
.40				10.160	.90				22.860
			.41	10.414				.91	23.114
		.42		10.668			.92		23.368
	.44			11.176		.94			23.876
			.45	11.430				.95	24.130
		.46		11.684			.96		24.384
			.47	11.938				.97	24.638
		.48		12.192			.98		24.892
.50				12.700	1.00				25.400